Science Networks · Historical Studies
Founded by Erwin Hiebert and Hans Wußing
Volume 27

Writing the History of Mathematics: Its Historical Development

Editors:
Joseph W. Dauben
Christoph J. Scriba

Birkhäuser Verlag
Basel · Boston · Berlin

Editors' addresses:

Joseph W. Dauben
Ph.D. Program in History
The Graduate Center, CUNY
365 Fifth Avenue at 34th Street
New York, NY 10016-4309
USA
email: jdauben@worldnet.att.net

Christoph J. Scriba
Fachbereich Mathematik
Universität Hamburg
Bundesstr. 55
D-20146 Hamburg
Germany
email: scriba@math.uni-hamburg.de

A CIP catalogue record for this book is available from the Library of Congress, Washington D.C., USA.

Deutsche Bibliothek Cataloging-in-Publication Data
Writing the history of mathematics: its historical development / ed.: Joseph W. Dauben ; Christoph J. Scriba. - Basel ; Boston ; Berlin : Birkhäuser, 2002
(Science networks ; Vol. 27)
ISBN 3-7643-6167-0
ISBN 3-7643-6166-2

ISBN 3-7643-6166-2 (Hardcover) Birkhäuser Verlag, Basel – Boston – Berlin
ISBN 3-7643-6167-0 (Softcover) Birkhäuser Verlag, Basel – Boston – Berlin

Member of the BertelsmannSpringer Publishing Group
Cover design: Micha Lotrovsky, Therwil, Switzerland
Cover illustrations: Portraits of Moritz Cantor (1829–1920), Gustaf Hjalmar Eneström (1852–1923), Mikami Yoshio (1875–1950), Li Yan (1892–1963), Adolf Pavlovich Youshkevich (1906–1993), Dirk Jan Struik (1894–2000)
Back Cover: Mathematisches Forschungsinstitut Oberwolfach, Germany;
Fotograf: Helmut Kastenholz
Typeset by the editors in LaTeX
Revision and layout by *mathScreen Online*, CH-4123 Allschwil

Printed on acid-free paper produced from chlorine-free pulp. TCF ∞
Printed in Germany
ISBN 3-7643-6166-2 (Hardcover)
ISBN 3-7643-6167-0 (Softcover)

9 8 7 6 5 4 3 2 1

www.birkhauser.ch

In memory of Kenneth O. May
and dedicated to the
Mathematisches Forschungsinstitut
Oberwolfach

KENNETH OWNSWORTH MAY (1915–1977)

Contents

Table of Portraits xv

Contributors xvii

Praescriptum xxi

Introduction xxix

Acknowledgements xxxiii

Glossary xxxv

Part I Countries

1 France

JEANNE PEIFFER 3

1.1 Introduction 3

1.2 Changing Appreciation of the Ancients, from the Renaissance to the Seventeenth Century 4

1.3 History of the Progress of the Human Mind (in the Enlightenment) 6

1.3.1 Fontenelle, the Initiator of a Tradition 7

1.3.2 The Historical Dimension of the Encyclopedia Project . . . 8

1.3.3 Montucla's Monumental Work 9

1.4 Historiography in Revolutionary Times 11

1.5 The Reform of Society Through the Sciences: Positivism 13

1.6 Oriental Studies and the History of Mathematics in the Nineteenth Century 14

1.6.1 Collaboration Between Scientists and Orientalists 15

1.6.2 Academic Controversies on the History of Arabic Mathematics 16

1.6.3 François Woepcke's Research 18

1.7 The History of Mathematics by Mathematicians: Chasles and his Successors . . . 20
1.8 The Spread of Historical Work in Journals and the Broadening of the Subjects Treated . . . 22
1.9 The History of Mathematics in Relation to the General History of Science: Paul Tannery . . . 25
1.10 Nationalist Tendencies . . . 30
1.11 Philosopher-Scientists of the Twentieth Century . . . 30
1.12 The History of Mathematics and the Institutionalization of the History of Science Between the Two World Wars . . . 33
1.13 The Historical Epistemology of Gaston Bachelard . . . 36
1.14 Towards an Autonomous History of Mathematics After the Second World War . . . 36
1.15 The Centre Alexandre Koyré and Research on the History of Mathematics . . . 38
1.16 History According to Bourbaki . . . 40
1.17 Further Developments . . . 41
1.18 Conclusion . . . 42

2 Benelux
Paul Bockstaele . . . 45
2.1 Geographical and Political Considerations . . . 45
2.2 Humanist-Inspired Return to the Sources . . . 46
2.3 Catalogues of Mathematicians . . . 47
2.4 The Eighteenth Century . . . 49
2.5 Historiography of Mathematics in the Kingdom of the Netherlands 1815–1830 . . . 49
2.6 Historiography of Mathematics in Belgium after 1830 . . . 50
2.7 Historiography of Mathematics in the Netherlands after 1830 . . . 53
2.8 Amateurs and Professionals, Journals and Societies . . . 56
2.9 Conclusion . . . 58

3 Italy
Umberto Bottazzini . . . 61
3.1 Introduction . . . 61
3.2 The Rediscovery of Classical Mathematics . . . 62
3.3 Interlude: Contributions of the Jesuits . . . 65
3.4 The History of Mathematics in the Enlightenment . . . 66
3.5 History as a Research Topic for Mathematicians . . . 69
3.5.1 Pietro Cossali . . . 69
3.5.2 Gregorio Fontana . . . 70
3.5.3 Giovanni B. Venturi and Pietro Franchini . . . 71
3.5.4 The Influence of Lagrange . . . 72

3.5.5 Guglielmo Libri 74
3.6 The Risorgimento and the Search for Italian Forerunners 77
3.7 Boncompagni's "Bullettino" and Its Influence 79
3.8 The History of Mathematics in the Early Twentieth Century 81
3.8.1 The School of Peano in Turin 82
3.8.2 The School of Enriques in Bologna 83
3.9 Second Interlude: Galileo and Leonardo 85
3.10 The Emergence of Professional Historians 87
3.10.1 Gino Loria 88
3.10.2 Ettore Bortolotti 90
3.11 Enriques and the Institute for the History of Science in Rome 91
3.12 After World War II 93
3.13 Conclusion 94

4 Switzerland
ERWIN NEUENSCHWANDER 97
4.1 Introduction 97
4.2 Humanism and Enlightenment 98
4.3 The Contribution of the "Naturforschende Gesellschaften" 100
4.4 The Major Editions 102
4.5 Further Developments 104
4.6 Conclusion 106

5 Germany
MENSO FOLKERTS, CHRISTOPH J. SCRIBA, and HANS WUSSING 109
5.1 Introduction 109
5.2 The Beginnings 110
5.2.1 The First Glimmer: Regiomontanus and Some Successors . 110
5.2.2 Two Extremes: Leibniz and Wolff 111
5.2.3 Three Additional Eighteenth-Century German Authors . . 111
5.2.4 Mathematician, Bibliographer and Epigrammatist: Kästner 113
5.3 First Half of the Nineteenth Century 113
5.3.1 From Watch-Maker to Professor of Technology: Poppe . . . 113
5.3.2 Handbooks and History: Klügel and Mollweide 114
5.3.3 Opposite Twins? Nesselmann and Arneth 114
5.4 From 1850 up to World War I 117
5.4.1 The "Philologists": Editions of Texts 117
5.4.2 Hermann Hankel and Moritz Cantor: First Comprehensive Studies 122
5.4.3 Günther and Braunmühl: On the Way to Professionalism . 125
5.4.4 Felix Klein and His "Vorlesungen." Editions of "Collected Works" 127

5.5 Between the Wars . . . 131
5.5.1 More Studies on Arabic Mathematics . . . 131
5.5.2 Toeplitz, Neugebauer, and Bessel-Hagen: The "Kiel-Göttingen-Bonn Group" . . . 134
5.5.3 Wieleitner, Tropfke, and Their Successors Vogel and Hofmann . . . 136
5.6 History of Mathematics Under the Third Reich . . . 140
5.7 The German Democratic Republic . . . 143
5.8 The Federal Republic of Germany . . . 145
5.9 History of Mathematics in a Reunited Germany . . . 148
5.10 Conclusion . . . 149

6 Scandinavia
Kirsti Andersen . . . 151
6.1 Introduction . . . 151
6.2 Early Publications on the History of Mathematics . . . 152
6.3 National Interests . . . 152
6.4 Professional History of Mathematics . . . 154
6.5 Eneström's Scientific History of Mathematics . . . 154
6.6 Zeuthen's Historical Mathematics . . . 156
6.7 Text Editions . . . 157
6.8 After Eneström and Zeuthen . . . 157
6.9 Conclusion . . . 160

7 The British Isles
Ivor Grattan-Guinness . . . 161
7.1 Introduction . . . 161
7.2 Prior to the Early Twentieth Century . . . 161
7.3 Studies of Special Topics . . . 164
7.4 Largely Historians of Newton . . . 166
7.5 Some Cambridge Historians . . . 169
7.6 The Royal Society Catalogue of Scientific Papers . . . 170
7.7 Historians of Greek Mathematics . . . 171
7.8 General Writing to the First World War . . . 172
7.9 Pearson and the History of Statistics . . . 174
7.10 Between the Wars . . . 175
7.11 After the Second World War . . . 176
7.12 Conclusion . . . 178

8 Russia and the U.S.S.R.
Sergei S. Demidov . . . 179
8.1 Introduction . . . 179
8.2 History of Mathematics Before 1917 . . . 179

8.3 The First Post-Revolutionary Years: Formation of the Soviet School 182
8.4 Research Schools After World War II 187
8.5 Dominant Postwar Research Themes and Publication Formats 191
8.6 Social Dimensions of the History of Mathematics After World War II 194
8.7 Conclusion 196

9 Poland
STANISŁAV DOMORADZKI and ZOFIA PAWLIKOWSKA-BROŻEK 199
9.1 Early Developments 199
9.2 Twentieth-Century Contributions 200
9.3 Conclusion 202

10 Bohemian Countries
LUBOŠ NOVÝ 205
10.1 Introduction 205
10.2 Beginnings 206
10.3 History of Mathematics 1750–1850 206
10.4 The Mid-19th Through the Mid-20th Century 207
10.5 History of Mathematics in Czechoslovakia Since 1945 210
10.6 Conclusion 212

11 Austria
CHRISTA BINDER 213
11.1 Introduction 213
11.2 Regiomontanus and Tannstetter 213
11.3 Decline and Revival 216
11.4 The “Encyklopädie” 217
11.5 Recent Developments 218
11.6 Conclusion 219

12 Greece
CHRISTINE PHILI 221
12.1 The Classical and Hellenistic Periods 221
12.2 The Byzantine Period 223
12.3 Voulgaris and the Athonian Academy 223
12.4 The Ionian Academy 225
12.5 The Greek National State (1822 to the Present) 226
12.6 Conclusion 229

13 Spain
Elena Ausejo and Mariano Hormigón 231
13.1 Introduction 231
13.2 Polemics With Respect to Sixteenth-Century Spanish Mathematicians 233
13.3 Mathematics in the Liberal State 234
13.4 Conclusion 236

14 Portugal
Luis M. R. Saraiva 239
14.1 Introduction 239
14.2 The Beginnings: Garção Stockler 239
14.3 The Second Half of the Nineteenth Century: Francisco de Castro Freire 241
14.4 The Golden Age of Portuguese Historiography of Mathematics: 1900–1940 243
14.5 The Modern Period: 1940–1970 246
14.6 Conclusion 246

15 The Americas
Ubiratan D'Ambrosio, Alejandro R. Garciadiego, Joseph W. Dauben, and Craig G. Fraser 249
15.1 Introduction 249
15.2 South America 249
15.2.1 Historiographical Remarks 249
15.2.2 Conquest and Early Colonial Times 250
15.2.3 The Established Colonies 250
15.2.4 Independent Countries 251
15.2.5 The Twentieth Century 252
15.2.6 Current Developments 254
15.2.7 Conclusion 255
15.3 Mexico 256
15.3.1 The Creation and Development of the Royal University 256
15.3.2 After Independence 257
15.3.3 The Emergence of Modern Mathematics in Mexico in the Twentieth Century: The Autonomous National University 258
15.3.4 The Faculty of Sciences 258
15.3.5 The Nightmare: The 1968 Student Movement 261
15.3.6 Conclusion 262
15.4 United States of America 263
15.4.1 Introduction 263

15.4.2 History of Mathematics in the United States: Early Efforts Through World War I . . . 264
15.4.3 Increasing Professionalization: History of Mathematics and History of Science . . . 273
15.4.4 Recent History of Mathematics in the United States . . . 276
15.4.5 Conclusion . . . 281
15.5 Canada . . . 285
15.5.1 Before 1966 . . . 285
15.5.2 Kenneth O. May . . . 286
15.5.3 The Situation Today . . . 287

16 Japan
Sasaki Chikara . . . 289
16.1 The Prewar Period, 1868–1945: The Flowering of the Study of the History of Japanese Mathematics . . . 289
16.1.1 Endō Toshisada and His Successor Mikami Yoshio . . . 289
16.1.2 The Tōhoku School . . . 291
16.2 The Postwar Period, 1945–1986: Beginnings of the Serious Study of the History of Western Mathematics . . . 292
16.3 Conclusion . . . 294

17 China
Liu Dun and Joseph W. Dauben . . . 297
17.1 The Decline of Traditional Chinese Mathematics . . . 297
17.2 Early Authors . . . 298
17.3 The "Qian-Jia School" and Its Successors . . . 300
17.4 The Modern Scholars . . . 304
17.5 Conclusion . . . 305

18 India
Radha Charan Gupta . . . 307
18.1 Introduction . . . 307
18.2 Beginnings of Indigenous Historiography . . . 308
18.3 Modern Historical Studies and Historiography of Indian Mathematical Sciences . . . 310
18.4 Indian Historians . . . 311
18.5 The Twentieth Century . . . 313
18.6 Conclusion . . . 315

19 Arab Countries, Turkey, and Iran
Sonja Brentjes . . . 317
19.1 Introduction . . . 317

19.2 Exchanges with Western Europe 318
19.3 A New Start for Historiography of Science and Mathematics 319
19.4 Diverse Attitudes towards History of Science and Mathematics 321
19.5 Trends in History of Mathematics in Iran and Turkey 323
19.6 The New Institutions 324
19.7 Conclusion 327

20 Postscriptum
Joseph W. Dauben, Jeanne Peiffer, Christoph J. Scriba, and Hans Wussing 329
20.1 The Character of Historiography 329
20.2 George Sarton's Views 329
20.3 Interrelations 330
20.4 On the History of Historiography 331
20.5 Functions of Historiography 332
20.6 Institutional Factors 334
20.7 History of Mathematics and Mathematics Education 335
20.8 History of Mathematics: Recent Trends 336
20.9 Electronic Resources 337
20.10 The Humanism of Mathematics 338

Part II Portraits and Biographies

Portraits 343

Biographies 351

Part III Abbreviations, Bibliography, and Index

Abbreviations 583

Bibliography 591

Index 645

Table of Portraits

1	Jean Etienne Montucla	(1725–1799)	343
2	Moritz Cantor	(1829–1920)	343
3	Hieronymus Georg Zeuthen	(1839–1920)	344
4	Paul Tannery	(1843–1904)	344
5	Zoel García de Galdeano	(1846–1924)	344
6	Walter William Rouse Ball	(1850–1925)	344
7	Francisco Gomes Teixeira	(1851–1933)	345
8	Gustaf Hjalmar Eneström	(1852–1923)	345
9	Johan Ludvig Heiberg	(1854–1928)	345
10	Florian Cajori	(1859–1930)	345
11	David Eugene Smith	(1860–1944)	346
12	Sir Thomas Little Heath	(1861–1940)	346
13	Gino Loria	(1862–1954)	346
14	Raymond Clare Archibald	(1875–1955)	346
15	Mikami Yoshio	(1875–1950)	347
16	Julio Rey Pastor	(1888–1962)	347
17	Kurt Vogel	(1888–1985)	347
18	Li Yan	(1892–1963)	347
19	Qian Baocong	(1892–1974)	348
20	Dirk Jan Struik	(1894–2000)	348
21	José Babini	(1897–1984)	348
22	Otto Neugebauer	(1899–1990)	348
23	Joseph Ehrenfried Hofmann	(1900–1973)	349
24	Adolf Pavlovich Youshkevich	(1906–1993)	349

Contributors

KIRSTI ANDERSEN
History of Science Department
University of Aarhus
Aarhus, Denmark
(K.A.)

ELENA AUSEJO
Departamento de Ciencias de la
Documentación e Historia de la Ciencia
Universidad de Zaragoza
Zaragoza, Spain
(E.A.)

LILIANE BEAULIEU
Centre de Recherches Mathématiques
Université de Montréal
Montréal, Québec, Canada
(L.B.)

KLAAS VAN BERKEL
Instituut voor Geschiedenis
Rijksuniversiteit Groningen
Groningen, Netherlands
(K.v.B.)

CHRISTA BINDER
Institut für Technische Mathematik
Technische Universität Wien
Vienna, Austria
(C.B.)

PAUL BOCKSTAELE
Wiskundig Instituut
Katholieke Universiteit te Leuven
Leuven, Belgium
(P.B.)

UMBERTO BOTTAZZINI
Dipartimento di Matematica e Applicazioni
Università di Palermo
Palermo, Italy
(U.B.)

SONJA BRENTJES
Institut für Geschichte der Naturwissenschaften
Johann Wolfgang Goethe-Universität
Frankfurt (Main), Germany
(S.B.)

IVOR BULMER-THOMAS †
Ancient Monuments Society
London, England
(I.B-T. †)

MAURICE CAVEING
Directeur de Recherche
Centre National de la Recherche Scientifique
Paris, France
(M.C.)

KARINE CHEMLA
REHSEIS
Centre National de la Recherche Scientifique
Paris, France
(K.C.)

UBIRATAN D'AMBROSIO
Rua Peixoto Gomide 1772, ap. 83
01409002 São Paulo SP
Brazil
(U.D'A.)

JOSEPH W. DAUBEN
Ph.D. Program in History
The Graduate Center
City University of New York
New York, NY, USA
(J.W.D.)

SERGEI S. DEMIDOV
Institute of History of Science
and Technology
Russian Academy of Sciences
Moscow, Russia
(S.S.D.)

Jean Dhombres
Centre Alexandre Koyré
Centre National de la Recherche Scientifique
Paris, France
(J.D.)

Stanisław Domoradzki
Institute of Mathematics
University of Pedagogy
Rzeszów, Poland
(S.D.)

Menso Folkerts
Institut für Geschichte der Naturwissenschaften
Universität München
Munich, Germany
(M.F.)

Craig G. Fraser
Institute for the History and Philosophy
of Science and Technology
University of Toronto
Toronto, Ontario, Canada
(C.G.F.)

Alejandro R. Garciadiego
Departamento de Matemáticas
Facultad de Ciencias
Universidad Nacional Autónoma de México
México DF, México
(A.R.G.)

Hélène Gispert
Institut Universitaire de Formation
des Maîtres de Versailles
Groupe d'histoire des sciences
Orsay, France
(H.G.)

Beat Glaus
Rütschistr. 28
Zürich, Switzerland
(B.G.)

Ivor Grattan-Guinness
Middlesex University at Enfield
Enfield, England
(I.G-G.)

Radha Charan Gupta
Gaṇita Bhāratī Academy
R-20, Ras Bahar Colony
Jhansi, U.P., India
(R.C.G.)

Mariano Hormigón
Departamento de Ciencias de la
Documentación e Historia de la Ciencia
Universidad de Zaragoza
Zaragoza, Spain
(M.H.)

LIU Dun
Institute for History of Natural Sciences
Chinese Academy of Science
Beijing, P.R.China
(L.D.)

Jesper Lützen
Department of Mathematics
University of Copenhagen
Copenhagen, Denmark
(J.L.)

Fritz Nagel
Bernoulli-Edition
Universitätsbibliothek
Basel, Switzerland
(F.N.)

Erwin Neuenschwander
Mathematisches Institut
Universität Zürich
Zürich, Switzerland
(E.N.)

Luboš Nový
Czechoslovak Academy of Sciences
Prague, Czech Republic
(L.N.)

Mario H. Otero
Universidad de la República
Montevideo, Uruguay
(M.H.O.)

Karen Hunger Parshall
Departments of History and Mathematics
University of Virginia
Charlottesville, Virginia, USA
(K.H.P.)

Zofia Pawlikowska-Brożek
Department of Mathematics
University of Mining and Metallurgy
Cracow, Poland
(Z.P-B.)

JEANNE PEIFFER
Centre Alexandre Koyré
Centre National de la Recherche Scientifique
Paris, France
(J.P.)

CHRISTINE PHILI
Department of Mathematics
National Technical University of Athens
Zographou Campus
Athens, Greece
(C.P.)

ROSHDI RASHED
Directeur de Recherche
Centre National de la Recherche Scientifique
Paris, France
(R.R.)

ADRIAN RICE
Department of Mathematics
Randolph-Macon College
Ashland, Virginia, U.S.A.
(A.R.)

LUIS M. R. SARAIVA
Centro de Matemática
e Aplicações Fundamentais
Universidade de Lisbon
Lisbon, Portugal
(L.M.R.S.)

SASAKI CHIKARA
University of Tokyo
Graduate School of Arts and Sciences
Department of History and
Philosophy of Science
Komaba, Tokyo, Japan
(S.C.)

CHRISTOPH J. SCRIBA
Fachbereich Mathematik
Schwerpunkt Geschichte der Naturwissenschaften,
Mathematik und Technik
Universität Hamburg
Hamburg, Germany
(C.J.S.)

JACQUES SESIANO
Département de Mathématiques
Ecole Polytechnique Fédérale
de Lausanne
Lausanne, Switzerland
(J.S.)

HOURYA SINACEUR
Institut d'Histoire et de
Philosophie des Sciences et des Techniques
Centre National de la Recherche Scientifique
Paris, France
(H.S.)

CHRISTIAN MARINUS TAISBAK
Institute for Greek and Latin
University of Copenhagen
Copenhagen, Denmark
(C.M.T.)

RENATE TOBIES
Fachbereich Mathematik
Universität Kaiserslautern
Kaiserslautern, Germany
(R.T.)

HANS WUSSING
Karl-Sudhoff-Institut für Geschichte
der Medizin und Naturwissenschaften
Universität Leipzig
Leipzig, Germany
(H.W.)

Editorial Assistance

ANKE JOBMANN
Institut für Geschichte der Naturwissenschaften,
Mathematik und Technik
Universität Hamburg
Hamburg, Germany

ELENA OPAYETS
Institut für Geschichte der Naturwissenschaften, Mathematik und Technik
Universität Hamburg
Hamburg, Germany

Praescriptum

JOSEPH W. DAUBEN, JEANNE PEIFFER, CHRISTOPH J. SCRIBA, and HANS WUSSING

The earliest written record that can be regarded as history of mathematics is to be found in PROCLUS's ancient *Commentary on Euclid.* This is also known as the *Eudemian Summary* because the *Commentary* may well have been copied, at least in part, from a now lost *History of Geometry* by EUDEMUS, a student of ARISTOTLE. Not only do we learn that geometry began with the Egyptians as the result of practical demands of land surveying, but more remarkably, that this is understandable "since everything which is in the process of becoming progresses from the imperfect to the perfect" [Thomas 1957, vol. 1, 147]. What Proclus recounts is the actual progress of mathematics from perception to reasoning to understanding, which as applied to the history of mathematics relates to the improvement of our knowledge of its origins and development — be it pure or applied, concrete or abstract, theoretical or practical. *Writing the History of Mathematics* likewise proceeds from the level of the facts of that history to evaluating their import, and, at its most revealing attempts, to understand the broader signifance of the history of mathematics for mathematicians, historians of science, and humanists in general.

The Enlightenment sought to chart the progress of the human spirit in the advance of science and technology, and later, in the hands of AUGUSTE COMTE, mathematics came to epitomize the best of positive knowledge. But if the Whiggish views of the 19th century embraced a vision of human progress, the 20th century has for the most part taken a darker view of history, and in the aftermath of two World Wars, world-wide depressions, the crash of stock markets and the disruption of international markets, global outbreaks of influenza, AIDS, environmental pollution, moral doubts related to genetic engineering and human cloning, it is no wonder that post-modernist historians have questioned the faith of earlier centuries in the inevitable progress of history, and of the sciences and technology as well.

Mathematics may seem immune to such concerns about progress, but only if mathematics is regarded as nothing more than the increase of a rather esoteric and technical kind of knowledge. But as this book makes clear, mathematics is much more than a body of theorems and proofs produced by mathematicians. It is not just a compilation of abstract results, but mathematics and its history are themselves part of larger cultural histories, and as we begin the 21st century, these can be understood and appreciated on an increasingly international scale.

As a result, the question of progress — and the progress of science and technology, as well as of mathematics — is an especially controversial one. Recently, a number of authors have written about the end of progress, the end of science, and even of the end of mathematics. The latter is not a new concern, for as early as the 18th century, a scientist like J. B. J. Delambre could lament that there was not much left for mathematicians to accomplish except ever better approximations.[1] In the modern world, however, just as some fear that the end of science is at hand, others fear that the impact of the sciences upon the world will be largely negative, increasingly disastrous. As a consequence, the dominant role that the sciences and mathematics have come to play in the modern world has made them increasingly objects of public concern. This in turn has prompted historians to consider the past and contemplate the future of the sciences more seriously. For historians of mathematics, as this book makes clear, the subject of their past and future is especially timely, since professionally it appears to have reached a certain maturity that deserves closer evaluation — with a consideration, at the end of this book, as to where its future may lead.

Answers to questions about the past and future of the sciences, however, including mathematics, depend in part on the resources at one's disposal, on philosophical or ideological views, on regional characteristics and local temperament, and on the purposes and intentions of particular scholars' points of view. Despite attempts to achieve scientific seriousness and objectivity, there still exist differences of opinion, competing views, and diverging interpretations — even when basic questions arise concerning the meaning and purpose or the form of historial studies and publications. In this sense, the historiography of mathematics is an open subject, one which offers a broad spectrum of possibilities and encourages thought-provoking discussions.

For clarity's sake, in the following text a distinction is made between "history" and "historiography," and in particular between "history of mathematics" and "historiography of mathematics." History is what has happened in the past; historiography is the analysis of "history" as a discipline, an account of its assumptions, methods, and the different approaches to which it has been subject in the hands of different historians writing in different places at different times under varying constraints including (but not limited to) economics, politics, philosophy, religion, and even health and psychological states of mind. Likewise, "history of

[1] J. B. J. Delambre: *Rapport historique sur le progrès des sciences mathématiques depuis 1789* (Paris, 1810), 99; quoted from [Sarton 1955, 12]: "It would be difficult and perhaps foolhardy to analyze the chances of further progress; in almost every part of mathematics one is stopped by unsurmountable difficulties; improvements in the details seem to be the only possibilities which are left."

mathematics" is concerned with the development, taking place in time, of the unfolding of mathematics; "historiography of mathematics," on the other hand, embraces the scholarly research, reconstruction, and description of the past development of the *history* of the subject. Historiography of mathematics involves the investigation, as exactly as possible, of this past evolution, to describe and interpret it for all relevant times and places in as critical and comparative a way as possible.

This has been the major aim of *Writing the History of Mathematics.* From the beginning, it was the goal of this project to provide more than a list of names with relevant biographical information about prominent historians of mathematics, and to do more than simply analyze the major works published by historians of mathematics from country to country. Each contributor was asked to go beyond the presentation of such data to provide a perspective on how and why history of mathematics has developed in the way it has, differently at different times in different countries under differing circumstances. Authors have sought to be especially critical in evaluating the forces that have shaped and directed the development of the history of mathematics in their own respective countries, and to conclude their analyses with characterizations of the position history of mathematics has come to play by the end of the 20th century in each of the countries surveyed here.

At the risk of over-simplifying, or overly-generalizing the main tasks of historiography, especially as the benefits as well as the perils of such research became clearer in the process of revising and discussing the many contributions to *Writing the History of Mathematics*, we outline here the preliminary thoughts of our working group in hopes of stimulating further discussion. We are well aware that such matters, and the answers to some of the questions we raise here, can only be given tentatively for now, and may in fact provoke disagreement. In spite of this, we offer the following in hopes that the aims, methods, and results of historiographic research may come to the fore more clearly. From any errors or rash conclusions in what follows, later authors may thereby be prompted, if not encouraged, to make further improvements of their own!

Considering one of the major innovations of this book, that the history of history of mathematics may be more easily compared from country to country, generation by generation, are there any general historiographic principles that emerge from these studies, ones that seem to transcend time and national boundaries? Or is it the case that national and/or regional characteristics prevail? What kind of relations are there — and which are most pertinent between the historiography of mathematics and related scientific disciplines? Can the study of the historiography of mathematics serve to sharpen the limits or boundaries of historiography of mathematics, in particular with respect to methodology? Is it possible to name areas in which it is especially urgent to investigate the history of the history of mathematics?

Of special interest historically are the interrelations between mathematics and historiography of mathematics. Often, as the contributions to this collection make clear, historiography of mathematics has been understood and carried out, without much reflection, as a part of mathematics. Where the historiography of mathematics for organizational reasons has been associated with mathematical departments or institutes, this has usually been for reasons of tradition, and may go back to issues related to basic scientific instruction, or may perhaps be the result of pragmatic considerations related to the institutions themselves. In addition, the image of the historiography of mathematics as a part of mathematics proper has been reinforced by its natural classification in bibliographies and abstracting journals like the *Jahrbuch über die Fortschritte der Mathematik* and *Mathematical Reviews*, where history of mathematics and biographies have had their special place from the beginning.

Mathematicians have often shown a supportive, interested, and pragmatic attitude towards historiography, especially with respect to chronology and where questions about priorities and the actual sequence of internal mathematical developments are concerned. Mathematicians, however, tend to be interested primarily in the history of concepts and methods. Some may also take an interest in outstanding personalities and important institutions, but others, like BOURBAKI, for example, may actually eschew any interest in individual mathematicians, preferring instead to emphasize the contributions rather than the contributors, insisting that mathematical ideas are what matter, not the individuals who have conceived them. As for the internal history of mathematics, there has been an astonishingly large number of mathematicians and teachers at secondary schools throughout Europe who, beginning in the 19th century, have contributed to this aspect of the discipline. The three hundred biographies in this book, and their numerous references to *Programmschriften* ([School] Program Writings), are clear testimony to the importance of such works for the history of mathematics.

The example of India, however, offers a somewhat different case, and shows that the existence of a highly developed mathematics in a Western sense is not a prerequisite for establishing a strong commitment to historiography. Mathematics, like other scientific disciplines, was originally considered in the earliest times to have been of divine origin and thus was highly revered. Consequently, mathematics and historiography of mathematics in India were thought to have arisen in more or less parallel ways, in contrast to historiographies in countries that only began to develop an historical sense of mathematics after it had achieved a certain standing. The situation is slightly different, yet again, in China, where mathematics was believed to have originated in the mythological mists of past time, but where an historiographic interest in the subject was the result of a well-established tradition of commentaries on early, often lost or fragmentary works, which of necessity required an intrinsic and critical view of the history of the manuscripts with which the commentators were concerned.

Thus no matter East or West, constructive collaboration between mathematicians and historians of mathematics will doubtless remain an indispensable part of the future development of the historiography of mathematics. It is especially important in connection with the evaluation and analysis of newer — and in fact the newest — results in mathematics, above all when the historical roots and the context of mathematical discoveries are being investigated. An example is the comparison of modern non-standard analysis with the early history of mathematical analysis of LEIBNIZ and CAUCHY, both in mathematics and in the historiography of mathematics. In Italy, there is a tradition of historically oriented research by mathematicians (consider, for example, BRIOSCHI and CASORATI). In some cases, important mathematicians have contributed in recent times to the history of mathematics. Among them, some of the most prominent are JEAN DIEUDONNÉ, A. N. KOLMOGOROV, DIRK J. STRUIK, B. L. VAN DER WAERDEN, and ANDRÉ WEIL (see for example [STRUIK 1980]).

Studies about individuals, their lives and fates as well as their historical influence, are generally considered to be an indispensable part of the historiography of mathematics. The same is true for publications about the history of institutions, mathematical schools, international collaborations at congresses, bibliographies, reference works, reviewing journals, etc. In the analyses presented here, such studies are exhibited as integral parts of extended historiographical activity.

In particular, following World War II there was an on-going debate for many years about "internal" versus "external" forces active in the development of the sciences. In some of the present studies this is at least touched upon, given the fact that historians of mathematics trained primarily as mathematicians, or whose primary occupation is the teaching of mathematics, tend to take the "internalist" approach to their subject, whereas historians of mathematics trained primarily as historians or as historians of science more often than not will take a broader, more "external" approach (it presumably goes without saying that the two are not mutually exclusive and that in most cases a satisfactory, more comprehensive history of complex subjects will likely benefit from if not require an amalgamation of both approaches). Here questions and opinions concerning theories of cognition and the philosophy of science must also play a fundamental role in determining the most appropriate and successful historiographic strategies one should adopt.

* * *

Writing the History of Mathematics represents an ambitious undertaking of a group of more than forty historians of mathematics from many countries representing diverse historiographic traditions as well. Conceived initially as a means of self-reflection by historians of mathematics about the changing perspectives of their own profession, this project grew out of meetings over several decades at the *Mathematisches Forschungsinstitut* in Oberwolfach, Germany. This is an international research and conference center primarily for mathematicians, and in an environment as conducive to reflection as to research on the frontiers of mathematics, this project quickly took on a life of its own.

The first meeting on history of mathematics at Oberwolfach was organized by J. E. HOFMANN in 1954. HOFMANN's approach to history of mathematics was in terms of *Problemgeschichte*, a fairly internalist approach to the subject that focused on the technical content of specific problems and methods. Over the years, as the participants at Oberwolfach meetings became increasingly international, approaches to the subject began to broaden considerably. By the 1990s, meetings were organized that reflected interests in the sociological and institutional dimensions of the history of mathematics, although the *Problemgeschichte* approach has never been entirely absent from meetings at Oberwolfach. But it was clear to regular participants that an historiographic change of considerable interest was underway in the discipline at large. It was this change that the International Commission on the History of Mathematics realized it was timely to investigate and evaluate.

As the result of a preliminary meeting at Oberwolfach in the early 1990s, during which several historiographic lectures were featured, it became clear that the history of the history of mathematics was not only one of impressive personalities, but that it had surprising cross-connections to other scientific and cultural currents. Subsequently, JOSEPH DAUBEN and CHRISTOPH SCRIBA agreed to accept editorial responsibility for a full-scale study of the history of history of mathematics, a project undertaken with the support of the International Commission on the History of Mathematics (ICHM), the International Union of History and Philosophy of Science, the International Mathematical Union, the Mathematisches Forschungsinstitut in Oberwolfach, and Birkhäuser Verlag of Basel, Switzerland.

Following the Oberwolfach meeting, the next step was a special ICHM symposium organized on the occasion of the XIXth International Congress for History of Science held in Zaragoza, Spain, in the summer of 1993. This was comprised of four sessions, where papers from a number of different countries were presented, and from which the current book began to take shape. Subsequently, with the help of more than forty collaborators, over the next few years chapters dealing with the history of history of mathematics in more than twenty-five different countries or regions were written, refereed, rewritten, and finally reconsidered in the context of the entire project by the team of scholars overseeing the project. The short biographies — three hundred in all — that comprise Part II of the book were also drafted and circulated for comment and revision.

Chronologically, the earliest historian of mathematics of whom record survives is the above-mentioned EUDEMUS who wrote of the origins of arithmetic in Phoenician commercial activity and of geometry in the land-surveying needs of the Egyptians. Modern history of mathematics, however, begins in the Renaissance, in Italy and Germany, with the first comprehensive history of mathematics that of MONTUCLA, written in French and first published in 1758. Historians of mathematics were soon active across Europe in the 19th century, and by the early 20th century, history of mathematics was also being pursued in Latin America, Mexico, the United States, Japan, and China. In fact, as the 20th century progressed, so too did interest in the subject.

Following World War II, the history of science matured as a discipline with its own journals, university positions, and institutes which helped to support a broad spectrum of activities on history of mathematics as well. Most contributions here bring the story down to at least the 1970s, with brief sketches of the most important subsequent developments completing the picture to the end of the century. The early 1970s serve as a convenient turning point of sorts for the history of mathematics in the last century because it represents the time when the International Commission on the History of Mathematics and the Commission's journal, *Historia Mathematica*, were founded, largely through the efforts of KENNETH O. MAY, to whom this book is dedicated.

Writing the History of Mathematics is also dedicated to the Mathematisches Forschungsinstitut in Oberwolfach, where the coordinating group of about a dozen historians of mathematics met on a number of occasions over the course of the last decade to guide this project to a successful conclusion. We are especially grateful to the two directors of the Institute during this period who were unhesitating in their support and cooperation, Prof. Drs. MARTIN BARNER and MATTHIAS KRECK.

Despite the effort and serious research that has gone into this project, as well as the many revisions and constant up-dating of material over the past decade, this book is still a first attempt. Readers will notice that the contributions remain to some extent inhomogeneous, perhaps to be expected in a project involving so many individuals and dealing with the historiography of a subject for which there is little precedent. Nevertheless, readers will find here an abundance of information about the individuals, publications, and historiographic emphases that in sum comprise various attempts to write the history of mathematics, especially over the past few centuries, but with certain notable precedents even earlier. We trust that our readers will be impressed by the richnesss of the story our colleagues have to tell, as were the authors and editors of these papers as the separate parts of this effort merged together, resulting in this truly international, cooperative account of *Writing the History of Mathematics*.

JOSEPH W. DAUBEN, JEANNE PEIFFER, CHRISTOPH J. SCRIBA,
and HANS WUSSING

Introduction

JOSEPH W. DAUBEN and CHRISTOPH J. SCRIBA

Writing the History of Mathematics is subdivided into three main parts, beginning with nineteen chapters on the different countries or regions covered in this book, and a brief postscriptum, followed by a second part devoted to biographies of historians of mathematics, concluding with a third part comprised of a bibliography of general works cited, and an index. Determining the order in which the countries should appear in Part I proved a special challenge, since neither an alphabetical nor a geographical sequence seemed satisfactory. Nor is there an appropriate chronological arrangement that provides a natural flow when reading from beginning to end, so that the final order of the chapters, country by country, has been in part a matter of arbitrary compromise between a number of possible solutions. After considerable discussion among members of the working group responsible for the final version of this book, it was decided to begin with France as the country in which JEAN ETIENNE MONTUCLA composed the first comprehensive history of mathematics. It then seemed natural to go on to the closely-related countries of Benelux, followed by Italy where GUGLIELMO LIBRI and BALDASSARRE BONCOMPAGNI were major contributors to the history of mathematics. The story then moves to Switzerland, and then in a natural way to the third major contributor to the subject at hand, Germany, where HERMANN HANKEL and MORITZ CANTOR were among the prominent early historians of mathematics. Then following a roughly chronological order with respect to their contributions in the 19th century, we come to Scandinavia, the British Isles, Russia and the USSR (and Russia again), Poland, the Bohemian countries, Austria, Greece, Spain, and Portugal. Attention then shifts to the New World, where South America, Mexico, the United States, and Canada are all treated in a single chapter, from which we proceed further East to consider Japan, China, India, and finally, the Arab Countries, Turkey, and Iran.

In making this journey from Europe to the Americas, the Far East, Asia, and then back to Europe, we have tried to be as consistent as possible in the spelling of names and places, and in the actual format followed throughout the book. In most cases, titles of books in foreign languages have been given English translations in parentheses when first mentioned, except where the meaning of the title is obvious, as in *Ars Magna* or *Histoire des mathématiques*. Names at first mention are accompanied by their relevant dates, i.e. JEAN ETIENNE MONTUCLA (1725–1799, **B**). The bold-faced **B** designates names for which additional biographical information has been provided in Part II of the book. This same convention is used in the index, where the bold-faced **B** indicates those names that also appear in Part II. For figures who are still alive, only dates of birth have been given, preceded by an asterisk *. Similarly, the dagger † preceeding a date notes the year of death. For some names it has not been possible to secure exact dates, and in such cases, approximate dates have been indicated wherever possible.

Throughout the convention of giving all dates according to the Gregorian calender has been followed. This pertains primarily to dates given for names and events in the chapter on Russia and the former Soviet Union, but is occasionally relevant elsewhere. Russian names are transliterated according to the convention adopted by *Mathematical Reviews*. Chinese and Japanese names are listed throughout the book using the now-usual East Asian format, where the family name is given first, followed by the given name(s). To avoid any confusion, these family names in the index are given entirely in upper case capital letters. Thus for QIAN BAOCONG and MIKAMI YOSHIO, QIAN and MIKAMI are the family names of these two historians of mathematics. In transliterating Chinese words and titles, we have adopted standard pinyin spelling.

Universities, archives, libraries, institutes, academies, and the like are indexed under the country in which the institutions in question reside. Institutions that have no permanent location, or events like International Congresses that move from place to place, are indexed according to their name or title rather than with respect to any city or country. There is also a separate listing for all journals mentioned in the text of Part I or II of the book, under "Journals & Series." Journal abbreviations are listed just before the bibliography.

The bibliography in Part III provides complete references for major primary and secondary works that appear in the main text of the book in Part I. Each of the individual biographies included in Part II provides additional bibliographic references usually not duplicated again in Part III. The primary sources listed include a given author's major historical works, but their publications devoted to mathematics or other fields are usually not mentioned unless they are of importance for the history of mathematics as well.

The references given for each of the biographies in Part II begin with articles to be found in the *Dictionary of Scientific Biography (DSB)*, KENNETH MAY's *Bibliography and Research Manual of the History of Mathematics (May)*, and J. C. POGGENDORFF's *Biographisch-literarisches Handwörterbuch zur Geschichte der exakten Naturwissenschaften (P)*. Then follow in chronological order further obituaries, biographical articles, and books. Where portraits of the individual historians of mathematics featured in Part II are known, this information has also been indicated.

Two collections have been especially useful in this regard: the *Table Générale des Tomes 1–35* of *Acta Mathematica*, RIESZ, MARCEL (Ed.), Berlin, Uppsala & Stockholm, Paris: Almqvist & Wiksell 1913 (with biographies and portraits of the authors 1882–1912), and *Das Fotoalbum für Weierstraß/A Photo Album for Weierstrass*, BÖLLING, REINHARD (Ed.), Braunschweig and Wiesbaden: Vieweg 1994.

The Working Group for *Writing the History of Mathematics*:

KIRSTI ANDERSEN (Aarhus, Denmark), ELENA AUSEJO (Zaragoza, Spain), UMBERTO BOTTAZZINI (Milan, Italy), JOSEPH W. DAUBEN (New York, NY, USA), SERGEI S. DEMIDOV (Moscow, Russia), MENSO FOLKERTS (Munich, Germany), HÉLÈNE GISPERT (Palaiseau, France), IVOR GRATTAN-GUINNESS (Bengeo, England), MARIANO HORMIGÓN (Zaragoza, Spain), ERWIN NEUENSCHWANDER (Zurich, Switzerland), KAREN H. PARSHALL (Charlottesville, VA, USA), JEANNE PEIFFER (Paris, France), CHRISTOPH J. SCRIBA (Hamburg, Germany), HANS WUSSING (Leipzig, Germany).

Acknowledgements

JOSEPH W. DAUBEN and CHRISTOPH J. SCRIBA

We are pleased to note our special appreciation here to a number of individuals and institutions, without whose help and continued support this project would not have been realized, certainly not in the form that it has achieved here. Among the institutions, most have already been noted, and we are especially grateful for grants in aid and other financial support provided by the International Commission on the History of Mathematics (ICHM), the International Union of History and Philosophy of Science (IUHPS), the International Mathematical Union (IMU), the Mathematisches Forschungsinstitut in Oberwolfach, the Volkswagenstiftung (Germany), the Hansische Universitätsstiftung, and Birkhäuser Verlag (Basel, Switzerland). The City University of New York (USA) and the University of Hamburg (Germany) provided on-going office and technical support. We are grateful to Academic Press and the Special Collections Department of the Tutt Library, Colorado College, Colorado Springs, Colorado, for permission to reproduce portraits of historians of mathematics that originally appeared in *Historia Mathematica* and the *Bulletin of the American Mathematical Society.*

Among the many individuals who contributed to this project, several deserve special mention here. As the book moved into its final phases, KAREN PARSHALL joined our working group, and was especially helpful in preparing translations of material in French and in providing on-going editorial consultation. She participated in the last meeting of the working group at Oberwolfach in 1998, and subsequently has continued to supply editorial consultation, reading every page of the final drafts and offering invaluable commentary and corrections throughout the various stages of production, from preliminary to final page proof. JEANNE PEIFFER and HANS WUSSING undertook the challenging task of marshaling the results of nearly a decade's worth of discussion about history of mathematics and historiography of the subject, and from their notes of the numerous meetings of our working group in Oberwolfach, produced a preliminary version of the *Praescriptum* to this book. Subsequently, the four of us joined forces to produce final versions of the *Praescriptum* as well as the *Postscriptum*, both of which appear over all four of our names. In Hamburg, ANKE JOBMANN and ELENA OPAYETS provided secretarial assistance and were responsible for much of the input and LaTeX formatting that the manuscript required, including the entries in Russian requiring the cyrillic alphabet as well as keying all names and subjects for the computer-generated index. Philip Beeley was also generous with his time in working with Christoph Scriba in proofreading English translations on a regular basis, especially as new text was written or substantial revisions were made in the process of preparing the final version of the manuscript for publication. Several colleagues helped to supply missing biographical information and undertook additional library and internet research in the final phase of our work, and for this we are pleased to thank in particular June Barrow-Green (Open University, Milton Keynes, UK), David Mascre (Paris, France), Carole Lee Sussman (New York, NY, USA), and XU Yibao (New York, NY, USA).

A special word of thanks goes to the Directors of the Mathematical Research Institute in Oberwolfach, already mentioned, Prof. Drs. Martin Barner and Matthias Kreck, who were always cooperative in scheduling meetings for our working group throughout the course of this venture; without their willingness to do so, this project would not have been possible. We also wish to acknowledge the support of our publisher, Birkhäuser Verlag (Basel, Switzerland), especially of Doris Woerner and Edgar Klementz, whose enthusiasm for our work and faith in our results were also a source of constant encouragement, and to Stevie Ammann and Regula Gysin Ammann for the final layout and careful handling of the proofs. It is thus especially gratifying that this book appears in Birkhäuser's series, *Science Networks.*

We also owe our thanks to all of the contributors to this volume, not only for the time and effort they all invested in writing the chapters that appear in Part I and the biographies that comprise Part II, but for agreeing to forego any royalties so that all proceeds from the production of this book might instead be contributed to the International Commission on the History of Mathematics, in support of its ongoing efforts to promote the history of mathematics.

Joseph W. Dauben and Christoph J. Scriba

Hamburg, April 2001

Glossary of Terms

Foreign Term	Translation or Corresponding Term in English
Abteilung	division, section (of an institute)
agrégation	competitive examination for posts on teaching in secondary schools
agrégé	(graduate) who has passed the *agrégation* examination
agrégé répétiteur	tutor, who has passed the agrégation; graduate assistant
Akademisches Gymnasium	classical, secondary school, upper level
alma mater	"fostering mother", the college or university from which one graduated
artes liberales examination	bachelor of arts graduation
Außerordentlicher Professor	associate professor
Außerplanmäßiger Professor	assistant professor; professor without paid position
baccalaureat	bachelor's degree
directeur d'études	director of studies
docent	lecturer, instructor
doctor honoris causa	honorary doctor
doctor medicinae	medical doctor
doyen	dean
Dozent	lecturer, instructor
école	school, institution
Extraordinariat = Außerordentliche Professur	associate professorship
Fachschule	institution at the post-*gymnasium* level, like schools for engineering, academies of music, etc.
Festschrift	publication celebrating an event or honoring a person
Gymnasium	classical, secondary school
Gymnasiallehrer	teacher at a classical, secondary school
Gymnasialprofessor	senior teacher at a classical, secondary school
Habilitation	qualification to obtain the *venia legendi*

Foreign Term	Translation or Corresponding Term in English
Habilitationsschrift	book, submitted to obtain the *venia legendi*
Hilfsarbeiter (in Bibliothek)	assistant librarian
Höhere Bürgerschule, Hogere Burgerschool	secondary school
Honorarprofessor	honorary professor
ibidem, ibid.	at the same place, in the same journal
in absentia	while absent
Inauguraldissertation	disseration for gaining a Dr. phil.
inter alia	among others
kand. fil.	candidate of philosophy
kandydat	candidate (for an academy position)
laurea	university degree
Lehramtsprüfung	teaching examination
Lehrauftrag	teaching assignment (for a specific course)
Leibarzt	physician in ordinary (to the king, etc.)
Leitartikel	leading article in a journal
libera docenza, venia legendi	qualification for teaching at a university
libero docente	university lecturer
licence	university degree after three years of study
licencié	holder of a diploma for teaching
licencié ès sciences	holder of a diploma for teaching science
liceo, lycée	classical, secondary school
literati	men of letters
madrasa	institution of higher learning for Islamic sciences
maître de conférences	non-professorial university lecturer
Nachlass	scientific estate
Naturforschende Gesellschaft	natural science asssociation
Oberlehrer	teacher at a secondary school
Oberschulrat	school inspector
ordentlicher Professor	full professor
Ordinariat = Ordentliche Professur	full professorship
polytechnicien	graduate of the Ecole Polytechnique
Polyteknisk Læreanstalt	polytechnical university
précis	sketch, outline
Privatdozent, private docent, privaatdocent	university lecturer without a salaried position
Professor Emeritus	emeritus professor, retired professor
professorat extraordinaire	associate professorship
Rechenmeister	reckoning master, private teacher of mathematics
répétiteur	examiner
savant	scholar
Schullehrer	schoolteacher
Schulprogramm	school program
secrétaire perpétuel	perpetual secretary

Foreign Term	Translation or Corresponding Term in English
Studienreise	research trip
Technische Hochschule, Technische Hogeschool	technical university
Textkritik	textual criticism
Töchterschule	girls school
venia legendi	certification for teaching at a university
Vielvölkerstaat	state of many nationalities, multi-national state
Zulassungsarbeit	thesis for certification for school-teaching

Part I

Countries

Chapter 1

France

JEANNE PEIFFER

1.1 Introduction

The first comprehensive history of mathematics written in the West was published in 18th-century France, namely MONTUCLA's *Histoire des mathématiques.* A lasting monument of the Enlightenment, it is related to the larger philosophical project of the Encyclopedists devoted to charting the progress of the human mind and to describing its history.

Since the 18th century, the history of mathematics has been written in diverse ways in France, often entangled with projects much broader in scope. These have involved, among other things, the universal history of the arts and sciences, questions of the origins of the sciences, comparative cultural history, oriental studies, positivism, the foundations of mathematics, philosophical theories of knowledge, and the history of philosophy. This entanglement complicates analysis of the history of mathematics in France and makes the subject difficult to isolate. Thus, in order to describe the history of mathematics as it emerges, develops, and professionalizes in France, it is necessary to place it in much larger contexts.

This chapter provides a first and rough outline of the history of the historiography of mathematics in France. Where factual material and thorough studies are lacking, the picture painted here needs dramatic refinement, enhancement, and correction. Nevertheless, the image given of the development of the history of mathematics in France points to major tendencies, opens paths for further research, describes different reasons for writing histories, and discusses the various historiographic practices of, for example, humanists, encyclopedists, historiographers of the major institutions, orientalists, mathematicians, philosophers, and historians of science.

1.2 Changing Appreciation of the Ancients, from the Renaissance to the Seventeenth Century

With the revival of antiquity in the Renaissance — the recovery of ancient authors and of the world in which they lived, worked, and wrote — a sense of history slowly developed. Humanists began to perceive something of the strangeness of the old authors as advances of the Scientific Revolution set them at a critical distance.

Renaissance editions and translations of mathematical classics from Greek antiquity offered the opportunity, in France as elsewhere, to revive, in the margins or in the prologue, the often mythical origins of the mathematical sciences as well as certain aspects of their past. Editors, philologists, commentators, and translators all engaged in mathematical activities. Familiarity with ancient mathematical texts was an assumed part of the mathematical humanist's general knowledge. For example, JACQUES LEFÈVRE D'ETAPLES (ca. 1455–1536), founder of the Parisian humanist school, assembled in a single, parallel edition (Paris 1516) the Latin version from the Arabic by CAMPANUS of EUCLID's *Elements* (which formed the basis of RATDOLT's 1482 edition) and ZAMBERTI's first Latin version from the Greek (Venice 1505). The text of the Euclidean *Elements* thus became an object of critical inquiry and historical study.

The French algebraists of the 16th century included in their treatises introductory chapters in which they cited as many ancient sources as possible. Although they inherited algebraic knowledge from the Arabs, they tried to hide that origin by attributing algebra to DIOPHANTUS (or even to EUCLID, like BORREL did). According to [CIFOLETTI 1996], they succeeded, while working on the *Arithmetica* of DIOPHANTUS (available in manuscript form and in the Basel edition of 1575) in transforming him from a mythical author into an object of critical analysis. Thus, JACQUES PELETIER (1517–1582) treated DIOPHANTUS in his introduction to *L'Algèbre*; JEAN BORREL or BUTEO (1492–1572) attempted to Latinize the term "algebra" by replacing it with the word "quadratura"; and GUILLAUME GOSSELIN (fl. 1577–1583) labored on an edition of the six then-known books of the *Arithmetica*, that never appeared.

The Arabist, GUILLAUME POSTEL (1510–1581), professor of mathematics at the *Collège Royal* in Paris, a research and teaching institution founded in 1530 by FRANÇOIS I, brought Arabic manuscripts, including the Euclidean *Elements* (in the version of NAṢĪR AL-DĪN AL-ṬŪSĪ (†1274)), back from the Levant. Nevertheless, when his contemporary, PIERRE DE LA RAMÉE (1515–1572) (or PETRUS RAMUS in the Latinized form), wrote his history of mathematics as the first book of his *Scholae mathematicae* (1569), he based it almost exclusively on Greek sources. His periodization follows the succession of the three great civilizations of the Chaldeans, Egyptians, and Greeks. RAMUS considered the last period, from THALES to THEON, the richest by far in mathematical discoveries.

Until the end of the 17th century, mathematicians maintained a close familiarity with their predecessors, looking for inspiration, searching for connections, and establishing parallels. Sometimes they even tried to restore lost ancient texts. Examples of such restorations include FRANÇOIS VIÈTE's *Apollonius gallus* (1600), CLAUDE-GASPARD BACHET DE MÉZIRIAC's Latin translation of the *Arithmetica* of DIOPHANTUS, and PIERRE DE FERMAT's restoration of APOLLONIUS' *Plane loci.*[1] The latter was based upon information contained in PAPPUS' *Mathematical Collection*, and eventually led to the invention of analytic geometry. In short, historical practice was part of scientific practice. The antiquity of mathematics conferred prestige, offered a legitimating context, or validated mathematicians' discoveries. Increasingly, this last motivation prevailed with the emphasis being on the progress of mathematics. The ancients were no longer insurpassable. The writings of illustrious modern mathematicians were considered worthy of being preserved and collected in prestigious volumes. Thus, FRANS VAN SCHOOTEN (ca. 1615–1660, **B**) published the works of FRANÇOIS VIÈTE (Leiden 1646); ADRIEN BAILLET (1649–1706) devoted a biography to RENÉ DESCARTES (1691, abridged in 1693); and CLAUDE CLERSELIER (1614–1686) edited DESCARTES's correspondence; among many other examples.

The progress of the mathematical sciences was documented in the second edition of CLAUDE FRANÇOIS MILLIET DESCHALES's *Cursus seu mundus mathematicus* (Course or World of Mathematics) (1690), which included a *Tractatus proemialis de progressu matheseos, et illustribus mathematicis* (Introductory Treatise on the Progress of Mathematics, and on Illustrious Mathematicians). MELCHISEDECH THÉVENOT (1620 or 1621–1692) began the publication of a volume of ancient mathematical texts entitled *Veterum mathematicorum ... opera* (Works of Ancient Mathematicians), which included, among others, works by HERO; this was completed by JEAN BOIVIN and PHILIPPE DE LA HIRE (1693).

Two events helped establish historical research as a scientific discipline toward the end of the 17th century. First, the quarrel between the so-called "ancients and moderns," started in the realm of literature, soon included the arts and sciences. At stake was how to write history, how to read and understand past authors. Second, the first principles of historical criticism were established in 1681 by the Benedictine JEAN MABILLON (1632–1707) in his *De re diplomatica* (On Diplomatic Matters), which MARC BLOCH termed "a big event indeed in the history of the human mind."[2] The flourishing of erudite historiographic studies contributed to a reevaluation of ancient scientific traditions, from which the history of mathematics benefitted.

[1]In a recent article, "Das Leben Fermats," KLAUS BARNER reported on his archival studies in France (*DMV Mitteilungen*, no. 3 (2001), 12–26): the year of birth 1601, which is traditionally quoted, is that of a half-brother PIERRE who died young; the mathematician PIERRE FERMAT was born between January 13, 1607, and January 12, 1608, but most likely in 1607.

[2]Quoted by BERNARD GROSPERRIN in *Dictionnaire du Grand Siècle*, ed. by FRANÇOIS BLUCHE, Paris: Fayard, 1990, 721.

1.3 History of the Progress of the Human Mind (in the Enlightenment)

This might well be the characteristic epithet of the historiographic project that took form and was realized throughout the century of the Enlightenment from FONTENELLE to CONDORCET, and that included the monumental *Encyclopédie* and the *Histoire des mathématiques* of MONTUCLA. PIERRE RÉMOND DE MONTMORT (1678–1719), an aristocrat who entered the republic of letters as author of the *Essai sur les jeux de hazard* (Essay on Games of Chance) (1708), stated the case in a letter to NICOLAS BERNOULLI on August 20, 1713:

> It would be desirable if someone wanted to take the trouble to instruct us how and in what order the discoveries in mathematics have followed themselves, one after another; and to whom we should be obliged for them. The history of painting, of music, of medicine have been written. A good history of mathematics, especially of geometry, would be a much more interesting and useful work. What a pleasure it would be to see the union, the connection between methods, the linkage of different theories, beginning from the earliest times up to our own, where this science has been brought to such a high degree of perfection. It seems to me that such a work, if done well, could be regarded to some extent as a history of the human mind since it is in this science [mathematics], more than in anything else, that man makes known the excellence of that gift of intelligence that God has given him to rise above all other creatures [MONTMORT 1713, 399].

Having explicitly recognized the superiority of modern science over that of the ancients, the historian of the human mind cannot be content with simply narrating things as they have occurred. Rather, it is necessary to describe the progress that has been made in understanding the natural world. The progress of the mathematical sciences reflects, more than that of the other sciences, the positive evolution of human intelligence and of humanity in the course of time. Thus, according to A. R. J. TURGOT (1727–1781), in his *Dissertation sur le progrès de l'esprit humain* (Dissertation on the Progress of the Human Mind) (1750), "only in mathematics was progress sure from the first steps," while all other areas of inquiry presented a "history of errors" [LAUDAN 1993, 5]. This is why the necessary succession of discoveries may also be seen as a genealogy of ideas, what MONTMORT termed "the union, the connection between methods, the linkage of different theories." It is this notion of progress that may serve as a framework for historiography in the 18th century.

1.3.1 Fontenelle, the Initiator of a Tradition

BERNARD LE BOVIER DE FONTENELLE (1657–1757) was the brilliant initiator and theoretician of this type of history.[3] For him, the method of writing history "resembles the way we make a system of philosophy. The philosopher has in front of him a certain number of effects of nature and of experiments. He must divine likely causes and, from what he sees and divines, he composes a well integrated whole: there is the system. The historian also has a certain number of facts of which he imagines the motive causes, and upon which he builds the best system of history he can, yet more uncertain and more subject to caution than a system of philosophy" [CROMBIE 1994, vol. 3, 1594]. The *Histoire de l'Académie royale des sciences — avec les mémoires de mathématiques et de physique* (History of the Royal Academy of Sciences with Mathematical and Physical Memoirs), of which FONTENELLE published 42 volumes between 1699 and 1740, exemplifies this approach to history.

Indeed, in 1697, FONTENELLE was named to replace JEAN-BAPTISTE DU HAMEL (1623–1706) as secretary of the *Académie Royale des Sciences.* As secretary, DU HAMEL had authored a history of the institution, *Regiae scientiarum academiae historia* (History of the Royal Academy of Sciences) (1696, 2nd enlarged edition, 1701), which drew on the verbal proceedings of its meetings as its principal source. When the Academy was reorganized in 1699, articles 40 to 42 specified the responsibilities of the secretary. These included giving "an extract of its register to the public," or a reasoned history ("histoire raisonnée") of the Academy's most remarkable achievements.[4]

Thus, while the memoirs of the academicians enumerated and described the particular circumstances related to their results, FONTENELLE, in the history part, endeavored to place them in an historical context, to establish the links between particular propositions, and to make clear their connections. Relative to mathematics, FONTENELLE was able to present technically complicated concepts and sophisticated procedures to a mixed public of scholars and the intellectually curious. He acted as an historian of mathematics in the preface to his *Eléments de géométrie de l'infini* (Elements of the Geometry of the Infinite) (1727). As a Cartesian, he took special care to put DESCARTES's genius and boldness in the proper light; DESCARTES had rebelled against the ancients using their own intellectual devices.

As secretary of the Academy, FONTENELLE wrote 69 éloges for academicians that remain, within certain limits, a good source of biographical information since they are based on authentic documents provided by the families concerned (as may be assumed from remarks in the Bernoulli correspondence). From these documents, FONTENELLE painted in an allegedly neutral style in fact idealized portraits of the savants. These éloges constitute what CHARLES B. PAUL has called an "unstable

[3] FONTENELLE included his reflections on history in his *Histoire des oracles* (1686), in *De l'origine des fables* (1724), and in his *Digression sur les anciens et les modernes* (1688).

[4] Article 41 of the Regulation of 1699, [COSTABEL 1988, 156–157].

combination of biographical veracity and hagiographic morality" [PAUL 1980, 7]. Note that an 18th-century reader like D'ALEMBERT appreciated above all the informative contents of FONTENELLE's éloges, describing them in the article "éloge" for the *Encyclopédie* as "historic" in style, whereas those of the French Academy were "merely oratorical" [*Encyclopédie*, art. "éloge"].

FONTENELLE accomplished this task with elegance and left his mark not only through his style but also through his philosophical conceptions about science and its history. To a certain degree, he served as a model for his 18th-century successors in the secretary's chair — JEAN-JACQUES DORTOUS DE MAIRAN (1741–1743), JEAN-PAUL GRANDJEAN DE FOUCHY (1744–1776), and JEAN ANTOINE CARITAT DE CONDORCET (1773–1791). PAUL sees in the collection of éloges written during the *Ancien Régime* (more than 200) "a collective biography of a new breed of savants" [PAUL 1980, cover].

These academic éloges were the principal sources for the *Histoire des philosophes modernes* (1760, 8 vols.), compiled by marine engineer ALEXANDRE SAVÉRIEN (born in Arles in 1720 or 1723). Even if the book was little more than a collection of somewhat anecdotal biographies, the underlying plan was significant and original. SAVÉRIEN grouped the biographies by class, with the mathematicians subdivided into geometers, algebraists, astronomers, opticians, and mechanicians. Each volume was prefaced with a "preliminary discourse," in which an overview of the history of the discipline in question was given. The sum of all these individual histories was meant to comprise a general history. SAVÉRIEN thus inaugurated a style, combining historical narrative with biography, that lasted until late in the 19th century. The *Histoire* went through three editions between 1760 and 1773, which reflects a certain success. In 1753, SAVÉRIEN had already published a *Dictionnaire universel de physique et de mathématiques* (Universal Dictionary of Physics and Mathematics), in which he had included short histories of geometry, the infinitesimal calculus, and algebra (for which he used WALLIS as a source). That same year, he had also produced a brief *Histoire critique du calcul des infiniment petits* (Critical History of the Calculus of Infinitesimals), in which English mathematicians assumed an important place. In 1766, he completed an *Histoire des progrès de l'esprit humain dans les sciences exactes* (History of the Progress of the Human Mind in the Exact Sciences), which amplified the historical notes given in the *Dictionnaire.* SAVÉRIEN saw in the "method of following the Sciences historically, ... one of the simplest means and the most certain to make them palatable to young people, and to the informed public" [BESSMERTNY 1934, 376]. His first aim seems thus to have been to popularize ancient and modern theories without distinction.

1.3.2 The Historical Dimension of the Encyclopedia Project

JEAN LE ROND D'ALEMBERT (1717–1783), in the *Discours préliminaire de l'Encyclopédie* (1751), considered the history of science to have no other aim than to determine the phases of the progress of knowledge. Even though he agreed with

Voltaire, Turgot, and others, that nature is static, he nevertheless accepted the idea of the historical dimension of human affairs (of knowledge and of worldly events), and reserved a not insubstantial place (in its ideal library) to the "study of history, which binds us to past centuries through the spectacle of their virtues and vices, of their knowledge and of their errors, and which transmits ours to future centuries" [d'Alembert 1751, p. x]. The synoptic unfolding that D'Alembert attributed to the study of history in the *Discours préliminaire* posited a network of connections and dependencies produced more often by the demands of rational reconstruction than by proven connections. The order of succession corresponded not so much to a true historical chronology of events as to the demands of distribution over time. This was evidenced by the *Discours*'s narrative form.

Condorcet's *Esquisse d'un tableau historique des progrès de l'esprit humain* (Sketch of a Historical Catalogue of the Progress of the Human Mind) (1794) represented the ultimate success of the history of the progress of human thought. This philosophical history, integrating the history of science centrally into cultural history, developed in an intellectual climate that favored comparative history. Western Europe was regarded as part of a wider world that had to be explored, described, and tamed. Through the writings of Voltaire, Montesquieu, and others, the philosophers of the Enlightenment were interested, despite the poverty of sources, in diverse aspects of Chinese, Indian, and Islamic cultures, including their arts and sciences (see Section 1.6 below).

1.3.3 Montucla's Monumental Work

Montucla's *Histoire des mathématiques, dans laquelle on rend compte de leurs progrès depuis l'origine jusqu'à nos jours* (History of Mathematics That Takes into Account Its Progress from Its Origins to Our Times) [Montucla 1758] mirrored the project of the Encyclopedists and approached, in its conceptions, ideas d'Alembert expressed in the *Discours préliminaire.* Born and educated in Lyon, Jean-Etienne Montucla (1725–1799, **B**) went to Toulouse to study law, then to Paris, where he was part of the literary circle around Jombert (and the Encyclopedists) and engaged in historical research.

The reception of the *Histoire*'s first edition was very positive, and this success encouraged Montucla to contemplate a third volume about the 18th century. Having, however, to earn his living in the Royal administration, he was unable to undertake this project before friends persuaded him to prepare a second enlarged edition of the *Histoire des mathématiques.* The first two volumes, in the revised and substantially expanded version, appeared in August of 1799. Montucla died four months later, before completing the proposed third volume devoted to the history of mathematics during the Enlightenment. Jérôme de Lalande (1732–1807, **B**), along with Charles Bossut (1730–1814, **B**) (who appears to have been educated by the same Jesuits at the same college in Lyon that Montucla attended [Sarton 1936c, 519]), completed the third volume and added a fourth, for which the help of other specialists was enlisted. Thus, Sylvestre François

LACROIX (1765–1843, **B**) revised the chapter on the integration of partial differential equations [SARTON 1936c, 552]. These two volumes (1528 pp.) eventually appeared in May of 1802.

MONTUCLA, in his preface, referred to FRANCIS BACON[5] who seems to have convinced him, like FONTENELLE and the Encyclopedists, of the importance of directing "a philosophical eye" toward the "spectacle" of the "development of the human mind and the different branches of knowledge" [MONTUCLA 1758, v]. His conception of the history of science included, as for MONTMORT, the attribution of results to their authors, the availability of sources, "the connection between methods," and the "linkage of different theories." The *Histoire des mathématiques*, as KURT VOGEL has said, is "the first attempt at a history of ideas and problems in mathematics" [VOGEL 1965, 183]. MONTUCLA was completely aware of the pioneering character of his work. He evaluated his predecessors and did not hesitate to make critical — sometimes severe — judgments. He took mathematics as the object of his history because it inspired respect in a philosophical mind for essentially two reasons: first, of all the sciences, mathematics is the most certain in the search for truth; second, its development is continuous, its progress constant. "His is very much a philosophic history — the progress of the human mind — but grounded upon a thorough technical knowledge of the original sources and the learned editions and studies of the érudits" [SWERDLOW 1993c, 303].

Encyclopedic in nature, the work included so-called "mixed" mathematics (mechanics, astronomy, optics, and music), qualified as such because of its "alliance with physics"; this comprised two-thirds of the whole volume. For the most part, the book followed a chronological order, but handled the history of some particular problems separately, such as the duplication of the cube. The mathematics of the Greeks, Arabs, and medieval Latin West are studied separately. Lacking the linguistic competence to study Arabic mathematics, MONTUCLA depended upon later Latin translations [GRATTAN-GUINNESS 1989, 13].

MONTUCLA attached great importance to the 17th century, devoting the entire second volume of the first edition to it. The mathematical discoveries of FERMAT, DESCARTES, HUYGENS, NEWTON, LEIBNIZ, the BERNOULLIS, and a

[5]The first paragraph of MONTUCLA's preface read: "One of the most worthy spectacles to interest a philosophical eye is without doubt that of the development of the human mind and the different branches of its knowledge. The famous Chancellor Bacon noted this more than a century ago, and this was why he compared history, such as it had been written until then, as a mutilated trunk separated from its most noble parts. I do not know, by what misfortune of fate, this branch of history [i. e. the history of mathematics] has always been the most neglected. Our libraries are overrun with prolix narrators of sieges, battles, revolutions; how many lives of supposed heroes are illustrated only by the amount of blood they have left in their path? It is almost impossible to find, as Pliny remarks with regret, any writers who have undertaken to preserve for posterity the names of these benefactors of humanity, some of whom have worked to relieve its burdens through useful inventions, while others have expanded the capacities of the intellect through their thought and research. Even less does one find someone who has thought about presenting the progress of these inventions, or of following the march and development of the human mind. Would such a picture be less interesting than that of the horrors and bloody scenes that engender the ambition and nastiness of mankind?"

multitude of other minor figures are described in minute detail. It is interesting to compare MONTUCLA's evaluation (in the second edition) of the priority dispute between LEIBNIZ and NEWTON over the invention of the infinitesimal calculus to the version BUFFON gave in his preface to the French translation of NEWTON's *La méthode des fluxions et des suites infinies* (The Method of Fluxions and Infinite Series) — which is clearly the source for MONTUCLA's chapter [MONTUCLA 1799/1802, vol. **3**, 103–109]. It followed the same structural and organizational format, cited events in the same order, repeated the linkages, but kept its distance from the evaluations made by BUFFON.

In fact, BUFFON's story followed that fashioned by the commissioners of the Royal Society of London from pieces of the *Commercium Epistolicum* (London 1712) (cf. Section 7.4). BUFFON sided with NEWTON and condemned LEIBNIZ, whom he regarded not as a second inventor but as a common plagiarist. The *Commercium*, in its false neutrality, claimed to make a certain number of censored documents available to the public, but, in fact, it imposed a one-sided reading that MONTUCLA denounced. In MONTUCLA's measured conclusion, "There can be no doubt that Newton was the first inventor of the calculus. The proofs are as clear as day. But was Leibniz guilty of publishing as his own a discovery that he had found in the writings of Newton himself? It is this we do not believe" [MONTUCLA 1799/1802, vol. **3**, 109]. The *Commercium Epistolicum* served as a source for both MONTUCLA and BUFFON, but whereas BUFFON accepted the reading it suggested, MONTUCLA maintained a critical distance.

1.4 Historiography in Revolutionary Times

The French Revolution led to a suppression or reorganization of all universities and academic institutions, but new ones were set in place by the mid 1790s.[6] The *Ecole Polytechnique* (1794), a new institution, rapidly became of central importance, while other engineering schools were remodeled. In 1808, a so-called "University" structure was set up by NAPOLÉON for the Empire; dealing mainly with school teaching, it included a small amount of science education in a few faculties. These arrangements led to a substantial number of publications in mathematics (and, indeed, in all sciences), at all levels from textbooks to treatises. Among this mass was a scattered but interesting collection of historical writings.

Many of the new figures in French mathematics were students at the *Ecole Polytechnique* and often, later, teachers or examiners there. The school thus became an institution of historical importance in its own right. This was recognized in 1828 by its librarian, AMBROISE FOURCY, who produced a detailed history of the school [FOURCY 1828]. He not only provided much institutional history but also accounts of some of the teaching (which did not include any courses in the history of mathematics) and a list of all students.

[6] The first six paragraphs of this section are due to IVOR GRATTAN-GUINNESS.

An important source for the history of mechanics predates the Revolution: JOSEPH LOUIS LAGRANGE's (1736–1813) *Méchanique analitique*, written during the period he spent in Berlin, but published in Paris in 1788, the year after he moved there. It presented mechanics in his own algebraic and variational style, supplemented with short paragraphs on appropriately chosen savants of the past such as GALILEO and COPERNICUS. While modest from a historical point of view, these passages had considerable (in fact, excessive) influence because of the renown of the book in which they appeared. The second edition (1811–1815 and re-editions) reinforced their influence during the 19th century. LAGRANGE's other textbooks and treatises also contained historical remarks.

LAGRANGE treated (his) basic principles and mostly terrestrial mechanics. PIERRE SIMON DE LAPLACE (1749–1827) brought a similar authority to celestial mechanics in his *Traité de mécanique céleste* (Treatise on Celestial Mechanics) (1799–1805), of which Book 11 in volume 5 (1823) was a short historical survey. His more popular prosodic account in *Exposition du système du monde* (Exposition on the System of the World) (editions from 1796) contained an historical *précis* in Book 5, and similar remarks in other Books. His *Essai philosophique sur les probabilités* (Philosophical Essay on Probability) (editions from 1814) performed the same function for the history of probability. As with LAGRANGE, LAPLACE's approach was rather Whiggish, his influence considerable.

The most important writing in the history of mechanics came later and from a much less renowned figure. The young ALFRED GAUTIER wrote two doctoral theses in the Paris Faculty of Sciences and expanded them both into a book, *Essai historique sur le problème des trois corps* (Historical Essay on the Three-Body-Problem) (1817). It is still one of the best accounts of its subject, especially for perturbation theory and the stability of the planetary system. After its publication, GAUTIER returned to his native Switzerland, where he wrote regularly (though not usually in an historical vein) for the Genevan journals *Bibliothèque britannique* and *Bibliothèque universelle*.

Of the major senior figures in French science, the textbook writer, SYLVESTRE FRANÇOIS LACROIX, took history most seriously. Although not an historian, LACROIX included much historical information in his textbooks and treatises. The most important is his three-volume *Traité du calcul différentiel et du calcul intégral* (Treatise on Differential Calculus and on Integral Calculus) (first edition 1797–1800, second edition 1810–1819); their tables of contents contained numerous references, and these books remain the principal source for the historian of late-18th-century analysis. Among LACROIX's other books, the *Essais sur l'enseignement* (Essays on Teaching) (editions from 1805–1838) includes an historical introduction on 18th-century mathematics in its cultural setting, and the first part a more detailed survey of mathematics education, including the first reforms after the Revolution.

The focus on ancient texts became strong about 1800 [DHOMBRES 1989, 549]. Thus LOUIS FRANÇOIS ANTOINE ARBOGAST (1759–1803), an Alsatian professor of mathematics who adopted revolutionary ideas and was very active in the reorganization of public education, gathered an important collection of manuscripts which he copied by hand from the originals of FERMAT, DESCARTES, JEAN BERNOULLI, PIERRE VARIGNON, the MARQUIS DE L'HÔPITAL, and others.[7] ARBOGAST also classified the papers left by MARIN MERSENNE. PTOLEMY was translated into French by the Abbot NICOLAS HALMA (1755–1828). EUCLID and ARCHIMEDES were made available in a new translation by FRANÇOIS PEYRARD (1760–1822), librarian at the *Ecole Polytechnique.* PEYRARD's edition of EUCLID's *Elements* [PEYRARD 1814/18] in itself is remarkable for providing the first Greek Euclidean text that is relatively independent of the version due to THEON OF ALEXANDRIA; previously all the manuscripts known in PEYRARD's day had been derived from that source. In fact, in 1808 PEYRARD had the singular opportunity of discovering and identifying, among manuscripts NAPOLÉON had had transferred from Rome to Paris, a Vatican codex (today *Codex Vaticanus Gr. 190*) that was a copy of an edition earlier than THEON's. This edition, based on the *editio princeps* of the Renaissance, took into account all the information provided by the Vatican codex.

At almost the same time that the second edition of MONTUCLA's *Histoire* appeared, the Abbot CHARLES BOSSUT published an *Essai sur l'histoire générale des mathématiques* (Essay on the General History of Mathematics) (1802), developing the outline he included at the beginning of the "dictionary of mathematics" in the *Encyclopédie méthodique* of 1784. His plan, according to the preface, was "to sketch a historical description of mathematics from the beginning to the present, and at the same time to honor the memory of great men who have extended its empire." He chose to be less technical than MONTUCLA, who had been criticized for being "understood only by professional mathematicians" [preface to his *Essai*]. Of the four periods he considered, the last, beginning with the discovery of infinitesimal analysis, occupies the major part of the second volume and is by far the most extensive and detailed. In particular, it followed the disputes of the two BERNOULLI brothers and the quarrel over the priority of the infinitesimal calculus. Astonishingly enough, within two years of publication, BOSSUT's book was available in English by JOHN BONNYCASTLE, in German by N. T. REIMER, and in Italian by ANDREA MOZZONI (1754–1842). BOSSUT (1730–1814) is also known for his edition of the works of PASCAL (1779). His younger colleague LOUIS BENJAMIN FRANCŒUR (1773–1849) wrote a general history of mathematics, but unfortunately it did not appear before his death in 1849 and now seems to be lost.

1.5 The Reform of Society Through the Sciences: Positivism

Was the history of mathematics written any differently, following the disruptions that the French Revolution of 1789 caused primarily in the institutions of math-

[7]Somehow LIBRI managed to secure ARBOGAST's collection, which has since been scattered. Part of it was deposited at the *Bibliothèque Nationale*, part in the *Laurenziana Library*, Florence.

ematics? The history of progress of the human mind as it was understood by Condorcet merged, in fact, into the positivist philosophy of Auguste Comte (1789–1857). Facing an unstable political situation, many wanted to reconstruct society on a rational basis. Comte, a graduate of the *Ecole Polytechnique*, proposed the reorganization of society depending on scientists, not only on narrow specialists, but on those who had a global view of their science and, indeed, whose knowledge was encyclopedic. In his *Cours de philosophie positive* (1830–1842, in six volumes), Comte introduced his famous law of the three stages, according to which the human mind progresses from theological (or fictive) through a metaphysical (or abstract) to a final, positive stage. The sciences, whose object is the most complex, were the last to attain the positive stage. Thus, mathematics was the most advanced of the sciences along the positivist route, followed by astronomy, then physics, chemistry, and physiology, to which Comte attributed a special value by virtue of their experimental character. Ultimately, Comte's "social physics" — or sociology — was regarded as the final outcome of the positive philosophy. An idea developed later by Tannery already appeared in Comte, namely, in order to understand correctly the whole of mathematics, a general knowledge of the other positivist sciences was required. According to Annie Petit, Comte's work underscored the need to approach the sciences from philosophical and pedagogical points of view. This undoubtedly explains the influence positivism had in France throughout the 19th century. Of course, positivism evolved and split into numerous warring factions; it also raised criticisms particularly at the turn of the 20th century. Still the high status it accorded to the sciences as an essential part of civilization and of collective intellectual development remained.

Toward the end of the century, Maximilien Marie (1819–1891, **B**), a tutor of mechanics and an examiner at the *Ecole Polytechnique*, was urged by Comte in a letter of April 15, 1841 "to study carefully the original works" as well as the creative ones that "teach spontaneously the historical links between the principal mathematical ideas" [Petit 1994, 68–69]. Marie went on to publish an *Histoire des sciences mathématiques et physiques* (1883–1887) in ten volumes, divided into 14 periods. Each of these periods opened with a brief introduction that traced the connections between ideas and methods. This was followed by the biographies of scientists of the period and by an analysis of their works.

1.6 Oriental Studies and the History of Mathematics in the Nineteenth Century

Under the pressure of colonial visions of the empire France was building in North Africa, a lively interest was awakened in the sciences of the East, notably of Islam. This gave rise, through interaction with mathematicians, to a series of high-level works.

The *Bibliothèque orientale* of Barthélémi d'Herbelot (1697, reprinted Maastricht 1776), according to [Charette 1995, 29], "signaled ... the birth of modern orientalism."[8] This first "Encyclopedia of Islam" contained succinct notices on the mathematical sciences, while the correspondence between Jesuit priests in China and India with the Royal Academy constituted a valuable source for the history of Chinese and Indian science. It was Indian astronomy above all, with its useful tables, that fascinated academicians like JEAN-DOMINIQUE CASSINI (1625–1712) [MARS 1666/99, vol. 8, 213–299], and JOSEPH-NICOLAS DELISLE (1688–1768), who put together a collection of manuscripts (deposited in the Naval archives). GUILLAUME LE GENTIL (1725–1792), following CASSINI, analyzed the methods of Hindu astronomical calculation, which he published as an appendix to his *Voyage dans les mers de l'Inde* (Travel in the Indian Oceans) (1779–1781, vol. **1**, p. 206–352). Thus, the first exact knowledge informed Europeans had about India concerned the traditions of mathematical astronomy. The history of the mathematical sciences and orientalism were intertwined from the beginning. The *Histoire de l'astronomie* (1775–1782) of JEAN-SYLVAIN BAILLY (1736–1793) drew upon the works cited above by CASSINI, LE GENTIL, and DELISLE, which were all characterized by their linguistic incompetence. BAILLY speculated on a possible oriental (Indian) origin of the sciences, which led to a lively controversy with VOLTAIRE. BAILLY ascribed to the Indians scientific results comparable to those obtained in the modern West, but made in time immemorial. The modern part of his *Histoire* was soon superceded by the *Histoire de l'astronomie* (1817–1827) of JEAN-BAPTISTE DELAMBRE (1749–1822, **B**).

1.6.1 Collaboration Between Scientists and Orientalists

During the *Ancien Régime*, astronomers like JÉRÔME DE LALANDE and orientalists continued to collaborate. Such contacts were also developed during the Napoleonic period, after NAPOLÉON decided to undertake the Egyptian expedition (1798–1801). He assigned such high caliber scientists as GASPARD MONGE (1746–1818), CLAUDE-LOUIS BERTHOLLET (1748–1822), and JOSEPH FOURIER (1768–1830), to work in the Egyptian Institute he founded in Cairo (on August 22, 1798) in order to propagate Western knowledge in Egypt, to study the past and present of that country, and to train experts to be at the disposal of the government.

At the Academy of Sciences, a commission composed of the astronomer DELAMBRE and the mathematician LAPLACE collaborated with JEAN-JACQUES CAUSSIN DE PERCEVAL (1759–1835), who held the chair of Arabic at the *Collège de France*, on the translation of a work by the 10th-century Arabic astronomer IBN-YUNUS. This translation appeared in the *Notices et extraits des manuscrits*

[8] The memoir by FRANÇOIS CHARETTE, "Orientalisme et histoire des sciences: l'historiographie européenne des sciences islamiques et hindoues, 1784–1900" (Orientalism and History of Science: The European Historiography of Islamic and Hindu Sciences, 1784–1900), presented to the Faculty of Advanced Studies at the University of Montreal in May, 1995, of which I only learned after completing my own research, confirms the conclusions drawn here. The memoir has been a great help in composing the final version of this section, and I thank SONJA BRENTJES for having made this work available to me.

de la Bibliothèque royale (Notices and Extracts from Manuscripts in the Royal Library) (1804).

The *Bureau of Longitudes* realized the benefits that astronomers could derive from knowledge of the mathematical and astronomical sciences of the Arabs and Persians. An assistantship for the history of oriental astronomy[9] was created in 1814 and occupied by JEAN-JACQUES SÉDILLOT (1777–1832, **B**), graduate of the *Ecole Polytechnique* and a student of ANTOINE-ISAAC SILVESTRE DE SACY (1758–1838), who occupied the chair for Arabic at the newly founded School for Living Oriental Languages (1795). SÉDILLOT collaborated actively with LAPLACE and DELAMBRE, who were among the founding members of the *Bureau of Longitudes.* His translations and analyses were used both by DELAMBRE in his *Histoire de l'astronomie au moyen âge* (History of Astronomy in the Middle Ages) (1819) and by LAPLACE in his "Précis d'histoire de l'astronomie" inserted into his *Exposition du système du monde* (1821). Contrary to SÉDILLOT, who ardently defended the originality and innovative power of the Arabs, DELAMBRE and LAPLACE reiterated the usual position reducing them to simple conservators of Greek theories. In his study of Indian science, for example, DELAMBRE took his inspiration from works in English, notably those of HENRY COLEBROOKE (1765–1837, **B**). DELAMBRE recognized that the Indians possessed advanced mathematics and that our system of numeration was derived from theirs. After examining ancient and medieval astronomy, including European contributions, he nevertheless concluded that there was a single source for scientific astronomy, the Greeks. Under the impulse of SÉDILLOT, whose works had but a weak echo within the milieu of orientalists, DELAMBRE and LAPLACE both became interested in Indian, Chinese, and Arabic mathematics and astronomy and integrated them into their histories.

1.6.2 Academic Controversies on the History of Arabic Mathematics

Around 1840, academic interests focused on Arabic culture. Indeed, the new colonies of North Africa offered a field of application of the knowledge of orientalists. LOUIS-AMÉLIE SÉDILLOT (1808–1875, **B**), son of JEAN-JACQUES, undertook through filial piety to pursue the work his father had failed to complete. His earliest studies concerned Arabic mathematical manuscripts in the *Bibliothèque Nationale*, among them the *Almagest* of ABŪ 'L-WAFĀ' (940–998). The *Journal asiatique*, founded in 1823, published his first articles, including a notice concerning a text of IBN AL-HAYTHAM. There he affirmed that Arabic mathematicians had successfully solved cubic equations by means of geometric constructions, but he also established a somewhat anachronistic parallel between the "geometry of position" developed by LAZARE CARNOT (1753–1823) and the concepts used by Arabic mathematicians. This brought criticism from such competing historians of

[9] G. BIGOURDAN, "Le Bureau des longitudes. Son histoire et ses travaux des origines (1795) à ce jour," *Annuaire du Bureau des longitudes pour l'an 1928*, Paris 1928, pp. A1–A72.

mathematics as GUGLIELMO LIBRI (1803–1868, **B**) and MICHEL CHASLES (1793–1880, **B**) (see Section 1.7).

In a letter addressed to the *Bureau of Longitudes*,[10] the younger SÉDILLOT announced the program that would occupy his entire life: "we are far from understanding all the facts relative to the history and progress of the mathematical sciences and astronomy of the peoples of Asia, especially of the Arabs and Persians." He divided the history of science into three periods linked to three distinct schools: the school of Alexandria, the Arab school, and the modern school. By insisting on the high level of algebraic knowledge of the Arabs, on their application of algebra to geometry, and on the geometrical solution of cubic equations by the Arabs, he strengthened the thesis his father had advanced concerning the originality of the Arabs. He reinforced it further in 1834 when he argued that ABŪ 'L-WAFĀ' had discovered, six centuries before TYCHO BRAHE (1546–1601), an inequality in the motion of the moon that modern astronomers call "variation" (namely, the deviation of the radius of the epicycle in relation to the point around which the mean motion is carried out).

This claim — as spectacular as it was false — was submitted in 1836 to a commission of the Paris Academy of Sciences comprised of JEAN-BAPTISTE BIOT, FRANÇOIS ARAGO, MARIE-CHARLES-THÉODORE DAMOISEAU, and GUGLIELMO LIBRI. A controversy spanning four decades followed between diverse mathematicians and orientalists, notably BIOT and SALOMON MUNK on the one hand, and the younger SÉDILLOT on the other, and in the next generation, between JOSEPH BERTRAND and MICHEL CHASLES [CHARETTE 1995, Chapter 4.2]. This is of interest here for two reasons. First and foremost, it forced the Academy of Sciences to recognize its incompetence on the subject of history of science (even if 30 notes on the question were published in the *Comptes rendus* between 1836 and 1873). In fact, it dissolved a commission composed of BIOT, ARAGO, and LIOUVILLE in 1843 on the grounds that this "historical question is not one on which the Academy is used to making a judgment, as a scientific body" [CHARETTE 1995, 125]. Second, this long and acrimonious controversy had harmful consequences for the history of Arabic mathematics.

JEAN-BAPTISTE BIOT (1774–1862, **B**), one of the protagonists in the controversy, had published as early as 1803 an *Essai sur l'histoire générale des sciences pendant la Révolution française* (Essay on the General History of Science during the French Revolution), which was still linked to the philosophical tradition of the progress of the human mind. Mathematics only occupied a few pages in BIOT's essay, however, and he characterized the influence of the French Revolution on the subject in negative terms. Moreover, in 1821 he edited what one recent author has called the first modern critical study of NEWTON's life and career for the *Biographie universelle* [GJERTSEN 1986, 83]. In collaboration with F. LEFORT, BIOT in 1856 published a new edition of the *Commercium Epistolicum*, which made obvious the innumerable shifts in meaning to be found in NEWTON's notes (see Section

[10] Published in 1834 in the journal, *Le moniteur*, as quoted in [CHARETTE 1995, 102].

7.4). Since his son EDOUARD-CONSTANT was a sinologist, the elder BIOT spent his old age studying the scientific heritage of India and China.[11] He thus took it upon himself to evaluate the ideas of the younger SÉDILLOT and to contribute to the controversy on lunar variation (*JS* 1841, 513–520; 602–610; 659–679; and *JS* 1843). Not only did BIOT lack the necessary linguistic competence to make such judgments, but in the process he also propagated racist stereotypes. SÉDILLOT was drawn into the debate, responding in the first volume of *Matériaux pour servir à l'histoire comparée des sciences mathématiques chez les Grecs et les Orientaux* (Materials for a Comparative History of the Mathematical Sciences of the Greeks and the Oriental Peoples) (1845). It should be emphasized that SÉDILLOT's defense of the originality of the Arabs relied upon a single fact: that the Arabs made original contributions to the sciences as exemplified by their discovery of lunar variation. As this latter "discovery" came more and more into question, SÉDILLOT's entire strategy was doomed to failure.

Nearly twenty years later, in May of 1862, MICHEL CHASLES resurrected the quarrel, taking SÉDILLOT's side. JOSEPH BERTRAND (1822–1900, **B**) succeeded BIOT at the *Collège de France* and defended his position. There followed an impassioned exchange at the *Académie* (1871–1873) that led to no new insights. The question was definitively settled by BERNARD CARRA DE VAUX (1867–1953, **B**) in the *Journal asiatique* in 1892; CARRA DE VAUX knew both the Arabic language and the historical context, which was not the case for any of the earlier protagonists. Historians who came after him were of the opinion that Arab astronomers had added nothing to the Ptolemaic theory of the moon. This controversy had a doubly negative influence on subsequent historiography. Paradoxically, SÉDILLOT's excessive zeal in supporting the originality of Arab mathematicians only reinforced old anti-Arab myths. Historical erudition was, of course, well represented in the scientific community of the 19th century, but under the influence of the polemics, it unfortunately focused on the Ptolemaic theory of the moon, to the exclusion of everything else.

1.6.3 François Woepcke's Research

A young German mathematician, FRANÇOIS WOEPCKE (1826–1864, **B**), having learned Arabic with a student of SILVESTRE DE SACY and having come, in 1850, to settle in Paris (then considered to be the capital of orientalism), carried out the program which the younger SÉDILLOT had already formulated, agreeing for the most part with SÉDILLOT's basic claims (see also Section 5.4.1). His *Algèbre d'Omar Alkhayyami publiée, traduite et accompagnée d'extraits de manuscrits inédits* (Algebra of ʿUmar al-Khayyām, Edited, Translated, and Accompagnied by Extracts

[11] Among the works of EDOUARD-CONSTANT BIOT (1803–1850, **B**) are two articles in the *Journal asiatique*, vols. 8 and 11 (of the third series), showing that the Chinese had at their disposal a place-value number system and claiming that they knew the Pythagorean theorem at least six centuries before PYTHAGORAS (which is not true). An article in *JS* 1835 was devoted to the so-called Pascal triangle.

of Unpublished Manuscripts) (1851) "must certainly be counted among the great products of French orientalism of the mid-century" [CHARETTE 1995, 146]. The *Algèbre d'Omar Alkhayyami* included results surpassing by far those which the Greeks had produced on the subject, and notably a general geometrical theory of the solution of equations up to the third degree. WOEPCKE was one of the first to be interested in diophantine analysis in Islam. He showed, moreover, that LEONARDO OF PISA had taken most of his results from the Arabs. The basic reason for WOEPCKE's relentless research in the collections of the *Bibliothèque Nationale* in Paris was, in his own words, "to destroy the prejudice that has prevailed for so long that the Arabs did nothing more than reproduce or comment on Greek works through which they had studied the sciences."[12] WOEPCKE published his original and meticulous research primarily in the *Journal asiatique*, which had been largely open to contributions on the history of mathematics since the work of SÉDILLOT.

CHASLES championed WOEPCKE's results at the *Académie*, and they were also discussed to a limited extent by mathematicians who, according to WOEPCKE, considered them to be "scholarly works that would do nothing to advance science" [CHARETTE 1995, 152]. In 1855, WOEPCKE was obliged, for financial reasons, to return to Germany, where he accepted a position as professor at a local school (lyceum). Thanks to the generous financial support of BALDASSARRE BONCOMPAGNI (see Section 3.7), WOEPCKE was able to return to Paris in 1858, where he began to study the history of arithmetic and the question of the transmission of science between India, the Islamic world, and medieval Europe. He took part, together with CHASLES, in the debate on the origin of numbers (see Section 5.4.1).

BONCOMPAGNI was largely responsible for publishing the work of his protégé. WOEPCKE, who died prematurely at the age of 38, exercised considerable influence through his writings. For example, ARISTIDE MARRE (1823–1918), a mathematician who specialized on Indonesia, completed part of WOEPCKE's unfinished work. The first issues of BONCOMPAGNI's *Bullettino di bibliografia e di storia delle scienze mathematiche e fisiche* (Rome, 1868–1887), the first specialized periodical for the history of mathematics, reflected the substantial number of French authors committed to the tradition described here, namely, SÉDILLOT, MARRE, LÉON RODET, and THÉODORE-HENRI MARTIN. In fact, the pages of the *Bullettino* bristled with the controversy then raging between BIOT and SÉDILLOT.

According to CHARETTE, and to my own conclusions arrived at independently, Eastern studies brought about a radical change in the humanities in France. Scholars like the SÉDILLOTS and WOEPCKE made mathematicians aware of their ignorance of the general history of science. Thus, in the second half of the century, the history of mathematics fell within the larger context of the general history of science, a trend which culminated in the work of PAUL TANNERY.

[12] FRANÇOIS WOEPCKE, "Notice sur une théorie ajoutée par Thabit Ben Korrah à l'arithmétique spéculative des Grecs," *Journal asiatique* **20** (1852, fourth series), 420–429; [CHARETTE 1995, 151]

1.7 The History of Mathematics by Mathematicians: Chasles and his Successors

Among the mathematicians who took part in the historical debates at the Academy of Sciences in Paris (1836–1873), some were well known for their historical works. BIOT, CHASLES, and his fellow students at the *Ecole Polytechnique*, NOËL-GERMINAL POUDRA, MAXIMILIEN MARIE, GUGLIELMO LIBRI, JOSEPH BERTRAND, and JULES HOÜEL were especially notable.

GUGLIELMO LIBRI (1803–1869, **B**), an Italian aristocrat forced into exile in France for his active involvement in democratic ideals, enjoyed, under the auspices of FRANÇOIS ARAGO, a brilliant career as a professor of mathematics in Paris. The author of an *Histoire des sciences mathématiques en Italie* (History of the Mathematical Sciences in Italy) in four volumes (1838), LIBRI collaborated first with the younger SÉDILLOT on the first volume devoted to antiquity and the Middle Ages. For the description of Chinese mathematics, he depended on the results of the sinologist, STANISLAS JULIEN, but his own strength was above all the interpretation of medieval Latin texts. A passionate bibliographer, LIBRI amassed an important collection of manuscripts, which eventually provoked a scandal that caused him to flee France in 1848. He was accused, in fact, of stealing documents from French libraries and of masterminding the forgery of fraudulent documents.

Among mathematicians who wrote on the history of mathematics, the most important representative was incontestably MICHEL CHASLES. He established his reputation as a geometer and historian of mathematics with the publication, in 1837, of an *Aperçu historique sur l'origine et le développement des méthodes en géométrie* (Historical Survey on the Origin and the Development of the Methods of Geometry), inspired by a question posed for a prize competition by the Royal Academy of Brussels in 1829. This called for a critical study of the methods of modern geometry. The memoir that CHASLES submitted on duality and homography won the prize and was published by the Brussels Academy after having been enriched with numerous historical and mathematical notes based on the latest research.

In writing his *Aperçu*, CHASLES had a precise objective in mind: to present geometrical discoveries in terms of methods, beginning from their origins and leading up to contemporary research. Thus, he claimed that he had not written a history of geometry, but "a general overview of the successive development of its methods, principally those with a connection to modern geometry" [CHASLES 1889, 94]. CHASLES took as his point of departure the contemporary state of geometry and, using a reverse teleological approach, tried to find its roots or predecessors in the past. Modern geometry had acquired the power of generalization, according to CHASLES; this was the process he wanted to clarify. Geometry had at its disposal methods capable of rivaling the tools of analysis "in a wide variety of questions" [CHASLES 1889, 2], and it thereby served to extend our knowledge of physical phenomena.

CHASLES distinguished five epochs in the history of geometry. The first ran from the birth of geometry in Mesopotamia and Egypt to the burning of the great library in Alexandria. The second began with the stereotype of a millennium of stagnation following the fire, the so-called Dark Ages. CHASLES then devoted a considerable amount of space to a discussion of the works of GIRARD DESARGUES and BLAISE PASCAL, ending just before DESCARTES, who opened the third epoch. The fourth saw the birth of the infinitesimal calculus of LEIBNIZ and NEWTON, along with "the progress and contributions of geometry in the course of the 18th century" [CHASLES 1889, 188]. The last epoch included the descriptive geometry of MONGE and the projective geometry of JEAN-VICTOR PONCELET. CHASLES, who, as noted above, was embroiled in the quarrel that rocked the Academy on the subject of the originality of Arabic mathematics, made abundant use of the work of the younger SÉDILLOT in his notes, which comprise more than half of the *Aperçu.* SÉDILLOT had introduced CHASLES to a world of unexpected riches and had thereby enabled him to rectify certain errors he had made in his historical introduction.

A good Latinist, CHASLES was interested in arithmetic and the origins of numerical systems. In his *L'histoire d'arithmétique* (The History of Arithmetic) (1843), he claimed (based on the descriptions of certain abacuses given by PSEUDO-BOETHIUS and GERBERT (ca. 940–1003)) that the origin of our numbers was Pythagorean, not Hindu, a view which he trumpeted loudly (see Section 5.4.1). Following ROBERT SIMSON (1687–1768), whom he criticized, CHASLES reconstructed the lost book of EUCLID's *Porisms* in 1860. This work not only revealed CHASLES's conceptualization of the history of mathematics but also raised numerous historiographic questions: how is it possible to know, in the absence of sufficient documentation, anything about the doctrine of porisms, its origins, the philosophical thought that created it, its purpose, its uses, its applications, and its transformation into modern theories? [CHASLES 1889, 274]. These sorts of questions guided CHASLES in his work.

Passionate about manuscripts like LIBRI, CHASLES was the victim of a forger, DENIS VRAIN-LUCAS. A gullible collector, CHASLES bought hundreds of manuscripts, some of which were presented to the Academy of Sciences and subsequently reproduced in the *Comptes rendus,* vols. 65–69, (1867–1869). The brouhaha that followed the publication of forged documents tarnished somewhat the reputation that the illustrious mathematician had previously enjoyed [FARRER 1907, 202–214].

At the insistence of the Minister of Public Instruction, and in the tradition of the *Ancien Régime* that NAPOLÉON had revived of using the expertise of the academies,[13] CHASLES wrote a *Rapport sur le progrès de la géométrie* (Report on the Progress of Geometry) (1870), which took as its point of departure the rupture

[13]The *Rapports à l'Empereur* (Reports to the Emperor) (1810–1815) served as a model. The *Institute* had been charged in 1802 to write "un tableau de l'état des progrès des sciences, des lettres et des arts depuis 1789 jusqu'au premier vendémiaire an X." DELAMBRE had been the author of the report on the mathematical sciences. Cf. [DELAMBRE 1989] for a recent edition.

of the French Revolution and the works of MONGE and CARNOT, and traced the subsequent evolution of methods in geometry. CHASLES's colleague, JOSEPH BERTRAND, the prestigious mathematician who had accumulated numerous key positions in the great scientific institutions, was the author of a similar report *Sur les progrès les plus récents de l'analyse mathématique* (On the Most Recent Progress in Mathematical Analysis) (1867).

CHASLES played a major role in establishing the history of mathematics as a subdiscipline of mathematics. Through his immense reputation, to which his election to the most important European academies attests, CHASLES exercised an influence on many mathematicians, French and foreign, including HIERONYMUS ZEUTHEN and MORITZ CANTOR, that is still under-appreciated. In particular, he urged his contemporaries to embark on historical research. One of the first to do so was NOËL-GERMINAL POUDRA (1794–1894, **B**). POUDRA had found a manuscript copy of DESARGUES's famous *Brouillon project* (Draft Project) of 1639, and CHASLES encouraged him to edit it ([POUDRA 1864a]). The senior professor of descriptive geometry at a military school, POUDRA based his book, *Histoire de la perspective ancienne et moderne* (History of Perspective, Ancient and Modern) [POUDRA 1864b], on his lecture course devoted to perspective. Similar to the study by MAXIMILIEN MARIE (recall Section 1.5 above), this consisted of a series of notes on the mathematicians who had contributed to perspective, from EUCLID to POUDRA, divided into two periods. In the first, "optics, perspective, and appearance all have the same meaning" [POUDRA 1864b, 2]; the second period, devoted to linear perspective, comprises most of the book.

BERTRAND also contributed in the second half of the 19h century to the reputation of the history of mathematics as a discipline. His numerous historical articles, some of which were popularizations, appeared in the *Journal des savants* and in the *Revue des deux mondes.* He also wrote a historical account, *Académie des sciences et les académiciens de 1666 à 1793* (The Academy of Sciences and the Academicians from 1666 to 1793) [BERTRAND 1869].

Moreover, at the beginning of the Third Republic, BERTRAND and JEAN-BAPTISTE DUMAS (1800–1884), the two perpetual secretaries of the Academy of Sciences, undertook to reconstitute and organize the Academy's archives. The aged BERTRAND exploited these rich archives for his biographies of D'ALEMBERT (1889) and PASCAL (1891).

1.8 The Spread of Historical Work in Journals and the Broadening of the Subjects Treated

The history of mathematics originated as a discipline during the second half of the 19th century in Germany and Italy. In France, JULES HOÜEL (1823–1886, **B**), professor at the University of Bordeaux (from 1859), played a decisive role as communicator and translator of foreign historical literature. He also contributed his own historical studies to the *Mémoires de la société des sciences physiques et* *naturelles de Bordeaux*, the journal that published the first writings of PAUL TANNERY. (TANNERY lived in Bordeaux from 1874 until 1877 and then again in 1888–1889.) For example, HOÜEL made the non-Euclidean geometries of LOBACHEVSKIĬ, BÓLYAI, BELTRAMI, and RIEMANN known in France by including translations of their works in the same *Mémoires ... de Bordeaux* (**8**, 1869) in which he published his translation of CARL A. BJERKNES's study of NIELS H. ABEL, *Tableau de sa vie et de son action scientifique* (Sketch of his Life and Scientific Activities). Most importantly, he reported at length on the history of mathematics in the *Bulletin des sciences mathématiques*, the journal he had founded together with GASTON DARBOUX (1842–1907) in 1870.

As the century progressed, the range of historical questions treated in France grew to include the more recent period. HOÜEL paved the way, largely through his patient work as a go-between, for the emergence of an autonomous history of mathematics. In reviewing foreign books, he sometimes took the opportunity to express his opinions on the history of mathematics and its status in France:

> The basic improvements which have been made independently in various parts of the history of mathematics following the appearance of the works of Montucla and Bossut, now make clear the urgency of a general treatise to replace the older ones. Of these, the most recent dates to the beginning of this century. The need for this became especially urgent for young French professors because the history of mathematics has recently been added to the subjects required for the qualifying examinations for teachers. A book on such a scale, edited to take into account as much chronological and bibliographical information as possible, would certainly be of the greatest help to readers who do not have the time or material resources to examine for themselves specialized publications or to search original documents. They would thus have both a general picture of the progress of science as well as a guide for the most detailed and profound studies.[14]

Despite this call, HOÜEL was extremely critical of the *Histoire des mathématiques, depuis leurs origines jusqu'au commencement du XIXe siècle* (History of Mathematics from Its Origins to the Beginning of the 19th Century) (1874) written by FERDINAND HOEFER. HOÜEL regarded HOEFER as "dangerous, because of the serious errors which he tends to propagate," and also "because of the discredit that would befall French science if the country of Chasles, the elder J.-J. E. Sédillot, and Th.-H. Martin were to accept M. Hoefer as their figurehead in the history of the exact sciences."[15] FERDINAND HOEFER (1811–1878), born in Thuringia, was a lively personality, a world traveler, and a military man. In addition to being a doctor of medicine (a degree he received in 1840) and the translator of KANT's *Kritik der reinen Vernunft* (Critique of Pure Reason) into French, he was also the

[14]See the review by JULES HOÜEL of HEINRICH SUTER's *Geschichte der mathematischen Wissenschaften*, First Part, Zurich 1873 in the *Bulletin des sciences mathématiques*, (1) 1874, 14–15.

[15]See HOÜEL's review in the *Bulletin des sciences mathématiques* (1) 1876, 136.

author of dilettantish histories of mathematics, astronomy, chemistry, and other sciences.

Until 1876, the third lunar inequality had been the focus of a heated and on-going controversy, evoking numerous French contributions to BONCOMPAGNI's *Bullettino*, which until then had been concerned almost exclusively with this topic. Interest in this subject declined as new authors began to appear. Among these were the number theorist EDOUARD LUCAS (1842–1891) and ADOLPHE DESBOVES (1818–1888), who taught in Parisian schools and who found in the works of the past a source of inspiration for their own research. CHARLES HENRY (1859–1926) examined new 17th-century archives, notably containing manuscripts of FERMAT, whose collected works he edited with LUCAS. In 1868, CLAUDE-ALPHONSE VALSON (1826–1901), professor of differential and integral calculus at the University of Grenoble, published a biobibliographical study of CAUCHY that was clearly more of a hagiography than a critical analysis. VALSON was also appointed by the Paris Academy of Sciences to oversee publication of CAUCHY's *Œuvres complètes*, of which some fifteen volumes appeared under his editorship. The edition was only finished in the second half of the 20th century, under the supervision of RENÉ TATON (*1915).

In addition to this activity, the number of mathematical journals was rapidly increasing in the second half of the 19th century, and some included historical papers. At a national level, the *Nouvelles annales de mathématiques*, edited by OLRY TERQUEM (1782–1862) and CAMILLE CHRISTOPHE GERONO (1799–1892), targeted students preparing their entrance examinations for the *Ecole Polytechnique* and the *Ecole Normale*. From 1855 to 1862, it had a section on bibliography and history, to which TERQUEM contributed notices of historical interest as well as biographies and book reviews [DAUBEN 1998, 2–6]. Internationally, GUSTAF ENESTRÖM founded *Bibliotheca mathematica* in 1887 in order "to promote the present great movement towards the study of history of mathematics." [*Bibl. Math.* (1) **1**, preface]. Among the French who supported his venture were HENRI BROCARD (Montpellier), ARISTIDE MARRE (Paris), and PAUL TANNERY (Tonneins). In 1899, yet another new journal, *L'enseignement mathématique*, was launched under the joint editorship of CHARLES LAISANT (1841–1920; French) and HENRI FEHR (1870–1954; Swiss) and proved to be exceptionally valuable for the history of mathematics.

By the turn of the century, international congresses also began to include the history of mathematics. It was in Paris, and under the presidency of HENRI POINCARÉ (1854–1912), that the International Congress on Bibliography of the Mathematical Sciences was held in 1889. The bibliography was to have a historical component listing mathematical works from 1600 until 1800. When the Second (after Zurich in 1897) International Congress of Mathematicians took place in Paris in 1900, several sessions were also devoted to the history of mathematics.

1.9 The History of Mathematics in Relation to the General History of Science: Paul Tannery

> When the history of our discipline is written, it will be necessary to give one of the most important places to PAUL TANNERY. Of course there were many historians of science before him, editors, translators, annotators of texts, etc. But he was one of the first to understand what these studies imply, and to combine in his own person philological exactitude, an erudition at once extensive and profound, and a philosophical sense that was always alert.[16]

PAUL TANNERY (1843–1904, **B**), to whom SARTON gave such homage, was the only historian of mathematics of the Third Republic who enjoyed immense intellectual recognition and who was, at the same time, part of the international network responsible for transforming the history of mathematics into an autonomous discipline. TANNERY was a good friend of the Danish scholars HEIBERG and ZEUTHEN, and translated their works into French. Moreover, his own research was praised by German critics, including CANTOR, GÜNTHER, and HULTSCH. Appointed in 1886 by the Italian Minister of Public Instruction to a commission mandated to study the manuscripts of DIOPHANTUS, TANNERY was also welcomed to Italy by BONCOMPAGNI. Along with LORIA and CANTOR, he was thus among the first to undertake cooperative research devoted to the history of mathematics on a truly international level.

Nevertheless, as a high functionary in the French Corps of Civil Engineers, and as a good linguist and devoted Hellenist, TANNERY was not a professional in the strict sense. He was not an academic fully devoted to the study of the history of the exact sciences. His interest in such history dated from 1876, during his stay in Bordeaux, where he may have met HOÜEL at meetings of the Society for the Physical and Natural Sciences. Later, TANNERY noted the influence that the positivist philosopher, AUGUSTE COMTE, had had on him. In 1903, he acknowledged explicitly, even though he had also come to criticize COMTE, that he had initially undertaken his studies with the aim of confirming COMTE's ideas:

> ...there is only one philosophy which I have ever really assimilated, and that is the philosophy of AUGUSTE COMTE, to which I have been devoted since the age of 22; it is his influence on me that has inspired my studies, wherein the aim has been to verify and to make more precise his ideas on the history of the sciences [TANNERY 1912/50, **10**, 134].

An indefatigable worker and a prolific writer, TANNERY concentrated on Greek and 17th-century science. In addition to his studies of Greek geometry and

[16] SARTON, GEORGE: "La correspondance de Paul Tannery et l'histoire de nos études." *Revue d'histoire des sciences* **7** (1954), 321–325; here: 321.

astronomy, he (in the words of his friend, ZEUTHEN) "looked for Greek arithmetic and algebra wherever it was to be found, and he found it" in the Pythagorean tradition, in the remaining traces of logistic, and in geometry [ZEUTHEN 1905, 265]. In keeping with this view, TANNERY considered DIOPHANTUS (whose manuscripts he had edited for the prestigious TEUBNER series) neither a student of the Hindus nor a creative genius, but a studious compiler of the arithmetic problems and their solutions that had long existed in logistic.

Related to this were the ongoing polemics on the concept of geometric algebra. TANNERY himself coined the phrase, and the concept was used extensively by ZEUTHEN, but above all, by HEATH. Geometric algebra refers to a reading of the Euclidean *Elements* that interprets in algebraic terms certain propositions said to have originated with the Pythagoreans. The most celebrated example of this is Proposition 11 of Book II, which is read as if it were a geometric solution to an equation of the second degree. It should be pointed out that, as early as 1939, ABEL REY (see below) raised questions about the algebraic interpretation of ancient mathematics.

As for TANNERY's research on the 17th century, ROBERT LENOBLE (1902–1959, **B**) observed that "There was one way of writing history of the 17th century before TANNERY, and another way after TANNERY" [*Rev. Hist. Sci.* **7**, 356] (see **B**). In short, TANNERY turned 17th-century science into an historical subject. He aimed, through meticulous study, to recapture scientific thought in all of its many branches regardless of how divergent they might be. He recognized that it was not exclusively the century of DESCARTES, and he thus gave a more correct and balanced view of the struggles and controversies of the period.

TANNERY made substantial contributions through systematic study of unpublished documents that he subsequently edited and interpreted. Together with CHARLES HENRY, he was entrusted with overseeing the official edition of the *Œuvres de Fermat*. The philosopher LOUIS LIARD (1846–1917), whom TANNERY had known in Bordeaux, also conceived of celebrating the tercentenary of the birth of DESCARTES, in 1896, with a new edition of his *Œuvres* to replace the defective one (1824–1826) done by VICTOR COUSIN (1792–1867). The Descartes project was administered by the Ministry of Public Instruction and entrusted jointly to TANNERY and CHARLES ADAM (1857–1940). In its subsequently revised edition, it is still a useful reference today. TANNERY also began, from the end of the 1880s, to collect material on the correspondence of MERSENNE. After his premature death, his widow MARIE TANNERY-PRISSET (1856–1945) took up the project with the help of CORNELIS DE WAARD (1879–1963, **B**). Continued by BERNARD ROCHOT (1900–1971, **B**), who edited vols. 6–12, this edition has only just been completed under the supervision of ARMAND BEAULIEU (*1909).

TANNERY did have the chance to teach the history of mathematics in the Paris Faculty of Sciences over the two-year period from 1884 to 1886. The course of lectures he offered was without doubt the first of its kind in France, even though the philosopher and professor of history and philosophy of science, ARTHUR HANNEQUIN (1856–1905), had pioneered the introduction of the general history of science to higher education. In 1884, the same year that TANNERY was teaching in Paris, HANNEQUIN offered a course in the Faculty of Letters in Lyon entitled "Critique of the Principles of the Mathematical Sciences."

Perhaps TANNERY's greatest contribution to the history of mathematics, however, was his introduction of a method to the field. Prior to his work, the critical evaluation of sources was notoriously insufficient in France. He was convinced that without rigorous documentation, there could be no legitimate conclusions. Thus, as soon as a new document was discovered and authenticated, the question of its date required systematic exploration. In the absence of a clear answer, he did not hesitate to discuss and, if necessary, revise hypotheses. He also tried to retrace the stages of thought which may have led to the writings under consideration, attempting to reason and to calculate as their authors did. When obscure passages resisted this approach, he relied upon his mathematical knowledge to suggest conjectural yet plausible reconstructions, without ever insisting that his might be a definitive solution. Subsequent research would either refine, confirm, or reject his hypotheses.

TANNERY's was not only an intellectual commitment to the field. He also gave considerable thought to the profession, just emerging, of the history of mathematics and to questions about how to write and teach the subject, how best to popularize its principal results, and how to appreciate them in the context of the history of science and intellectual history. In fact, an entire volume of TANNERY's *Mémoires scientifiques* — volume 10, entitled "Historical Generalities" — concerns his thoughts on these matters. According to TANNERY, the history of science could only be treated by someone who was in total command of the science in question, or, at the very least, who was capable of understanding on his own all the scientific questions with which he had to deal. As examples TANNERY cited the works of LAPLACE and CHASLES. But it is also clear that he believed that to be a good historian,

> it does not suffice to be a scientist. It is necessary, above all, to be devoted to history, that is, to have good historical sense; it is imperative that the historian develop such a sensibility for history, which is essentially different from a scientific sensibility; it is necessary to acquire as well a number of special skills, additional talents that are indispensable for history, even if they are of absolutely no use to the scientist who is only interested in the progress of science [TANNERY 1912/50, **10**, 165].

When the protagonists of the young discipline got together during the first international congresses, TANNERY did not even consider the possibility of establishing a special section for the history of mathematics as part of the quadrennial International Congress of Mathematicians, even though he had participated in the Heidelberg Congress in 1904. He also ruled out the idea of an association with congresses of philosophy, but strongly advocated the creation of a section for the history of science in the Congresses of Comparative History, because, as he said, "we make use of historical methods, not philosophical ones" [TANNERY 1912/50,

10, 107]. TANNERY's position was unique and was not shared by other European historians of mathematics [SCRIBA 1993]. His views, it must be said, were strictly dependent on his idea of connecting the special histories (*histoires particulières*), one of which was the history of mathematics. Their synthesis was what he called the general history of science. It should also be noted that HENRI BERR (1863–1954, **B**) focused his intellectual efforts on the idea of synthesis at the same time. He made the pages of his newly founded *Revue de synthèse historique* immediately available to TANNERY. It was important, in TANNERY's eyes, not to isolate the different branches of the history of science, which in turn came to fill in "the picture of intellectual development of a given civilization or epoch" [TANNERY 1912/50, **10**, 167]. He envisioned a comprehensive course for teaching the history of science with the histories of the individual sciences taught by professors of the corresponding sciences at the secondary level and the general history of science taught at an advanced level and capped by a university diploma.[17] This course was discussed at the Congresses of Comparative History held in Paris in 1900 and in Rome in 1903.

If a desire for synthesis came to light in Paris, objections to the general history of science were raised in Rome. In a letter to ZEUTHEN of January 10, 1904, TANNERY wrote: "In a period when the trend is advocated instead for the isolated study of the history of each of the particular sciences, I believe I am the only one in Europe who is capable of representing seriously the general point of view of the founder of positivism, and at the same time of showing that alongside special histories, a general history represents its interests even from the practical point of view of historical progress" [TANNERY 1912/50, **10**, 137]. For TANNERY, the only effort of historical synthesis worth mentioning was that of COMTE, however critical he may have been otherwise.

TANNERY was just as involved in diffusing and popularizing results concerning the history of mathematics. Thus, he wrote the chapters on the "Histoire des sciences en Europe depuis le XIVe siècle jusqu'à 1900" (History of the Sciences in Europe from the XIVth Century to the Present) in the *Histoire générale du IVe siècle à nos jours* (General History from the IVth Century to the Present) (12 vols., Paris 1893–1901) edited by ERNEST LAVISSE (1842–1922) and ALFRED NICOLAS RAMBAUD (1842–1905). Most important, he contributed historical notes to the first volume of the French edition of the *Encyklopädie der mathematischen Wissenschaften* conceived by FELIX KLEIN (see Section 5.4.4) and published in France between 1904 and 1916 by GAUTHIER-VILLARS (partly in cooperation with TEUBNER). JULES MOLK (1857–1914, **B**), a friend of TANNERY, was in charge of the publication in France of the *Encyclopédie*. The French edition is notable because the historical treatment is more extensive, and often more precise (thanks

[17]For more details about this course and what TANNERY meant by the general history of science, see his contribution to volume **8** (February, 1904) of the *Revue de synthèse historique*, and [COUMET 1981].

to the collaboration of TANNERY and ENESTRÖM) than the original German version.[18]

In 1892 a chair was established for the general history of science at the *Collège de France.* The matter of this chair prompted "the first explicit acknowledgment of our studies," in the words of GEORGE SARTON.[19] "The chief significance of the chair of the general history of science is that it provoked a set of statements about the nature of the young discipline by a remarkable group of practitioners at the turn of the century in France" [PAUL 1976, 396]. Since the history of this chair is well known (see [SARTON 1947], [PAUL 1976], [COUMET 1981], and [PETIT 1995]), it suffices here to summarize the situation briefly. Such a position, called for as early as 1832 by AUGUSTE COMTE, was finally established in 1892. The first chair was held by PIERRE LAFFITTE (1823–1903), then one of the leading exponents of positivism. When the chair became vacant in 1903, TANNERY proposed himself as a candidate and was ranked first in the voting of both the *Collège de France* and the *Académie des Sciences.* The Minister, however, departed from usual custom (which would have been to approve this double vote), and chose instead the second person on the list, GRÉGOIRE WYROUBOFF (1843–1913), a chemist, crystallographer, and positivist close to EMILE LITTRÉ. (LITTRÉ, a disciple of COMTE, was opposed to the political and mystical developments of positivism.) Unfortunately, the chair was abolished in 1913 after WYROUBOFF's death. This defeat, deeply resented by TANNERY and the international community of historians of science, prevented the formation of a school of historians of mathematics. ANNIE PETIT has discussed the intellectual and philosophical stakes involved in this complex history [PETIT 1995]. In her view, positivism became the official philosophy of the Third Republic, and one of its components actually triumphed. It seems clear that the construction of the discipline of history of science was far from independent of politics. The history of mathematics is part of the general history of science, the institutionalization of which was intimately related to diverse positivist movements and to battles between them.

By considering the other candidates for the chair at the *Collège de France*, including LOUIS COUTURAT (1868–1914, **B**), GASTON MILHAUD (1858–1918, **B**), and ARTHUR HANNEQUIN, another phenomenon peculiar to France emerges. Although TANNERY insisted on the separation of philosophy and the history of science, in fact, the history of mathematics was led by what may be called, following DOMINIQUE PARODI, "philosopher-scientists" [NADAL 1959, 101]. But first we have to deal with PIERRE DUHEM, for whom TANNERY paved the way.

[18] This has been analyzed in detail in [GISPERT 1999].

[19] Review of PAUL TANNERY, *Mémoires scientifiques* **10**, Toulouse-Paris 1930, in *Isis* **16** (1931), 155.

1.10 Nationalist Tendencies

The works of TANNERY, by focusing on Greek antiquity and the so-called Scientific Revolution, overshadowed the results of orientalists on the mathematics of Islam. By projecting algebra into EUCLID and DIOPHANTUS, TANNERY served to divert attention away from, and consequently to diminish the luster of, the algebraic research of the Arabs. As for the European Middle Ages, the history of mathematics had painted the period in somber colors. Thanks to the influence of BONCOMPAGNI, for whom the Latin Middle Ages constituted a principal center of interest, this period increasingly attracted the attention of historians of science, including PIERRE DUHEM (1861–1916, **B**). DUHEM exploited long-neglected sources and enlarged the body of knowledge concerned with scholastic mathematics and philosophy. He defended the thesis that, through an uninterrupted sequence of barely perceptible improvements, modern science arose from doctrines taught in the medieval schools like those of JORDANUS. As NOEL SWERDLOW has put it, "[DUHEM]'s role may really have been to narrow the focus of a previously universal (multicultural?) subject to the confines of Northern Europe" [SWERDLOW 1993c, 309]. DUHEM was responsible to a considerable extent for overlooking — and with him historians of mathematics in the first half of the 20th century and since — some of the most original mathematics of the Middle Ages.

Exaggerated nationalism at the end of the 19th century encouraged celebration of the great Republican institutions that had emerged from the turmoil of the French Revolution, including the *Ecole Normale Supérieure (ENS)* and the *Ecole Polytechnique*. DOMINIQUE PESTRE has shown that the self-histories commissioned for the centenaries of these institutions at the end of the 19th century — as well as for their bicentenaries at the end of the 20th century — reflected the traits of glorious and heroic histories that often resort to self-congratulation and glorification of national prowess in the techno-scientific domain [BELHOSTE/DAHAN/PICON 1994, 333]. Nevertheless, these histories constituted unique testimonies to the ways in which the actors themselves experienced certain events. Such accounts provide indications, often difficult to find elsewhere, of the milieu, the personal qualities, and the relations between different individuals.

1.11 Philosopher-Scientists of the Twentieth Century

At the beginning of the century, under the influence of positivism and in light of the place that COMTE's system gave to mathematics, there was increased interest in France on the part of philosophers in mathematics and of mathematicians in philosophy. The alliance of philosophy and mathematics also affected French institutions for research. As ENRICO CASTELLI GATTINARA has convincingly shown, the French response to the crisis of reason, which characterized a large part of mid-European culture in the first half of the 20th century, assumed the form of an intersection "of a double articulation between history and epistemology" [CASTELLI 1998, 16] that was nourished by the exact sciences (above all physics).

Some mathematicians, confronted with epistemological questions, turned to philosophy in an effort to scrutinize the principles or to criticize the value of their own research. GASTON MILHAUD is one of the best representatives of this point of view. His program consisted of an examination of fundamental notions, including results from many different positive disciplines, and of the relations between certainty and truth. He discussed the origins, the role, and the range of proof in mathematics and in physics. Among his *Etudes sur la pensée scientifique chez les Grecs et chez les modernes* (Studies on the Scientific Thought of the Greeks and the Moderns) (1906), a study entitled "La géométrie grecque considérée comme œuvre personnelle du génie grecque" (Greek Geometry Considered as the Personal Œuvre of Greek Genius) held that the demonstrative character of arguments was a historical peculiarity of Greek science. MILHAUD also popularized the idea of a Greek miracle. Like TANNERY, he worked on the 17th century as well and especially on DESCARTES as scientist.

MILHAUD exercised considerable influence through his teaching. A local school teacher, CHARLES BIOCHE (1859–1950/51), encouraged by MILHAUD (and HENRI BERGSON), edited an *Histoire des mathématiques* (History of Mathematics) (1914) available and accessible to a wide public. Only after MILHAUD's death, in 1920, was a chair for the general history of science again established at the *Collège de France*. Unfortunately, it too disappeared soon after the death of its first occupant, the mathematician PIERRE BOUTROUX (1880–1922, **B**), son of EMILE BOUTROUX. LÉON BRUNSCHVICG (1869–1944, **B**) and PIERRE SERGESCU (1893–1954, **B**) took the opportunity to follow BOUTROUX's course on the history of mathematics.

Meanwhile, philosophers were becoming increasingly interested in studying the exact sciences and in adopting their methods. During the last quarter of the 19th century, positivism coexisted with a strong reaction against it. Criticisms on both sides were founded on affirmations of a basic rationalism. According to JACQUES REVEL, the historicization of statements and of scientific procedures made it possible to substitute an evolutionary and dynamic perspective for the idea of system: "If the sciences are really as important in France as the rationalist philosophy maintains, then in the wake of Descartes and Kant, it is necessary to understand what they do and what they are, it is necessary to expose the rational model and structures that support them" [BIARD 1997, 171]. Here the mathematical sciences occupied a central place, with notable interest in Greek mathematics, where the origins of rationalism or, in the formulation of ABEL REY, the origins of scientific thought could possibly be discovered.

The works of LEIBNIZ were also the subject of profound studies on the part of philosophers. What was the impulse? ARTHUR HANNEQUIN devoted his Latin thesis (1895) to LEIBNIZ. For HANNEQUIN, a deep understanding of scientific techniques was indispensable for the historian, who could not expect to understand the development of science in its entirety except through the history of a particular science. Like COUTURAT, he acquired a solid mathematical background by taking courses taught by JULES TANNERY (1848–1910), the brother of PAUL, at the *Ecole Normale Supérieure*, as well as those of PICARD (1856–1941), JORDAN (1838–1922), and POINCARÉ (1854–1912). In his famous thesis on the mathematical infinite, COUTURAT drew on the ideas of GEORG CANTOR (1845–1918), whose work he thus helped to diffuse. His research on the collections of the Library of Hannover resulted in the publication of his important study, *La logique de Leibniz* (The Logic of Leibniz) (1901), and the *Opuscules et fragments inédits de Leibniz* (Unpublished Works and Fragments of Leibniz) (1903). At the initiative of JULES LACHELIER (1832–1918), an international association devoted to the works of LEIBNIZ began to meet in Paris in 1901. Meanwhile, the mathematicians DARBOUX (1842–1917) and POINCARÉ, along with the philosophers BOUTROUX and LIARD, were all members of the Leibniz Committee of the Institut de France. The project of producing an edition of the works of LEIBNIZ by an international group of scholars was interrupted by the First World War, but at least resulted in a catalogue compiled by the historian and philosopher, ALBERT RIVAUD (1876–1956), among others.

The radical transformations experienced by the exact sciences at the beginning of the century — problems with the foundations of set theory, questions as to the principles of classical mechanics, etc. — contributed to the concerns of philosophers, including LÉON BRUNSCHVICG, author of *Etapes de la philosophie mathématique* (Stages of Mathematical Philosophy) (1912). As a result of contact with PIERRE BOUTROUX, BRUNSCHVICG (who was opposed to the logicism of COUTURAT), began to take an interest in mathematics. It was his hope, as JEAN-TOUSSAINT DESANTI noted in the preface to his new edition of the *Etapes* (1981), to highlight the formative moments in the emergence of the mathematical mind. DESANTI counted BRUNSCHVICG among those "philosopher-readers" who, with respect to published mathematics, charted the evidence of the processes of productive reasoning. PIERRE BOUTROUX, on the other hand, distanced himself from BRUNSCHVICG's project, which had received support from mathematicians like POINCARÉ, EMILE BOREL, and RENÉ BAIRE.[20] BOUTROUX contrasted historical reconstructions advanced by philosophers, like his student BRUNSCHVICG, with those put forward by scientists. He held that the two kinds of reconstructions are not of the same type. The scientist is interested in science as it is, whereas the philosopher wants to create systems and is interested in stable and closed theories.

BRUNSCHVICG greatly influenced epistemologists like GASTON BACHELARD (1884–1962, **B**), JEAN CAVAILLÈS (1903–1944, **B**), and ALEXANDRE KOYRÉ (1892–1964, **B**). This tradition, which made the history of science a laboratory of examples for philosophy, continues in France to this day, notably with DESANTI as its latest representative. One of the most profound of this group between the two World Wars was JEAN CAVAILLÈS. A student of BRUNSCHVICG, CAVAILLÈS combined

[20] See BOUTROUX's review published in the *Revue de métaphysique et de morale* **21** (3–5) (1913), 307–328, and his *L'idéal scientifique des mathématiciens* (1920).

profound philosophical interests with solid mathematical training. He believed it was necessary to understand from within the development of mathematical thought and the mechanisms whereby mathematicians create their ideas.

CAVAILLÈS wrote at a time when not only the question of the truth of mathematics arose within mathematics itself (prompted, for example, by the antinomies of set theory), but also questions concerning necessity or the contingency of the development of mathematics. These concerns appeared in sharp relief in the responses LEOPOLD KRONECKER (1823–1891), HENRI POINCARÉ, and L. E. J. BROUWER (1881–1966) made to the crisis in foundations, namely, to exclude infinite recursive procedures from mathematics. CAVAILLÈS, in his thesis of 1938, presented an historical method which he applied to the fundamental question related to set theory: "was its appearance necessary or, and to what extent, was it a historical fantasy; was it an autonomous structure or a plurality covered by a contingent system?" [CAVAILLÈS 1962, 29]. CAVAILLÈS concluded that: "There are certain problems of analysis which give rise to fundamental concepts, and which have engendered certain methods already guessed by Bolzano or Lejeune-Dirichlet, which became fundamental methods perfected by Cantor. Autonomy, so necessity" [FICHANT/PÊCHEUX 1974, 158].

For CAVAILLÈS, it was important to follow "mathematics in progress" (*mathématiques en acte*) through all its convolutions and "unpredictable developments." By retracing its multifaceted route on its own terms, the philosopher thus rediscovers, in terms of their genesis, the links ideas have with specific problems at a given time. The nature of mathematics is taken, in this sense, to be identical with its progress [SINACEUR 1994, 15].

1.12 The History of Mathematics and the Institutionalization of the History of Science Between the Two World Wars

During the years between the two World Wars, the history and philosophy of science were organized in France in two distinct ways and in two different places: at the university, where an epistemological point of view focusing on history was adopted, and at the *Centre de Synthèse* founded by HENRI BERR, where close foreign contacts nurtured a more scholarly approach to history. These two Parisian groups, one at the *Sorbonne*, the other at the *Nevers Mansion* at 12 rue Colbert, periodically developed relatively strong antagonisms. Most often, however, these two centers of activity were frequented by the same individuals who enjoyed getting together and discussing their ideas with considerable collegiality.

HENRI BERR founded the *Revue de synthèse historique* in 1900; in 1930, it became the *Revue de synthèse.* His pronounced interest in the history of science was decisive for the progressive institutionalization of the discipline in France, even if the discipline itself seems to have benefited only slightly as a result. The collaboration with historians of science came to an end after the death of PAUL TANNERY, whose opening lecture written for a history of mathematics course at the *Collège de France* (never delivered) was published by BERR.[21] As a result of international "synthesis" weeks, as well as through the editorial project "Evolution de l'humanité" (began in 1920), BERR maintained an interdisciplinary dialogue with scientists. The discussions, however, focused more frequently on matters of current science, usually physics, than on historical matters.

At the *Centre International de Synthèse (CIS)* — created in 1925/26 by BERR with the support of PAUL DOUMER (1857–1932), President of the Republic from 1931 — a section for the history of science was established after ALDO MIELI (1879–1950, **B**) arrived in France in 1929. PIERRE BRUNET (1893–1950) and HÉLÈNE METZGER (1889–1944) were members; foreign members included such historians of mathematics as FLORIAN CAJORI (1859–1930, **B**), GINO LORIA, (1862–1939, **B**) and QUITO VETTER (1881–1960, **B**). The *CIS* was involved (along with an International Committee on the History of Science) in organizing the first International Congress of History of Science, presided over by MIELI in Paris, May 20–24, 1929. This organizing committee later became the nucleus of the International Academy of the History of Science *(AIHS)*, which MIELI served as Permanent Secretary. BERR, also a founding member of the Academy, decided to offer the mezzanine of the *Nevers Mansion* to serve as the Academy's headquarters; the Academy's library and files on history of science from the *CIS* were also located there. "From this time on," wrote PAUL CHALUS in 1955, "the library became the meeting place for French historians of science: HÉLÈNE METZGER, PIERRE BRUNET, ALEXANDRE KOYRÉ, PAUL MOUY (1888–1946, **B**), and CHARLES SERRUS (1886–1946), as well as for foreign scholars visiting Paris."[22] *Archeion*, the journal for the history of science which MIELI had founded in Italy and which he continued to edit in Paris, along with the *Revue de Synthèse*, reflected discussions which were held there. Indicative of this concentration of activity was the creation of a French Committee of Historians of Science in 1931.

Meanwhile, in 1932 ABEL REY, a philosopher associated with the *CIS*, established an Institute for History of Science and Technology with a positivistic and internalist perspective at the University of Paris (13, rue du Four).[23] The creation of this institute, which drew from five faculties, signaled the first serious representation of the history of science in French university education, although, as mentioned, isolated courses resulting from individual initiatives had been taught earlier. At present, the course of study at the Institute for History and Philosophy of Science, for which a certificate is awarded, includes "mathematics and its progress" as an explicit part of the curriculum throughout different periods, beginning with the 16th century.

[21] PAUL TANNERY, "De l'histoire générale des sciences," *Revue de synthèse historique* **8**, no. 1, févr. 1904, p. 1–16.

[22] PAUL CHALUS, "Henri Berr (1863–1954)." *Rev. Hist. Sci.* **8** (1955), 73–80.

[23] REY edited four volumes devoted to the role Greece played at the beginning of scientific thought for the series "Evolution de l'humanité." The last volume was devoted to mathematics.

The project of the *Encyclopédie française*, directed after 1932 by historian LUCIEN FEBVRE (1878–1956) (founder of the *Annales* in 1929), was linked to the activities of the Institute. An entire volume treated mathematics and opened with an historical introduction. Edited by JACQUES HADAMARD (1865–1963), this was intended primarily to classify then-current mathematical problems.

In 1934 another journal, *Thalès*, was created by REY. Published from 1934 to 1968, it aimed to provide an annual record of work done at the Institute on the rue du Four, although, in fact, it only appeared irregularly. Among contributors in the history of mathematics, the most notable included BRUNET, PIERRE SERGESCU, PAUL SCHRECKER (1889–1963), and RENÉ DUGAS (1897–1957). After the War, RENÉ TATON (*1915), FRANÇOIS RUSSO (1909–1998), and LÉON AUGER (1886–1964) also contributed to the journal. HENRI LEBESGUE (1875–1941), who became interested in history at the end of his life, published a study of the mathematical works of VANDERMONDE (**4**, 28–42), while MAURICE FRÉCHET's (1878–1973) biography of the Alsatian mathematician, ARBOGAST (1759–1803), appeared there as well. The absence of any connection between MIELI and *Thalès*, however, seems to indicate that the Institute wanted to maintain a certain distance from the evolution of the *CIS* and its section on the history of science.

As ABEL REY noted in his foreword to the first issue of *Thalès*, the history of science plays a fundamental role in the history of civilizations. At its best, it should teach us what the sciences, including mathematics, have done for mankind and the place of the sciences in the evolution of humanity. The history of science should be neither a mere "chronological presentation of formulas and the advance of techniques" nor an "anecdotal record." It should show how the history of thought and scientific ideas are related in general.

The Italian, FEDERIGO ENRIQUES (1871–1946, **B**), who was in Paris in 1939 after losing his position at the University of Rome, should also be mentioned in relation to French epistemologists. It was ENRIQUES who offered, at the Third International Congress for the History of Science held in Coimbra in 1934, the notion of the history of scientific thought that was to characterize the works of GASTON BACHELARD and ALEXANDRE KOYRÉ.[24] ENRIQUES was responsible for submissions to the section on "History of Scientific Thought" for the series, *Actualités scientifiques et industrielles*, published by HERMANN. It was in this series that KOYRÉ's famous *Etudes galiléennes* (Galileian Studies) was published in 1939.

The period of the Second World War, 1940–1945, was a low point for the history of mathematics in France. Not only was there a lack of promoters and researchers, but MIELI left, REY died, and METZGER, who was deported, subsequently died in a concentration camp.

[24]By the "history of scientific thought," ENRIQUES meant the history "of ideas arising from science strictly speaking, which have come to play a role in the larger domain of culture, where they are associated with different social and political trends" [*Archeion* **16** (1934), 346].

1.13 The Historical Epistemology of Gaston Bachelard

After ABEL REY, the philosopher GASTON BACHELARD became the director of the Institute for History of Science at the *Sorbonne*, a position he held from 1940 until 1955. "French historians of science are greatly in his debt," wrote PIERRE COSTABEL, "even if he was not a great historian of science."[25] According to JEAN HYPPOLITE, in his eulogy of BACHELARD published in the *Revue d'histoire des sciences* in 1964, BACHELARD actually made epistemology a domain of original research that was not only conceived but, in part, explored by him [HYPPOLITE 1964, 6]. To begin with, BACHELARD recounted as exactly as possible the irreversible development of the natural sciences and insisted upon the gulf that separated these sciences from naive experience. He also described the obstacles that prevented the scientific spirit from thriving. He thought of truth as "historical rescue from persistent error," experience and thought as "rescue from common and early illusions" [BACHELARD 1975, 177]. It was the continuous correction of errors that led to knowledge. BACHELARD devised an original conception of history that followed no necessary order. It is the result — the science of the present — that makes it possible to reorganize and to rethink the past. On these terms, the ancient should be thought of as a function of the new.

BACHELARD was one of those philosophers of great historical erudition who knew perfectly well "how to use ancient scientific texts to make a point," and who therefore used the history of science to support a theory of knowledge [METZGER 1987, 190]. He developed "a philosophy of creative intelligence" [HYPPOLITE 1964, 3] that he applied to mathematical physics, drawing some of his examples from the history of mathematics. Ultimately, BACHELARD significantly influenced the work of historians of the exact sciences. He provided incentives and encouragement on the intellectual as well as the institutional level; he supported young mathematicians like RENÉ TATON, who wanted to write an historical thesis. From 1958 until 1968, TATON also had the opportunity to teach the history of mathematics at the Institute of the History of Science at the *Sorbonne*, then under the directorship of GEORGES CANGUILHEM (1904–1995). BACHELARD also inspired an editorial enterprise of great breadth, which can be seen as inspired by TANNERY, *L'histoire générale des sciences* (The General History of Sciences) overseen by TATON and published in four volumes (Paris 1957–1964).

1.14 Towards an Autonomous History of Mathematics After the Second World War

The Romanian mathematician, PIERRE SERGESCU, who immigrated to France in 1946, played a role in the separation, relatively speaking, of the history of mathematics from the history of science. In particular, in the 1950s he founded, with

[25] PIERRE COSTABEL, *Rev. Hist. Sci.* **16** (1963), 258.

the support of MAURICE FRÉCHET, the Seminar for the History of Mathematics (which still exists) at the *Institut Henri Poincaré (IHP)* in Paris.

Following the First World War, SERGESCU had visited Paris regularly and had turned, under MIELI's influence, to the history of mathematics. In 1939, SERGESCU was commissioned by the French government to present a picture of contemporary French mathematics at the International Exposition in New York. His style, although relatively technical, was nevertheless marked by a concern for setting the history of mathematics within a panorama of the general history of science [TATON 1987, 107]. As a specialist in the history of 19th- and 20th-century French mathematics, SERGESCU successively enlarged the subjects of his research to include medieval science, 17th-century mathematics, and science generally during the Scientific Revolution. He thus refused, following DUHEM, to ascribe the Scientific Revolution to the modern period, and as a result, the birth of modern science was no longer considered the exclusive and unique result of the 17th century [SERGESCU 1951]. SERGESCU regarded the Middle Ages as "the period of gestation of modern science, without which the latter cannot be properly understood in all of its true significance" [TATON 1987, 108–110].

After the Second World War, SERGESCU actively participated in the creation of the International Union of the History of Science, along with PIERRE BRUNET, ARNOLD REYMOND (1874–1958), and ARMANDO CORTESAO (1891–1977). Appointed to the General Secretariat in October of 1947, he played an essential role in the reorganization and international collaboration of historians of science (for details, see [TATON 1987, 108–110]).

During the 1950s, in addition to a limited number of researchers (including ROBERT LENOBLE (1902–1959, **B**), RENÉ TATON, PIERRE COSTABEL, and FRANÇOIS RUSSO), there were also teachers interested in the history of mathematics like JEAN ITARD (1902–1979, **B**), LÉON AUGER (a specialist on Roberval), and CHARLES NAUX (who wrote a history of logarithms). Others such as PAUL-HENRI MICHEL (1894–1964, **B**), librarian at the Mazarine Library, and retired engineers like RENÉ DUGAS (1897–1957), also worked in the history of mathematics. DUGAS, in particular, edited the *Histoire abrégée des sciences mathématiques* (Abridged History of the Mathematical Sciences) by MAURICE D'OCAGNE (1862–1938, **B**).

This small Parisian community of historians of mathematics met at the seminar of the *Institut Henri Poincaré (IHP)* and published in the *Revue d'histoire des sciences* (created in 1947). PIERRE BRUNET was the journal's first editor, but when he left this position due to illness, HENRI BERR entrusted SUZANNE DELORME (*1913) and RENÉ TATON jointly with the journal's editorship. In accordance with BERR's wishes, the *Revue* had the double aim of studying the genesis of discoveries and of considering the collective needs and effects of science on society. The articles on the history of mathematics took up a considerable amount of the journal's space.

Among those who taught mathematics just after the Second World War and who came to history basically for pedagogical reasons, JEAN ITARD (1902–

1979, **B**) is perhaps the only one to have acquired an international reputation. His *Mathématiques et mathématiciens* (Mathematics and Mathematicians) (1959), written in collaboration with PIERRE DEDRON, was translated into English in 1973. In the tradition pioneered by PAUL TANNERY at the turn of the century, ITARD worked on Hellenistic mathematics and on mathematics in France of the 16th and 17th centuries. At KOYRÉ's request, ITARD taught the history of Greek mathematics at the *Ecole Pratique des Hautes Etudes (EPHE).*

1.15 The Centre Alexandre Koyré and Research on the History of Mathematics

In 1958, at the suggestion of ALEXANDRE KOYRÉ, a Center for Research on the History of Science and Technology was created by FERNAND BRAUDEL (1902–1985) at the *EPHE.* KOYRÉ became its first director. This center, like the International Academy for the History of Science, was organized around the library of the history of science at the *Nevers Mansion* in Paris. In 1966, two years after KOYRÉ's death, the Center was named in his honor.

"If the history of science has reached a viable maturity as a discipline, it owes this maturity in large measure to Alexandre Koyré" (MARSHALL CLAGETT, cited in [REDONDI 1986, ix]). It is in this capacity that KOYRÉ deserves to be mentioned in this chapter. Historian of religion, he frequented the *CIS* beginning in the 1930s, where he had a somewhat ambiguous connection with MIELI [TATON 1987, 40]. His influence on the milieu of French historians of science prior to 1940 was limited: ROBERT LENOBLE paid homage to him in his thesis on *Mersenne et la naissance du mécanisme* (Mersenne and the Birth of Mechanism) (1943), and PIERRE COSTABEL reviewed his *Etudes galiléennes* (Galilean Studies) (1939) in 1947. From this moment on, KOYRÉ enlivened the history of science in France, along with BACHELARD and SERGESCU. At the *Centre National de la Recherche Scientifique (CNRS)* (The National Center for Scientific Research), created in 1948, KOYRÉ fought along with BACHELARD and RAYMOND BAYER (1898–1959) to give the history of science an honorable place among the philosophical disciplines.

Beginning in the 1930s, KOYRÉ had progressively distanced himself from the history of religious thought to become more closely associated with that of scientific thought. He considered the latter closely connected with the evolution "of trans-scientific, philosophical, metaphysical, and religious ideas" (KOYRÉ in 1951, cited by [REDONDI 1986, 127]. KOYRÉ radically transformed the frameworks of historical understanding by showing that the histories of astronomy and mechanics involved not only erudite collections of "scientific" facts, but also religious and ontological conceptions, cosmologies, and mathematical thoughts. He also modified the traditional connections between epistemology and the history of science. No longer would these serve as the vanguard for a philosophical theory concerning the physical world, the structure of the human mind, or social organization.[26] KOYRÉ focused primarily on modern science from KEPLER to NEWTON, and in collaboration with I. BERNARD COHEN (*1914), he produced a critical edition of NEWTON's *Principia.* On a methodological level, KOYRÉ placed great importance on the conceptual analysis of texts in their proper intellectual and religious context. He viewed it as essential "to interpret them as a function of mental states, of the preferences and aversions of their authors." This is why he argued against translating their thoughts, for the sake of greater clarity, into modern languages. And finally, he considered the study of errors and failures as well as successes, because they "reveal the difficulties which one had to overcome, the obstacles which one had to surmount."[27]

In the 1960s and 1970s, the *Centre A. Koyré*, headed by RENÉ TATON beginning in 1964, was an important meeting point for historians of mathematics who had been merged into the history of science as inherited from BRUNET and MIELI. TATON taught "History of the Exact Sciences" (1964–1983) whereas KOYRÉ entitled his course "History of Scientific Thought" (1954–1962). From the beginning of the Center, PIERRE COSTABEL was closely associated with its work and, from 1962 until 1981, he offered a regular seminar on the history of mechanics. TATON and COSTABEL both had strong mathematical backgrounds and guided the formation of the next generation of historians of mathematics. As a result, dynamic and more professional research was promoted at the *Centre Koyré*, where a number of collective projects were undertaken, including a study of the teaching of science in the 18th century. The Center also made important contributions to editions of complete works of MARIN MERSENNE, NICOLA DE MALEBRANCHE, the BERNOULLIS, EULER, CAUCHY, and others. The tradition, begun by BERR at the *CIS*, of organizing major international colloquia in an interdisciplinary spirit has also continued.

In the 1970s, through the energetic efforts of ROSHDI RASHED (*1936), there has also been a reawakening of interest in the history of Arabic mathematics. These studies were pursued within the Center for the History of Arabic Science and Philosophy, first directed by JEAN JOLIVET (*1925) and then by RASHED.

Institutionally, historians of mathematics are linked above all to the *CNRS*, where most are affiliated with the Section on Philosophy and History of Science, and a few with the Section on Mathematics. At the same time French mathematicians, under the influence of BOURBAKI, have given a warm reception to the historical works of their colleagues.

[26]Consider here the two facets of the dilemma described in 1935 by HÉLÈNE METZGER in *Archeion* [METZGER 1987, 27].

[27]These quotations are taken from a *curriculum vitae*, which A. KOYRÉ had composed in February 1951. Cf. "Orientations et recherches," *Etudes d'histoire de la pensée scientifique*, Paris: Gallimard, 1973, 14.

1.16 History According to Bourbaki

In the 1930s, a group of mathematicians under the pseudonym of (NICOLAS) BOURBAKI (20th century, **B**) undertook to produce a unified reconstruction of all of mathematics in a famous series of books devoted to the *Eléments des mathématiques.* Almost from the beginning, these included historical notes which were later collected together in a single volume: *Eléments d'histoire des mathématiques* (Elements of the History of Mathematics) (1960). The series was organized according to the "fundamental structures of analysis," not according to the traditional mathematical disciplines, and historical developments were placed at the ends of chapters, following the same logic. What resulted was an internal history of concepts selected by virtue of a structuralist conceptualization of mathematics. BOURBAKI, as he said in the preface, sought to make clear how mathematical ideas "are developed and react one against the others." This pure history of ideas excludes any biographical or institutional information, and depends only on a reading of original texts, translated into modern terms and interpreted with respect to the current state of mathematical understanding. This kind of history was accessible only to readers with a solid background in classical mathematics.

Consider, for example, the relatively extensive treatment BOURBAKI gave to the infinitesimal calculus. In his words,

> the development of the infinitesimal calculus in the 17th century can be thought of as if it were the gradual and inevitable development of a symphony, where the positive *Zeitgeist* is both the composer and the conductor; taking up the baton, everyone plays his part with his own stamp, but no one controls the themes which are heard, themes which a knowing counterpoint has almost inextricably entangled. It is in the form of a thematic analysis that history ought to be written.[28]

BOURBAKI chose the following themes: mathematical rigor, kinematics, algebraic geometry, the classification of problems, interpolation and the calculus of differences, the progressive algebrization of infinitesimal analysis, and the notion of function. Underlying all this was the idea that the very nature of mathematical objects and relations inspires the mathematician to forge the necessary tools for problem-solving. The accent was placed on the structural elements and not on connections with philosophy, for example, or with physics.

The successive editions and translations of this collective work have played a noticeable role in the rebirth of interest in the history of mathematics among mathematicians which was apparent beginning in the 1970s. JEAN DIEUDONNÉ (1906–1992), one of the founding members of BOURBAKI, wrote historical works of his own. Convinced that it was not possible to understand contemporary mathematics without an idea of its history, he provided an historical preface for his course

[28] NICOLAS BOURBAKI: "Calcul infinitésimal," in: *Eléments d'histoire des mathématiques*, 2nd ed., Hermann, Paris 1969, 207–249; here: 215.

on algebraic geometry. DIEUDONNÉ also brought together a group of mathematicians willing to write a collective work called *Abrégé d'histoire des mathématiques, 1700–1900* (Abridgment of the History of Mathematics, 1700–1900) (1978). Moreover, he regarded LAGRANGE as the first mathematician to produce an internal history.[29] This was like COMTE, who despite recognizing the "irrational disdain" of mathematicians for intellectual history, had earlier made an exception for "the great Lagrange (who) alone foresaw its great possibilities" [PETIT 1994, 68–69].

DIEUDONNÉ supported a number of projects to prepare the collected works of French mathematicians for publication. The French National Committee of Mathematics, in conjunction with the *CNRS* and with the patronage of the Academy of Sciences, is also involved in this work. Thus far, the complete works of HENRI POINCARÉ and ELIE CARTAN, among others, have already appeared.

1.17 Further Developments

Interest in the history of mathematics is reflected at the university level in various individual initiatives. Even though there are no positions devoted specifically to the history of mathematics, some courses have been given, for example, by PIERRE DUGAC at *Paris VI* and by JEAN-LUC VERLEY at *Paris VII* (with support from the Rector of the University, FRANÇOIS BRUHAT). In the *CNRS*, new research groups in the history of science have been created with strong interests in the history of the mathematical sciences, such as the *REHSEIS*, or the Research Group for Epistemological and Historical Studies on the Exact Sciences and Scientific Institutions, founded by CHRISTIAN HOUZEL, MICHEL PATY, and ROSHDI RASHED in 1983.

Since 1980, lectures presented at the seminar of the *IHP* have been published in the *Cahiers du séminaire d'histoire des mathématiques* founded by DUGAC. In 1995, this was transformed into an international journal for the history of mathematics with support from the French Mathematical Society, the *Revue d'histoire des mathématiques.*

Interest in the history of mathematics has also had pedagogical (didactic) ramifications in the Institutes for Research on Mathematics Education *(IREM)* where there is a very active Commission on Epistemology and History that publishes collections intended for secondary school teachers. Among these are the sourcebook *Mathématiques au fil des âges* (Mathematics Throughout the Ages) (1987) and *History of Mathematics — Histories of Problems* (1998).

Beginning in the 1980s, lively interest in the universities has resulted in the creation of small research groups, seminars, and editorial projects, even though institutional recognition for the history of mathematics remains problematic.

[29]Personal communication from JEAN-LUC VERLEY.

1.18 Conclusion

The history of mathematics as a separate field of study in France was the creation of the "philosophes" of the Enlightenment. Their legacy has essentially been two-fold. First, the history of mathematics benefitted during the period of colonial expansion in North Africa from a strong interest in the history of oriental, especially Arabic, science.[30] Thus, the first position created in France for historical studies in the mathematical sciences was a position associated, thanks to the active support of DELAMBRE and LAPLACE, with the *Bureau des Longitudes* (in 1814) and devoted to the history of oriental astronomy. Second, the history of mathematics has since then been associated in France with philosophy. Its intellectual and institutional development was, during the 19th century, linked to a certain extent with Comtian positivism.

Simultaneously, with the professionalization of mathematics after the French Revolution, mathematicians began to take charge of their past by referring in their own research to historical developments (LAGRANGE) and by developing a history of mathematical methods and problems (CHASLES). The *Encyclopédie des sciences mathématiques pures et appliquées* (Encyclopedia of the Pure and Applied Mathematical Sciences), aiming at an adequate representation of the state of mathematics around 1900, reflects nevertheless a teleological interest in the historical development of mathematical methods during the 19th century.

These three different approaches, embodied by three different groups, the orientalists, the philosophers, and the mathematicians, sometimes succeeded in and somtimes failed in collaborating. The discussions during the second half of the 19th century at the *Académie des Sciences* in Paris between orientalists and mathematicians made the latter aware of their lack of knowledge of the history of their discipline. HOUËL's efforts to make German and Italian historical works available in France resulted in the slow emergence of a more autonomous history of mathematics, of which the work of PAUL TANNERY is an outstanding example. This in turn brought new interests to the fore. Historical research on 17th-century French mathematics, notably on DESCARTES, FERMAT, and MERSENNE, developed mostly owing to TANNERY. Focusing on the achievements of classical antiquity and of what the French call the *Grand Siècle* (Great Century), historical research no longer took Arabic mathematics into account. With the rehabilitation of the Latin Middle Ages under DUHEM's ideological pen, the mathematical achievements of the Arabs were not only neglected but dramatically denied.

At the beginning of the 20th century, the complementary interest of mathematicians like GASTON MILHAUD and PIERRE BOUTROUX in philosophy, and likewise of philosophers like BRUNSCHVICG in mathematics, led to an alliance between the two fields that left its marks on the later institutionalization of the discipline. In response to the remarkable advances of the exact sciences, epistemol-

[30] ROSHDI RASHED stresses the anthropological foundations of later 19th-century orientalist approaches. See his "La notion de science occidentale," *Entre arithmétique et algèbre. Recherches sur l'histoire des mathématiques arabes*, Paris: Les Belles Lettres, 1984, 301–318.

ogists and philosophers of science (under the influence of REY and BACHELARD) used examples from the history of science to devise a theory of knowledge. This entanglement eventually drew the history of science, and with it much of the history of mathematics, into the academic realm of philosophy. The emergence of an autonomous history of mathematics, if such an autonomy were possible or desirable, would have been difficult under such circumstances.[31] In the field of mathematics proper, BOURBAKI favored the development of historical studies, provided they were compatible with his own theoretical framework. It should be acknowledged that most recently positions for history have been created in the mathematical division of the *CNRS*. The French Mathematical Society *(SMF)* has also shown interest in history by providing financial support for publications in the history of mathematics. Historians proper are relatively absent from the story told here, despite PAUL TANNERY's plea at the turn of the century for a rapprochement of the two disciplines. Even if historians like LUCIEN FEBVRE and ROBERT MANDROU were interested in the history of science and included some aspects in their research, the dialogue between the two disciplines has only recently become more intense. This has led to a wider array of questions on the one hand and to more profound contextual research on the other. The history of mathematics, located as it is at the convergence of historical, mathematical, and philosophical research, should continue to benefit from these interactions.

[31] ROGER, JACQUES, "Pour une histoire historienne des sciences," *Pour une histoire des sciences à part entière*, Paris: Albin Michel, 1995, 43–73. ROGER stresses the monopolization that at times philosophy, at times science, but rarely history, have had on the history of science in France.

Chapter 2

Benelux

Paul Bockstaele

2.1 Geographical and Political Considerations

The following pages offer a survey of what has been accomplished in the historiography of mathematics since the sixteenth century in the region that corresponds roughly with the current territory of Belgium, the Netherlands, and the Grand Duchy of Luxembourg. The reference to "Benelux" in the title is, of course, anachronistic. In what follows, reference will be made instead to the "Low Countries" or the "Northern" and "Southern Netherlands," or related terms. In the early part of the 16th century, these regions came under Habsburg rule. They were initially controlled by the Austrian branch of the dynasty, before passing to Spanish sovereignty in 1556, when Philip II was crowned king of Spain. Political conflicts, aggravated by religious disputes stemming from rising Protestantism, resulted in the separation of the South and the North in 1585. The northern regions went the way of complete independence, while the South remained Spanish until 1714, when it was returned to Austrian rule. French troops occupied the entire Low Countries in 1795, and the Austrian Netherlands were officially subsumed into France. The North retained a degree of independence until 1810, when it too was annexed. Following the defeat of Napoleon at Leipzig and Waterloo, the Southern and Northern provinces were rejoined in 1815 into the Kingdom of the United Netherlands. The new union was short-lived, however, as the Southern provinces soon asserted their independence and established the Kingdom of Belgium in 1830.

2.2 Humanist-Inspired Return to the Sources

Although not appointed by the University, the physician JUSTUS VELSIUS (ca. 1510–ca. 1581, **B**) started giving public lessons on EUCLID at Leuven in 1542. He stated in an inaugural lecture that anyone who wanted to gain a solid knowledge of mathematics would best avoid the recent *compendia* like the plague, and turn instead to the authors of antiquity. He referred to EUCLID, ARCHIMEDES, APOLLONIUS, PTOLEMY, NICOMACHUS, and PROCLUS among the Greeks, followed by BOETHIUS, JORDANUS NEMORARIUS, and WITELO, and, among recent authors, PEUERBACH and REGIOMONTANUS [VELSIUS 1544, f. C iv *v*°]. VELSIUS himself published the Greek text of PROCLUS's *De Motu* (On Motion), along with a Latin translation [VELSIUS 1545].

It is not surprising that VELSIUS did not mention DIOPHANTUS, as his work was unknown at the time in the West. It was first introduced to the Low Countries by SIMON STEVIN (1548–1620), in his *Arithmétique* [STEVIN 1585]. There, STEVIN provided a French translation, or rather a paraphrase into the notation of *L'Arithmétique*, of the first four books of DIOPHANTUS's *Arithmetica*, drawing on the Latin translation by XYLANDER, published in Basle in 1575.

In his *In Archimedis circuli dimensionem expositio et analysis*, a defence of ARCHIMEDES against attacks of JOSEPH SCALIGER (1540–1609), ADRIANUS ROMANUS (1561–1615, **B**) included the Greek text of ARCHIMEDES's *On the Measurement of the Circle*, plus a Latin translation and a double commentary [ROMANUS 1597, 1–18]. Unfortunately, another of ROMANUS's ventures, which might have secured him an honorable place among the commentators on key mathematical texts, remained unfinished. A friend of his in Prague, THADDÄUS HÁJEK, had given him a copy of ROBERT OF CHESTER's translation of AL-KHWĀRIZMĪ's *Algebra*. ROMANUS decided to publish the work, accompanied by a detailed commentary, but, for some reason, only 72 pages were printed [ROMANUS ca. 1602].[1] The fragment opens with ROMANUS's introduction, entitled *In Mahumedis Arabis Algebram Prolegomena* (Introduction into Mohammed's Algebra). It also includes the beginning of his commentary on the preface of AL-KHWĀRIZMĪ's work. The original plan to publish the whole of the *Algebra* came to naught.

Meanwhile, WILLEBRORD SNELLIUS (1580–1626, **B**), in his *Apollonius Batavus* [SNELLIUS 1608], attempted to restore lost works by APOLLONIUS OF PERGE. Half a century later, FRANS VAN SCHOOTEN JR. (ca. 1615–1660, **B**) published a reconstruction of APOLLONIUS's treatise on *Plane Loci* [VAN SCHOOTEN 1657, book III: *Apollonii Pergæi Loca plana restituta*, 191–292].

An annotated Latin edition of EUCLID's *Elements* was published in Antwerp in 1645 by the French Jesuit CLAUDE RICHARD (1588/89–1664, **B**), professor of mathematics at Madrid [RICHARD 1645]. Added are treatises by ISIDORUS OF

[1] The original printed fragment no longer exists. Two copies survive, however, one in the Jesuit Archive in Brussels, produced by H. BOSMANS, and one in the University Library in Stockholm, which was made for Prince BONCOMPAGNI.

MILETUS and HYPSICLES, and propositions of PROCLUS. Ten years later the same author published, also at Antwerp, *Apollonii Pergæi Conicorum libri IV* [RICHARD 1655], with commentary on the first four books of APOLLONIUS's *Conics*, the only ones to survive in the original Greek. He added a reconstruction of the four final books, the Arabic version of which was still unknown at the time. The first Arabic manuscripts, including books five, six, and seven of the *Conics*, had, however, been brought back from the Levant a good fifteen years earlier by JACOB GOLIUS (1596–1667, **B**), professor of oriental languages and mathematics at Leiden. Nothing came, unfortunately, of the latter's plan to publish and translate APOLLONIUS's work. Nevertheless, GOLIUS did publish the Arabic text and a Latin translation of *The Elements of Astronomy* by the ninth-century astronomer AL-FARGHĀNĪ [GOLIUS 1669]. Also noteworthy is the publication, with Latin translation, of IAMBLICHUS's commentary on NICOMACHUS OF GERASE's *Introduction to Arithmetic* by SAMUEL TENNULIUS (*1635, **B**), lecturer in history at Nijmegen [TENNULIUS 1668].

2.3 Catalogues of Mathematicians

Some of ADRIANUS ROMANUS's works include interesting historical elements. He went furthest in this regard in his *Prolegomena* (Introduction) to AL-KHWĀRIZMĪ's *Algebra*. The work begins with his thoughts on the nature, purpose, and value of algebra, including a short history of the methods for solving equations of the first to the fourth degree and of equations of any degree. Appended is a list of the *Inventores artis resolutoriae* or the inventors of algebra, beginning with the Arabs, represented by AL-KHWĀRIZMĪ, who was "said to have been the first to write about algebra." Then came the Greeks with DIOPHANTUS. The list continues with the Italians, French, Portuguese and Spanish, Germans, and, finally, the Belgians and Dutch [ROMANUS ca. 1602, 6-10]. ROMANUS did not just list the names, but also cited the precise titles of the works of these authors, together with the year of publication. For AL-KHWĀRIZMĪ, he mentions that his *Algebra* had been translated from Arabic into Latin by ROBERT OF CHESTER, although he admits not knowing when AL-KHWĀRIZMĪ had lived.

A further piece of history is to be found in *De mathematicis disciplinis libri duodecim* (Twelve Books [= Chapters] on the Mathematical Disciplines), which was published in Antwerp in 1635 by the Scottish Jesuit HUGH SEMPLE or SEMPILIUS (1596–1654, **B**), who taught mathematics in Madrid. SEMPLE began his account with ADAM and his son SETH. According to FLAVIUS JOSEPHUS, SETH was the first to practice mathematics. Among the ancients he cited THALES, ARCHYTAS, APOLLONIUS, DIOPHANTUS, PTOLEMY, and HIPPARCHUS, but he also mixed in mythical figures like ORPHEUS, AESCULAPIUS, and ATLAS. He then went straight to the *recentiores*, citing NEPER and ANDERSON in Scotland, BRIGGS and GUNTER in England, AGUILON, STEVIN, VAN CEULEN, and SNELLIUS in the Low Countries, TYCHO BRAHE in Denmark, KEPLER, SCHEINER, and CLAVIUS as the most important in Germany, GALILEO, MAGINI, and DEL MONTE in Italy, and for France, VIÈTE, HENRION, and BACHET. SEMPLE also stated the fields in which they had distinguished themselves [SEMPILIUS 1635, 5-6]. He rounded off his work with a highly detailed, alphabetical list of authors who had written about the various branches of mathematics [SEMPILIUS 1635, 262–310].

In 1654, the Jesuit ANDREAS TACQUET (1612–1660, **B**) published at Antwerp his *Elementa geometriae planae ac solidae* (Elements of Plane and Solid Geometry) for school use. The work begins with an *Historica narratio de ortu et progressu Matheseos* (Historical Report on the Origin and Progress of Mathematics) [TACQUET 1654, ff. a[6] r^o – b3 r°]. In order to make his students aware of the value of mathematics, TACQUET considered it important to show how wise men had committed themselves to this science in every era. He acknowledged that in so doing he was following the example of RAMUS (1515–1572), who had devoted the first book of his *Scholae mathematicae* (Mathematical Lectures) to the history of mathematics. Like RAMUS (and SEMPLE), TACQUET also began with ADAM and his son SETH, although he did not attach much credence to this story. Also like RAMUS, he limited himself to the Greeks and their predecessors. Only in the part on DIOPHANTUS, whom he called the inventor of algebra, did he also mention VIÈTE and DESCARTES. He did not concur, however, with RAMUS's attribution of the *Elements* to THEON OF ALEXANDRIA, and he also rejected his criticism of EUCLID. TACQUET's work remained one of the most successful manuals of geometry until well into the eighteenth century. It was reprinted many times throughout Europe, including editions by P. VAN MUSSCHENBROEK (1692–1761) and, most notably, by WILLIAM WHISTON (1667–1752), Lucasian Professor at Cambridge. TACQUET's *Historica narratio* appeared unchanged in all these reprints.

By far the most important 17th-century contribution to the history of mathematics in the Low Countries is *De universae mathesios natura et constitutione liber* (Book on the Nature and Constitution of the Universe of Mathematics) by the humanist and philologist GERARDUS JOANNES VOSSIUS (1577–1649, **B**). At the time of his death he left an almost complete manuscript that his nephew FRANCISCUS JUNIUS (1624–1678) published in 1650 [VOSSIUS 1650]. In it, VOSSIUS dealt with the various disciplines considered as belonging to the mathematical sciences. He provided an historical review of each one in the form of a chronological list of authors whose writings had contributed to each discipline from the Greeks and Romans through to his contemporaries. His book says nothing about the development of the science of mathematics as such, but is concerned rather with the people who devoted themselves to mathematics over the centuries. MONTUCLA, in his *Histoire des mathématiques* (see Section 1.3.3), rightly noted that VOSSIUS's work "ne présente guère que des divisions et des subdivisions des mathématiques, et une énumération d'auteurs avec les titres de leurs ouvrages" (presents scarcely more than the divisions and subdivisions of mathematics, and a list of authors with the titles of their works) [MONTUCLA 1799/1802, Vol. 1, v–vi]. Nevertheless, he admitted that it had been useful to him on more than one occasion.

2.4 The Eighteenth Century

In 1759, on the occasion of his acceptance of the post of lecturer in mathematics at Leiden University, PIBO STEENSTRA (†1788, **B**) gave an inaugural lecture on the rise and development of geometry [STEENSTRA 1759]. It comprised little more than a list of the same names given by TACQUET, supplemented with details on mathematicians from the 16th and 17th centuries. Following TACQUET's example, STEENSTRA began his *Grondbeginselen der Meetkunst* (Elements of Geometry), a schoolbook on geometry published at Leiden in 1763, with a fairly lengthy introduction on the development of geometry since antiquity. It was a reworking and continuation of his inaugural lecture of 1759, and drew extensively on MONTUCLA's *Histoire des mathématiques* [MONTUCLA 1758].

The publication of a Dutch translation by ARNOLDUS BASTIAAN STRABBE (1741–1805, **B**) of MONTUCLA's *Histoire* might have had some influence on the historiography of mathematics in the Netherlands. The translation, based on the first edition [MONTUCLA 1758], was published in four parts, the first in 1782, and the remaining three in 1787, 1797, and 1804 [MONTUCLA 1782/1804]. It does not appear to have been very highly rated; a report dating from 1813 describes the translation as faulty in numerous respects.

A modest contribution to the history of mathematics was made by CORNELIS EKAMA (1773–1826, **B**). Upon his appointment in 1809 as professor of philosophy, physics, and astronomy at Franeker University, Friesland, he gave an address on the contribution of Frisians to the development of mathematics [EKAMA 1809]. It is a rather overdone paean of praise to Frisians who as surveyors, amateurs, teachers, or in some other manner had been involved in the field of mathematics, in Friesland or elsewhere. In 1823, EKAMA devoted a separate study to the best-known of them: GEMMA FRISIUS (1508–1555) [EKAMA 1823]. This was the first serious study of the life and work of this scholar.

2.5 Historiography of Mathematics in the Kingdom of the Netherlands 1815–1830

In the meantime, the political situation in the Low Countries had altered fundamentally. The Southern and Northern provinces were joined in the Kingdom of the Netherlands in 1815. Three new universities were set up in the South — in Ghent, Leuven and Liège — to join the ones that already existed in the North. Following the French example, each university had a separate faculty for mathematical and physical sciences. Some sporadic interest was displayed in the history of these fields, probably encouraged by a Royal Decree of 2 August 1815, which prescribed that a short history ought to be given for each science.

The country's universities organized annual competitions, and occasionally a subject from the history of mathematics was chosen. In 1824, for instance, the Faculty of Mathematics and Physics at Ghent University asked for a survey of pure mathematical studies since TSCHIRNHAUS (1651–1708) on caustics arising from reflection and refraction. In a competition announced by Leiden University in 1822, the winning entry, dealing with the problem of the quadrature of curves, was largely historical in character, drawing extensively on MONTUCLA's *Histoire des mathématiques* [MONTUCLA 1799/1802]. Winning entries of this kind may have been little more than solid student work, but they nevertheless illustrate how the history of mathematical ideas and problems began to prevail over the purely biographical and bibliographical approach.

In 1815, GERARD MOLL (1785–1838, **B**) became lecturer in physics at Utrecht University. His published works dealt mainly with astronomy, physics, and technology, but he was also interested in the history of mathematics and astronomy. He wrote a concise history of the mathematical sciences in the Low Countries in the eighteenth century entitled *Bijdragen tot de geschiedenis van de wiskundige wetenschappen in de Nederlanden in de achttiende eeuw* (Contributions to the History of the Mathematical Sciences in the Netherlands during the 18th Century). It was published in volume three of N. G. VAN KAMPEN's *Beknopte geschiedenis der letteren en wetenschappen in de Nederlanden* (Short History of Literature and Science in the Netherlands) [VAN KAMPEN 1826, 289–312].

In 1826, a *Musée des Sciences et des Lettres* was founded in Brussels. Public courses were inaugurated to foster interest in science, one of which was on the history of science. It was entrusted to ADOLPHE QUETELET (1796–1874, **B**), who gave his first lecture on March 6, 1827.

2.6 Historiography of Mathematics in Belgium after 1830

Following the break-up of the Kingdom of the United Netherlands in 1830 and the proclamation of Belgian independence, a movement arose in the new kingdom to explore what Belgians had achieved in the arts and sciences in the past in order to foster a sense of national identity. Scholars like GERARD MERCATOR (1512–1594), GEMMA FRISIUS, and STEVIN were rescued from obscurity. It is in this romantic and patriotic tradition that we must situate ADOLPHE QUETELET's historical work. As far as he was concerned, the history of science was first and foremost the history of the scientific practice of a particular people or era, rather than the history of the development of the science itself. This spirit is clearly apparent in his *Histoire des sciences mathématiques et physiques chez les Belges* (History of the Mathematical and Physical Sciences in Belgium), the first edition of which was published in 1864 [QUETELET 1864]. The work is predominantly biographical and bibliographical and was based mostly on secondary sources. Little attention is paid to the mathematical content of the cited works. All in all, QUETELET showed himself to be much more a fervent patriot than a good historian [SARTON 1935].

In the decades after QUETELET, three professors of mathematics displayed an interest in the history of their subject: PHILIPPE GILBERT (1832–1892, **B**) in

Leuven, PAUL MANSION (1844–1919, **B**) in Ghent, and CONSTANTIN LE PAIGE (1852–1929, **B**) in Liège. GILBERT's historical writings were largely biographical, the most original being those concerned with the trial of GALILEO. MANSION, too, was interested in GALILEO. His primary interest, however, was in mathematics itself and its history. One of his favorite topics was the body of mathematical and philosophical problems associated with Euclidean and non-Euclidian geometry. He also participated in the disscussion between M. CANTOR (1829–1920, **B**) and G. ENESTRÖM (1852–1923, **B**), carried out in the journal *Bibliotheca mathematica*, about problems of the history of mathematics (see Section 6.5).

Among the achievements of MANSION was the pioneering role he played in the foundation of the first Belgian university course devoted to the history of mathematics. Indeed, he was the first lecturer in the "History of the Physical and Mathematical Sciences," a course that the Belgian government set up in 1884 at the *Ecole Normale des Sciences* associated with the Faculty of Science at Ghent University.[2] When the *Ecole Normale* was abolished in 1889, the Higher Education Act of 1890 made the history of physical and mathematical sciences a compulsory course in the doctoral programme for mathematics and physics at the two state universities of Ghent and Liège, as well as at the universities of Leuven and Brussels. MANSION took charge of the course in Ghent, LE PAIGE in Liège, CHARLES J. DE LA VALLÉE POUSSIN (1866–1962) in Leuven, and EUGÈNE BRAND (1861–1936) in Brussels. The latter two lecturers were seemingly less enthusiastic about their appointments, as neither had ever worked on the history of mathematics. LE PAIGE, by contrast, had been directing his research increasingly towards the history of mathematics and astronomy. As early as 1884, in BONCOMPAGNI's *Bullettino* (see Section 3.7), he published the correspondence of RENÉ FRANÇOIS DE SLUSE (1622–1685) with PASCAL (1623–1662), HUYGENS (1629–1695), OLDENBURG (ca. 1620–1677), WALLIS (1616–1703, **B**), and others — the fruits of his research into the practice of mathematics in the old prince-bishopric of Liège [LE PAIGE 1884]. He published further findings in his *Notes pour servir à l'histoire des mathématiques dans l'ancien pays de Liège* (Notes Contributing to the History of Mathematics in the Ancient Provinces of Liège) [LE PAIGE 1888].

When MANSION retired in 1910, he asked if he might continue to teach his course on the "Elements of the History of the Physical and Mathematical Sciences." His request was denied, and the course was entrusted instead to ARTHUR CLAEYS (1875–1949), who taught it until his own retirement in 1945. In Liège, LE PAIGE retired in 1922 and was succeeded by JOSEPH FAIRON (1863–1925). Upon the latter's death, the course — now combined with a course on methodology of mathematics — was taken over by LÉON MEURICE (1866–1943). DE LA VALLÉE POUSSIN handed over the Leuven course to GEORGES LEMAÎTRE (1894–1966) in 1926. BRAND was succeeded in Brussels by THÉOPHILE DE DONDER (1872–1957), who was given permission in 1931 to pass on his teaching to JEAN PELSENEER

[2]For the program and intent of the course, see [MANSION 1888] and [MANSION 1900]. As a student at Ghent University, GEORGE SARTON took MANSION's course in 1908–09 and in 1909–10 [GILLIS 1973].

(1903–1985, **B**). Even though some of these lecturers were internationally famous mathematicians or astronomers, none of them, apart from PELSENEER, ever carried out any research in the history of science. As a consequence, in 1928, when the "Elements of the History of Physical and Mathematical Sciences" was no longer an obligatory course for mathematics and physics students, it withered except at the University of Brussels. After World War II, the torch was picked up in 1945 when the course was once again entrusted to PELSENEER. PAUL BOCKSTAELE (*1920) lectured on the history of mathematics to fourth-year mathematics students at the Dutch-speaking part of Leuven University from 1962 onwards. Nevertheless, the history of mathematics has yet to be recognized as an academic discipline in Belgium.

That Belgium enjoyed a solid reputation among historians of mathematics in the first quarter of the twentieth century is due primarily to the work of Father HENRI BOSMANS, S.J. (1852–1928, **B**). Between 1897 and his death in 1928, he published over 200 articles on the history of mathematics, especially on the Renaissance in the Southern Netherlands in the 16th and 17th centuries. His study of unpublished sources and careful scrutiny of original documents enabled him to revive entire areas of the history of mathematics.

FERNAND VAN ORTROY (1856–1934, **B**), professor of geography and history of geography at the University of Ghent, was a contemporary of BOSMANS. His historical work focused primarily on early cartography. Among his descriptive and critical bibliographies of various cartographers and mathematicians are those of GERARD (1521–1591) and CORNELIUS DE JODE (1568–1600), GEMMA FRISIUS, GERARD MERCATOR, PETRUS PLANCIUS (1552–1622), PETRUS APIANUS (1495–1552), and ABRAHAM ORTELIUS (1527–1598).

The engineer PAUL VER EECKE (1867–1959, **B**) undertook a remarkable project. An interest in the science of classical antiquity directed him to the study of the works of ARCHIMEDES. This led in 1921 to the publication of a French translation of ARCHIMEDES's surviving works. He continued his research in this field, publishing in succession translations of APOLLONIUS, DIOPHANTUS, THEODOSIUS, SERENUS, PAPPUS, EUCLID, DIDYMUS, ANTHEMIUS, and PROCLUS. By contrast, Belgian philologists and classicists showed little or no interest in the science of classical antiquity for some time. ADOLPHE ROME (1889–1971, **B**), however, professor at Leuven University, was a pioneer in this field. He focused principally on astronomy, more especially on PTOLEMY's *Almagest.* A small group of philologists formed around him, including ALBERT LEJEUNE (1916–1988, **B**) and JOSEPH MOGENET (1913–1980, **B**), who devoted themselves to the study of early Greek and Latin scientific texts. Thanks to the efforts of MOGENET, a course on the history of science in antiquity was added to the curriculum for students in classical philology at Leuven University. In 1965, he established the basis for a center that is still active in Louvain-la-Neuve, the present *Centre d'Histoire des Sciences Grecques et Byzantines.*

LUCIEN GODEAUX (1887–1975, **B**), professor of mathematics at Liège, was also interested in the history of his profession. He wrote several articles on the status of mathematical research in Belgium in the first half of the twentieth century, and produced an *Esquisse d'une histoire des sciences mathématiques en Belgique* (Essay on the History of the Mathematical Sciences in Belgium) [GODEAUX 1943]. Meanwhile, the history of mathematical and astronomical instruments is dealt with in the work of the engineer HENRI MICHEL (1885–1981, **B**). He published many books and articles on astrolabes, sundials, and other measuring instruments.

The absence of institutions of higher education in the Grand Duchy of Luxembourg meant that the teaching of the history of mathematics could not develop. That is not to say, however, that there was no historiography at all in this field. Between 1939 and 1966, ALBERT GLODEN (1901–1966, **B**) published a series of historical articles on subjects including Luxembourg mathematicians and topics relating to analysis and the theory of numbers.

2.7 Historiography of Mathematics in the Netherlands after 1830

Articles on the history of mathematics, primarily on Dutch mathematicians, were published sporadically in the Netherlands in the decades after 1830 [BIERENS DE HAAN 1891]. The most interesting of these were written by JOHANNES JACOBUS DODT VAN FLENSBURG (1800–1847, **B**) and GEORGE AUGUSTE VORSTERMAN VAN OYEN (1836–1915, **B**). Above all, however, DAVID BIERENS DE HAAN (1822–1895, **B**) gave a new impetus in the final quarter of the 19th century to the study of the mathematical past. His immense knowledge of Dutch mathematical literature gave birth to his *Bibliographie néerlandaise historico-scientifique sur les sciences mathématiques et physiques* (Historical-Scientific Dutch Bibliography on the Mathematical and Physical Sciences). This work, which is still frequently consulted today, was initially published in BONCOMPAGNI's *Bullettino* between 1881 and 1883 before appearing in book form [BIERENS DE HAAN 1883]. In the meantime BIERENS DE HAAN published numerous studies on 16th- and 17th-century mathematicians in his series *Bouwstoffen voor de Geschiedenis der Wis- en Natuurkundige Wetenschappen in de Nederlanden* (Building Stones for the History of the Mathematical and Physical Sciences in the Netherlands), on which he commenced work in 1874. All of these strongly biographical and bibliographical studies were based on original documents and are noteworthy for their accurate and detailed description of the works employed.

BIERENS DE HAAN's greatest merit, however, was that he helped establish a long tradition of editing historically important documents. In 1879, he published two early Dutch treatises on life insurance. These were followed in 1884 by two previously unpublished small works by STEVIN. That same year, he reprinted ALBERT GIRARD's (1595–1632) *Invention nouvelle en l'algèbre* (New Invention in Algebra) and two small mathematical works attributed to SPINOZA (1632–1677).

In the meantime, the Royal Academy of Science in Amsterdam had expressed a desire to see the publication of the collected works of CHRISTIAAN HUYGENS

(1629–1695). BIERENS DE HAAN was entrusted with the project and devoted his energies to this enterprise for the remainder of his life. The first volume of the *Œuvres complètes de Christiaan Huygens* (Complete Works of Christiaan Huygens) was published in 1888. The venture was continued after BIERENS DE HAAN's death by JOHANNES BOSSCHA (1831–1911, **B**) and then by DIEDERIK JOHANNES KORTEWEG (1848–1941, **B**) and JOHAN ADRIAAN VOLLGRAFF (1877–1965, **B**), who published the twenty-second and final volume of this exemplary edition in 1950. KORTEWEG, who succeeded BOSSCHA as editor-in-chief in 1911, wrote numerous articles on HUYGENS's life and works. VOLLGRAFF took responsibility for the HUYGENS project in 1927. He lectured on the history of mathematics and physics as *privaatdocent* at Leiden University between 1911 and 1916.

An historian of science of international standing who must be mentioned here is CORNELIS DE WAARD (1879–1963, **B**). In 1914, he discovered unpublished documents relating to FERMAT in the Groningen University library. He published them in 1922 as a *Supplément aux tomes I–IV* of FERMAT's *Œuvres*, which had been edited by PAUL TANNERY (1843–1904, **B**) and CHARLES HENRY (1859–1926) (see Section 1.9).

The research carried out over several decades on the work of HUYGENS and others had a profound influence on attitudes towards the historiography of science in the Netherlands. In addition to biographical, cultural, and historical aspects, the history of mathematical ideas and problems began to occupy an increasingly important place. This shift is illustrated by the historical studies of NICOLAAS L. W. A. GRAVELAAR (1851–1913, **B**), lecturer at the teacher training college in Deventer. His publications on PITISCUS's (1561–1613) *Trigonometria* and the works of JOHN NAPIER are characterized by great accuracy and scrupulous study of the sources.

Even greater emphasis was placed on the historical development of mathematical sciences in the historical writings of HENDRIK DE VRIES (1867–1954, **B**), professor at the Municipal University of Amsterdam. His lectures took in algebra and analysis, but from 1921–22 onwards, he focused increasingly on his preferred field, giving public lectures on the development of geometry. These culminated in a series of articles in the *Nieuw Tijdschrift voor Wiskunde* (New Journal for Mathematics), which were later collected, together with other items, in a three-volume publication entitled *Historische Studiën* [DE VRIES 1926/40]. DE VRIES wrote in the introduction that he wanted to focus attention on the historical development of very precisely defined topics, even of specific problems or theorems. He pointed out the didactic benefits that the historical approach to mathematical problems could offer.

In 1923 the young mathematician DIRK JAN STRUIK (1894–2000, **B**) became a *privaatdocent* at Utrecht University. His inaugural lecture dealt with the development of differential geometry. It was during a stay at Rome and Göttingen in 1925–26 that he first began to take a serious interest in the history of mathematics. From this period date his first historical papers. Shortly after returning to Holland he accepted a visiting appointment at the Massachusetts Institute of Technology

(MIT) and left for the United States in November of 1926. He remained on the faculty there until his retirement in 1960 (see Section 15.4.4).

Interest in the history of science also arose around 1920 among secondary school teachers. HERMANUS J. E. BETH (1880–1950, **B**) and EDUARD JAN DIJKSTERHUIS (1892–1965, **B**) played an inspirational role in this process. The latter gave a series of lectures on the history of mathematics in 1927 and 1928 as a kind of continuing education for teachers. DIJKSTERHUIS stressed the importance of the history of science as an element in teacher-training and was given the opportunity to put his ideas into practice in 1930, when he was appointed external lecturer in history of mathematics at Amsterdam Municipal University. Two years later, he was offered the corresponding post at Leiden.

An initiative that was to have a positive impact on the historiography of mathematics in the Netherlands was the foundation by DIJKSTERHUIS and BETH of the *Historische Bibliotheek voor de Exacte Wetenschappen.* The aim was to publish a series of monographs on historically important topics as well as annotated editions of classic works. First to appear were DIJKSTERHUIS's study of EUCLID's *Elements* [DIJKSTERHUIS 1929/30] and BETH's *Inleiding in de niet-Euclidische meetkunde op historischen grondslag* (Introduction to Non-Euclidian Geometry on an Historical Basis) [BETH 1929]. These were followed in 1932 by BETH's study on NEWTON's *Principia* [BETH 1932] and in 1938 by the first volume of DIJKSTERHUIS's *Archimedes* [DIJKSTERHUIS 1938]. The final work in the series was published after the war: BARTEL LEENDERT VAN DER WAERDEN's *Ontwakende Wetenschap* (Science Awakening), a history of Egyptian, Babylonian, and Greek mathematics [VAN DER WAERDEN 1950]. Like STRUIK, VAN DER WAERDEN (1903–1996, **B**) only worked briefly in the Netherlands. In 1951 he was appointed professor of mathematics at Zürich (see Section 4.5).

In 1944, during the latter part of World War II, DIJKSTERHUIS accepted the invitation of Amsterdam Municipal University to teach a course on the history of mathematics, a position he lost after the war under unpleasant circumstances. On several occasions before the war he had cited the need for a modern edition of SIMON STEVIN's works. He approached the Netherlands Royal Academy of Science about this in 1946, which resulted in the first volume of *The Principal Works of Simon Stevin*, containing STEVIN's writings on mechanics, published in 1955. DIJKSTERHUIS wrote the general introduction and the commentaries on the individual works. Volume two, which contains the mathematical writings, followed in 1958, with introduction and commentaries by STRUIK. Volumes three, four, and five were published in 1961, 1964, and 1966, respectively.

Recognition of the history of science as an academic discipline only came after the war. The first Dutch chair for the history of science was established at the Free University of Amsterdam in 1945. Its first occupant was REIJER HOOYKAAS (1906–1994), who held the post until his retirement in 1976. Between 1967 and 1976, he performed the same task at Utrecht University. He chose to focus, however, on the history of the natural sciences, particularly chemistry and geology, rather than on mathematics. In 1953, DIJKSTERHUIS was appointed professor of history of the "exact sciences" at Utrecht, a post he also took at Leiden two years later. Ill health obliged him to stop teaching in 1963, and he died two years later. Former students of DIJKSTERHUIS include ALPHONSE J. E. M. SMEUR (*1925), who investigated the 16th-century arithmetics printed in the Netherlands, and HUBERTUS L. L. BUSARD (*1923) whose researches helped to disentangle the complex medieval Euclid tradition.

An extraodinary professorship in history of mathematics was founded in 1969 at Amsterdam Municipal University especially for EVERT M. BRUINS (1909–1990, **B**), who had built up an international reputation as a versatile historian of mathematics. His teaching, however, which he continued until his retirement in 1979, was less successful. He largely limited himself to private tutorials and to directing working groups. He had virtually no students. Meanwhile, the mathematician HANS FREUDENTHAL (1905–1990, **B**) gained a special place in the historiography of mathematics thanks to his original studies on both early and more recent mathematics.

A few words on the development of academic teaching of the history of mathematics since 1970 will serve to complete this survey. A course on the history of mathematics was offered as an option at Groningen in 1971, but was dropped the following year due to illness on the part of the lecturer. It was, however, revived in the 1990s. An attempt was made at Delft Technical University in 1979 to found a "Cultural History of Mathematics" course. Utrecht University began a course on the history of mathematics in 1973. Initially intended to be given every two years, it was taught by HENK J. M. BOS (*1940). The University's Institute of Mathematics began to pay increasing attention to the history of mathematics from 1977 onwards, culminating in the foundation of a department devoted to the History and Social Function of Mathematics under BOS's direction. This has since developed into an important center for both teaching and research in the history of mathematics. Among staff members, ANTONIE FRANS MONNA (1909–1995) deserves special mention for his research into the history of contemporary mathematics, as does JAN HOGENDIJK (*1955) for his contributions to the history of Arabic/Islamic mathematics.

2.8 Amateurs and Professionals, Journals and Societies

Although works in the Low Countries devoted to the history of science in general, or to the history of mathematics in particular, were limited in the first half of the 19th century, other enterprises with wider or different objectives, such as cultural historical studies or occasional publications honoring deserving scientists, also reflect serious interest in the history of science.

For example, the seven-volume *Holland's Roem in Kunsten en Wetenschappen* (Holland's Fame in the Arts and Sciences) by HENDRIK COLLOT D'ESCURY (1773–1845) [COLLOT D'ESCURY 1824/44] gave ample consideration to mathematics and the natural sciences. The sixth volume, which appeared in 1835, is devoted entirely to the development of the mathematical sciences and their applications. It treats, in turn, mathematics (pp. 1–135), astronomy, geography and navigation (pp. 137–424), the art of fortification (pp. 427–539), and hydraulic engineering (pp. 541–700).

The dedication in 1846 of a statue in honor of SIMON STEVIN in Bruges, the city of his birth, occasioned the publication of a number of studies on his life and work. Alongside well-intentioned amateur appraisals, there are also contributions that represent serious historical research. The effort to celebrate the cartographer and geographer GERARD MERCATOR was primarly the work of the physician JAN HUBERT VAN RAEMDONCK (1817–1899). He wrote about prehistory, Gallo-Roman archaeology, and the history of public health, but his most striking achievement is his biography *Gerard Mercator, sa vie et ses oeuvres* (Gerard Mercator, his Life and Works) [VAN RAEMDONCK 1869], which rests on solid scientific research. In the Netherlands, PIETER HARTING (1813–1885), professor of medicine first in Franeker and then in Utrecht, campaigned for a national commemoration of CHRISTIAAN HUYGENS. In 1868, he published the first biography of HUYGENS [HARTING 1868].

The authors of these studies set the tone for the 19th century, but their approaches were predominantly biographical, usually nationalistic, and rarely focused on or devoted much consideration to purely scientific or mathematical details.

The history of science began to be professionalized in the last quarter of the 19th century. In the Low Countries, the Netherlands took the lead. Initially, primarily physicians were involved; CAREL EDOUARD DANIËLS (1839–1920) and HENDRIK PEYPERS (1853–1904) launched the journal *Janus* in 1896, subtitled *Archives internationales pour l'histoire de la médecine et la géographie médicale* (International Archives for the History of Medicine and Medical Geography). Publication of *Janus* was suspended in 1941 due to the War, but started again in 1957 with a new subtitle: *Revue internationale de l'histoire des sciences, de la médecine, de la pharmacie et de la technique* (International Review of the History of Science, Medicine, Pharmacy, and Technology). EVERT M. BRUINS became a co-editor of the journal, and assumed full editorship in 1963. He saw to it that ample space was devoted to the history of mathematics. *Janus* ceased publication with volume 73 in 1990. Meanwhile, at the beginning of the century, thanks to the efforts E. C. VAN LEERSUM (1862–1938), professor of pharmacology in Leiden, and J. A. VOLLGRAFF, a society for the history of science was founded in 1913 — known today as the *Genootschap voor de Geschiedenis der Geneeskunde, Wiskunde, Natuurwetenschappen en Techniek* (Society for the History of Medicine, Mathematics, Science and Technology), abbreviated as *Gewina.* In 1978, the Society commenced publication of the *Tijdschrift voor de Geschiedenis der Geneeskunde, Natuurwetenschappen, Wiskunde en Techniek* (Journal for the History of Medicine, Science, Mathematics, and Technology).

In Belgium, organizations for the history of science arose more slowly than in the Netherlands. In 1913, from his home in Wondelgem near Ghent, GEORGE SARTON (1884–1956, **B**) launched the international journal *Isis.* With the outbreak of World War I, however, he left Belgium. After a short stay in England, he emigrated in 1915 to the United States (see Section 15.4.3). In 1933, a *Comité Belge d'Histoire des Sciences* was formed, headed successively by PAUL VER EECKE (1867–1959, **B**) (president from 1946 to 1950), ADOLPHE ROME (1889–1971, **B**), (president from 1951 to 1955), and JOSEPH MOGENET (1913–1980, **B**) (president from 1961 to 1965). JEAN PELSENEER (1903–1985, **B**) was the secretary from 1933 until 1955. The object of the *Comité* was to further the history of the sciences in Belgium, but it was not very successful in this endeavor. From the 1970s on, its activities declined. An attempt to revive the *Comité* in 1989 produced little. Somewhat earlier, in 1959, a group of historians of science launched the journal *Scientiarum Historia*, devoted to the history of medicine, mathematics, and the natural sciences. After 15 volumes, because of financial problems, publication was suspended in 1973. In the meantime, the *Zuidnederlands Genootschap voor de Geschiedenis van Geneeskunde, Wiskunde en Natuurwetenschappen* was founded in 1960. Thanks to this Society, publication of *Scientiarum Historia* was resumed in 1990, after an interruption of sixteen years. Meanwhile, the National Committee for Logic, History, and Philosophy of Science was formed in 1973. Its task is to provide official contact between Belgium and the two Divisions of the International Union of the History and Philosophy of Science, and has assumed most of the activities of the *Comité Belge d'Histoire des Sciences*.

2.9 Conclusion

The history of history of mathematics in Belgium and The Netherlands has until recently remained an interest of scattered individuals, mostly mathematicians, who have studied the history of mathematics either because they were required to teach the subject, or because they wished to honor and advance their own native nationalistic agendas. Early interest was typically bibliographic and antiquarian, and was devoted either to recovering important texts of the past or to recounting the lives of major mathematicians.

The first concerted and serious interest in history of mathematics is reflected in the numerous publications of HENRI BOSMANS early in the 20th century. This Belgian Jesuit was especially concerned with Renaissance mathematics and works produced in the Southern Netherlands in the 16th and 17th centuries, and he drew substantially on previously unpublished archival materials in the course of his research.

Meanwhile, even more substantial efforts to promote the history of mathematics were made in The Netherlands, thanks at first to interest stemming from the official publication by the Dutch Royal Academy of Sciences of the works, largely mathematical, of CHRISTIAAN HUYGENS. Especially influential in promoting history of mathematics among high school teachers were lectures given by H. J. E. BETH and E. J. DIJKSTERHUIS beginning in the 1920s. They were also instrumental in launching a series of historical monographs and edited texts, the last of which was B. L. VAN DER WAERDEN's study of ancient mathematics and astronomy, *Science Awakening* (1950).

In 1945, after World War II, the first chair for history of science was established at the University of Amsterdam, followed in 1953 by the appointment of DIJKSTERHUIS as professor of history of the exact sciences at the University of Utrecht. There, especially thanks to the historical interests of HANS FREUDENTHAL, a strong center for history of mathematics has emerged, with the founding of a Department for the Study of the History and Social Function of Mathematics.

Of various journals for history of science, including mathematics, founded in Belgium and The Netherlands, two survive. The oldest, *Scientiarum Historia*, was begun in Belgium in 1959, but stopped publishing in 1973. The journal was revived several decades later, in 1990, but this time with the sponsorship of the *Zuidnederlands Genootschap voor de Geschiedenis van Geneeskunde, Wiskunde en Natuurwetenschappen* (South Netherlands Society for the History of Medicine, Mathematics, and Science). Meanwhile, in 1978, the Dutch *Genootschap voor Geneeskunde, Wiskunde, Natuurwetenschappen en Techniek (Gewina)* (Dutch Society for the History of Medicine, Mathematics, Science and Technology) had begun publication of its own journal, *Tijdschrift voor de Geschiedenis der Geneeskunde, Natuurwetenschappen, Wiskunde en Techniek.*

At present, the history of mathematics appears to receive only slight encouragement in Belgium, but it has flowered substantially in The Netherlands. This is due in particular to those working at the University of Utrecht, where an especially active circle of younger historians of mathematics is producing a new generation with excellent training in both the technical and historical skills needed to pursue history of mathematics in all its diversity.

Chapter 3

Italy

UMBERTO BOTTAZZINI

3.1 Introduction

In referring to Italy in the preface to the Italian edition of his book, *A Concise History of Mathematics*, DIRK J. STRUIK (1894–2000, **B**) emphasized that "no country except China has a longer tradition of mathematical activity, much of it of basic importance. We may begin with BOETIUS (if not with the Roman agrimensores) and continue to the present day; a period of more than a thousand years" [STRUIK 1981, 9]. Naturally, there were ups and downs, but one is struck by the continuity of creative interest in mathematics. STRUIK concluded his preface by recalling his stay in Rome from 1924 to 1925 when, as a young researcher, he worked with TULLIO LEVI-CIVITA (1873–1941). It was on "the historical soil of Italy," and through ETTORE BORTOLOTTI (1866–1947, **B**) and GIOVANNI VACCA (1872–1953, **B**), that he became interested in the history of mathematics and published his first papers in the field. Even STRUIK's wife, RUTH, became attracted to history; she contributed historical notes and commentaries to volume X of EUCLID's *Elements* edited by FEDERIGO ENRIQUES (1871–1946, **B**; see Section 3.11).

STRUIK's impression of mathematics in Italy also rings true for the writing of history of mathematics, which has a long tradition beginning in the Middle Ages and Renaissance, and continuing to the present. Of course, the term "history" has to be understood in a rather broad sense. Indeed, as mathematics has changed over the course of the centuries, so has the history of mathematics. Although rather far from the modern understanding of what constitutes the field, the editorial work of medieval and Renaissance scholars who translated and edited ancient texts of classic Greek mathematics seems an appropriate starting point for the history of mathematics in Italy. For more than two centuries, historical work went hand-in-hand with the development of original mathematical research, and contributed significantly to creating the sense of historical continuity mathematicians felt with antiquity.

During the 17th and 18th centuries, the history of mathematics found its place in Italy mainly in encyclopedias and in general works on the history of literature and culture. At the end of the 18th century, the work of PIETRO COSSALI (1748–1815, **B**) on the history of algebra marks the beginning of a more modern approach, and some forty years later, GUGLIELMO LIBRI (1803–1869, **B**) published his influential *Histoire des sciences mathématiques en Italie* (History of the Mathematical Sciences in Italy) (1838–1841). The works of both COSSALI and LIBRI are indicative of the distinctive character of Italian studies on the history of mathematics. On the one hand, by emphasizing the roles of Italians — and particularly of GALILEO (1564–1642) — in the development of mathematics, LIBRI's *History* served to promote patriotic sentiments among Italian scientists. From the middle of the 19th century to World War II, the history of mathematics was not simply considered a mathematical discipline; it played a political role in giving Italian mathematicians a sense of belonging to a scientific community which had its own historical tradition and cultural unity. During the 1920s and 1930s this often turned into a form of political nationalism, and in some cases writing the history of mathematics in Italy was simply reduced to the history of Italian mathematics.

On the other hand, by the end of the 18th century, the history of mathematics had become a research subject for mathematicians, but without notable connections to either philosophers or historians of science. In other words, one could say that in Italy history was studied and written by mathematicians for mathematicians. Most Italian historians of mathematics began as practicing mathematicians who later became interested in history. This well-established tradition, which dates back at least to LAGRANGE (1736–1813), has continued to the present century and until well after World War II.

In recent times, the historiography of mathematics in Italy has become the subject of an increasing number of studies. These make it possible to suggest a general picture which is, however, still far from complete. It involves not only the development of mathematics but also the cultural and political history of Italy. Taking this into account, what follows is a preliminary and provisional outline of a long and complex history, one that invites further historical research.

3.2 The Rediscovery of Classical Mathematics

The first traces in Italy of an interest in the history of mathematics is reflected in the rediscovery of the most important mathematical texts of antiquity, although the editorial results of medieval scholars can hardly be equated with history in a modern sense. More accurately, their work re-established a body of mathematical knowledge which had been nearly lost for centuries and which was to serve as the basis for the modern development of mathematics. However, the rediscovery of classical Greek mathematics also reinforced the sense that mathematics must be understood to a certain extent as an historical discipline, one which has its roots far in the past.

During the Middle Ages, a number of translations and editions of ancient texts became available throughout Western Europe. In the 12th century the most important figures were PLATO OF TIVOLI and GERARD OF CREMONA. Although they were born in Italy, they were active in Spain, where towns like Seville and Toledo were among the liveliest centers of Western culture and science. PLATO and GERARD translated many ancient mathematical texts from Arabic and Hebrew into Latin. PLATO's translations included ARCHIMEDES's *De mensura circuli* (Measure of the Circle) and THEODOSIUS's *Spherica* (Spherics). GERARD is credited with translations of some fifty works, including EUCLID's *Elements*, PTOLEMY's *Almagest*, and AL-KHWĀRIZMĪ's *Algebra.* (An extensive study of GERARD OF CREMONA's life and work was published by BALDASSARRE BONCOMPAGNI (1821–1894, **B**) in 1851).

In the 13th century, an edition of EUCLID's *Elements*, largely based on an Arabic translation, was due to CAMPANUS OF NOVARA. This became the *editio princeps*, the standard edition of the *Elements*. After the invention of printing it was reissued numerous times during the Renaissance, even after the first direct translation from the Greek by BARTOLOMEO ZAMBERTI (1473?–?) had appeared in print in 1505. Some years later, in 1509, LUCA PACIOLI (1445?–1514?) edited an improved version of CAMPANUS's Latin translation. Unfortunately, PACIOLI's own Italian translation of the *Elements*, although circulated in manuscript, was eventually lost. The first Italian version of the *Elements* (mistakenly credited to EUCLID of Megara), was published by NICCOLÒ TARTAGLIA (1499/1500–1557) in 1543. That same year TARTAGLIA also published a Latin edition of ARCHIMEDES's *Floating Bodies.*

All this editorial activity anticipated the work of FEDERICO COMMANDINO (1509–1575) and FRANCESCO MAUROLICO (1494–1575). COMMANDINO was by far the most important Renaissance editor of ancient mathematical texts. He made Latin translations and editions with commentaries of ARCHIMEDES's major texts, including *Measure of the Circle*, *Spirals*, *Quadrature of the Parabola*, *Conoids and Spheroids*, and the *Sand Reckoner*. In addition, he translated and edited works by ARISTARCHUS, EUTOCIUS, SERENUS, PTOLEMY, and HERON, as well as PAPPUS's *Collections* (books III-VIII). His critical edition of EUCLID's *Elements*, in Latin (1572), was long the standard reference. A resident of the court of the Duke of Montefeltro in Urbino, COMMANDINO was able to create a "school" which attracted some of the leading Italian and foreign mathematicians of his day, including GUIDOBALDO DEL MONTE (1545–1607), JOHN DEE (1527–1608), and BERNARDINO BALDI (1533–1617, **B**), who later served as mathematician to the Court of Gonzaga in Mantua.

BALDI coupled his editorial work with a tremendous amount of original research. He was the author of some one hundred works of a mathematical and literary character. He translated ARATUS's *Phenomena* into Italian, and edited with both mathematical and historical commentaries the pseudo-Aristotelian *Questions of Mechanics* which, according to STILLMAN DRAKE ([DSB 1970/90, **1**, 419]), was one of "the most important works of this kind to appear up to that time." BALDI also translated HERON's *Automata* into Italian, prefacing it with a history of mechanics. He also wrote the *Vite de' matematici* (Lives of Mathematicians), probably between 1587 and 1596, inspired by PLUTARCH's *Lives* and the *Lives of Philosophers* by DIOGENES LAERTIUS, as well as by the *Scolarum mathematicarum libri unus et triginta* (Thirty-one Books on Schools of Mathematics) (1569), in which RAMUS provided a general, historical introduction to the subject.

BALDI's *Lives* is an impressive collection of biographies amounting to some two thousand handwritten (and largely unpublished) pages wherein he describes the lives of 201 mathematicians from antiquity to the Renaissance [BILINSKI 1977]. An abbreviated, chronologically ordered version of the *Lives*, written by BALDI himself, was eventually published in 1707. Many other manuscripts of BALDI's *Lives*, including the lives of Arabic and Italian mathematicians, as well as biographies of PYTHAGORAS and LUCA PACIOLI, were eventually edited by ENRICO NARDUCCI (1832–1893) and published in BONCOMPAGNI's *Bullettino* (see Section 3.7). (A critical edition of BALDI's *Lives* of medieval and Renaissance mathematicians has recently been published (in 1998) by ELIO NENCI; see [Baldi 1998]). In 1714 BALDI's *Vita di Federico Commandino* (Life of Federico Commandino) was published in the *Giornale de' Letterati d'Italia.* This journal, a review of general culture, is a valuable source for the history of mathematics in the 17th and 18th centuries [LORIA 1899]. Indeed, many mathematicians, including DOMENICO GUGLIELMINI (1655–1710), TOMMASO CEVA (1649–1736), and GAETANO FONTANA (1645–1719), contributed to it. The *Giornale* also published advanced research papers in mathematics, among them a paper in 1692 by LEIBNIZ (1646–1716) on the differential calculus, which was the first on the subject to appear in Italy.

In contrast to COMMANDINO, MAUROLICO worked in Messina, far from the centers of Italian Renaissance culture (see [MOSCHEO 1988]). He edited a number of Greek texts, including EUCLID's *Phaenomena*, AUTOLYCUS's *Sphaera*, ARCHIMEDES's *De quadratura parabolae* (Quadrature of the Parabola), *De mensura circuli* (Measure of the Circle), *De spiralibus* (Spirals), and *De conoidibus et spheroidibus* (Conoids and Spheroids), as well as the *Conics* of APOLLONIUS for which he reconstructed books V and VI. In fact, in MAUROLICO's day only the first four books of this work were known. After the discovery in the 17th century of an Arabic manuscript containing books V to VII, the *Conics* was eventually edited again by GIOVANNI ALFONSO BORELLI (1608–1679).

MAUROLICO, sometimes called the "Archimedes of the Renaissance," was primarily interested in original mathematical research rather than philological restorations. He regarded ancient texts as sources of inspiration for his own mathematical work. Many of his editions were only published posthumously and a number of his manuscripts have been lost. (An edition of his collected works has recently been planned by an international group of historians of mathematics.)

MAUROLICO's fame attracted a number of scholars to Messina, including the German Jesuit CHRISTOPH SCHLÜSSEL (1537–1612), better known by the latinized name of CLAVIUS. A professor of mathematics at the *Collegium Romanum* in Rome, CLAVIUS was one of the most eminent mathematicians of his day. His celebrated edition of EUCLID's *Elements* (1574) — inspired by MAUROLICO's work — was reprinted several times during the 17th century.

This tradition of editing classic texts continued. Mathematicians like MARINO GHETALDI (1566–1626), VINCENZO VIVIANI (1622–1703), and VITALE GIORDANI (1633–1711) edited and "reconstructed" works of APOLLONIUS, ARISTAEUS THE ELDER, and EUCLID. As for EUCLID's *Elements* in particular, in addition to the edition by CLAVIUS, BORELLI's *Euclides restitutus* (Euclid Restored) (1658) and GIORDANI's *Euclide restituto* (Euclid Restored) (1680) are also worth mentioning.

Although these works belong more properly to mathematics than to the history of mathematics, they show the lasting interest of Italian mathematicians in the classical sources of their science. During the 17th century, interest in reconstructing ancient texts began to decrease and eventually disappeared with the concomitant increase of new, original mathematics. The "*imitatio*" and "*emulatio*" of the authors of antiquity, which had served to guide Renaissance mathematicians, ceased to be goals of mathematical research. For mathematicians, the long quarrel over ancients versus moderns eventually came to an end with the proud affirmation of the superiority of modern over ancient authors.

3.3 Interlude: Contributions of the Jesuits

CLAVIUS made the Jesuit *Collegium Romanum* in Rome one of the outstanding mathematical centers of his day. From the last decades of the 16th century until the second half of the 18th century, Jesuit schools and colleges played an important role in the development and transmission of mathematical knowledge, not only throughout Europe but also in the Far East, most notably China. Aside from CLAVIUS, there were many other Jesuits who did original mathematical work, or who contributed to the history of mathematics by writing chronologies and encyclopedic works which included substantial historical material.

An early example of mathematical chronology is the *Clarorum mathematicorum chronologia* (Chronology of Illustrious Mathematicians) appended by the mathematician and astronomer GIUSEPPE BIANCANI (1566?–1624) to a treatise *Aristotelis loca mathematica ex universis ipsius operibus collecta et explicata* (Aristotle's Mathematical Topics Extracted from His Collected Works and Explained) (1615). Another example is the *Almagestum novum* (New Almagest) (1651) by the astronomer GIOVAMBATTISTA RICCIOLI (1598–1671). This work is a treatise in two volumes on astronomy, greatly enriched by historical notes and commentaries. In particular, the first volume begins with a *Chronicon* (Chronicle), wherein astronomers, astrologers, and geographers are listed by RICCIOLI in both chronological and alphabetical order along with a substantial amount of biographical information.

By far the most important 17th-century encyclopedia of mathematics was due to the Jesuit scientist CLAUDE-FRANÇOIS MILLIET DESCHALES (1621–1678), who was born in Savoy, a region that became part of France in 1860. His *Cursus seu mundus mathematicus* (Course or World of Mathematics) (1674; second edition in four volumes, 1690), includes 31 "treatises." The opening treatise, *Tractatus proemialis de progressu matheseos, et de illustribus mathematicis* (Introductory Treatise on the Progress of Mathematics and on Illustrious Mathematicians), is entirely devoted to the history of mathematics, which is understood explicitly in terms of progress. The development of each branch of the mathematical sciences of DESCHALES's time is described on the basis of biographies of the most important mathematicians from antiquity up to the 17th century.

3.4 The History of Mathematics in the Enlightenment

In tune with the philosophy of the Enlightenment, Italian scholars of the 18th century regarded the history of mathematics as a branch of human knowledge. Therefore it was included along with general histories of human activities and culture, and was cultivated by erudite scholars rather than by mathematicians. Throughout this period a number of histories of "literature" were published in Italy, most of which included mathematics and its history. (The term "literature" had a more general sense then, and included any kind of general cultural activity.) RICCARDI (1828–1898, **B**) listed more than thirty histories of this kind, published from the beginning of 18th century up to the early 19th century, which were of interest for the history of mathematics.

An outstanding example of such works was *Gli scrittori d'Italia* (Italian Writers) (1753–1763) by GIOVANNI MARIA MAZZUCCHELLI (1707–1765), wherein writers are listed in alphabetical order. In 1737 MAZZUCCHELLI published *Notizie storiche e critiche intorno alla vita, alle invenzioni e agli scritti d'Archimede siracusano* (Historical and Critical Notes on the Life, Inventions, and Writings of Archimedes of Syracuse), which includes historical details on writings by TARTAGLIA concerning ARCHIMEDES. As for MAZZUCCHELLI's major work, he was only able to publish a part of *Italian Writers* prior to his death in 1765. Some 35 volumes of unpublished manuscripts were eventually given as a present to Pope PIUS IX in 1865, and are now kept in the Vatican library [NARDUCCI 1867]. At BONCOMPAGNI's suggestion, from this collection of manuscripts NARDUCCI extracted the "Memorie concernenti il marchese G. C. de' Toschi di Fagnano" (Papers Concerning the Marquis G. C. de' Toschi di Fagnano), as well as a paper on BENEDETTO CASTELLI, both of which were published in BONCOMPAGNI's *Bullettino* (vols. III and XI, respectively).

Perhaps the most remarkable example of these general histories is provided by the *Storia della letteratura italiana* (History of Italian Literature) (1772–1795) in 11 volumes (and 14 parts) by the former Jesuit GIROLAMO TIRABOSCHI (1731–1794). (The order of Jesuits was suppressed in 1773). TIRABOSCHI's *History* was reprinted many times, beginning with an edition in 9 volumes (and 16 parts) published posthumously in Venice in 1795–96. Each volume of this work, very popular in its day, opens with a chapter on "Philosophy and Mathematics" that includes historical commentaries and biographical details of mathematicians from antiquity up to the early 18th century. TIRABOSCHI included in his *History* the culture of the ancient Greek colonies of Southern Italy (Magna Graecia). Indeed, according to him, "the literary history of any country is the history of the peoples who lived in that country whether it was their ancient homeland or whether they came there from elsewhere" [TIRABOSCHI 1772–1795, **1**, 34]. On the basis of this rather questionable historical criterion, TIRABOSCHI pointed with pride to such Italians as PYTHAGORAS and ARCHIMEDES. This became typical of Italian historiography until the present century [CASINI 1998], and was one of the many issues over which BORTOLOTTI and GINO LORIA (1862–1954, **B**) would disagree (see Section 3.10).

In addition to his *History*, TIRABOSCHI published the *Biblioteca modenese o Notizie della vita e delle opere degli scrittori nati negli Stati del Duca di Modena* (Modena Library or Notices on the Lives and Works of Writers Born in the States of the Duke of Modena) (1781–1786) in 6 volumes. Five additional volumes of *Notizie bibliografiche in continuazione della Biblioteca modenese* (Bibliographical Notices Continuing the Modena Library) by the same author eventually appeared between 1833 and 1837. These volumes stemmed from TIRABOSCHI's post as the "First Librarian" of the library of the Duke of Modena. In 1814 the engineer ANTONIO LOMBARDI (1768–1847), who had been TIRABOSCHI's assistant since 1790, was appointed to succeed him. Three years later, in 1817, LOMBARDI was elected secretary of the *Società italiana delle scienze, detta dei XL* (Italian Society of Sciences, Called of the XL). In addition to mathematical work, LOMBARDI also did historical research. His first important historical study was a biography of RUFFINI, the *Notizie sulla vita e su gli scritti di Paolo Ruffini* (Notices on Paolo Ruffini's Life and Works) (1824). His *Storia della letteratura italiana del secolo XVIII* (History of Italian Literature in the 18th Century) (1827–1830), in 4 volumes, was intended to be a continuation of TIRABOSCHI's *History*. Some 170 pages of the first volume are devoted to "pure and mixed mathematics." There LOMBARDI presented a short history of mathematics in Italy in the 18th century, from SACCHERI (1667–1733) to the Venetian mathematician ANTONIO COLLALTO, who died in 1820. Nationalism motivated LOMBARDI's work. Indeed, he intended "to vindicate the honor which is due Italians" [Lombardi 1827–1830, vol. 1, 351]. At the same time he wanted to improve the "very valuable" *Histoire des mathématiques* of MONTUCLA and LALANDE, which was "rather superficial" on the subject of Italian mathematicians, who were "usually dismissed by briefly hinting at the content of their works" [Lombardi 1827–1830, **1**, 352].

Among general histories, an important example is *Dell'origine, progressi e stato attuale d'ogni letteratura* (On the Origin, Progress, and Present Status of Every Literature) (1782–1799) by GIOVANNI ANDRES (1740–1817), in seven volumes (a new edition in 23 volumes was published in 1821). ANDRES was a Jesuit born in Spain (actually his name was JUAN ANDRÈS). After the expulsion of the Jesuits from his native country, he settled in Italy. In the introduction to his monumental work, ANDRES referred to D'ALEMBERT's *Preliminary Discourse* to the *Encyclopédie* in order to support his claim that science played a central role in the origin and development of "every literature." Many chapters are devoted to the history of exact sciences, of mathematics and astronomy in particular. In addition, ANDRES was the author of a *Saggio della filosofia di Galileo* (Essay on the Philosophy of Galileo) (1775).

Another work of this kind was *Vitae Italorum doctrina excellentium qui saeculis XVII et XVIII floruerunt* (Lives of Distinguished Italian Scholars who Lived in the 17th and 18th Centuries) (1778–1805), in 20 volumes by ANGELO FABBRONI (1732–1803). According to LORIA [LORIA 1946, 159], this is "a fundamental work for the knowledge of Italian intellectual life during the 17th and 18th centuries." It includes biographies of mathematicians like GALILEO, CASTELLI, CAVALIERI, TORRICELLI, CASSINI, GUGLIELMINI, MANFREDI, GRANDI, RICCATI, and many others.

Along with the works of these scholars, the *Storia dell'astronomia dalla sua origine fino all'anno MDCCCXIII* (History of Astronomy from Its Origins up to the Year 1813) by GIACOMO LEOPARDI (1798–1837) should also be mentioned. This work was written by LEOPARDI (one of Italy's greatest poets) when he was 15 years old. It remained unpublished until 1880 (repr. 1997), and therefore exerted virtually no influence on LEOPARDI's contemporaries. Nonetheless, this impressive work is of interest because it shows not only the tremendous erudition of its incredibly young author, but also the cultural climate of the Enlightenment, when the Newtonian "System of the World" (and its related astronomy) became a very popular topic of intellectual conversation in the Italian cultural milieu.

According to the new *esprit* of the age of the Enlightenment, Greek mathematics was no longer considered the benchmark for comparison or the insuperable model of mathematical knowledge. In the Age of Reason, the Middle Ages was considered a period of religious fanaticism and scientific decline compared to the Renaissance with its later innovations such as algebra. Thus, for instance, medieval contributions to mechanics were completely ignored, to the point that LAGRANGE in his *Mécanique analytique* (Analytical Mechanics) could claim that "the interval which has separated these two great geniuses [ARCHIMEDES and GALILEO] disappears in the history of mechanics" [LAGRANGE 1788, *Œuvres*, **12**, 265].

However, in spite of these historical flaws, it was in the last decades of the 18th century that the history of mathematics in a modern sense first made its appearance in Italy.

3.5 History as a Research Topic for Mathematicians

3.5.1 Pietro Cossali

At the very end of the 18th century, PIETRO COSSALI (1748–1815, **B**) published his *Origine, trasporto in Italia, primi progressi in essa dell'algebra* (Origins, Transmission to Italy, and Early Progress of Algebra There) (in two volumes, 1797–99). This may be considered the first truly professional work on the history of mathematics written in Italy. It still makes for interesting, although difficult, reading. COSSALI was a mathematician and a physicist who taught physics at the University of Parma from 1787 up to 1805, when NAPOLEON named him a professor of higher calculus at the University of Padua. Although COSSALI's name is remembered today only for his *History of Algebra* — as COSSALI himself used to call his book — he also published some other historical works including an *Elogio di Luigi Lagrange* (Eulogy of Luigi Lagrange) (1813).

As RAFFAELLA FRANCI has remarked [FRANCI 1989, 207], in spite of the "origins" suggested by its title, COSSALI's *History of Algebra* first deals with the transmission of algebra to Italy and then with its origins. The book opens with the figure of FIBONACCI (ca. 1180–ca. 1250), whom COSSALI emphasizes as important for his role in the history of modern algebra by crediting him with the "transmission to Italy" of the Arabic science of algebra in the early 13th century. Rightly dating back FIBONACCI's *Liber abaci* to 1202 — as clearly stated on the heading of almost all manuscripts of this work — COSSALI avoided a major mistake made by MONTUCLA, who dated FIBONACCI's work to the early 15th century. COSSALI was right in saying that algebra spread first from FIBONACCI's hometown of Pisa to Tuscany (Florence in particular), then to the rest of Italy, and eventually throughout Europe. By correcting a mistake which, following MONTUCLA, was often repeated by other 18th-century historians, COSSALI's work also revealed that in fact the "progress" of algebra had been much slower than generally thought. This raised a major historical question concerning the development of mathematics during the Middle Ages. A lack of primary sources, however, prevented COSSALI from "filling the gap" of nearly three centuries which separated FIBONACCI's work from PACIOLI's *Summa de arithmetica, geometria, proportioni et proportionalità* (All that is Known about Arithmetic, Geometry, Proportions, and Proportionality) (1494). As FRANCI has noted [FRANCI 1989, 208], this led COSSALI mistakenly to credit PACIOLI with "all that is new in the *Summa* with respect to the *Liber abaci*." Looking for FIBONACCI's sources, COSSALI was unable to find any original Arabic texts, and so his work was based on second-hand sources. In separate chapters COSSALI also expounded the content of DIOPHANTUS's *Arithmetic* and FIBONACCI's *Liber quadratorum* (Book of Squares).

COSSALI devoted a great deal of his book to the history of Renaissance algebra, particularly to the resolution of cubic and quartic equations. He presented a balanced view, based on original sources, of the celebrated priority dispute between TARTAGLIA and CARDANO (1501–1576) over discovery of the formula for solving cubic equations. COSSALI also carefully described CARDANO's and BOMBELLI's (1526–1572/73) contributions to the solution of fourth-degree equations.

In order to vindicate the role played by Italians in the development of algebra in the Renaissance, COSSALI corrected some historical mistakes made by GUA DE MALVES (ca. 1712–1786), WALLIS (1616–1703, **B**), and, once again, MONTUCLA. "It seems a crude, and harsh charge, which is however too often confirmed," he stated, "that these historians wrote without bothering to read. What has one to say about those who transcribed their own narratives from them?" [FRANCI 1989, 211].

The concluding chapter of COSSALI's book was devoted to a detailed historical account of the *casus irriducibilis* for the cubic equation. COSSALI also devoted a lengthy manuscript to this subject, the *Storia del caso irriducibile* (History of the Irreducible Case), which has only recently (1996) been edited by ROMANO GATTO. According to GATTO, this work was ready for printing when COSSALI died. It is divided into five chapters. The first two present a critical analysis of CARDANO's *Ars Magna* and *Regula Aliza*. In GATTO's view, "there is no comparably acute and detailed analysis even in the more recent historiography of mathematics" [COSSALI 1996, 10]. COSSALI again pointed out historical mistakes made by WALLIS, GUA DE MALVES, and MONTUCLA, but failed to mention GIRARD's (1595–1632) relevant contribution to the subject as presented in *L'invention nouvelle en algèbre* (New Invention in Algebra) (1629). In the remaining part of his *History*, COSSALI critically analyzed more recent contributions to the subject, including papers by CLAIRAUT (1713–1765), D'ALEMBERT (1717–1783), and VINCENZO RICCATI (1707–1775), all of which reflect interest in the irreducible case of the cubic equation throughout the 18th century. COSSALI did not limit himself to merely describing the content of the papers in question, but also pointed out incidental mistakes. In order to clarify their historical significance, he developed methods and results of these authors in original ways as well.

Along with this work, COSSALI left a number of unpublished manuscripts concerning the history of mathematics, some of which were published by BONCOMPAGNI in a volume of *Scritti inediti del chierico regolare teatino P. D. Pietro Cossali* (Unpublished Writings of the Regular Theatine Clerk P. D. Pietro Cossali) (1857). In addition to information about COSSALI's life and work, BONCOMPAGNI's volume includes papers by COSSALI on LEONARDO PISANO, LUCA PACIOLI, NICCOLÒ TARTAGLIA, and preliminary material for a history of arithmetic.

3.5.2 Gregorio Fontana

To a lesser extent than COSSALI, the mathematician GREGORIO FONTANA (1735–1803) combined original mathematical work in analysis and applied mathematics (mechanics, optics, and astronomy) with an interest in history. FONTANA published a number of his historical works as notes and additions to the Italian translation of BOSSUT's *Saggio sulla storia generale delle matematiche* (Essay on the General History of Mathematics), which appeared in four volumes in 1802–03.

In eighteen appendices to BOSSUT's *Essay*, FONTANA demonstrated his historical competence. The appendices include such topics as ARCHIMEDES's *Sphere and Cylinder*, HIPPOCRATES's "quadrature of lunes," the dispute between LAGRANGE and ALEXIS FONTAINE DE BERTINS over the tautochrone, the tides, and various questions concerning optics and instruments for measuring longitude. As MARIA TERESA BORGATO has stated [BORGATO 1992, 133], these themes mainly deal with FONTANA's own mathematical research. In the appendices, he "gives full details and more recent contributions, with mathematical insight and ability, and with an accurate historical investigation."

Even MARIANO FONTANA (1746–1808), a physiscist without any apparent relationship to GREGORIO FONTANA, was interested in history, and particularly in MAUROLICO's works which he planned to edit. In the same year 1808 when a paper by FONTANA on MAUROLICO's arithmetic appeared, the naturalist DOMENICO SCINÀ (1764–1837) published in Palermo an historical *Elogio di Francesco Maurolico* (Eulogy of Francesco Maurolico) (1808; repr. 1994) with notes and mathematical commentaries.

3.5.3 Giovanni B. Venturi and Pietro Franchini

By the turn of the century, along with COSSALI and both of the FONTANAS, many other Italian mathematicians were contributing to the history of their science. Among them were GIOVANNI BATTISTA GUGLIELMINI (1763–1817), GIOVANNI BATTISTA VENTURI (1746–1822), and PIETRO FRANCHINI (1768–1837). GUGLIELMINI, who taught mathematics for many years at the University of Bologna, published historical papers including in particular an *Elogio di Leonardo Pisano* (Eulogy of Leonardo of Pisa) (1813).

A professor of physics at the University of Modena from 1776, VENTURI had a strong and lasting interest in the history of science. His historical works include an *Essai sur les ouvrages physico-mathématiques de Leonardo de Vinci* (Essay on the Physico-Mathematical Works of Leonardo da Vinci) (Paris 1797; repr. Paris 1937). According to [LORIA 1946, 52], "this work inaugurates the literature about Leonardo as a proper scientist." In addition, VENTURI published an essay on the history of optics, *Commentarj sopra la storia e le teorie dell'ottica* (Commentaries on the History and Theories of Optics) (1814), which included the first edition of HERON OF ALEXANDRIA's *Dioptra*, a treatise devoted to a device for surveying and geodesic measurement. VENTURI also edited unpublished letters and memoirs of GALILEO in a two-part volume, *Memorie e lettere inedite finora o disperse di Galileo Galilei, ordinate e illustrate con annotazioni* (Memoirs and Hitherto Unpublished or Missing Letters by Galileo Galilei, Ordered and Explained with Annotations) (1818–1821).

FRANCHINI was also a professor of mathematics but at the *Liceo* of Lucca beginning in 1802. He combined mathematics with his religious and political interests. In addition to a treatise and papers on calculus, algebra, and mechanics, he published a popular, comprehensive history of mathematics, *Saggio sulla sto-*

ria delle matematiche corredato di scelte notizie biografiche ad uso della gioventù (Essay on the History of Mathematics Accompanied by Selected Biographical Notices for the Use of Young People) (1821). FRANCHINI's *Essay* is divided into two parts. In the first, he presents an historical overview of the main mathematical results in mechanics, hydrodynamics, and optics; the second is devoted to biographical sketches of mathematicians from antiquity to 1821 (see [RIVOLO 1989]). The *Essay* was accompanied by a *Supplemento al Saggio sulla storia delle matematiche* (Supplement to the Essay on the History of Mathematics) (1824) devoted to such topics as plane and solid loci, continuous fractions, probability, geography, gnomonics (the doctrine of sundials), and nautics (the art of navigation). As a second supplement, FRANCHINI published *La storia dell'algebra e dei suoi principali scrittori sino al secolo XIX* (History of Algebra and its Main Writers up to the 19th Century) (1827). This work has, as its motto, COSSALI's statement quoted above about the historical errors promulgated by MONTUCLA and all too often simply copied by subsequent historians without any attempt to ascertain the truth. About one third of FRANCHINI's *History* is devoted to DESCARTES's *Geometry*, "because the analysis of this geometry is overlooked" in works on the history of mathematics and of astronomy by DELAMBRE (1749–1822, **B**), and only sketched briefly in BOSSUT's and MONTUCLA's histories.

3.5.4 The Influence of Lagrange

At the turn of the century, Italian mathematicians and historians of mathematics were greatly influenced by the works of JOSEPH LOUIS LAGRANGE (1736–1813) (for LAGRANGE's biography, see [BURZIO 1942], repr. 1993). After his death, to counter claims that LAGRANGE was a major French mathematician, Italians firmly vindicated his Italian nationality. Although it is true that he never renounced his nationality, LAGRANGE was a typical representative of the Enlightenment, and in fact was a citizen of Europe. Born in Turin, where he spent the first part of his life, LAGRANGE taught mathematics at the local military school and was among the founders of a learned society which eventually became the Academy of Sciences of Turin. In 1766 he left his hometown, never to return to Italy. After more than twenty years at the Berlin Academy, he moved in 1788 to Paris where he played a major role in French mathematics until his death in 1813.

Although LAGRANGE was not a professional historian, he nevertheless exerted a significant influence on the field of the history of mathematics. According to [BORGATO 1989, 108], "it is evident that for Lagrange historical investigation was a necessary preliminary to any original work or systematization." LAGRANGE provided excellent examples of serious historical research in many of his publications, and in his major treatises in particular.

LAGRANGE's concern for history was already apparent in his early papers, for instance in his *Recherches sur la nature et la propagation du son* (Research on the Nature and Propagation of Sound) (1761). In this paper he took up the problem of the vibrating string, one of the most controversial questions of the day.

Before presenting his own solution, he discussed at length relevant earlier contributions to the problem: NEWTON's *Principia*, papers by JOHANN I BERNOULLI and BROOK TAYLOR (1685–1731), and more recent studies by D'ALEMBERT, EULER, and DANIEL I BERNOULLI. LAGRANGE also considered the history of music, referring to works of MERSENNE (1588–1648), WALLIS, RAMEAU (1683–1764), and TARTINI (1692–1770), among others.

In his paper, *Réflexions sur la résolution algébrique des équations* (Reflections on the Algebraic Solution of Equations) (1772), the historical investigation appears not only to have provided the background but also to have served as a guideline for LAGRANGE's own research. New mathematical concepts like the "resolvent" of an algebraic equation arose in a natural way as the result of his detailed, historical, and critical examination of Renaissance methods proposed by CARDANO, TARTAGLIA, and FERRARI (1522–1565), as well as more recent approaches by BEZOUT (1730–1783), EULER, CRAMER (1704–1752), and WARING (1736–1798). The theory of equations was one of LAGRANGE's favorite subjects. A number of relevant historical remarks are scattered throughout various papers, and these were eventually collected in *Notes sur plusieurs points de la théorie des équations algébriques* (Notes on Various Points of the Theory of Algebraic Equations), which was appended to LAGRANGE's *Traité de la résolution des équations numériques* (Treatise on the Resolution of Numerical Equations) (1798). In the *Notes*, which constitute two-thirds of the book, LAGRANGE analyzed the historical development of various topics, including methods of approximation, interpolation, and the fundamental theorem of algebra. Valuable historical information about indeterminate analysis can also be found in the *Additions*, published as an appendix to the French translation of EULER's *Eléments d'algèbre* (Elements of Algebra) (1798).

LAGRANGE's great treatises, his *Mécanique analytique* (Analytical Mechanics) (1788), *Théorie des fonctions analytiques* (Theory of Analytic Functions) (1797), and *Leçons sur le calcul des fonctions* (Lectures on the Calculus of Functions) (1806), clearly show how much his philosophy (or "metaphysics" as he put it) of mathematics involved history. Each section of the *Analytical Mechanics* (Statics, Hydrostatics, Dynamics, Hydrodynamics) opened with historical introductions, in order, as he said, "to please those who like to follow the progress of the mind in the sciences, and to know the paths actually taken by the inventors as well as the more direct ones they could have followed" [LAGRANGE 1788, *Œuvres*, **12**, 11]. In the section on "Statics," he sketched the historical development of the three major principles of this branch of mechanics — the principle of the lever, the composition of forces, and virtual velocities — discussing as well the evolution of these concepts and their proofs since the time of ARCHIMEDES. This culminated in LAGRANGE's own statement of the principle of virtual velocities, which he regarded as the basic principle of mechanics. Similarly, in the long historical introduction to the section on "Dynamics," he presented "the sequence and gradual advance of ideas which most contributed to enlarge and to bring this science to perfection" [LAGRANGE 1788, *Œuvres*, **11**, 238].

LAGRANGE emphasized (for the first time in history) GALILEO's pioneering role (the principle of inertia and composition of movements), and gave a fair account of various principles and theorems due to NEWTON, HUYGENS, the BERNOULLIS, and EULER. His historical reconstruction culminated in D'ALEMBERT's principle and its analytical formulation by LAGRANGE himself. Although the introductions to the various sections of his *Analytical Mechanics* do not constitute a complete history of mechanics, they nevertheless provide an important example of critical, historical analysis which exerted a lasting influence on the field (see [TRUESDELL 1984, 254]).

As he had done in his *Analytical Mechanics* for the history of the foundations of mechanics, in the introduction to his *Theory of Analytic Functions*, LAGRANGE discussed the history of foundations of the calculus from its origins in LEIBNIZ and NEWTON through the attempts of EULER, D'ALEMBERT, and LANDEN to make it rigorous. He resumed this story in the introduction to his *Lectures on the Calculus of Functions*, where Lecture 18 provides an historical and critical account of the invention of the calculus which has largely withstood the test of time.

Even today, the pages devoted to historical material scattered among LAGRANGE's papers and treatises make for interesting reading. There is no doubt that he had a genuine interest in history, as is also shown by the historical annotations found among his papers at the time of his death. These shed considerable light on his views and on his way of doing original research. Indeed, for LAGRANGE the history of mathematics was not of interest *per se*; it was intimately related to his own mathematical development. His historical approach is an early example of a style that was to become popular among mathematicians. Many great mathematicans (BOURBAKI is a case in point) create their own historical tradition in the sense that they examine past mathematical results as anticipations of their own work.

3.5.5 Guglielmo Libri

In contrast to the historical character of LAGRANGE's critical reconstructions, the approach to history taken by GUGLIELMO LIBRI (1803–1869, **B**), an Italian refugee in Paris whose mathematical skills were apparently overestimated in his day, was entirely different. "History is facts" ("l'histoire ce sont les faits"), he once wrote, and he believed that "historical facts" should be established through documented evidence. This justified LIBRI's tireless search for rare books, manuscripts, and unpublished material.

As LORIA aptly remarked [LORIA 1937, 350], in light of LIBRI's mediocre mathematical ability, the fame he gained in France as a mathematician was "really incredible." LIBRI coupled a true bibliomania with tremendous erudition, and eventually he emerged as a first-rate historian of mathematics. His major work was the *Histoire des sciences mathématiques en Italie depuis la Renaissance des lettres jusqu'à la fin du XVIIIe siècle* (History of Mathematical Sciences in Italy from the Revival of Letters up to the End of the 18th Century) (in four volumes,

1838–1841; repr. 1989). In fact, his *History* begins with the introduction of algebra in the West and comes to an end with the death of GALILEO. "Because of his somewhat rhetorical style and a display of passionate devotion to his native land," Loria regarded LIBRI's work as "rather antiquated" [LORIA 1946, 42]. However, in spite of the fact that "it is not free from inaccuracies, it is still one of the more remarkable works among those devoted to the history of positive sciences" (according to LORIA as quoted by [PROCISSI 1989, 179]). Even today LIBRI's work remains an extremely valuable resource for the history of Italian mathematics.

Remarkably, the first comprehensive history of Italian mathematics was published in French while its author was in political exile from Italy. (An Italian translation of this work was provided by LUIGI MASIERI in 1853). By its emphasis on the role of Italians, and particularly of GALILEO, in the development of the mathematical sciences, LIBRI's *History* contributed to increasing national pride among Italian scientists in the period of the political unification of Italy. LIBRI himself trumpeted his role (which in fact was a minor one) as a patriot in the struggle for Italian independence, and during his stay in London he boasted of his acquaintance with GIUSEPPE MAZZINI (1805–1872), one of the leaders of the Italian *Risorgimento* (see Section 3.6) who at the time was in exile there.

LIBRI's *History* opens with a *Discours préliminaire* (Preliminary Discourse) which spans some 190 pages, wherein LIBRI offered an "Account of the progress of science by the various peoples of the world, beginning from the most ancient antiquity." In the *Discourse* he rejected TIRABOSCHI's claim that PYTHAGORAS was Italian, as well as the claim that the Indian numerals had a Pythagorean origin [LIBRI 1838/41, Vol. I, Note II]

According to LIBRI, the introduction of Indian numerals as well as of algebra to Italy dated back to the work of LEONARDO PISANO (Fibonacci) in the early 13th century (see [LIBRI 1838/41, Vol. I, Note II]). LIBRI's statement gave rise to a lively polemic with CHASLES (see Section 1.7). Contrasting LIBRI's (and COSSALI's) opinions, on the basis of a dubious passage credited to BOETHIUS, CHASLES mistakenly claimed that Arabic numerals and arithmetic were not transmitted from India, but that the Arabs had received all this knowledge from the Greeks. This debate also involved NESSELMANN (see Section 5.3.3), who in his *Geschichte der Algebra der Griechen* (History of the Algebra of the Greeks) (1842) shared CHASLES's opinion. In fact, LIBRI was wrong about FIBONACCI's *Liber quadratorum* (Book of Squares) being lost; the work was discovered and published by BONCOMPAGNI in 1856.

In his *History* LIBRI joined VENTURI (see Section 3.5.3) in calling attention to LEONARDO DA VINCI as a scientist, and in vindicating the role GALILEO played as a founder of modern science. He published excerpts from LEONARDO's scientific manuscripts, which were at that time almost completely unexplored, and devoted about one half of the concluding volume of his *History* to GALILEO's life and works.

Earlier the first biographical sketch of GALILEO, the *Racconto istorico della vita di Galileo* (Historical Account of Galileo's Life), had been written by VIVIANI

("Galileo's last pupil" as he called himself) at the behest of the Prince (later Cardinal) LEOPOLDO DE' MEDICI (1617–1675) and of the Grand Duke of Tuscany FERDINANDO II DE' MEDICI (1621–1670). Both were interested in science and were responsible for founding the *Accademia del Cimento* (Academy of Experiments) at Florence in 1657. Viviani's *Account* became better known after it was reprinted in the edition of the *Opere di Galileo Galilei* (Works of Galileo Galilei) (1718), in three volumes edited by TOMMASO BONAVENTURI (1675–1731) and BENEDETTO BRESCIANI (1658–1740), and enriched with mathematical notes by GUIDO GRANDI (1671–1742).

Thanks to special permission from the Holy Office *(S. Uffizio)*, GALILEO's celebrated *Dialogue* was reprinted for the first time in 1744 in a new edition of his *Collected Works* edited in Padua by Abbot GIUSEPPE TOALDO (1719–1797). By the end of the 18th century an erudite nobleman of Florence, GIOVANNI BATTISTA CLEMENTE DE' NELLI (1725–1793), published the two-volume *Vita e commercio letterario di Galileo Galilei* (Life and Correspondence of Galileo Galilei) (1793). LIBRI provided a great deal of new information going beyond all earlier references. He gave particular emphasis to GALILEO's trial and stressed the obscurantism of the Catholic Church (and of the Jesuits as well) in preventing the diffusion of Copernicanism by putting GALILEO's *Dialogue* on the *Index of Prohibited Books.*

LIBRI's *History* is written in a discursive style and avoids all mathematical detail. About one half of each volume is devoted to "Notes" which provide the lasting interest of this work. These include an enormous amount of unpublished material that added significantly to knowledge of the history of medieval and Renaissance mathematics. More than anything else, the "Notes" also reflect one of LIBRI's main interests which eventually led him to disgrace — namely his passion for collecting (sometimes in a dubious way) rare books and manuscripts. A devoted bibliophile, in 1837 LIBRI happened to buy a set of FERMAT's manuscripts from an antiquary. These had been collected by the mathematician LOUIS-FRANÇOIS ARBOGAST (1759–1803), and were sold after his death. LIBRI communicated his discovery to the French Academy, and some time later he published an important paper on the manuscripts (1839). This aroused great interest in FERMAT's work, to the point that the French government decided to entrust LIBRI with editing FERMAT's collected works. In 1845, in the *Revue des deux mondes*, LIBRI published a paper on FERMAT which is one of his best pieces of work. But this was all that LIBRI was able to produce, and the following year he was dismissed. (The project was revived in 1882, and eventually led to the edition of FERMAT's collected works by PAUL TANNERY and CHARLES HENRY.)

LIBRI's *History* invited others to continue historical research on the development of mathematics in Italy after GALILEO. A first step in this direction was taken by GABRIO PIOLA (1791–1850), who read (and published) an *Elogio di Bonaventura Cavalieri. Con note e postille matematiche* (Eulogy of Bonaventura Cavalieri: With Mathematical Notes and Commentaries) (1844). This coincided with the sixth meeting of Italian scientists held in Milan, when a monument in honor of CAVALIERI was unveiled. PIOLA was a devout Catholic, and through his writings he sought to promote complete agreement between science and religious faith. While preparing his paper, he asked SILVESTRO GHERARDI (1802–1879, **B**) for information about CAVALIERI's teaching in Bologna from 1629 to 1647. Some forty years later, in 1888, the same subject was again treated extensively by ANTONIO FAVARO (see Section 3.9) in a paper on *Cavalieri nello studio di Bologna* (Cavalieri at the University of Bologna).

GHERARDI was one of the most representative Italian scientists of his day. A professor of mechanics at the University of Bologna beginning in 1827, he became increasingly concerned with the history of mathematics at his own university. In 1846 he published a book, *Di alcuni materiali per la storia della Facoltà Matematica nell'antica Università di Bologna* (Some Materials for the History of the Mathematical Faculty at the Old University of Bologna), which was later translated into German by CURTZE in 1871.

As a result of both his political activity and his historical interests, GHERARDI also contributed to a better knowledge of GALILEO. GHERARDI's views differed from those of PIOLA concerning religion and science. At the time of the War of Independence in 1849 he took part in the uprising against the Pope and served as Minister of Public Education during the subsequent, short-lived Roman Republic. This position gave him an opportunity to examine the Archives of the Holy Office (*S. Uffizio*), where he found a number of documents concerning GALILEO's trial which he eventually published in 1869–70.

3.6 The Risorgimento and the Search for Italian Forerunners

The political unification of Italy in 1861 marked a turning point not only in the development of mathematics but also for the history of mathematics in Italy. In the second half of the 19th century, the number and quality of Italian historians of mathematics increased dramatically, as reflected for instance in the updated bibliography published by PIETRO RICCARDI (1828–1898, **B**) at the end of the century [RICCARDI 1897].

Many of the leading Italian mathematicians of the time — including ENRICO BETTI (1823–1892), FRANCESCO BRIOSCHI (1824–1897), and LUIGI CREMONA (1830–1903) — had strong patriotic sentiments, and in the 1840s and 1850s they participated in the wars against the Austro-Hungarian empire — the Italian *Risorgimento*. After the unification of Italy they were directly involved in the political life of the new state, and a number of them were elected (or named) as members of the new Italian Parliament. As for mathematical research and teaching, they gave up old traditions and made a serious effort at reform by taking French and German universities as their models. An essential task was to provide students and younger mathematicians with surveys of contemporary research, as well as with historical studies of the development of recent mathematics.

The best example of this new spirit is Felice Casorati (1835–1890). A professor of analysis at the University of Pavia, he prefaced his treatise *Teorica delle funzioni di variabili complesse* (Theory of Functions of Complex Variables) (1868) with a historical introduction amounting to 143 pages, where he expounded the development of complex function theory from Gauss to the early publications of Weierstrass. His detailed, careful historical analysis of Cauchy's many contributions to the field is particularly noteworthy. As a matter of fact, Casorati's *Notizie storiche* (Historical Notices) is still one of the best introductions to the history of the theory of complex functions, including the theory of elliptic functions and Abelian integrals.

The endeavor made by mathematicians of the *Risorgimento* to create an Italian school of mathematics — which eventually succeeded in establishing internationally recognized schools in analysis, geometry, and logic — prompted the rediscovery of a virtually forgotten Italian tradition in mathematics. These mathematicians viewed history as closely related to mathematical research, and this in turn motivated them to undertake their historical studies. Thus, for instance, Brioschi (1824–1897), in editing a posthumous paper of Piola on the movement of fluids in channels for publication, prefaced it with an historical introduction "on those parts of hydraulics which Piola had made the object of his own studies," in order to "classify his works in the history of this important part of mathematical physics" [Brioschi 1852, 119–120]. Some ten years later, while working on the solution of the quintic equation by means of elliptic functions, Brioschi called attention to a paper by the 18th-century Italian mathematician Gianfrancesco Malfatti (1731–1807), in which a particular kind of "resolvent" of the fifth-degree equation was introduced. In his own paper on Malfatti's resolvent, Brioschi suggested that the history of mathematics could play a role in building the consciousness of a national, mathematical tradition, which might inspire the new generation of Italian mathematicians. He began by complaining that Lagrange's early works and the works of 18th-century Italian mathematicians like Ruffini, Paoli, Malfatti, Gregorio Fontana, Saladini, and Lorgna were little known in Italy and almost unknown abroad. As Brioschi continued:

> The history of mathematics of that period is closely related to the history of our Academies, to the point that only by carefully rummaging in the *Proceedings* of the Italian Society of Science, as well as in the *Memoirs* of the Academies of Turin, Bologna, the Physiocrats in Siena, and so forth, are we able to construct a clear picture of how these studies flourished among us at that time. There is hardly any doubt that a historical work about the progress of mathematics, in which evidence is given of the great part Italians have played, would not only be useful on the scientific side, but it might be regarded as discharging of a debt of gratitude to Italy, which has regained its place as a nation [Brioschi 1863, 39].

In the same vein, EUGENIO BELTRAMI (1835–1900), celebrated for his *Saggio di interpretazione della geometria non euclidea* (Essay on the Interpretation of non-Euclidean Geometry) (1868), joined BRIOSCHI by publishing a historical paper [BELTRAMI 1889] on GEROLAMO SACCHERI's *Euclides ab omni naevo vindicatus* (Euclid Vindicated of Every Defect) (1733), which by then had been completely forgotten. BELTRAMI presented a detailed analysis of the first 39 propositions of this work, showing that a number of SACCHERI's results — the theorems mistakenly named by HILBERT (1862–1943) after LEGENDRE (1752–1833) and the concept of the angle of parallelism — clearly established him as a forerunner of (hyperbolic) non-Euclidean geometry. This prompted the translation of SACCHERI's text into German by ENGEL and STÄCKEL (1895). Inspired by BELTRAMI's paper, CORRADO SEGRE (1863–1924) later showed that there was evidence to support the conjecture that LEGENDRE, LOBACHEVSKIJ, and JÁNOS BÓLYAI all could have known SACCHERI's work [SEGRE 1902/03].

Related to the pre-history of non-Euclidean geometry and BELTRAMI's interpretation was ANGELO GENOCCHI's (1817–1889, **B**) historical analysis of a forgotten paper by the 18th-century Italian mathematician DAVIET DE FONCENEX on the theory of the lever, where he identified the relationship between the principles of mechanics and EUCLID's fifth postulate. As an acute analyst and number theorist, GENOCCHI also did substantial work on the history of both medieval number theory and 18th-century analysis. His interest in history had been stimulated by BALDASSARRE BONCOMPAGNI's publication of FIBONACCI's *Liber quadratorum* (Book of Squares). In the 1850s they exchanged an intensive correspondence of more than 1500 letters, mostly related to problems concerning medieval mathematics [ISRAEL 1990]. GENOCCHI also produced some papers on FIBONACCI's work and on his *Liber quadratorum* in particular. Some of BONCOMPAGNI's further publications prompted GENOCCHI to publish papers of historical content. For example, GENOCCHI published an interesting historical paper on the addition of elliptic integrals in BONCOMPAGNI's *Bullettino* in 1870 (see Section 3.7).

To be sure, neither BRIOSCHI nor CASORATI (nor BELTRAMI nor GENOCCHI) claimed to be professional historians of mathematics. Nonetheless, the extent to which they engaged in historically oriented work suggests an important link between the history of mathematics and their original mathematical research. History was not only important in itself; they believed it was also of interest for working mathematicians who could find both the origins and motivations for their own research through historical studies. By the end of the century, this attitude was common among Italian mathematicians (see Section 3.8).

3.7 Boncompagni's "Bullettino" and Its Influence

Prince BALDASSARRE BONCOMPAGNI's appearance on the scene marked a turning point in Italian historiography. He may be considered not only the founder of the

Italian school of historiography of mathematics, but also one of the most prominent and influential figures in this field.

While mathematicians conceived of history as only occasionally related to their own research, BONCOMPAGNI was a genuine historian. In his works, the history of mathematics acquired the standards of a rigorous discipline. He paid special attention to critical apparatus and carefully avoided quoting from second-hand sources. BONCOMPAGNI's early contributions to the history of mathematics were scholarly editions of medieval manuscripts. Of particular importance was his edition of the writings of LEONARDO PISANO, which became an exemplary reference. However, there is hardly any doubt that his major contribution to the history of mathematics was his creation of the celebrated *Bullettino di bibliografia e storia delle scienze matematiche* (1868–1887). To produce the highest quality publications possible, BONCOMPAGNI established his own printing press and published the journal at his own expense. In all, he edited twenty volumes of the *Bullettino* (reprinted in 1964 by Johnson Reprint, New York, and in 1968 by Forni, Bologna), which remain a standard reference and an extremely useful research tool. According to [LORIA 1946, 70], BONCOMPAGNI's series "is a collection that must be kept on the bookshelves of every scholar of the history of mathematics. The *Bullettino* is of great importance not only for the indisputable value of the papers it published but also because it shows by its own example the necessity of both direct searches for sources and precision in documentation." LORIA warmly recommended beginners to study BONCOMPAGNI's *Bullettino* so that they could observe for themselves "the highest degree of rigor and precision" [Loria 1946, 75].

BONCOMPAGNI's style of writing the history of mathematics is exemplified in his edition of the works of LEONARDO PISANO, and in his edition of a treatise on arithmetic published in 1478 at Treviso. This so-called *Aritmetica di Treviso* (Arithmetic of Treviso) is a booklet of 62 sheets which BONCOMPAGNI edited, adding some 740 pages of notes and commentaries. Another example of BONCOMPAGNI's style is reflected in his paper "Intorno ad un trattato di aritmetica del P. D. Smeraldo Borghetti lucchese, canonico regolare della congregazione del SS. Salvatore" (On a Treatise of Arithmetic by P. D. Smeraldo Borghetti of Lucca, Canon Regular of the Congregation of the Holy Redeemer), which appeared in vol. 13 of the *Bullettino*. As LORIA remarked, the "essential part" of the latter paper (280 pages long) is given in footnotes wherein BONCOMPAGNI displayed immense historical and bibliographical erudition ([LORIA 1946, 75]).

Through the *Bullettino*, BONCOMPAGNI contributed greatly to the promotion of the history of mathematics in Italy. He himself published some 130 papers there, including a number of notes, additions, and remarks on papers by other authors. Apart from BONCOMPAGNI, the most prominent Italian contributor to the *Bullettino* was ANTONIO FAVARO (1847–1922, **B**), who published nearly fifty papers there, including his early articles on GALILEO.

Other scholars published notable works in the *Bullettino*. Among these was PIETRO RICCARDI (1828–1898, **B**), for example, who later published a celebrated *Bibliografia matematica italiana dalle origini della stampa ai primi anni del secolo*

XIX (Bibliography of Italian Mathematics from the Origins of Printing to the Early Years of the 19th Century) (1893; repr. 1952). Corrections and additions made by RICCARDI himself to his *Bibliography* were published posthumously in 1928.

GIAMBATTISTA BIADEGO (1850–1923), an engineer with a strong interest in history, wrote a long essay on the 18th-century Italian mathematician GIANFRANCESCO MALFATTI (along with a list of his publications). BIADEGO also translated MORITZ CANTOR's essay, "Euclid and His Century,"into Italian with commentaries, as well as ADOLF MAYER's (1839–1908) paper on the history of the principle of least action, both of which were published in the *Bullettino.* Although the primary languages of the *Bullettino* were Italian and French (plus, of course, Latin), in order to acquaint Italians with the results of international (and particularly German) historical research, BONCOMPAGNI chose to publish translations. Thus, along with papers in French by LOUIS P. E. A. SEDILLOT, VON BEZOLD, BIERENS DE HAAN, C. HENRY, MARRE, and others, the *Bullettino* published Italian translations of papers by German scholars like CURTZE, S. GÜNTHER, and H. HANKEL, as well as French translations by MANSION of papers (mainly obituaries) originally published in German (for instance SCHERING's *éloge* of RIEMANN, KLEIN's of HESSE, and CLEBSCH's of PLÜCKER).

BONCOMPAGNI's secretary, ENRICO NARDUCCI (1832–1893), also contributed to the *Bullettino.* In addition to editing some of BALDI's *Lives* (see Section 3.2), NARDUCCI also composed commentaries on a 14th-century Italian translation of ALHAZEN's *Optics*, and on two treatises devoted to the abacus kept and preserved in the Vatican Library. He was also a bibliographer, and in 1862 produced a *Catalogo dei manoscritti ora posseduti da Baldassarre Boncompagni* (Catalog of Manuscripts Now in the Possession of Baldassarre Boncompagni) (2nd edition 1892). BONCOMPAGNI's celebrated library amounted to some 600 manuscripts and 20,000 volumes, and according to FAVARO was "unique in the world" [ARRIGHI 1989, 25]. In spite of this, soon after BONCOMPAGNI's death his magnificient collection was dispersed and sold partly at auction.

3.8 The History of Mathematics in the Early Twentieth Century

Following the tradition inaugurated by mathematicians of the *Risorgimento*, by the end of the 19th century many outstanding Italian mathematicians occasionally published historically-oriented papers. Among these were contributors to the German *Encyklopädie der mathematischen Wissenschaften* (Encyclopedia of Mathematical Sciences), including SALVATORE PINCHERLE (1853–1936), GUIDO CASTELNUOVO (1865–1952), FEDERIGO ENRIQUES (1871–1946, **B**), ORAZIO TEDONE (1870–1922), and LUIGI BERZOLARI (1863–1949). In the spirit of the German *Encyklopädie*, they produced up-to-date overviews of their respective fields of mathematics, presented in terms of the relevant historical context. In the same vein, the Italian *Enciclopedia delle matematiche elementari* (Encyclopedia of Elementary Mathematics), in three volumes and seven parts, was edited in the 1930s by BERZOLARI, GIUSEPPE VIVANTI (1859–1949, **B**) and DUILIO GIGLI (1878–1933). VIVANTI also contributed to the (posthumous) vol. 4 of CANTOR's *Vorlesungen über Geschichte der Mathematik* (see Section 5.4.2) with an essay of 230 pages on the history of the calculus from 1755 to 1799.

Moreover, plenary lectures given by Italian mathematicians at International Congresses of Mathematicians were often historical in content. At the *ICM* in Paris in 1900, VOLTERRA (1860–1940) gave a celebrated lecture on the roles BETTI, BRIOSCHI, and CASORATI had played in the rebirth of mathematics in Italy after political unification. Eight years later, VOLTERRA lectured at the *ICM* held in Rome on the development of mathematics in Italy in the second half of the 19th century. VOLTERRA also joined GINO LORIA and DIONISIO GAMBIOLI (1858–1941) in editing the three volumes of collected works of GIULIO CARLO FAGNANO (1682–1766), *Opere matematiche del marchese G. C. de' Toschi di Fagnano* (Mathematical Works of the Marquis G. C. de' Toschi di Fagnano) (3 vols., 1911–12). (Volumes I and II include FAGNANO's *Produzioni matematiche* (Mathematical Results), while Volume III contains other writings of both a scientific and polemical character, as well as his correspondence and a biography). GAMBIOLI also edited the Italian translation of W. W. ROUSE BALL's *Short Account of the History of Mathematics* (1901), which he provided with a long appendix of more than 150 pages on recent Italian mathematicians, including biographies and bibliographical information concerning some fifty 19th-century Italian mathematicians. The second edition (1927), revised by LORIA, included more biographies and a further appendix on Italian historians of mathematics.

By the turn of the century, a number of mathematical "schools" were flourishing in Italy. Of particular interest for the history of mathematics was the school of FEDERIGO ENRIQUES, which first developed in Bologna and then, after 1922, in Rome. There was also the school of mathematicians and logicians gathered around GIUSEPPE PEANO (1858–1932) in Turin. In both groups the history of mathematics was cultivated in more than an amateurish way.

3.8.1 The School of Peano in Turin

As reflected in his *Formulario matematico* (Mathematical Formulary) (5th ed. 1908), PEANO himself had a true interest in history. To be sure, the *Formulario* was not intended as an historical work, but it did include a number of accurate historical references appended to the relevant mathematical results. These were intended to identify as carefully and accurately as possible the first discoverer of a formula or a "mathematical truth." PEANO's view of history was reflected in the works of his pupils, notably GIOVANNI VAILATI (1863–1909, **B**), GIUSEPPE VACCA (1872–1953, **B**) and, later on, UGO CASSINA (1897–1964, **B**).

"What is the history of a science?" VACCA asked in a letter to LOUIS COUTURAT in 1901. One might think, he continued, that the answer would be a "fair exposition" of the scientific ideas of authors in the past. But, according to VACCA, this was just a preliminary step to history. "The only means that makes it possible to decide [what is scientific] among the works of the ancients is to begin from our point of view. To study the history of scientific truths means to look for and expound all of the attempts (*essais*) made in the past which have step-by-step led to the truths that we now know. One example of this type of history is the history of Taylor's formula in [Peano's] *Formulario*" (quoted in: [DI SIENO 1998, 769]).

Similar views were held by VAILATI, who was primarily a philosopher. VAILATI served as PEANO's assistant before becoming an honorary assistant to VOLTERRA, who by then was professor of rational mechanics at the University of Turin. In this capacity VAILATI taught courses on the history of mechanics for three years beginning in 1897. "History shows us a series of successes each of which surpasses and overshadows the previous ones," he asserted in the opening lecture to his courses. Were ARCHIMEDES or APOLLONIUS still alive, they could easily see that no proposition has since been found to contradict their conclusions. Similarly, EUCLID could verify that "his rigor is still our rigor, and his starting point is still our starting point" (quoted in: [GUERRAGGIO 1989, 265]). In this connection, VAILATI's paper on a proof of the principle of the lever credited to EUCLID is worth mentioning. This paper as well as VAILATI's opening lectures to his courses on the history of mechanics appeared in the *Bollettino di Bibliografia e Storia delle Scienze Matematiche*, a new journal edited by LORIA in 1897 as a supplement to BATTAGLINI's *Giornale di Matematiche* (see Section 3.10.1).

VAILATI's philosophy of history emphasized continuity. In particular, he held that mathematical symbolism played an essential role in providing historical continuity in mathematics. In this respect PEANO's logical symbolism appeared as the conclusion of a long path which could be traced back to LEIBNIZ and, even further in the past, to ARISTOTLE. In the field of history of logic, VAILATI produced an important paper on SACCHERI's "remarkable conclusion" (*conclusio mirabilis*) as presented in the book *Logica demostrativa* (Demonstrative Logic) (1697). SACCHERI's reasoning was a particular case of *reductio ad absurdum*, which he used in his attempt to prove EUCLID's fifth postulate, but this can already be found earlier in CARDAN's writings.

3.8.2 The School of Enriques in Bologna

As early as 1894, when ENRIQUES began teaching in Bologna, he became interested in the history of geometry and gave lectures on it, including the development of non-Euclidean and many-dimensional geometries. The similarity between the approaches of KLEIN and ENRIQUES to the foundations of geometry is striking. This became especially evident when an article on "Prinzipien der Geometrie" (Principles of Geometry), written by ENRIQUES for the German *Encyklopädie der mathematischen Wissenschaften*, appeared in print in 1907. However, as early as 1896, when the Italian translation by FRANCESCO GIUDICE (1855–1936) (with a preface by LORIA) of KLEIN's lectures on elementary geometry appeared, ENRIQUES took these as a source of inspiration for his own book, *Questioni riguardanti la geometria elementare* (Questions Concerning Elementary Geometry), which he edited in 1900 (German translation 1907).

This collection of expository papers, largely historical in content, included a thorough study of the history of non-Euclidean geometry by ENRIQUES's assistant, ROBERTO BONOLA (1874–1911, **B**). Some years later, an extended version of BONOLA's monograph was published as the well-known book, *La geometria non euclidea. Esposizione storico critica del suo sviluppo* (Non-Euclidean Geometry: A Critical and Historical Study of Its Development) (1906). This was soon translated into both English and German, and still represents the standard introduction to the history of this subject. In 1912–14 ENRIQUES edited a revised and extended version (in two volumes) of his *Questions* under the title *Questioni riguardanti la matematica elementare* (Questions Concerning Elementary Mathematics). A third edition in four volumes appeared in 1924–27 (repr. 1983). In this work ENRIQUES intended to present "a comprehensive view of the progress of mathematical ideas," joining as he said "scientific theory with a historical basis" [ENRIQUES 1924–27, **1**, preface]. The *Questions* opened with an introduction by ENRIQUES devoted to the evolution of geometrical ideas in Greek thought. In addition to many papers by ENRIQUES himself, the *Questions* included contributions from colleagues like VAILATI (the theory of proportions) and CASTELNUOVO (solvability of geometrical problems), and from ENRIQUES's students, including UGO AMALDI (1875–1957) (concepts of straight line and plane, equivalence theory), BONOLA (non-Euclidean geometry), GIUSEPPE VITALI (1875–1932) (the postulate of continuity in elementary geometry), and OSCAR CHISINI (1889–1967) (area, length, and volume in elementary geometry, and the elementary theory of isoperimeters).

By the turn of the century, ENRIQUES had become increasingly involved in both philosophy and history of science. In 1906 he published his first book on philosophy, *Problemi della scienza* (Problems of Science), and the following year he founded the *Rivista di scienza* (later *Scientia*). As a consequence of the editor's conceptions in philosophy, this international review of philosophical synthesis published numerous papers by such professional historians as CAJORI (1859–1930, **B**), ZEUTHEN (1839–1920, **B**), HEIBERG (1854–1928, **B**), and by the Italians BORTOLOTTI, LORIA, MIELI (1879–1950, **B**), VACCA, and, of course, ENRIQUES himself. Not only did ENRIQUES's historical interests influence his conception of science, but they also affected his teaching of mathematics, as he explained in the introduction to his four-volume *Lezioni sulla teoria geometrica delle equazioni e delle funzioni algebriche* (Lectures on the Geometric Theory of Algebraic Equations and Functions) (1915–1934; repr. 1985), edited by ENRIQUES together with his former student OSCAR CHISINI.

According to ENRIQUES, the actual division of the exact sciences into different branches and apparently disconnected domains made it necessary to adopt what he called a "dynamic conception" of knowledge in order to provide a unified view of mathematics and, more generally, of science. "A dynamic vision of science leads naturally to the domain of history," he wrote [ENRIQUES 1915–1934, **1**, xi]. "The sharp distinction which is usually drawn between science and history of science is founded on the latter's concept as pure, literary erudition. When conceived in this way, history enriches theory with a complementary amount of chronological and bibliographical information" [ENRIQUES 1915–1934, **1**, xi].

The concept of history ENRIQUES had in mind was quite different. In his opinion, history was the only possible route to follow in order to achieve a satisfactorily deep understanding of theory. "Such a history becomes an integral part of science and finds its place in the exposition of theories [*dottrine*]" [ENRIQUES 1915–1934, **1**, xi]. This point of view pervaded ENRIQUES's *Lectures*, which have become a basic reference for the study of Italian algebraic geometry. Needless to say, his approach to history was quite different from that of BONCOMPAGNI and FAVARO, or even from professional historians like VACCA, BORTOLOTTI, and LORIA (see Section 3.10.1). ENRIQUES's approach characterized his historical work between the two World Wars, when history became one of his major research interests (see Section 3.11).

3.9 Second Interlude: Galileo and Leonardo

LIBRI's *History* had strongly emphasized the importance for Italian culture and science of historical research on LEONARDO and GALILEO. By 1835 GALILEO's name had been removed by the Roman Catholic Church from the *Index* of "prohibited authors." Thus, at the Third Congress of Italian Scientists held in Florence in 1841, a forthcoming edition of GALILEO's collected works could be announced. In fact, the first volume of the works of GALILEO GALILEI, *Le opere di Galileo Galilei, prima edizione completa condotta sugli autentici manoscritti palatini* (The Works of Galileo Galilei, First Complete Edition Based on the Authentic Palatine Manuscripts), jointly edited by EUGENIO ALBÈRI (1807–1878), VINCENZO ANTINORI (1792–1865), and CELESTINO BIANCHI (1817–1885), was published in 1842. The 15th and last volume appeared in 1856, together with a supplement. No less than five volumes were devoted to GALILEO's correspondence. However, as LORIA remarked, "the swiftness with which this edition was produced prevented it from achieving the expected perfection" [LORIA 1938, 126].

By that time GALILEO was a figure whose significance far exceeded his scientific achievements and their historic importance. Indeed, during the *Risorgimento* the Vatican opposed the political unification of Italy, and in the 1860s refused to accept Rome as the capital of the new state (although this eventually happened in 1870). The political climate perpetuated the political relevance of the GALILEO affair.

On the scientific side, FAVARO published an extensive, two-volume study, *Galileo Galilei e lo studio di Padova* (Galileo Galilei and the University of Padua) (1883; repr. 1966). An appendix to the second volume included an outline of the project to produce a *National Edition of Galileo's Works*. In addition, FAVARO began to publish papers on GALILEO in BONCOMPAGNI's *Bullettino*. Eventually, in 1887, the Italian government accepted FAVARO's proposal to sponsor a *National Edition of Galileo's Collected Works*, which was to be edited according to the most rigorous historiographical standards. FAVARO aimed at "the greatest philological rigor in the presentation of the texts." Accordingly, as he emphasized, the new "truly complete" edition "would not include a single line which did not exist either in the autograph, the original document, or a copy of the time, without a careful collation having been made" [FAVARO 1888, 32].

As BUCCIANTINI has aptly remarked, "FAVARO shared [with BONCOMPAGNI] that new manner of doing history, in which erudition and bibliography [play] integral and special parts" [BUCCIANTINI 1985, 441]. This "positivist historiography" was based on a scrupulous respect for data and sources, and relied upon BONCOMPAGNI's *Bullettino* as its main reference. FAVARO was one of the leading representatives of this approach. In his opinion, the direct knowledge of sources and the elimination of any interpretation not supported by documented evidence revealed the ultimate truths of the facts themselves.

The *National Edition of Galileo's Collected Works* was an impressive enterprise which took years to complete. The first volume of *Le opere di Galileo Galilei, edizione nazionale sotto gli auspici di sua maestà il re d'Italia* (The Works of Galileo Galilei; National Edition under the Auspices of H. M. the King of Italy), edited by FAVARO, appeared in 1890, and the last (the 20th), in 1909. Nine volumes are devoted to GALILEO's correspondence. (Family letters were also published by FAVARO in *Galileo Galilei e suor Maria Celeste* (Galileo Galilei and Sister Maria Celeste) (Firenze 1891)). Reprinted in 1964–66, this work still represents the starting point for scholarly research on GALILEO.

The publication of the *National Edition* renewed interest in GALILEO. A *Bibliografia galileiana (1568–1895)* (Galileian Bibliography (1568–1895), 1896), published by FAVARO and ALARICO CARLI (1824–1900), contains more than two thousand entries. Its continuation as the *Bibliografia galileiana (1896–1940)* (Galileian Bibliography (1896–1940), 1943), edited by GIUSEPPE BOFFITO (1869–1944), added some 1700 new entries. Included in this immense secondary literature are a number of GALILEO biographies. Indeed, with all of its scientific and religious implications, the GALILEO affair continued to attract (and still does) the attention of historians. Among Italian scholars, the most prominent devoted to Galileo include ANTONIO BANFI (1886–1957) [works 1930, 1949], LORIA [1938], GIORGIO DE SANTILLANA (1902–1974) [1955], LUDOVICO GEYMONAT (1908–1991, **B**) [1957], and ATTILIO FRAJESE (1902–1986, **B**) [1964]. This is only a small sample of a flourishing literature that even in recent times shows no sign of abating.

In addition to his involvement with the *Galileo National Edition*, FAVARO was also involved in the analogous project to produce a *National Edition of the Works of Leonardo da Vinci*. Perhaps because LEONARDO's manuscripts and codices are dispersed in various (public and private) libraries in different countries, this was not as successful as the Galileo project. As a member of the Royal Vinciana Commission (named after LEONARDO's hometown) entrusted with producing the *National Edition*, FAVARO was only able to edit (with ENRICO CARUSI (1878–1945))

LEONARDO's manuscript *Del moto e misura del'acqua* (On the Movement and Measure of Water), which appeared posthumously in 1923, one year after FAVARO's death.

ROBERTO MARCOLONGO (1862–1945, **B**) succeeded FAVARO as a member of the Commission. For some twenty years MARCOLONGO studied LEONARDO's contributions to mathematics and physics, completing the picture provided by LIBRI's *History* and the account given by RAFFAELLO CAVERNI (1837–1900) in his *Storia del metodo sperimentale in Italia* (History of the Experimental Method in Italy) (1891-1898). (It should be noted that LORIA had a low opinion of CAVERNI. With respect to GALILEO, for instance, LORIA did not hesitate to criticize CAVERNI's *History*, noting that CAVERNI's "disparaging attitude towards the great scientist makes it necessary to be extremely cautious in accepting as true what is written therein" [LORIA 1946, 42]. In this connection, see also [MIELI 1920]).

Modern historical research on LEONARDO DA VINCI was inaugurated by PIERRE DUHEM (see Section 1.10). Note that an updated bibliography on the life and works of LEONARDO was produced in 1919 by LUCA BELTRAMI: *Bibliografia vinciana, 1885–1919* (Vinciana Bibliography, 1885–1919). The study of newly-discovered material allowed MARCOLONGO to point out that LEONARDO's knowledge of mathematics was much broader and deeper than DUHEM had thought. MARCOLONGO focused primarily on the *Codex Arundel* and the *Codex Forster*, wherein LEONARDO showed his mastery of geometry and mechanics. MARCOLONGO published some thirty papers on this subject, which are still worth studying. After his retirement, these were collected in *Memorie sulla geometria e meccanica di Leonardo da Vinci* (Memoirs on the Geometry and Mechanics of Leonardo da Vinci) (1937). In addition to MARCOLONGO's work, [BELTRAMI 1919] and [LEONARDO DA VINCI 1942] are also worth mentioning.

3.10 The Emergence of Professional Historians

"As a true Cinderella," the history of science "does not find its place in any faculty" (quoted in [GUERRAGGIO 1989, 260-261]). According to VAILATI, this was the institutional plight of the history of science in Italy at the end of the 19th century. However, by that time the history of mathematics was being taught there at a serious academic level. The first official courses in Italian universities were offered by FAVARO at Padua and by FEDERICO AMODEO (1859–1946, **B**) at Naples, although they had very different conceptions of history.

In the opening lecture of his first course in 1905, AMODEO argued that history should be neither a mere collection of facts and information about books and theorems, nor a chronological list of propositions with the names of those mathematicians who first stated or proved them (compare with FAVARO's ideas above). All such work was only preliminary to history proper. In his opinion, the history of mathematics should "make use of all these methods as a basis for reaching the conclusions to be sought. Why did a given theory develop in such and such a manner? Why has a viable method been forgotten? Why has a particular theorem not been appreciated? Why has a given theory remained unnoticed for centuries and then suddenly developed in a quick and surprising way? This is history of mathematics" (quoted in [PALLADINO 1989, 265]).

AMODEO and FAVARO taught the history of mathematics as adjuncts (*incaricati*) in their respective mathematical faculties. When the Minister of Public Education decided in 1910 to eliminate all courses taught by adjuncts, the history of mathematics disappeared from the academic curricula. Later, the mathematical faculties at a few universities accepted "free courses" on the history of mathematics usually taught by members of the faculty. This explains why Italian historians of mathematics tended to be working mathematicians, and why they had at least a mathematical degree (*laurea*). Looking at those who taught "free courses" on the history of mathematics thus provides a picture of the community of historians active in the first decades of the 20th century in Italy. In addition to FAVARO and AMODEO, scholars like GINO LORIA and ETTORE BORTOLOTTI contributed the most among Italians to the development of the history of mathematics.

3.10.1 Gino Loria

In his day, LORIA was arguably the preeminent historian of mathematics in Italy. A full professor of higher geometry at the University of Genoa beginning in 1891, LORIA wrote the history of mathematics as a mathematician writing for other mathematicians. He emphasized this approach repeatedly in his works. For instance, in the introduction to his *Storia delle matematiche dall'alba della civiltà al tramonto del secolo XIX* (History of Mathematics from the Dawn of Civilization to the End of the 19th Century) [LORIA 1929/33], he stated that general history of mathematics was written "by a mathematician for mathematicians" [LORIA 1950, v]. In his *Le scienze esatte nell'antica Grecia* (Exact Sciences in Ancient Greece) (1914), he explained that the book had been written "by a mathematician for persons who, although with modest scientific background, are interested in mathematics" [LORIA 1914, vi]. Concerned only with exact sciences, LORIA felt himself compelled to justify his incursions into "foreign fields" like philosophy, geodesy, and astronomy, forced as he said by "the indissoluble links" among the various branches of his subject matter [LORIA 1914, vii].

LORIA's main works are devoted to the history of geometry, his area of research and teaching as a mathematician. His first important historical work was an essay on NICOLÒ FERGOLA (1753–1824) [LORIA 1892], which marked the rediscovery of a virtually forgotten school of geometry that flourished in Naples during the early decades of the 19th century. LORIA himself first heard of the existence of FERGOLA's school when he read in CHASLES's *Aperçu historique sur l'origine et le développement des méthodes en géométrie* (Historical Essay on the Origin and Development of Methods in Geometry) (1875) that "we owe several important works that reestablish ancient geometrical analysis in its original purity to the celebrated FERGOLA and his students" [CHASLES 1837, 46]. Following BON-

COMPAGNI's example, LORIA founded the *Bollettino di bibliografia e storia delle scienze matematiche* in 1898. LORIA, however, adapted editorial policies different from those of BONCOMPAGNI. In LORIA's view, subsequent historians "slightly modified his (BONCOMPAGNI's) principles" [LORIA 1946, 75]. On the one hand, they tried to pay attention to influential mathematicians; on the other hand, they presented quotations in a clear and exact form, though reduced to their essentials, in order to save space and spare the time of scholars by not "inflicting them with such boredom that drives readers away" [LORIA 1946, 76].

LORIA's *Bollettino* thus sought to avoid the overly rigorous, erudite style of BONCOMPAGNI. It included generally short though well-documented papers, and gave ample space to reviews. Among historians who contributed to the 21 volumes of LORIA's journal were BONOLA, BORTOLOTTI, VAILATI, and MARCOLONGO (see biographies). In addition to LORIA himself, the major contributor was VACCA, who published papers on manuscripts of LEIBNIZ and HARRIOT, as well as articles on the principles of the calculus according to LAGRANGE's unpublished lectures at the Academy of Turin, and on the curve usually credited to MARIA GAETANA AGNESI (1718–1799). As VACCA proved, the latter was actually due to GUIDO GRANDI (1671–1742). In the same vein, VACCA established the pioneering role MAUROLICO played in stating the principle of mathematical induction.

In 1922 the *Bollettino* was reduced in size but continued to be published for some twenty more years as the "historical and bibliographical section" of the *Bollettino di Matematica*, a journal edited by ALBERTO CONTI (1873–1940) and mainly addressed to teachers of mathematics in secondary schools. This became *Archimede* in 1948, and is still published today.

LORIA did his best to help make history an internationally recognized branch of mathematics. He did so in part by trying to organize the production of a textbook for training in the history of mathematics, a project he launched at the Fourth International Congress of Mathematicians held in Rome (1908). When this international project failed because of World War I, LORIA published material he himself had collected as a *Guida allo studio della storia della matematica* (Handbook for the Study of the History of Mathematics) (1916; second expanded edition 1946). Although outdated, LORIA's *Handbook* is still a valuable reference tool.

Continuity and progress characterized LORIA's view of the historical development of mathematics. As he admitted in a lecture on "The Uninterrupted Continuity of Italian Mathematical Thought" [LORIA 1932], although he never missed an opportunity to make clear the merits of Italians, at the same time he did not share the exaggerated nationalistic views expressed by BORTOLOTTI, which led the latter to over-emphasize the roles of Italians in the history of mathematics. LORIA (1922) also believed that the constant criticism of foreign historians which pervaded the works of COSSALI and LIBRI was largely the result of the unfortunate political conditions which existed in Italy at that time. This criticism was also meant as a reply to BORTOLOTTI. Indeed, in reviewing LORIA's edition of TORRICELLI's works, BORTOLOTTI (1922a) pointed out flaws and mistakes, above all LORIA's failure to emphasize the fact that in TORRICELLI's works "one finds the foundations and the origins of modern analytical theories" [Bortolotti 1922b, 369].

Enriques, editor of the *Periodico di Matematiche*, put an end to this debate among Italians by insisting that the important thing to consider in historical research was "the unity and continuity of the progress of ideas" independently of the languages in which they were written. Accordingly, infinitesimal analysis begins with the emergence of the concept of infinity in the Eleatic school, and continues through Democritus, Eudoxus, Archimedes, and later Galileo, Cavalieri, Torricelli, Fermat, Roberval, and so on until it culminates in the works of Leibniz, Newton, and Euler.

Loria found such sweeping generalizations unacceptable, and refused to accept claims about the early origins of Italian mathematics in the Pythagorean school, which had been made by Tiraboschi (see Section 3.4) and resurrected by Bortolotti. Should Archytas and Archimedes be ranked among Italian mathematicians, as Tiraboschi and Bortolotti argued, or should the origins of Italian mathematics only be dated back to Fibonacci in the 13th century, as Loria maintained? However unfounded this question may seem today, it was nevertheless hotly debated at the time.

3.10.2 Ettore Bortolotti

Like Loria, Bortolotti occupied a chair in mathematics before devoting himself to history. He contributed primarily to the history of Renassaince and 17th-century Italian mathematics, focusing on the history of mathematics at the University of Bologna. This was the subject of a number of his papers, whereby he established Scipione dal Ferro's (1465–1526) priority in solving algebraic equations of the third degree, clarified the controversial story of the discovery of the relevant formula by Tartaglia and Cardano, and threw new light on Bombelli's work in algebra. In particular, Bortolotti's tireless search for sources and documents concerning Bolognese mathematics led him to find and edit the unpublished books IV and V of Bombelli's *Algebra* in 1929. In addition, he called attention to the pioneering papers on continuous fractions by Cataldi (1548–1626), and vindicated Ruffini's priority in proving that fifth-degree equations cannot be solved algebraically. He was also the editor of Ruffini's collected works.

As already mentioned, Bortolotti sought "to vindicate for Italy the glory of her mathematicians, not for shabby national spirit but for the pure love of truth and justice" [Bortolotti 1931, preface]. This would then re-establish what he called "the primacy of Italy in the field of mathematics" [Bortolotti 1939, 598]. (This was actually the title of a paper he presented at a meeting of the Italian Society for the Advancement of Science in 1938.) According to Agostini (1947), Bortolotti urged his students to keep in mind that in doing historical research, it was best "to reason as far as possible with the mentality of the time of the author one is studying, and not from the point of view of a modern mathematician" [Agostini 1947, 88]. Despite his own recommendation, Bortolotti's claims for priority were often based on modern interpretations of methods and results from the past. This is not surprising. Indeed, in order to search for "forerunners" of modern theories and methods, and to make claims for priorities, it is often necessary to interpret old results in "modern" terms. This frequently led BORTOLOTTI to engage in polemics and arguments over priorities with both Italian and foreign historians. As was the case for LORIA, BORTOLOTTI worked largely in isolation. His only student was AMEDEO AGOSTINI (1892–1958, **B**), who (beginning in 1927) gave a free course on the history of mathematics at the University of Pisa.

3.11 Enriques and the Institute for the History of Science in Rome

ENRIQUES tried to create a "school" of historians of mathematics (and of science) after his move to the University of Rome at the end of 1922. Two years earlier he had been elected president of *Mathesis*, the Italian Society for Teachers of Mathematics, and editor of its journal, *Periodico di Matematiche.* ENRIQUES began a new series of the *Periodico* which gave considerable room to historical papers, and many Italian historians of mathematics, BORTOLOTTI in particular, regularly contributed papers and reviews. On the occasion of its annual meeting in 1921, *Mathesis* encouraged the creation of an institute to promote the history of mathematics. Accordingly, a National Institute for the History of Mathematical and Physical Sciences was founded in Rome in 1923 with financial support in the amount of 3,000 lire from *Mathesis*, and substantially more — 20,000 lire — from the *Istituto Nazionale delle Assicurazioni* (INA, the National Insurance Institute). In addition, the University of Rome helped to establish the Institute's library. Indeed, maintaining a collection of books and manuscripts was among the goals of the Institute, as were the promotion of scientific culture and the publication of classical texts.

The Institute's Scientific Committee was chaired by ENRIQUES and included BORTOLOTTI, LORIA, MARCOLONGO, and VACCA, among others (VOLTERRA was an honorary member; see [NASTASI 1990]). At the same time ALDO MIELI was involved in promoting a "Federation" (*Federazione*) of all Italian societies and individuals interested in the history of science. MIELI was editor of the *Archivio di storia della scienza,* which he founded in 1919 (after 1928, the journal was known as *Archeion*). According to MIELI, from the time it was founded the *Archivio* had played a significant part "in organizing and promoting our studies in Italy, as well as in pointing them increasingly in a rigorously scientific direction" [Mieli 1926, 42]. History of mathematics was one of the leading subjects of papers published in MIELI's journal (for a list, see [Loria 1946, 90-92]). The *Archivio* soon became an internationally recognized journal, and beginning in 1927, it published papers in foreign languages with abstracts in PEANO's international language, *Latino sine flexione* (Latin Without Inflexion).

As for the role and scope of the history of science, however, ENRIQUES and MIELI had diverging views — ENRIQUES was much more concerned with relations between historians of science and philosophers, whereas MIELI was primarily interested in addressing scientists. In spite of their differences, the two succeeded in 1926 in creating a new center, and what had formerly been ENRIQUES's National Institute for the History of Mathematical and Physical Sciences became simply the National Institute for the History of Science. This was jointly headed by ENRIQUES and MIELI, and adopted as its official journals ENRIQUES's *Periodico di Matematiche* as well as MIELI's *Archivio*. However, their joint project was destined to have a short life. Because of MIELI's opposition to the Fascist regime, he emigrated to Paris in 1928. In addition to MIELI's exile, increasing economic and political troubles served to reduce the Institute's entire range of activities to the publication of classical texts. Nevertheless, ENRIQUES's "dynamic conception" of science and the importance he placed on understanding its historical development were reflected in the editorial work he did in connection with the Institute. As a related enterprise he founded a series "For the History and Philosophy of Mathematics," which aimed at presenting classic texts in forms accessible to students and teachers in secondary schools.

The series included not only a four-volume edition of EUCLID's *Elements* (1925–1936) edited by ENRIQUES with the help of many collaborators, but also the works of ARCHIMEDES, BOMBELLI, GALILEO, NEWTON, and DEDEKIND (1831–1916), among others. This project involved ENRIQUES's colleagues (BORTOLOTTI and CASTELNUOVO, in particular), and students like UMBERTO FORTI (*1901), ENRICO RUFINI (1890–1924), and later GIORGIO DE SANTILLANA and ETTORE CARRUCCIO (1908–1980, **B**).

In addition, foreign students who came to Rome attracted by the fame of Italian mathematicians were fascinated by ENRIQUES's work and joined him in editing texts. This was the case for RUTH STRUIK (1896–1995), who contributed to the edition of EUCLID's *Elements*, and OSCAR ZARISKI (1899–1986), who edited the Italian translation with historical notes of DEDEKIND's essays on the foundation of numbers in 1926. There were also volumes by ENRICO RUFFINI providing an Italian edition (1926) of *The Method* of ARCHIMEDES discovered by HEIBERG, and CASTELNUOVO's study of the origins of the infinitesimal calculus (1938).

By the late 1920s ENRIQUES was also involved in the *Enciclopedia Italiana Treccani* (Italian Encyclopedia Treccani) edited by GIOVANNI GENTILE (1875-1944). This was a monumental enterprise which GENTILE intended to represent Italian culture independently of the political ideas of its contributors. As editor-in-chief for the mathematical entries, ENRIQUES invited many of his colleagues to contribute articles of historical content to the *Enciclopedia*.

ENRIQUES combined direction of the Institute for the History of Science with direction of a Postgraduate School in the History of Science, which he founded at the University of Rome in 1924. There he lectured on the history of scientific concepts, while VACCA gave courses on the history of mathematics and MIELI taught history of chemistry. The Postgraduate School attracted an increasing number of young students interested in the history of mathematics, including DE SANTILLANA, CARRUCCIO, and FRAJESE (see biographies). All of these later served as "volunteer assistants" to ENRIQUES (FRAJESE was also entrusted with editing ENRIQUES's *Le matematiche nella storia e nella cultura* (Mathematics in History and Culture) (1938)). In particular, ENRIQUES and DE SANTILLANA worked together on a joint project to produce a comprehensive history of science, of which they were only able to complete a part (see [ENRIQUES/SANTILLANA 1932] and [ENRIQUES/SANTILLANA 1937]).

Like most historians of his generation, the approach ENRIQUES took to history emphasized the continuity of the development of science from antiquity to the present. Accordingly, in ENRIQUES's view the study of ancient Greek science (and mathematics) was of particular importance. A striking example of this was given in his interpretation of the history of relativity. Introducing ALBERT EINSTEIN (1879–1955) during his visit to Bologna, ENRIQUES asserted that "the philosophical revolution achieved by EINSTEIN appears to be the result of an evolution of thought which began in the 5th century B.C. with PARMENIDES OF ELEA, who first maintained the relativity of motion" [ENRIQUES 1922, 79]. In discussing relativity from a philosophical point of view, ENRIQUES argued that the history of science cannot be "reduced to a collection of texts and erudite notices, but it should be constructed by order, and connect the data of the written record by means of hypotheses and explain them with reasons" [ENRIQUES 1936, 45–46]; see also [ENRIQUES 1934]. ENRIQUES thus regarded history as the search for continuity in the progress of ideas. In his own words, "a discovery corresponds to an idea, and an idea supposes, in general, an entire preceding development of ideas" [ENRIQUES 1936, 46].

Both ENRIQUES's Institute for the History of Science and the Postgraduate School in History of Science ceased to operate when he was forced to leave the university in 1938 because of Fascist racial laws. In 1940, on the occasion of the Second Congress of the Italian Mathematical Union, at BORTOLOTTI's, VACCA's and AGOSTINI's suggestion, the Congress officially expressed its hope that a new journal on the history of mathematics and astronomy might be founded, and the publication of classical texts on the history of science might continue. Because of the outbreak of World War II, however, this hope remained unfulfilled. After the War, ENRIQUES and CASTELNUOVO launched a new series of books, but were only able to publish the first one, GALILEO's *Dialogue*, before ENRIQUES's death in 1946. A new journal for the history of mathematics was not founded until 1981 (see Section 3.13).

3.12 After World War II

With the deaths of ENRIQUES (1946), BORTOLOTTI (1947), and LORIA (1954), the history of mathematics in Italy began to change. On an academic level, the tradition they had begun continued in the works of AGOSTINI, CARRUCCIO, and CASSINA. The history of mathematics also attracted officials of the Ministry of Public Education like FRAJESE and LUIGI TENCA (1877–1960) (who was concerned primarily with post-Galileian mathematics). A lasting interest in the history of mathematics was also cultivated by ARNALDO MASOTTI (1902–1989), a professor of rational mechanics at the Polytechnic in Milan who published two volumes on TARTAGLIA in 1974 (including a reprint of the *Cartelli di matematica disfida* concerning the mathematical challenge that TARTAGLIA exchanged with CARDANO). Medieval mathematics was the primary research field of GINO ARRIGHI (*1906), an engineer who taught at the Naval Academy of Livorno (see [GEMIGNANI 1993]).

Courses on the history of mathematics almost disappeared from the mathematical faculties in Italy following World War II. Among the few exceptions were courses offered by LUDOVICO GEYMONAT (1908–1991, **B**) at the University of Turin [GEYMONAT 1947], as well as the "free course" taught for a few years, beginning in 1948, at the University of Rome by ALPINOLO NATUCCI (1883–1975). NATUCCI, a former student of the *Scuola Normale Superiore* in Pisa, spent more than forty years as a secondary school teacher. His major historical work was a book on the concept of number [NATUCCI 1923]. Courses were also taught at the University of Pisa by AGOSTINI, at the University of Pavia (and later Milan) by PEANO's former student CASSINA, and by CARRUCCIO.

Along with CASSINA, CARRUCCIO was perhaps the most active historian of this period. Beginning in 1955, he taught courses on the history of mathematics at the universities in Turin, then Modena, and eventually Bologna. He obtained a chair in the history of mathematics at the latter in 1977. This was actually the first full professorship for the history of mathematics at an Italian university.

CARRUCCIO, who combined historical research with a deep interest in the philosophy of mathematics and logic, devoted most of his works to the history of mathematical logic and the foundations of mathematics. A self-styled disciple of ENRIQUES, CARRUCCIO also insisted that all too frequently "the history of science, of which the history of mathematics is a branch, is still often considered as a collection of scientific enquiries through the ages," little more than "a museum of documents and scientific curiosities." Rejecting this view, CARRUCCIO followed ENRIQUES in maintaining that "the history of science is identified with science itself in terms of its development, the thought of those who constructed it, and its relation to the various aspects of culture and human life" [CARRUCCIO 1958, 9]. In studying history, he continued, "what is required is not so much erudition as an effort to enliven the texts by interpretation and by relating them to their own times" [CARRUCCIO 1958, 9].

3.13 Conclusion

In considering the long historiography of mathematics in Italy from the Middle Ages to the mid-20th century, it is possible to distinguish a number of distinct periods. In its earliest phase, from medieval times through the Renaissance, the mathematical achievements of antiquity were rediscovered and a body of knowledge that for centuries had nearly been lost was reestablished. In the process, mathematics came to be regarded as an historical discipline in so far as it had its origins far in the past. Moreover, editing ancient texts eventually led to original research inspired by those same texts. From the early 17th century until the end of the 18th century, the emergence of new mathematical concepts and techniques not only promoted a clear sense of the superiority of modern achievements over ancient mathematics, but contributed substantially to the idea of progress in mathematics as well. Historical interest was likewise stimulated in the biographies of great mathematicians. The idea that mathematics is a fundamental part of human knowledge was directly reflected in the general histories of human activities and culture which were so popular in the 18th century.

By and large, the history of mathematics in Italy from the end of the 18th century may be characterized as a subject of research primarily for mathematicians. As GINO LORIA once said, the history of mathematics is done by mathematicians for mathematicians. This view was largely shared as well by FEDERIGO ENRIQUES (and his school), even though ENRIQUES promoted historical research as an integral part of his project to establish a scientific philosophy wherein mathematics and its history played major roles. But due to the general failure of ENRIQUES's efforts and the inability of either LORIA or BORTOLOTTI to create a school of followers, directly after World War II the history of mathematics in Italy was cultivated by only a handful of scholars. Indeed, by the time of the Second World War, ENRIQUES, BORTOLOTTI, and LORIA were all retired, and died soon after the end of the War (ENRIQUES and BORTOLOTTI in 1947, and LORIA in 1954, at the age of 92). This helps to explain why the history of mathematics following World War II almost disappeared from the curricula of courses taught in mathematical faculties at Italian universities.

Renewed interest in the history of mathematics was stimulated by GEYMONAT's teaching at the University of Milan, where he attracted a number of students concerned with logic, philosophy, and the history of mathematics in the 1960s. GEYMONAT promoted the diffusion of these disciplines in Italian universities and combined his academic work with intense editorial activity, founding and editing a new series of books devoted to the history and philosophy of science. At the same time, LUCIO LOMBARDO RADICE (1916–1982), a former student of ENRIQUES, also promoted historical research at the University of Rome. The history of mathematics has developed considerably in the last few decades in Italy. In addition to CARRUCCIO's chair in Bologna, since the end of the 1970s official courses on the history of mathematics are now taught by professional historians of mathematics at most universities in Italy. Institutionally, the history of mathematics has been supported by the *Unione Matematica Italiana* (Italian Mathematical Union), which has devoted a section of its *Bollettino dell'Unione Matematica Italiana* to didactic and historical questions since 1940. In 1981 the Union founded a new journal in support of the history of mathematics, the *Bollettino di storia delle scienze matematiche*, which by the end of the 20th century had published eighteen volumes.

Chapter 4

Switzerland

Erwin Neuenschwander

4.1 Introduction

Although Switzerland is a relatively small country, it has produced many important mathematicians. The first outstanding center of mathematics was the University of Basel, where Leonhard Euler studied with Johann I Bernoulli. Somewhat later, Cramer and L'Huillier were active in Geneva, where the *Académie* made it another center of some importance. In 1855, the *Eidgenössische Polytechnische Schule* (today the *Eidgenössische Technische Hochschule (ETH) Zürich*) was founded. It immediately attracted numerous foreign mathematicians, and soon overshadowed the older centers. Today it is the home of the only Swiss research institute for mathematics. Moreover, Zürich is the only city in the world that has hosted the quadrennial International Congress of Mathematicians on three occasions: in 1897, 1932, and 1994.

Despite these favorable conditions, the history of science, and in particular the history of mathematics, have not fared so well in Switzerland — in contrast to the history of medicine — and enjoy little institutional support. Research on the history of mathematics, for example, has traditionally been left to professors or teachers with historical interests who, for the most part, have carried out their work in addition to their teaching at local secondary schools or universities.

In the 19th century, the most important source of encouragement for historical studies came from the flourishing scientific societies. Later, when the government began to promote science, it did so with respect to historical studies by supporting editions of the collected works of famous Swiss mathematicians. However, general research devoted to the history of mathematics had to be pursued largely without support from the government. Hence it is not surprising that most publications have a strong local orientation; a general history of mathematics in

Switzerland remains a desideratum that would have to be compiled from numerous dispersed sources. Therefore this chapter begins with a survey of the general development of mathematics in Switzerland followed by an account of individual contributions to the history of mathematics.

4.2 Humanism and Enlightenment

Accompanying the economic growth of cities and towns during the Renaissance, and in direct response to the Reformation, several theological seminaries were founded in Switzerland. These were later raised to the status of academies and, in the 19th century, as a consequence of the general upheaval after the French Revolution, to universities. The universities of both Zürich and Geneva date back to theological schools founded by the Protestant reformers ZWINGLI (1484–1531) and CALVIN (1509–1564), in 1525 and 1559, respectively. The universities of Bern and Lausanne also developed from two theologically oriented academies, whereas the universities in Fribourg (Freiburg) and Neuchâtel (Neuenburg) are partly based on later secular academies, founded in the 18th and 19th centuries.

Only in Basel, due to favorable circumstances created by the Council of Basel (1431–1449), was an early associated university *(Konzilsuniversität)* established. With the approval in 1460 of the Council's former secretary, ENEA SILVIO PICCOLOMINI (Pope PIUS II since 1458), the present University of Basel was officially founded. Favorable political, economic, and cultural conditions facilitated construction of paper mills and attracted the settlement of printers and independent scholars. Due in great measure to ERASMUS OF ROTTERDAM (ca. 1466–1536), by the beginning of the 16th century, Basel was a major center of scholarship and printing, and consequently was one of the most significant cities for intellectual activity north of the Alps.[1]

Among the many editions of Greek texts in the mathematical sciences printed in Basel were several *editiones princepes* by such notable scholars as SIMON GRYNAEUS, XYLANDER, and ERASMUS, including among others the *Elements* of EUCLID (1533), PTOLEMY's *Geography* (1533) and *Almagest* (1538), PROCLUS's *Hypotyposis of Astronomical Hypotheses* (1540), the works of ARCHIMEDES (1544), the then six books of the *Arithmetica* of DIOPHANTUS (1575), and numerous writings from or on ARISTOTLE ([HARLFINGER 1989] [HIERONYMUS 1992, 404–442]).

Basel was also the only city in Switzerland where (as early as the 16th century) there was a professorship for mathematics, originally meant to cover all four parts of the *quadrivium* (arithmetic, geometry, astronomy, and music). The first mathematician to become widely known at the University of Basel was HEINRICH LORITI, called GLAREANUS (1488–1563), a scholar of universal learning who was

[1]The importance of Basel for academic publishing was reflected in 1989 in a comprehensive exhibition, entitled *Graecogermania*, of early editions of Greek works printed in German-speaking countries, in the *Herzog August Bibliothek Wolfenbüttel*. This exhibition included nearly 200 incunabula, about 50 of which were printed in Basel.

also a musician and geographer. Among his successors, CHRISTIAN WURSTISEN (1544–1588) and PETER MEGERLIN (1623–1686) were especially eminent. The latter composed mathematico-historical tables of world history, wherein major events were arranged according to the great conjunctions of Jupiter and Saturn. With his successor, JAKOB I BERNOULLI (1654–1705), the so-called Golden Age of mathematics in Basel commenced. For the next hundred years, the professorship of mathematics remained in the hands of the Bernoulli family (see [BONJOUR 1960], [MERIAN 1860], [SPEISER 1939], [STAEHELIN 1957], and [VISCHER 1860]).

Throughout this period there was intensive exchange of ideas between mathematicians in Basel and their colleagues in Geneva, Lausanne, and Bern. Mathematics in the French-speaking part of Switzerland was greatly stimulated when GABRIEL CRAMER (1704–1752) and JEAN-LOUIS CALANDRINI (1703–1758) were called to share a newly created professorship of mathematics at the Academy in Geneva in 1724. Later this chair was occupied by LOUIS NECKER (1730–1804), LOUIS BERTRAND (1731–1812), and SIMON-ANTOINE L'HUILLIER (1750–1840). Active in Geneva at that time were also NICOLAS FATIO DE DUILLIER (1664–1753), JEAN JALLABERT (1712–1768), JEAN-ANDRÉ DELUC (1763–1847), JEAN-ROBERT ARGAND (1768–1822), and JACQUES-ANDRÉ MALLET (1740–1790), the latter of whom became the first professor for astronomy at the newly opened observatory in 1771.

At the same time mathematics was also furthered at the Academy of Lausanne, where it was represented by JEAN-PIERRE DE CROUSAZ (1663–1750), FRANÇOIS-FRÉDÉRIC DE TREYTORRENS (1687–1737), LOUIS DE TREYTORRENS (1726–1794), and JEAN-PHILIPPE LOYS DE CHESEAUX (1718–1751). The latter is famous for predicting the comet of 1744 by the methods of Newton. For more information on the early development of mathematics in the French-speaking part of Switzerland, see [ISELY 1901], [GRAF 1888/90] and [GRAF 1906], [METHÉE 1991], and [TREMBLEY 1987]; for a general survey of the early period, see [FUETER 1941] and [WOLF 1858/62].

Among the increasing number of mathematicians working in Switzerland during this period, several published contributions to the history of mathematics. Above all, GABRIEL CRAMER (of Cramer's rule for determinants) provided detailed historical notes on the development of the theory of algebraic curves in the preface to his best-known work, *Introduction à l'analyse des lignes courbes algébriques* (Introduction to the Analysis of Algebraic Curves) (1750). He also edited collected works of famous mathematical contemporaries, including the *Elementa matheseos universae* (Elements of Universal Mathematics) of CHRISTIAN WOLFF (1679–1754), the *Opera* of JOHANN I (1667–1748) and JAKOB I BERNOULLI (1654–1705), and the correspondence between LEIBNIZ (1646–1716) and JOHANN I BERNOULLI. All in all, he edited the impressive number of 13 volumes. In addition, he wrote several minor historical works on the "invention" of Arabic number symbols (1739), the burning mirrors of ARCHIMEDES (1741), mathematicians from antiquity (1748), and the quadrature of lunes by HIPPOKRATES OF CHIOS (1750) ([P I, 493], [ISELY 1901, 126–138], and [SPEZIALI 1959]). Apart from CRAMER, during the 18th century numerous scholars occasionally did work in the history of mathematics (for a detailed list, see [SUTER 1890]).

4.3 The Contribution of the "Naturforschende Gesellschaften"

Towards the end of the 18th century, scientists in the major cities of Switzerland joined together to found societies for the promotion of science which later came to bear the name *Naturforschende Gesellschaften* (scientific societies). The oldest of these is the *Naturforschende Gesellschaft in Zürich*, founded by JOHANNES GESSNER (1709–1790) in 1746. Similar societies were established in Bern (1786), Geneva (1790), Aargau (1811), Vaud (1815), Basel (1817), etc. This led in 1815 to the foundation of the national *Schweizerische Naturforschende Gesellschaft*, to which the local societies were subordinated. The national society meets annually in different Swiss towns. In two steps it was raised in 1983 and 1988 to the rank of the *Schweizerische Akademie der Naturwissenschaften* (Swiss Academy of Sciences).

These societies played an important role in the development of scientific life in Switzerland. They often maintained their own libraries and scientific collections, sometimes supported botanical gardens or an observatory, published scientific journals and edited collected works of important Swiss scientists, and fostered the study of local history of science. Among the latter contributions, one finds historical reports occasionally issued at annual meetings ([MERIAN 1838], [BAYS 1927a], and [BAYS 1927b]), and at special anniversary celebrations ([CANDOLLE 1915], [MERIAN 1867], [GRAF 1886], [RUDIO 1896], etc.); these represent one of the most important stimuli for the early historiography of mathematics in Switzerland (see [GRAF 1888/90, vol. 1, v]).

Most of the well-known Swiss historians of mathematics from the last century held important offices within these *Naturforschende Gesellschaften*, and in many cases presided over an often associated local library. One of the first scholars who linked the history of mathematics, the activities in the *Naturforschende Gesellschaft*, and library directorship in a really exemplary manner was DANIEL HUBER (1768–1829), the Basel astronomer, mathematician, and head librarian. At first, like his father, HUBER occupied himself with astronomy, but in 1792 he succeeded his teacher JOHANN II BERNOULLI (1710–1790) as professor of mathematics at the university. Moreover, beginning in 1802, HUBER was director of the university library (to which he later bequeathed his own rich private collection). As a result, Basel University Library has a nearly complete collection of the most important publications in the exact sciences of the 17th and 18th centuries. In keeping with his interests, in 1817 HUBER inspired the creation of the *Naturforschende Gesellschaft in Basel*, over which he presided until his death. In the history of mathematics his publications on JOHANN HEINRICH LAMBERT (1728–1777) are of special importance, above all his monograph *Johann Heinrich*

Lambert nach seinem Leben und Wirken (J. H. Lambert: His Life and Work) (1829) ([Graf 1902], [Merian 1830], [Wolf 1858/62, I 441–462, III 317 *et passim*]).

One of the following presidents of the *Naturforschende Gesellschaft in Basel* was Peter Merian (1795–1883), who continued the traditions started by Huber. In 1838 he became annual president of the *Schweizerische Naturforschende Gesellschaft* and, at its Basel meeting of that year, gave the opening address, wherein he summarized the development of the sciences in Switzerland from the Renaissance up to his time [Merian 1838]. On the occasion of the 400th anniversary of the University of Basel he wrote the "Festschrift" *Die Mathematiker Bernoulli* [Merian 1860], and in 1867 published a history of the *Naturforschende Gesellschaft in Basel* [Merian 1867].

Following this tradition, the physicist and historian of science Fritz Burckhardt (1830–1913), also an active member of the *Naturforschende Gesellschaft in Basel*, edited the autobiography of Johann II Bernoulli, produced a genealogy of the Euler family, and wrote several articles on the Basel mathematicians [Schneider 1914, with list of works]. At the same time Hermann Kinkelin (1832–1913), a Basel mathematician, published some historical studies on medieval mathematics and a paper on Euler in the society's transactions [Fäh 1913, with list of works].

But the best-known of all these 19th-century figures in Switzerland who combined work in the three areas of activity mentioned above (*Naturforschende Gesellschaft*, local library, and history of science) was the extraordinarily productive astronomer and historian of science, Johann Rudolf Wolf (1816–1893; **B**). Director of the *ETH* library in Zürich from 1855 until his death and founder of the journals of the *Naturforschende Gesellschaft* in Bern and in Zürich, he published several books and about 500 short articles on the history of the exact sciences.

Wolf's successor in Bern was the mathematician and historian of science, Johann Heinrich Graf (1852–1918), who again contributed to the same three areas. From 1883 to 1910 he edited the journal of the *Naturforschende Gesellschaft in Bern* and from 1895 to 1918 presided over the *Schweizerische Bibliothek-Kommission* (Swiss Library Commission) through which he promoted the foundation of the *Schweizerische Landesbibliothek* (Swiss National Library). Noteworthy among his numerous contributions to the history of mathematics, astronomy, and cartography is his four-volume *Geschichte der Mathematik und der Naturwissenschaften in bernischen Landen* (History of Mathematics and the Sciences in the Territories of Bern) (1888–1890) ([Graf 1888/90], [Graf 1886]–[Graf 1906], [Andres 1918], and [Crelier 1918, with list of works]).

Wolf's successor in Zürich was the German (by birth) Ferdinand Rudio (1856–1929, **B**), who continued to work again along the same lines. In addition to Huber, Merian, Burckhardt, Wolf, Graf, and Rudio, many minor contributors also wrote in the same way on various aspects of the history of mathematics. These include the Fribourg mathematician Séverin Bays (1885–1972), who as annual president of the *Schweizerische Naturforschende Gesellschaft* in 1826 gave an opening address on Swiss mathematicians [Bays 1927a]. He also published studies specifically devoted to Fribourg mathematicians [Bays 1927b], as well as on Fermat (1601–1665), Sadi Carnot (1796–1832), magic squares, and the philosophy of mathematics [Piccard 1973, with list of works].

During this period lecture courses were occasionally given on the history of mathematics at various Swiss universities. Beginning in 1845, for example, such lectures were announced at the University of Zürich by the former librarian of the University of Heidelberg, Anton Müller (1798–1860), who later became the first holder of a full professorship in mathematics at the University of Zürich. Johann Caspar Hug (1821–1884) also lectured on history of mathematics, beginning in 1861, as did Rudolf Wolf at the *ETH*. Later, lecture courses on the history of mathematics were also announced from time to time at Bern, for example by Friedrich Graefe (1855–1918) in 1879 and 1881, and by Georg Joseph Sidler (1831–1907) in 1896. At the University of Basel history of mathematics was first announced relatively late, in the winter term of 1894/95 by Robert Flatt (1863–1955). From 1906 onwards, the well-known historian of mathematics Otto Spiess (1878–1966, **B**) took over this task, although only from 1916 on with regular success. After World War I history of mathematics and history of science flourished in Basel, with lectures on these subjects being quite often the best attended in their respective subject areas. Around 1920 Spiess's historical lectures attracted far more students than any of the courses in pure mathematics.

4.4 The Major Editions

The *Naturforschende Gesellschaften* (scientific societies) also played an important role in connection with the publication of collected works of famous Swiss mathematicians. The first and most prestigious of these major editions is that of the collected works of Leonhard Euler (1707–1783) which did not begin to appear until 125 years after his death, and was so extensive that no one individual could have been responsible for editing it single-handedly. It is well known that on several occasions Euler himself claimed to have produced so many papers that it would take the St. Petersburg Academy more than 20 years after his death to complete their publication. In this, he was to be proven over-correct; it took even longer than 40 years before the last of his manuscripts was ready for publication and sent to press in 1830. Various unsuccessful proposals to publish Euler's collected works came from Belgium (1838), St. Petersburg (1844) and Berlin (1903). Finally, due to the tireless efforts of Rudio and the *Schweizerische Naturforschende Gesellschaft*, an edition of *Leonhardi Euleri Opera omnia* began to appear early in the 20th century that as late as 2000 has yet to be completed ([Rudio 1911], [Biermann 1983]; see also Section 5.4.4).

Already at the commemoration of the centenary of Euler's death in Basel and Zürich in 1883, Rudio agitated without success for a Swiss edition of Euler's collected works. He renewed his call at the First International Congress of Mathematicians in Zürich in 1897, in hopes of gathering support from foreign mathematicians and societies. Eventually, in connection with the bicentenary of EULER's birth in 1907, RUDIO succeeded in convincing the *Schweizerische Naturforschende Gesellschaft* to appoint a commission to determine the feasibility of an EULER edition. The idea was finally approved in 1909 at the annual meeting of the *Gesellschaft* in Lausanne. Based on a plan by PAUL STÄCKEL (1862–1919), the edition was divided into three series with a planned total of 43 volumes (series I: pure mathematics, 18 vols.; series II: mechanics and astronomy, 16 vols.; series III: physics, and volumes of mixed contents and correspondence, 9 vols.). RUDIO was chosen as the first general editor and oversaw 22 volumes that were published prior to his death in 1929.

ANDREAS SPEISER (1885–1970), professor of mathematics in Zürich and Basel, was an energetic successor to RUDIO as general editor. He was particularly able in persuading representatives from major Swiss businesses to support the EULER edition financially. This became necessary as the plan for the edition had been repeatedly enlarged after the completion of the ENESTRÖM inventory [ENESTRÖM 1910/13]. In 1947 SPEISER submitted a new plan for 72 volumes, not including the correspondence and unpublished manuscripts (under SPEISER's editorship 35 new volumes were published). Since 1965, when WALTER HABICHT (*1915) was named general editor, eleven further volumes have been issued, nearly completing all three series.

In 1967 the *Euler Kommission* decided to crown the edition with a fourth series, subdivided into two parts: *IV A* is to comprise EULER's correspondence (in ten volumes) and *IV B* to reproduce his scientific notebooks, diaries, and hitherto unpublished manuscripts (in 7 or 8 volumes). Since EULER's scientific papers and correspondence are preserved in St. Petersburg, this fourth series is to be published jointly with the Russian Academy of Sciences. The responsibility for this series rests with EMIL A. FELLMANN (*1927), while the Basel mathematician HANS-CHRISTOPH IM HOF (*1944) inherited editorial responsibility for the entire project. So far, four volumes have been published in series *IV*. It is, however, uncertain whether this enormous project will be finished before the next great EULER anniversary, the tercentenary of his birth in 2007 ([RUDIO 1911], [BURCKHARDT 1983], and [FELLMANN/IM HOF 1993]).

The second major edition being produced in Basel is the BERNOULLI edition. This was mainly started on the initiative of the Basel mathematician OTTO SPIESS. After the publication of his book on EULER [SPIESS 1929], SPIESS began to search for the BERNOULLIS' scientific estate, of which only a small portion was then kept in Basel. In 1935 SPIESS succeeded in obtaining some of JOHANN I BERNOULLI's papers from Stockholm. He also won support from the industrialist, JOHANN RUDOLF GEIGY, as well as from the *Naturforschende Gesellschaft in Basel*. As a result, the *Bernoulli Kommission* was founded. SPIESS was soon able to arrange for the acquisition of the rest of JOHANN I BERNOULLI's papers, including those of JOHANN HEINRICH LAMBERT (1728–1777), which had been in the possession of the *Herzogliche Bibliothek* in Gotha. Over the next few decades

SPIESS inspected and sorted this very rich material, and in 1955 published the first volume of the BERNOULLI edition ([SPIESS 1955], [NAGEL 1993]).

When SPIESS died in 1966, JOACHIM OTTO FLECKENSTEIN (1914–1980, **B**), his assistant and later professor for history of exact sciences at the Technical University in Munich, assumed responsibility for the BERNOULLI edition. SPIESS bequeathed his estate for the creation of an Otto-Spiess-Foundation to support the editorial work financially. Nevertheless, the project progressed slowly, and by 1980 only two further volumes had appeared. After the death of FLECKENSTEIN, the general editorship was transferred to DAVID SPEISER (*1926), professor of theoretical physics at the University of Louvain-la-Neuve, who, together with his collaborators, produced a new plan for the edition. According to this, the edition will now comprise ca. 45 volumes divided into several series, i. e., the works and correspondence of the various members of the BERNOULLI family. In 1988, thanks to support from the Swiss National Science Foundation, two research positions were created. By expanding the editorial team and bringing in numerous foreign scholars, the frequence of publication has been considerably increased, so that since 1987 nearly every year a new volume has appeared ([STÄHELIN/SPEISER 1989/90], [NAGEL 1993])).

Among smaller editions supported by the *Naturforschende Gesellschaften* was the production of the *Collected Works of Ludwig Schläfli* (3 vols., 1950–1956), edited by the *Steiner-Schläfli Committee* of the *Schweizerische Naturforschende Gesellschaft.* The edition was greatly aided by preparatory work done by SCHLÄFLI's successor, GRAF, as well as by JOHANN JAKOB BURCKHARDT (*1903), who examined SCHLÄFLI's *Nachlass.*

Mention should also be made of the *Kurze Mathematiker-Biographien*, a series of short biographies of mathematicians issued between 1947 and 1980 as *Beihefte* to the journal *Elemente der Mathematik.* Of the 15 issues dealing with themes from the history of mathematics, five were devoted to Swiss mathematicians (STEINER, EULER, SCHLÄFLI, BÜRGI, JAKOB I, and JOHANN I BERNOULLI). Since 1987 this series has been replaced by the larger *Vita Mathematica* series edited by E. A. FELLMANN and likewise published by *Birkhäuser Verlag* in Basel — the publisher of numerous works on the history of mathematics alongside the EULER and BERNOULLI editions.

4.5 Further Developments

With the increasing specialization and institutionalization of the sciences, the *Naturforschende Gesellschaften* gradually lost their former importance and were replaced in part by societies devoted to specific disciplines. Similarly, scientists preferred to discuss their mutual interests at ever more specialized international meetings, rather than at the traditional general assemblies of the local *Naturforschende Gesellschaften.* The *Schweizerische Mathematische Gesellschaft* (Swiss Mathematical Society), for example, was founded in 1910. Thanks to the general development and institutionalization of science, the number of full professorships in mathematics during the last hundred years has increased by about a factor of ten in Switzerland. The history of mathematics, unfortunately, has not profited from this development (in contrast to the history of medicine, which is relatively well-represented in Switzerland with five professorships). Consequently, it is still pursued by only a few "outsiders" and by some interested mathematicians. Hence, it has not attained sufficient institutional backing that would have resulted from adequate support, and its practitioners therefore have been unable to study and evaluate properly the rich sources in Switzerland for the history of mathematics.

The first Swiss historian of mathematics who reflects this new level of specialization was HEINRICH SUTER (1848–1922, **B**), professor at the Kantonsschule in Zürich. Despite the general lack of institutional support, he set a high standard for the history of mathematics in Switzerland with his works on the history of Arabic mathematics. Among others who made contributions at this new, more specialized and internationally oriented level were LOUIS ISELY (1854–1916) and LOUIS GUSTAVE DU PASQUIER (1876–1957). ISELY wrote an *Histoire des sciences mathématiques dans la Suisse française* (History of the Mathematical Sciences in French Switzerland) (1901) and a small study, *Les femmes mathématiciennes* (Mathematical Women) (1894), while DU PASQUIER published *Le développement de la notion de nombre* (The Development of the Concept of Number) (1921), *Le calcul des probabilités, son évolution mathématique et philosophique* (Probability Theory, its Mathematical and Philosophical Development) (1926), and *Léonard Euler et ses amis* (Leonhard Euler and His Friends) (1927).

Of recent scholars in the French-speaking part of Switzerland, PIERRE SPEZIALI (1913–1995), FRANÇOIS LASSERRE (1919–1989), and FERDINAND GONSETH (1890–1975) should be mentioned. SPEZIALI worked above all on the history of mathematics in Geneva, and LASSERRE published several books on Greek mathematics. GONSETH was a philosopher with historical inclinations in the French tradition (cf. Section 1.11). His writings on the philosophy of mathematics and on relativity theory were informed by historical studies, and his work as a journal editor and organizer of conferences made him a noteworthy entrepreneur. For additional information about the development of mathematics in the French-speaking part of Switzerland, see [BAYS 1927b], [ISELY 1901], [METHÉE 1991], and [TREMBLEY 1987].

In Zürich, LOUIS KOLLROS (1878–1959), MICHEL PLANCHEREL (1885–1967), and RUDOLF FUETER (1880–1950) made noteworthy contributions to the history of mathematics, followed somewhat later by BARTEL LEENDERT VAN DER WAERDEN (1903–1996, **B**) and JOHANN JAKOB BURCKHARDT. For example, the Fribourg-born mathematician PLANCHEREL, who worked at the *ETH*, published a survey article on the history of mathematics of the preceding one hundred years in Switzerland [PLANCHEREL 1960]. After World War II, the history of mathematics in Zürich was associated most closely with BURCKHARDT and VAN DER WAERDEN. The former published his *Lesebuch zur Mathematik. Quellen von Euklid bis heute* (Readings in Mathematics: Sources from Euclid until Today) (1968) and *Die*

Mathematik an der Universität Zürich 1916–1950 (Mathematics at the University of Zürich, 1916–1950) in 1980. For VAN DER WAERDEN's many notable contributions to history of mathematics, see the biographical notice (**B**); for further information on the development of mathematics in Zürich, see [BURCKHARDT 1980] and [FREI/STAMMBACH 1994].

History of mathematics in Basel has always reflected the legacy of EULER and the BERNOULLIS. Among the most significant recent contributors was the mathematician OTTO SPIESS, known especially for his biography, *Leonhard Euler. Ein Beitrag zur Geistesgeschichte des XVIII. Jahrhunderts* (Leonhard Euler. A Contribution to the Intellectual History of the 18th Century) (1929), and for his efforts on behalf of the BERNOULLI edition project. Exceptional by virtue of his regular delivery of lectures on the history of mathematics and the exact sciences was JOACHIM OTTO FLECKENSTEIN (assistant to SPIESS and his successor in the Bernoulli project). He produced a steady stream of historical publications, including *Johann und Jakob Bernoulli* (1949), *Der Prioritätsstreit zwischen Leibniz und Newton* (The Priority Dispute Between Leibniz and Newton) (1956), and the major study, *Gottfried Wilhelm Leibniz. Barock und Universalismus* (Gottfried Wilhelm Leibniz: Baroque and Universalism) (1958) [GASSER 1979].

Among those attached to the Euler Edition was ANDREAS SPEISER, who also edited the *Opera mathematica* of JOHANN HEINRICH LAMBERT and the scientific writings on optics of JOHANN WOLFGANG VON GOETHE (1749–1832) in the Artemis edition. His major contributions to the history of mathematics include the source book *Klassische Stücke der Mathematik* (Classical Pieces of Mathematics) (1925), *Die mathematische Denkweise* (Mathematical Thinking) (1932, ^{3}1952), *Ein Parmenideskommentar. Studien zur platonischen Dialektik* (A Parmenides Commentary. Studies in Platonic Dialectics) (1937, ^{2}1959), and *Die Basler Mathematiker* (The Basel Mathematicians) (1939) [BURCKHARDT 1980, with list of works]. For more information on the development of mathematics in Basel, see [BONJOUR 1960], [SPEISER 1939], and [STAEHELIN 1957]; for more information on the history of mathematics in Switzerland in general, see [NEUENSCHWANDER 1999].

4.6 Conclusion

In summary one may say that research in the history of mathematics has a long tradition in Switzerland. In the beginning, its most important institutional support came from the *Naturforschende Gesellschaften*, a feature seemingly unique to Switzerland. Most of the well-known Swiss historians of mathematics from the last century such as WOLF, GRAF, and RUDIO held important offices within these *Naturforschende Gesellschaften*, and in many cases presided over an often associated local library. Later encouragement also came from the edition of collected works and from renowned mathematicians or philosophers with historical interests, chiefly RUDIO, SPEISER, GONSETH, and VAN DER WAERDEN. Often these influential figures inspired doctoral dissertations on the history of mathematics. At present, regular lectures on the history of mathematics, however, are offered only at the universities in Geneva and Zürich, and at the *Ecole Polytechnique Fédérale de Lausanne*.

Chapter 5

Germany

MENSO FOLKERTS, CHRISTOPH J. SCRIBA, and HANS WUSSING

5.1 Introduction

The history of mathematics as an academic discipline is one of the youngest branches of the historical sciences. Even in Germany, where the history of science has a rather long tradition (the first professional history of science society, the *Deutsche Gesellschaft für Geschichte der Medizin und Naturwissenschaft* (German Society for History of Medicine and Science), was founded in 1901 in Hamburg), only a few of the numerous universities have as yet established a chair — or even an institute — for the history of science, and positions expressly devoted to the history of mathematics are extremely rare.

On the other hand, the *Vorlesungen über Geschichte der Mathematik* (Lectures on History of Mathematics) by MORITZ CANTOR (1829–1920, **B**), published in four stately volumes between 1880 and 1908, are testimonies to the great interest the history of mathematics already enjoyed among mathematicians more than a century ago. The work has been reprinted several times, and although outdated, it still remains a valuable source of information [CANTOR 1880/1908]. Other testimonies for the growing interest of mathematicians in the history of their discipline around the turn of the century are also apparent: for example, the 1903 dissertation by CONRAD H. MÜLLER (1878–1953, **B**), supervised by FELIX KLEIN (1849–1925, **B**), on the history of mathematics in Göttingen during the 18th century [MÜLLER 1903], which reflects growing interest in historical research by the end of the 19th century [SCRIBA 1967].

Considering their style and scope, CANTOR's *Vorlesungen* have never been superceded by any similar undertaking. (Moreover, given the current tendencies and main objectives of the history of science and mathematics, it will doubtless never be replaced in a similar form either.) The *Vorlesungen* thus represent a milestone in the development of writing the history of mathematics. They were not, however, without some interesting forerunners. From today's point of view, it is difficult to understand how one of the greatest advocates of the history of science and of mathematics, GEORGE SARTON (1884–1956, **B**), in his introduction to *The Study of the History of Mathematics* [SARTON 1936a, 22], could claim that "the main reason for studying the history of mathematics, or the history of science, is purely humanistic. Being men, we are interested in other men, and especially in such men as have helped us to fulfil our highest destiny." He thus seemed to assume that biographical interests would provide the main motivation for all historians of mathematics. At the same time, in Sarton's *Bibliography of the History of Mathematics* appended to this book (p. [41]), he argued that of early treatises only one deserved to be mentioned, namely that by JEAN ETIENNE MONTUCLA (1725–1799, **B**), *Histoire des mathématiques* [MONTUCLA 1799/1802].

The magnificent efforts of MONTUCLA and JOSEPH JÉRÔME DE LALANDE (1732–1807, **B**), as well as the contemporaneous publications of CHARLES FRANÇOIS MAXIMILIEN MARIE (1819–1891, **B**), to name but three French authors (see Section 1.3.3), are indeed comparable in size to CANTOR's *Vorlesungen*. The Italian author GUGLIELMO LIBRI — actually: LIBRI-CARUCCI DALLA SOMMAJA (1803–1869, **B**), who, strangely enough, is not mentioned by SARTON, though generally well known (see Section 3.5.5) — should also be remembered. MONTUCLA in particular aimed at writing a history of ideas and problems in mathematics (one of the main concerns of the historian of mathematics), and in this he surpassed by far all earlier histories of mathematics. As for CANTOR, he had some predecessors in Germany who should not be totally overlooked in an historiographical survey, for these, too, have both a "humanistic interest" and an historical interest as well.

5.2 The Beginnings

5.2.1 The First Glimmer: Regiomontanus and Some Successors

Although it is problematic to assign REGIOMONTANUS, i.e. JOHANNES MÜLLER (born in Königsberg in Franconia, 1436–1476) — or any other scholar of that period — to a single country, his importance for writing history of mathematics in Germany is considerable. REGIOMONTANUS (whom DIRK J. STRUIK has called "a figure of flesh and bones" [STRUIK 1980, 4]), delivered an inaugural lecture at Padova in 1464 which contains valuable historical remarks, based on a good knowledge of primary sources. The humanistic touch reflected in his approach served as a model for his successors, and many followed his example and incorporated historical remarks or brief historical introductions into their works. Among those who did so were MICHAEL STIFEL (ca. 1487–1567), JOACHIM JUNGIUS (1587–1657), and JOHANN CHRISTOPH STURM (1635–1703). To call their works "histories," however, would be a great exaggeration.

5.2.2 Two Extremes: Leibniz and Wolff

The philosopher and mathematician, GOTTFRIED WILHELM LEIBNIZ (1646–1716), was doubtless the first German scholar to see clearly the need for what would now be called "internal history of mathematics," i.e., a description of the progress of mathematical ideas, problems, and methods. In his *Historia et origo calculi differentialis* (History and Origin of the Differential Calculus) of 1714, he emphasized as well the value of an impartial history for the augmentation of the "ars inveniendi," the heuristic art:

> It is not just that history may give everyone his due and that others may look forward to similar praise, but also that the art of discovery be promoted and its method known through illustrious examples [LEIBNIZ 1714, 392–410; here: 392].

An apt description of this ideal can be found in a letter of August 20, 1713, by PIERRE RÉMOND DE MONTMORT (1670–1719) to NICOLAUS I BERNOULLI (1687–1759). Therein MONTMORT expresses the hope that someone will write a chronological account of the discoveries in mathematics, attributing each to its author. He thought that "there could be no greater pleasure than to see the liaison, the connection of methods, the linkage of different theories," and he regarded such a presentation as an important contribution to a history of the human mind.[1]

The very influential German representative of early Enlightenment philosophy and, to some extent, pupil of LEIBNIZ, CHRISTIAN WOLFF (1679–1754), was professor of philosophy and mathematics at Halle and Marburg. He supplemented his elementary and very popular introductions to mathematics, entitled *Mathematische Anfangsgründe* (Elementary Foundations of Mathematics), with a fifth volume, *Kurtzer Unterricht von den vornehmsten mathematischen Schriften* (Brief Guide to the Most Important Mathematical Literature), in 1710/11 (new edition 1750). His aims were much more modest than what LEIBNIZ and MONTMORT had called for, and were motivated by didactical concerns: he wanted to provide some guidance to the mathematical literature for the novice and did not, at this point, intend to write a full history of mathematics. The Latin edition of WOLFF's work, published in 1741 and entitled *De Praecipuis Scriptis Mathematicis Brevis Commentatio* (Brief Commentaries on Distinguished Mathematical Writings), again was no more than a dry annotated guide to the literature.

5.2.3 Three Additional Eighteenth-Century German Authors

The 18th century saw the appearance of at least three additional works that can be classified as "histories of mathematics," namely studies by HEILBRONNER, KRAFFT, and FROBESIUS. JOHANN CHRISTOPH HEILBRONNER (1706–1747, **B**), a teacher in Leipzig, published his *Versuch einer mathematischen Historie* (An

[1] For the complete quotation, see the beginning of Section 1.3.

Attempt at a Mathematical History) in German (1739), followed by a Latin publication, *Historia matheseos universae* (History of Universal Mathematics) (1742). About HEILBRONNER and his Dutch forerunner, the polyhistorian GERARD JOHANN VOSSIUS (1577–1649, **B**) (on VOSSIUS, see Section 2.3), KURT VOGEL (1888–1985, **B**) wrote in the *Dictionary of Scientific Biography* [DSB 1970/90, vol.9, 501] that they "give only a jumble of names, dates, and titles." And with respect to HEILBRONNER's Latin work which had a very presumptuous title, VOGEL complained [VOGEL 1965, 182]: "Truly an enormous project! ... This book was at the same time a veritable chaos." JOSEPH E. HOFMANN (1900–1973, **B**), although he did not find much of value, at least appreciated the emphasis HEILBRONNER gave to excerpts from early writers and references to manuscripts and rare books of his time [BECKER/HOFMANN 1951, 255]. As early as 150 years ago, however, the historically sensitive G. H. F. NESSELMANN (1811–1881, **B**) leveled this harsh judgment against HEILBRONNER's Latin work [NESSELMANN 1842, 16]: "There is not the faintest sense of history to be found in this book."

Only slightly more historical, perhaps, was the approach of GEORG WOLFGANG KRAFFT (1701–1754). Professor of mathematics and physics at Tübingen beginning in 1744, KRAFFT had formerly taught at the *Academic Gymnasium* in St. Petersburg and was professor of mathematics and physics at the Petersburg Academy. His *Institutiones geometriae sublimioris* (Principles of Advanced Geometry) of 1753 contains some valuable historical selections on specialized topics, yet the book is far from constituting a continuous history of mathematics — or of geometry, for that matter.

A third figure from the 18th century is JOHANN NIKOLAUS FROBESIUS (1701–1756, **B**), a pupil of WOLFF and later professor of logic, metaphysics, and (beginning in 1741) of mathematics and physics at the University of Helmstedt. He wrote an *Historica et dogmatica ad mathesin introductio* (Historical and Dogmatic Introduction to Mathematics) (1750) and a collection, *Rudimenta biographiae mathematicorum* (A First Attempt at Biographies of Mathematicians), and left behind an unfinished history of mathematics at Helmstedt. FROBESIUS subdivided the history of mathematics proper into three sub-disciplines: (i) Biography, (ii) Bibliography and organography (from the Greek "organon," meaning here description of the contents of important mathematical works), and (iii) History of mathematics in its strictest meaning: the treatment of the origins and progress of mathematics. As examples he mentioned authors such as PETRUS RAMUS (1515–1572), G. J. VOSSIUS, ANDRÉ TACQUET (1612–1660), CLAUDE FRANÇOIS MILLIET DESCHALES (1621–1678), JOHN WALLIS (1616–1703, **B**), and JOHANN CHRISTOPH HEILBRONNER. Of particular interest is the periodization FROBESIUS provides of the history of mathematics: barbarian or oriental mathematics, Greek, Alexandrino-Roman, Arabic, and occidental mathematics. While he devoted 17 pages to the first and 20 to the second period, he offered a survey of *mathesis occidentalis* from the Middle Ages to the present in just two pages (i.e. to the time of Tzar PETER I of Russia)!

5.2.4 Mathematician, Bibliographer and Epigrammatist: Kästner

The Göttingen professor of mathematics and physics, ABRAHAM GOTTHELF KÄSTNER (1719–1800, **B**), well known for his epigrams and his literary works, published a four-volume *Geschichte der Mathematik* (1796–1800). In his introduction (vol. 1, 22–23) he mentions that HEILBRONNER, whom he knew from Leipzig, owned a considerable library. After HEILBRONNER's death this was sold at auction, and KÄSTNER acquired a number of HEILBRONNER's books. It has been said that KÄSTNER wrote his *History of Mathematics* on the book-ladder, for he too owned a substantial library. His *History* for the most part consists of remarks upon and excerpts from many of his books. In 1980 DIRK J. STRUIK characterized this as "an artless compilation of titles and descriptions of ancient books without an attempt at a readable narrative" [STRUIK 1980, 7]. Nevertheless, KÄSTNER called attention to the parallel problem, and in doing so was influential in bringing it to the attention of CARL FRIEDRICH GAUSS (1777–1855), JÁNOS BÓLYAI (1802–1860), and NIKOLAI IVANOVICH LOBACHEVSKIJ (1792–1856). As STRUIK acknowledged: "This is one example showing how research in the history of a science can lead to vital new discoveries."

5.3 First Half of the Nineteenth Century

5.3.1 From Watch-Maker to Professor of Technology: Poppe

JOHANN HEINRICH MORITZ (VON) POPPE (1776–1854), who started out as a clock-maker, held various positions as professor of mathematics and physics, ending his career as professor of technology at the University of Tübingen (beginning in 1818). He is a late representative of the 18th-century school of "technologists" that in part superceded the traditional school of cameralists. (The traditional academic training of future public servants under the "cameralists" had centered around public finance and administration. The "technologists" emphasized instruction in mechanical, technical, and manufacturing procedures in order to enable public administrators to understand and promote industrial activities.) At the age of 21, POPPE had already published a history of clock-making, and between 1807 and 1811, he issued a valuable three-volume *Geschichte der Technologie* (History of Technology). In 1828 (when he was *Hofrat* and Professor in Tübingen), he edited his *Geschichte der Mathematik seit der ältesten bis auf die neueste Zeit* (History of Mathematics from the Oldest Times until the Present Time), a book-length study that had grown out of a prize essay on the history of curved lines in the mechanical arts and architecture. His *Geschichte der Mathematik* also included a large section of 400 pages on applied mathematics (the mechanical, optical, and astronomical sciences), and a selected bibliography. In his preface, POPPE justified his publication by arguing that there was as yet no continuous chronological history of mathematical inventions and discoveries in German that was not constantly interrupted by learned quotations. Moreover, his work was a kind of handbook of pure and applied mathematics; it had the advantage of being more concise than MONTUCLA's voluminous study [MONTUCLA 1799/1802], and it was not preoccupied with concern primarily for French authors. As for CHARLES BOSSUT's (1730–1814, **B**) *Histoire des mathématiques* (translated into German by NIKOLAUS THEODOR REIMER in 1804), POPPE found it too fragmentary and, in a number of applied sections like mechanics and optics, too incomplete. He was equally critical of KÄSTNER's *Geschichte* of 1796. This, said POPPE, could hardly be called a "history" because it was no more than a collection of material for a future history of mathematics.

Yet such a universal history of mathematics, as POPPE envisaged it, which would put equal emphasis on pure as well as on applied mathematics — an 18th-century ideal, it must be said — was never realized. That many authors of mathematical monographs, or textbooks, were at least conscious of the development of their particular field of research, is revealed by historical surveys often contained in an introductory section. The reform of the German university system inspired by WILHELM VON HUMBOLDT (1767–1835), however, accentuating academic research, resulted in the advance of pure mathematics in Germany during the 19th century at the cost of applied areas. The consequences for comprehensive German publications on the history of mathematics will become apparent in the following sections.

5.3.2 Handbooks and History: Klügel and Mollweide

Although not constituting an actual history, the *Mathematisches Wörterbuch* (Mathematical Dictionary) in five volumes by GEORG SIMON KLÜGEL (1739–1812), and continued by CARL BRANDAN MOLLWEIDE (1774–1825), contains useful historical information. Following the example of CHRISTIAN WOLFF's *Mathematisches Lexicon* of 1716, which included historical remarks as part of its various entries ([WOLFF 1716]), the KLÜGEL/MOLLWEIDE *Wörterbuch* (published between 1803 and 1831, with two supplementary volumes in 1833/1836) included articles such as "Abacus," "Algebra," "Analysis," "Geometry," "Logarithms," "Parallels," and numerous others, sometimes with extensive historical and literary comments. Even if these fail on the whole to represent a complete history, they reveal that the authors had a strong historical interest and included historical information on pure as well as applied mathematics. While there is not such a general encyclopedic tradition in Germany as is described in Section 7.3 for the British Isles, the extent to which WOLFF, KLÜGEL, and MOLLWEIDE included rich historical information in mathematical dictionaries reveals a trend along similar lines.

5.3.3 Opposite Twins? Nesselmann and Arneth

With NESSELMANN and ARNETH, the spirit of educational reform, initiated in Prussia by WILHELM VON HUMBOLDT and spread from there to other German states, began to penetrate into the historiography of mathematics. GEORG HEIN-

RICH FERDINAND NESSELMANN (1811–1881, **B**), a student under CARL GUSTAV JACOB JACOBI (1804–1851) at Königsberg (East Prussia) and later professor of Sanskrit and Arabic there, was an author trained in mathematics as well as in the classical languages. His important *Versuch einer kritischen Geschichte der Algebra. Erster Theil: Die Algebra der Griechen* (Attempt at a Critical History of Algebra. Part 1: Algebra of the Greeks) (1842) clearly reflects the "kulturphilosophische" ideas of JOHANN GOTTFRIED VON HERDER (1744–1803). The influential German historical school under BARTHOLD GEORG NIEBUHR (1776–1831) and LEOPOLD VON RANKE (1795–1886) also began to influence the historiography of mathematics.

NESSELMANN, who in 1843 edited the *Essenz der Rechenkunst* (Essence of Arithmetic) of BAHĀʾ AL-DĪN AL-ʿĀMILĪ in Arabic and German, explained his historical point of view and his intentions in the preface to his *Algebra der Griechen* this way:

> It was not my intention to excerpt masses of historical data from older historical works and offer these to the public as a pleasant book for entertainment ... Even less was it my plan, to put together a hasty conglomeration of superficially-grasped facts, arbitrarily stretched and modeled on the procrustean bed of a philosophical system. As the title says, I intended to write a *critical* history. I wanted to investigate the history of algebra not as it is taught traditionally, but as it results from a prolonged and conscientious study of the sources. But the basic critical element is doubt. For this reason, I did not accept any fact as such from older historical works until my own inspection had convinced me of their truth and reliability. Understandably I have often had to tear down various edifices in this undertaking, and unfortunately, I have not always been in a position to erect a new structure on the ruins. But whether I have only been destructive or not may be judged by the expert and well-intended readers [NESSELMANN 1842, Vorrede, pp. ix–x].

NESSELMANN suggested a periodization of the history of mathematics and originally intended to devote one volume to each of the first four periods, i.e., up to the end of the 17th century. Unfortunately, only the first volume appeared. Of its nearly 500 pages on Greek algebra, 200 are devoted to number systems, logistics, and arithmetic, and about 250 to DIOPHANTUS and number theory, in particular to indeterminate problems. Throughout the work, NESSELMANN strove for a contextualization of the mathematics under scrutiny. In his words, "It is one of the most difficult tasks of the historian to understand and convey the conceptual background not only of each author but also of each period of time in its individual character, and at the same time to keep the modern viewpoint in mind, without lifting the ancient author from his peculiar sphere of thinking into our present one" [NESSELMANN 1842, p. 38].

In contrast to NESSELMANN's painstaking approach to sources, ARTHUR ARNETH (1802–1858, **B**) pursued broader, global questions. Professor of mathematics and physics at the Lyceum in Heidelberg and *Privatdozent* at the university, ARNETH made a book-length contribution to the *Neue Encyklopädie für Wissenschaften und Künste*, his *Die Geschichte der reinen Mathematik in ihrer Beziehung zur Geschichte der Entwicklung des menschlichen Geistes* (History of Pure Mathematics in Its Relation to the History of the Development of the Human Mind) of 1852. Having become aware of great differences between Indian and Greek mathematics, ARNETH, when asked to contribute to this general encyclopedia, wanted to give more than a simple description of *how* mathematics had risen to the present level. He wanted to look for the reasons *why* at different times and in different large populations it developed in characteristic ways.

ARNETH, like MORITZ CANTOR later, was influenced by the cultural philosopher EDUARD M. RÖTH (1807–1858). In Paris, RÖTH had studied science with DOMINIQUE FRANÇOIS JEAN ARAGO (1786–1853), JEAN BAPTISTE BIOT (1784–1862), PIERRE LOUIS DULONG (1785–1838), and JEAN-BAPTISTE-ANDRÉ DUMAS (1800–1884), as well as oriental languages. He began lecturing at Heidelberg in 1840, and became professor of philosophy and Sanskrit there in 1850. He believed that the roots of modern knowledge could be traced back to ancient Egypt and to the teachings of ZOROASTER [LÜTZEN/PURKERT 1994, 3].

In the introduction to his work, ARNETH argued that when definite knowledge about the earliest history of mathematics is not available, the historian has to deduce the missing information from laws of history which themselves have to be deduced from the intellectual history of a people as a whole. Abstract science, he continued, is the result of a long development during which mathematics was closely related to reality. A complete history of mathematics would be a description of the processes which led to the emergence of abstract thinking. Such a history, according to ARNETH, is precisely a "history of the development of the human mind" ("Geschichte der Entwickelung des menschlichen Geistes") (see Section 1.3). The whole life of a people (*Volk*) — its intellectual and material culture, its public and private state of affairs, and, in particular, its religion — form the basis for everything else. The inference therefore must be that in cases where we have insufficient information, the historian must fill in the gaps by drawing conclusions from this general intellectual and cultural background in order to construct a consistent unity.

ARNETH's description of the development of mathematics is thus based on his belief that general laws and conditions determine the development of the world and of the human species. This led him to sketch (in 60 pages) the laws of the development of all life on our planet. One of his principles is that a contrast of opposites is needed to produce something of a higher order (thesis/antithesis, in Hegelian terms), and indeed, ARNETH was most likely influenced by HEGEL's philosophy in holding such a view.

ARNETH's central theme then focuses on the rivalry between Greek and Indian mathematics. Indian mathematics shows a complete lack of proofs in favor of descriptive rules. Mathematical theorems in Indian mathematics do not state that something exists, but rather how something can be constructed or determined. Indian mathematics stresses not the properties of numbers as such, but operational schemes. Such arithmetical tendencies never fully developed in Greece, while in India the geometrical approach was not completely exploited (BHASKARA and DIOPHANTUS are seen as two exceptions). According to ARNETH, a new synthesis of both the algebraic and geometric approaches would have to occur in yet another cultural setting, where neither was emphasized over the other. A true amalgamation of algebra and geometry would eventually prove especially powerful for posterity ([ARNETH 1852, 179]). It must be admitted that the correctness of ARNETH's point, in light of considerable hindsight, is obvious.

ARNETH's work represents an early attempt in the German-speaking world to write a history of mathematics based on general philosophical principles. Doubtful as these may have been, and daring as this simplistic construct must appear, a comparison with NESSELMANN reveals that, by the middle of the 19th century, writing the history of mathematics no longer involved merely the compilation of facts or the naked chronicling of events.

5.4 From 1850 up to World War I

5.4.1 The "Philologists": Editions of Texts

In the second half of the 19th century, the history of mathematics in Germany was mostly written by interested high school teachers. Reforms in universities and higher education in Germany during the early decades resulted in the training of professional teachers with excellent scholarly qualifications — often in both the classics and the mathematical sciences — for the German *Gymnasium*. Thus, it was natural that they took as their main goal the editing of mathematical texts. This required first locating manuscripts and libraries where the works in question could be found. It then involved careful comparison of various copies to reconstruct the original texts. The *Gymnasium* teachers used the methods of *Textkritik* (textual criticism) expounded by KARL LACHMANN (1793–1851), professor of German at the University of Berlin. LACHMANN had demonstrated the advantage of his method not only in his editions of Old German texts, but in his edition (1848–1852) of the Roman *Corpus agrimensorum* (a collection of texts concerning land measurement), which contained not only theological and legal treatises but also practical texts on mathematics of import for Roman land surveyors. Because all *Gymnasium* teachers, even those going into mathematics teaching, were well trained in Latin and Greek, they were also prepared to edit mathematical texts written in the classical languages. In fact, most of those who undertook such editions were mathematics teachers, and only a small fraction actually taught Latin and Greek. Most were concerned with mathematical writings from Greek antiquity and medieval Latin texts, but some were interested in Arabic and Hebrew texts and in the mathematics of the 17th century, especially the works in Latin by GOTTFRIED WILHELM LEIBNIZ (1646–1716).

It should be noted that this interest in editing classical or later Latin texts was not solely a German phenomenon; there were scholars with similar interests in other countries: PAUL TANNERY (1843–1904, **B**) in France; NICHOLAI BUBNOV (1858–1943, **B**) in Russia; JOHAN LUDVIG HEIBERG (1854–1928, **B**) and AXEL ANTHON BJØRNBO (1874–1911, **B**) in Denmark; BALDASSARRE BONCOMPAGNI (1821–1894, **B**) in Italy, although in most cases these individuals started somewhat later. But the *Schulprogramme*, i.e., annual reports on the activities of local high schools, were typically German, and usually contained at least one academic article by one of the teachers. From time to time, some used this opportunity to publish editions of mathematical texts or other research on the history of mathematics.

Among the most notable scholars of this sort were Arabists and Hebraists. FRANZ WOEPCKE (1826–1864, **B**) was one of the first to edit Arabic texts from manuscripts and to reconstruct Greek works based upon Arabic translations. Fully aware of the importance of the Arabs in the transmission of mathematical and astronomical texts and of the poor state of Arabic studies — at the time only one Arabic text on mathematics was available in a modern edition in the West, namely AL-KHWĀRIZMĪ's *Algebra* edited by FREDERIC ROSEN (1831) — WOEPCKE decided to learn Arabic. At the time Paris was the capital of oriental studies in Europe. Since WOEPCKE lived there almost continuously from 1850 until his early death at age 38, he was able to learn Persian and Sanskrit as well, and even improved his knowledge of mathematics by studying with JOSEPH LIOUVILLE (1809–1882). WOEPCKE was a pioneer and an outstanding expert on Eastern contributions to the history of mathematics. Influenced by the ideas of ALEXANDER VON HUMBOLDT (1769–1859), whose treatise on different numeral systems *Über die bei verschiedenen Völkern üblichen Systeme von Zahlzeichen* (On the Various Number Systems Used by Different Peoples) he translated into French (1851), WOEPCKE focused his studies on Arabic algebra and the influence of Indian and Arabic mathematics on the West, especially the transmission of the Hindu-Arabic numerals and the sources of the *Liber abbaci* by LEONARDO OF PISA (FIBONACCI) (ca. 1180–ca. 1250). Among the more than thirty texts WOEPCKE published are the *Algebra* (1851) by AL-KHAYYĀM (1048?–1131?) and another *Algebra* (1853) by AL-KARAJĪ (ca. 1000). He also reconstructed lost Greek texts of EUCLID (ca. 365–ca. 300) and APOLLONIUS (ca. 260–ca. 190) by using Arabic manuscripts (see also Section 1.6.3).

Just as WOEPCKE was a pioneer in the study of Arabic literature, MORITZ STEINSCHNEIDER (1816–1907, **B**) did ground-breaking work on the history of mathematics and astronomy in Hebrew. In addition to his position as director of the Jewish *Töchterschule* (Girls School) in Berlin, he held a menial position ("Hilfsarbeiter") at the *Königliche Bibliothek* (Royal Library) in Berlin. There, he had access to one of the world's important collections of Arabic and Hebrew manuscripts. STEINSCHNEIDER was a typical representative of Jewish culture that flourished in Germany and in the Austro-Hungarian Empire in the second half of the 19th century. He published many philological articles on the Hebrew language, but he was mainly interested in mathematics and astronomy. His research began in earnest in 1844 when he started to work on an article devoted to "Jüdische Literatur" (Jewish Literature) for the *Realencyclopädie* edited by JOHANN SAMUEL ERSCH (1766–1828) and JOHANN GOTTFRIED GRUBER (1774–1851). Realizing then that there was almost no material available on the mathematical sciences, he started collecting information about Hebrew mathematical texts and their authors. His main results were published in a nine-part series entitled *Mathematik bei den Juden* (Mathematics of the Jews) [STEINSCHNEIDER 1893/1901]. These articles comprise a chronological list of authors of mathematical (and astronomical) texts in Hebrew up to 1500. Based on all of the material available to STEINSCHNEIDER, including numerous unpublished manuscripts, these articles are still useful today. Another helpful reference work is STEINSCHNEIDER's *Die europäischen Übersetzungen aus dem Arabischen bis Mitte des 17. Jahrhunderts* (The European Translations from the Arabic until the Middle of the 17th Century) (1904–1905, reprinted 1956), which lists (in alphabetical order) the names of the authors or translators who produced translations from Arabic into Latin and other languages. His *opus magnum* was *Die hebräischen Übersetzungen des Mittelalters und die Juden als Dolmetscher* (Medieval Hebrew Translations and the Jews as Translators) (1893), which contains valuable information on Arabic as well as Hebrew manuscripts. Besides these bibliographical works, STEINSCHNEIDER also edited the *Mishnat ha-Middot* (1864) (cf. SOLOMON GANDZ, in *Quellen und Studien* **A 2** (1932)), the earliest geometrical treatise in Hebrew, and he wrote long articles on ABRAHAM IBN EZRA (1867), ABRAHAM SAVASORDA (1880), and the so-called "middle" books of the Arabs (1865).

Another teacher who worked on Hebrew mathematics was GUSTAV WERTHEIM (1843–1902, **B**). His main historical work was a detailed analysis with partial translation of an arithmetical treatise of ELIA MISRACHI (1893; 2nd edition 1896), which was first published in Constantinople in 1534. He also brought out a translation of the works of DIOPHANTUS (ca. 250 A.D.) into German (1890) with additions and remarks by PIERRE DE FERMAT (1601–1665).

The most important German secondary school teachers of the 19th century who dealt with Greek and Roman mathematics were FRIEDLEIN and HULTSCH. GOTTFRIED FRIEDLEIN (1828–1875, **B**) edited some important Greek and Latin mathematical texts: the commentary by PROCLUS (410–485) on EUCLID (1873), the writings of BOETHIUS (ca. 480–524) on the *quadrivium* (1867), the *Geometry* of the Byzantine author PEDIASIMOS (1866), and the *Calculus* of VICTORIUS (1871). Other important contributions to the history of mathematics include FRIEDLEIN's book on elementary arithmetic in Greek and Roman antiquity as well as in the medieval West [FRIEDLEIN 1869], and a series of articles and books on the history of Arabic numerals in the West beginning with [FRIEDLEIN 1861]. In these works, FRIEDLEIN argued that the so-called *Geometry* of BOETHIUS was not a genuine work, but that it originated in its present form in the Middle Ages. Therefore, the Arabic numerals in the manuscripts of this text could not have originated with BOETHIUS but were introduced sometime after GERBERT (ca. 940–1003). This problem had been discussed earlier in France, especially by GUGLIELMO LIBRI (1803–1869, **B**) and MICHEL CHASLES (1793–1880, **B**) (see Section 1.7). FRIEDLEIN's arguments were later supported by NICHOLAI BUBNOV, but they were heavily criticized by MORITZ CANTOR (1829–1920, **B**) — wrongly, as we now know. Hindu-Arabic numerals seem not to have been known in the West before the end of the 10th century.

Although FRIEDRICH HULTSCH (1833–1906, **B**) never studied mathematics, he was interested in its history, especially in Greek antiquity. He published many articles, mostly on philological topics. HULTSCH was led to Greek mathematics through his interest in Greek and Roman metrology. In 1862 he wrote a fundamental work on this subject and edited the writings of the classical authors in this field (1864–1866). Although his edition of the geometrical and stereometrical works (1864) of HERO (ca. 75 A.D.) — the first modern edition of these texts — was superseded four decades later by JOHAN LUDVIG HEIBERG, HERMANN SCHÖNE (1870–1941), and WILHELM SCHMIDT (1862–1905), HULTSCH's edition of the *Collection* (1876–1878) of PAPPUS (ca. 320 A.D.) has never been replaced. HULTSCH was also interested in the writings on spherical geometry by THEODOSIUS (ca. 100 B.C.) and AUTOLYCUS (about 330 B.C.), and in the square root approximations in the *Measurement of the Circle* by ARCHIMEDES (287?–212). Moreover, HULTSCH contributed extensive articles, notably "Abacus," "Apollonios von Perge," "Archimedes," "Arithmetica," "Diophantos," "Dioptra," and "Eudoxos," to the *Real-Encyclopädie des classischen Alterthums*, founded by AUGUST PAULY (1796–1845), the revised edition of which by GEORG WISSOWA (1859–1931) was just beginning to appear at that time.

Another group of *Gymnasiallehrer* also studied the history of mathematics in the Latin West during the Middle Ages and the Renaissance, and it is no exaggeration to say that fundamental insights in this field came from research done by German scholars such as M. CANTOR and MAXIMILIAN CURTZE (1837–1903, **B**). This is also true of less important figures like PETER TREUTLEIN and HERMANN EMIL WAPPLER. TREUTLEIN (1845–1912, **B**) edited some texts on the abacus in the tradition of GERBERT (1877) and the *De numeris datis* (On Given Numbers) (1879) by JORDANUS NEMORARIUS (first half of 13th century). Especially worth mentioning are TREUTLEIN's articles on arithmetic in the 16th century (1877) and on algebra in Southern Germany in the 15th and 16th centuries (1879). This topic, the so-called *Coss*, was investigated at the same time by other teachers such as HERMANN EMIL WAPPLER (1852–1899, **B**), who went to school in Annaberg where the German *Rechenmeister* ADAM RIES (1492–1555) had lived and taught in his own private school. Later, when WAPPLER was a teacher at the *Gymnasium* in Zwickau (1879–1899), he became acquainted with the school's extraordinary collection of mathematical books and manuscripts. His articles on German algebra in the 15th century published between 1887 and 1900 were fundamental contributions to our understanding of the German *Coss* and the arithmetic of the *Rechenmeister*, and they continue to represent valuable secondary sources. Together with RUDOLF

PEIPER (1834–1898), WAPPLER rediscovered the *Rithmimachia*, a medieval board game based on Pythagorean number theory.

By far the most active and productive historian of mathematics of this group was MAXIMILIAN CURTZE (1837–1903, **B**). From the 1860s onwards, he edited many medieval mathematical texts, including works by NICOLE ORESME, BANŪ MŪSĀ, (PS.) JORDANUS NEMORARIUS, JOHANNES DE SACROBOSCO, AL-NAYRĪZĪ, SAVASORDA, LEONARDO MAINARDI, the so-called "INITIUS ALGEBRAS" (a fictitious author, probably of the early 16th century), and the correspondence of REGIOMONTANUS. With the exception of CANTOR, who was his friend, CURTZE was the most important German historian of mathematics in the 19th century. Thanks to his activities, many formerly unknown texts from the Middle Ages came to light, and during his sojourns in various European libraries, he found many important manuscripts. In his time CURTZE was the outstanding expert on medieval mathematical texts. But it should be noted, too, that CURTZE did not try to establish critical editions. In general, he was content to print the Latin text from one manuscript without trying to find more copies or to check the quality of the manuscript upon which his edition was based. He usually wrote only very short introductions and almost never provided commentary. Apparently, he had inadequate paleographic knowledge, and especially in his early editions, there are many incorrect resolutions of abbreviations and other paleographic mistakes. Therefore, all of his editions must be read carefully for such errors, although the general value of CURTZE's work should not be underestimated.

The last *Schullehrer* (schoolteacher) to be mentioned here is CARL IMMANUEL GERHARDT (1816–1899, **B**). While teaching in Salzwedel, he continued work on the history of the calculus which he had begun in Berlin. In the Salzwedel *Schulprogramm* (annual school report) for 1840, he treated the historical development of the calculus up to LEIBNIZ. Later, he obtained access to LEIBNIZ's *Nachlass* in Hannover, and following a thorough, preliminary survey of its contents, he was able to reconstruct LEIBNIZ's route to the discovery of the calculus. GERHARDT published the most important documents he found in the *Nachlass* in two series: *Mathematische Schriften von Leibniz*, 6 vols., 1849–1863, and *Philosophische Schriften von Leibniz*, 7 vols., 1875–1882.

GERHARDT edited not only the writings of LEIBNIZ, but also the *Arithmetica* of the Byzantine mathematician MAXIMOS PLANUDES (1865) and books 7 and 8 of PAPPUS's *Collection* (1871). Another important contribution to the history of mathematics was his *Geschichte der Mathematik in Deutschland* (History of Mathematics in Germany) (1877), which was part of the series *Geschichte der Wissenschaften in Deutschland* (History of the Sciences in Germany) promoted by the influential German historian LEOPOLD VON RANKE (1795–1886). Almost half of its 300 pages are devoted to mathematics up to 1650 with primary emphasis given to the 16th century, although the second chapter deals largely with the calculus in the 17th century. This remarkable book gives detailed information about the writings of some of the most important mathematicians up to the 17th century. Although CANTOR criticized the general idea of writing a "history of

German mathematics" because "with such an international science as mathematics it is scarcely possible to make distinctions in very early history on geographical grounds" [CANTOR 1900, 30], GERHARDT's book nevertheless retains much of its value.

5.4.2 Hermann Hankel and Moritz Cantor: First Comprehensive Studies

GERHARDT's *Geschichte der Mathematik in Deutschland* was an attempt to write a comprehensive study of a large part of the history of mathematics. At almost the same time another book appeared: *Zur Geschichte der Mathematik in Alterthum und Mittelalter* (Towards a History of Mathematics in Antiquity and the Middle Ages) (1874). At the early age of 28, its author HERMANN HANKEL (1839–1873, **B**) had become professor of mathematics in Leipzig, then (in the same year) in Erlangen, and in 1869, at the age of 30, in Tübingen. In his inaugural lecture at Tübingen, he dealt with the development of mathematics during the preceding centuries. In his unfinished and posthumously published book, some details are not correct, and the chapter on medieval Western mathematics is especially weak. CANTOR, as well as other authors, criticized numerous details, although HANKEL would certainly have corrected many of these had he lived longer. Yet at the same time CANTOR emphasized that the leading ideas, the vast historical landscape against which HANKEL presented the development of mathematics, was an entirely original creation.

Of particular interest is HANKEL's idea of dividing the history of mathematics into three periods. The first period, geometrical, runs up to the first century A.D. Then follows an arithmetical-algebraical period, which includes Hindu and Arabic mathematics as well as developments in Western mathematics in the 16th century. A new third period begins with RENÉ DESCARTES (1596–1650), who unified the two previous aspects of mathematics. HANKEL stressed throughout the similarity in development between philosophy and mathematics.

In his 35-page review in BONCOMPAGNI's *Bullettino* of 1875, PAUL MANSION (1844–1919, **B**) neatly expressed one of HANKEL's central arguments as follows:

> To summarize, Indian mathematics distinguishes itself from that of the Greeks by two principal characteristics: 1. It relies much more on intuition than on logical deduction; 2. Abstract speculations in arithmetic and algebra dominate significantly. One cannot deny that modern mathematics has a greater resemblance to that of the Indians than to that of the Greeks [MANSION 1875].

When HANKEL elaborated on this subject, he pointed to the importance of the role of Arabic authors in the transmission of mathematics to medieval Europe. Although he regretted the decline of Greek culture with its great mathematical talent, in his opinion the strictly logical, constructive synthesis in Greek mathematics had reached its peak. It was thus no accident that Greek mathematics began to decline and that Indian mathematicians took the lead.

HANKEL did not refer to his forerunner ARNETH when he made these general judgments. Yet he quoted one work by ARNETH in connection with Indian trigonometry. On the other hand, he referred to NESSELMANN four times — not to his book on Greek algebra but to the edition of BAHĀʾ AL-DĪN AL-ʿĀMILĪ, which he quoted. It seems unlikely that he did not know NESSELMANN's important *Versuch einer kritischen Geschichte der Algebra* (Attempt at a Critical History of Algebra), too. In fact, HANKEL's *Zur Geschichte der Mathematik in Alterthum und Mittelalter* may be regarded as a kind of synthesis of the approaches taken by NESSELMANN and ARNETH.

Without doubt, MORITZ CANTOR (1829–1920, **B**) was the most outstanding historian of mathematics in 19th-century Germany. Although his contributions to mathematics are insignificant, his reputation is due entirely to his historical studies. It seems that CANTOR was led to the history of mathematics by ARTHUR ARNETH, who emphasized the relationship between history of mathematics and history of the development of the human mind [ARNETH 1852]. In 1860, CANTOR visited Paris and met MICHEL CHASLES (1793–1880, **B**) (see Section 1.7), who encouraged him to continue his historical work. From that time on, CANTOR gave lectures on the history of mathematics at the University of Heidelberg. Beginning in 1875, he offered a three-semester course on the entire history of mathematics, from its origins to his own time. One of his students was SIEGMUND GÜNTHER (1848–1923, **B**), who himself later became an historian of science.

CANTOR started his historical publications with an article on the introduction of the Hindu-Arabic numerals into Europe (1856). This question was also central to his first extensive study on the history of mathematics, *Mathematische Beiträge zum Kulturleben der Völker* (Mathematical Contributions to the Cultural Life of the Peoples) (1863). CANTOR tried to show that the Hindu-Arabic numerals were Pythagorean and had been transmitted to the Middle Ages through BOETHIUS. Although he was criticized by FRIEDLEIN and later by BUBNOV, he held this incorrect position up to his death. In his second book, *Euclid und sein Jahrhundert* (Euclid and His Century) (1867), he summarized the contents of the main works of the three great Greek mathematicians, EUCLID, ARCHIMEDES, and APOLLONIUS, supplying an uncritical mass of facts. In 1875 he published *Die römischen Agrimensoren und ihre Stellung in der Geschichte der Feldmesskunst* (Roman Surveyors and Their Place in the History of Surveying). In this book, CANTOR described the writings of the Roman surveyors, raised questions about their sources and their influence on later scholars, and edited one important mathematical text which was missing from LACHMANN's edition (1848): that of EPAPHRODITUS and VITRUVIUS RUFUS (erroneously attributed by CANTOR to NIPSUS). CANTOR was very interested in the Roman surveyors because their writings were central in the transmission of methods from Egyptian and practical Greek geometry to the Latin Middle Ages.

As a result of his lectures and contemporary research on the history of mathematics, CANTOR published his monumental *Vorlesungen über Geschichte der Mathematik* (Lectures on History of Mathematics). The first volume appeared in 1880, the second in 1892, and the third in several parts between 1894 and 1898 [CANTOR 1880/1908]. These volumes cover the history of mathematics from its beginning up to 1758, and subsequently appeared in several revised editions. At the Second International Congress of Mathematicians in Heidelberg (1904), a fourth volume was planned that would go up to 1799. Written by nine historians of mathematics, among them ANTON VON BRAUNMÜHL (1853–1908, **B**) and SIEGMUND GÜNTHER, with CANTOR as editor-in-chief, this volume was published in 1908.

CANTOR tried to include virtually all information available to him in his *Vorlesungen*. It was without rival and remains to this day the most extensive work on the history of mathematics from antiquity to modern times. Therefore, it is not surprising that when it first appeared, it was received with admiration and enthusiasm. As the later volumes began to appear, however, they were also criticized because of many inaccuracies in detail. GUSTAF ENESTRÖM (1852–1923, **B**), who had reviewed the first volume positively, later devoted a special section of his *Bibliotheca mathematica* solely to correcting errors in CANTOR's *Vorlesungen*. Many well-established historians of mathematics, such as VON BRAUNMÜHL, CURTZE, FERDINAND RUDIO (1856–1929, **B**), HEINRICH SUTER (1848–1922, **B**), TANNERY, and WERTHEIM, contributed to these *Kleine Bemerkungen* (Short Comments). FLORIAN CAJORI (1859–1930, **B**) was correct when he wrote:

> The Herculean efforts of Cantor and the extremely penetrating criticisms of Eneström clearly point out two lessons to scholars of today: (1) The need for a more accurate general history of mathematics, prepared on the scale of that of Cantor's *Vorlesungen* and embracing the historical researches of the last twenty years; (2) the impossibility of this task for any one man. A history of the desired size and accuracy can be secured only by the cooperative effort of many specialists ([CAJORI 1920/21, 26]).

MORITZ CANTOR's monumental achievement can be compared with MONTUCLA's *Histoire des mathématiques* [MONTUCLA 1799/1802]. But in contrast to MONTUCLA, CANTOR restricted himself to so-called "pure mathematics," in keeping with the attitude towards mathematics prevalent at German universities in the 19th century. CANTOR's approach to history of mathematics was also somewhat similar to that taken by LIBRI in his *Histoire des sciences mathématiques en Italie* (History of the Mathematical Sciences in Italy) [LIBRI 1838/41] (cf. Section 3.5.5).

CANTOR not only contributed to the history of mathematics through his *Vorlesungen*, but he was also responsible for making the *Zeitschrift für Mathematik und Physik*, founded by OSKAR SCHLÖMILCH (1823–1901), a periodical increasingly sympathetic to the history of mathematics. CANTOR was a member of its editorial board from 1859, and was the editor of its *Historisch-literarische Abtheilung* (established in 1875) and of the *Abhandlungen zur Geschichte der Mathematik*, which appeared from 1877 as a supplement to the *Zeitschrift für Mathematik und Physik* and later became a separate publication. This periodical, BALDASSARRE BONCOMPAGNI's *Bullettino*, and GUSTAF ENESTRÖM's *Bibliotheca mathematica*, were the most important places for publishing articles and monographs prior to World War I.

5.4.3 Günther and Braunmühl: On the Way to Professionalism

The first attempts to establish the history of mathematics in universities were made in the 1880s. The two leading advocates, SIEGMUND GÜNTHER (1848–1923, **B**) and ANTON VON BRAUNMÜHL (1853–1908, **B**), were both professors at the *Technische Hochschule* in Munich. Thanks to their efforts, Munich soon became one of the major centers for the history of mathematics (and of the sciences) in Germany.

GÜNTHER was a prolific author; the number of his publications runs to four figures, with 200 of them books or substantial articles. His earliest writings were devoted to pure mathematics and geography, but his mathematical works never overlooked the historical side of the subjects he treated. His first major historical work was the *Vermischte Untersuchungen zur Geschichte der mathematischen Wissenschaften* (Diverse Researches on the History of Mathematical Sciences) (1876), in which he considered such varied topics as star polygons, continued fractions, magic squares, Jewish astronomy, and the pendulum, all from an historical point of view. In 1887, he published his *Geschichte des mathematischen Unterrichts im deutschen Mittelalter bis zum Jahre 1525* (History of Mathematical Instruction in the German Middle Ages Until 1525), a useful and comprehensive description of how mathematics was taught in the medieval cathedral schools, universities, and private schools. His most comprehensive book on the history of science is the *Geschichte der anorganischen Naturwissenschaften im 19. Jahrhundert* (History of the Inorganic Sciences in the 19th Century), which appeared in 1901 as the fifth volume of *Das 19. Jahrhundert in Deutschlands Entwicklung* (The 19th Century in the Development of Germany). In almost a thousand pages, GÜNTHER tried to provide a survey of the physics, chemistry, earth sciences, and astronomy of the previous century. In 1908, he wrote the *Geschichte der Mathematik. I. Teil. Von den ältesten Zeiten bis Cartesius* (History of Mathematics. Part I. From the Oldest Times Until Descartes).

It is clear that GÜNTHER's writings are basically compilations. They contain many names, dates, and facts, but reveal little of motives, interconnections, or results. GÜNTHER's knowledge of scientific matters and the secondary literature resulted, however, in readable historical accounts. Although HEINRICH WIELEITNER (1874–1931, **B**) rightly complained that GÜNTHER "published a bit too much and some things a bit too quickly" [WIELEITNER 1923, 2], his voluminous works contributed much to history of mathematics and are still worth reading. GÜNTHER's influence also extended beyond his writings. He was an excellent teacher and had numerous students; typically, his lectures on topics in the history of science attracted up to 27 students.

At the same time GÜNTHER was giving occasional lectures on the history of science, his colleague BRAUNMÜHL offered the first regular lectures on the history of science and especially of mathematics. As a mathematician, BRAUNMÜHL was not important, but around 1890 he became increasingly interested in the history of mathematics and astronomy. His first major work, a biography of CHRISTOPH SCHEINER (1573–1650), appeared in 1891. There followed further biographical sketches of GALILEO GALILEI (1564–1642), NICOLAUS COPERNICUS (1473–1543), LEONHARD EULER (1707–1783), KARL WEIERSTRASS (1815–1897), LUDWIG SEIDEL (1821–1896), SOPHUS LIE (1842–1899), and others. In 1895 he announced his plan to write a history of trigonometry. After a series of studies on individual topics related to Arabic and medieval Western trigonometry, the first part of his *Vorlesungen über Geschichte der Trigonometrie* (Lectures on History of Trigonometry) appeared in 1900 and treated the subject as far as the beginning of the 17th century. Later developments, until the end of the 19th century, were described in a second part (1903). This two-volume treatise is notable for its thorough investigation of sources. It was the first comprehensive history of trigonometry, and its treatment of the Christian Middle Ages has not yet been surpassed. BRAUNMÜHL intended to write a sequel to the *Geschichte der Mathematik*, to complete the work of GÜNTHER, but he died before he could finish this. WIELEITNER eventually completed the project (1911, 1921).

As of 1893, in addition to his mathematical lectures, BRAUNMÜHL also began to offer courses on the history of mathematics. Beginning in the winter semester 1893/94, he held a "mathematical-historical seminar" as well — perhaps the first such at any university. Whether he lectured himself, or his assistants or advanced students gave lectures, it was expected that primary sources would be treated [BRAUNMÜHL 1895], [BRAUNMÜHL 1897], [BRAUNMÜHL 1902], [BRAUNMÜHL 1905]. Among seminar participants were WILHELM END (1864–1922), WILHELM KUTTA (1867–1944), CARL RAIMUND WALLNER (1881–ca.1933), and especially AXEL ANTHON BJØRNBO, whose dissertation on the trigonometry of MENELAUS was suggested by BRAUNMÜHL.

Meanwhile interest in the history of mathematics persisted among the group of *Gymnasiallehrer*. MAX SIMON (1844–1918, **B**) analyzed the six planimetric books of EUCLID's *Elements* (1901) and published a survey of Greek mathematics in cultural perspective (1909), while EDMUND HOPPE (1854–1928, **B**) studied the works of HERON. In contrast to SIMON's overview, HOPPE's survey of Greek astronomy and mathematics (1911) concentrated on what much later would be called "internal history."

5.4.4 Felix Klein and His "Vorlesungen." Editions of "Collected Works"

Apart from university professors like CANTOR, GÜNTHER, and BRAUNMÜHL, who devoted most of their time to historical studies, at the beginning of the 20th century there were professors who, in addition to their teaching duties in mathematics, were interested in the recent development of mathematics. The most important figure in this category was FELIX KLEIN (1849–1925, **B**) who was not only interested in problems of teaching mathematics, but also had a keen interest in its historical development. In order to provide the mathematical layman with "a picture of what mathematicians are doing, of the boundless diversity of the problems which our science is always trying to include in the compass of its power" [KLEIN 1926/27, **1**, 1–2], KLEIN considered an historical approach as particularly appropriate. He argued that "the natural interest in the growth of the subject will spontaneously pull the reader along ... Finally, emphasis on the influence of mathematics upon its neighboring sciences, together with a lively description of its relations to our cultural life as a whole, will serve as a starting-point for every educated reader" [KLEIN 1926/27, **1**, 2].

KLEIN's most important contributions in this respect were his *Vorlesungen über die Entwicklung der Mathematik im 19. Jahrhundert* (Lectures on the Development of Mathematics in the 19th Century), which he had given at the University of Göttingen. These circulated in typescript during his lifetime and were edited and published after his death by RICHARD COURANT (1888–1972) and OTTO NEUGEBAUER (1899–1990, **B**) [KLEIN 1926/27]. In these lectures, KLEIN did not intend to give a systematic account of the development of the main areas of mathematics. Instead, he restricted himself to "putting together selected sketches, in which I sometimes present the life-work of exceptional individuals, and sometimes the goals and results of certain schools" [KLEIN 1926/27, **1**, 5–6]. Indeed, his account is historically unbalanced: geometry, the theory of functions, mathematical physics, and mechanics are treated at length; other areas such as algebra, number theory, and the theory of sets are hardly considered. Although KLEIN (as one who was actively involved in its development) naturally gives a subjective account of the history of mathematics in the 19th century, his book, the first attempt to give a summary account of the subject, was a distinguished piece of historical writing.

In addition, KLEIN stimulated others to edit and to comment on the collected works of famous mathematicians, among them PAUL STÄCKEL (1862–1919, **B**) on EULER and GAUSS, HEINRICH LIEBMANN (1874–1939, **B**) on NIKOLAĬ LOBACHEVSKIĬ, and FRIEDRICH ENGEL (1861–1941, **B**) on HERMANN GRASSMANN and SOPHUS LIE. It was ENGEL's edition of the mathematical and physical papers of GRASSMANN that rescued his important *Ausdehnungslehre* (Theory of Extension) from oblivion. In 1895, ENGEL and STÄCKEL published *Die Theorie der Parallellinien von Euklid bis auf Gauss* (The Theory of Parallels from Euclid to Gauss). There are also special books and articles on the history of non-Euclidean geometry by ENGEL (especially on LOBACHEVSKII, including the first translation of his works into a Western European language), and by STÄCKEL (especially on the two BÓLYAIS, WOLFGANG and JOHANN).

The most important undertaking of this kind was the edition of the *Werke* (Collected Works) of GAUSS (see [SCRIBA 1977, 45–51]). When GAUSS died in 1855, the government of Hanover bought his scientific estate from his heirs and entrusted one part of it to Göttingen's *Akademie der Wissenschaften* (Academy of Sciences), formerly the *Königliche Gesellschaft der Wissenschaften* (Royal Society of Sciences), and another part to the astronomical observatory. The idea of editing the collected works eventually led to twelve volumes published from 1863 to 1935 under sponsorship of the Göttingen *Akademie*. The first editor, ERNST C. J. SCHERING (1833–1897), had known GAUSS as a student and had been made professor of mathematics at Göttingen in 1858. Between 1863 and 1874, he edited volumes 1–6 which were devoted primarily to papers already published by GAUSS. After SCHERING's death, FELIX KLEIN (who had been called to Göttingen in 1886) took over the edition. Under his supervision, volumes 7 to 10,1 (i.e. volume 10, part 1) were published (1900–1917). These distinguished themselves from the earlier volumes in that material from GAUSS's correspondence with HEINRICH CHRISTIAN SCHUMACHER (1780–1850), ALEXANDER VON HUMBOLDT (1769–1859), FRIEDRICH WILHELM BESSEL (1784–1846), WOLFGANG (FARKAS) BÓLYAI, and WILHELM OLBERS (1758–1840) was included, as well as excerpts from unpublished papers in GAUSS's estate. Of special importance was GAUSS's scientific diary edited by KLEIN in 1901. As was his custom, KLEIN distributed the task of actually editing and publishing this material among several of his collaborators.

For volumes 10,2 and 11,2, KLEIN suggested a collection of articles on various aspects of GAUSS's scientific activities, to be written by competent scientists in these areas. In carrying out this plan, KLEIN, together with MARTIN BRENDEL (1862–1939) and LUDWIG SCHLESINGER (1864–1933), founded a series entitled *Materialien für eine wissenschaftliche Biographie von Gauss* (Materials for a Scientific Biography of Gauss), in which eight essays were published between 1911 and 1920. Here specialists investigated the development of GAUSS's research in different areas: PAUL BACHMANN (1837–1920, **B**) on the theory of numbers; SCHLESINGER on the theory of the arithmetic-geometric mean and the theory of functions; ANDREAS GALLE (1858–1943) on GAUSS's methods of computing with numbers; STÄCKEL on geometry; PHILIPP MAENNCHEN (1869–1945) on the interrelation between GAUSS's methods of computation and number theory; BRENDEL on astronomy; ABRAHAM FRAENKEL (1891–1965) on number theory and algebra; and ALEXANDER OSTROWSKI (1893–1986) on the interconnections between two of the proofs given by GAUSS for the fundamental theorem of algebra. In revised form these articles were re-issued between 1922 and 1933 as volumes 10,2 and 11,2 of the GAUSS *Werke*, augmented by papers on the calculus of variations (OSKAR BOLZA); mechanics and potential theory (HARALD GEPPERT); geodesy (ANDREAS GALLE); and magnetism, electrodynamics, and optics (CLEMENS SCHAEFER). After KLEIN's death in 1925, editorial responsibilities passed on to his collaborators

BRENDEL and SCHLESINGER. In 1927 and 1929, respectively, volumes 11,1 (supplementary material on physics, chronology, and astronomy) and 12 (*varia*) were published. Consequently, the *Works* covered all of the main areas in which GAUSS worked. No index volume has yet been prepared, nor have those papers as yet unpublished been systematically examined.

In connection with preparation of the *Works*, the editors constantly tried to collect further parts of GAUSS's correspondence. Through donations and purchases, the GAUSS archives in Göttingen now include additional letters from GAUSS to SCHUMACHER, OLBERS, and BESSEL, as well as letters to CHRISTIAN LUDWIG GERLING (1788–1864), KARL LUDWIG HARDING (1765–1834), and to GAUSS's son JOSEPH (1806–1873). Subsequently, some of these correspondences have also been published.

Based in part on these resources in Germany, and additional materials he collected for his own archive of GAUSS memorabilia in Natchitoches (Louisiana), the American professor of German literature, GUY WALDO DUNNINGTON (1906–1974, **B**) later wrote what at present remains the definitive biography of GAUSS in English: *Carl Friedrich Gauss – Titan of Science* (1955).

German mathematicians not only made fundamental contributions to the edition of the works of GAUSS, but also to the EULER edition (see [RUDIO 1911], [BIERMANN 1983], [FELLMANN/IM HOF 1993]; see also Section 4.4). As early as 1841–1849, JACOBI, who was very eager to publish a complete edition of the works of EULER, corresponded with PAUL HEINRICH FUSS (1798–1855) about such a project. Although he made very detailed recommendations, nothing came of his proposal at the time. Nor was it possible to realize another EULER project, planned in 1903 by members of the academies in Berlin and St. Petersburg. Finally, a proposal which RUDIO began to develop in 1883 was realized in 1907, the bicentenary of EULER's birth. On this occasion, the *Schweizerische Naturforschende Gesellschaft* (Swiss Scientific Society) (since 1983/88: *Schweizerische Akademie der Naturwissenschaften*) (Swiss Academy of Sciences) appointed a commission to investigate the possibility of producing a complete edition. In 1909 work on the *Leonhardi Euleri Opera omnia* officially began. Its first general editor was RUDIO; under his direction 22 volumes appeared through 1928. Besides RUDIO, credit above all for the EULER edition is due to STÄCKEL (see [Rudio 1923]), chair of the *Euler Kommission* which had been formed at the annual meeting of the *Deutsche Mathematiker-Vereinigung* in Dresden in 1907. The Commission was charged with the task of finding ways and means to complete the EULER edition. STÄCKEL had submitted his own plan about how best to group EULER's books and papers according to common themes within this edition. STÄCKEL's proposal was based upon suggestions made in the course of correspondence between JACOBI and FUSS, which STÄCKEL had edited with the help of WILHELM AHRENS (1872–1927) [STÄCKEL/AHRENS 1908]. STÄCKEL was also closely involved as ENESTRÖM compiled a comprehensive bibliography of EULER's publications. This reference work, published by the *Deutsche Mathematiker-Vereinigung* (1910–1913), remains the standard reference on EULER's works. Two volumes of EULER's *Opera omnia* were edited by STÄCKEL himself: the *Mechanica* and (with RUDIO) the *Commentationes algebraicae.*

The last decades of the 19th century also witnessed the rise of another collective undertaking that was to become of great importance for the further development of mathematics: the founding of the *Jahrbuch über die Fortschritte der Mathematik* (Yearbook on the Advances of Mathematics). Begun in 1869 by two mathematics teachers at *Gymnasia* in Berlin, CARL ORTHMANN (1839–1885) and FELIX MÜLLER (1843–1928, **B**), who adopted the model of the *Fortschritte der Physik* (founded in 1845), the *Jahrbuch* is of considerable importance as a source for historians of mathematics. In particular, from the beginning it included a section on the history of mathematics; its reviews therefore provide insight into the scope and understanding as well as into the main areas of mathematico-historical research between 1870 and 1945.

In the beginning the journal suffered from lack of active support by first-rank mathematicians, although KLEIN praised its general conception and tried to interest his colleagues in active collaboration. The *Jahrbuch*, as it was generally called, because its reviews covered mathematical publications (books and articles) year by year in separate volumes, was the only comprehensive reviewing journal of mathematics for its first six decades of existence. Among regular contributors, there was more than a proportional share of well-known historians of mathematics. All of the following, for instance, are represented in the present volume with a biography in Part II (listed here roughly in the order in which they began to contribute reviews to the *Jahrbuch*):

MAXIMILIAN CURTZE (1837–1903)
PAUL MANSION (1844–1919)
PETER TREUTLEIN (1845–1912)
SIEGMUND GÜNTHER (1848–1923)
GUSTAF ENESTRÖM (1852–1923)
ANTON VON BRAUNMÜHL (1853–1908)
PAUL STÄCKEL (1862–1919)
GINO LORIA (1862–1954)
JOHANNES TROPFKE (1866–1939)
WILHELM LOREY (1873–1955)
KARL BOPP (1877–1934)
CONRAD H. MÜLLER (1878–1953)
PHILIP JOURDAIN (1879–1919)
KURT VOGEL (1888–1985)
JOSEPH E. HOFMANN (1900–1973)
HANS FREUDENTHAL (1905–1990)

However, due to the rapid increase of mathematical production and the disturbances caused by World War I, the *Jahrbuch* fell increasingly behind schedule (see [SIEGMUND-SCHULTZE 1993]). This was one of the reasons why the *Zentralblatt für Mathematik* was founded in 1931 (see Section 5.4.6.), which eventually superseded the *Jahrbuch.*

Increased interest in history of mathematics is also reflected in the appearance and success of the series *Ostwalds Klassiker der exakten Wissenschaften* (Ostwald's Classics of the Exact Sciences) [*Ostwalds Klassiker* 1889]. Founded by the famous physico-chemist WILHELM OSTWALD (1853–1932), the series aimed at producing inexpensive editions (in German translation, where necessary) of important works that had become "classics" in their field. The series was to include mathematics, astronomy, physics, chemistry, and physiology. As co-editor (and editor beginning with vol. 43), OSTWALD chose his former physics professor ARTHUR VON OETTINGEN (1836–1920), who moved from Dorpat to Leipzig in order to be near to OSTWALD and the publishing house WILHELM ENGELMANN.

By 2001 the series *Ostwalds Klassiker* has grown to almost 300 volumes. Among its titles are the standard German translations of the works of EUCLID and ARCHIMEDES as well as important publications by about two dozen mathematicians, from NEWTON and LEIBNIZ to KLEIN and HILBERT. Several of these volumes have been reprinted more than once, a testimony to the great interest in these annotated, moderately priced editions (for bibliographical details, see [DUNSCH/MÜLLER 1989]). During the years after World War II, while Germany was divided, the series was continued in Leipzig (East Germany). Among the volumes published in East Germany are one on HILBERT's problems (vol. 252, with commentaries, based on the Russian edition, Moscow 1969) and a republication of GAUSS's mathematical diary (vol. 256, also with detailed annotations); both have been reprinted several times. In West Germany six volumes of a *Neue Folge* (New Series) appeared between 1965 and 1970. Among them, KURT VOGEL's German translation of the *Chiu Chang Suan Shu. Neun Bücher arithmetischer Technik* (1968) made the ancient Chinese text *Jiuzhang suanshu* (Nine Chapters on the Mathematical Arts) (1st century A.D.) (see chapter 17) available for the first time in a Western language (apart from the Russian translation by EL'VIRA I. BERËZKINA that appeared in 1957).

5.5 Between the Wars

5.5.1 More Studies on Arabic Mathematics

Another focus of research in Germany during the first half of the 20th century was mathematics in the Arabic-Islamic culture. A number of individuals are responsible for the fact that interest in this subject began to develop elsewhere as well, stimulated by the work of German scholars. Among the earliest was JULIUS RUSKA (1867–1949, **B**), who first studied mathematics and science, but then turned to Semitic languages, and from 1911 taught Semitic philology at the University of Heidelberg. All of RUSKA's books and articles focused on one problem: identifying the role of Arabic science in the transmission of Greek thought to medieval Europe and tracing its historical evolution. He always sought to avoid traditional opinions by going back to primary sources. To do this, it was necessary to make editions of the original texts. In his dissertation he wrote (as quoted directly from the *éloge* in English for RUSKA written by PAUL KRAUS):

> For a real history of medicine and the natural sciences in the domain of Islamic civilization almost everything is still lacking ... In order to build it up, we need the texts themselves; from these we have to draw the fundamental facts of history; with their help we have to

> substantiate the theories of the scholars and their transformations, to establish the relations of the single branches to each other, to determine the contributions of the great authors and their influence upon the development of our knowledge: tasks enough to occupy many generations of students, even if more of them will devote themselves to these questions (RUSKA, according to [KRAUS 1938, 12]).

RUSKA contributed substantially to filling these lacunae. Although his main areas of research were alchemy and mineralogy, he also published studies on Arabic mathematics. His *Zur ältesten arabischen Algebra und Rechenkunst* (On the Oldest Arabic Algebra and Arithmetic) (1917) is still considered a basic work on the development of Arabic arithmetic and algebra.

RUSKA also had a considerable hand in the establishment of the history of science as a discipline in German universities. As vice-chair of the *Deutsche Gesellschaft für Geschichte der Medizin und Naturwissenschaft*, he used his connections with KARL SUDHOFF (1853–1938), the society's chair and director of the *Institut für Geschichte der Medizin* in Leipzig (founded in 1906), to establish an *Institut für Geschichte der Naturwissenschaften* in Heidelberg in 1922. Initially financed by a private foundation, this was the first institute for the history of science in Germany. In 1927 RUSKA was called to Berlin as director of the newly founded *Forschungsinstitut für die Geschichte der Naturwissenschaften.* In 1931, this institute became part of an *Institut für Geschichte der Medizin und Naturwissenschaften* at the University of Berlin, and RUSKA was made *Abteilungsvorsteher für Geschichte der anorganischen Naturwissenschaften* (Section Head for History of the Inorganic Sciences) [RUSKA 1925], [RUSKA 1926], [RUSKA 1928], [RUSKA 1929], [RUSKA 1930]. Together with the Institute's historian of medicine, PAUL DIEPGEN (1878–1966), RUSKA edited the journal *Quellen und Studien zur Geschichte der Medizin und Naturwissenschaften.*

RUSKA also had close connections with EILHARD WIEDEMANN (1852–1928, **B**), professor (from 1886) of physics in Erlangen and director of the Institute of Physics. Beginning in 1900, WIEDEMANN gave up his experimental laboratory work and increasingly concentrated on his research related to the history of Arabic-Islamic science, translating and commenting on the works of numerous scientists, mathematicians, and engineers. His publications were devoted primarily to physics, astronomy, and scientific instruments. Many of WIEDEMANN's publications were issued under the collective title "Beiträge zur Geschichte der Naturwissenschaften" (Contributions to History of the Sciences) in the *Sitzungsberichte der Physikalisch-Medizinischen Sozietät zu Erlangen* (1902–1928; 78 parts). It was thanks to his inspiration and support that the series *Abhandlungen zur Geschichte der Naturwissenschaften und der Medizin* was founded (1922–1925). HEINRICH SUTER's translations of PAPPUS's commentary to book X of EUCLID, a treatise on geometrical constructions by ABŪ 'L-WAFĀ' (940–998) (1922), and BJØRNBO's edition of THĀBIT's work on MENELAUS's theorem (1924, published with comments by H. SUTER and with additional material by HERMAN BÜRGER and KARL KOHL), were among the important contributions about astronomy and mathematics of the Arabs published in the *Abhandlungen.*

Great credit for pioneering research and publication of sources on the history of astronomy and trigonometry of the Arabs is due to CARL SCHOY (1877–1925, **B**). His historical interests date to his student days in Munich, where he studied with GÜNTHER and BRAUNMÜHL. Unfortunately, SCHOY never realized his intention of becoming a university professor. After his *Habilitation* in the history of mathematics and astronomy (1919), under the post-war conditions in the occupied *Ruhrgebiet* he was prevented from lecturing as *Privatdozent* for political reasons (see **B**). SCHOY died shortly after he had been made *Dozent* for the history of the exact sciences in the Near East at the University of Frankfurt. Nevertheless, between 1911 and 1925, SCHOY made substantial contributions to the history of Arabic mathematics and mathematical astronomy through a series of excellent articles. His research focused on Arabic methods of determining coordinates and on gnomonics. Thanks to SCHOY, the achievements of the then almost forgotten AL-BĪRŪNĪ (973–1048) were rediscovered. In connection with his research on Arabic trigonometry, he found the treatise by ARCHIMEDES on the heptagon and showed that the so-called "Heronic formula" for a triangle was in reality first discovered by ARCHIMEDES.

PAUL LUCKEY (1884–1949, **B**), too, made important contributions to the study of Arabic mathematics. Originally interested in nomography, he turned to Arabic science in the 1930s. LUCKEY argued that Muslim mathematicians of the 10th century had furthered Chinese and Indian methods for extracting square roots, applying algorithms equivalent to those known as the "Ruffini-Horner method." LUCKEY made clear the numerical ability of the Arabs by editing and analyzing the works of AL-KĀSHĪ (†1429) on the computation of π and on arithmetic. His research focused as well on the development of spherical trigonometry by ABŪ 'L-WAFĀ' (940–998) and others.

Finally, MAX KRAUSE (1909–1944, **B**) also undertook important research on Arabic mathematics. He obtained his Ph.D. in 1933 with an excellent edition of an Arabic redaction of the *Sphaerica* of MENELAUS. His catalogue, *Stambuler Handschriften islamischer Mathematiker* (1936), which covers an extraordinary collection of Islamic manuscripts, remains a masterpiece of scholarship. Unfortunately, KRAUSE's plan to edit AL-BĪRŪNĪ's *Qānūn al-Mas'ūdī* was never realized, as he lost his life in Russia in 1944.

The War also interrupted the research of KARL GARBERS (1898–1990) who, after he had obtained his Ph.D. with an edition of THĀBIT B. QURRA's treatise on plane sundials (1936), worked for a while in RUSKA's department in Berlin and in the Sudhoff Institute in Leipzig. After World War II, GARBERS became a schoolteacher and turned his main attention to Arabic texts in chemistry. Among the very few individuals in Germany who continued to do some research on the history of Arabic mathematics was the patent lawyer HEINRICH HERMELINK (1920–1978, **B**) whose studies included publications by Muslim scholars on magic squares.

5.5.2 Toeplitz, Neugebauer, and Bessel-Hagen: The "Kiel-Göttingen-Bonn Group"

Important stimulation for the history of mathematics after World War I came from a group of mathematicians teaching at the universities of Kiel, Göttingen, and Bonn. The central figure was OTTO TOEPLITZ (1881–1940, **B**), who had studied in Breslau, Berlin, and Göttingen, and who became professor first in Kiel (1913–1928) and then in Bonn (1928–1935). TOEPLITZ was convinced that historical considerations could be useful in bridging the gap between mathematics at the *Gymnasium* level and at the universities, resulting in what he called the "genetic method." This did not aim at recreating what had happened in history (the task of the historian), but sought to select from history the origins and motives for those mathematical results and methods that managed to withstand the test of time. Thus, his concern was not simply history for its own sake but reflected an interest in the genesis of problems, theorems, and proofs in their historical development. TOEPLITZ presented these ideas in a lecture given in 1926 in Düsseldorf [TOEPLITZ 1927], but they may also be seen most clearly in his book, *Die Entwicklung der Infinitesimalrechnung. Eine Einleitung in die Infinitesimalrechnung nach der genetischen Methode* (The Development of the Infinitesimal Calculus. An Introduction to the Infinitesimal Calculus Following the Genetic Method), published posthumously in 1949.

At the University of Kiel, TOEPLITZ — jointly with the philosopher, HEINRICH SCHOLZ (1884–1956, **B**; from 1928 in Münster), and the classical philologist, JULIUS STENZEL (1883–1935) — established a seminar for the history of mathematics which focused on Greek mathematics [BEHNKE 1949, 91]. One of the participants was SIEGFRIED HELLER (1876–1970, **B**), who after an active career in teaching and administration, wrote fundamental papers on the mathematics of the PYTHAGOREANS, PLATO, ARCHIMEDES, and the history of number theory [HOFMANN 1970, 215].

When TOEPLITZ was offered a position at the University of Bonn in 1928, he succeeded in establishing a division for pedagogy and the history of mathematics within the *Mathematisches Seminar* [SCHUBRING 1985, 160–162]. Because he was Jewish, his career was cut short in 1935, when he was forced to give up his professorship. TOEPLITZ retired to Palestine, where he died in 1940.

At TOEPLITZ's instigation, a salaried *Lehrauftrag* for "Mathematics with special consideration of its history and pedagogy" was established in Bonn and given to ERICH BESSEL-HAGEN (1898–1946, **B**) for the winter term 1928/29. After obtaining his Ph.D. at the University of Berlin in 1920, BESSEL-HAGEN spent seven years in Göttingen where he met, among others, OTTO NEUGEBAUER (1899–1990, **B**) and CARL LUDWIG SIEGEL (1896–1981). FELIX KLEIN had an exceptionally strong influence upon him, for not only did BESSEL-HAGEN live in KLEIN's house, but he was KLEIN's private assistant from 1921 to 1923. BESSEL-HAGEN also played a major role in establishing the historical division, including an important library, as a part of the mathematical seminar in Bonn [NEUENSCHWANDER 1993,

384–385]. Apart from TOEPLITZ and BESSEL-HAGEN, the mathematical-historical seminar in Bonn was also directed by OSKAR BECKER (1889–1964, **B**). BECKER had been an *Assistent* of the renowned philosophers EDMUND HUSSERL (1859–1938) and MARTIN HEIDEGGER (1889–1976) in Freiburg, and had become an *ordentlicher Professor* of philosophy at Bonn in 1931. (Among his publications is a *Geschichte der Mathematik* (1951, French ed. 1956), written jointly with J. E. HOFMANN, containing a sketch of the development of the historiography of mathematics.) Other participants in this seminar included the astronomer FRIEDRICH BECKER (1900–1985) and the assyriologist ALBERT SCHOTT (*1901, † probably between 1941 and 1945) [NEUENSCHWANDER 1993, 386–387]. As in Kiel, the Bonn seminar concentrated primarily on themes from mathematics and astronomy in antiquity.

BESSEL-HAGEN was also involved in editing the works of GAUSS. He had particularly close contacts with OTTO NEUGEBAUER, founder (in 1931) of the *Zentralblatt für Mathematik*, which soon established itself world-wide as the leading review journal for all areas of mathematics, including the history of mathematics. BESSEL-HAGEN wrote more than 200 reviews for NEUGEBAUER's *Zentralblatt.* In 1929 NEUGEBAUER, STENZEL, and TOEPLITZ collaborated in founding a new series of publications for the history of mathematics: *Quellen und Studien zur Geschichte der Mathematik, [Astronomie und Physik].* In their preface, they stressed that original sources were a necessary prerequisite for serious historical research. Moreover, they held that primary sources should be edited according to standards of philologists but satisfying the interests of mathematicians at the same time, thus leading to co-operative work between both disciplines. *Quellen und Studien* was not only addressed to specialists; it aimed to demonstrate that mathematical thinking is closely interwoven with culture in general and its historical development. The editors believed that "in considering historical growth of mathematical thought, a bridge can be constructed between the so-called 'humanities' and the seemingly a-historical 'exact sciences' " [*Quellen und Studien*, Abt. B, Band 1, Heft 1, 2]. *Quellen und Studien* appeared in two series: *Abteilung A* contained editions of sources; *Abteilung B* consisted of articles "which might have a direct or general connection with material derived from the sources" [*Quellen und Studien*, Abt. B, Band 1, Heft 1, 2]. From 1929 to 1938, four volumes of *Abteilung B*, fundamental research papers, were published; in *Abteilung A* numerous substantial editions appeared, among them the *Mathematische Keilschrifttexte* (Mathematical Cuneiform Texts) by NEUGEBAUER and the edition of the *Moscow Mathematical Papyrus* by VASILIĬ V. STRUVE (STRUWE) (1889–1965).

With the departure of NEUGEBAUER and TOEPLITZ from Germany as a result of the politics of the National Socialists, and after the untimely death of STENZEL, the demise of *Quellen und Studien* followed quickly. Activities in the history of mathematics in Bonn were drastically curtailed during World War II owing to severe damage to the library of the mathematical-historical seminar. They ceased totally when BESSEL-HAGEN died in 1946.

A similar fate halted the activities of a seminar on the history of mathematics which had been inaugurated in 1922 by MAX DEHN (1878–1952, **B**) (see also Section 15.4.3) at the University of Frankfurt (Main). Regular participants included his colleagues PAUL EPSTEIN (1871–1939), ERNST HELLINGER (1883–1950, **B**), and SIEGEL. The main object of the seminar was the study of classical mathematical works in their original languages [SIEGEL 1965], which served to inspire the study by ADOLF PRAG (*1906) of JOHN WALLIS [PRAG 1929]. PRAG, like DEHN, HELLINGER, and SIEGEL, was forced by the Nazi terror to emigrate.

5.5.3 Wieleitner, Tropfke, and Their Successors Vogel and Hofmann

Between the wars the leading scholars in Germany were WIELEITNER, TROPFKE, VOGEL, and HOFMANN. They all taught in *Gymnasien*, none of them having a normal university post, although some taught university-level courses. And despite the fact that none was a student of any of the others, they did not work entirely independently: WIELEITNER extended BRAUNMÜHL's program — in particular his efforts to introduce the history of mathematics into the universities as a regular subject. WIELEITNER and TROPFKE, who had no direct connection with universities, were old friends. VOGEL and HOFMANN worked in close association with WIELEITNER, and VOGEL continued WIELEITNER's lectures at the University of Munich. He also continued TROPFKE's *Geschichte der Elementarmathematik* (History of Elementary Mathematics). There was some friction between this group and the professors mentioned in the previous section. While there was a tendency among the Kiel-Göttingen-Bonn group to take modern mathematics as its starting point and to construct, by selecting suitable pieces from the historical development, a "genetic approach" as a teaching aid, WIELEITNER, TROPFKE, and his younger colleagues started their research from the sources.

HEINRICH WIELEITNER (1874–1931, **B**) was one of the first professional historians of mathematics in Germany. Although he lived only three years after completing his *Habilitation* (1928), he had considerable influence. His first publications were on advanced geometry and mathematical education. Around 1910, however, he turned to the history of mathematics through the indirect influence of BRAUNMÜHL. The publishing house of Göschen planned to update the four volumes of CANTOR's *Vorlesungen*, adapted to the needs of students and teachers; GÜNTHER was to cover the period up to DESCARTES, and BRAUNMÜHL the history from DESCARTES to the end of the 18th century. GÜNTHER published his volume in 1908, but BRAUNMÜHL died that same year without having completed more than some preliminary drafts. GÜNTHER persuaded WIELEITNER to continue BRAUNMÜHL's work, the first part of which came out in 1911, the second in 1921.

Besides these books, WIELEITNER also published articles on several special problems, among others the prehistory of analytic geometry, perspective, infinite series, and the law of free fall in the later Middle Ages. He was especially interested in NICOLE ORESME (1323?–1382) and the latitude of forms, which he regarded as the forerunner of analytical geometry. In 1909 WIELEITNER came into contact with RUSKA and, as a result, published a lengthy article (1922) on mathematical problems related to inheritance that were treated in AL-KHWĀRIZMĪ's *Algebra*. HUGO DINGLER (1881–1954), whose *Habilitation* at the University of Munich in 1912 for history, methodology, and teaching of mathematics was the first in Germany that included the history of mathematics, was also in close contact with WIELEITNER and encouraged him to pursue his historical research.

WIELEITNER wrote more than 150 publications. He was accurate and always went back to original sources. Biography did not interest him: he preferred to write the history of ideas, usually with the notion of their development as a *Leitmotiv*, and introduced this approach to the history of mathematics. WIELEITNER was a world-famous historian of mathematics who, according to BORTOLOTTI, after the death of TANNERY, ENESTRÖM, and ZEUTHEN, was the world's best historian of science [BORTOLOTTI 1932].

Like WIELEITNER in Munich, JOHANNES TROPFKE (1866–1939, **B**) taught mathematics at secondary schools in Berlin. But unlike WIELEITNER, he never succeeded in obtaining a position as *Privatdozent* or professor at a university. TROPFKE wished to introduce historical aspects into secondary school mathematics. He first presented his ideas in a *Schulprogramm* [TROPFKE 1899] and followed this in 1902–1903 with the publication of the first edition of his two-volume *Geschichte der Elementarmathematik in systematischer Darstellung* (Systematic Presentation of the History of Elementary Mathematics). There, he described the history of those individual parts of mathematics that he believed were most important for mathematics as taught in secondary schools. He intended his history to inform teachers about the origin of special problems, terms, and methods in school mathematics. Although he based the first edition of this work mainly on secondary literature, TROPFKE later improved his analysis by examining original texts, and in this he was supported by positive criticism from ENESTRÖM and WIELEITNER. The result was a seven-volume second edition (1921–1924) and a later third edition, of which only volumes 1–3 appeared in TROPFKE's lifetime (1930–1937; with the assistance of J. E. HOFMANN and K. VOGEL).

TROPFKE's approach to history of mathematics at this time was new and even now is not yet out of date. The only comparable work is the second volume of D. E. SMITH's *History of Mathematics*, first published in 1925, but which gives far less detailed information.

KURT VOGEL (1888–1985, **B**) continued, with a short break, the tradition started by WIELEITNER in Munich, giving regular lectures on the history of mathematics from 1936 into the 1960s. Before the War he had already begun to institutionalize the subject in the university: his efforts then and later resulted in the present *Institut für Geschichte der Naturwissenschaften* (Institute for the History of Science), which was founded in 1963. A substantial part of VOGEL's research dealt with pre-Greek mathematics. His dissertation was on the $(2 : n)$-table of the *Rhind Papyrus* (1929). This papyrus, the most important mathematical text from the Egyptians, was first published by the German egyptologist AUGUST EISEN-

LOHR (1832–1902, **B**) in 1877; new editions appeared just before VOGEL's dissertation (by THOMAS ERIC PEET (1882–1934), 1923, and by ARNOLD B. CHACE (1845–1932, **B**), 1929). Simultaneously with OTTO NEUGEBAUER, VOGEL was the first to investigate Babylonian mathematics on cuneiform tablets. In 1958/59 he published a readable, general account of Egyptian and Babylonian mathematics.

VOGEL's philological gifts enabled him to contribute to many fields of the history of mathematics. His interests comprised Greek, Byzantine, and Chinese mathematics as well as Western European mathematics in the Middle Ages and the Renaissance. Together with HANS GERSTINGER (1885–1971), he edited a Greek mathematical papyrus from the first century B.C. which established a connection between pre-Greek mathematics and the works of HERO OF ALEXANDRIA. VOGEL, who produced numerous scholarly editions, placed special emphasis on practical mathematics. His *Habilitationsschrift* was on logistics, the practical arithmetic of the Greeks, but given how much of the ancient Greek sources had been lost, VOGEL instead had to concentrate primarily on the analysis of Byzantine texts which proved to be essential for the reconstruction of Greek mathematics. Later, he published masterful editions of two Byzantine books of practical arithmetic from the 14th and 15th centuries (1963, 1968). After learning Chinese when he was over 70 years old, he translated the oldest known Chinese book on practical mathematics, the *Jiu Zhang Suan Shu (Jiuzhang suanshu)* (Nine Chapters on the Art of Mathematics), which had previously been available to Western scholars only in a Russian edition by E. I. BERËZKINA (see Section 8.5).

VOGEL also worked extensively on Western mathematics in the Middle Ages and Renaissance, particularly on questions of transmission from the East. He edited the reworking of the Latin translation of AL-KHWĀRIZMĪ's arithmetical treatise (1963), produced several studies of Italian mathematicians (LEONARDO FIBONACCI, PEGOLOTTI, *Columbia Algorismus*), and analyzed the mathematics of Southern Germany in the 15th century. He edited the *Algorismus Ratisbonensis* (1954), a German algebra of 1481 (1981), the so-called *Bamberger Rechenbuch* (1950), the *Bamberger Blockbuch* (1980), and three shorter texts from South-East Germany (1973), among other texts.

In addition to numerous editions and studies of the mathematics of particular regions or cultures, VOGEL wrote a series of articles on the history of specific problems. In the commentaries and appendices to his editions, he worked to establish connections between problems and methods of solutions at different times and in different places. He also edited volume four of TROPFKE's *Geschichte der Elementarmathematik* in 1940 and, together with HELMUTH GERICKE and KARIN REICH, completely revised the first three volumes (on arithmetic and algebra) into a single volume in 1980. In the latter, he included for the first time a special section on applied arithmetic and geometry, with a systematic description of individual problems of mercantile and recreational mathematics.

Like VOGEL, JOSEPH E. HOFMANN (1900–1973, **B**) taught mathematics at high schools, but he devoted all his spare time to writing his many publications on the history of mathematics. One question that repeatedly arose in his work was: how could a mathematical genius be recognized? Accordingly, many of his publications concern the 17th century, a time rich in discoveries by such luminaries as GOTTFRIED WILHELM LEIBNIZ (1646–1716), ISAAC NEWTON (1642–1727), PIERRE DE FERMAT (1601–1665), JAMES GREGORY (1638–1675), and NICOLAUS MERCATOR (1620–1687). As director of the group editing the works of LEIBNIZ at the Academy of Sciences in Berlin (1939–1945), HOFMANN prepared the first volume of LEIBNIZ's mathematical correspondence, covering the Paris years from 1672 to 1676, i.e., the time of his discovery of the calculus. Although almost complete in 1946, the difficult post-war conditions in divided Germany resulted in a delay of publication until 1976, when it was finally published posthumously.

LEIBNIZ and his works were central to HOFMANN's numerous (several hundred) publications. Particularly important are his book on LEIBNIZ's discovery of the calculus, *Die Entwicklungsgeschichte der Leibnizschen Mathematik während des Aufenthaltes in Paris 1672–1676* (1949), translated into English by ADOLF PRAG as *Leibniz in Paris 1672–1676. His Growth to Mathematical Maturity* (1974), and the volume of correspondence already mentioned. HOFMANN also focused on FERMAT, analyzing especially his contributions to the theory of numbers. He also edited (in 1951) the mathematical writings of NICHOLAS OF CUSA (1401–1464), with introduction and comments. The German translation of the text had been prepared by his wife JOSEPHA HOFMANN.

HOFMANN published two general histories of mathematics, one with OSKAR BECKER (1951) and another, crammed with facts, in three concise volumes (1953–1957; vol. 1: 2nd edition 1963). Like TROPFKE, HOFMANN tried to introduce the history of mathematics into the mathematics curriculum of secondary schools, suggesting that mathematics could be brought to life through its history. He interpreted the subject exclusively as the study of problems (*Problemgeschichte*). By *Problemgeschichte der Mathematik* HOFMANN meant a history of mathematics that emphasized what has also been termed "internalist" history, whereby the great lines of mathematical theoretical development might be traced. This point of view colored not only his own works but also his tireless efforts on behalf of the institutionalization of the history of mathematics.

In addition to these four major figures, other scholars were also active and testify to the continued interest in the field in those decades. Again, most were teachers at *Gymnasia*. Attention to ethnographical and pedagogical aspects inspired professors of mathematical didactics, like WALTER LIETZMANN (1880–1959, **B**) and EWALD FETTWEIS (1881–1967, **B**) to write on pre-historic aspects of the development of mathematics. KARL MENNINGER (1898–1963, **B**) in his standard work, *Zahlwort und Ziffer* (1934; 3rd ed. 1979; Eng. trans. 1969), described the development of number words and number symbols in connection with the rise of various cultures.

Greek mathematics remained an attractive area of occupation: CLEMENS THAER (1883–1974, **B**) prepared German translations of EUCLID's *Elements* and *Data*, while ARTHUR CZWALINA (1884–1964, **B**) translated the works of ARCHIMEDES, APOLLONIUS, DIOPHANTUS, and of several other mathematicians into German. Subsequently, these translations have proved indispensible as primary texts for seminars in the history of mathematics. GUSTAV JUNGE (1879–1959, **B**) collaborated with an edition of the Arabic translation of EUCLID's *Elements* by AL-HAJJĀJ (early 9th century) and investigated several special themes in Greek mathematics. KURT REIDEMEISTER (1893–1971, **B**), a creative mathematician with strong philosophical inclinations, published several books on Greek mathematics, logic, and philosophy in the 1940s, thereby contributing to an area which otherwise, in contrast to France, was not much cultivated in Germany.

During the first half of the 20th century the productivity of historians of mathematics in Germany was affected not only by the lack of academic positions but also by the limited number of journals devoted to the history of science. A glance at the bibliographies of WIELEITNER, VOGEL, or HOFMANN will make this evident. Their articles appeared in a wide range of journals, including Academy journals which offered possiblities of publishing detailed investigations. A few mathematical journals sometimes accepted historical articles, provided these had a strong mathematical flavor. Other periodicals, especially those addressed to teachers of mathematics or to an interested learned public, were open to more general articles, such as biographies of famous mathematicians, or survey articles. NEUGEBAUER's *Quellen und Studien zur Geschichte der Mathematik* had tried to remedy this situation, but ceased publication all too early due to the political disaster of Nazism in Germany in the 1930s. The journal *Deutsche Mathematik*, launched in 1936 by the mathematician and fervid National Socialist functionary THEODOR VAHLEN (1869–1945), also published (in line with VAHLEN's intention "to present a lively picture of the entire mathematical production of German *Volksgenossen* (compatriots)") a number of papers submitted by historians of mathematics between 1936 and 1944 (see Section 5.6). — A summary of publications in the history of mathematics that appeared in Germany during the war is to be found in [HOFMANN 1948a].[2]

5.6 History of Mathematics Under the Third Reich

History of Mathematics could not escape the consequences of the National Socialists when HITLER came to power in Germany in 1933. The effects of the Nazi period on mathematics in particular and the sciences generally have recently been the focus of increased interest and study: [PINL 1969–1973], [PINL/FURTMÜLLER 1973], [MEHRTENS 1987a], [MEHRTENS 1987b],

[2]Together with similar reports about various mathematical disciplins, HOFMANN's survey is contained in the first volume, *Pure Mathematics*, of the *FIAT Review of German Science 1939–1946* (German edition: *Naturforschung und Medizin in Deutschland 1939–1946*, vol. 1: Reine Mathematik). This series was published upon the instigation of the *Field Information Agency (Technical); (FIAT)* of the post-war Military Government of the Allies. The director of the *Mathematisches Forschungsinstitut* in Oberwolfach, WILHELM SÜSS, had been commissioned to edit the first volume (in two parts) on pure mathematics.

[Mehrtens 1994], [Renneberg/Walker 1994], [Meinel 1998], and [Siegmund-Schultze 1998].[3]

The effect of the Nazi period specifically on the history of mathematics has not been studied, but its greatest consequence was upon scholars who either chose or were forced to leave in response to the racial laws enacted by the Nazis. The most prominent was Otto Neugebauer, who was appointed head of the Mathematical Institute in Göttingen in April of 1933, to replace Richard Courant (1888–1972) who was removed from his position as Director when the racial purges began at universities throughout Germany. Neugebauer served as Director for only one day, since he refused to take the oath of loyalty to the Nazi regime [Segal 1980, 48]. Neugebauer, who was not Jewish, went first to Copenhagen and then, somewhat later, to the United States where a professorship for history of mathematics was created for him, along with an Institute, at Brown University. Among mathematicians who were Jewish and forced to give up their positions, several emigrated to the United States where they later contributed to the history of mathematics, including Salomon Bochner (1899–1982), Max Dehn, and Ernst Hellinger. Otto Toeplitz emigrated to Palestine, while Adolf Prag, an active participant in the seminar on history of mathematics offered by Dehn and Hellinger at the University of Frankfurt, had to leave Germany before he was able to submit his thesis. Only many years later, having relocated to Great Britain where he became a school teacher in London, was Prag able to return to work on the history of mathematics. Among his post-war contributions, he assisted D. T. Whiteside in editing the *Mathematical Papers* of Isaac Newton (see Section 7.11).

History of mathematics as a discipline was not affected as much as mathematics under the Third Reich, due primarily to the fact that there were only a few serious scholars who might be called professional historians of mathematics at the time. But Nazi ideology as it related to history of mathematics did have an effect, especially in the hands of those like the Berlin mathematician Ludwig Bieberbach (1886–1982). Bieberbach may not have been an historian of mathematics, but he certainly made use of it to promote his racial theories. In a controversial and widely-debated lecture published as "Persönlichkeitsstruktur und mathematisches Schaffen" (Personality Structure and Mathematical Creativity), he distinguished between Aryan or purely German mathematics and its opposite, abstract or Jewish mathematics ([Bieberbach 1934]; cf. [Hardy 1934]). In doing so he drew inspiration from Felix Klein and Henri Poincaré (1854–1912), among others, to promote his views that there was a distinctive Jewish mathematics as opposed to a purely Aryan mathematics, the S-type versus the J-type in the terminology Bieberbach employed.[4]

[3]The *Deutsche Mathematiker-Vereinigung* underwent a purge of its non-Aryan members in 1938–1939 (see [Gericke 1966]). On the foundation of the *Reichsinstitut für Mathematik* in the Black Forest by Wilhelm Süss in 1944 see [Gericke 1968] and [Remmert 1999].

[4]As Herbert Mehrtens explains, "Positive, German-type would be intuitive, realistic, close to *Volk* and life, while the 'counter-type' would be a highly sophisticated juggler of formal con-

Aryan mathematics was associated with concrete, objective origins and results, and as examples BIEBERBACH offered such German mathematicians as C. F. GAUSS and FELIX KLEIN. The opposite, S-type, he linked to abstract, formalist mathematics of the sort he attributed to C. G. J. JACOBI, GEORG CANTOR (1845–1918), and "the Jew" EDMUND LANDAU (1877–1938) [LINDNER 1980, 97–101]. These were ideas that found their most direct expression in articles, some of them historical, that appeared in *Deutsche Mathematik*, a journal "devoted to the publication of 'Aryan mathematics'," [Segal 1980, 54]. In so far as its authors' views on mathematics conformed to Nazi ideology, the idea that there was a typically Aryan mathematics could not help but affect the history of mathematics in Germany, at least in the hands of those who sought to justify such claims as BIEBERBACH and VAHLEN made through historical research and examples drawn from the history of mathematics.

As for BIEBERBACH's general influence, or that of *Deutsche Mathematik*, HERBERT MEHRTENS summarizes their significance as follows:

> Bieberbach attempted, so to speak, a scientific counter-revolution with the help of the new posers. His attempt to dominate mathematics in Germany failed, however. The Mathematicians' Association managed to exclude him in early 1935. He started a journal *Deutsche Mathematik* to create his own, competing organization, but did not find the support he needed from the Nazi Minister of Education. By 1937 Bieberbach and his group were an ideological residue in the system of mathematics without substantial influence [MEHRTENS 1994, 299].

For the most part, those who allowed ideology to drive their views of history were minor figures. These are now mostly forgotten in the annals of the history of mathematics. The only historians of mathematics of any note to publish in *Deutsche Mathematik* were JOSEPH E. HOFMANN and KURT VOGEL. HOFMANN published a number of articles related to internalist questions of mathematics on such topics as NICOLAUS MERCATOR's *Logarithmotechnia* of 1668, the study of the logarithmic series in England, and the Theodoros section of PLATO's *Theaetetus.* He even wrote an article jointly with his wife JOSEPHA on the first quadrature of the cissoid.

All of these contributions were very much in the style of "Problemgeschichte der Mathematik," an approach to the history of mathematics that was typical of HOFMANN's usual focus primarily on the internal development of mathematical problems viewed historically. HOFMANN did write one expository article for *Deutsche Mathematik*, "Über Ziele und Wege mathematikgeschichtlicher Forschung" (On the Aims and Means of History of Mathematics Research) in which he called specifically for research devoted to German mathematicians, and

cepts with no bonds to reality and no other motive than his own success" [MEHRTENS 1994, 299].

discussed mathematics in terms of *Volk* and race, factors which he took to determine the kinds of problems a mathematician chose to work on, and consequently were to be considered as "a unique characteristic and thus very important for the historian" [HOFMANN 1940, 153]. But aside from such phraseology reflecting National Socialist rhetoric, HOFMANN kept to the major point of his article, that the best approach for the history of mathematics was through the study of individual problems. (For a brief analysis of this article, including questions about Nazi ideology and history of mathematics, see [EPPLE 2000].)

VOGEL published, apart from a short summary of his *Habilitationsschrift* of 1936, "Beiträge zur griechischen Logistik I" (Contributions to Greek Logistics I), an article "Zur Geschichte der linearen Gleichungen mit mehreren Unbekannten" (On the History of Linear Equations with Several Unknowns) in *Deutsche Mathematik*, also a contribution to internalist history.

It may be that one result of the extreme politicization of scholarship under the Third Reich was to encourage just the sort of concentration upon what HOFMANN called *Problemgeschichte* rather than biographical, sociological, or other aspects of the history of mathematics. As a contemporary study of German universities during the Nazi period concluded, "One unanticipated effect of the deliberate politicization of knowledge may be the encouragement of interest in recondite subjects beyond the ken or concern of dictators" [HARTSHORNE 1938, 222].

5.7 The German Democratic Republic

At the end of the Second World War in what was to become the German Democratic Republic (GDR) in 1949, there were only a few isolated enterprises where there was evidence of any interest in the history of science. These included, for example, the publisher B. G. TEUBNER (*Teubner Verlag*), which had a program in the history of science, the editorial staff of *Poggendorff's literarisch-biographisches Handwörterbuch*, and the collection of scientific instruments in the *Mathematisch-Physikalischer Salon* in Dresden, which survived despite the destruction of its buildings. There were also three Institutes for the History of Science at universities in Berlin, Leipzig, and Jena, as well as one at the *Deutsche Akademie der Naturforscher Leopoldina* in Halle.

In keeping with the country's socialist/communist principles, the ideology of Marxism also affected the history of science. The goal of constructing a suitable Marxist history of science in the GDR is associated, above all, with the name of the physicist GERHARD HARIG (1902–1966). As a member of the communist party, he had to emigrate to the Soviet Union in 1933. In 1937 he was among the German emigrants handed over by STALIN to HITLER, and was imprisoned (from 1937 to 1945) in the concentration camp at Buchenwald. After World War II, he served as Secretary of State in the ministry for universities and specialized high schools *(Ministerium für das Hoch- und Fachschulwesen, MHF)*. In 1957, HARIG was appointed to head the division for the history of science at the *Karl*

Sudhoff-Institut für Geschichte der Medizin und Naturwissenschaften in Leipzig (founded in 1906 for the history of medicine, and before World War II enlarged to incorporate the history of science). There was also an assistantship for the history of mathematics. In 1959 HARIG became director of the Sudhoff Institute, and gradually the number of assistantships was increased.

Independently and spontaneously, the history of mathematics was also taught in Greifswald by FRANZ VON KRBEK (1898–1984) in the 1950s, and the subject soon began to gain ground elsewhere. The history of mathematics was further strengthened in the GDR through research in Berlin at the *Alexander-von-Humboldt-Forschungsstelle*, directed for many years by KURT-R. BIERMANN (*1919); by the founding of the journal *Schriftenreihe zur Geschichte der Naturwissenschaften, Technik und Medizin (NTM)* (continued since 1992 as a new series by the publisher Birkhäuser), and of the series *Teubner-Archiv zur Mathematik*; and with the revival of the series, *Ostwalds Klassiker der exakten Wissenschaften.* Thanks to the publication of new research by rising scholars, including HANS WUSSING (*1927), HANS REICHARDT (1908–1991), and HANS SALIÉ (1902–1978), along with translations, notably of works by DIRK STRUIK (1894–2000, **B**) and ADOLPH P. YOUSHKEVICH (1906–1993, **B**), the field was greatly enriched. Meanwhile, all of this activity was supported by a substantial number of interested mathematicians, especially by REICHARDT.

In 1962 the so-called "mathematics decree" of the *Politbüro* of the *SED (Sozialistische Einheitspartei Deutschlands)* also brought the history of mathematics into the limelight. Following the socialist principles advanced in the GDR, among other things, special emphasis was placed on cultivating the history of science and technology, including the history of mathematics, as part of the popularization of a Marxist historical worldview. Consequently, new courses and lectures on the history of mathematics for future teachers and, later, for all mathematicians, were made obligatory at universities and special institutions for higher education. (Historical courses were, in fact, introduced for all disciplines in the natural sciences, including physics, chemistry, and biology). As a result, the *Beirat für Wissenschaftsgeschichte* of the *MHF* supported a working-group of historians of mathematics which drew up recommendations for syllabi in the history of mathematics. Textbooks, lecture notes, and teaching supplements, including biographies of mathematicians, slide collections, and other teaching aids, were regularly made available. For the history of mathematics, too, teachers were trained at an accelerated pace. In general, historical lectures were readily accepted by the students, although their positive response depended directly on their own previous training and on the personal skill of their lecturers. Greifswald, Erfurt, and Leipzig were the main centers for such training in the history of mathematics. Concurrent with the increased level of teaching, accessible biographies of mathematicians (GAUSS, NEWTON, RIES, LOBACHEVSKIĬ, EUCLID, SOF'YA KOVALEVSKAYA (1850–1891), and EULER) were published in large editions by the *Akademische Verlagsgesellschaft* in its series *Biographien hervorragender Naturwissenschaftler,*

Techniker und Mediziner (Biographies of Leading Scientists, Technicians, and Physicians).

As a result of the considerable expansion of teaching, a number of new positions were created and staffed (in part) with younger mathematicians. Consequently, an increasing number of dissertations and *Habilitationen* in history of mathematics were produced, the majority of which were submitted to faculties for mathematics and the sciences.

Within the *Mathematische Gesellschaft der DDR* (Mathematical Society of the GDR) a *Fachsektion für Geschichte, Philosophie und Logik der Mathematik* (Division for the History, Philosophy, and Logic of Mathematics) was established, and beginning in 1975 it organized annual meetings, intended (in part) to offer further training for teachers. With the reunification of Germany in 1989, in the following year this *Fachsektion* was incorporated into the *Deutsche Mathematiker-Vereinigung* (German Society of Mathematicians) where it continues its work successfully. In collaboration with mathematicians and historians of mathematics at the universities, the *Akademie der Wissenschaften* in Berlin has organized several international commemorations of mathematicians, among others GAUSS, EULER, KARL WEIERSTRASS (1815–1897) and (especially for victims of National Socialism), LEON LICHTENSTEIN (1878–1933), and FELIX HAUSDORFF (1868–1942).

Due to the emphasis placed on teaching the history of mathematics, the GDR successfully trained a considerable number of historians of mathematics who, in turn, began to publish works of their own. Among the areas which received special attention are Islamic mathematics, functional analysis, non-Euclidean geometry, the history of mathematical concepts (groups, fields, algebra), biographical research, and institutional histories. Not only has this research gained international recognition, but researchers outside the GDR have published their work through GDR presses (see, for example, vols. 1–4 of *Science Networks*, vol. 1 of which was published in 1989 by VEB Deutscher Verlag der Wissenschaften, Berlin; subsequently the series has continued to be published by Birkhäuser Verlag Basel).

5.8 The Federal Republic of Germany

Most influential for the further development of the history of mathematics in Germany in the post-war period was J. E. HOFMANN's instigation of regular conferences on the history of mathematics at the *Mathematical Research Institute* in Oberwolfach (Black Forest). From small beginnings in 1954, these annual meetings, which for many years HOFMANN called *Tagungen zur Problemgeschichte der Mathematik* (Conferences on Internal History of Mathematics), grew to an international meeting place for historians of mathematics [SCRIBA 1985].

After HOFMANN's death in 1973, the meetings were organized jointly by historians of mathematics from Germany and various other countries under more general themes, reflecting the broadening of interest of the members of the profession. Interrelations between mathematics and society (1975), problems of founda tions of mathematics (1984), mathematics and the arts (1985), oriental and non-European mathematics (1987), mathematical schools (1992), and innovation and transmission of mathematics (1994), are examples of such meetings concentrating on special themes. Others were devoted to the development of mathematics in certain historical periods, in specific subdisciplines, or to the interrelations between mathematics and various areas of application. Repeatedly, and increasingly as time passed, problems of methodology and of modern trends in the historiography of mathematics were also discussed.

Perhaps because relations with philosophy as well as history were rather weak during the 20th century in Germany, related questions did not play so important a role at first for historians of mathematics. Neither VOGEL nor HOFMANN had much to say about methodological problems in their autobiographies (see **B**). VOGEL, at the XIth International Congress of the History of Science in Poland in 1965, gave a lecture on the historiography of mathematics prior to MONTUCLA ([VOGEL 1965]), while HOFMANN published two articles in which he outlined his understanding of historiography of mathematics ([HOFMANN 1940], [HOFMANN 1948b]). The section on the development of the discipline in [BECKER/HOFMANN 1951] is a rather matter-of-fact description (in no way comparable to the series of articles ENESTRÖM had published in the third series of his journal *Bibliotheca Mathematica* at the beginning of the century). Likewise, GERHARD KROPP gave a short bibliographically oriented survey in 1948 [KROPP 1948]. Some aspects of the tools of an historian of mathematics, in part referring back to ENESTRÖM's articles, were later discussed in [SCRIBA 1967], [SCRIBA 1968], and [SCRIBA 1970].

Beginning in the late 1960s and throughout the 1970s, a growing historical consciousness emerged. Institutionalization of the history of mathematics at universities in the Federal Republic of Germany began after the country had begun to recover from the destructions of World War II. Apart from the *Karl-Sudhoff-Institut* for the history of medicine and science founded in Leipzig in 1906, and after the division of Germany at the end of World War II, situated in the German Democratic Republic (DDR), newly-created West German institutes were established in Frankfurt, Hamburg, Munich, West Berlin (at the Technical University), and in other cities in the 1960s and 1970s. While all are devoted to the history of science in general, history of mathematics is only one of many branches of the history of science cultivated by these centers. Indeed, their very existence has provided a strong institutional basis that was sorely lacking for many decades. Without these institutions, the growing number of publications in the field would not have been possible. Several series, publishing mostly monographs, were created by the first generation of historians of mathematics, science, and technology during this period: *Boethius* (1963), *Arbor scientiarum* (1978), *Studien zur Wissenschafts-, Sozial- und Bildungsgeschichte der Mathematik* (1985), *Dokumente zur Geschichte der Mathematik* (1985), and *Algorismus* (1988).

HELMUTH GERICKE, who continued the strong tradition at the University of Munich published a two-volume *Geschichte der Mathematik* (History of Mathematics) up to Descartes, taking into account the results of recent scholarship in

Greek, oriental and medieval mathematics ([GERICKE 1984], [GERICKE 1990]). At the University of Mainz NICOLAI STULOFF, who paid special attention to mathematics in Byzantium, represented the history of mathematics (now succeeded by DAVID ROWE) in a division that was created within the mathematics department. The department later published his lectures in four volumes, *Die Entwicklung der Mathematik* (The Development of Mathematics; from prehistory to BOURBAKI) ([STULOFF 1988/91]). At the Technical University of Berlin, after CHRISTOPH J. SCRIBA and MENSO FOLKERTS accepted positions in Hamburg (1975) and Munich (1980), respectively, EBERHARD KNOBLOCH continues the tradition of offering lectures and seminars on the history of mathematics.

At the same time international contacts have proliferated, not only on a personal basis with foreign historians of mathematics at the rapidly increasing number of conferences and congresses (cf. Section 8.4), but also in response to the significant increase in the publications of historians, sociologists, and historians of science writing in English and French. The questions of how, why, and for whom history of mathematics should be researched and written, among other pressing historiographic issues, have become increasingly important and more widely discussed. Debate over the so-called "Kuhnian model" — to give but one example — was an international phenomenon that had repercussions in Germany, too. To isolate specific German contributions, however, to the history of mathematics is no longer reasonable at a moment when writing the history of mathematics has become a truly international undertaking. Suffice it here to call attention, *pars pro toto*, to some recent publications that testify to the present awareness of the importance of methodological concerns among the current generation of German historians of mathematics: [MEHRTENS 1990], [SCHNEIDER 1992a], [SCHNEIDER 1992b], [SIEGMUND-SCHULTZE 1993], and [SCHUBRING 2000].

Apart from university institutes for the history of science, there are other institutions in Germany which support work at least partly related to the history of mathematics. The increasing number of pedagogical departments for mathematics offers expanded opportunities for studying the relevance of history of mathematics for mathematical instruction, as, for instance, at the University of Bielefeld. At offices for major projects like the COPERNICUS, KEPLER, LEIBNIZ, HILBERT, or HAUSDORFF editions, historians of mathematics are among those preparing the collected works of these scientists for publication.

The *Institut für Geschichte der Arabisch-Islamischen Wissenschaften*, Frankfurt is directed by FUAT SEZGIN, who founded it after having received the King Feisal Prize for Islamic Sciences in 1979. Besides publishing the *Geschichte des arabischen Schrifttums* (History of Arabic Publications), volume V of which is devoted to mathematics, and the *Zeitschrift für Geschichte der Arabisch-Islamischen Wissenschaften*, which includes *inter alia* articles on the history of mathematics, the Institute re-issues facsimiles of manuscripts (e.g. of ABŪ KĀMIL's *Algebra*) and numerous texts, books, and articles in several series: e.g. NAṢĪR AL-DĪN's *Traité du quadrilatre* (1891), several volumes of articles about EUCLID in the Arabic tradition, the collected papers of F. WOEPCKE, J. M. MILLÁS VALLICROSA's *Es-*

tudios sobre Azarquiel (1943–1950), and several volumes of "Miscellaneous Texts and Studies on Islamic Mathematics and Astronomy." It also publishes new books.

5.9 History of Mathematics in a Reunited Germany

One of the first developments to affect history of mathematics in Germany after its reunification in 1989 was the creation of the *Fachsektion Geschichte der Mathematik* (Section for History of Mathematics) of the *Deutsche Mathematiker-Vereinigung* (German Society of Mathematicians). The establishment of this *Fachsektion* in 1991 was initiated by PETER SCHREIBER (*1938) of the University of Greifswald in connection with the reunification of Germany and the subsequent merger of the two professional societies, the *Mathematische Gesellschaft der DDR* (Mathematical Society of the GDR) and the *Deutsche Mathematiker-Vereinigung* (of West Germany) in 1990. The stated aim of the *Fachsektion*, to further cooperation between historians of mathematics and mathematicians, has perhaps had the greatest resonance thus far among educators. Indeed, pedagogy (or didactics) of mathematics is a discipline that has gained considerable strength in Germany during the past half-century. Moreover, in recent decades German historians of mathematics have taken up a far wider spectrum of themes and research topics than the earlier, more restricted emphasis on "Problem-" and "Ideengeschichte." Many new areas of interest have been stimulated by shifts in the general history of science; in fact, a stronger mutual interaction between history of science and history of mathematics is one of the characteristic features of much recent research in both fields, in Germany as elsewhere. On the other hand, cooperation between general historians and historians of mathematics still leaves a great deal to be desired.

A direct reflection of increased commitment to interest in history of mathematics in Germany is the fact that in recent years the number of conferences on special themes has grown considerably. In addition to the regular conferences at the *Mathematische Forschungsinstitut Oberwolfach*, every year several well-attended meetings are organized, partly through individual initiatives, and partly by the section for history of mathematics of the German mathematicians union. Such meetings are often attended by mathematicians showing a particular interest in certain aspects of their discipline, or in important mathematicians of the past. An example is DETLEF LAUGWITZ (1932–2000) and his recently published biography of RIEMANN in the Birkhäuser series *Vita Mathematica*.

Lastly, the founding of the Max Planck Institute for the History of Science in Berlin in 1994, five years after the reunification of Germany, has made it possible for many young historians of mathematics whose futures would otherwise have been quite uncertain, to continue their research with stipends providing several years of support.

5.10 Conclusion

In this short survey it has only been possible to treat the most important contributors to the history of mathematics in Germany; many other scholars could also have been named who have worked in this field. In retrospect, it does not seem surprising that interest in the historical study of mathematics and the sciences advanced slowly during the 19th century, tending towards positivistic, ahistorical interpretations. Nevertheless, idealistic, neohumanistic developments in the humanities that helped to shape German university reforms in the early 19th century, eventually exerted their influence on the interpretation of the sciences. This is not the place to describe in detail the unfolding of new branches of history in the second half of the 19th century (archaeology and the histories of law, art, music, etc.), nor to investigate the motives behind such developments, yet the historicization of the seemingly ahistorical natural sciences, including mathematics, must be seen as one stage in this general development.[5]

As in other countries, the history of mathematics in Germany has, until recently, consisted largely of the contributions of many individuals, most of them high-school teachers of mathematics. There have also been university professors who, in addition to teaching mathematics, published works on its history, mostly in connection with the editing of collected works of famous mathematicians.

Towards the end of the 19th century, MORITZ CANTOR was the key figure in developing the field. He not only tried to introduce history of mathematics into universities, but provided an outlet for the publication of historical articles by transforming a mathematical periodical into an historical one — namely, the *Abhandlungen zur Geschichte der mathematischen Wissenschaften*. Unfortunately, this disappeared — like *Bibliotheca Mathematica* — before World War I. CANTOR did succeed, however, in stimulating further research through publication of his monumental *Vorlesungen zur Geschichte der Mathematik*.

History of mathematics as a serious discipline at the university level was a somewhat later phenomenon. After HUGO DINGLER's *Habilitation* in 1912, devoted to history, methodology, and the teaching of mathematics, it was not until 1928 that there was an *Habilitation* specifically in history of mathematics (HEINRICH WIELEITNER). This development continued with the *Habilitationen* of VOGEL in 1936, and HOFMANN in 1939. After World War II, and especially since the late 1960s, the number of *Habilitationen* in history of mathematics (or at least including it as a partial field of specialization) has been increasing.

One question naturally arises: is there a typical German style of history of mathematics? In general, the answer is probably no. A characteristic feature, however, is certainly the strong interest among German scholars in a thorough investigation of the development of special mathematical problems, *i.e.* "Problemgeschichte der Mathematik." More generally, apart from biographical studies,

[5]Considering the influence of the philosopher EDUARD RÖTH (1807–1858), however, a further, more detailed study may well establish unequivocally the import of ideas adapted from French philosopher-historians of science on the development of "histories" in Germany.

including obituaries and contributions to encyclopedias (from MOLLWEIDE to F. KLEIN and to J. E. HOFMANN'S numerous articles in [NAAS/SCHMIDT 1961]), it is the so-called "internal history" of mathematics that has been most widely cultivated. This is not surprising, since most authors in the field have been mathematicians by training. There has also been, perhaps, a greater tendency in Germany than elsewhere to study and edit source materials. Here there may well be a direct influence from the long-established German tradition of preparing editions of biblical and classical literary texts as well as documents from political history. In contrast to the development of the historiography of mathematics in France, however, in Germany there has never been a strong interaction between the history and philosophy of mathematics — OSKAR BECKER, a philosopher with a lively interest in foundational problems, being a rare exception.

At present, throughout Germany there are more well-trained historians of mathematics teaching and undertaking their own research than at any other time in the discipline's history. They are also lecturing and writing for a much wider audience, both locally and internationally, than at any previous time. Given the substantial interest shown in history of mathematics by pedagogues and mathematicians alike, history of mathematics should continue to play a strong role in university curricula, and more broadly, in reaching both specialist and generalist audiences alike.

Chapter 6

Scandinavia

Kirsti Andersen

6.1 Introduction

In this chapter, the term Scandinavia refers to Denmark, Finland, Iceland, Norway, and Sweden. Culturally, these countries are rather close, owing partly to political interrelations. During the Renaissance, Finland became a Swedish province and Norway a Danish province. After the Napoleonic wars, however, Finland came under the rule of Russia and remained under Russian control until the revolution in 1917, while Norway won self-governance under the Swedish Crown and obtained complete independence as a kingdom in 1905. Until 1944 Iceland was connected to Denmark and did not have its own institutions of higher education.

These connections aside, it is impossible to distinguish a particular Scandinavian development of the history of mathematics. Thus, to a large extent this chapter consists of different stories for the various countries organized nevertheless in common sections.[1] Denmark, Norway, and Sweden receive the most attention, because only a recent interest in the history of mathematics in Iceland is documented, and because not much is known about the situation in Finland. For the three first-mentioned countries there are useful lists of publications on the history of mathematics before 1889 in [Christensen/Heiberg 1889], [Holst 1889], and [Eneström 1889].

While fighting for their independence, the Norwegians managed to open a university in Christiania (Kristiania, Oslo) in September 1813. The young university soon received its first outstanding student of mathematics, Niels Henrik Abel (1802–1829). He was, in fact, the first exceptionally talented Scandinavian mathematician. Earlier, universities had been founded in Uppsala (1477), Copen-

[1] Otto Bekken offered valuable help in providing information on Norway, as did Erkka Maula in commenting upon history of mathematics in Finland.

hagen (1479), Åbo or Turku (1640), and Lund (1668); at these — like at most other European universities — mathematics was taught as a part of the *artes liberales* to an audience of generally uninterested students. When the discipline itself only appealed to a few, it is not surprising that its history was little cultivated.

6.2 Early Publications on the History of Mathematics

The earliest signs of a Scandinavian interest in the history of mathematics occur in rather brief dissertations *(disputatser)* which students defended at the universities or at gatherings in their living quarters. Some of the dissertations were written by the students themselves, but most were authored by their professors. In the period from the mid-17th century to the end of the 18th century, a number of dissertations took up themes from the history of mathematics — especially mathematics in antiquity.

Scandinavia created its first learned societies in the 18th century. Sweden opened its Royal Academy of Sciences in Stockholm in 1739, Denmark followed suit and inaugurated its Royal Academy of Sciences and Letters in Copenhagen in 1742, and Trondheim became the home-town of the first Norwegian academy (Royal Norwegian Academic Society of Sciences and Letters in Trondheim) in 1767. The proceedings of these academies show that some of the academicians found topics for their lectures and papers in the history of mathematics. Antiquity was still the favorite subject, but there are also examples of papers on more recent mathematics. As more journals came to exist, more Scandinavian papers on the history of mathematics appeared.

Most of the early Scandinavian publications on the history of mathematics fall into three categories: 1) mathematical problems — like the duplication of the cube; 2) concepts — such as logarithms; and 3) aspects of the history of a mathematical discipline — for instance the calculus. There are no biographies, and only one bibliographical survey seems to have been compiled, namely, a Swedish thesis, *Specimen historiae literariae de arte conjectandi* (Example of a history of the literature on the theory of probability). Sixteen pages long and published in Uppsala in 1731, it was written by M. STRÖMER, who defended it in a disputation supervised by SAMUEL KLINGENSTIERNA (1698–1765).

Much of this early work found Scandinavian authors largely repeating material known to European scholars. This situation changed around 1800 as national history started to define a focus.

6.3 National Interests

In 1799, the associate professor at Greifswald University, JOHANN FRIEDRICH DROYSEN (1770–1814, **B**), gave a birthday speech for the head of state who at that time was the Swedish king. As his subject he chose the contributions Swedish scholars had made to mathematics and physics [DROYSEN 1800]. Three quarters of a century later ERNST MAURITZ DAHLIN (1843–1929, **B**) made mathematics in Sweden until 1670 the theme of his doctoral thesis [DAHLIN 1875]. At the end of the 20th century, the Swedish mathematician LARS GÅRDING (*1919) published the history of mathematics in Sweden up to 1950, concentrating on the period after 1800 [GÅRDING 1994].

In Norway, ABEL became a national historical subject. Indeed, papers and books on his life and works have dominated Norwegian literature on the history of mathematics until well into the 20th century. In 1839 — ten years after ABEL's death — BERNT MICHAEL HOLMBOE (1795–1850, **B**) edited the works of his famous pupil. Another edition with commentaries was provided by SOPHUS LIE (1842–1899, **B**) and LUDVIG SYLOW (1832–1918, **B**) in 1881. Later SYLOW collaborated with ELLING HOLST (1849–1915, **B**) and CARL STÖRMER (1874–1957, **B**) in preparing the publication of ABEL's letters which appeared in 1902. In 1880 CARL ANTON BJERKNES (1825–1903, **B**) brought out the first biography of ABEL. Among ABEL's other biographers is the Norwegian mathematician, ØYSTEIN ORE (1899–1968, **B**), who contributed much to the history of mathematics while on the faculty at Yale University.

One could have expected that the two famous mathematicians, LIE and SYLOW, would have received similar attention in Norway, but that has not yet been the case, although SYLOW was the subject of three papers published by the librarian, HELGE BERGH KRAGEMO (1897–1968, **B**), in the 1930s.[2]

Another special Norwegian interest — not among mathematicians but among historians and philologists — was the mathematical content of old Nordic literature. A number of papers on this theme was published in the mid-19th century [HOLST 1889].

Mathematics in Norway in general was treated by the Danish teacher of mathematics, SOPHUS ANDREAS CHRISTENSEN (1861–1943, **B**), who wrote a doctoral thesis on the development of mathematics in Denmark and Norway in the 18th century [CHRISTENSEN 1895]. As a result of this work LIE, among others, became aware of the historical significance of the studies of CASPAR WESSEL (1745–1818) on complex numbers. Much later the Norwegian number theorist, VIGGO BRUN (1885–1978, **B**), described the early history of arithmetic in Norway [BRUN 1962].

Mathematical activities in Finland up to the beginning of the 18th century were described by J. DAHLBO [DAHLBO 1897]. A century later, GUSTAV ELFVING (1908–1984) treated the period from 1828 to 1918 [ELFVING 1981].

During the 19th century, the account of the history of mathematics in Denmark was limited to what CHRISTENSEN included in his above-mentioned book. At the beginning of the 20th century, the Danish mathematician NIELS NIELSEN (1865–1931, **B**) provided an extremely valuable source for treating the subject by listing Danish publications on mathematics from the period 1528–1908

[2]While the last proofs of this chapter were read, a biography of LIE has recently appeared: [STUBHAUG 2000b].

[NIELSEN 1910/12]. He did not draw many conclusions from this material himself, focusing instead on French mathematicians, about whom he published two books. Partly based on NIELSEN's work, the history of mathematics in Denmark has been described in [ANDERSEN 1980] and [ANDERSEN/BANG 1983].

6.4 Professional History of Mathematics

Not all new contributions from Scandinavia to the history of mathematics concerned national roots. Actually, a change of interests occurred during the last decades of the 19th century in Norway, Sweden, and Denmark, and at the same time the historiographical level increased considerably. ELLING HOLST, GUSTAF HJALMAR ENESTRÖM (1852–1923, **B**), HIERONYMUS GEORG ZEUTHEN (1839–1920, **B**) and JOHAN LUDVIG HEIBERG (1854–1928, **B**) were mainly responsible for this change. Although developments in the three countries were to a large extent independent and different, the publication of profound papers on non-national themes was a common feature. It was also significant for the new period that lectures on the history of mathematics were given at the universities, by ZEUTHEN in Copenhagen and by HOLST in Christiania. OLE JACOB BROCH (1818–1889, **B**) had lectured in Christiania as early as 1852–1853 on the development of the mathematical sciences and in 1858 on the history of mathematics until NEWTON and LEIBNIZ. Among many other things, BROCH actively engaged in reorganizing mathematical education at all levels in Norway.

Together with professionalism in Scandinavia came internationalism; ENESTRÖM, HEIBERG, and ZEUTHEN published a large part of their papers in French or German. In fact, these three scholars had quite an influence on the international history of mathematics, each in his own way. Details on their lives, their contributions to the history of mathematics, and their relations with colleagues can be found in their biographies in Part 2. This chapter concentrates on their styles, their views on how to treat the discipline, and their influences.

6.5 Eneström's Scientific History of Mathematics

One of ENESTRÖM's contributions to the history of mathematics was to provide the field with its own journal, *Bibliotheca Mathematica* (for its history see the biography of ENESTRÖM). Each of the fourteen volumes of the third series of this journal (1900–1914) opens with a "Leitartikel" by ENESTRÖM, in which he dealt with the history of mathematics as a science. He reemphasized his ideas over the years, becoming ever more outspoken. He stated as an axiom that the aim of research in the history of mathematics is to work out "eine rein fachmässige Darstellung der Geschichte der Mathematik" (a purely professional exposition of the history of mathematics) [vol. 4, 230]. Opposite the "rein fachmässige" history of mathematics ENESTRÖM placed the "Kulturgeschichte der Mathematik" (cultural history of mathematics), which he considered to be adequate in principle but unobtainable in reality. He thus set so high a standard for an ideal cultural history of mathematics that it could not possibly be fulfilled, and he used MORITZ CANTOR's four-volume *Geschichte der Mathematik* (History of mathematics) [CANTOR 1880/1908], to illustrate his point. ENESTRÖM was willing to accept a form of cultural history of mathematics, but only if it was viewed as an interesting diversion and not mistaken for real history [vol. 4, 2]. ENESTRÖM first approved of CANTOR's work as a diversion, but later criticized it strongly (see the biography of ENESTRÖM).

ENESTRÖM divided "rein fachmässige" history into two categories: an *Entdeckungsgeschichte* (history of discoveries) and an *Entwicklungsgeschichte* (history of development), and considered the second the true aim. The first, he held, did not require great expertise or sophistication, but could be useful by providing good material for the *Entwicklungsgeschichte.* He required that the latter be treated exactly and scientifically and include recent developments. He often emphasized the point that the history of mathematics should provide an exact and scientific presentation of the development of mathematics up to recent times. He found that only in this way would it win the acceptance of mathematicians — a goal in itself for ENESTRÖM. He was aware that it would be difficult to produce a work of such high standard, and that to do so would take a long time. In his reviews, he often criticized authors for not having learned the scientific approach to the history of mathematics and so labeled their work preliminary. But to reach the goals ENESTRÖM had set seemed nearly impossible: "eine wirkliche Sisyphusarbeit muss die mathematisch-historische Arbeit werden, solange es Personen mit ungenügenden Kenntnissen gibt, die sich vornehmen, die Geschichte der Mathematik zu bearbeiten ..." (the mathematico-historical work will remain Sisyphean as long as persons with insufficient knowledge take it upon themselves to treat the history of mathematics) [Bibl. Math. (3) **12**, 11]. Nevertheless ENESTRÖM was optimistic. He hoped that his ambitions could be realized by educating young, talented students and by schooling them well in mathematics and languages.

It is remarkable that ENESTRÖM's ideas concerning an exact and scientific approach to the history of mathematics went so far as to imply that there is *one and only one history of mathematics.* Concerning the methodology for producing this one history, ENESTRÖM did not offer much positive advice. He made some remarks about how to write the history of mathematics, but they are rather superficial. He had, however, much to say about what the methodology should *not* be. To illustrate his views, he used the works of his colleagues — and he criticized them all. He showed some respect for PAUL TANNERY (1843–1904, **B**) and ZEUTHEN, but he found shortcomings in their works as well. JULES MOLK (1857–1914, **B**) was one of the scholars whose work ENESTRÖM seems to have accepted; he called him a promoter of exact mathematico-historical research in an obituary [Bibl. Math. (3) **14**, 336–340].

Having written a great number of papers on history of mathematics himself, ENESTRÖM was quite courageous in publishing so much criticism of the art. No one seems to have been tempted to reply by analyzing ENESTRÖM's works and pointing out that they also failed to live up to his own criteria. The reason is presumably that, while respecting his enormous contribution to the history of mathematics, his colleagues more or less ignored ENESTRÖM's views on historiography.

6.6 Zeuthen's Historical Mathematics

ZEUTHEN's interest in the history of mathematics was most likely awakened when he stayed with MICHEL CHASLES (1793–1880, **B**) in Paris in 1863. However, as JESPER LÜTZEN has argued, ZEUTHEN's own work on mathematics, in particular on conic sections, also naturally led him to history (see **B**). In his mathematical research, ZEUTHEN applied synthetic methods, and these gave him a good background for understanding the history of Greek geometry, especially APOLLONIUS's works. He also recognised the importance of Books Two and Six of EUCLID's *Elements* for the geometrical calculations for which ZEUTHEN used TANNERY's expression "geometrical algebra" (see Section 1.9). This term has been much criticized and debated in recent decades, one of the arguments being that the Greeks did not do algebra. It should be remarked, however, that ZEUTHEN himself was fully aware of the difference between geometrical and usual algebra.

As noted above, ZEUTHEN found the history of mathematics a suitable topic for his lectures, and he actually taught courses on the subject from 1876 onwards. History *per se* was not what he wanted to teach his students first and foremost, rather he wanted to expose them to different ways of mathematical thinking. ENESTRÖM and MORITZ CANTOR commented upon ZEUTHEN's approach to the history of mathematics, and he himself agreed that it was rather ahistorical (on discussions between ZEUTHEN and CANTOR see [LÜTZEN/PURKERT 1994]). For ZEUTHEN, the important thing was to unravel the splendour of deductions in a mathematical text from the past, keeping to the techniques of the time. He also tended to restrict his investigations to texts written by first rank mathematicians. Readers who want to know about the contexts in which mathematical ideas were born, or to trace interactions and influences in the development of mathematics, will not find ZEUTHEN's work useful, but those who want to get a deeper mathematical understanding of a given historical text can still benefit much from his penetrating analyses.

Independently of ZEUTHEN, another kind of historical mathematics was taught in Denmark outside the university at the Askov School, one of the Danish so-called folk high schools attended by young people with a practical bent. The main themes taught there were Danish history and literature. In 1878 the physicist POUL LA COUR (1846–1908, **B**) started to teach mathematics and physics at the Askov School, where he was inspired to present mathematics through its history. He also wrote some very entertaining books which became very popular, but they do not reflect the new professionalism of history of mathematics in Scandinavia.

6.7 Text Editions

With a strong interest in classical texts rather than in the history of mathematics, the Danish philologist, JOHAN LUDVIG HEIBERG, gave an important impulse to the study of Greek mathematics. In his doctoral thesis, *Quaestiones Archimedeae* (1879), he treated the transmission of ARCHIMEDES's texts. He naturally followed this with an edition of the works of ARCHIMEDES, after which he undertook the enormous task of editing the works of EUCLID, APOLLONIUS, SERENUS, PTOLEMY, HERON, and THEODOSIUS, and of composing the anthology, *Mathematici graeci minores* (Minor Greek Mathematicians). His editions contain the Greek texts and translations into Latin (for details see the biography of Heiberg).

HEIBERG's work attracted attention abroad as well as in Denmark. Inspired by his editions, several new translations of works by Greek mathematicians appeared in modern languages. In recent years, criticism has been raised about some of HEIBERG's text solutions, but no one denies a debt to him. In Denmark, the first woman who graduated in mathematics from Copenhagen University, THYRA EIBE (1866–1955, **B**), was stimulated by HEIBERG's work to translate his edition of EUCLID's *Elements* into Danish (1897–1912). Presumably due to EIBE's translation, parts of the *Elements* were reintroduced into the university curriculum for mathematics.

Another of the Copenhagen students of mathematics, AXEL ANTHON BJØRNBO (1874–1911, **B**), combined his interest in mathematics and classical languages by studying Greek mathematics in Munich. Like ENESTRÖM, BJØRNBO earned his living as a librarian. While serving the Royal Library in Copenhagen, he edited a series of medieval texts on mathematics, mathematical astronomy, optics, and cartography.

ZEUTHEN followed HEIBERG's work closely, and over the years they had many stimulating discussions on Greek mathematics. An exceptional occasion for demonstrating the fruitfulness of their collaboration arose in 1906 when HEIBERG discovered a manuscript of the work by ARCHIMEDES known as *The Method.*

6.8 After Eneström and Zeuthen

Neither ENESTRÖM nor ZEUTHEN had students who continued their work in the history of mathematics. They nevertheless left very different scenarios for the future of the subject when they died in the early 1920s.

ENESTRÖM had worked in isolation. When he edited the first series of *Bibliotheca Mathematica* (1884–1887), he was in touch with the circle around GÖSTA MITTAG-LEFFLER (1846–1927), but the latter never took an interest in his work, nor did any other outstanding Swedish mathematician. Consequently, the history of mathematics in Sweden almost died with ENESTRÖM, only to be revived much later.

A chair in *lärdomshistoria* (history of learning) was created at Uppsala University in 1932, and this field became quite strong in Sweden. It is a rather broad field and includes the history of ideas and of science, but neither mathematics nor in general the technical aspects of the mathematical sciences. Thus, the history of mathematics has remained an individual activity in Sweden. One of the mathematicians who took up the subject in teaching and research, CLAS-OLOF SELENIUS (1922–1991, **B**), had a special interest in Indian mathematics.

Aside from NIELS NIELSEN, ZEUTHEN's colleagues and students did not show a great enthusiasm for the history of mathematics either, but because they respected ZEUTHEN highly as a mathematician, they took an interest in his historical work and were prepared to acknowledge history of mathematics as a possible field of research for students. It also happend that some professors worked on historical themes. JOHANNES HJELMSLEV (1873–1950, **B**), for instance, paid much attention to GEORG MOHR (1640–1697) after one of his students found a previously unknown work by MOHR; HJELMSLEV also published on Greek mathematics. Another of the professors, NIELS ERIK NØRLUND (1885–1981), edited a work on 18th century French geometers, left by NIELS NIELSEN. At the philosophical institute, JØRGEN JØRGENSEN (1894–1969, **B**) made a valuable contribution to the history of logic in his *A Treatise of Formal Logic*, published in three volumes in 1931.

The positive attitude towards the history of mathematics at the Institute of Mathematics in Copenhagen was stimulated when OTTO NEUGEBAUER (1899–1990, **B**) joined the group of professors there in January of 1934. NEUGEBAUER lectured on the history of mathematics and mathematical astronomy, and intended to publish a work in three volumes called *Vorlesungen über Geschichte der antiken mathematischen Wissenschaften* (Lectures on the History of the Mathematical Sciences in Antiquity). The first volume, *Vorgriechische Mathematik* (Pre-Greek Mathematics), appeared in 1934, but the project was never completed; NEUGEBAUER realized that a precise presentation of Babylonian astronomy required further research, and became absorbed in that quest [SWERDLOW 1993a, 146–147] and [SWERDLOW 1993b, 290–292]. During the time he spent in Copenhagen, NEUGEBAUER also published *Mathematische Keilschrift-Texte* (Mathematical Cuneiform Texts), and (assisted by his wife), the *Zentralblatt für Mathematik.* Growing tensions between NEUGEBAUER and FERDINAND SPRINGER (1881–1965) over political aspects of the editorial work led to NEUGEBAUER's resignation from the *Zentralblatt* in December of 1938. In the summer of 1939, NEUGEBAUER left Copenhagen for a professorship at Brown University and the task of starting the *Mathematical Reviews* (see Section 15.4.4). One of his Danish students, OLAF SCHMIDT (1913–1996, **B**), later became professor of the history of science at Copenhagen University. After NEUGEBAUER had settled at Brown there was still so much of his spirit left in Copenhagen that the mathematician ASGER AABOE (*1922) took up a study of Babylonian science and later became professor of the history of science at Yale University.

When SCHMIDT retired from Copenhagen University his position was preserved for the history of science. During the 1960s, OLAF PEDERSEN (1920–1997) created a department for the history of science at Aarhus University, where a position for the history of mathematics was also established. To facilitate international contacts was considered important, and hence research results in the history of mathematics were generally published in English. In recent decades, a positive attitude towards the history of mathematics has manifested itself among teachers at Danish secondary schools, with the result that the topic is now a part of the basic curriculum in mathematics.

In Norway, the history of mathematics was kept alive by a few mathematicians. The number theorist, VIGGO BRUN, was particularly eager to promote the history of mathematics, arguing that every large university ought to have a chair in the discipline [SCRIBA 1980, 3]. Besides the above-mentioned book on the history of arithmetic in Norway and some brief biographies, he published a book in 1964 based on his lectures given at Oslo University. Called *Alt er tall* (Everything is Number), this treats mathematics from antiquity to the Renaissance. BRUN spent a large part of his career at the *Norges Tekniske Högskole* in Trondheim. Another of the mathematicians there, OLE PEDER ARVESEN (1895–1991, **B**), also had an interest in the history of mathematics. As the theme of his inaugural lecture in 1927, he chose MONGE's creation of descriptive geometry. In the period 1940–1973, ARVESEN published three books in Norwegian on the history of mathematics, centered on biographies of important mathematicians. Beginning in the 1960s, JOHANNES AUGUST LOHNE (1908–1993, **B**), a teacher of mathematics at Flekkefjord high school, regularly published on the history of mathematics in international journals, his main interests being geometrical optics and THOMAS HARRIOT (ca. 1560–1621).

In Finland, the influential mathematician ROLF NEVANLINNA (1895–1980, **B**) became interested in how mathematics developed and wrote some papers on this subject. He also encouraged students who wanted to work on the history of mathematics or logic.

Several other Scandinavian mathematicians in addition to those mentioned in this section occasionally engaged in historical investigations and published their results in one of the Scandinavian journals of mathematics. The international journal devoted to the history of mathematics, *Bibliotheca Mathematica*, ceased publication at the beginning of the First World War. Efforts to revive the journal after the war progressed so slowly that, by the time the revival seemed possible, ENESTRÖM had died and the plan was abandoned, a loss for the field. Later, publication outlets did increase owing to the creation of a number of international journals for the history of science. One of these new journals, *Centaurus*, was founded in 1950 as a result of the initiatives of a circle of Danish historians of science. Some of these also founded the book series *Acta historica scientarum naturalium et medicinalium*, the first book of which was published in 1942.

6.9 Conclusion

Apart from text editions, almost all work on the history of mathematics in Scandinavia has, until recently, been done by mathematicians. Moreover, the fate of the history of mathematics in Scandinavia has depended to a large extent on how mathematicians have reacted to the field.

It seems that the situation in Denmark was more favorable than in the other Scandinavian countries. The respect that ZEUTHEN and NEUGEBAUER enjoyed as mathematicians translated not only into a general approval of the history of mathematics, but also into actual university positions. The esteem for NEUGEBAUER, combined with the fact that he was engaged in Babylonian mathematics, had the added consequence that it became acceptable to work on other aspects of the history than the development of highbrow mathematics.

This chapter shows that because the field in Scandinavia was so small, the progress of the history of mathematics was very dependent on individuals.

Chapter 7

The British Isles

IVOR GRATTAN-GUINNESS

7.1 Introduction

This chapter traces a roughly straightforward chronology of historical writing from the earliest examples in the late 17th century to the early 1960s.[1]

Some preliminary observations are useful, for they guide the design. Despite the reasonable importance of the country in the development of mathematics, the measure of historical writing is, and always has been, quite modest: "good old" British empiricism and Baconianism (as its adherents call it) have dictated not only dominating empiricist philosophies but also an historiography which has always deemed the history of mathematics to be irrelevant to current knowledge and understanding, or at best a marginal activity. Thus the quantity of work is far less than for Germany, Italy and France, even than for the USA (which became prominent in mathematics only about a century ago); so, for example, no schools, networks, organizations, traditions or journals were formed. Most historians have been mathematicians by training (though not necessarily by profession); several contributions have been made by classicists; no specimens of *historien-philosoph* (see Section 1.11) are to be found.

7.2 Prior to the Early Twentieth Century

The earliest historical writing on mathematics took place as part of an interest in the history of ancient sciences (not only mathematical ones) which developed in the 16th and 17th centuries. English translations included that of PROCLUS's *Sphere*.

[1]Comments on an earlier draft of this chapter were generously supplied by JOHN FAUVEL, KIRSTI ANDERSEN, and JEANNE PEIFFER.

More significantly, HENRY SAVILE (1549–1622) lectured on Greek astronomy at Oxford and collected many manuscripts, which he placed in the Bodleian Library. When he founded two chairs in the university for geometry and astronomy in 1619 he expected that their holders would pay due attention to their classical heritage (see [MADDISON 1977], and A. GRAFTON and others in [FEINGOLD 1990]).

One holder did pursue this aim, but he also brought patriotic partiality into his historical scholarship on more recent developments. *A Treatise of Algebra, Both Historical and Practical* (1685) of JOHN WALLIS (1616–1703, **B**) provided a valuable survey of this developing branch of mathematics (and related topics in its various addenda), but after a reasonable discussion of the Italian Renaissance there occurs an "historical" account of equations that advocated THOMAS HARRIOT (ca. 1560–1621) over DESCARTES (1596–1650) to an absurd degree (pp. 125–207, as against pp. 208–212). HARRIOT's posthumous book of 1631, which in fact (as we now know) does not contain all of his best work, hardly rivals DESCARTES to the extent that, as WALLIS claims, DESCARTES's *Géométrie* (1637) "adds very little of his own, if any thing, (as to *pure Algebra*)" (p. 208). Some of WALLIS's other remarks indicate the non-availability of sources: in particular, "Leonardus Pisanus" was "not now extant," but "Vossius placeth around the year 1400, or somewhat sooner" (pp. [iii], 61–64). WALLIS's judgement against DESCARTES aroused some controversy, to which he responded with a more referenced Latin edition of his book in 1693.

WALLIS also discussed some pre-calculus methods, but omitted DESCARTES's. By then LEIBNIZ (1646–1716) was just in print, but NEWTON (1642–1727) not at all, and nationalism became even more marked in the priority row over the infinitesimal calculus confected in the 1710s by the ageing NEWTON, with hair and perhaps brain saturated with mercury.[2] The historical refutation of these claims needs no discussion here, but their historical influence must be stressed; although they became steadily less recent history, their tenets remained in place for the rest of the century and thereby contributed to the stagnation of British mathematics by blocking it from achieving a better appreciation of the significant advances of the Leibnizian calculus.

WALLIS provided some of the fuel for this controversy when in 1693 he included NEWTON's account of the invention of the calculus in the second volume of his own *Opera mathematica.* Continental reactions led him to publish correspondence between NEWTON and LEIBNIZ in the third volume (1699) and insinuate that some of LEIBNIZ's procedures were already in NEWTON. This volume also contained WALLIS's own translations into Latin of various classic Greek texts on mathematics and astronomy. One author was APOLLONIUS whose main works were rendered into Latin in 1706 and 1710 by WALLIS's younger contemporary, EDMUND HALLEY (1656–1743).

[2]The most substantial single source is [NEWTON 1967/81]. Among texts of the time other than NEWTON's own was J. RAPHSON, *History of Fluxions* (1715). Newton's nephew by marriage, JOHN CONDUITT, collected information and anecdotes, which were used by BERNARD LE BOVIER FONTENELLE in his "Eloge de Monsieur Neuton" (1728).

Historical research in the 18th century was rather intermittent. The stream of dictionaries and encyclopedias which was to characterize British contributions to a considerable extent began with JOHN HARRIS's (1667–1719) *Lexicon technicum* (1704) and continued with a *New Mathematical Dictionary* (1726) by EDMUND STONE (1700?–1768), a *Cyclopaedia* (1728) by EPHRAIM CHAMBERS (1680?–1740), and further publications [WALLIS 1975]. The level of historical scholarship in these works, however, was not profound.

More serious work is evident from the 1770s. For example, SAMUEL HORSLEY (1733–1806) put out a five-volume edition of NEWTON (1779–1785), including not only the *Principia* and the *Opticks* but also some fluxional texts. In Scotland a *History of Gunnery* (1776) by JAMES GLENNIE (1730–1817) dealt with theories of air resistance, while the *System of Chronology* (1784) by JOHN PLAYFAIR (1748–1819) considered earlier versions, including NEWTON's. Then in 1787 the EARL OF BUCHAN (1742–1829) and WALTER MINTO (1753–1796) produced *An Account of the Life, Writings and Inventions of John Napier*, which contained a good survey of the production of logarithmic tables. Later WILLIAM TRAIL (1746–1831) recalled his Glasgow teacher in *An Account of the Life and Writings of Robert Simson* (1812), including manuscripts related to SIMSON's reconstruction of PAPPUS and of Euclidean porisms.

Signs of further change appear in the new generation of mathematicians who at last were aware of Continental calculus. CHARLES HUTTON (1737–1823), professor at the Royal Military Academy [JOHNSON 1989] and from 1779 Foreign Secretary of the Royal Society, produced a massive *Mathematical and Philosophical Dictionary* in two large volumes (1796–1795: sic), in which many entries on topics as well as on mathematicians looked back in some way. HUTTON's appreciation of EULER (1707–1783) and of LAMBERT (1728–1777) (though not LEIBNIZ) was particularly warm. His work drew substantially on his magnificent library, which unfortunately was dispersed after his death.

HUTTON also produced in 1785 an edition of some mathematical tables, and this surprisingly stimulated the largest and strangest single production in the history of mathematics in Britain. FRANCIS MASERES (1731–1824) held the splendidly named legal post of "Cursitor Baron of the Exchequer," but he was a talented mathematician and devoted much of his time to an edition of reprints and new articles (especially by himself) of material on logarithms and their functions. In the end, six vast volumes of *Scriptores logarithmici* appeared between 1791 and 1807, providing reappearances for many original tables by (among others) NAPIER (1550–1617), BRIGGS (1561–1630), and KEPLER (1571–1630), as well as papers on series by NEWTON and his followers; volume 4 (1796) also covered navigation and volume 5 (1801), interest, and insurance. The whole rather overemphasized British texts, but despite its strange design, it deserves to be used more, especially for making available several extremely rare items. Near the end of his long life, HUTTON prepared a smaller collection on optics (1823), in which he was helped by the young CHARLES BABBAGE (1792–1871).

Encyclopedias became a great speciality in Britain at this time, partly due to the economic opportunities made available by publishers but also for the competence of authors. Table 1 lists the principal encyclopedias that appeared during the first half of the 19th century. The fourth edition of the *Encyclopaedia Britannica* included a concise summary of the "Progress of Mathematical and Physical Science since the Arrival of Letters in Europe." Of particular note is the *Encyclopaedia Metropolitana*, which contained a remarkable range of lengthy articles on most branches of pure and applied mathematics (see Section 3 below). HUTTON's younger colleague PETER BARLOW (1776–1862) filled most of volume 3 (1829) with a range of excellent articles on mechanics, with extensive references to contributions from the 18th century.[3] Similarly, when the British Association for the Advancement of Science (BAAS) was founded in 1831, it commissioned survey articles on developments in various sciences; mathematics was well represented, and some of the pieces are durable historical contributions to the recent history of their time.

7.3 Studies of Special Topics

For all their considerable merits, these encyclopedias and dictionaries were not intended as historical texts. Books of the latter kind did begin to appear in the early 1800s. Typically, the first was a translation; CHARLES BOSSUT's (1730–1814, **B**) superficial *A General History of Mathematics* (1803) (see Section 1.4) was prepared by HUTTON's colleague JOHN BONNYCASTLE (1750–1821). No English translation of MONTUCLA/LALANDE [MONTUCLA 1799/1802] was ever prepared.

However, a real book appeared in 1810 from a Cambridge mathematician who was aware of Continental developments to the extent of summarizing one of the main ones: *A Treatise on Isoperimetrical Problems and the Calculus of Variations* by ROBERT WOODHOUSE (1773–1827). Although the word "history" does not appear in its title, it *was* such a study, comprising a careful account of the principal sources. Moreover, he made no effort to dress the story in official British fluxionese: he had already shown his allegiance to LAGRANGE's algebraic style (indeed, this was doubtless his original inspiration), and he showed himself quite capable of handling Continental theories.

He was soon to play a role in the main (but not the first) reform, effected by BABBAGE and JOHN HERSCHEL (1792–1871) in founding the Analytical Society in 1812. The preface which they wrote to the only volume of the Society's *Memoirs* in 1813 is itself a valuable introduction to Continental calculus over the previous thirty years.

[3] A bibliographical difficulty with the *Encyclopaedia Metropolitana* and other encyclopedias of this time arises from the publishers' practise of issuing title pages for each volume when the whole edition was completed. Thus most copies are dated this way to 1845, whereas, for example, Barlow's articles (which also covered optics and magnetism) were written in the middle-to-late 1820s; the volume in which they were printed first appeared in 1829.

Short title	Vols	Place and date of publication	Principal editor(s)	Comments
The British Encyclopaedia	6	London, (1809)	W. Nicholson	Often called *'Nicholson's encyclopaedia'*
The Cabinet Encyclopaedia	133	London, 1829–1846	D. Lardner	61 titles in all; some appeared in 2nd eds. 12 titles for the cabinet on 'natural philosophy'
The Cyclopaedia	45	London, 1802–1820	A. Rees	Edited by the revisor of E. Chambers's *Cyclopaedia.* — Sometimes called *'The new cyclopaedia'*
The Edinburgh Encyclopaedia	18	Edinburgh, 1808–1830	D. Brewster	Contains several important articles for mathematical physics
The Encyclopaedia Britannica; 4th edition	20	Edinburgh, 1801–1810	J. Millar	Scots responsible for organising articles on science (J. Playfair, J. Robison, D. Stewart, T. Thomson)
5th edition	20	Edinburgh, 1810–1817	J. Millar	First five volumes reprinted from 4th ed.
6th edition	20	Edinburgh, (1823)	C. MacLaren	Not a substantial revision of the 4th ed.
Supplement to the 4th, 5th and 6th editions	6	Edinburgh, 1815–1824	Mac. Napier	Several lengthy articles relating to mathematical physics
7th edition	21	Edinburgh, 1830–1842	Mac. Napier	A substantial revision of the 6th ed., using material from the supplement
Encyclopaedia Edinensis	6	Edinburgh, (1827)	J. Millar	A more popular work, though still scholarly
Encyclopaedia Mancuniensis	2	Manchester, 1813		Principally concerned with science and engineering
The Encyclopaedia Metropolitana	29	London, 1817–1845	T. Curtis; E. Smedley, H. J. Rose	Contains several important articles for mathematical physics. 2nd ed. 1848–1858 (40 vols.)
Encyclopaedia Perthensis	23	Perth, 1796–1808	A. Aitchison	2nd ed. 1807–1816 (24 volumes)
The Oxford Encyclopaedia	7	Oxford, 1828–1831	Various	
Fantologia	12	London, (1813)	J. M. Good, O. Gregory, N. Bosworth	
The Penny Encyclopaedia	29	London, 1833–1856	E. G. Long	Associated with The Society for the Diffusion of Useful Knowledge
The Popular Encyclopaedia	7	Glasgow, (1841)	A. Whitelaw	Includes a survey of science by T. Thomson

Table 1: The principal encyclopaedias that appeared during the first half of the 19th century

At this time appeared the first substantial history of mathematics written by a Briton — from a chemist. THOMAS THOMSON (1773–1852), also well known as a journal editor, wrote a *History of the Royal Society, from its Institution to the End of the Eighteenth Century* (1812). Historical studies of the Society had appeared before, indeed THOMAS SPRAT (1635–1713) wrote a history of the Society in 1667, shortly after the time of its launching, but mathematics had not been a prominent concern in its early years. THOMSON, however, not only treated all sciences but even covered developments from before the Society's time. Thus his chapter of 72 quarto pages "Of Mathematics: Number and Extension" began with around 30 on Egyptian, Greek, and medieval mathematics. For the 17th century the calculus controversy dominated his treatment: the common error persisted of thinking that NEWTON's $\dot{x}$ and LEIBNIZ's dx were equivalent, as did the claim that the two theories "are absolutely the same in principle, and differ only in the method," although LEIBNIZ was "far inferior to NEWTON, both as a philosopher and as a man" (pp. 291–296).

The final figure of this time is one of the most interesting British figures, HENRY T. COLEBROOKE (1765–1837, **B**). His father, chairman of the East India Company, sent him to a post there in 1782 in his 17th year; his own intellectual curiosity and mathematical ability drew him as a hobby into a career as one of the first major European scholars of Sanskrit. He focussed especially on Indian algebra in his book, *Algebra, with Arithmetic and Mensuration from the Sanskrit of Brahmegupta and Bhaskara* (1817), published soon after his return to England. He pointed out the superiority over others of Sanskrit authors with regard to algebraic procedures, and in this and other ways he pioneered the serious Western study of non-Western mathematics using original sources [WHITROW 1975] (see Section 18.3).

7.4 Largely Historians of Newton

SIR DAVID BREWSTER (1781–1868) produced a *Life* in 1831, covering NEWTON's work in a manner which corresponded more closely to the interests of the author rather than of his subject; 90 pages on optics came first, followed by less than 70 on mechanics and astronomy, less than 40 on the calculus (including the controversy), and only a trace on NEWTON's "supposed attraction to alchemy." The book was reprinted in 1858, with a discourse by LORD BROUGHAM on the unveiling of a statue of NEWTON in his home town of Grantham. Three years earlier BREWSTER had produced a much larger two-volume study of *Memoirs* on NEWTON; the structure and balance of this work was much the same, but a good deal of new information was provided.

BREWSTER had been helped by interim work effected by two scholars. STEPHEN PETER RIGAUD (1774–1839) produced in 1838 a short *Historical Essay on the First Publication of Sir Isaac Newton's Principia*, including correspondence with HALLEY and the first (partial) publication of NEWTON's manuscript

"De motu." Soon afterwards his son completed a two-volume edition of 250 letters of *Correspondence of Scientific Men of the Seventeenth Century* (1841); the collections for ISAAC BARROW (1630–1677), JOHN FLAMSTEED (1646–1719), JOHN WALLIS and NEWTON were the most significant. Also of value was the edition by JOSEPH EDLESTON (1816–1895) of *The Correspondence of Sir Isaac Newton and Professor Cotes* (1850), for not only was an important exchange made available (mainly concerning the second edition of the *Principia* in 1713), but the editor also prefaced it with a fine synopsis of NEWTON's life, lists of his lecture courses, and "De motu."

Shortly afterwards ROBERT GRANT (1814–1892) published *A History of Physical Astronomy ... Comprehending a Detailed Account of the Establishment of the Theory of Gravitation by* NEWTON ... (1852), running from ancient times up to current work. While he did not include many (enough, even) mathematical details, he clearly characterized them in prosodic descriptions, and his book remains a most distinguished aid to the historian of mathematics.

RIGAUD's Oxford colleague BADEN POWELL (1796–1860) produced in 1834 a *History of Natural Philosophy From the Earliest Periods to the Present Time*, for DIONYSIUS LARDNER's "Cabinet Cyclopaedia" book series. Although derivative, the book drew upon a wide range of sources, and devoted a good deal of space to descriptive accounts of mathematics, especially its applications. He also included a useful chronology (ending oddly with "1833 At Cambridge"); but he never mentioned LEIBNIZ or the BERNOULLIS, and so showed that British historical perspective remained warped.

Among general historians, HENRY HALLAM (1777–1859) deserves to be better remembered. His large *Introduction to the Literature of Europe, in the Fifteenth, Sixteenth, and Seventeenth Centuries* (1838–39), which became well enough known for a 7th edition in 1864, was extraordinarily wide-ranging, to the extent of treating science and mathematics; his surveys are short and derivative, but serious.

By the 1840s the Cambridge renaissance had led to the appearance of a distinguished cohort of mathematicians, of whom a few took an interest in history. The two outstanding figures were WILLIAM WHEWELL (1794–1866, **B**) and AUGUSTUS DE MORGAN (1806–1871, **B**).

WHEWELL, a major influence on the teaching of mathematics at Cambridge [BECHER 1980], philosophized himself almost to excess into erudite historical and mathematical corners. One of the most sophisticated philosophers of the century, he united historical and philosophical thinking, most notably in his *History of the Inductive Sciences* (1837) and *Philosophy of the Inductive Sciences, Founded upon Their History* (1840). Mathematics occupied quite a prominent place in the first volume and the first half of the second volume of the *History*. Although most of his statements were derivative, they included sensitive comments about the achievements of his historian predecessors. Regarding methodology, the adjective "inductive" of his title shows that the BACON-like canons of induction were flourishing. However, he was far removed from the simple-minded empiricist tradition of most of his compatriots. On the contrary, he was an early enthusiast for IM-

MANUEL KANT (1724–1804), and was influenced by the emphasis that KANT had given to mental objects and processes in epistemology. He also stressed the need to divide history into epochs. In *Philosophy* he used history largely for general encouragement rather than a formalized historiography, but he treated pure and applied mathematics in some detail in the first part of the book.

WHEWELL also defended NEWTON against the charges of misuse of FLAMSTEED's astronomical data made in the *Account of the Rev. John Flamsteed* (1835) by FRANCIS BAILY (1774–1844). The significance to us of the charges lies in the finesse, uncommon for the time, with which BAILY used manuscripts to prepare his case.

BAILY's revisions of NEWTON were continued by DE MORGAN. After graduating from Cambridge in the mid 1820s, he soon began his non-stop career of writing, mainly in algebra and education, but also with much historical work. His most significant historical finding concerned the controversy between NEWTON and LEIBNIZ of the early 18th century over the invention of the calculus. In Britain LEIBNIZ's guilt had been taken for granted, but by drawing on documents in the archives of the Royal Society (examined on his behalf as he was not a Fellow), DE MORGAN revealed some of NEWTON's manipulations and deceit (cf. Section 1.3.3). DE MORGAN's main writings on NEWTON appeared in 1846: a long biographical article in the book *Cabinet Portrait of British Worthies*, and a short paper in the *Philosophical Transactions of the Royal Society.*

DE MORGAN's historical side was first revealed, however, in book reviews for the *Quarterly Journal of Education* (1831–1835), a very interesting serial published by the Society for the Diffusion of Useful Knowledge. The Society also sponsored a large-scale encyclopedia, the *Penny Cyclopedia* (1833–1856), to which he contributed a large number of entries, many with historical content. One source of information was his own enormous library, which must be one of the finest collections ever assembled by a mathematician. It was essential for the compilation of his valuable bibliography of *Arithmetical Books from the Invention of Printing to the Present Time* (1847), which also contained a reasonable summary of the development of arithmetic. He dedicated the book to GEORGE PEACOCK (1791–1858), whose article on this subject for the *Encyclopaedia Metropolitana* in 1829 was a major contribution in its own right.

DE MORGAN's many historical writings were well known until the early 20th century but are now largely forgotten (see **B**, [RICE 1996]). He also put historical remarks in many of his writings, including his research papers on mathematics and logic, but he spoiled his extraordinary erudition by incoherence of purpose and poor organisation of material. Over 20 years, for example, DE MORGAN published works on the history of mathematics and astronomy buried in the *Companion of the British Almanac.* On occasion he touched on the history of probability, to an aspect of which EDWARD J. FARREN contributed an *Historical Essay on the Doctrine of Life Contingencies in England* (1844), a branch of actuarial mathematics for which England has been uncommonly important.

7.5 Some Cambridge Historians

The dominance of British (especially English) mathematics by Cambridge University carried over into historical work, since most main figures of the second half of the 19th century were trained there. An early practitioner was ISAAC TODHUNTER (1820–1884, **B**). TODHUNTER was inspired to historical work by early exposure to DE MORGAN at University College London, and upon his appointment to Cambridge University he produced books on the progress of various branches of mathematics [JOHNSON 1996]. The first dealt with the calculus of variations, which was soon succeeded by the book by ROBERT WOODHOUSE; its publication in 1861 led to TODHUNTER's election to the Royal Society the following year. There followed books on the history of probability theory (1865, and this at a time when the subject itself was not prominent in mathematics), *Theories of Attraction and Figure of the Earth* (1873), and a work on elasticity and strength of materials (1886–1893, partly posthumous: see **B**). In all these works TODHUNTER tried to maintain chronology at almost any cost to subject matter; as a result his texts are difficult to read, although the information presented is exceedingly valuable, and his books remain in reprint today.

The same judgement can be passed on his extensive two-volume life and letters of WHEWELL (1876). A similar style of the time for biographies in Britain is also evident in *The Life of James Clerk Maxwell* (1882) by LEWIS CAMPBELL (1830–1908) and WILLIAM GARNETT (1850–1932), in *The Life of Sir William Rowan Hamilton* (three volumes, 1882–1891) by ROBERT GRAVES, and later in the biography of LORD KELVIN (1910) by SYLVANUS PHILLIP THOMPSON (1851–1916). Three autobiographies may also be noted: BABBAGE (1864), interesting but rather fabulous; GEORGE B. AIRY (posthumous, 1896), and as dead as its late author; and J. J. THOMSON (1936), rather good and in places surprising.

JAMES W. L. GLAISHER (1848–1928) built up a library of De Morganian proportions, which he deployed in historical articles on various branches of mathematics. Many of them, and also some articles by others, appeared in the two journals that he edited for many years, and which ceased publication after his death: the *Quarterly Journal of Pure and Applied Mathematics* (1855–1927), and *Messenger of Mathematics* (1862–1928). In 1890 he even opined to the BAAS that mathematics suffered the most of all disciplines from being separated from its history, but he could not awaken his colleagues to this fact.

GLAISHER's colleague at Trinity College, WALTER W. ROUSE BALL (1850–1925, **B**), made his career largely as an historian. His best remembered work is *A Short Account of the History of Mathematics*, which first appeared in 1889 as the written version of a lecture course of the previous year; a further edition came out in 1893, with slight changes wrought for later editions. A potboiling *Primer* was first published in 1895. ROUSE BALL worked in the era of MORITZ CANTOR, and in the preface to his *Account* he stated that he had relied much on CANTOR as well as other historians. Indeed, his text often shows the character of second-hand reading. Still, ROUSE BALL provided the most substantial general account of the history of mathematics then available in English.

ROUSE BALL's best book is *A History of the Study of Mathematics in Cambridge* (1889), wherein the quality of first-hand acquaintance makes a contrasts with the inevitably derivative character of his general history. Another book, *Essays on the Genesis, Contents and History of Newton's "Principia"* (1893), contributed to the NEWTON industry. In this work ROUSE BALL continued the efforts of BREWSTER and others (Section 4) who considered manuscripts and also published some letters for the first time. By then, enough work had been done on NEWTON for GEORGE JOHN GRAY (1863–1934) to compile a *Bibliography of the Works of Newton* and of writings on him (1888, 1907).

Important historical research on the history of symbolic logics came from PHILIP JOURDAIN (1879–1919, **B**), a graduate from Cambridge in 1902. Until his early death in his 41st year from a crippling illness, he devoted himself largely to the history of set theory and logic. In his history of recent logicians in the early 1910s, JOURDAIN had the happy idea of sending his essays to those figures still alive and incorporating their comments in the published versions, making them of permanent importance. He also worked (without distinction) in set theory itself and on the history of mechanics. In 1915 he produced an edition of DE MORGAN's writings on NEWTON. He was succeeded in his work by ARCHIBALD E. HEATH (1887–1961) and JAMES MARC CHILD (1871–1960), who, with ARTHUR T. SHEARMAN (1866–1937) and ALFRED E. TAYLOR (1869–1945), maintained a modest tradition of historical interest in logic and related subjects until after the Second World War.

7.6 The Royal Society Catalogue of Scientific Papers

A valuable aid to historical research is worth mentioning here. In mid-century the Royal Society undertook a major bibliographical task. The initiative came from the American JOSEPH HENRY (1797–1878), who suggested at the 1855 meeting of the BAAS a collaboration with the Smithsonian Institution to produce a catalogue of scientific papers. After some delay and inertia, the Royal Society took over the project in the form of a catalogue of papers in all sciences (but not technology) published in journals since 1800. The first six volumes, covering the period 1800–1863, appeared in 1867, and a succeeding sextet up to 1883 was out by 1902. Then index volumes for pure mathematics, mechanics, and physics for the whole century were published in 1908–1914. The remaining seven volumes of the catalogue came out between 1914 and 1925. By then the Royal Society had taken the lead in a continuation, the *International Catalogue of Scientific Literature* (with mathematics and mechanics as the first of 17 branches of science), which produced annual volumes until 1916, covering the period 1901–1913. However, World War I ended this venture.

Of the various parts of these two projects the index volumes of the first *Catalogue* are the most remarkable, constituting one of the best guides to 19th-century science (not least in the citation of historical papers themselves). The modes of classification of subjects also give guidance as to the conception of mathematics at that time, and the lengths of the entries indicate the relative popularity of the various subjects. Curiously, although mathematics (pure and applied) was well represented in these projects, especially in the indexing, mathematical Fellows of the Royal Society played little role in their preparation.[4]

7.7 Historians of Greek Mathematics

The fetish for editions of EUCLID in Britain is extraordinary: over 200 by the end of the 19th century. However, this "work" did not lead to much significant scholarship on Greek mathematics (the most valuable were attempts to reconstruct porisms by ROBERT SIMSON (1687–1768) and MATTHEW STEWART (1717–1785)). This point was well stated in the preface to his refreshing *A Short History of Greek Mathematics* (1884) by the Cambridge scholar JAMES GOW (1854–1923):

> The work, as usual, has been left to Germany and even to France, and it has been done there with more than usual excellence! It demanded a combination of learning, scholarship and common sense which we used, absurdly enough, to regard as peculiarly English.

GOW naturally drew much on CANTOR, but covered material which many historians did not treat, including gematria, and he wrote throughout in a nice style. His volume was soon followed by *Greek Geometry from Thales to Euclid* (1889), published in Dublin by the Irish mathematician GEORGE JOHNSTON ALLMAN (1824–1904). Also basing his book on CANTOR, ALLMAN handled this early period in more detail than GOW. Soon afterwards Scotland joined in, with *Early Greek Philosophy* (1892) from the classicist JOHN BURNET (1863–1928), who took suitable note of mathematics.

The most noteworthy scholar-historian of Greek mathematics of this time was professionally employed elsewhere. SIR THOMAS LITTLE HEATH (1861–1940, **B**) gained both Fellow of the Royal Society (1912) and Fellow of the British Academy (1932) — an unusual double honour — for his historical researches into Greek mathematics. However, his knighthood came for his career as a Treasury official, which he pursued after graduating from Cambridge (where he came under the influence of GOW), although for some years he held a Fellowship at Trinity College. His spare-time achievements are remarkable: starting with a study of DIOPHANTUS (1885, second edition 1910), he continued with a modernized edition and

[4]See the editorial introductions to the opening volumes of both catalogues and to the index volumes. A summary is given in [ROYAL SOCIETY 1940, 180–182]; the Society Archives hold files at ms. 539, and in MC. 9 and MM. 14. A study of the history of these catalogues (not yet undertaken by anyone) would constitute a fine doctoral dissertation.

translation of APOLLONIUS in 1896 (an interesting contrast with the intentions of WALLIS 180 years earlier), ARCHIMEDES (1897, with a supplement in 1912), and EUCLID (1908, second edition 1926). He pulled together his thoughts in *A History of Greek Mathematics* (1922), and in his later years he prepared a survey of Greek astronomy (1932) and a study of ARISTOTLE (1949, posthumous).

HEATH largely aimed to highlight Greek figures who had been eclipsed by the British EUCLID fetish, to update British readers with discoveries and other developments made abroad since the time of GOW, and to prepare scholarly English translations of their works. Historiographically speaking, his method was anachronistic, for he rewrote the mathematics of several figures in modern notation. Despite this mathematician-friendly approach and his academic accolades, his work had little impact on the mathematical community and only marginal influence on historians in general (the acknowledgements in his books were addressed to classicists). However, most of his books are still in print.

7.8 General Writing to the First World War

The German-born historian JOHN THEODORE MERZ (1840–1922) published a four-volume *History of European Thought in the 19th Century* between 1896 and 1914. While the breadth of his vision prevented penetrating or original interpretation of developments, he paid attention to science to a remarkable degree, with well-judged comparisons of major European countries and a good deal of accurate information given on mathematics and its institutions in its first part. A similar volume for its field was *A Short History of Astronomy* (1898) by ARTHUR BERRY (1862–1929) (Cambridge trained but then at University College London). BERRY's *History* provided coverage from ancient to current times; the mathematical component is perhaps too slight, but the book is a valuable adjunct to the history of mathematics.

The *Encyklopädie der mathematischen Wissenschaften* was being generated from the mid 1890s under FELIX KLEIN's general direction, with an extended French edition prepared alongside (see Sections 1.9, 5.4.4, and 11.4). Nevertheless, no interest was excited in Britain in a comparable English version, despite the efforts of WILLIAM HENRY YOUNG (1863–1942) and GRACE CHISOLM YOUNG (1868–1944), British-born mathematicians living and working on the Continent.[5] The German project used around 150 authors in all over its forty years of preparation, but only eight were British (in various parts of applied mathematics, which was the strongest area of British mathematics itself). However, their contributions

[5]The YOUNGs started to translate articles on analysis in the first volume in the 1900s, but they gave up when ANDREW R. FORSYTH (1858–1942) told them that it would be professionally much more advantageous to write a textbook. No manuscripts of such translations survive in their *Nachlass* (Liverpool University Archives). The papers are in Liverpool because there from 1913 to 1916, YOUNG was part-time Professor of the History and Philosophy of Mathematics; he seems not to have done any historical research at all.

were significant, and constitute some of the best British writing of recent history which is mostly available only in German translation.[6]

The generally high level of writing, which we saw as a British speciality, was renewed in encyclopedias at the end of the century, this time especially in the *Encyclopaedia Britannica.* From mid century this publication had included some fine articles in all sciences, but it excelled in the 10th (1890) and especially the 11th (1910–1911) editions. The latter, which was reprinted until the 1920s as the bulk of the 12th and 13th editions, is among the greatest of all encyclopedias in any language, and its mathematical articles are often first rate. There are too many to be presented even in tabular from here, but as a whole they constitute the equivalent of several books on the then-current state of mathematics, often with considerable historical background. ALFRED NORTH WHITEHEAD (1861–1947) wrote a rather short (5 pages) general article on mathematics, but included a short passage on the recent history of its history.

A surprising omission from the contributors to this encyclopedia, however, is SIR EDMUND WHITTAKER (1873–1956, **B**), by far the most distinguished British mathematician to write substantially on the history of mathematics. The wide range of interest in pure and applied mathematics allowed him to turn out centenary articles of uncommon quality on scientists. WHITTAKER's deep religiosity may have drawn him to his main historical effort, *A History of the Theories of Æther and Electricity* (1910). The book covered the period of classical physics; after the relativistic revolution he revised it and published a new volume on "the modern theories 1900–1926" in the early 1950s. Both volumes are standard reference works, the first one being a pioneering work, although in some ways they are disappointing from an author of his quality. Quite apart from his well-known denigration of ALBERT EINSTEIN (1879–1955) in the second volume, the first is disappointing in its treatment of the (crucial) mathematical aspects of the story: the notations are modernized (compare his contemporary HEATH), and discussion of modes of mathematical modelling and reasoning is largely absent.

WHITTAKER spent the bulk of his career in Edinburgh; another figure with Scottish connections was SIR THOMAS MUIR (1844–1934, **B**). After studying at Glasgow under KELVIN and teaching there, he spent his career from 1892 until retirement as Cape Superintendent (of Education) in South Africa. Nevertheless, he managed to write a considerable amount of mathematics and history, especially a gigantic study of *The Theory of Determinants in the Historical Order of Development* (1906–1923), with a succeeding volume in 1930 which told the story to 1920 and provided an index to the whole work. His penchant for synchrony matched TODHUNTER's, with the same results for unreadability, but his attention to bibliography allowed him to notice, and criticize, repeated "discoveries" of sev-

[6] Articles by British authors in the *Encyklopädie der mathematischen Wissenschaften* are G. T. WALKER (article IV 9 (1900), 36 pages), A. E. H. LOVE (IV 15–16 (1901), 100), H. LAMB (IV 26 (1906), 96), G. H. BRYAN (V 3 (1903), 90), E. W. HOBSON (V 4 (1904), 48), S. S. HOUGH and G. H. DARWIN (VI 1,6 (1908), 83: English original in DARWIN's *Scientific Papers* (1911)), and E. T. WHITTAKER (VI 2,12 (1912), 45).

eral results [May 1968]. His volumes remain standard reference sources in an area of history which is still seriously understudied; indeed, they go beyond their title, for parts of the history of matrix theory are also recorded.

At Glasgow the mathematician George Gibson (1858–1930) wrote some valuable articles, mainly in the 1890s. A lecture is still held at the University in his memory, occasionally devoted to historical themes. At St. Andrews Duncan M. Y. Sommerville (1879–1934) extended his interest in non-Euclidean geometry to produce an excellent *Bibliography* in 1911. John Sturgeon Mackey (1843–1914) wrote several articles, especially on the history of geometry. Computational mathematics of various kinds received attention from the Scots in these years. In 1905 the Chartered Accountants of Scotland celebrated their 50th anniversary as the first such national association with a valuable collective volume, *A History of Accounting and Accountants*, edited by Jack Brown, which covered material such as double-entry bookkeeping often not discussed in histories. A decade later it was time to celebrate 300 years since the creation of logarithmic tables by Napier and Briggs, with a conference organized by the Royal Society of Edinburgh. This led to a substantial *Napier Tercentenary Volume* (1915) edited by Cargill G. Knott (1856–1922), which included valuable historical articles by authors from various countries. An exhibition of tables and calculating machines was also held, for which Ellice M. Horsburgh (1870–1935) edited a valuable *Handbook* (1914).

7.9 Pearson and the History of Statistics

A different tradition stemmed from Karl Pearson (1857–1936, **B**), who infused his writing with Marxist and atheistic preferences (he had changed the spelling of his Christian name "Carl" to the given name "Karl"). Initiated into history by his completion of Todhunter's *History of the Theory of Elasticity* around 1890, he then turned to his life's work of mathematical statistics. In 1911 he accepted the Chair of Eugenics at University College London, financed by Francis Galton (1822–1911). Pearson's historical interest continued with a magnificent biography of Galton (done in the usual life-and-letters style) which appeared in three volumes between 1914 and 1930. Profusely illustrated throughout, it also contained massive genealogies (back to the Darwin and Wedgewood families), as well as facsimiles of some of his finger-prints, all produced by his Francis Galton Laboratory for National Eugenics. In the span of 1400 pages he not only described Galton's introduction of statistics into meteorology, anthropology, heredity, and psychology, but also commented on the probity of his methods. It is most fortunate that the project was completed: the first volume appeared just before the First World War (like the Napier Conference) and so lost money, and its successors appeared only with the help of private support.

Pearson also gave a lecture course on the history of statistics from 1921 onwards, and came to encourage historical articles in *Biometrika*, the statistical journal which he had helped to found in 1901. He was the initial (and occasional) historical author there from 1920; but the inclusion of historical articles increased from the 1940s and then became a regular feature of the journal — the sole case of a continuing tradition of history within mathematics in the British Isles. Under his influence MAJOR GREENWOOD (1880–1949) and FLORENCE N. DAVID (*1909) made important contributions.

7.10 Between the Wars

After World War I, and partly because of it, much of the previous impetus was lost world-wide. Most writing was of a secondary or derivative kind, though often quite well done.

One of the speakers at the NAPIER conference had been ROBERT T. GUNTHER (1869–1940), who was Curator of the Museum of the History of Science in Oxford after the War. He produced 14 volumes on *Early Science in Oxford* (1922–1945), which included mathematics in places: in general in vol. 2 pt. 2 (1922) for 100 pages, *passim* in vol. 11 (1937) on Oxford colleges, and in parts of the five volumes devoted to ROBERT HOOKE (1635–1703). The most valuable (because unusual) parts of these studies, as also of his companion single-volume *Early Science in Cambridge* (1937), treat mathematical models and mathematical instruments such as slide-rules and astrolabes (on which he also published *The Astrolabes of the World* in 1932).

Similar general works included *A History of Science and Its Relations with Philosophy and Religion* (1929) by WILLIAM C. D. DAMPIER-WHETHAM (1867–1952), where mathematics was excessively neglected. The large *A History of Science, Technology and Philosophy in the 16th and 17th Centuries* (1935) by ABRAHAM WOLF (1876–1948), with a sequel analysis of the 18th century three years later, gave an unoriginal treatment (and rather few references) but was excellent in its wide coverage (including, for example, building and cartography). The first edition of *Mathematics for the Millions* (1936) by LANCELOT HOGBEN (1895–1975) sold very well, perhaps due in part to an excellent title and its then-popular leftwing slants. *Science since 1500* (1939) by HUMPHREY THOMAS PLEDGE (1903–1960) scampered through everything but provided good historical context for each science among the others. The history of classical thought, including some mathematics, was treated by BENJAMIN FARRINGTON (1891–1974) (University College, Swansea) in books such as *Science and Politics in the Ancient World* (1939).

NEWTON studies largely stopped, but the bicentenary of his death in 1927 inspired a "memorial volume edited for the Mathematical Association." The coverage of mathematics and physics was reasonable, although the predominance of scientists among the authors led to a watered-down historiography. Five years later EUGENE F. MACPIKE (* 1870) produced a scholarly edition of the *Correspondence and Papers of Edmund Halley*. The Scots attended to some of their Newtonians, in the style of TRAIL on SIMSON (Section 2): in 1922 CHARLES TWEEDIE (1867/68–1925) published a *Sketch* of the life and work of JAMES STIRLING (1692–1770), including much correspondence for the first time; similarly, in 1939 HERBERT W. TURNBULL (1885–1961, **B**) edited a substantial *James Gregory Tercentenary Memorial Volume*, including many letters and manuscripts.

During the 1930s the Royal Society increased its attention to historical matters. In 1932 it inaugurated its *Obituary Notices of Fellows*, to extend the notices published annually in the *Proceedings*; some obituaries (such as WHITTAKER's) are of high historical value. Five years later the *Notes and Records* was started as an historical journal, largely but not exclusively devoted to scientific (including mathematical) developments in which the Society had played a role. In 1940 the Society also published the fourth and most important edition of its book-length *Record* of its history, publications, and membership, to which a supplementary volume for the years 1940 to 1989 was issued in 1992. None of these sources is explicitly devoted to the history of mathematics, but all of them are of value for it.

A significant event of another kind occurred in 1934, when a privately financed research institute in Renaissance studies in Hamburg, the *Kulturwissenschaftliche Bibliothek Warburg*, escaped the Nazi attentions upon its personnel by transferring to London as the Warburg Institute. In 1944 it was incorporated into the University of London. It has always been an important center for the history of medieval and Renaissance thought, mathematics well included, with a remarkable library.

Various other quite worthy authors could be mentioned; for example, D. K. WILSON on *The History of Mathematical Teaching in Scotland to the End of the 18th Century* (1935), or FLORENCE A. YELDHAM (* 1877), whose study of *The Teaching of Arithmetic through Four Hundred Years* (1936) is limited to Europe. But no general increase in historical appreciation was evident, not even among the teaching community, by books such as these, and despite the interest of figures such as WILLIAM J. GREENSTREET (1861–1930), the editor of the NEWTON volume in 1927. One contribution was of greatest use to classicists: HEATH's follower IVOR THOMAS (1905–1993) produced *Selections Illustrating the History of Greek Mathematics* (1939) for the Loeb series, with the Greek original set opposite his translation; in his two volumes he followed almost exactly the design of HEATH's *History*.

7.11 After the Second World War

The British Society for the History of Science was founded soon after World War II, and within its modest scale of activities the history of mathematics had some place; but it did not rise within its realm to the extent that other sciences such as chemistry and biology were to flourish. Posts in the field were not given to those with a speciality in mathematics. Activity was confined, for example, to teaching at University College London by J. F. SCOTT (1892–1971), who produced non-durable studies of WALLIS (1938) and DESCARTES (1952).

Growing out of a seminar on the transition from school to university mathematics, a seminar on the history and teaching of mathematics at Imperial College London was founded by G. J. WHITROW (1912–2000) and CECILY TANNER (1900–1992) in 1958. The seminar continued until the latter's retirement in 1969. TANNER, the daughter of the YOUNGs mentioned in Section 8, had been inspired by the example of seminars run at Göttingen by her parents' mentor FELIX KLEIN, and by his follower, the historian and educator WILHELM LOREY (1873–1955, **B**). Among her other activities, TANNER financed the founding of the Thomas Harriot Seminar, which began formally in 1967, in his birthplace Oxford. The Seminar has become a forum for discussion of mathematics and science in Elizabethan Britain, at the beginning under the nominal directorship of ALISTAIR CROMBIE (1915–1996). CROMBIE, an Australian-born scholar, originally trained as a chemist, later made notable contributions to the history of early modern European science in general. However, the dream of an edition of HARRIOT's manuscripts, which goes back to RIGAUD in the 1830s, remains unfulfilled.

In the history of applied mathematics a substantial body of work was produced by EVA G. R. TAYLOR (1879–1966). In *The Haven-Finding Art* (1956) she traced the development of navigation from early times to that of Captain COOK, including some details of mathematical methods. The word "mathematicians" applies to her figures in the full historical sense of lower-than-"geometers" which was to obtain for so long. She advanced awareness of this neglected side of the history of mathematics with a survey of *The Mathematical Practitioners of Tudor and Stuart England* (1954), covering the period 1420–1715. To this she added in 1966 a successor study, *The Mathematical Practitioners of Hanoverian England (1714–1840)*. In each case her narrative was followed by capsule biographies of many figures, most of whom were little- or un-known before. While her data suffer on details to a notable extent, she opened up a largely forgotten area of the history of science, which has been admirably deepened since by RUTH and the late PETER WALLIS (1918–1992, **B**).

During the early 1960s the history and philosophy of science grew rapidly. (The British Society for the Philosophy of Science had formed in 1950, out of a philosophy group within the British Society for the History of Science). But mathematics was given very little room within this development. For example, the eight-volume edition of *The Mathematical Papers of Isaac Newton* [NEWTON 1967/81] was very much the personal effort of D. T. WHITESIDE (*1932), with the help of the German immigrant ADOLF PRAG (*1906) (see Section 5.4.6). The full edition of NEWTON's correspondence [NEWTON 1959/77], which began to appear in the late 1950s under the editorship of H. W. TURNBULL, was of broader import and so gained more interest.

7.12 Conclusion

As in most other countries, the history of mathematics began to (re)develop early in the 1970s. Most recruits were mathematicians and teachers, driven by deep dissatisfaction with mathematics teaching deprived of historical or cultural context. The British Society for the History of Mathematics was launched in 1971 by JOHN DUBBEY (*1934), with WHITROW as founding President, and it has continued to this day [BSHM 1992]. Since the mid 1980s its membership has increased considerably and become more international. The connection with education has become a speciality, with annual meetings drawing relatively large audiences.

Another innovation of the mid 1970s was a course on the history of mathematics, launched at the Open University around special series of television programs. In recent years radio broadcasts or cassettes have been introduced, and recruitment to the course across the country is quite steady. Outside these two centers, however, activity is often confined to individuals. To this extent the history is continuous. In the past many historians, such as COLEBROOKE, HEATH, and MUIR, were professionally employed elsewhere. Even those with academic positions either functioned principally in other areas (THOMSON, ALLMAN, GOW, DE MORGAN, WHEWELL, WHITTAKER), or could not generate a general impact (TODHUNTER, ROUSE BALL, HEATH). Some authors were writing for a wider audience, such as historians of science in general (THOMSON, WHEWELL, MERZ, WOLF), or neighboring ones such as classicists (BURNET, FARRINGTON, THOMAS), or astronomers (GRANT, BERRY).

The most favored topics for detailed work were, unsurprisingly, Greek mathematics and NEWTON. The best products were general encyclopedia articles, unequalled in any other language, in many cases of exceptional quality, that have appeared in a variety of reference works. Among these, the most recent is the *Companion Encyclopedia of the History and Philosophy of the Mathematical Sciences* [GRATTAN-GUINNESS 1994]; the British presence among the authors was substantial.

Chapter 8

Russia and the U.S.S.R.

SERGEI S. DEMIDOV

8.1 Introduction

The history of mathematics was not seriously cultivated in Russia until the 1880s, although important contributions were made by earlier Russian authors. Among these was a Russian translation of the first part (which ends at the close of the 17th century) of MONTUCLA's *Histoire des Mathematiques* [MONTUCLA 1758] (see Section 1.3.3) published in 1789–1791 in *Akademicheskie Izvestiya* (Academic Proceedings). Historical notes also appeared in the mathematical works (1812) of PËTR A. RAKHMANINOV (†1813), in the works (1815) of SEMËN E. GUR'EV (1766–1813), and in the *Leksikon Chistoĭ i Prikladnoĭi Matematiki* (Lexicon of Pure and Applied Mathematics) (1839) of VIKTOR YA. BUNYAKOVSKIĬ (1804–1889). Old Russian sources were also published in connection with histories of the old Russian culture, for example, EVGENIĬ BOLKHOVITINOV's 1828 publication of the chronology of KIRIK FROM NOVGOROD (1136) [DEMIDOV 1992]. In fact, it is possible to distinguish two periods in the development of the history of mathematics in Russia corresponding to the two periods of Russian civil history, before and after the Bolshevik Revolution of 1917. The second period is marked above all by works of the "Soviet school in the history of mathematics" which flourished in the 1960s and 1970s.

8.2 History of Mathematics Before 1917

In 1880 MIKHAIL E. VASHCHENKO-ZAKHARCHENKO (1825–1912, **B**), professor of mathematics at Kiev University and an active advocate of teaching geometry in the *Gymnasium* according to EUCLID, translated EUCLID's *Elements* into Russian with historical commentary. Three years later he published the first volume of a *History of Mathematics* which was devoted mainly to geometry, from antiquity to the Renaissance. Despite the fact that VASHCHENKO-ZACHARCHENKO's translation of EUCLID was free and sometimes inaccurate, and that his *History of Mathematics* was little more than a compilation of the works of western European authors, especially MORITZ CANTOR (see Section 5.4.2), both works were of considerable importance. They marked the beginning of the history of mathematics in Russia as a discipline, and not simply as a field of sporadic research. In Russia (and later in the U.S.S.R.), the history of mathematics was developed mainly by mathematicians and so was regarded as a mathematical discipline. This perception has persisted to the present, although recently there has been a tendency to regard the field as a part of the history of science.

At the end of the 19th, and well into the 20th century the most important scholar in the history of mathematics in Russia was VIKTOR V. BOBYNIN (1849–1919, **B**), who taught at Moscow University. Beginning in 1888, his papers appeared regularly in GUSTAF ENESTRÖM's journal, *Bibliotheca Mathematica* (see Section 6.5), and BOBYNIN was invited to contribute the chapter on 18th-century elementary geometry for the fourth volume of MORITZ CANTOR's *Vorlesungen über Geschichte der Mathematik* (1908). BOBYNIN was also the first to teach the history of mathematics in Russia. Beginning in 1882 (and each year thereafter until 1919), he gave his lectures in courses varying in length from detailed surveys designed to last four years to elementary one-year courses ranging from antiquity to the modern period.

One of BOBYNIN's greatest contributions to the professionalization of the history of science in Russia was his founding and editing of the first journal devoted to the history of exact sciences, *Fisiko-Matematicheskie Nauki v ikh Nastoyashchem i Proshedshem* (Physico-Mathematical Sciences, Present and Past) (1885–1898). In 1898, he transformed the journal (which existed until 1904) into *Fisiko-Matematicheskie Nauki v Khode ikh Razvitiya* (Physico-Mathematical Sciences in the Course of Their Development). He also published a three-volume *Russkaya Fiziko-Matematicheskaya Bibliografiya* (Russian Physico-Mathematical Bibliography) (1886–1908). In all of his works BOBYNIN combined a truly historical attitude towards sources with a philosophical approach to the developmental process of mathematical ideas. Moreover, as the editor of a scientific journal, and through his many contributions to different scientific meetings and conferences, he greatly helped to generate interest in the history of mathematics.

At Kiev University, historian NIKOLAĬ M. BUBNOV (1858–1943, **B**) devoted his research to medieval European mathematics. In 1899 he published an edition of *Gerbert's Opera Mathematica* (in Berlin). He followed this, in 1908, with *The Arithmetical Autonomy of European Culture*, in which he presented his original idea concerning the origin of Arabic numerals, namely, that they had been independently invented by Europeans.

If BOBYNIN and BUBNOV approached questions of the history of mathematics primarily as historians, IVAN YU. TIMCHENKO (1863–1939, **B**) of Odessa University, arguably the second most important historian of mathematics after BOBYNIN in pre-revolutionary Russia, wrote primarily from the perspective of a mathematician. He was, however, well versed in basic historical and sound philological scholarship. TIMCHENKO authored, for example, a very important, mathematically oriented treatise on the history of analytic functions up to the publication in 1825 of CAUCHY's *Mémoire sur les intégrales définies prises entre des limites imaginaires* (Memoir on Definite Integrals Taken between Imaginary Limits) [TIMCHENKO 1899]. In this work, TIMCHENKO emphasized the diverse historical connections between analytic function theory and different branches of mathematical analysis. TIMCHENKO also added excellent historical explanations to the second edition of the Russian translation of FLORIAN CAJORI's book, *History of Elementary Mathematics with Hints on Methods of Teaching* [TIMCHENKO 1917].

From the 1860s through 1900, Odessa experienced active economic development and served as a very important cultural center. Its citizens made sgnificant contributions to literature, art, and science, as well as (like TIMCHENKO) to the expansion of the history of mathematics in pre-revolutionary Russia. VENIAMIN F. KAGAN (1869–1953, **B**), also primarily a mathematician but one deeply interested in historical problems, published, for example, the influential study, *Foundations of Geometry. Part II. Historical Essay on the Development of the Theory of the Foundations of Geometry.* This detailed account takes the history of this subject up to the beginning of the 20th century [KAGAN 1907]. KAGAN is also well known for his studies of NIKOLAĬ I. LOBACHEVSKIĬ (1792–1856), most of which were published after the October Revolution of 1917.

Another prominent mathematician and historian of mathematics, ALEKSANDR V. VASIL'EV (1853–1929, **B**) of Kazan' University, was the first to focus serious attention on LOBACHEVSKIĬ's results in algebra and analysis. Under VASIL'EV's direction, the first edition of the geometrical works of LOBACHEVSKIĬ was published in two volumes in 1883–1886. VASIL'EV also published the first scientific biography of LOBACHEVSKIĬ in 1914. All of VASIL'EV's works are characterized by a profound interest in the mathematical roots of technical questions and reflect VASIL'EV's sensitivity to historical and mathematical as well as philosophical questions.

Other achievements related to the history of mathematics during this early period include editions of the works of important Russian and foreign mathematicians, such as PAFNUTIĬ L. CHEBYSHEV (1883–1886), JAKOB BERNOULLI (1913), and ISAAC NEWTON (notably ALEKSEĬ N. KRYLOV's translation of the *Principia*) [KRYLOV 1916]. V. P. SHEREMETIEVSKIĬ's essay on the history of mathematics in the Russian edition (1898) of HENDRIK A. LORENTZ's book, *Lehrbuch der Differential- und Integralrechnung* (Textbook of Differential and Integral Calculus), also deserves to be mentioned.

8.3 The First Post-Revolutionary Years: Formation of the Soviet School

The Bolshevik Revolution of 1917 forced radical changes throughout Russia, which affected the development of science as well as mathematics and the history of mathematics. In particular, the Revolution and ensuing civil war had serious consequences for the lives and work of scientists of the pre-revolutionary generation. BOBYNIN, for example, died in 1919. Although his death was not directly connected with political events, he and others might have lived longer had the general standard of living been better. TIMCHENKO lived until 1939, but published nothing more of importance. The only person who was active during the early years following the Revolution was VASIL'EV. In 1919 he published a substantial historical monograph on *The Integer* [VASIL'EV 1919] and, in 1921, the first systematic survey of the history of mathematics in 18th- and 19th-century Russia. As VASIL'EV noted, however, his research for these works began before the Revolution, and the difficulties of the post-revolutionary years did not facilitate their completion.

If the Revolution adversely affected research in the history of mathematics, it nevertheless encouraged historical analysis of both the past and the present and so sparked new historical studies. By the mid-1920s, when life began to return to normal, a renaissance of scientific and pedagogical activity — including activity in the history of mathematics — was well underway, owing to the fact that communist ideology placed a special emphasis on the sciences. In addition to VASIL'EV, the distinguished mathematician of the Petersburg school, VLADIMIR A. STEKLOV (1864–1926), published *Mathematics and its Significance for Humanity* in 1923, and authors like VSEVOLOD K. BELLYUSTIN (1865–1925), STEPAN A. BOGOMOLOV (1877–1965), GEORGIĬ N. POPOV (1878–1930), and MIKHAIL N. MARCHEVSKIĬ (1884–1974) also produced popularized historical accounts. By the end of the 1920s, DMITRIĬ D. MORDUKHAĬ-BOLTOVSKOĬ (1876–1952, **B**), a mathematician who moved in 1914 because of the War (together with the University) from Warsaw to Rostov-on-Don, had also begun to work on the history of mathematics. Among his publications, *Research Concerning the Origins of some Principal Ideas of Modern Mathematics*, published in 1928 in *Izvestiya Severo-Kavkazskogo Universiteta* (Proceedings of the North-Caucasian (Rostov-on-Don) University), was an important collection of essays on the origins of mathematical concepts like number, limit, etc.

In the 1930s, after the Bolsheviks had abandoned their ideal of an international communistic revolution and had begun in earnest to build a communist state at home (in a single country only), overall scientific activity increased even more. The doctrine of Marxism-Leninism — upon which Soviet orthodoxy was based — set itself the task of creating a scientific theory for the development of society. Naturally, this meant that official Marxist ideology (adapted to the momentary needs of the policies of the party) came to dominate the social sciences, to the detriment of social sciences themselves, while the physical sciences and mathematics benefited and developed under more favorable circumstances. Although the authorities tried to exert a certain amount of ideological control over the latter two areas, these efforts never proved as suffocating or harmful for them as they did for the life sciences (in the case of Soviet biology) and humanities. Despite various attempts, the mathematical sciences were not readily susceptible to ideological pressures. Moreover, the physical and mathematical sciences had the advantage that the technological and military power of the state depended directly upon them.

Throughout the Soviet period, then, the mathematical sciences were encouraged and serious studies in the history of mathematics were pursued with increased support in Moscow, in Leningrad, and in other scientific centers as well. In addition to Rostov-on-Don, Kazan' was a significant center, owing to the presence there of NURI YUSUPOV (1888–1937?), who published his *Essays on the History of Development of Arithmetic in the Middle East* (1932). This was the first book in Russian to contain a detailed exposition of Arabic mathematics. Also in Kazan', NIKOLAĬ G. CHEBOTARËV (1894–1947) studied the works of NEWTON and GALOIS. At Kiev University, MIKHAIL F. KRAVCHUK (1892–1942) undertook a study in 1935 of the development of mathematics at Kiev University during the preceding hundred years, and DMITRIĬ A. GRAVE (1863–1939) published a book on the history of algebra. Meanwhile, in Odessa, NIKOLAĬ A. CHAĬKOVSKIĬ (1887–1970) compiled a mathematical bibliography for the Ukraine covering the period 1834–1929, and in Kharkov in 1940 ANTON K. SUSHKEVICH (1889–1961) gave lectures on the history of mathematics.

The main achievements of Leningrad historians were connected with mathematics in antiquity — in 1930 BORIS A. TURAEV (1868–1920) and VASILIĬ V. STRUVE (1889–1965, **B**) published their decipherment of the Moscow papyrus in Berlin, and in 1937 SOLOMON YA. LUR'E published, also in Berlin, his *Die Infinitesimaltheorie der antiken Atomisten* (The Infinitesimal Theory of Atomists in Antiquity).

The 1930s witnessed early steps towards the professionalization of the history of science and mathematics. In 1933, the Institute for the History of Science and Technology was founded in Leningrad and under its auspices, ten volumes of the *Arkhiv Istorii Nauki i Tekhniki* (Archive for the History of Science and Technology) appeared, which included papers on the history of mathematics. The Institute also published collections devoted to LEONHARD EULER (1935) and JOSEPH LOUIS LAGRANGE (1937). NIKOLAĬ I. BUKHARIN (1888–1938), director of the Institute and a famous Soviet statesman, fell prey to STALIN's terror, however, and the Institute closed immediately following BUKHARIN's arrest in 1938.

The tendency toward centralization characteristic of Russian social and state life in the Soviet period also manifested itself very powerfully relative to science in the 1930s. In 1919, the state capital moved from Petrograd to Moscow, and in 1934, the Academy of Sciences of the U.S.S.R. moved to Moscow as well. Not surprisingly, Moscow then became the center of activity in the history of mathematics. The so-called "Soviet school of the history of mathematics" was created there in the 1930s as a result of the efforts of SOF'YA A. YANOVSKAYA (1896–1966, **B**), MARK YA. VYGODSKIĬ (1898–1965, **B**), and ADOLF P. YUSHKEVICH (1906–1993, **B**).

Ideology played an important role in the foundation of this school. Since dialectical methods obliged anyone studying a given subject, including mathematics (which was an important part of the education of students in many specializations), to explore its history, the history of mathematics as a discipline had an ideological justification. However, the history of science generally also assumed special significance within the framework of Marxist doctrine. As LENIN in his "Filosofskie tetradi" (Philosophical Notebooks, published in 1933) explained: "The continuation of the work of Hegel and Marx will include the dialectical analysis of the history of human thought, of science and technology" [LENIN 1963, 314]. The history of science was thus inherently an ideological discipline, and although this might have been advantageous relative to political support, it was dangerous in that work on the history of science was continuously subject to the control of party ideology. From its very beginning, however, the Soviet school of the history of mathematics allied itself with the mathematical community and considered itself an integral part of that field. As a result, it developed the history of mathematics from a highly technical point of view for an audience of mathematicians. By thus aligning itself with the relatively ideology-free discipline of mathematics, the history of mathematics — as opposed to the history of science — successfully escaped the extreme and often devastating scrutiny of the ideologues. This was primarily achieved through the efforts of YANOVSKAYA, VYGODSKIĬ, and YUSHKEVICH.

YANOVSKAYA had become a professor at Moscow University in 1925, with VYGODSKIĬ joining her first as a post-graduate student in 1926 and then as a professor in 1931. When Bolshevik scientists there began a campaign against the older faculty in hopes of transforming the curriculum to conform to the new spirit of Marxism, YANOVSKAYA organized a seminar on the history and philosophy of mathematics. Many enthusiastic students and young professors attended, including IGOR' V. ARNOLD (1900–1948), PËTR K. RASHEVSKIĬ (1907–1983), and YUSHKEVICH.

The seminar focused on a Marxist analysis of problems concerning the foundations of mathematics, including discussions of formalism, intuitionism, and logicism, all of which were especially fashionable at this time. While consonant with the overall political and ideological climate, this had a highly negative side in the hands of YANOVSKAYA and VYGODSKIĬ. Together with ERNST KOLMAN (1892–1979) — one of the most reprehensible representatives of Soviet science in this period — they organized an attack on so-called "reactionary" professors, whose ideology they regarded as alien to Marxism [DEMIDOV 1993a], [DEMIDOV/FORD 1996]. This led to the persecution of many intellectuals, including some distinguished mathematicians like DMITRIĬ F. EGOROV (1869–1931), who was arrested in 1931. It cannot be denied, however, that YANOVSKAYA and VYGODSKIĬ were also prominent scientists who favored the expansion of science in the U.S.S.R. In particular, YANOVSKAYA exerted her influence within the party and in philosophical circles to protect the history of mathematics in the U.S.S.R., just when it was entering a period of active development.

YANOVSKAYA and VYGODSKIĬ were committed to promoting a Marxist interpretation of mathematics. Their interests — and philosophy — are reflected in their publications. For example, YANOVSKAYA wrote "The Category of Quality in Hegel's Works and the Essence of Mathematics" (1931), while VYGODSKIĬ published an article on "The Problems of History Viewed from a Marxist Methodology" (1930). Their own ideological positions aside, YANOVSKAYA and VYGODSKIĬ, both professional mathematicians, had excellent historical intuition. In their historico-mathematical studies they proceeded from the mathematical nature of their subject, which they understood well, and from the historical context, which they investigated in detail. Above all, they were careful never to make the historical facts fit preconceived philosophical schemes (in contradistinction to the majority of their contemporary militant ideologues in the Soviet Union).

Whereas YANOVSKAYA and VYGODSKIĬ exercised their influence ideologically, YUSHKEVICH was primarily an historian of mathematics. He was, moreover, a universal historian with interests ranging from calculating methods in primitive cultures to the ingenious mathematical theories introduced at the beginning of the 20th century. Along with VYGODSKIĬ and YANOVSKAYA, YUSHKEVICH served to set the agenda — and the approach — that the Soviet school of the history of mathematics would follow in the early stages of its existence.

In the 1930s, YANOVSKAYA and VYGODSKIĬ both taught courses on the history of mathematics, reviving the series that BOBYNIN had offered up until 1919. In 1933 they organized a new seminar on the history of mathematics that exists to this day. YUSHKEVICH was soon invited both to participate and to assume a leadership role. As a result of the efforts of this seminar, an entire generation of historians of mathematics became well known after World War II. Among those who attended were IVAN N. VESELOVSKIĬ (1892–1977, **B**) (mathematics in antiquity), GAREGIN B. PETROSYAN (1902–1998) (mathematics in medieval Armenia), KONSTANTIN A. RYBNIKOV (*1913) (early history of calculus of variations), and ESFIR' YA. BAKHMUTSKAYA (1916–1972) (infinitesimal methods in the 17th century).

Interest in the foundations of mathematics grew considerably after 1925 when the Marx-Engels Institute in Moscow received photocopies of the mathematical manuscripts of KARL MARX (1818–1883). In these manuscripts, MARX discussed the foundations of analysis, and YANOVSKAYA headed the group responsible for publishing these works. A great part of the manuscripts was published in 1933 in the book *Marksizm i Estestvoznanie* (Marxism and Natural Sciences). A complete version was published only in 1968 [YANOVSKAYA 1968], and made the history of mathematical analysis (and especially its foundations) a respectable subject of scientific research, as well as an essential subject from an ideological point of view. It was in this political climate that YUSHKEVICH began his studies of the history of mathematical analysis and its foundations, which became a primary focus of research undertaken by Soviet historians of mathematics (see below).

YUSHKEVICH's first papers were devoted to problems concerning the foundations of mathematical analysis, among them "Filosofiya Matematiki L. Karno" (The Philosophy of Mathematics of L. Carnot) (1929) and "Angliĭskaya Filosofiya Empirizma i Teoriya Flyuksiĭi" (English Empirical Philosophy and the Theory of Fluxions) (1934). Historical studies devoted to aspects of the foundations of analysis were also published by NIKOLAĬ N. LUZIN (1883–1950) (on LEONHARD EULER (1933) and on the function concept (1935)), and by VALERIĬ I. GLIVENKO (1897–1940) on the differential concept in MARX's mathematical manuscripts (1934). In this same period, ALEKSANDR O. GELFOND (1906–1968) wrote a paper on the history of transcendental numbers (1930), and at the end of the 1930s RYBNIKOV began his studies on the early history of the calculus of variations.

In addition to these original publications, many translations of classic texts began to appear in Russian, including works by LAZARE CARNOT, CAVALIERI, DESCARTES, EULER, DE L'HÔPITAL, KEPLER, MONGE, and NEWTON. All of these translations were accompanied by substantial commentaries that represented important contributions to the history of mathematical analysis. Editions of the works of EVARISTE GALOIS, PETER G. LEJEUNE-DIRICHLET, JOHANN B. LISTING, and EGOR I. ZOLOTARËV (1847–1878) rounded out the classic mathematics texts published in the U.S.S.R. in the 1930s.

Another principal topic of interest to the Soviet school was mathematics in Russia and in the U.S.S.R. YUSHKEVICH methodically organized a series of studies on the history of mathematics in Russia in the 18th century; KAGAN produced works — on LOBACHEVSKIĬ (1934), ALEKSANDR YA. KHINCHIN (1937), and SERGEĬ N. BERNSHTEĬN (1940) — related to the history of probability in Russia; VYGODSKIĬ published a major study of Babylonian mathematics in *Uspekhi Matematicheskikh Nauk* (Progress of the Mathematical Sciences) in 1940–41, as well as his principal work, *Arithmetic and Algebra in Antiquity* [VYGODSKIĬ 1941]; and a variety of other studies by VYACHESLAV V. STEPANOV (1899–1950), PAVEL S. ALEKSANDROV (1896–1982), and BORIS V. GNEDENKO (1912–1995, **B**), to name but a few of the most prominent, came out in the late 1930s and into the 1940s. The principal unifying element of all of this work was the seminar on the history of mathematics at Moscow University.

One final work of note from the pre-World War II period was written by ANDREĬ N. KOLMOGOROV (1903–1987, **B**), one of Russia's leading mathematicians. His article on mathematics for the first edition of the *Bol'shaya Sovetskaya Entsiklopediya* (Great Soviet Encyclopedia) (1936) contained a well-known periodization of the history of mathematics with very concise descriptions of each period. KOLMOGOROV divided the history of mathematics into four periods: the birth of mathematics (6th to 5th century B.C.); the period of elementary mathematics (up to the 16th century); the establishment of the mathematics of variables (to the middle of the 19th century); and the period of modern mathematics. This periodization has been widely discussed by historians of mathematics (see for example [YOUSHKEVICH 1994]) and became universally adopted in Soviet historiography. During all of his creative life KOLMOGOROV was actively interested in history of mathematics. As a result, he wrote several remarkable works (for example on NEWTON's researches [KOLMOGOROV 1946]). In 1978–1987 together with YUSHKEVICH he edited three volumes on the history of mathematics of the 19th century [KOLMOGOROV/YOUSHKEVICH 1978/87]. He also devoted considerable energy to improving mathematical education in secondary schools, and was especially active in the reform of mathematics education [PETROVA 1996].

8.4 Research Schools After World War II

Like the October Revolution, World War II deeply affected life in the U.S.S.R. It substantially curtailed the publication of scientific research, although it did not come to a complete stop. After the War, mathematics in the U.S.S.R. rebounded dramatically. The geographical distribution of scientists and research institutions grew significantly. Relocation of numerous scientific institutions and establishments during the War from war-ravaged territories of the U.S.S.R. resulted in an increased scientific potential of many regions, especially in the East.

Moscow, however, remained the principal focus of scientific research in the U.S.S.R. In 1944, when studies at Moscow University resumed, the Seminar on the History of Mathematics reconvened and quickly reestablished itself as the country's center for such study. As before, YANOVSKAYA, VYGODSKIĬ, and YUSHKEVICH led these efforts. (They had all returned to Moscow — YANOVSKAYA from Perm', VYGODSKIĬ from Kazakhstan, and YUSHKEVICH from Izhevsk.) Somewhat later, they were joined by RYBNIKOV and IZABELLA G. BASHMAKOVA (*1921). Not only were all of the most important results obtained by Soviet historians of mathematics first presented at this seminar, but the early volumes of *Istoriko-Matematicheskie Issledovaniya (IMI)* (Research on the History of Mathematics), founded in 1948 by GEORGIĬ F. RYBKIN (1903–1972) and YUSHKEVICH, were also published as its transactions. YUSHKEVICH oversaw this journal, which is the most influential Russian serial in the history of mathematics. At the time of his death in 1993, 35 volumes had been published [RYBKIN/YOUSHKEVICH 1948/66], [YOUSHKEVICH 1973/94]. (The second series of *IMI* began in 1995.)

Actually, in post-war Moscow, two groups of historians of mathematics emerged — one at the University and a second at the Institute for the History of Science and Technology (founded in 1946) of the Soviet Academy of Sciences. The university group (YANOVSKAYA, RYBNIKOV, BASHMAKOVA, including later arrivals SVETLANA S. PETROVA (*1933) and ALLA V. DOROFEEVA (*1935)), delivered lectures on the history of mathematics within the faculty of mathematics and mechanics. In 1956, a Department for the History of Mathematics and Mechanics was created, headed by RYBNIKOV. At Moscow University, a number of mathematicians also conducted historical research, including KAGAN, VLADIMIR V. GOLUBEV (1884–1954), BORIS N. DELONE (1890–1980), PAVEL S. ALEKSANDROV (1896–1982), LAZAR' A. LYUSTERNIK (1899–1981, **B**), KOLMOGOROV, GELFOND, ALEKSEĬ I. MARKUSHEVICH (1908–1979), GNEDENKO, IGOR' R. SHAFARE-

VICH (*1923), ALEKSANDR D. SOLOV'ËV (*1927), VLADIMIR M. TIKHOMIROV (*1934), VLADIMIR I. ARNOLD (*1937), YURIĬ I. MANIN (*1937), and ALEKSEĬ N. PARSHIN (*1942).

Although this group at Moscow University covered many areas of the history of mathematics in its work, three principal fields serve to characterize its main interests. A number of YANOVSKAYA's students specialized in the history of mathematical logic and foundations of arithmetic, among them NIKOLAĬ I. STYAZHKIN (1932–1986), who authored *The Formation of the Ideas of Mathematical Logic* [STYAZHKIN 1967], and ZINAIDA A. KUZICHEVA (*1933). In analysis, RYBNIKOV and his students — like DOROFEEVA — pursued the history of the calculus of variations [RYBNIKOV 1949], [DOROFEEVA 1961]. RYBNIKOV's group went on to study the pre-history and early development of functional analysis. PETROVA, for example, worked on DIRICHLET's principle [PETROVA 1965], while DOROFEEVA wrote on the history of the theory of integral equations. Finally, in the history of algebra, BASHMAKOVA's school concentrated on algebra, algebraic geometry and number theory. Although BASHMAKOVA's primary interest was initially the history of algebraic numbers and the theory of algebraic equations, in the mid-1960s her focus shifted to the history of Diophantine analysis. In 1972, she published her groundbreaking study, *Diophantus and Diophantine Equations* [BASHMAKOVA 1972]. In this book and in the book *History of Diophantine Analysis from Diophantus to Fermat*, written with EVGENIĬ I. SLAVUTIN (*1948) [BASHMAKOVA/SLAVUTIN 1984], she analyzed the history of Diophantine analysis from DIOPHANTUS to POINCARÉ and began to reconsider the role of Diophantine analysis in the history of algebra. She later argued that Diophantine analysis was a decisive factor in the development of algebra up to the end of the 18th century.

Unlike its counterpart at Moscow University, the group at the Academy's Institute of the History of Science and Technology — VASILIĬ P. ZUBOV (1899–1963, **B**) (medieval European mathematics), YUSHKEVICH, IOSIF B. POGREBYSSKIĬ (1906–1972, **B**) (mathematical analysis and analytical mechanics from the 18th to the 20th centuries), BORIS A. ROZENFELD (*1917) (medieval Arabic mathematics and geometry), LEONID E. MAĬSTROV (1920–1982, **B**) (probability theory and computer science), FËDOR A. MEDVEDEV (1923–1993) (set theory, theory of functions, foundation of analysis), AL'GIRDAS B. PAPLAUSKAS (1931–1984, **B**) (theory of series) — had no teaching duties and so was able to engage exclusively in research. Like the university group, however, the academy group had several main focal points of research. The principal personality inspiring and encouraging this work was YUSHKEVICH. ZUBOV and YUSHKEVICH, for example, worked on the history of mathematics and the exact sciences in medieval Europe. The results of YUSHKEVICH's many years of research appeared in his monumental book, *History of Mathematics in the Middle Ages* [YOUSHKEVICH 1961]. In this work, he presented views on medieval mathematics that he had begun to formulate as early as the 1950s. He was convinced that medieval mathematics formed a coherent whole, the common features of which lay in the medieval mathematical cultures of Europe and Asia. His book was translated into a number of languages, and each subsequent edition was updated in light of the most recent scholarly research in the field. YUSHKEVICH also inaugurated the study of medieval Arabic mathematics in the U.S.S.R. In this case, his research was officially supported primarily because it was consistent with the government's interest in illuminating the roots of the cultural legacies of the peoples of Central Asia and Transcaucasia. BORIS A. ROZENFELD, with encouragement from YUSHKEVICH, worked on Arabic medieval mathematics, and together they organized a substantial and very active school [ROZENFELD/YOUSHKEVICH 1965].

The most popular subject studied by Soviet historians of mathematics, however, was mathematical analysis. As already noted, prior to World War II considerable attention was devoted to foundations, which remained one of the favorite topics pursued after the War by YUSHKEVICH and his students. YUSHKEVICH concentrated primarily on aspects of the foundations of analysis in works by mathematicians of the 18th and 19th centuries, and on the history of the principal concepts of analysis, including functions (see his well-known article about the development of the notion of function [YOUSHKEVICH 1971]) and the CAUCHY integral. His pupil, FËDOR A. MEDVEDEV, published *The Development of the Theory of Sets in the Nineteenth Century* [MEDVEDEV 1965], after which he began to study the history of the concept of the integral and the history of real functions [MEDVEDEV 1974], [MEDVEDEV 1975]. Another of YUSHKEVICH's pupils, PAPLAUSKAS, studied the history of the theory of series and wrote a book, *Trigonometric Series from Euler to Lebesgue* [PAPLAUSKAS 1966].

Another, and perhaps the most popular area of the history of mathematical analysis, was ordinary and partial differential equations, along with difference equations. Among the most important works published on this topic were YUSHKEVICH's own essay on the development of the theory of differential equations (1953); studies by NIKOLAĬ I. SIMONOV (1910–1979, **B**) on differential equations in the works of EULER [SIMONOV 1957]; a monograph by NIKOLAĬ D. MOISEEV (1902–1955) on the development of the theory of stability (1949); a book by VYACHESLAV A. DOBROVOL'SKIĬ (*1919) on the history of the analytical theory of differential equations; studies on boundary problems in mathematical physics by FELIKS I. FRANKL' (1905–1961), MOISEĬ G. SHRAER (1918–1983), and VARVARA I. ANTROPOVA (1924–1991); and a work by VLADIMIR S. SOLOGUB (1927–1982, **B**) on the history of elliptic partial differential equations.

History of the exact sciences has become an important focus of research, especially in studies by IOSIF B. POGREBYSSKIĬ on the development of analytical mechanics [POGREBYSSKIĬ 1966], ALEKSEĬ N. BOGOLYUBOV (*1911) on the history of mathematical physics, and VLADIMIR P. VIZGIN (*1936) on the role of mathematics in the development of physics in the 19th and 20th centuries.

Next to Moscow, Leningrad — where the Soviet Academy's Institute for the History of Science and Technology maintained a branch — was the second most important center for the history of mathematics in the U.S.S.R. This community was led by the well-known mathematician, VLADIMIR I. SMIRNOV (1887–1974, **B**), and included ALEKSANDR D. ALEKSANDROV (1912–1999) (history of geometry), IVAN YA. DEPMAN (1885–1970, **B**) (mathematics in Russia and history of arithmetic), EFIM M. POLISHCHUK (1914–1988) (mathematics in the 19th and the first half of the 20th century), ANDREĬ A. KISELËV (1916–1994) (history of number theory, of operational calculus, and of constructive function theory), IL'YA G. MEL'NIKOV (1916–1979) (history of number theory), and ELENA P. OZHIGOVA (1923–1994, **B**) (history of number theory and mathematics in Russia).

Kiev, with its special seminar in a Department for the History of Mathematics in the Institute of Mathematics, ranked third in importance. This department, headed by IOSIF Z. SHTOKALO (1897–1987), was later transformed into the Department of the History of Science and Technology, a division of the Institute for the History of the Ukrainian Academy of Sciences, and its seminar became the Republican Seminar on the History of Mathematics. The principal research areas pursued here were the history of mathematics in Russia and the history of mathematical analysis in the 18th and 19th centuries (primarily works of BOGOLYUBOV), and also of NAUM I. AKHIEZER (1901–1980) (history of the theory of approximations), YURIĬ M. GAĬDUK (1914-1993) (mathematics in Russia), VYACHESLAV A. DOBROVOL'SKIĬ (*1919) (history of the theory of differential equations and of mathematics in Russia), YURIĬ A. BELYĬ (*1925) (mathematics in Russia and mathematics of the Renaissance), SERGEĬ N. KIRO (1926–1990) (mathematics in Russia), and VLADIMIR S. SOLOGUB (1927–1982) (history of the theory of partial differential equations).

A fourth center was Tashkent, where research focused on medieval Arabic mathematics. The works of TASHMUKHAMED N. KARY-NIYAZOV (1896–1970, **B**) on ULUG BEG's astronomical school (15th century) marked the beginning of such research, which was later continued under the leadership of GALINA P. MATVIEVSKAYA (*1930), originally from Leningrad, and SAGDY KH. SIRAZHDINOV (1921–1988) [MATVIEVSKAYA 1962], [MATVIEVSKAYA 1967].

Systematic study of the history of mathematics also developed in geographically dispersed regions throughout the U.S.S.R. — from Lithuania, Latvia, and Estonia to Kazakhstan and Tadzhikistan — with the areas of historical research pursued centering on medieval Arabic mathematics and developments in all areas of Western mathematics from the 17th through the 20th centuries, as well as on individualized studies of mathematics in the various Soviet states.

L. EULER, who stimulated interest in contemporary mathematics in Russia, still serves as a kind of link between Russia and science in the rest of the world, even now. For example, following the death of STALIN in 1953, when the iron curtain began to part and scientific relations with the West — which almost completely had been broken off — were gradually restored, one of the first scientific events to bring scientists from both parts of the previously divided world together was the celebration of the 250th anniversary of EULER's birth. In the spring of 1957, in both Berlin and Leningrad, special colloquia dedicated to this event made it possible for scientists from Germany and the Soviet Union to renew their previous contacts. Even more important for the development of international scientific cooperation has been the truly international effort of mathematicians and historians of mathematics from many countries (Switzerland, both parts of Germany, France, and the USSR) who have contributed to the editing of EULER's works, among them the historians of mathematics EMIL FELLMANN, EDUARD WINTER, and ADOLF YOUSHKEVICH. YOUSHKEVICH, in fact, played a leading role in promoting scientific collaboration between Soviet scholars and their foreign colleagues. This began first through contacts with scholars from the communist block, especially the German Democratic Republic (KURT-R. BIERMANN, HANS WUSSING, and others), and Czechoslovakia (QUIDO VETTER, LUBOŠ NOVÝ, and others), and extended further throughout Europe, including above all France (ALEXANDRE KOYRÉ, RENÉ TATON, and others) and West Germany (KURT VOGEL, JOSEPH E. HOFMANN, WILLY HARTNER, CHRISTOPH J. SCRIBA, and others). Of major European centers for study and scholarly research, several have been especially important in promoting international contacts for historians of mathematics, including the *Centre Koyré* in Paris and the Mathematical Research Institute in Oberwolfach (Germany). In addition to the participation of Soviet scholars in international meetings and collaborative research projects, the value of efforts in the Soviet Union to promote research and publications of a very high quality in history of mathematics has also been reflected in the election of YOUSHKEVICH in 1956 and of VASILIĬ ZUBOV in 1958 to membership in the International Academy of the History of Science.

8.5 Dominant Postwar Research Themes and Publication Formats

As the preceding section makes clear, a number of themes in the history of mathematics emerged as dominant after the close of World War II. Thereafter, several postwar directions of research have been especially noteworthy.

Studies on the history of mathematics in antiquity continued. Here the most important results were works by AIZIK A. VAIMAN (*1922) on deciphering Sumerian-Babylonian texts, *Sumerian-Babylonian Mathematics* (1961); a new Russian translation of EUCLID's *Elements* undertaken by D. D. MORDUKHAĬ-BOLTOVSKOĬ, with the help of M. YA. VYGODSKIĬ and I. N. VESELOVSKIĬ [MORDUKHAĬ-BOLTOVSKOĬ 1948/50], which stimulated considerable interest in the history of ancient mathematics in the U.S.S.R.; and translations of ARCHIMEDES's works [VESELOVSKIĬ 1962], as well as DIOPHANTUS's *Arithmetica* [BASHMAKOVA/VESELOVSKIĬ 1974]. Finally, BASHMAKOVA and her pupils continued to produce studies on the history of mathematics in ancient Greece — in 1953 her works on ARCHIMEDES's differential methods were published, and in 1958 she edited her *Lectures on the History of Mathematics in Ancient Greece.* Subsequently, the main object of her research focused on Greek mathematics of the first centuries A.D., first of all on DIOPHANTUS's *Arithmetica.* BASHMAKOVA showed that DIOPHANTUS's methods were not merely sets of separate "clever tricks," but that they were modifications of general methods, the geometrical meaning of which was only discovered later [BASHMAKOVA 1972], [BASHMAKOVA/SLAVUTIN 1984].

As for studies dedicated to medieval mathematics (besides those mentioned above on Arabic themes by YUSHKEVICH, ROZENFELD, MATVIEVSKAYA, and later by MIRIAM M. ROZHANSKAYA (*1928) [ROZHANSKAYA 1976]), important contributions were also made to the history of the mathematical culture of ancient China, especially in YUSHKEVICH's work (1955), and through numerous studies and translations by ELVIRA I. BERËZKINA (*1931), most notably her translation of the classic text, *Nine Chapters on the Art of Mathematics* (Jiuzhang Suanshu) [BERËZKINA 1957] — the first in a modern European language. Additionally, research by ALEKSANDR I. VOLODARSKIĬ (*1938) on the history of mathematics in ancient and medieval India should be mentioned as well.

Among Soviet studies on the mathematics of 18th–20th centuries, the most popular subject was the history of mathematical analysis, specifically the history of the variational calculus which was regarded as one of the principal sources of functional analysis. The study of another source, of the symbolic calculus, was begun in the 1960s by LYUSTERNIK and PETROVA, whereas the history of operational methods was studied by KISELËV, OZHIGOVA, BOGOLYUBOV, and his pupils.

Other studies on the history of mathematical analysis include MARKUSHEVICH's *Essay on the History of Analytic Functions* [MARKUSHEVICH 1951], the works of VIKTOR V. GUSSOV (*1911) on the history of special functions, and the studies of AKHIEZER and ALEKSEĬ A. GUSAK on the history of approximation theory. VLADIMIR N. MOLODSHIĬ (1906–1986) (1953) and MATVIEVSKAYA [MATVIEVSKAYA 1967] continued their studies on the evolution of the number concept, while FEODOSIĬ D. KRAMAR (1911–1980) and NADEZHDA V. ALEKSANDROVA (*1932) were occupied with the history of different generalizations of that notion, including the vector calculus and the quaternion calculus.

Soviet studies on the history of geometry mainly concerned questions of the foundations of geometry, especially LOBACHEVSKIĬ's ideas and the development of non-Euclidean geometry. Still worth consulting are KAGAN's works on the history of the foundations of geometry (1949), his studies on LOBACHEVSKIĬ's life and works (1948), and his publication of the complete works of LOBACHEVSKIĬ (1946–1951). There is also research by BORIS L. LAPTEV (1905–1989) on LOBACHEVSKIĬ, as well as the well-known book by ROZENFELD on the history of non-Euclidean geometry [ROZENFELD 1976]. Works on the history of differential geometry, including the history of the Moscow geometrical school founded by K. M. PETERSON, were published by DEPMAN and YULO G. LUMISTE (*1929).

It is not surprising that in the U.S.S.R., with its famous school of probability theory, studies on the history of probability were of considerable interest. In 1954 GNEDENKO published an historical essay on its development in his well-known textbook on probability theory. In the 1960s LEONID E. MAĬSTROV (1920–1982) produced a series of articles which were summarized in his book, *Probability Theory. An Historical Essay* [MAISTROV 1967]. At about the same time OSKAR B.

SHEĬNIN (*1925) also began to study the history of probability theory and mathematical statistics [SHEYNIN 1965], [SHEYNIN 1966].

The history of computing machines was presented in a number of works by IGOR A. APOKIN (*1936) and MAĬSTROV, and summarized in their book, *Development of Computing Machines* (1976). Other studies on the history of computing have been written by ABRAM B. SHTYKAN (1906–1985), NIKIFOR D. BESPAMYATNYKH (1910–1987), and RAFAIL S. GUTER (1919–1978).

One of the main (if not the main) focus of post-war historico-mathematical studies has been the history of mathematics in Russia and the U.S.S.R. This is not surprising because official ideology attached great importance to the promotion of "Soviet patriotism." In this connection, historians were primarily interested in the pre-revolutionary period. Although not as politically charged as research devoted to the Soviet period, this work had to be approached according to accepted Soviet historiography. This usually led historians to avoid questions related to social history and to study only the pure history of ideas instead.

Most works devoted to the Soviet period popularized the achievements of the Soviet epoch. Among these, for example, are surveys that appeared in such well-known compendia as *Thirty Years of Mathematics in the U.S.S.R.* [KUROSH/MARKUSHEVICH/RASHEVSKIĬ 1948], and *Forty Years of Mathematics in the U.S.S.R.* [KUROSH 1959]. Both of these included very informative essays written by YUSHKEVICH on the development of research devoted to the history of mathematics.

Systematic research on the mathematics of old Russia was taken up again in a series of studies by DEPMAN, ZUBOV, YUSHKEVICH, YURIĬ A. BELYĬ (*1925), and REM A. SIMONOV (*1929). These represented more detailed analyses than had been done before on themes like the relations between the Russian and Byzantine traditions [SIMONOV 1977], mathematical practice in pre-Mongol Russia (in particular ZUBOV's analysis of the works on chronology by KIRIK OF NOVGOROD [ZUBOV 1953]), and geometrical manuscripts of the 16th and 17th centuries.

A great number of studies were also devoted to the 19th and 20th centuries. The works of LOBACHEVSKIĬ, MIKHAIL V. OSTROGRADSKIĬ (1801–1862), CHEBYSHEV (1821–1894), SOF'YA V. KOVALEVSKAYA (1850–1891), and ALEKSANDR M. LYAPUNOV (1857–1918) received attention. So did the activities of the major mathematical schools, including the Chebyshev school [DELONE 1947], the Moscow school of differential geometry, and the Moscow school of the theory of functions (MEDVEDEV, LYUSTERNIK), as well as the activities of the main university centers in St. Petersburg [DEPMAN 1960], Moscow [VYGODSKIĬ 1948], [YOUSHKEVICH 1948b], [LIKHOLETOV/YANOVSKAYA 1955], Kazan', Kharkov, Warsaw, and Odessa. Developments in certain special branches of mathematics, such as the history of the theory of numbers (see [OZHIGOVA 1972]) and of probability theory, were also explored.

Emphasis on the more recent period also prompted attempts to present a global view of the development of mathematics in the U.S.S.R. Although GNEDENKO published *Essays on the History of Mathematics in Rus-*

sia [GNEDENKO 1946], the definitive studies in this field were YUSHKEVICH's well-known book, *The History of Mathematics in Russia to 1917* [YOUSHKEVICH 1968], and SHTOKALO's *The History of Mathematics in Our Country* [SHTOKALO 1966/70]. The latter contains an extensive bibliography and numerous valuable references.

Finally, biographies of famous scientists and editions of classical works were regarded as important publication projects, since they were seen to contribute to the formation of a scientific world view. After World War II the works of AL-KHWARIZMI, L. EULER, N. I. LOBACHEVSKIĬ, M. V. OSTROGRADSKIĬ, P. L. CHEBYSHEV, B. RIEMANN, N. E. ZHUKOVSKIĬ, S. V. KOVALEVSKAYA, H. POINCARÉ, A. M. LYAPUNOV, P. BOHL, G. F. VORONOĬ, D. F. EGOROV, N. N. LUZIN, and others were published. For a list of biographies and editions of classical works see [DEMIDOV 1993b].

Study of EULER's scientific publications became a separate focus of research. Many scientists participated in this work, headed by YUSHKEVICH, among them SMIRNOV, N. SIMONOV, OZHIGOVA, MATVIEVSKAYA, and GLEB K. MIKHAĬLOV (*1929) [YOUSHKEVICH/WINTER 1959/76], [SMIRNOV/YOUSHKEVICH 1974].

Most of this historico-mathematical literature was designed primarily for mathematicians who were interested in the history of their science (many of whom contributed to history of mathematics as professionals), as well as for professors and students of the mathematical faculties. A considerable part of the historico-mathematical literature (books of DEPMAN, GERSH I. GLEIZER (1904–1967), and others) was intended for teachers at the high school and lower levels. Such historico-mathematical books were printed in large editions and could even be found in the libraries of small villages.

8.6 Social Dimensions of the History of Mathematics After World War II

The successful development of historico-mathematical studies in the U.S.S.R. in the post-war years (culminating in the 1960s and 1970s), as well as the establishment of obligatory courses on the history of mathematics at universities and pedagogical institutes, led to a substantial increase in the number of historians of mathematics in the Soviet Union. For example, sessions for the history of mathematics at the All-Union Conferences on the History of Physico-Mathematical Sciences regularly attracted more than one hundred participants from different parts of the country (the file of active historians of mathematics included an even larger number of names). In the same period, there was a correspondingly dramatic increase in both research and publishing activity.

As more and more authors turned to serious study of the history of mathematics, the journal *Istoriko-Matematicheskie Issledovaniya* could neither accept nor publish all of the articles submitted for consideration. Thus, numerous papers were published in other journals, including *Voprosy Istorii Estestvoznaniya i Tekhniki*

(Problems on the History of Science and Technology), *Matematika v Shkole* (Mathematics in School), *Istoryko-Matematychnyĭ Zbirnyk* (Collection on the History of Mathematics, edited in Ukrainian in 1959–1963) [SHTOKALO 1959/63], and *Narysy z Istorii Pryrodoznavsta i Tekhniky* (Essays in the History of Natural Science and Technology), also edited in Ukrainian since 1962. The mathematical volumes in the series *Istoriya i Metodologiya Estestvennykh Nauk* (History and Methodology of Science) [RYBNIKOV 1966/89], published by Moscow University, also provided key publication outlets.

From a broader institutional point of view, the activities of historians of mathematics throughout the U.S.S.R. were coordinated through the mathematical section of the Soviet National Committee of Historians of Science and Technology, which organized meetings twice a year in which historians from all parts of the country participated. The official seminar of the section was the Moscow University seminar on the history of mathematics. The majority of dissertations in the field, however, were presented in the Academy's Institute for the History of Science and Technology in Moscow. The Institute also organized seminars in various cities — beginning in Tartu in 1973 and later in Liepaya, Odessa, and Kamenetz-Podolskyĭ — that served as schools where the history of mathematics could be taught and promoted throughout the U.S.S.R.

The majority of leading Soviet historians of mathematics began their postgraduate studies and prepared their theses mainly in Moscow, but to a lesser degree in Leningrad, Kiev, and Tashkent. Returning to their homes, they maintained scientific and personal contacts with their teachers and colleagues, but also participated regularly in national conferences and summer schools. Moreover, lectures and courses on history of mathematics were given in the major universities across the country, all of which tended to be based on the same texts, RYBNIKOV's *History of Mathematics* [RYBNIKOV 1960/63, with several later editions], and *A History of Mathematics from Ancient Times to the Beginning of the 19th Century* [YOUSHKEVICH 1970/72]. The latter was a collective work published in 1970–72 under YUSHKEVICH's direction, and was later continued under the editorship of KOLMOGOROV and YUSHKEVICH in a series of books, *Mathematics in the 19th Century* [KOLMOGOROV/YOUSHKEVICH 1978/87]. In the mid-1970s, another collective work produced under YUSHKEVICH's direction, *Source Book on the History of Mathematics* [YOUSHKEVICH 1976/77, published in two volumes], was also widely adopted. This resulted in a certain "common culture" in the history of mathematics throughout the U.S.S.R.

With the dissolution of the Soviet Union in the 1990s, the subsequent political muddle and economic difficulties have had very negative consequences for science and education throughout all the territories of the former U.S.S.R. The entire system of scientific connections and academic networks was destroyed and the organization of the larger scientific community was seriously compromised. Research activities in many centers which had worked successfully before (as for example, Tashkent) almost came to a stop. Many leading scientists, above all professional mathematicians, left the country as a result of economic difficulties.

Simultaneously, the spirit of market reforms changed the scientific priorities of students — not as many, for example, now choose the history of mathematics as their scientific field. Economic difficulties have also created great obstacles to normal academic publishing. However, life in the leading scientfic centers (Moscow and Saint-Petersburg in Russia, and Kiev in the Ukraine) continues, and there is promising activity in some provincial centres, for example in Perm, Orenburg.

Recent transformations of the educational system at the university level (the creation of many new universities, including universities for the humanities, and a general shift of education to the humanities) has led to the appearance of new disciplines in new colleges, including the appearance of new courses for the history of mathematics aimed not at mathematicians but at students who choose non-scientific careers, among them journalism. As a result, the history of mathematics has left its traditional (for Russia) place allied with the mathematical disciplines and has now come to be related much more broadly to the history of culture. Consequently, historians of mathematics have begun to cooperate more closely with historians and philosophers of science.

The best evidence that there is a viable future ahead for historico-mathematical studies in post-Soviet Russia is the continuing publication of *IMI*, founded by YUSHKEVICH, and the recent appearance of such substantial and scholary editions as the first Russian translation of PTOLEMY's *Almagest* (1998), and two annotated volumes of works of DAVID HILBERT (1998).

8.7 Conclusion

Early studies on the history of mathematics in Russia and in the U.S.S.R., closely connected to and therefore largely determined by interests of the mathematical community, were first of all mathematical studies connected with the problems and needs of this community. Although history of mathematics had already been pursued before the revolution of 1917 (BOBYNIN, TIMCHENKO), it began to flourish with the activities of the so-called Soviet school in the history of mathematics (YANOVSKAYA, VYGODSKIĬ, YUSHKEVICH); for studies related to the historiography of the history of mathematics in Russia and in the U.S.S.R., see [YOUSHKEVICH 1948a], [YOUSHKEVICH 1959], [YOUSHKEVICH 1979].

For several reasons (including ideological ones) the history of mathematics in Russia and in the U.S.S.R. was, strictly speaking, the history of ideas. Questions related to social history were rarely studied or even ignored. The central theme of research was the history of mathematics in Russia and in the U.S.S.R. — priorities derived from ideological directives. Special attention was given to the scientific works of EULER, LOBACHEVSKIĬ, and CHEBYSHEV (Eulerian studies even led to the formation of a special scientific school headed by YUSHKEVICH).

Active research was devoted to history of mathematics from the 18th century to the beginning of the 20th century, above all to the history of mathematical analysis and its different branches, with a special emphasis on foundations, on the theory of functions of real and complex variables, on the theory of ordinary and partial differential equations, on the calculus of variations, and on the operational calculus. As for other branches of mathematics since 1700, Soviet historians produced works on the history of algebra and algebraic number theory (especially on the history of Diophantine analysis), on the history of geometry, particularly on the history of non-Euclidean geometry, and on the theory of probability.

As far as the history of mathematics in antiquity is concerned, valuable contributions were made to the history of Babylonian mathematics (deciphering Sumerian-Babylonian mathematical texts in particular), to the mathematics of ancient Egypt, and to Greek mathematics with works connected with EUCLID's *Elements*, with ARCHIMEDES's infinitesimal methods and with DIOPHANTUS's ideas. Medieval mathematics in Europe was studied (YUSHKEVICH), with additional research devoted to ancient and medieval mathematics in Arabic countries, China and India.

The history of mathematics in the 20th century has not been studied so thoroughly. As yet, there have been no investigations (as fashionable elsewhere) of "women in science." Nevertheless, the mathematical works of female mathematicians have been studied, above all those of SOF'YA KOVALEVSKAYA [for example, research by PELAGEYA YA. POLUBARINOVA-KOCHINA (1899–1999)], but only along traditional lines. Similarly, ethnomathematics has not been studied systematically either, although one of the pioneers in this area (known as "people's mathematics") was BOBYNIN.

Previously, if ideological concerns (prompted by the government) affected research in the history of mathematics in the U.S.S.R., nowadays even the slightest hint of ideology arouses suspicion. At the same time, the association of studies on the history of science with dialectical materialism (as was compulsory in the Soviet period) led to a close connection between philosophers and historians of mathematics. This connection has been maintained, although in the last decade the philosophical community and its ideology has changed drastically. For Russian historians of mathematics, their priority has more recently become the understanding of the history of mathematics as a part of the history of ideas. This does not preclude studies on the social history of mathematics (primarily in Russia and in the U.S.S.R.). However, the principal goal of all such studies has been to help identify and analyze the significance of social and cultural contexts on the development of mathematics.

Chapter 9

Poland

STANISŁAV DOMORADZKI and ZOFIA PAWLIKOWSKA-BROŻEK

9.1 Early Developments

Prior to the 17th century, the history of mathematics had only a few representatives in Poland, the earliest of which was JAN BROŻEK (1585–1652). The first Ministry of Education in Europe — *Komisja Edukacji Narodowej* (Commission on National Education) (1773–1794) — hoping to improve and broaden mathematical knowledge, recommended that students be acquainted with the history of mathematics beginning from antiquity. In an algebra textbook of 1782 there is even a short outline of the development of mathematics written in Polish [LHUILIER 1782].

In 1818, according to a *Proposal for a Statute of the Royal University of Warsaw,* provision was made for a course on the history of the mathematical sciences. Unfortunately, due to a lack of staff, nothing came of this idea, nor was a proposed Department for the History of Mathematics established either. Nevertheless, despite the lack of such institutional foundations, works devoted to the history of mathematics sparked considerable interest in Poland, as is clear from school reports of Polish grammar and secondary schools in the 19th century. Such publications, primarily based on the famous work by J. E. MONTUCLA [MONTUCLA 1799/1802] (see Section 1.3.3), also dealt with geometry and the history of mathematics in Poland.

Especially important in this regard are works by JAN ŚNIADECKI (1756–1830). His were the first independent treatises written on history of mathematics in Poland. These were devoted primarily to the achievements of J. L. LAGRANGE [ŚNIADECKI 1815], the history of probability theory [ŚNIADECKI 1837a], and the special character and nature of argumentation in mathematics throughout the ages [Śniadecki 1837b]. Equally valuable is a major compilation due to Teofil Żebrawski (1800–1887), *Bibliografia piśmiennictwa polskiego z działu matematyki i fizyki oraz ich zastosowań* (Bibliography of Polish Literature on Mathematics and Physics and their Applications) [Żebrawski 1873/86]. Żebrawski's *Bibliografia* surveys scientific achievements made by Poles from the earliest (preserved) handwritten works of the 13th century to publications up to 1830. This served to stimulate interest in historical and mathematical studies, as evidenced by the writings of Jan N. Franke (1848-1918) [Franke 1879] [Franke 1884], Aleksander M. Baraniecki (1848-1895) [Baraniecki 1884], Józef Bieliński (1848–1926) [Bieliński 1890], and others.

The one person who contributed most to the development and popularization of the history of mathematics in Poland in the decades around 1900, however, was Samuel Dickstein (1851–1939, **B**). Dickstein's strengths were both technical and organizational. He founded two journals, *Prace Matematyczno-Fizyczne* (1888) and *Wiadomości Matematyczne* (1897), both of which included reviews, news, and articles devoted to the history of mathematics. Dickstein's scientific achievements as an historian of mathematics were also noteworthy. His most important work dealt with Józef Hoene-Wroński (1776–1853), including the monograph [Dickstein 1896]. For many years, Dickstein also collected materials for a monograph on Adam Kochański (1631–1700), some of which he published, for example Kochański's correspondence with Leibniz [Dickstein 1901/02].

Another outstanding representative of the history of mathematics in the 19th and the early years of the 20th century was Ludwik Antoni Birkenmajer (1855–1929), who devoted a considerable part of his research to Nicolaus Copernicus's personality and achievements. From 1897 Birkenmajer was the first to lecture on the history of mathematics at the Jagiellonian University in Cracow. University publications, in fact, are an important source for historical research, including history of mathematics; among these are *Programy Wykładów i Składy Osobowe Wydziałów* (Lecture Reports and List of Lectures at Faculties) for the universities of Cracow, Warsaw, Lwów, Wilno, and Poznań. *Sprawozdania Dyrekcji Gimnazjów i Liceów* (Reports of Grammar and Secondary School Head Masters) are also worth consulting.

9.2 Twentieth-Century Contributions

The period 1918–1939 saw extraordinary developments not only in Polish science in general, but also in the history of mathematics, as Poland regained its political sovereignty during this period. In 1918 Birkenmajer and his son Aleksander Birkenmajer (1890–1967) issued an unprecedented study: *The Most Important Desiderata of Polish Science in the Field of History of Mathematical Sciences* [Birkenmajer/Birkenmajer 1918]. Many of the goals they set were accomplished over the decades that followed, for example, Zofia Pawlikowska compiled a history of Polish mathematical terminology from the beginning of the 16th century until the start of the 20th century [Pawlikowska 1964/66]. Monographs by Stanisław Dobrzycki (1905–1989) and Jadwiga Dianni (1886–1981) covered the history of mathematics in universities [Dobrzycki 1971], while Zdzisław Opial (1930–1974) and Stanisław Gołąb (1902–1980) also produced comprehensive papers on mathematicians connected with the Jagiellonian University [Opial 1964/65], [Gołąb 1964]. Other works covered the history of mathematics at Cracow University, the oldest Polish university, from its creation in 1364 until the end of the 19th century [Dianni 1963], [Opial 1964/65], [Gołąb 1964].

General research on the history of mathematics in Poland is reflected in papers by Edward Marczewski (1907–1976), Opial, Pawlikowska-Brożek, and J. Dianni and Adam Wachułka (1909–1991) [Marczewski 1948], [Opial 1966], [Pawlikowska-Brożek 1983], [Dianni/Wachułka 1963]. Among studies of historical periods is a monograph on the history of Polish mathematics of the 17th century written by Edward Stamm (1886–1940) [Stamm 1935]. A history of Polish mathematics over the period 1919–1970 was published by Kazimierz Kuratowski (1896–1980) [Kuratowski 1973], and Tadeusz Iwiński (1906–1993) has written a history of the Polish Mathematical Society (1919–1973) [Iwiński 1975].

Numerous works on history of mathematics, including biographical studies and articles on the scientific achievements of outstanding Polish mathematicians, have been published in *Wiadomości Matematyczne*, *Colloquium Mathematicum*, *Matematyka*, and in prefaces to various collected papers. The authors of the majority of such works have often been specialists dealing with pure mathematics, for example Karol Borsuk (1905–1982), St. Gołąb, K. Kuratowski, E. Marczewski, Jacek Szarski (1921–1980), Władysław Ślebodziński (1884–1972), Andrzej Schinzel, Jerzy Mioduszewski, and Andrzej Pelczar.

Noteworthy among these scholars is Jadwiga Dianni, who amassed a unique bibliography on the history of mathematics in Poland, which is now kept in the archives of *PAN* (Polish Academy of Sciences) in Cracow. Many of Dianni's papers, which dealt primarily with the history of mathematics in Poland, were coauthored with A. Wachułka and published in *Kwartalnik Historii Nauki i Techniki* (Quarterly Journal of the History of Science and Technology).

History of mathematics has also been popularized more recently by The Commission on History of Mathematics of the Polish Mathematical Society (founded in 1978) under the leadership of Z. Pawlikowska-Brożek. Every year the Commission organizes conferences lasting several days for mathematicians and historians of mathematics called Schools of History of Mathematics (the 10th *School* took place in 1998). Each year the Commission also publishes books (about 200 copies every year) reporting the proceedings of each *School.* These include articles written by well-known mathematicians: Julian Musielak on functional analysis, Józef Siciak on complex analysis, Roman Duda on the first Polish renowned mathematical periodical, and Krzysztof Tatarkiewicz on mechanics in Poland. Stanisław Domoradzki has even presented the history of Polish periodicals publishing mathematical works [Domoradzki 1992a]. Thanks to the activities of the Commission, it has been possible to present Polish achievements in the history of mathematics against the background of activities elsewhere in the world, as historians of mathematics from abroad (the Czech Republic, Germany, Slovakia, Russia, and the Ukraine) have also participated in the conferences.

At universities, history of mathematics has been included as a regular subject since the 1960s. The lectures given by Marek Kordos have been published as a book [Kordos 1994]. In addition, a specific course on the history of mathematics in Poland has also been regularly presented at Warsaw University. Similar lectures on history of mathematics are delivered at the Jagiellonian University in Cracow by Zdzisław Pogoda, at the Wrocław University by Witold Więsław and Władysław Narkiewicz, and at the Adam Mickiewicz University in Poznań by Roman Murawski, who is the author of an anthology of classical texts [Murawski 1986].

9.3 Conclusion

Historiography of the history of mathematics in Poland has been conditioned largely by political events. A century-long occupation by Austria, Germany, and Russia greatly influenced the development of science in Poland. A first, futile attempt had been made early in the 19th century to introduce a course on the history of mathematics and to create a department for this discipline at the University of Warsaw. For a long time, school reports were the main publications in which articles on the history of mathematics can be found.

Following some forerunners in the second half of the 19th century, two scholars must in particular be mentioned. Samuel Dickstein, well-known as an author and a founder of journals, fostered increasing interest in this field, while Ludwik Antoni Birkenmajer's research on Nicolaus Copernicus received international recognition.

After liberation uprisings state institutions that had made research possible were closed under the Russian occupation (1795–1918). Subsequently, private foundations were established, among which the Józef Mianowski Fund sponsored many scientific enterprises and publications such as the multi-volume project *Poradnik dla Samouków* (Handbook for Autodidacts), edited in 1898–1911. S. Dickstein was the author of the mathematical part of the mathematical and scientific volume (1898). The mathematical volume beginning the second series (1915) was written by many authors: Jan Łukasiewicz (1878–1956), Zygmunt Janiszewski (1888–1920), Stefan Mazurkiewicz (1888–1945), Wacław Sierpiński (1882–1969), Stanisłw Zaremba (1863–1942), and Stefan Kwietniewski (1874–1940), who contributed an entry on "History of Mathematics in Poland."

In 1884 Marian A. Baraniecki (1848–1895) had launched the *Biblioteka Matematyczno-Fizyczna* (1884) (Mathematical and Physical Library), and this too contains publications important for the history of mathematics. Dickstein edited privately, with his own money, *Prace Matematyczno-Fizyczne* (Mathematical and physical works) and *Wiadomości Matematyczne* (Mathematical News), both of which contained many articles on history of mathematics. *Wiadomości Matematyczne* reappeared after World War II (in a second series), and continues to publish important works on history of mathematics in Poland.

Chapter 10

Bohemian Countries

Luboš Nový

10.1 Introduction

The Czechs and the Slovaks are two closely related Slavic peoples. Historically, they have constituted the great majorities in their countries, although politically they have tended to live separately. The Czechs, in the kingdom of Bohemia (under the rule of the Habsburg monarchy from 1526 to 1918), dominated in the second half of the 19th century both economically and politically. This kingdom was created mainly from two Czech countries: Bohemia and Moravia. The Slovaks, since the second half of the 19th century, have lived under the rule of the Hungarian Kingdom and have thus been subjected to cultural and national pressures. Czechoslovakia was created after World War I; since 1993 it has consisted of two autonomous republics.

The history of mathematics has a long tradition in the former Czechoslovakia, although this is little known outside the country.[1] One of the main reasons is the relative isolation of Czech and Slovakian science. Until 1918, despite many difficulties, it developed slowly in the *Vielvölkerstaat* (multi-national state) of the Habsburg monarchy. Another reason stems from the fact that most works were published in the national languages. Here the main periods in the development of the history of mathematics in Czechoslovakia are analyzed by characterizing selected publications and by describing the work of important historians of mathematics.

[1]A bibliography devoted to the history of mathematics in Bohemia was published by Quido Vetter [Vetter 1924]. Throughout this chapter, the term Czechoslovakia is used to refer to the 20th-century nation as it existed prior to 1993.

10.2 Beginnings

Until the middle of the 18th century, the history of mathematics was limited primarily to historical remarks contained in mathematical publications. Whereas mathematics was the focus, history served to explain the context of problems the authors studied, by tracing their origins most often to ancient authors. One exception, and doubtless the first historico-mathematical work of Bohemian provenance, was the 1557 *Oratio de laudibus geometriae* (Lecture in Praise of Geometry) by the well-known mathematician, astronomer, and physician Tadeáš Hájek a Hájek (1525–1600). Hájek a Hájek, whose patients included the emperors Maximilian II and Rudolf II, delivered this *Oratio* as an introductory lesson for his academic explanation of mathematics. The *Oratio* not only contains remarks about the usefulness of mathematical knowledge, but also about the history of mathematics (beginning with ancient Egypt), as well as important information concerning mathematics in Bohemia, especially the professors at the University of Prague.

10.3 History of Mathematics 1750–1850

Although isolated remarks on the history of mathematics in Bohemia may be found in 17th- and early 18th-century sources, the earliest serious development of interest in the history of mathematics in the Bohemian countries began in the second half of the 18th century, and was closely tied to the rise of the Czech national movement. The proponents of this movement sought to emphasize the most significant contributions of the Czech nation and the kingdom of Bohemia in all cultural and intellectual realms.[2] In mathematics, as in so many areas, these contributions had been interrupted at the beginning of the Thirty Years' War, and especially after the suppression of the Bohemian uprising in 1620. Among the clearly nationalistic works published during the second half of the 18th century was the four-volume work (in German), *Abbildungen böhmischer und mährischer Gelehrten und Künstler nebst kurzen Nachrichten von ihrem Leben und Wirken* (Portraits of Bohemian and Moravian Scholars and Artists, with Brief Reports of their Lives and Works), consisting of brief biographies of Bohemian scholars [Pelzel 1773/82]. These included descriptions of their scientific work and ranged from the earliest times until about 1770. At almost the same time, Stanislav Vydra (1741–1804), professor of elementary mathematics at the University of Prague, published his *Historia matheseos in Bohemia et Moravia cultae* (History of Mathematics Cultivated in Bohemia and Moravia) [Vydra 1778].

It was from such beginnings that serious research concentrating on the history of mathematics in the Bohemian countries began to develop. A summary of the major contributions of Bohemian mathematicians made in this period is given

[2]This had already been done in the work of Bohuslav Balbín (1621–1688). His research, however, remained unpublished until the second half of the 18th century; see especially his *Bohemia docta* (vols. I–III, Prague 1776–1780).

in JOSEF SMOLÍK's (1832–1915) *Matematikové v Čechách od založeni university* (Mathematicians in Bohemia from the Founding of the University) (1864). This work was intended to cover the period from the creation of the University of Prague by KARL IV in 1348 to the beginning of the 19th century. In fact, when published, the first part only covered the period from 1348 to 1622 [SMOLÍK 1864] and was based primarily on printed sources.[3] (It is unfortunate that even modern scholars interested in the early development of Bohemian mathematics often fail to make full use of the rich manuscript holdings available). Mathematics at the University of Prague is also comparatively well documented, thanks to the efforts of V. V. TOMEK (1818–1905). He collected and edited the most important sources and thus provided fundamental information about teaching and the professors of mathematics [TOMEK 1849]. To date, TOMEK's research has been supplemented by only a few additional sources.

10.4 The Mid-19th Through the Mid-20th Century

The second distinct period in the development of the history of mathematics in Czechoslovakia extends from the second half of the 19th through the first half of the 20th century and witnessed the creation of a self-sustained, scholarly discipline. Maintaining a focus on historical research, especially on Czech national history, active mathematicians increasingly began to undertake serious historical research. Their expertise as mathematicians was also reflected in a growing emphasis on recent mathematical developments. The well-established *Karls-Universität* of 1348 (named *Karl-Ferdinand-Universität* since 1622) was divided into the Czech (Bohemian) and the German university in 1882.

Among Czech historians of mathematics, FRANTIŠEK JOSEF STUDNIČKA (1836–1903), professor of mathematics at the Czech University in Prague, was the most important figure of the second half of the 19th century. An active mathematician with numerous creative interests, STUDNIČKA produced editions (with commentaries) of the minor manuscripts of TYCHO BRAHE [STUDNIČKA 1886]. He also published BRAHE's sketches of trigonometry [STUDNIČKA 1903]. The latter contains an explanation of the *prostaphairesis method* which replaces the product (or quotient) of two numbers by their sum (or difference) by means of formulas involving goniometric functions (i.e. those that measure angles). One of STUDNIČKA's most influential papers was "A. L. Cauchy als formaler Begründer der Determinantentheorie" (A. L. Cauchy as Formal Founder of the Theory of Determinants) [STUDNIČKA 1876]. In his own mathematical research he was concerned with the theory of determinants and wanted to ascertain the origins of the

[3]The manuscript of the second part, assuming it was ever written, seems to have been lost. However, the work of SMOLÍK was taken up by other authors. Later, based on further research, QUIDO VETTER gave a more complete picture of the earlier development of Czech mathematics, regularly publishing the results of his research in foreign countries. Most complete is the Russian version [VETTER 1958/61]

theory (which, at the time, was rather overestimated in importance).[4] STUDNIČKA was also one of the first Czech mathematicians to appreciate the importance of BERNARD BOLZANO (1781–1848), and contributed, if only to a limited extent, historical remarks related to BOLZANO's life and work in Bohemia [BOLZANO 1882].

In the 20th century, the history of mathematics has become increasingly involved with mathematics education. On the whole, the scholars in the Bohemian countries have followed the same thematic trends, and at comparable levels of sophistication, as historians of mathematics elsewhere in the world.

The further development of the history of mathematics in Czechoslovakia is connected with the scientific and organizational activities of QUIDO VETTER (1881–1960, **B**). VETTER studied mathematics at the Czech University of Prague, where he received the first *Habilitation* (permission to teach at the university) in Czechoslovakia for the history of mathematics in 1919. Later he was promoted to an extraordinary professorship, but due to complex circumstances, he did not teach as a professor but instead served as the director of a middle school outside Prague. After the Second World War, he resumed his lectures on the history and didactics of mathematics at the University of Prague.

VETTER's numerous papers in a variety of areas on the history of mathematics were published in Czechoslovakian and foreign journals. Among his many works, his *Habilitationsschrift*, "O metodice dějin matematiky" (On the Methodology of the History of Mathematics), is outstanding [VETTER 1919]. It still inspires Czech students and remains of interest even when read from a more contemporary point of view. VETTER's work is based on a sound knowledge of the historico-mathematical literature of his day. Above all, he valued the works of GUSTAF ENESTRÖM (1852–1923, **B**) and GINO LORIA (1862–1954, **B**), without favoring any one point of view.[5] In his knowledgeable survey, VETTER pointed to the diversity of publications in the history of mathematics, beginning with an analysis of mathematical publications on a given theme, and continuing to an interpretation of mathematics as a general component of cultural history. He had the utmost respect for the importance of *all* serious studies in the history of mathematics. Philosophically, however, he believed that the history of mathematics must be viewed as part of the history of culture.[6] He also argued that the discipline had to accept historical method as it was understood by contemporary historians, but this in turn had to be adjusted to the needs of its special subject matter, namely, mathematics. In his opinion, without this acceptance, a scientific standard for the

[4]For an alternative to STUDNIČKA's views on this subject, see [MUIR 1906, 131]. MUIR placed special emphasis on the importance of VANDERMONDE.

[5]VETTER, however, was not familiar with LORIA's *Guida allo studio della storia matematiche* (Milano 1916). Although VETTER submitted his *Habilitationsschrift* on November 14, 1917, scientific contacts with Italy had already been interrupted due to World War I. His knowledge was thus based upon LORIA's papers published in *Bibliotheca Mathematica*, and in *Bollettino di bibliografia e storia delle scienze matematiche* published by GINO LORIA since 1898.

[6]Naturally, VETTER's understanding of the aims of the history of mathematics was influenced by the state of research in cultural history of his time. Thus, he could not avoid accepting the hypothesis that differences in culture are based, in part, on differences between nationalities.

history of mathematics could not be secured. Many of the ideas promoted by VETTER can be found in papers by other historians of mathematics writing at virtually the same time, but they are most inspiring when considered in connection with the inner coherence of his own writings.

VETTER's publications range from pre-Greek mathematics to analyses of the most recently published sources, from the *Method* of ARCHIMEDES to numerous articles about mathematicians working in Bohemia, in which he compared their work with contemporary international tendencies in mathematics. A summary of his evaluations was published between 1958 and 1961 in a survey of the development of mathematics in the Bohemian countries up to the end of the 17th century.[7]

VETTER also played an important part in the creation of national and international organizations for the history of science. He had numerous international contacts and was among the small circle of scholars who founded the *Académie internationale d'histoire des sciences*. VETTER was actively involved in the work of the Academy, becoming, in 1934, vice-president and serving as its (fourth) president from 1934 to 1937. In Czechoslovakia, he founded the *Volné sdružení pro dějiny reálnych věd* (Free Association for the History of Science) in 1928; it was later renamed the National Comittee for the History of Science. During his presidency in 1937, the Fourth International Congress of the History of Science was organized in Prague [FOLTA/NOVÝ 1973].

During the first half of the 20th century, QUIDO VETTER was not the only specialist who took a scholarly interest in the history of mathematics. Numerous historical articles were also published, mostly by teachers at *Gymnasia*. These often focused on the works of BERNARD BOLZANO. For example, VOJTĚCH JARNÍK (1897–1970), professor of mathematics at the Czech University in Prague, and KAREL RYCHLÍK (1885–1968), professor of mathematics at the Technical University of Prague, produced fundamental studies of BOLZANO's contributions to mathematical analysis.[8] In 1924, a Bolzano Commission was established to prepare an edition of BOLZANO's unpublished papers. Between the Wars, however, only four volumes were published.[9]

[7]See note 3.

[8]After JAŠEK defined a continuous, nondifferentiable function (*Věstník Královské české společnosti nauk, Třída mat. přírodověd* (1921), pp. 1–32), JARNÍK published an analysis of BOLZANO's method of construction of these functions in 1922. JARNÍK's papers about BOLZANO were translated into English on the occasion of the bicentenary of BOLZANO's birth. See: [JARNÍK 1981]. — RYCHLÍK informed the international mathematical public of BOLZANO's theory of functions at the Sixth International Congress of Mathematicians in Bologna in 1928. Later he studied BOLZANO's theory of real numbers [RYCHLÍK 1962].

[9]*Spisy Bernarda Bolzana — Bernard Bolzano's Schriften*: Bd. 1: *Functionenlehre*, Praha 1930; Bd. 2: *Zahlentheorie*, Praha 1931; Bd. 3: *Von dem besten Staate*, Praha 1932; Bd. 4: *Der Briefwechsel B. Bolzano's mit F. Exner*, Praha 1935.

10.5 History of Mathematics in Czechoslovakia Since 1945

The third period in the history of mathematics in Czechoslovakia began after the end of the Second World War. It is characterized, at least in part, by a continuation of tendencies, and even personalities, that had been dominant between the Wars. To those, however, were added younger collaborators and newly founded, financially better equipped scientific institutions. Greater attention paid to the history of mathematics at the universities not only enabled VETTER to resume his lectures at Prague but also resulted in new instructors for the history of mathematics, like FRANTIŠEK BALADA (1902–1961) in Brno, who successfully passed his *Habilitation* in 1960.

When the Czechoslovakian Academy of Sciences was founded in 1952, followed by the Slovakian Academy of Sciences in 1953, both included working groups for the history of science and technology. Within these groups, historians of mathematics pursued their studies, too. Similarly, improved working facilities encouraged mathematicians to give lectures on the history of their own particular disciplines. Meanwhile the following specialized publications and journals for the history of science were founded:

- *Sbornik pro dějiny přirodních věd a techniky,* Praha, Academia, 1954–1967, 12 vols.;
- *Acta historiae rerum naturalium necnon technicarum, Special Issues*, Praha, Československá akademie věd, since 1965, 25 vols.;
- *Dějiny věd a techniky*, Praha, since 1968;
- *Z dejín vied a techniky na Slovensku*, Bratislava, Slovenská akadémie vied, since 1962.

An important characteristic of the comparatively late development of the history of mathematics in Czechoslovakia is its close association with history of science. As a result of these strong ties, considerable emphasis has been placed on the use of sophisticated historical methods and tools in the history of mathematics.

More recently, Czechoslovakian historians of mathematics have followed the ideals and inspiration that VETTER articulated at the end of World War I. This is reflected, for instance, in the collective work, *Dějiny exaktních věd v českých zemích do konce 19. století* (History of Science in the Bohemian Lands up to the End of the 19th Century) [NOVÝ 1961]. Here, the development of mathematics, astronomy, physics, and chemistry is described with emphasis upon social and cultural interrelations.

In Slovakia, JÁN TIBENSKÝ (*1923) published *Dejiny vedy a techniky na Slovensku* (History of Science and Technology in Slovakia) (1979), which includes chapters on the history of mathematics. Somewhat earlier, KAREL KOUTSKÝ (1897–1964), professor at *Gymnasia* in Slovakia, and then professor of mathematics at Brno University, compiled a list of the lives and works of Slovak mathematicians.

Very useful for the history of Slovak science is the two-volume biobibliographical work edited by JÁN TIBENSKÝ [TIBENSKÝ 1976].

Although the leading historians of mathematics in Czechoslovakia have been expected to pursue general problems in the history of science, in addition to their more specialized research, the history of mathematics is flourishing. Many recent selective bibliographies reflect the extent of this work, although they do not include all publications; in particular, they omit more specialized investigations published in Czech or Slovakian languages.[10]

Most articles on the history of mathematics have been written by mathematicians employed at universities or high schools, rather than by scholars who systematically work in the field of history. Only a very few specialists are occupied with history of mathematics.

Among them, LUBOŠ NOVÝ (*1929), now retired, worked at the Academy of Sciences in Prague. He focused on the origins of modern algebra [NOVÝ 1973] and the development of Czech mathematics [NOVÝ 1961], as well as theoretical and methodological problems in the history of science. JAROSLAV FOLTA (*1933), first at the Academy of Sciences in Prague and now at the National Technical Museum in Prag, investigated the modern development of geometry and published *Česká Geometrická Škola, Historická Analýza* (The Czech Geometric School, Historical Analysis) [FOLTA 1982]. JINDŘICH BEČVÁŘ (*1947), lecturer of mathematics at the University of Prague, concentrates mainly on Czech mathematics since the second half of the 19th century. His most important publication, *Eduard Weyr, 1852–1903*, is a biography of a professor of mathematics at the Czech Polytechnic in Prague [BEČVÁŘ 1995].

Beginning in 1980, summer schools for the history of mathematics have been offered. An even more important step that may positively influence the future development of the history of mathematics as a discipline in the Czech Republic was the foundation of a division for the history of mathematics within the Institute of Mathematics at the Charles University in Prague in 1988. This division, besides having positions for historians of mathematics, also provides for associate external collaborators and visiting scholars.

Following VETTER's example, Czechoslovakian historians of science have been active in international collaboration. They regularly participate in international congresses and have organized international symposia in Czechoslovakia. In 1967, for instance, a symposium, *La révolution scientifique du 17ième siècle et les sciences mathématiques et physiques* [NOVÝ 1968], was organized on the occasion of the 300th anniversary of the death of MARCUS MARCI (1595–1667). In 1981, a symposium on the *Impact of Bolzano's Epoch on the Development of Science* took place [NOVÝ 1982] to celebrate the bicentary of BOLZANO's birth.

[10]See, for example, HANA BARVÍKOVÁ and MÁRIA HROCHOVÁ, "Czechoslovak History of Science: Selected Bibliography 1970–1980," *Acta hist., Special Issue* **15** (1981), 214 pp. (the history of mathematics is covered on pp. 41–50), and HANA BARVÍKOVÁ, LUDMILA CUŘÍNOVÁ, and MÁRIA HROCHOVÁ, "Czechoslovak History of Science: Selected Bibliography 1980–1988," *Ibid., Special Issue* **22** (1989), 213 pp.

10.6 Conclusion

Some general interest in the history of mathematics in the Bohemian countries arose in the second half of the 18th century, in connection with the Czech national movement. This stimulated significant contributions of the kingdom of Bohemia and the Czech nation in culture and science, including mathematics. From these beginnings, serious research on the history of mathematics in the Bohemian countries began to develop during the 19th century. Mostly pursued by active mathematicians, they increasingly turned their attention to more recent mathematical developments. BERNHARD BOLZANO became one of the main figures of historical research.

During the first half of the 20th century, GUIDO VETTER assumed a leading role as an historian of mathematics. He was not only of great importance for the professionalization of this discipline on a national scale, he was also active and influential in international organizations for the history of science. In 1937, he organized the Fourth International Congress for History of Science in Prague.

The third period in the history of mathematics in Czechoslovakia began after the Second World War. Ties to history of science were strengthened when both the Czechoslovakian and the Slovakian Academies of Sciences established working groups in the history of science and technology. A number of journals devoted to these areas were founded, regularly publishing articles in the history of mathematics — mostly contributed by mathematicians teaching at universities or high schools. Only very few specialists were fortunate to obtain positions devoted entirely to the history of mathematics.

Chapter 11

Austria

CHRISTA BINDER

11.1 Introduction

History of mathematics has never been a major occupation among Austrian mathematicians. There have been no major figures or professors of history of mathematics like MORITZ CANTOR (1829–1920; **B**). There have nevertheless been singular studies by mathematicians and historians on their predecessors and on the development of mathematical ideas. As these studies are scattered among various books, journals and other manuscripts, and as the meaning of "Austria" has varied over the course of time, the story which follows is complex, and completeness can in no way be guaranteed. In what follows, the term "Austrian" is used in a broad sense and includes all those who either lived in Austria as it was understood in their day, or those who were born or educated there.

11.2 Regiomontanus and Tannstetter

The first records of mathematical activity within the boundary of Austria date to 9th-century Salzburg, where the books of EUCLID and BOETHIUS were studied in the *Schola Sancti Petri*, founded by ST. RUPERT in the sixth century. During the following centuries mathematical knowledge was very scarce, but in every monastery at least one person was able to calculate the date of Easter and was familiar with the basic statics of buildings. From 1157 on, monks were no longer allowed to build churches; this led to the foundation of so-called *Bauhütten*, guilds that accumulated much knowledge of applying geometry, but that kept this knowledge secret. After completing a building, guild members even destroyed their plans, surely *not* an historical attitude

Austria, or "Ostarrichi," was first mentioned in 996. But until the founding of the University of Vienna in 1365 (the third German University, after Prague (1348) and Cracow (1364)), we cannot seriously claim that any mathematics was done in the region (apart from the activities in the monasteries).

One exception to this was HERMANN OF CARINTHIA who lived in the early 12th century and translated books of PTOLEMY and EUCLID, as well as other astronomical and astrological texts, from Arabic into Latin. No known document relates him to the region from which his name is derived (today: Kärnten), but he was probably born in Carinthia. (He may have been of Slavic origin.) He was educated at Chartres and worked in Spain.

The first activities which can truly be considered as constituting the history of mathematics date from the 15th century at the University of Vienna. This institution had gained a worldwide reputation through the work of JOHANNES VON GMUNDEN (ca. 1380–1442) and GEORG VON PEUERBACH (1423–1461).

Undoubtedly, the most important figure not only for mathematics but for its history was JOHANNES MÜLLER (REGIOMONTANUS) (1436–1476) (see [METT 1996]). Although he was not Austrian, he did spend 11 years in Vienna, from 1450 until the death of PEUERBACH in 1461. In Vienna, he learned astronomy, mathematics and Greek and became acquainted with classical literature. He also began to work on the *Problemata almagesti* (a commentary to PTOLEMY now lost) in collaboration with PEUERBACH. In 1463 in Venice he found (and immediately recognized the value of) six books of DIOPHANTUS's *Arithmetica*. REGIOMONTANUS intended to translate them himself, and he urged others to look for the missing books.

His next activity in the history of mathematics was a famous lecture on astrology given at Padua in 1464. In it he outlined the development of mathematics from the flooding of the Nile to PEUERBACH and Cardinal BESSARION (1403–1472), thus giving a clear picture of what was known of Greek and Arabic authors. He also singled out missing books and bad translations (see [ZINNER 1990, 70–74]).

His most important contribution to history, however, was yet another enterprise. He clearly recognized the deficiencies of many of the available manuscripts and books and sought to improve the situation. To do so he established his own printing press in Nuremberg in 1472. He intented — as he explicitly stated in a famous *Tradelist*, i.e., an announcement of texts scheduled for publication which he sent to several universities — to provide his successors with the best possible sources for further studies. Nuremberg, the center of book-printing, was certainly a good place for starting this enterprise. There REGIOMONTANUS could find typesetters and woodcutters whom he trained personally for the difficult job of printing tables and mathematical manuscripts. In fact, he personally supervised the whole printing process. The *Tradelist* consists of one page and includes the following prayer:[1]

[1]His prayer was not answered, since REGIOMONTANUS died under mysterious circumstances before completing this work.

> May Almighty God look with favor upon it. — Then after completion — death will not be bitter, if its organizer should soon die, if he has left such a gift behind for later generations [ZINNER 1990, 111].

Unfortunately, only a few of the books announced on the *Tradelist* ever appeared. The first one was PEUERBACH's *Theoricae novae planetarum* (New Theory of the Planets) (1453/54), then some calenders and almanacs (always a guaranteed success). REGIOMONTANUS printed his own *Ephemerides* for the years 1475–1506 in 1459, which ultimately played an important role during COLUMBUS's travels some twenty years later (for more details see [ZINNER 1990, 47]). In 1475, REGIOMONTANUS was called by Pope SIXTUS IV to participate in calender reform, whereupon he left for Italy, taking some of his manuscripts with him. He died in Rome in 1476.

The first Austrian — if REGIOMONTANUS is considered more a European — to write a history of mathematics was GEORG TANNSTETTER (COLLIMITIUS) (1482–1535) ([GRÖSSING 1983], [BINDER 1999]). TANNSTETTER was the leading figure of the so-called Second Viennese Mathematical School [BINDER 1998] (the first consisting of GMUNDEN, PEUERBACH, and REGIOMONTANUS [BINDER 1996]). Born in Rain am Lech, he studied in Ingolstadt with JOHANNES STABIUS († 1522) and ANDREAS STIBORIUS (ca. 1470–ca. 1515), and followed them to Vienna in 1502, where he taught from 1503 to 1528 in various positions. He belonged to the famous *Collegium poetarum* (founded by KONRAD CELTIS († 1508)). The *Collegium* was an academic society comprised of scientists and poets from the Danube region; later it became a university with professors for astronomy, mathematics, and poetry. TANNSTETTER was a mathematician and astronomer. He had the royal privilege to publish calenders (copies exist for the years from 1504 to 1526), and was consulted in connection with calender reform. He was also a physician (*Leibarzt* of Emperor MAXIMILIAN I), an astrologer (famous for his prediction of the date of MAXIMILIAN's death and of a great flood), a cartographer, and poet. Although he edited many works of his predecessors, his intention was not humanistic (as with REGIOMONTANUS who wanted to publish the classics). Instead, he was a teacher of mathematics and astronomy, who wanted to provide his students with reasonably priced reliable editions of the theories to be learned. This is best seen from a collection of the five most important books on higher mathematics — the *Arithmetic* of JOANNES DE MURIS (1290?–1360?), the *Theory of Proportions* of THOMAS BRADWARDINE (1290?–1349), the *Latitudines* of NICOLAUS HOREM (NICOLE D'ORESME) (1323?–1382), the *Algorithm of Integers* of PEUERBACH, and the *Algorithm of Sexagesimal Fractions* of GMUNDEN, which he published together in one volume in 1515.

The year before, TANNSTETTER published the *Tabulae Eclypsium* (Table of Eclipses) of PEUERBACH and the *Tabulae primi mobilis* (special trigonometric tables) of REGIOMONTANUS. He closed the latter with a chapter, the *Viri Mathematici* (Lifes of Mathematicians). This was the first historical study of mathematics in Austria. A fundamental work, it remains the main source for our knowledge of scientific activity in Vienna from the late 14th century until the early 16th century. It contains biographies of 26 astronomers and mathematicians in chronological order, starting with HEINRICH VON LANGENSTEIN (ca. 1340–1397) and ending with TANNSTETTER himself. Some of them (GMUNDEN, PEUERBACH, REGIOMONTANUS, STABIUS, STIBORIUS) are extensive and contain a list of their works; some are rather short and record little more than the name. (The *Viri Mathematici* has been reprinted in Latin and German in [GRAF-STUHLHOFER 1996, 156–171].)

STIBORIUS also reputedly wrote a *Büchlein über mathematische Schriftsteller* (Booklet about Mathematical Authors), but it has not survived.

11.3 Decline and Revival

In the course of the political changes of the 16th century — including the wars against the Turks and various plagues — universities in Austria lost their competence in mathematics. Moreover, schools (in monasteries and in cities) and universities came under the influence of the Jesuits, and the teaching of mathematics declined in importance.

By the second half of the nineteenth century, however, there was an increase in historical research in central Europe, although this did not extend to Austria. There were, however, isolated studies, done mostly by well-educated high school teachers.

For example, FRANZ VILLICUS, director of a school of commerce and a high official, wrote a carefully researched *Geschichte der Rechenkunst vom Alterthume bis zum XVIII. Jahrhundert* (History of Reckoning from Antiquity to the 18th Century) (1883). This went through three editions and was positively received not only among high school teachers but also by the general public. ([VILLICUS 1897] is the third edition.)

The famous analyst OTTO STOLZ (1842–1905), a student of WEIERSTRASS and professor of mathematics at the University of Innsbruck, had strong historical interests. His extensive knowledge permitted him to enrich his courses with remarks both on history and on recent developments. He always recommended a study of the masters, and urged his pupils to read original texts. Given that STOLZ's calculus books were the first and most profound textbooks in German on Weierstrassian analyis, it is not surprising that STOLZ also studied in detail the earlier contributions to analysis of BOLZANO, comparing them to the work of CAUCHY.

Since BOLZANO's manuscripts were kept in the library of the *Österreichische Akademie der Wissenschaften* (Austrian Academy of Sciences), STOLZ submitted his paper "B. Bolzano's Bedeutung in der Geschichte der Infinitesimalrechnung" (B. Bolzano's Relevance for the History of the Infinitesimal Calculus) for publication in the *Sitzungsberichte* of the Academy in March 1880. This met with unexpected difficulties, however. The paper was rejected because it did not give a full history of every previous attempt to define limits. This seemed to STOLZ an overwhelming task — how could one ever be sure that something had not been overlooked? His paper was eventually published in the *Mathematische Annalen* [STOLZ 1881].

In Hungary history of mathematics has not been neglected. There have been many studies — mainly on FARKAS (WOLFGANG) and JÁNOS (JOHANN) BÓLYAI — and on the emergence of science in the 19th century, most of which are written in Hungarian. The first detailed history is [SZÉNÁSSY 1992] which describes mathematics in Hungary up to 1900, including extensive biographical data ([BINDER 1994]).

11.4 The "Encyklopädie"

One of the largest collaborative enterprises ever attempted in mathematics was the *Encyklopädie der mathematischen Wissenschaften mit Einschluss ihrer Anwendungen* (Encyclopedia of the Mathematical Sciences, Including their Applications), initiated by FELIX KLEIN during a session of the *Deutsche Gesellschaft der Naturforscher und Ärzte* (German Society of Scientists and Physicians) in Vienna in 1894. This is not the place for a history of the *Encyklopädie* (see Sections 1.9 and 5.4.4), but it deserves to be mentioned here because the Austrian Academy of Sciences played a major role in it from the beginning. Although the articles in the *Encyklopädie* vary in quality, all are based on serious historical research, giving references to many otherwise forgotten contributions and putting them into a general perspective. Even now the study of late 19th- and early 20th-century mathematics is greatly enriched thanks to the relevant chapters in the *Encyklopädie.*

GUSTAV VON ESCHERICH (1849–1935) belonged to the first team of editors, and he urged his best pupils to contribute articles on their fields. Thus WILHELM WIRTINGER (1865–1945) wrote "Algebraische Funktionen und ihre Integrale" (Algebraic Functions and their Integrals) (1901), HANS HAHN (1879–1934) (together with E. ZERMELO (1871–1953)) contributed "Weiterentwicklung der Variationsrechnung in den letzten Jahren" (Progress of the Variational Calculus in Recent Years) (1904), and later HEINRICH TIETZE (1880–1964) and LEOPOLD VIETORIS (*1891) coauthored "Beziehungen zwischen den verschiedenen Zweigen der Topologie" (Interrelations Between the Various Branches of Topology) (1929).

Other Austrian mathematicians who wrote for the *Encyklopädie* included EMANUEL CZUBER (1851–1925) on "Wahrscheinlichkeitsrechnung" (Probability Theory) (1900) (one of the first articles to appear), PHILIPP FURTWÄNGLER (1869–1940) on "Die Mechanik der einfachsten physikalischen Apparate in Versuchsanordnungen" (The Mechanics of the Simplest Physical Equipment in the Arragement of Experiments) (1904), GUSTAV KOHN (1859–1921) on "Spezielle ebene algebraische Kurven" (Special Plane Algebraic Curves) (with GINO LORIA (1862–1954, **B**)) and on "Ebene Kurven dritter und vierter Ordnung" (Plain Curves of Third and Fourth Order) (1908), EMIL MÜLLER (1861–1927) on "Die verschiedenen Koordinatensysteme" (The Various Coordinate Systems) (1910), and Richard von Mises (1883–1952) on "Dynamische Probleme der Maschinenlehre" (Dynamical Problems of the Theory of Machines) (1911).

Wirtinger followed Escherich as editor and contributed articles on "Elliptische Funktionen" (Elliptical Functions) (prepared by J. Harkness (1864–1923) in Montreal and Wirtinger, published by R. Fricke (1861–1930) in 1913) and on "Abelsche Funktionen und allgemeine Thetafunktionen" (Abelian Functions and General Theta-Functions) (with A. Krazer (1858–1926) in Karlsruhe) (1920). Since Wirtinger did not really like to write such survey articles (he preferred research), it took a long time and the help of coauthors to complete them. Nevertheless he had a strong interest in history, as is shown, for instance, by a seminar on Archimedes in Greek which he offered regularly from 1922 until 1927.

Some later contributions by Austrian mathematicians to the *Encyklopädie* were: Konrad Zindler (1866–1934) on "Algebraische Liniengeometrie" (Algebraic Line Geometry) (1921), and Roland Weitzenböck (1885–1955) on "Neuere Arbeiten der algebraischen Invariantentheorie: Differentialinvarianten" (Recent Publications on the Algebraic Theory of Invariants: Differential Invariants) (1921).

11.5 Recent Developments

Since World War II, a number of mathematicians have maintained an active interest in the history of mathematics. Paul Funk (1886–1969) had new translations prepared of a number of letters of Regiomontanus. Edmund Hlawka (* 1916) whose historical knowledge exceeds that of many historians, has published many obituaries and autobiographical articles. Wilfried Nöbauer (1928–1988) regularly lectured on the history of mathematics [Kaiser/Nöbauer 1984], and Friedrich Katscher (*1923), a journalist and book collector, also did research on the history of π. Last but not least, Auguste Dick (1910–1993), a school teacher, started serious research after her retirement. Her contributions to Max Pinl's series, "Kollegen in einer dunklen Zeit" (Colleagues in a Dark Period) [Pinl 1974], [Pinl 1976], are indispensable for any investigation on mathematics in the middle of the 20th century, especially during the Nazi period. Her articles and books on Emmy Noether (1882–1935) [Dick 1970] are an essential contribution to the study of this famous algebraist. Auguste Dick took her studies very seriously. She went to cemeteries and churches to study their registers. She traced family connections and maintained an extensive international correspondence. Her archive, deposited with the Archive of the Austrian Academy of Sciences, awaits scholarly study.

In the field of institutional history, for a long time the dissertation *Geschichte des Studienfaches Mathematik an der Universität Wien* (History of the Mathematics Curriculum at the University of Vienna) [Peppenauer 1953], which was written as part of a history of the entire University of Vienna, was the only study of its kind. More recent investigations are [Einhorn 1985] and [Ottowitz 1992].

The history of mathematics finally became institutionalized in Austria in the 1960s. The *Kommission für Geschichte der Mathematik, Naturwissenschaften und Medizin der Österreichischen Akademie der Wissenschaften* (Commission for the History of Mathematics, Sciences and Medicine of the Austrian Academy of Sciences) was founded in 1961, followed by the *Österreichische Gesellschaft für Wissenschaftsgeschichte* (Austrian Society for the History of Science) in 1980. The latter society has its own journal, *Mitteilungen der Österreichischen Gesellschaft für Wissenschaftsgeschichte* (Notices of the Austrian Society for the History of Science), and regularly organizes meetings on the history of mathematics in Neuhofen an der Ybbs (1986, 1989, 1992, 1995, 1999).

The new "Wiener Kreis" (Vienna Circle) should also be mentioned. In contrast to its famous predecessor during the 1920s and 1930s, whose main tendencies were philosophy and foundations of mathematics, the new Circle has strong historical interests; these focus mainly but not exclusively on the historic Vienna Circle ([STADLER 1997]).

11.6 Conclusion

After isolated studies, beginning in the Renaissance with REGIOMONTANUS, historical research was greatly stimulated by the publication of the *Encyklopädie der mathematischen Wissenschaften mit Einschluss ihrer Anwendungen* towards the end of the 19th century, to which a number of Austrian mathematicians contributed. Although it has never been institutionalized by a chair at a university, study of the history of mathematics in Austria has nevertheless achieved modest gains in the past few decades, and is supported by individuals and by institutions with wider interests. There are regular lectures on various topics, and increasingly students have worked on themes devoted to the history of mathematics for their diplomas or their dissertations, for example [HOFER 1985], [FAUSTMANN 1994], and [GRUBER 1996]. Since 1986, regular meetings on the history of mathematics, attracting international attendance, have been organized in Neuhofen an der Ybbs.

Chapter 12

Greece

CHRISTINE PHILI

12.1 The Classical and Hellenistic Periods

Early Greek mathematics developed in close connection with the formation of abstract ideas, theories, and philosophy starting about 600 B.C. in Greece and parts of the eastern Mediterranean. Owing to a lack of original sources, very few details about this fundamental beginning of European thought can definitely be established. Yet the descriptions of historical events — at first embodied in mythical poetry, then increasingly taking on the form of factual, even rationalized historical reports — were an early part of Greek intellectual activity. During the classical period of PLATO (429–348) and ARISTOTLE (384–322), philosophical and scientific thinking was developed and refined in confrontation with the thought of prior schools, the Ionian philosophers, the Pythagoreans, etc.

Similarly, mathematicians were most likely aware of the importance of historical tradition and transmission. THALES and PYTHAGORAS reportedly acquired important mathematical knowledge during travels to neighboring countries, especially Egypt and perhaps Mesopotamia, although such reports cannot be verified. Occasionally, later mathematicians refer to "the ancients," indicating thereby that they were conscious of their indebtedness to the achievements of previous generations.

Due to the introspective nature of ancient Greek thought, ancient commentaries on Greek mathematics, and surviving fragments about its early history convey a picture that is both selective and colored by certain intentions. While the practical side of mathematics is largely neglected, its theoretical, deductive nature is strongly emphasized, as is the "disinterested" nature of Greek mathematics which stresses its abstract character as a primarily theoretical body of learning. The commentaries thus reflect a specific normative vision of mathematics (one of whose features is the necessity of proofs) which has influenced how mathematics and its historiography have been understood through all of European history (cf. [VITRAC 1996]).

The first writer on the history of mathematics, whose work survives only in fragments and citations, is EUDEMUS (4th century B.C.) of Rhodes, a disciple and friend of ARISTOTLE. Besides philosophical and other treatises, he wrote three histories of the mathematical sciences: on arithmetic, geometry, and astronomy. The largest fragments are to be found in PROCLUS's commentary on EUCLID and in SIMPLICIUS's commentary on ARISTOTLE's *Physics*.

Thanks to EUDEMUS, many important elements of the history of mathematics are known from the early period of its development. He held that arithmetic arose from practical matters: the needs of Phoenician merchants. Similarly, he believed geometry began with land surveying in Egypt, which was necessary after the flooding of the Nile. THALES then brought mathematics from Egypt to Greece and made his own contributions. To PYTHAGORAS and the Pythagoreans he attributed the pedagogy that enabled mathematics to spread in Greece. In addition to his historical remarks, EUDEMUS mentioned many mathematicians and philosophers up to his own time, including PLATO, ARCHYTAS, and EUDOXUS.

One of the fragments described HIPPOCRATES's treatment of the quadrature of lunes, others recounted the squaring of the circle (ANTIPHON's and BRYSON's methods), the application of areas (παραβολή, parabole), their exceeding (ὑπερβολή, hyperbole) and their falling short (ἔλλειψις, elleipsis), and several results attributed to the Pythagoreans. From EUDEMUS's history of arithmetic only one fragment (in PORPHYRY) remains; it concerns Pythagorean musical theory. For EUDEMUS's fragments see [WEHRLI 1955].

After this early period, no mathematicians of the stature of EUCLID or ARCHIMEDES and no historian of the caliber of EUDEMUS followed. However, many of the mathematicians of late antiquity included historical material in their summaries and compendia.

PAPPUS (fl. 300–350 A.D.) quoted liberally from the mathematical heritage of his epoch in his principal work, *Synagoge* (Mathematical Collection) in eight books. This was not so much a collection of his own contributions to mathematics as it was a presentation of several branches of mathematics that particularly interested him. Almost the whole of the treatise is extant in a twelfth-century manuscript. Particularly valuable from an historical point of view is book VII, since it gives an account of the *Treasury of Analysis*, a collection of Greek works, some lost.

PROCLUS (410–485) wrote commentaries on numerous Greek mathematical and philosophical classics, such as Plato's *Republic*. Having studied the texts of Greek mathematicians and prior commentaries, he wrote his own commentary on the first book of EUCLID's *Elements*, which was more philosophically than mathematically oriented. PROCLUS also utilized EUDEMUS's history of geometry in his work, while his pupil and biographer, MARINUS OF NEAPOLIS (Flavia Neapolis in Palestine), wrote a preface to EUCLID's *Data*.

This tradition continued with EUTOCIUS (6th century A.D.), a pupil of ISIDORUS OF MILETUS, who composed commentaries on the works of ARCHIMEDES, and with JOHANNES PHILOPONUS (490–566), who wrote on the *Arithmetic* of NICOMACHUS.

12.2 The Byzantine Period

As in late antiquity, conservation of the Greek heritage characterized Byzantine mathematics. Even more than in previous periods, the emphasis was on the writing of commentaries, and many such works resulted.

The tenth century encyclopedia known as the *Suda* is an invaluable source of information, including mathematics. A century later, MICHAEL PSELLOS (1018–ca. 1078) wrote encyclopedic scientific works and letters containing some additions and commentaries on DIOPHANTUS's *Arithmetic.* In the thirteenth century, GEORGIOS PACHYMERES (1242–ca. 1310) presented a paraphrase of the first book of DIOPHANTUS and some extracts of EUCLID and NICOMACHUS in a treatise on the quadrivium. In the 14th century MAXIMOS PLANUDES (1255?–1310) wrote a commentary to the first books of DIOPHANTUS. In this study, the symbol for zero and the nine numerals appeared for the first time in the Byzantine culture. PLANUDES's pupil and friend MANUEL MOSCHOPOULOS lived under the emperor ANDRONICUS (1282–1328) and wrote many commentaries and a treatise on magic squares. The monk, ISAAK ARGYROS (1310?–1371), was a disciple of NIKEPHOROS GREGORAS (1295–1359) and wrote commentaries on Euclidean geometry. All these works, however, were composed with the aim of explaining mathematical texts; they were not intended to make otherwise new contributions to the development of mathematics.

Throughout this period of commentary, the Byzantine Empire fought constantly for its survival. Following the Schism of the Eastern Orthodox Church (1054), first the Seljuks and the Normans attacked an Empire already weakened considerably by interior revolt. The taking of Constantinople in 1204 during the fourth crusade and the establishment of the Latin Empire of Constantinople (1204–1261) marked the beginning of the end for the Empire, despite its partial reconstruction under the PALAEOLOGUS dynasty. Constantinople fell to the Turks in 1453.

Nevertheless, Byzantium succeeded in maintaining its culture and its tradition; paradoxically, the times of turmoil also produced great cultural achievements. Learning flourished during and even after the "Palaeologian renaissance" of the 14th and 15th centuries.

12.3 Voulgaris and the Athonian Academy

After the conquest of Constantinople in 1453, some Greek scholars fled to the West, but others remained under the Ottoman regime [POULITSAS 1957]. Although most of their works may be characterized as compilations, they occasionally include useful historico-mathematical information.

After a centuries-long period of stagnation, members of the Greek orthodox clergy led a revival of Greek scholarship [CHRYSOSTOMOS 1821]. One of the main figures of this "Greek Enlightenment" (1750–1821) was EUGENIOS VOULGARIS (1716–1806), a widely esteemed scholar, teacher, and church leader. He had spent a few years in Venice, had studied mathematics with GIUSEPPE SUZZI (1701–1764), and is also supposed to have studied philosophy at the University of Padua (ca. 1739–ca. 1742). Having an extraordinary ability for languages, he mastered Italian, French, Latin, Hebrew, and later even Russian. Besides Aristotelian logic, Greek and Latin classics, Byzantine and later Greek history and theology, VOULGARIS was also interested in Newtonian physics, and in astronomy. He returned to Greece from Italy in 1742. At the newly established school of Ioannina (in Epire) he began his career of twenty years as a teacher in important educational centers of Hellenism. Many of VOULGARIS's translations of Western philosophers and mathematicians date from this period.

In 1749, MELETIOS, the abbot of the Vatopediou convent in Athos, established a school propagating Greek culture and the teaching of philosophy, logic, and theology at Mount Athos, thereby realizing an idea dating back to 1726. Financial problems, particularly the provision of an enormous building, were largely solved through the generosity and patronage of the patriarch of Constantinople, CYRIL V. In 1753 ,VOULGARIS was appointed as the first director of the *Athonias* (Athonian Academy). In his patriarchal edict of July 1753, CYRIL described VOULGARIS as a scholar "who can educate his students not only in grammar and logic, but also in philosophy and mathematics." In another edict that same year, the patriarch explicitly allowed the entrance of "young unbearded," i.e. secular, students to the Academy. In the five years from 1753 to 1758, the number of students rose from 70 to 200, with many coming from Greek communities in the Balkans and Asia Minor.

In his almost six years as director, VOULGARIS made enormous contributions to the intellectual rebirth of his country, by imparting his enthusiasm to others and even by contributing financially to the Academy. Imitating PLATO, he had inscribed over the Academy's door the words "Only those who know geometry may enter here ...." The mathematical sciences taught included arithmetic, geometry, physics, and astronomy. For the first time in the Ottoman period, Greece supported an institution of great intellectual standing and cultural ambition. When VOULGARIS left for Constantinople in 1759, however, the Athonian Academy entered a decline that ended in its closing in 1761. Despite this, many students — among them the distinguished scholar BENJAMIN OF LESBOS (1762–1824) — continued to travel to Mount Athos for their scientific education.

In 1764 VOULGARIS went to Germany, principally to publish his books. He spent over seven years there, mainly in Leipzig and Halle, where he met J. A. SEGNER and undertook a translation of SEGNER's *Elements of Mathematics* (Leipzig 1767).

Like VOULGARIS's translation of SEGNER's text into Greek, many mathematical books written in the 18th century by Greek scholars contain important elements of the history of mathematics in their introductions. Some, like VOULGARIS's, treat aspects of the history of ancient Greek mathematics. Others, such as KONSTANTINOS KOUMAS's *Elementary Series of Mathematics* (Vienna 1807), underscore the important intermediary role of Arab scientists. Finally some, like SPYRIDON ASSANIS's translation of ABBÉ DE LA CAILLE's book, *Leçons élémentaires des mathématiques*, address issues in the then contemporary development of mathematics.

12.4 The Ionian Academy

The members of the Ionian Academy — among them IOANNIS CARANDINOS (1784–1834) and OTTAVIO MOSSOTTI (1795–1863) — made important contributions to the intellectual rebirth of Greece. Established on the island of Corfu in 1824, the Academy lasted until 1864.

The Ionian islands had a turbulent history marked by numerous conquerors who left their marks, both politically and culturally. The first academy in Corfu — chief town of the Ionian islands — was established in 1695, during the Venetian occupation. Called *Assured*, it survived until 1716. At almost the same time another Academy was founded under the name of *Fertiles*, with a third, *Errants*, following in 1732. Unfortunately, none of these institutions has left any written traces.

During the first period of French occupation after the treaty of Campoformio from 1797 to 1798 (Corfu and Paxi until 1799), the French amassed an important library of 4,000 volumes by accumulating books from various monastic libraries on the island. They also set up a modern press (the first French document was published in Greece in May 1798) and offered 30 scholarships for students to study in Paris.

The treaty of Tilsit (1807) changed the political status of the Ionian islands; from 1800 to 1807 they constituted the Republic of Heptanesos which was created by the Russians under Turkish sovereignty. The French returned in 1807 and stayed until 1811 (Corfu and Paxi until 1815). They founded an association of cultivated persons under the name of the Ionian Academy in 1808 that aimed to combat the ignorance so prevalent on the island. According to the statutes drawn up by CHARLES DUPIN (1784–1873), the Academy was to consist of regular and corresponding members. Public courses, competitions and presentations of communications were all defining parts of its mission. Divided into three sections (Sciences, Ethics, Humanities), the Academy offered many subjects (botany, physiology, political economy, penal and civil law, etc.) and served to exemplify the methodical organization of useful teaching for the future of the island's residents.

DUPIN showed evidence of his multiform abilities during his mandate (1808–1811) in Corfu as a researcher in geometry, as an engineer in shipyards (of the islands), and as an educator through the creation of the Academy. (Unfortunately, all documents concerning the French period of the Ionian Academy seem to be lost, both in Greece and in France.) It seems that DUPIN's aim was to mark his sojourn in Corfu by creating an institution. Not only was he one of the founders of the Ionian Academy, he also served as its secretary. This Academy had the same status as the *Institut d'Egypt*. Thus it is possible that he tried to follow, or imitate, MONGE's creation, since MONGE had been his teacher. DUPIN was also a gifted teacher. He gave free public lectures on physics and chemistry. He taught analysis and mechanics privately to an excellent and talented young student, IOANNIS CARANDINOS. Later he continued to teach in Paris, at the *Conservatoire des Arts et Métiers*. (For a study of DUPIN in Paris, see the thesis of FERNAND PERRIN, supervised by RENÉ TATON [PERRIN 1983]).

When the Ionian islands became a British protectorate in 1815, FREDERIC NORTH, fifth Earl of Guilford (1766–1827), was nominated chancellor. He was the British representative on matters of education. A great admirer of ancient Greece, he served as the official representative of the British crown beginning in 1820 and immediately set about to realise his dream of reviving the Academy. In 1823 he succeeded in doing so when a new *Ionian Academy* was founded on Corfu. (University teaching officially began in 1824.) Many students from Corfu also studied in Western Europe on scholarships from Lord GUILFORD. CARANDINOS, for instance, so impressed him with his enthusiasm for the sciences that the Earl sent him to study in the prestigious *Ecole Polytechnique* in Paris. There CARANDINOS attended the lectures of A. L. CAUCHY and studied his *Cours d'Analyse* (1821).

Returning to Corfu in 1823, CARANDINOS resolved to teach what he had learned and to organize the Academy. He taught mathematics and mechanics there until 1832, and served as the Academy's rector. Though little remembered today, CARANDINOS's activities deserve to be stressed. In 1826, he published as his first work a Greek translation of a section of LACROIX's *Complément des Eléments d'Algèbre* (Paris 1798). It contained LAGRANGE's results and methodology on the resolution of numerical equations, which LAGRANGE had first published more than fifty years earlier, in 1770–1771, in the *Memoirs* of the Academy of Berlin. Through his translations, CARANDINOS made an important contribution to the development of a Greek mathematical terminology and provided an intellectual climate favorable for an increasing interest in the history of mathematics.

12.5 The Greek National State (1822 to the Present)

After the re-establishment of the Greek state in 1822, institutions to support scientific research were founded: the University of Athens (1837) and the School of Arts (1837), which later became the National Technical University of Athens. By the end of the 19th century, although the history of mathematics did not exist as an autonomous discipline, many re-editions of classical texts produced by classical scholars included material related to the history of mathematics. When, upon an initiative of the French Mathematical Society, at the International Congress on Bibliography of the Mathematical Sciences held in Paris in 1889, an International Committee was set up under the presidency of HENRI POINCARÉ (1854–1912), KYPARISSOS STEPHANOS (1857–1917), internationally known for his mathematical research, became the representative of Greece [PHILI 1999]. There is also a study by a Greek author of an historical character: a survey of the history of non-Euclidean geometry; its author had studied mathematics with FELIX KLEIN [KARAGIANNIDES 1893].

The first scholar responsible for serious study of the history of science in modern Greece was MICHAEL STEPHANIDES (1868–1957). Although not an historian of mathematics (his main interest was the history of natural sciences), STEPHANIDES did write some papers on the history of mathematics, e.g. "The Mathematics of the Byzantines" (*Athena* (1923), 206–218). Following his studies at the University of Athens, STEPHANIDES taught mathematics and physics in high schools from 1896 to 1912 and, for the first time, introduced a weekly course devoted to the history of science in high schools, while continuing his own research. When, in 1924, he was appointed to the first chair for the history of science at the University of Athens, GEORGE SARTON noted:

> The National University of Greece has taken the initiative of devoting a chair to the History of Sciences .... As far as we know, the chair of the University of Athens is at present the only chair in the world devoted to our studies.... How long will it take before other university chairs are founded in the nobler phases of human evolution? [*Isis* **8** (1926), 158]

Similarly, FRIEDRICH DANNEMANN wrote:

> The University of Athens is to be praised. A first chair for the history of the sciences has been founded there, and efforts are being made to derive modern scientific culture from classical Greek culture.[1]

STEPHANIDES taught at the University of Athens until 1939. He was a member of the Academy of Athens, of the International Academy of the History of Science, and was one of the founders of the (American) History of Science Society. He wrote many books and numerous papers, including *The Sciences in Greece before the Revolution* (1926), *L'histoire des sciences en Grèce* (1932), and [STEPHANIDES 1932].

One family of note for history of mathematics in Greece is that of IOANNIS HADJIDAKIS (1844–1921). HADJIDAKIS, who studied at the University of Athens, continued his education in France at the University of Paris, and in particular in Germany, where in the course of five years at the University of Berlin (1869–1873) he attended the lectures of (among others) KRONECKER, KUMMER, and

[1] "Rühmend ist die Universität Athen zu nennen. Man hat dort zuerst einen Lehrstuhl für die Geschichte der Wissenschaften gegründet und bemüht sich die neuere wissenschaftliche Kultur aus der griechischen abzuleiten" [*FF* **7**, Nr. 27 (20. September 1931), 368].

WEIERSTRASS. Upon returning to Greece, HADJIDAKIS taught mathematics, wrote textbooks, and helped to found a kind of academy named the Scientific Society in 1888. The Society published an annual journal, *Athena*.

HADJIDAKIS devoted an Inaugural Lecture (May 15, 1880) as an instructor (Maître de Conférences) at the University of Athens to the "History of Mathematics in Ancient Greece" [HADJIDAKIS 1885], which reflected his deep erudition and demonstrated his qualifications as an historian of mathematics. When shortly thereafter he gave his Inaugural Lecture as Professor of Mathematical Analysis, it was devoted to the development of the infinitesimal calculus, which he covered down to the time of its establishment by NEWTON and LEIBNIZ [HADJIDAKIS 1884]; for details, see [PHILI 2000].

Meanwhile, NIKOLAOS HADJIDAKIS (1873–1941), son of IOANNIS, had become a distinguished mathematician. Having studied in Paris, Göttingen, and Berlin, he was the first in Greece to teach a weekly course on the history of mathematics, beginning in 1905. His wife being Danish, HADJIDAKIS had a special interest in Scandinavian mathematics, and he wrote an article devoted to the subject [HADJIDAKIS 1899]. In another article, he reported about the state of higher mathematics in Greece [HADJIDAKIS 1901/02]. Somewhat later, in a survey article, he also considered contemporary mathematics, mentioning the major figures at the turn of the century. He not only described their research, but highlighted as well the international mathematical movement and its indispensable role for the development of mathematics in Greece [HADJIDAKIS 1910].

However, it was EVANGELOS STAMATIS (1898–1990, **B**) who inaugurated serious research on the history of mathematics in modern Greece. STAMATIS is especially well known for his research on ancient Greek mathematics. Following his studies at the University of Athens, he went on to the University of Berlin in 1931 and then returned to Berlin again for further study from 1936 to 1940. When he went back to Athens, he began a systematic analysis of the ancient Greek mathematical heritage. STAMATIS presented numerous translations of classical Greek texts into modern Greek, including the works of ARCHIMEDES, EUCLID's *Elements*, APOLLONIUS's *Conics*, and DIOPHANTUS's *Arithmetica*. Especially important was STAMATIS's collaboration with the publishing house B. G. Teubner, which reissued HEIBERG's editions (with minor corrections and additions by STAMATIS) of EUCLID's *Elements* (1969,1970, 1972, 1973, 1977) and the *Opera Omnia* (Vols. I–II, 1972) of ARCHIMEDES. STAMATIS wrote more than ten books and numerous papers. He was also a member of the International Academy of the History of Science.

Although there has never been a chair for the history of mathematics, nor a department or special course for history of mathematics in any Greek university, the subject has been studied sporadically by some mathematicians who have published on the history of mathematics (including CONSTANTIN CARATHÉODORY: "Einführung in Eulers Arbeiten über Variationsrechnung" (Introduction to EULER's Publications on the Calculus of Variation) in L. EULER: *Opera Omnia* (I) **24**

(1952), viii–lxii, and PANAGIOTIS ZERVOS: *Mathematics and its Relations to other Sciences and to Philosophy*, Inauguraldissertation, University of Athens, 1918).

In the last twenty years there has been increasing activity in the history of mathematics, and a number of historians of mathematics have started to work in different fields of research, related both to the history of ancient Greek mathematics and to modern Western mathematics in general.

12.6 Conclusion

In ancient Greece, reflections about the origin and progress of mathematics seem to have arisen almost as soon as mathematics began to develop as a discipline. Historical narratives have been transmitted only in fragments, but are supplemented by a number of commentaries on specific mathematical texts. These all testify that, in Greece, philosophical, epistemological, and methodological considerations accompanied the production of mathematical knowledge from a very early stage, and that they left remarkable traces in historical reports as well. In addition to providing a chronology of important events, the texts transmit an ideal conception of mathematics as a pure, deductive science. This image became not only decisive for the self-understanding of mathematics up to modern times, but also normative for the future development of the historiography of mathematics, when it was re-awakened in European humanism.

To the Byzantine and Renaissance traditions of producing editions and commentaries on ancient texts, the Greek Enlightenment (1750–1821) began to stimulate interest in the history of mathematics through the efforts of (among others) EUGENIOS VOULGARIS, who translated the works of many Western philosophers and mathematicians into modern Greek. After establishment of the Greek state in 1822, the University of Athens was founded not long thereafter, in 1837. Towards the end of the century, IOANNIS HADJIDAKIS gave occasional lectures on history of mathematics, both ancient Greek and modern, but it was his son, NIKOLAOS HADJIDAKIS, who was the first to offer a course of weekly lectures on history of mathematics at the University of Athens, beginning in 1905.

Several decades later, in 1924, the University of Athens created a chair for the history of science, which has also served indirectly to promote history of mathematics. Its first occupant, MICHAEL STEPHANIDES, wrote occasionally on the history of mathematics. It was EVANGELOS STAMATIS, however, who began serious study of history of mathematics, especially of ancient Greek mathematics, and he produced numerous editions of texts in modern Greek as well as his own research. His legacy continues in the work of a few scholars, whose interests range more broadly over the spectrum of the history of mathematics, from ancient Greece to modern times.

Chapter 13

Spain

Elena Ausejo and Mariano Hormigón

13.1 Introduction

The historiography of mathematics in Spain has a long and uneven history. For example, as early as 1068, Sa^c^id al-Andalusi, a scholar from Toledo, wrote his *Book on the Categories of Nations* [Sa^c^id Al-Andalusi 1935], which includes discussion of mathematicians and their works. This book in particular reflects a serious level of attention given to the historiography of mathematics in a part of the Iberian Peninsula already regarded (although from afar) as Spain. The rise of Christianity, however, eventually succeeded in suppressing much of the previous Islamic cultural presence, reducing significantly the place of science and historiography in al-Andalus.[1]

Through the 16th century, this loss was not considered important; Spain, thanks to its conquests in the Americas, could simply import whatever mathematics it needed. As Spanish wealth began to decline with the depletion of South American silver, it was necessary to consider new alternatives. This prompted an evaluation of who had articulated earlier concepts and ideas.

The conflicts that accompanied the formation of the liberal state in 19th-century Spain did not overlook the exact sciences. Generally speaking, those who overvalued national science were the crown, the aristocracy, the clergy, and the defenders of the Old Regime. Their basic assumption was that the Spanish empire had once been a world superpower, controlling an enormous territory and promoting numerous intellectual efforts, all of which were exceptional (the *Golden Age*). Due to the envy of Spain's adversaries, the argument went, the merits of Spanish geometricians had been obscured. In support of such claims, lengthy catalogues

[1] The bibliography of Spanish-Muslim science is extensive; the most recent synthesis produced by the so-called Spanish school of Millás is [Samsó 1992].

of authors and titles were produced the mere existence of which was accepted as a sign of the inherent quality of the works included. As a result, it was taken as self-evident that everything — even science — had been good in Spain under an absolute monarchy that should therefore be maintained.

On the other hand, certain intellectuals were unhappy about what had happened to Spain, despite their personal prosperity. They found notable successes in philosophy and theology, literature and art. But when they looked at international lists of scientists and mathematicians, they took the names as evidence that, in Spain, the absolute state had failed in science. Assuming that science was indeed an indispensable part of the modern world, Spain had to overcome the results of its delayed development in order to survive. As for mathematics, as JOSÉ ECHEGARAY (1833–1916, **B**) observed in 1866: "Mathematics does not owe us anything: it is not ours; there is no name that Spanish lips can pronounce without an effort;"[2] there are only "books of accounts and geometries of tailors"[3] in Spanish mathematics [ECHEGARAY 1866, 185, 176]. Thus he expressed his dismay about the fact that all the important mathematicians in history were foreigners (not Spanish) and that Spanish mathematicians only produced elementary books on arithmetic and geometry.

In the 19th century, Spanish histories of mathematics were basically of three types. General histories of mathematics were written by authors like ZOEL GARCÍA DE GALDEANO (1846–1924, **B**) or VENTURA REYES PRÓSPER (1863–1922, **B**). There was also a focus on the scientific merits of Spanish mathematicians in the 16th and 17th centuries. Finally, the progressive expansion of mathematical knowledge was also examined from the tentative beginning of industrial development, starting in the second third of the 19th century, up to the beginning of the last civil war (1936–1939).

Some studies were published by such important institutions as the early *Real Academia de Ciencias Exactas, Físicas y Naturales (RACEFNM)* (Royal Academy of the Exact, Physical and Natural Sciences) in Madrid, which published the journal *Revista de los Progresos de las Ciencias Exactas, Físicas y Naturales* (Review of the Progress of the Exact, Physical and Natural Sciences). Authors like GALDEANO and REYES PRÓSPER also published on topics in the general history of mathematics, but these two authors did not publish their works in the Academy's journal but in journals like *El Progreso Matemático*, *Revista de la Sociedad Matemática Española*, or *Revista Matemática Hispano-Americana*.

[2] "La ciencia matemática nada nos debe: no es nuestra; no hay en ella nombre alguno que labios castellanos puedan pronunciar sin esfuerzo."

[3] "Libros de cuentas y geometrías de sastres."

13.2 Polemics With Respect to Sixteenth-Century Spanish Mathematicians

As in other European countries during the Renaissance, mathematics was developed in Spain specifically for its applications to navigation [Cano Pavón 1993] as well as to artillery and engineering at the *Academia de Matemáticas* (Academy of Mathematics) founded by Philip II (1527–1598) in 1572 [Vicente 1991]. Mercantile arithmetics were also written, some books of Euclid were translated, instruments were manufactured and some original works were produced, although these were of uneven quality. One especially indispensable reference work that includes the majority of Renaissance texts is *Apuntes para una biblioteca científica española del siglo XVI* (Notes for a Spanish Scientific Bibliography in the 16th Century) [Picatoste 1891] by Felipe Picatoste Rodríguez (1834–1892, **B**).

By the end of the 19th century, the *regenerationist* climate sought to look beyond the era of Spain's great exploits and to begin a new period emphasizing everyday work and improvements in the standard of living for the majority of Spain's citizens. The translation of such ideals into mathematics was made by Galdeano and Reyes Prósper, among others. In short, appreciating the level of Spanish Renaissance mathematics was less important than raising the level of contemporary knowledge of the sciences in Spain, and this was no less true of works on the history of mathematics.

At the turn of the century, the Spanish mathematical community and those who were most interested in its history, following the leadership of Luis Octavio de Toledo (1857–1934) — redirected their objectives towards research.[4] By the 1910s it looked as if this new initiative had produced its first concrete results in the work of the mathematician Julio Rey Pastor (1888–1962, **B**). When Rey Pastor was invited to deliver the inaugural lecture for the 1912–1913 academic year at the University of Oviedo, he chose as his topic "Spanish Mathematicians of the 16th Century" [Rey Pastor 1913], and launched his arguments against traditionalist positions. By examining those works regarded as key for the history of mathematics, he attempted to end debate by disclaiming the merits of all Renaissance mathematicians — except for half a dozen — including, of course, the Spanish ones. The crudeness of the conclusion only contributed to further exclusion from historical analysis of figures who could have helped to shed light on the historical situation.[5] Often during the first third of the 20th century, when

[4]This resulted, in part, from the military defeat of Spain (and the accompagnying loss of the last Spanish colonies of Cuba, Puerto Rico, and the Philippines) during the Spanish-American War in 1898. It meant the end of the Spanish imperial dream (or nightmare). In this sense, a new period of regeneration (i.e., modernization according to European standards) started in all domains, including mathematics. The Nobel Prizes in Literature were awarded to Echegaray in 1904 and in Medicine to Ramón y Cajal in 1906, and the foundation of the *Junta para Ampliación de Estudios e Investigaciones Científicas (JAE)* (Council for Improving Study and Scientific Research) occurred in 1907.

[5]Such as Reyes Prósper [Reyes Prósper 1911].

the subject of 16th-century Spanish mathematicians arose, strong views were expressed in opposition to those of REY PASTOR.[6] In turn, REY PASTOR tried to prove he was right by requiring some of his students to carry out further studies on authors he had already condemned.[7] Unfortunately, the history of Spanish Renaissance and Baroque mathematics is still largely unknown, even in Spain. REY PASTOR had become the most important mathematician in 20th-century Spain by the 1960s; for this reason his negative views about Spanish Renaissance and Baroque mathematics discouraged interest in these periods.

13.3 Mathematics in the Liberal State

Modern mathematics in Spain may be said to coincide with the rise of liberal economy and politics (around 1833). The reform of the Spanish education system, advancing from the Old Régime to the new one, lasted almost one century. It began in the reigns of FERDINAND VI (1746–1759) and CHARLES III (1759–1788) and achieved official legal status through the first Law of Public Instruction enacted under the minister CLAUDIO MOYANO in 1857.

In addition to trying to recruit more and better mathematicians in all fields, science needed social prestige. The public, however, found it more difficult to appreciate the benefits of mathematics than, say, those of life-saving vaccines. Mathematics thus required a subtle approach in its public relations efforts in order to take its place among the emerging sciences. The use of the history of mathematics as part of this strategy can be explained as part of the first stage of the institutionalization of mathematics in the liberal framework that developed in Spain during the second third of the 19th century. Another significant factor is the foundation of the RACEFNM in Madrid in 1847. The Academy required that all lectures delivered there reflect original research. Perhaps for this reason, or perhaps because some academicians did not dare touch on more technical subjects, a certain number of historical works appeared in the Academy's early years. However, the historicist affirmation of science can be traced back in Spain at least to the Enlightenment. Examples of such efforts include a history of the Spanish navy and its connection with mathematics (1846) by MARTÍN FERNÁNDEZ DE NAVARRETE (1765–1844, **B**), who was a moderate, enlightened, liberal navy man with links to 18th-century mathematics.

Artillerymen and military engineers played crucial roles in helping to establish and develop modern mathematics in Spain.[8] Two military engineers, ANTONIO REMÓN ZARCO DEL VALLE (1775–1866) and MANUEL MONTEVERDE († 1868), both of whom were members of the founding core of the RACEFNM, included

[6]The most significant case is FRANCISCO VERA (1888–1967, **B**), a republican mathematician whose work remained unfinished due to the Spanish Civil War (1936–1939).

[7]The most important of these is the doctoral thesis of JOSÉ Mª LORENTE PÉREZ [LORENTE PÉREZ 1921].

[8]On this subject, see the synthesis and further references offered in [VELAMAZÁN 1994].

themes related to the history of science and mathematics in their speeches on the occasion of the Academy's incorporation. ZARCO DEL VALLE's speech, on "Condiciones favorables que España reúne, por su posición geográfica y su topografía física, para el cultivo de las ciencias" (Favorable Conditions that Spain Embodies, by its Geographical Position and Physical Topography, for Cultivation of the Sciences), was a sort of military sermon directed to all sectors of society in hopes of dignifying the study of science. MONTEVERDE's work, later published as a report by the Academy [MONTEVERDE 1853], was directly intended to give social prestige to mathematics. However, these speeches could not escape the dry, scholarly framework within which they were presented, and therefore they were almost totally lacking in originality.

The University was not completely left out of this process. For example, the professor of mathematics on the Faculty of Philosophy at Madrid, FRANCISCO DE TRAVESEDO (1786–1861), defended his doctoral thesis in 1855 — when he was nearly seventy — on *Los progresos de las matemáticas entre los antiguos y el obtenido por los modernos* (The Progress of Mathematics among the Ancients and the Moderns).

At the beginning of the second half of the 19th century, however, Spain's scientific structures were extremely weak. In this context, general overviews were important because they fulfilled two functions: to place before the public scientific themes that were as modern as possible, and to publicize topics in the history of science that were not included in more specialist works. However, except for the above-mentioned studies on Renaissance themes, and for the debate on the advancement or stagnation of science in the 19th century, there was no original research, despite the appearance of material of great historiographic importance — such as *Los libros del saber de astronomía* (Books on the Knowledge of Astronomy) by ALPHONSE X the Wise (1221–1284).

Although slowly and despite financial problems, the number of professional mathematicians began to increase, and this in turn encouraged retrospective reflection on their activity. The visibility of the history of mathematics also increased with the appearance in 1891 of the first Spanish mathematical journal, *El Progreso Matemático* (The Progress of Mathematics). Its founding editor, GARCÍA DE GALDEANO, as well as some of the journal's contributors, sometimes used an historical approach to present mathematical themes (the geometry of triangles or projective geometry, for instance). Both in GALDEANO's journal and in the subsequent journals, the history of mathematics — whether generally or in Spain — was always represented.

The professional life of Spanish mathematicians was thus progressing at the beginning of the 20th century when REY PASTOR sought to evaluate the evolution of mathematics in 19th-century Spain [REY PASTOR 1915]. REY PASTOR, the most important Spanish mathematician of his day, had received numerous honours, but did not enjoy commensurate influence due to a domineering, controlling nucleus in the Faculty of Science at Madrid that opposed him. REY PASTOR reacted violently, criticizing 19th-century Spanish mathematicians, and claiming that for mathematics in Spain, the 19th century did not begin until 1865. Even then, he maintained, there were only four mathematicians above contempt: to protect his rear guard, he named the influential ECHEGARAY and EDUARDO TORROJA Y CABALLÉ (1847–1918), REY PASTOR's thesis director and head of the Madrid clan; but to be just he added GALDEANO and REYES PRÓSPER because they deserved to be held in high esteem. REY PASTOR likened the rest to bats because they promoted darkness and delay. The consequence was not a liberating self-criticism, but a rigid conformity throughout most of the mathematical community. As was true in the Renaissance, only a few with strong personalities — and sufficient maturity — were able to assert their independent views. Among these was GALDEANO, who delivered objective lectures on the recent evolution of mathematics [GARCÍA DE GALDEANO 1921].

Luckily, the flexibility of Spanish life encouraged some scholars to pursue their own interests, independently from their normal responsibilities. Their work dealt with many aspects of the history of mathematics, including the controversial 16th and 19th centuries. Shortly before the Fascists came to power in 1936, a group of these historians constituted the *Asociación de Historiadores de la Ciencia Española* (Association of Historians of Spanish Science), which the new regime soon suppressed.

During more than 35 years of dictatorship (1939–1975), only the history of medicine received any significant institutional support and development. Nevertheless, there were still valuable isolated studies produced by individuals and by a few noteworthy collective groups of scholars.[9]

Beginning in 1976, contemporary history of mathematics in Spain has been seriously pursued, in spite of institutional difficulties, to a remarkable and significant degree. Above all, the *Sociedad Española de Historia de las Ciencias y de las Técnicas* (SEHCYT) (Spanish Society for the History of Science and Technology), with its congresses, symposia, and its journal, *Llull*, has provided a foundation upon which a growing number of faculty and students, along with serious writers and popularizers, now comprise a substantial community of historians of science in Spain.

13.4 Conclusion

Despite sporadic interest in the history of mathematics in the Iberian Peninsula from medieval times — and especially in the early Renaissance, Spain's "Golden Age" — a tradition of sustained research only began in the late 19th century. Initially, this involved a critical reassessment of 16th-century contributions by

[9]NORBERTO CUESTA DUTARI (1907–1989), with his works on 18th-century mathematics in Spain, is the most outstanding individual figure. JOSÉ AUGUSTO SÁNCHEZ PÉREZ (1882–1958, **B**) published a number of studies on the history of mathematics in Muslim Spain. As for collective efforts, the above-mentioned Spanish school of MILLÁS VALLICROSA (footnote 1) is the most outstanding.

Rey Pastor. It soon broadened, however, to include the more recent history of the 19th century. One characteristic of 19th-century Spanish mathematics was the centralization of academic political power and influence within the University of Madrid. Nevertheless, more forward-looking mathematicians in the provinces, like Galdeano, promoted competing initiatives not only in mathematics but also in the history of mathematics. While these initiatives seemed promising, the Spanish Civil War and its aftermath witnessed a period of diminished activity.

The situation began to change in the last quarter of the 20th century. One reflection of the vitality of the growing community of historians of mathematics in Spain is the series of Symposia organized at the University of Zaragoza in honor of García de Galdeano, of which four have been held to date (2000) on a wide range of topics, and drawing an international list of participants. These have included "Mathematical Journals" (1991), "Paradigms and Mathematics" (1994), "Science and Ideology" (1996), and "The Hilbertian Paradigm: Mathematics at the Turn of the Twentieth Century" (1999).

Chapter 14

Portugal

LUIS M. R. SARAIVA

14.1 Introduction

There has never been a school of history of mathematics in Portugal. Works on the history of mathematics have come rather from isolated researchers working with specific goals determined by individual circumstances. In particular, five mathematicians — FRANCISCO DE BORJA GARÇÃO STOCKLER (1759–1829, **B**), FRANCISCO DE CASTRO FREIRE (1811–1884, **B**), FRANCISCO GOMES TEIXEIRA (1851–1933, **B**), RODOLFO FERREIRA DIAS GUIMARÃES (1866–1918, **B**), and PEDRO JOSÉ DA CUNHA (1867–1945, **B**) — authored works that served to define a strand of interest in the history of mathematics in Portugal from the late 18th through the 20th centuries. An analysis of these men and their works largely shapes this essay.[1]

14.2 The Beginnings: Garção Stockler

Until the 18th century, there were no relevant writings by Portuguese authors on the history of mathematics. From the middle of the 16th century to the middle of the 18th century, the Portuguese cultural and scientific world suffered a period of general stagnation and isolation. Under the leadership of CARVALHO E MELO (1699–1782), the Prime Minister of King D. JOSÉ and future MARQUIS OF POMBAL, a strategy was adopted to bring Portugal up to the level of the most cultured countries of the European Enlightenment. The University Reform of 1772, aiming to modernize the Portuguese University and bring it into line with its European counterparts, represented the most striking cultural and scientific change due to

[1] All Portuguese titles of papers and books are given in English translations.

this initiative. In 1779, moreover, the Lisbon Academy of Sciences was founded. Its two main instigators were the naturalist and diplomat, José Francisco Correia da Serra (1750–1823), and D. João de Bragança de Sousa e Ligne (1719–1806), the second Duke of Lafões, who had lived in England during the major part of Pombal's government and was a member of the Royal Society of London. Until the middle of the 19th century the Academy was the only Portuguese institution that supported work in the history of mathematics, publishing books and papers in its series of *Memoirs*.

Among early Academy members, António Ribeiro dos Santos (1745–1818) and Garção Stockler made noteworthy contributions to the history of mathematics. Ribeiro dos Santos, a Doctor of Law and head of the Library of Coimbra University beginning in 1777, wrote several papers on the history of Portuguese mathematics in the *Memorias de Litteratura da Academia das Sciencias de Lisboa.* In 1806, for example, he published two papers in volume 7: "Memoir on the Life and Writings of Pedro Nunes" [Santos 1806b] and "Memoir on the Life and Writings of Don Francisco de Mello" [Santos 1806a]. Pedro Nunes (1502–1578) was Portugal's leading mathematician of the 16th century, and Don Francisco de Mello (1490–1536) was a clergyman who graduated in mathematics from the University of Paris and wrote several papers on mathematics and physics. In 1812, his *Historical Memoirs on Some Portuguese and Foreign Mathematicians Based in Portugal* appeared in volume 8 [Santos 1812].

In contrast to the librarian Ribeiro dos Santos, Francisco de Borja Garção Stockler was a mathematician who later served as a general in the army and as secretary of the Academy of Sciences. He published mathematical works on analysis as well as on the history of mathematics. In 1805 the first volume of his *Works* appeared, and contained several of his historical papers: "Eulogy of D'Alembert" [Stockler 1805a], "Eulogy of José Joaquim Soares de Barros e Vasconcellos" (a Portuguese astronomer and member of the Berlin Academy of Sciences) [Stockler 1805b], and a "Memoir on the Originality of Portuguese Naval Discoveries during the XVth Century" [Stockler 1805c].

In 1819 Stockler published in Paris his *Historical Essay* [Stockler 1805d], the earliest history of mathematics devoted to a single European country. The work grew from Stockler's sense of the moral duty of the cultured citizen. Faced with a general indifference to Portugal's cultural tradition and worried about the scarcity and fragility of documents, Stockler felt obliged to do what he could to preserve the nation's scientific heritage. His book thus includes transcriptions of documents, or brief accounts of their contents, which are not always commensurate with their mathematical relevance.

On a methodological level, Stockler followed the generalistic approach of the *Histoire des Mathématiques* by Jean Etienne Montucla (1725–1799, **B**) [Montucla 1799/1802]. Like Montucla, whom he quoted extensively, Stockler believed that the methods used to analyze topics in general history sufficed for the historical analysis of mathematics. He thus viewed the historian's role to be:

1. The analysis of the nature of the country, combining its physical character with the successive changes in its moral and political situation;

2. The compilation of data from historians and public archives;

3. The comparison of contemporaneous works and of contemporaneous works with earlier ones;

4. The formulation of hypotheses when there is not enough information to decide on a definite historical interpretation.

Assuming these roles in writing his *Historical Essay*, STOCKLER divided the history of Portuguese mathematics into three main periods, a periodization that later historians have perpetuated. According to STOCKLER, the first period opened in the middle of the 15th century with works of HENRY THE NAVIGATOR (1394–1460), and closed with the end of of the reign of King D. JOÃO III (1502–1557). The leading mathematician in this period was PEDRO NUNES (1502–1578), whose works STOCKLER described in some detail. A period of decadence followed, which lasted for some two centuries, until the middle of the 18th century. STOCKLER isolated two main causes for this: the uncontrolled power of the Inquisition and the Jesuit monopoly on public instruction. This view was adopted by all major historians of Portuguese mathematics until PEDRO JOSÉ DA CUNHA (who had accepted it at first) abandoned it in the 1940s [CUNHA 1940c]. JOSÉ VICENTE GONÇALVES (1896–1985) also questioned STOCKLER's interpretation of the second period [GONÇALVES 1966]. Finally, STOCKLER defined a third period in the history of Portuguese mathematics that began in the middle of the 18th century, when King D. JOSÉ came to power. This period witnessed the recovery spearheaded, in STOCKLER's view, by the outstanding work of JOSÉ ANASTÁCIO DA CUNHA (1744–1787).

14.3 The Second Half of the Nineteenth Century: Francisco de Castro Freire

During the first half of the 19th century, Portugal was ravaged by the Napoleonic invasions, by extended periods of civil unrest, and by actual Civil War. During this period there was almost no mathematical activity. Eventually, there were signs of social and economic recovery, and by mid-century the country had achieved some degree of stability. The Academy Reform of 1851 led to the founding of two new Academy publications: a *Buletim da Academia Real das Sciencias de Lisboa* to report the proceedings of its meetings, and a periodical (named *Jornal de Sciencias Mathematicas, Physicas e Naturaes*) for scientific works that might not ordinarily have been included in the *Memorias da Academia Real das Sciencias de Lisboa*.

In 1852, a scientific and literary academy, *O Instituto*, was founded in Coimbra. Comprised of three classes — Moral and Social Sciences, Physical and Mathematical Sciences, and Literature and Arts — these directly reflected the initial structure of the Lisbon Academy of Sciences. *O Instituto* published a governmentally subsidized journal as well, also called *O Instituto*, that published scientific and literary papers by members of the university's faculties, including many important papers on Portuguese history of mathematics.

Within this period, the mathematician FRANCISCO DE CASTRO FREIRE turned his attention to the history of mathematics. He had joined the Faculty of Mathematics at Coimbra after graduating there in 1830 and later became a sub-dean of the university and president of *O Instituto*. His noteworthy *Historical Memoir on the Faculty of Mathematics* was published in Coimbra in 1872 [FREIRE 1872]. Chapter I, the most extended and important, recorded a wealth of data on syllabi, laws governing the university, texts adopted, and the development of the teaching staff. Extended notes on the life and works of JOSÉ MONTEIRO DA ROCHA (1734–1819) [FREIRE 1872, 32–33] and JOSÉ ANASTÁCIO DA CUNHA [FREIRE 1872, 33–37] preceded an account of the students of these two mathematicians [FREIRE 1872, 47–56]. The third and final chapter analyzed the influence of the Faculty of Mathematics in Portugal by focusing on the works and teaching posts assumed by its graduates. Some went on to teaching posts in the naval academies, and a number of the faculty were admitted as members of the Academy of Sciences. CASTRO FREIRE also emphasized the work of his colleagues: General PHILLIPE FOLQUE (1800–1874) was one of the pioneers of geodesy in Portugal, and a teacher in the Faculty of Mathematics, while DANIEL AUGUSTO DA SILVA (1814–1878) and GOMES TEIXEIRA were together the most important Portuguese mathematicians of the 19th century [FREIRE 1872, 115–116]. In thus singling out his colleagues, CASTRO FREIRE demonstrated awareness of the historical importance of his contemporaries.

While the amount of information contained in this work is considerable and its fifty-page bibliography of Portuguese mathematics (including history of mathematics) was compiled with great care, the same cannot be said for its historical analysis. CASTRO FREIRE essentially limited himself to a compilation of data. In the few places where he sketched an analysis, he offered no critical insights and displayed no real historical sensitivity. CASTRO FREIRE essentially composed a eulogy to the Faculty of Mathematics from which all critical elements were missing. This should not however, detract from the value of a work which, above all, compiled data that provide a detailed picture of the evolution of Portuguese mathematics in the century from 1772 to 1872.

14.4 The Golden Age of Portuguese Historiography of Mathematics: 1900–1940

The last quarter of the nineteenth century witnessed a great increase in the number of mathematical works published in Portugal. Many of these appeared in the *Jornal de Sciencias Mathematicas e Astronomicas.*[2] This was founded by GOMES TEIXEIRA, an energetic mathematical activist who created a wide network of contacts within Portugal as well as abroad. The production of papers in the history of mathematics also increased, and three main figures were largely responsible for this: RODOLFO GUIMARÃES, PEDRO JOSÉ DA CUNHA, and GOMES TEIXEIRA.

RODOLFO FERREIRA DIAS GUIMARÃES pursued a career in the military, attaining the rank of colonel. A member of the Academy of Sciences, he published mainly on topics in geometry, although from 1900 on his main research field became history of mathematics. In that year, he published his ground-breaking study on *Les Mathématiques en Portugal au XIXe Siècle*, which was distributed by the Portuguese section of the 1900 Paris Universal Exhibition. The main feature of this book was a 122-page catalogue of 19th-century Portuguese mathematical works, compiled according to the norms of the 1889 *Congrès International de Bibliographie des Sciences Mathématiques.*

GUIMARÃES followed the advice of GUSTAF ENESTRÖM (1852–1923, **B**) (in a review of GUIMARÃES's book in *Bibliotheca Mathematica* [ENESTRÖM 1901, 169]), and transformed a four-page note on Portuguese mathematics he had written for the book into a series of 12 articles, published in *O Instituto* between 1904 and 1909. These papers combined to form a 96-page *Aperçu historique*, a summary of Portuguese mathematics up to the beginning of the 20th century. This summary, together with a bibliography of all Portuguese mathematical works, was a great improvement on the earlier book and was titled, more inclusively, *Les Mathématiques en Portugal* ([GUIMARÃES 1909]).

Like STOCKLER, GUIMARÃES neither analyzed the works of Portuguese mathematicians nor tried to follow the evolution and transformation of mathematical concepts. For him, the historian of mathematics should describe the contents of a work in a strictly objective way, leaving analysis of the mathematics and historical connections to the reader. GUIMARÃES accepted STOCKLER's periodization of Portuguese mathematics and built on the information STOCKLER provided but also augmented STOCKLER's account in many places. For example, GUIMARÃES introduced a completely new section, on mathematics in Portugal during the Spanish occupation from 1580 to 1640 [GUIMARÃES 1909, 26–30], a topic barely mentioned in STOCKLER's book. GUIMARÃES aimed, moreover, to go beyond the period stud-

[2]From 1905 this was superceded by another journal founded by GOMES TEIXEIRA in Oporto under the title *Annaes Scientificos da Academia Polytechnica do Porto.* Following the Republican Revolution of 1910, the mathematical part of this journal was published separately as the *Anais da Faculdade de Ciencias do Porto.*

ied by STOCKLER. Thus, more than half of his *Aperçu* (58 of its 96 pages) dealt with the period 1779–1905.

GUIMARÃES published numerous other works on the history of Portuguese mathematics, principally on PEDRO NUNES. He also published an *Historical Outline on Historiography of Mathematics* [GUIMARÃES 1915b], which includes a list of 28 histories of mathematics.

Like GUIMARÃES, PEDRO JOSÉ DA CUNHA had a military background, graduating from Lisbon's Army School in military engineering in 1891. In that same year, he became a teacher of astronomy at the Polytechnic School of Lisbon, and after 1929 focused his research on the history of mathematics. DA CUNHA wrote an *Historical Outline of Mathematics in Portugal* (69 pages) for the 1929 Seville International Exhibition [CUNHA 1929a], for which he relied primarily on the works of STOCKLER and GUIMARÃES for mathematics, and, for the related history of navigation, on the works of the early 20th-century historians LUCIANO PEREIRA DA SILVA (1864–1926), JOAQUIM BENSAÚDE (1859–1952), and ERNEST GEORGE RAVENSTEIN (1834–1913). The latter sources allowed DA CUNHA to correct both STOCKLER and GUIMARÃES on several points connected to Portuguese navigation (for example, emphasizing the minor role of MARTIM BEHAIM (ca. 1459–1507) in the Portuguese discoveries, and the importance of ABRAÃO ZACUTO's *Almanach Perpetuum* in the compilation of Portuguese nautical tables). DA CUNHA's awareness of contemporary research in the history of mathematics and related sciences thus allowed him to augment and to complement previous work in the history of mathematics. In a similar spirit, he offered a reappraisal of DANIEL DA SILVA, as did GUIMARÃES in the second edition of his book, following GOMES TEIXEIRA's call to bring this mathematician to the attention of both the national and international communities of scientists. DA CUNHA's willingness to engage published texts critically also came to the fore in his 1940 paper, "Mathematics in Portugal in the XVIIth Century" [CUNHA 1940c]. There he challenged the accepted explanation due to STOCKLER of the Portuguese period of decline from the middle of the 16th century to the middle of the 18th century.

FRANCISCO GOMES TEIXEIRA, unlike the two previously-mentioned mathematicians, did not have a military background. He is the best-known Portuguese mathematician of his time. He wrote over 140 research papers on analysis and on geometry. He also had active interests in the history of mathematics, and published a variety of studies over his long career. In 1890, for example, TEIXEIRA published a study "Sur les écrits d'histoire des mathématiques publiés au Portugal" [TEIXEIRA 1890] in *Bibliotheca Mathematica.* This was the first paper on Portuguese historiography of mathematics published in an internationally respected journal for the history of mathematics. In keeping with the guidelines set by the journal's editor, GUSTAF ENESTRÖM, TEIXEIRA limited his account to the history of works in pure mathematics. Even so, he omitted important details about some works and failed to include some of the most important studies, despite the fact that they were by no means difficult to obtain. For instance, he quoted one author as F. A. MARTINS, although his complete name was FRANCISCO ANTÓNIO

MARTINS BASTOS (1799–1868), and while ANTÓNIO JOSÉ TEIXEIRA's book on *Studies on Proportion Theory* [TEIXEIRA 1865] was mentioned, he did not include TEIXEIRA's important *On an Argument between J. Anastácio da Cunha and J. Monteiro da Rocha*, which is mentioned both by CASTRO FREIRE and GUIMARÃES in their above-mentioned books. All of this suggests either that GOMES TEIXEIRA was not greatly motivated to write this paper or, more probably, that he was at the beginning of his career as an historian and was still a bit unsure of what was relevant and important for historians of mathematics.

GOMES TEIXEIRA continued to publish papers and books on the history of mathematics [TEIXEIRA 1902], [TEIXEIRA 1905/06], [TEIXEIRA 1923], [TEIXEIRA 1925], culminating with his posthumous, 295-page book, *History of Mathematics in Portugal* [TEIXEIRA 1934]. Unfortunately, the revision of this book remained incomplete at the time of his death; the many notes promised in the main text to clarify mathematical and other discussions were lost or never written. Nevertheless, the study provided a new look at Portuguese history of mathematics from its beginnings to the 1850s. Although the general interpretation of the evolution of Portuguese mathematics and its major characteristics are the same as STOCKLER's, GOMES TEIXEIRA's methodological approach was completely new. Unlike previous authors, GOMES TEIXEIRA placed Portuguese mathematics in the broader context of European mathematics, thereby underscoring the multiplicity of relations between different mathematical traditions. More important, he analyzed mathematical works from the point of view of a mathematician, guided by an historical perspective. Whereas Portuguese historians of mathematics had tended to avoid technical, mathematical analyses in favor of purely non-mathematical assessments, this changed radically with GOMES TEIXEIRA. He set an example for a more technically sophisticated approach to the history of mathematics for future Portuguese researchers.

FERNANDO DE ALMEIDA LOUREIRO E VASCONCELLOS (1874–1944, **B**), who combined a military career with an academic one, also belongs to the period under discussion. In 1925, while still teaching at the Faculty of Sciences of Lisbon, he published a *History of Mathematics in Antiquity*, a 637-page book that deserves a study of its own. It opens with an analysis of the history of mathematics in ancient eastern civilizations, among them Egypt and Mesopotamia. Then Greek mathematics is covered from Ionia and the Pythagoreans to the Second Alexandrian School, with extensive chapters on PTOLEMY, PAPPUS, and DIOPHANTUS. The final part is devoted to the transmission of Greek and Indian mathematics, with chapters on the schools of Cairo and Bagdad, the Arab schools in Spain and Morocco, and on Church and Monastic schools up to the 12th century, ending with a discussion on the teaching of mathematics in medieval universities.

14.5 The Modern Period: 1940–1970

From the mid 1930s to the mid 1940s, the so-called "Portuguese scientific generation of the 40s" created an internal network for cultural and scientific exchange. It organized events, seminars, and public meetings; it created cultural and scientific societies and journals; it even attempted to create a university for the people. With the end of World War II and the defeat of Fascism, many believed that the two Iberian dictatorships would not last long. Thus, many cultural and scientific figures dared to demand more openly what seemed only their right as citizens. Many petitions were signed on behalf of various causes, in hopes of a peaceful transition to a democratic regime. When these hopes went unfulfilled, many intellectuals and scientists found themselves barred from public service, and were persecuted and imprisoned. Many had to emigrate, and the country's loss was the benefit of others. Nevertheless, research concerning the history of mathematics continued throughout this period, but the number of publications dropped drastically. A few interesting studies on the history of Portuguese mathematics did appear, although they were isolated singularities in more general research devoted to the history of science in Portugal.

During this difficult period the most prominent historian of mathematics was José Vicente Gonçalves. His work, which focused primarily on the history of 20th-century Portuguese mathematics, included an "Analysis of Book VIIII [= 9] of José Anastácio da Cunha's *Mathematical Principles*" [Gonçalves 1940], an "Historical Eulogy of Pedro José da Cunha" [Gonçalves 1966], and a "Preface" to the first (and as yet only) volume of the collected works of Aureliano de Mira Fernandes (1884–1958) [Gonçalves 1971], in addition to an earlier obituary [Gonçalves 1958]. More recently, the 1970s and 1980s witnessed the revival of serious interest in the history of mathematics with the publications of José Joaquim Dionísio (1924–1999), José Tiago da Fonseca Oliveira (1928–1992), Fernando Roldão Dias Agudo (*1925), Luis de Albuquerque (1917–1992), and José Morgado (*1921). Their studies have served as a foundation for contemporary research.

14.6 Conclusion

Individuals in unique contexts have essentially been responsible for advancing the history of mathematics in Portugal. Many of the major works analyzed here, for example, resulted from the relations of their authors to external circumstances. Stockler felt a moral obligation to help save Portugal's cultural inheritance, namely its mathematics and its history; the hundredth anniversary of Coimbra University moved Castro Freire to undertake his *Historic Memoir*; Guimarães undertook *Les Mathématiques en Portugal* because a catalogue of 19th-century Portuguese mathematical works was needed for the Paris Universal Exhibition of 1900; and Pedro José da Cunha wrote his *Historical Outline* in order to present the history of Portuguese mathematics at the Seville International Exhibition of 1929. Only Gomes Teixeira pursued the history of mathematics independently of external events, and in a more careful and considered way. The time and thought that went into his ideas are clearly reflected in his texts, which demonstrate the value of combining the understanding of a good mathematician with an equally good sense of history.

From the mid-1980s there has been renewed interest in the history of mathematics, focused at the Universities of Coimbra, Lisbon, Minho, and Oporto. A National Seminar for the History of Mathematics was founded in 1988, and since then it has been holding regular meetings.

Chapter 15

The Americas

Ubiratan D'Ambrosio, Alejandro R. Garciadiego, Joseph W. Dauben, and Craig G. Fraser

15.1 Introduction

History of mathematics as an autonomous discipline is comparatively recent in the Americas. Particularly in regions with previously limited mathematical activity of their own, as is the case throughout North and South America, history of mathematics as an independent subject naturally appeared quite late. Nonetheless, examples of historical interest are reflected in a select but limited number of mathematical works that appeared as early as the colonial period. Beginning with a survey of South America in general, this section proceeds to examine specifically the cases of Mexico, the United States, and Canada.

15.2 South America

Ubiratan D'Ambrosio

15.2.1 Historiographical Remarks

In what follows, historical periods are defined according to the general chronology associated with the conquest and colonization of the Americas. Geographic divisions are defined following the administrative organization in Viceroyalties: New Spain (roughly what is today Mexico and upper Central America), New Granada (southern Central America, approximately Costa Rica, Colombia, Venezuela, and Ecuador), Peru (roughly Peru and Bolivia), La Plata (roughly present day Chile,

Paraguay, Argentina, and Uruguay), and, under Portugal, the Viceroyalty of Brazil.

Since the independence movements began early in the 19th century, the cultural map has remained largely the same. The chronology adopted here is based on five major divisions: pre-Columbian era; conquest and early colonial times; the era of established colonies; the rise of independent countries; the 20th century. When we refer to South America we are concerned primarily with these divisions, with the exception of the Viceroyalty of New Spain.

It is important to mention that this section is not on the history of the development of mathematics in South America, but rather on South American mathematicians who have made some contributions to the field of history of mathematics. Thus there is no reference to several important South American mathematicians.

15.2.2 Conquest and Early Colonial Times

This period begins in about 1500, when Pedro Alvares Cabral claimed what became Brazil for the Portuguese King Dom Manuel, and 1510, when Vasco Nuñez de Balboa founded Antigua in what is now Panama, which became a springboard for the Spanish conquests of South America.

In early colonial times, the Spanish and the Portuguese tried to establish schools, mostly run by Catholic religious orders. The demand for mathematics in these schools was essentially related to the economic purposes of trade, but there was also an interest in mathematics related to astronomical observations. Reliance on indigenous knowledge was limited, but there was some concern for the nature of native knowledge.

Above all there was modest interest in pre-Columbian mathematical knowledge among the chroniclers, mainly those who reported on the Peruvian quipus, although these were hardly identified and barely recognized as a form of mathematical knowledge. A basic reference, however, was written by Bernabé Cobo: *Historia del Nuevo Mundo* (History of the New World) (1653) [Cobo 1964]. Yet to be explored are the archives of the Jesuit missionaries, as well as other religious orders.

15.2.3 The Established Colonies

From the middle of the 18th century, a number of expatriates and *criollos* played an important role in creating a scientific atmosphere in the colonies. This happened under the influence of the *Ilustración* (Enlightenment), the important intellectual revival that began in Spain under Charles III, who reigned from 1759 to 1788, and in Portugal under José I, who reigned from 1750 to 1777, and his strong minister Sebastião José de Carvalho e Melo (1699–1782), the Marquis of Pombal. A number of intellectuals well versed in a variety of areas of knowledge were responsible for introducing mathematics to the colonies. Among these, Juan

ALSINA (†1807) and PEDRO CERVIÑO (†1816) in Buenos Aires lectured on infinitesimal calculus, mechanics, and trigonometry. In Peru, COSME BUENO (1711–1798), GABRIEL MORENO (1735–1809), and JOAQUÍN GREGORIO PAREDES (1778–1839) are best known. In Brazil, JOSÉ FERNANDES PINTO ALPOIM (1695–1765) wrote two books, *Exame de Artilheiros* (Questions and Answers for Artillerymen) (1744) and *Exame de Bombeiros* (Questions and Answers for Bombardiers) (1748), both of which focused on what might be called military mathematics, but with an historical tone.

Among South Americans of the pre-independence period, the most important relative to history of mathematics was JOSÉ CELESTINO MUTIS (1732–1808), who not only translated NEWTON but was also responsible for introducing modern mathematics to Colombia. However, the works of all these early colonial mathematicians do not offer much in the way of history of mathematics. They exhibit no special concern for facts of the past as important in their own right or in the contexts in which they occurred. This can hardly be called history.

15.2.4 Independent Countries

The independence of the Viceroyalties of New Granada, Peru, La Plata, and Brazil was achieved in the first quarter of the 19th century. However, the ensuing process of modernization in the newly independent countries did not change the prevailing attitude towards mathematics, nor was there any immediate evidence of concern with the history of mathematics.

The establishment of university centers after independence generated more liberal attitudes toward the sources of knowledge which would guide the newly-established countries of Latin America in their development. Formerly restricted to influences from Spain and Portugal, the new, independent countries attracted considerable attention throughout Europe, and numerous scientific expeditions were sent to South America. Particularly important were those of ALEXANDER VON HUMBOLDT (1769–1859) from Germany, and GEORG H. VON LANGSDORFF (1774–1852) from Russia. They tried to explain nature and the way natives made use of local resources, and in doing so greatly stimulated new intellectual climates throughout the region. This in turn helped to spark greater interest in building up large and diversified libraries.

Particularly interesting is the case of JOAQUIM GOMES DE SOUZA (1829–1863), known as "Souzinha," the first Brazilian mathematician with a European reputation. He published in the *Comptes rendus de l'Académie des Sciences de Paris* and in the *Proceedings of the Royal Society.* His works, devoted to partial differential equations, were permeated with interesting historical and philosophical remarks, revealing his familiarity with the most important literature then available. This was possible thanks to the existence of important private collections in Maranhão, his home state in the Northeast of Brazil.

In Buenos Aires, the private library of BERNARDINO SPELUZZI (1835–1898) contains the main works of NEWTON (1642–1727), D'ALEMBERT (1717–1783),

EULER (1707–1783), LAPLACE (1749–1827), CARNOT (1796–1832), and several other modern classics. VALENTIN BALBIN (1851–1901), while rector of the *Colegio Nacional de Buenos Aires*, proposed in 1896 a new study plan that included history of mathematics as a distinct discipline. This laid the groundwork for history of mathematics in Argentina, which ultimately developed only in the early 20th century.

15.2.5 The Twentieth Century

In 1917 the Spanish mathematician JULIO REY PASTOR (1888–1962, **B**) visited Argentina, and remained there for the rest of his life, except for frequent trips to Spain. He was a Professor at the *Universidad de Buenos Aires*. In addition to making important contributions to mathematics, mainly in projetive geometry, REY PASTOR is well known for his contributions to the history of mathematics, especially to the history of 16th-century Iberian mathematics. REY PASTOR also marked new directions in historiography by drawing attention to the mathematical achievements that made possible the great age of navigation. What little history of mathematics had been written in the 19th century was largely concerned only with developments that could be traced back from current mathematics. REY PASTOR, however, by exploring mathematics less relevant to contemporary developments, helped to elucidate what was essential to the practical achievements of the period in question. Thus REY PASTOR pioneered a modern, sophisticated, academic approach to the history of mathematics. Indicative of this new approach is his *La Ciencia y la Técnica en el Descubrimiento de América* (Science and Technology in the Discovery of America) [REY PASTOR 1942]. In this small book, REY PASTOR criticizes historians who only look into the facts which have been recognized scientifically, but fail to evaluate their significance or place them in the context in which the discoveries were made. In a brief preliminary note he clarifies his historiographic approach. He claims that it is premature to expect to find "science" as an organic and well-defined set of ideas in the Middle Ages. Particular importance is given in the book to astronomy, the nautical sciences, and metallurgy.

A disciple of REY PASTOR in Argentina, JOSÉ BABINI (1897–1983, **B**) became one of the first and most distinguished historians of science and mathematics in Latin America. The special issue of the *Boletin Informativo de la FEPAI*,[1] *Historia de la Ciencia*, Año 4, N° 7, 1° Semestre 1985, gives an incomplete list of Argentinian publications in the history of science; 246 entries are for JOSÉ BABINI. His career as a driving force of mathematics in Argentina is significant. He was a founder of the *Unión Matemática Argentina* and in 1920 he became professor at the *Universidad Nacional del Litoral*. In addition to his many books and articles in non-specialized periodicals, BABINI added considerably to scholarship on medieval Jewish contributions to mathematics. His major work, *Historia de la Matematica* (History of Mathematics), coauthored with REY PASTOR

[1] FEPAI: *Fundación para el Estudio del Pensamiento Argentino e Iberoamericano.*

[Babini/Rey Pastor 1951], is one of the truly formative books in the history of mathematics. Unfortunately it has not as yet been translated into other languages. The book, which provides an overall coverage of the history of mathematics, includes a substantial chapter on Hindu, Chinese, and Arabic mathematics. In the Preface, the authors mention that the novel aspect of the book is its historical approach based on the cultural atmosphere surrounding the development of mathematics.

Aldo Mieli (1879–1950, **B**), a distinguished historian of science from Italy (see Section 3.11), was invited in 1939 to become the Director of the recently founded *Instituto de Historia y Filosofia de la Ciencia* at the *Universidad Nacional del Litoral.* Babini had initiated courses on History of Science at the University and was responsible for the creation of the Institute and for the invitation to Mieli. The official journal of the International Academy of the History of Science, *Archeion*, was published in Rosario, from 1939 through 1943. Babini lost no time in arranging a most productive collaboration with Mieli, including their joint work with Desiderio Papp (1895–1993, **B**), a *Panorama general de historia de la ciencia* (General Panorama of History of Science), 10 vols. [Babini/Mieli/Papp 1945].

In the 1930s, some European mathematicians emigrated to Argentina. The high cultural and economic level of the country since the late 19th century made Argentina a very attractive option for European immigrants. Cooperative programs, particularly with Germany, were noteworthy. This is clearly discussed in [Pyenson 1985]. Among them, the distinguished Italian mathematician, Beppo Levi (1875–1961), established an important research center in Rosario along with an influential journal, *Mathematica Notae.* Well-known for his seminal theorem in mathematical analysis, Beppo Levi devoted much of his research to the history of mathematics. Of special note is his book *Leyendo a Euclides* (Reading Euclid) [Levi 1947], which offers a critical analysis of the general organization of the *Elements.*

Luis Santaló (1911–2001) emigrated to Argentina during the Spanish Civil War to accept a professorship at the *Universidad de Rosario*; later he transferred to the *Universidad de Buenos Aires.* Santaló was a distinguished specialist in integral geometry who subsequently became a very influential scholar in mathematics education and the history of mathematics. He made especially important contributions to the history of geometric probabilities and published a number of substantial studies on Buffon (1707–1788).

In neighboring Uruguay, a major tradition of mathematical research was established early in the 20th century. An important figure here, particularly devoted to the history of mathematics, was Eduardo García de Zuñiga (1867–1951, **B**). Above all he succeeded in creating a library for the history of mathematics in the *Facultad de Ingenería de la Universidad de la Republica* (Engineering Faculty of the University of the Republic) in Montevideo. His actual research was mainly devoted to Greek mathematics (see his collected works [García de Zuñiga 1992]).

In Brazil, the beginning of the 20th century marked a break from the dominant trend of positivism. An important figure at the beginning of the century was Manoel Amoroso Costa (1885–1928). A graduate of the *Escola Politécnica do Rio de Janeiro*, Costa worked on the history of mathematics, but primarily on the philosophy of numbers and of analysis. Most of his publications are collected in *As Idéias Fundamentais da Matemática* (Fundamental Ideas of Mathematics) [Costa 1920], a book largely philosophical in character. Among the basic notions of mathematics, Costa was interested in the evolution of the concept of infinity.

More recently, Francisco M. de Oliveira Castro (1902–1993) contributed a chapter on "A Matemática no Brasil" (Mathematics in Brazil) [Oliveira Castro 1994] to the book by Fernando de Azevedo (1894–1974), *As Ciências no Brasil* (The Sciences in Brazil) [Azevedo 1994], a major reference tool for the history of Brazilian science. In the same book, the mathematical physicist Abrão de Morais (1916–1970) wrote on "A Astronomia no Brasil" (Astronomy in Brazil) [Morais 1994], a chapter crucial for understanding early colonial mathematics in South America. Two of Brazil's most distinguished scientists, theoretical physicist Mario Schemberg (1916–1990) and mathematician Leopoldo Nachbin (1922–1993), included useful reflections on the history of mathematics, if in a non-systematic way, in their lectures and writings. Also, Guilherme M. de la Penha (1943–1996) studied Euler, mainly analyzing the *Lettres à une princesse d'Allemagne* (Letters to a German Princess). Rolando Chuaqui (1935–1994), a prominent logician at the *Pontificia Universidad Católica de Chile*, in Santiago, also published and promoted history of mathematics and was the author of several noteworthy articles on the subject.

15.2.6 Current Developments

In the last several decades interest in the history of mathematics has grown considerably throughout South America. The *Sociedad Latinoamericana de Historia de las Ciencias y la Tecnologia (SLHCT)* was founded in Mexico in 1982. This stimulated the organization of national societies devoted to the history of science, including sections for the history of mathematics.

The *SLHCT* meets every three years in different countries, when the directive board is elected. The Society has no permanent site. The Mathematics Section of the Society is quite active. The official journal of the Society is *Quipu. Revista Latinoamericana de Historia de las Ciencias y la Tecnología.* It can be assessed through the electronic website ≪http://www.smhct.org≫.

Some specific activities should also be mentioned. There is an electronic discussion group for history of mathematics, based in Montevideo: ≪historia-matematica@chasque.apc.org≫. This is one of several promising indicators of the increasing interest for history of mathematics in the region, which benefits from the excellent library of the *Universidad de la Republica*, in Montevideo. Together with the library of the *Sociedad Científica Argentina* and other research assets in different countries, such institutions provide scholars with the resources necessary for intensive study of the history of mathematics in Latin America.

In university developments, Doctoral and Masters Programs for the history of mathematics should also be mentioned. Particularly in Brazil these programs have attracted students from other countries of the region. Also in Brazil, the *Seminario Nacional de Historia da Matematica* takes place every odd year from Monday through Wednesday preceding Easter, and the *Sociedade Brasileira de História da Matemática (SBHMat)* founded in 1999, publishes the *Revista Brasileira de História da Matemática.*

15.2.7 Conclusion

Although it is possible to identify various activities like courses, conferences, and sessions for history of mathematics at scientific meetings in just about every country in South America, neither history of mathematics nor history of science was institutionalized, with the exception of Argentina, before the 1980s, when the institutionalization, under the influence of the *SLHCT*, took place in many countries of the region. Meanwhile, young mathematicians from South America have obtained doctorates in the history of mathematics from universities in Europe and North America, a hopeful sign for the continuing professionalization of the subject throughout the countries of the region.

The choice of research subjects is also a promising sign of the creation of strong local groups. There has been increasing interest in the history of early universities of South America, of astronomical observations, and of scientific expeditions. Particularly important is research on the history of the teaching of mathematics. The school systems in South America, with access restricted to an elite, were well advanced for their time. Many important innovations in curricula are found in each of the various countries. This constitutes an important and very rich topic for research. There is also an increasing recognition of the importance of mathematical developments in pre-Columbian cultures. Some of their mathematical knowledge has been lost or transformed during the colonial regime. The search for new and diverse primary sources is a major historiographical challenge. Extant traditional knowledge may help to identify specific mathematical ideas of pre-Columbian cultures, but historians will require new competencies in order to deal with the cultural dynamics implicit in mathematical knowledge and practice. Thus study of ethnomathematics is a field very active in Latin America, which calls for a new historiography. In the last two decades, it has become clear that there are many and very interesting themes to be studied in Latin America, which offers very rich material to historians of mathematics for future research.

15.3 Mexico

ALEJANDRO GARCIADIEGO[2])

15.3.1 The Creation and Development of the Royal University

Mexico, as it is usually described in modern history or geography textbooks and general encyclopedias, is a product of the 19th century.[3] As its geographical boundaries were negotiated with its neighbors, its first political constitution was also drawn up, establishing Mexico as an independent republic in 1821. Native regions of the Olmec, Mayan, and Aztec cultures, among others, which had flourished before the local populations were conquered by the Spanish, fell within its boundaries. The Spanish then ruled "Mexico" for 300 hundred years, from 1521 until 1821.

Within 30 years of the Spanish conquest, following the structure and image of the Spanish University of Salamanca, the *Universidad Real y Pontificia de México* (Royal and Pontifical (1595) University of Mexico) was established in 1551, the first to be founded on the American continent.[4]

Two years later, when the university officially opened its doors, the following professorships had already been assigned: Theology, Sacred Scriptures, Canon Law, Law, Arts, Grammar, and Civil Law. Other "chairs" (e.g. Medicine and Moral Theology, among others) were subsequently added. It was another hundred years, however, until a Chair for "Astrology and Mathematics"was established.[5] A "Mathematics Chair" proper was officially recognized only in the second half of the 18th century. This was largely the result of a debate provoked by the physician JOSÉ IGNACIO BARTOLACHE (1739–1790). Although some intellectuals in New Spain were interested in following the recent "scientific" and mathematical

[2]Given the fact that the professionalization of the history of mathematics in Mexico is comparatively recent, it has been impossible to support every argument with specific references; some of the most recent history covered here is necessarily of a subjective and personal character, although historical references have been provided whenever possible.

[3]A few isolated references containing "historical" accounts of the evolution of "Mexican" science (including mathematics) — written between the 17th and 19th centuries — have survived [TRABULSE 1984, 9–10, notes 1 and 2]. The analysis here, of historians of mathematics active in Mexico, and of mathematicians who incorporated significant amounts of historical material in their writings, will concentrate primarily on the first half of the 19th century and on the personality of SOTERO PRIETO (1884–1935), tutor and mentor of the first generation of professional mathematicians.

[4]The founding of the Royal University stimulated the development of other schools of advanced studies, among them the *Colegio Mayor de Santa María de Todos los Santos* (1573), the *Colegio de San Pablo* (1575), where it was possible to study the trivium, quadrivium, and theology, and the *Colegio de San Gregorio* (1575).

[5]FRIAR DIEGO RODRÍGUEZ (ca. 1596–1668) was the first to hold this chair; he taught mathematics to students of the faculty of medicine where it was a compulsory subject, along with astrology [FERNÁNDEZ DEL CASTILLO 1953, 39].

developments on the European continent, their efforts were largely thwarted by the Sacred Inquisition.[6]

Following its "duty" to prevent the corruption of readers, the Inquisition elaborated a long list of forbidden books, including, among others, scientific works by GIORDANO BRUNO (1548–1600), GALILEO GALILEI (1564–1642), ISAAC NEWTON (1642–1727), and CHRISTIAN WOLFF (1679–1754).[7] To add a book to the list, only an anonymous accusation was necessary, arguing that the book in question contained some statement against Catholic principles. In New Spain (i.e. Mexico), the index of prohibited works was taken very seriously. It is difficult to assess, for example, the possible influence in Mexico of the *Histoire des Mathématiques* (History of mathematics) [MONTUCLA 1799/1802] by MONTUCLA (1725–1799, **B**) — since it was black-listed in 1808, only four years after its arrival in the new world.[8]

15.3.2 After Independence

Throughout much of the 19th century, the university suffered several decades of turmoil due (in great part) to the persistent rivalry between liberal and conservative parties in their efforts to control the country, once the rebel Mexican army officially signed a treaty recognizing independence from Spain in 1821. Although the conservative party was able briefly to impose another foreign government, the Republic was reestablished when emperor FERDINAND MAXIMILIAN JOSEPH (1832–1867) of Austria was executed on June 19, 1867. Hoping to control the country politically, the new liberal leaders sought to establish the basis for a durable and peaceful social order. One of their main doctrines, in opposition to the ideology of the Inquisition and the Catholic clergy, was freedom of conscience, which was firmly rooted in three different forms of emancipation — scientific, religious, and political. GABINO BARREDA (1818–1881), spokesman for a new educational policy, adapted the positivist thought of AUGUSTE COMTE (1789–1857) in support of his ideas.[9] The original positivist motto, "love, order, and progress," was later replaced by "freedom, order, and progress." (Some years later, the motto was abbreviated simply to "order and progress," as if "love" and "freedom" no longer mattered.)

---

[6] FRIAR DIEGO RODRÍGUEZ included readings of COPERNICUS, TARTAGLIA, CARDANO, BOMBELLI, BRAHE, and KEPLER, among others, in his courses on mathematics and astrology.

[7] However, a positive consequence of this prohibition is that we now possess listings of some of the books that arrived from Europe during the Colonial period. In trying to fulfill their commitments, the inquisitors drew up detailed descriptions of some private libraries. Thus it is not only possible to list some of the authors (EUCLID, PTOLEMY, KEPLER, and NEWTON, among others), but to date precisely when they first arrived in Mexico. See, for example, [FERNÁNDEZ DEL CASTILLO 1982], [MORENO CORRAL 1992], and [O'GORMAN 1939].

[8] Unfortunately, when lists of forbidden books were published, the Inquisition did not provide reasons or explanations. The list simply gave the complete references for such books. On the other hand, the private record shows that the Inquisition had received an "oral report" that MONTUCLA's book contained "heretic propositions" [*Catálogo* 1992, slip 1147, page 228].

[9] The most complete and serious attempt to explain the origins, development, and decline of positivist doctrines in Mexico is [ZEA 1968].

The positivist ideal envisioned a complete reform of the Mexican educational system, and one of its main goals was the popularization of the exact and natural sciences. This reform, on the other hand, explicitly excluded studies promoting religious debates; some of the latter were actually replaced by the study of experimental sciences. Nevertheless, some reforms were more easily described on paper than put into practice. In fact, by the turn of the century (after nearly 35 years of a new dictatorship), the study and development of technology and the sciences (both natural and exact) had fallen well behind prevailing European standards.

15.3.3 The Emergence of Modern Mathematics in Mexico in the Twentieth Century: The Autonomous National University

The University of Mexico attracted the most cultured, educated, and, at times, outspoken members of the community. Consequently, on various occasions the university was closed, due in large measure to the political instability of the country.

In 1910, still under the dictatorship of PORFIRIO DÍAZ (1830–1915), a newly transformed National University reopened its doors. JUSTO SIERRA (1848–1912), one of the main advocates of the "scientific party" — so called because its members justified a reelection of DÍAZ using "scientific" arguments — regarded the university as the most important cultural institution of the nation. This presumed a new role for the teaching and development of the sciences. The *Escuela Nacional Preparatoria* (National High School) — where students were prepared to enter the university — and the newly-created *Escuela de Altos Estudios* (School of Advanced Studies) began to offer specialized courses in biology, physics, chemistry, and mathematics. These schools promoted a favorable academic climate in which those interested in the teaching and study of higher mathematics could discuss their findings. A non-exhaustive list of those who did includes ÁNGEL DE LA PEÑA (1837–1906), SOTERO PRIETO (1884–1935), ALFONSO NÁPOLES (1897–1992), and FRANCISCO CÁRDENAS (1898–1969). Among the most distinguished of these was PRIETO, Mexico's first serious historian of mathematics. Trained by his father, an elementary mathematics teacher, PRIETO was a natural leader and a highly admired professor according to reminiscences of some of his students. He taught various branches of higher mathematics and is credited with the introduction (in Mexico) of the theory of relativity (see, for example, [PRIETO 1921/23]).

15.3.4 The Faculty of Sciences

Between 1920 and 1940, political and academic conditions changed dramatically regarding the role and value of the mathematical sciences. By then, political spokesmen envisioned the transformation of Mexico, anticipating its modernization through industrialization and application of new technologies. The study of branches of engineering was to be supported (see [VASCONCELOS 1926]). Mathematics, as a foundation for engineering, was thus viewed much more favorably than it had been in the past. PRIETO happened to be at the right place at the right time.

He taught mathematics at the *Escuela Nacional de Ingenieros* (National School of Engineers), where he was able to unite a small but highly motivated group of students wishing to pursue the study of mathematics. In a relatively brief period of time, the status of this group was formalized and established as a "section" (or department) in the Faculty of Philosophy. Between 1931 and 1933, PRIETO lectured on the history of mathematics [PRIETO 1991]. His original manuscript lecture notes — in a very poor facsimile edition — do not reveal origins, methodology, pedagogic goals, or the significance of mathematics. PRIETO does not mention the sources he read or the influences on his historical views. The notes only seem to reflect specific data needed to lend substance to the lectures, and unfortunately never mention possible hypotheses, theses, major concepts, trends, or traditions. In his notes, PRIETO constantly lists (in modern notation) problems mathematicians were trying to solve, but he does not illustrate reasons why particular questions were important, why there was any interest in attempting to solve them, or what tools were available to attack them. Basically, each lesson was prepared independently of the others following a chronological order.[10] Nevertheless, from the comments of some of PRIETO's students, it seems clear that the lecture notes do not reflect the character of the courses as delivered or the influence they may have had on PRIETO's students.[11] The notes, however, are admittedly dry and unappealing.

The opening phrase of the lecture notes suggests that PRIETO was interested in doing more than presenting a chronological outline of mathematics to his students. He stated: "The history of a science clarifies the origins of its fundamental concepts and exhibits the evolution of its methods," [PRIETO 1991, 1]. That is to say, it was not simply a matter of memorizing hundreds of names, dates, and titles. History as a discipline, for PRIETO, had a much deeper meaning. He had read GIOVANNI BATISTA VICO (1668–1744), for example, and believed that history played a major role in the understanding of culture. This, in turn, meant that it was necessary to know how a given culture evolved. History was about more than collecting data; it also included laws explaining trends and clarifying how history — and the history of mathematics — developed as they did. Although some of his followers (e.g., AGUSTÍN ANFOSI (1889–1966), CARLOS GRAEF (1911–1988), and FRANCISCO ZUBIETA (*1911)) subsequently lectured on historical topics, none seems to have shared PRIETO's identification with VICO or any other school of historical thought.

A few years later, in 1939, the Department of Mathematics was officially established as part of the Faculty of Sciences of the now renamed *Universidad Na-*

[10] Originally, PRIETO planned two yearly academic courses, each divided into two terms. The first course ran from October 1, 1931, to June 7, 1932, and from July 19 to November 1, 1932, and covered (in 71 lectures) the history of mathematics from the Egyptians up to the 16th century, including some comments on Portugal and Spain. The second course covered parts of the 17th and 18th centuries (in only 30 lectures), and was offered from February 14 to June 8, 1933.

[11] In fact, in a personal communication, Dr. ALBERTO BARAJAS, a former student of PRIETO, has questioned the authenticity and authorship of the facsimile notes ascribed to PRIETO.

cional Autónoma de México (UNAM) (National Autonomous University of Mexico).[12] This "autonomous" character meant that, although the university received most of the necessary funds to operate from the federal government, the university had complete freedom to govern itself, both politically and economically.

Academic conditions were substantially improved in both the teaching and transmission of mathematics.[13] Well-established scholars from abroad regularly visited the *UNAM*, and offered substantial support to the efforts of their Mexican colleagues.[14] Students trained abroad started to return to Mexico, bringing with them new ideas and methods. Meanwhile, the *Instituto de Matemáticas* (Institute of Mathematics) and the *Sociedad Matemática Mexicana* (Mexican Mathematical Society) were founded in 1942 and 1943, respectively.[15]

It was not long before the profession began to enjoy a new level of recognition, in part because some mathematicians had developed very close ties to major political figures. Most all of the sciences went through a powerful renaissance at this same time. By 1951, when the university celebrated its four-hundredth birthday, part of the festivities included the organization of the *Congreso Científico Mexicano* (Mexican Scientific Congress). When the proceedings were published, they occupied 15 thick volumes containing more than 500 works, some of which were historical in character [*Memoria* 1953/54]. The first volume contains the mathematical papers. Some of which (see RECILLAS AND NAPOLES, among others) are survey works describing research developed at Mexican institutions.

[12]Unfortunately, PRIETO was no longer alive to appreciate this moment. Four years earlier, he had committed suicide by shooting himself in the head.

[13]This transformation was not limited to mathematics. Other disciplines, like physics, chemistry, and astronomy to mention just a few, enjoyed similar metamorphoses. Such transfiguration also took place within the humanities.

[14]DIRK STRUIK, for example, visited Mexico for six weeks in 1934. Thereafter, he visited Mexico repeatedly; on the occasion of his most recent visit (1976), he delivered a set of six lectures on the history of mathematics. As a direct consequence of this visit, ALEJANDRO GARCIADIEGO enrolled at the University of Toronto (Toronto, Ontario, Canada) to become the first trained professional historian of mathematics in Mexico. He received his masters and doctoral degrees between 1979 and 1983. Visits of GEORGE DAVID BIRKHOFF, GARRET BIRKHOFF, SOLOMON LEFSCHETZ, and NORBERT WIENER, among others, proved to be of immense value for the development of mathematics in Mexico in the present century.

[15]The Federal Government also supported the foundation of other major institutions of research and training. President LAZARO CÁRDENAS (1895-1970), in his wish to support higher education, created the *Instituto Politécnico Nacional (IPN)* (National Polytechnical School) in 1939. This institution offered a bachelor's degree in physics-mathematics and, later, provided an alternative for the doctoral degree in mathematics. In 1943, some of the most prestigious intellectuals in Mexico founded *El Colegio Nacional* (The National School). Members of the Spanish academic elite, refugees from the atrocities of the Civil War, established *El Colegio de México* (The Mexican College), the most influential research and teaching school for historical and social studies. Unlike *El Colegio de México*, the *Colegio Nacional* does not prepare students. Its members offer conferences and lectures for the interested public and publish, generally, their collected works. Somewhat earlier, the foundation of the *Sociedad Mexicana de Historia Natural* (Mexican Society of Natural History) (1936) heavily promoted historical studies on scientific themes through its official journal.

When a newly built University City opened in 1954, the "Tower of Sciences" housed most of the scientific institutes. This building provided the opportunity for colleagues from different scientific disciplines to gather on a daily basis. Some shared philosophical interests and an inquisitive attitude towards nature and the importance of science in general. On February 21, 1955, SAMUEL RAMOS (philosophy), GUILLERMO HARO (astronomy), and ELI DE GORTARI, a logician, founded an interdisciplinary *Seminario de Problemas Científicos y Filosóficos* (Seminar on Scientific and Philosophical Problems). In a relatively short period of time, this seminar attracted a large number (225) of professors and researchers, some of whom lived abroad. Every month, anyone who wished to do so was free to participate in the seminar, and in the course of its first few years, DE GORTARI oversaw publication of more than 74 pamphlets and 33 books.[16]

15.3.5 The Nightmare: The 1968 Student Movement

In mid-1968, an apparently insignificant street fight between two schools (administered by *UNAM* and *IPN*), was suppressed by city police and subsequently escalated into one of the most important social and political events in Mexico in the second half of the 20th century — surpassed in its extent and consequences only by the Mexican Revolution (ca. 1911–1920). University and polytechnic students, emotionally and ideologically motivated by similar political student movements elsewhere — especially those in France that erupted in Paris during May of that same year — along with support from the intellectual community in general, carried out public rallies on the main streets of Mexico City. The government (1964–1970) of GUSTAVO DÍAZ chose not to negotiate or compromise with the students. Instead, to end the conflict, DÍAZ decided to use brute force to repress a mass meeting held in Tlaltelolco — the cradle of Mexican culture in Mexico City — on October 2, 1968. When the Mexican army invaded the university (violating its "autonomous" character) as well as the *IPN*, the university was forced to close its doors. Meanwhile, many students and professors, among them DE GORTARI, were illegally detained by the police and the army. With them went the full support of the *Seminario de Problemas Científicos y Filosóficos*.

It was nearly a year before the university reopened, but most students and professors did not obtain their freedom immediately. Most were accused of different offenses under common law and not for their political convictions. From time to time over the next few years, the government offered amnesty to most of those who had participated in the 1968 student movement. Nevertheless, the damage

[16] These included original contributions, as well as translations from both classic sources (LOBACHEVSKIĬ, PAVLOV, and PLANCK, among others), and contemporary thinkers (e.g., JACQUES HADAMARD, KURT GÖDEL, ALEXANDER ALEXANDROV, and GORDON CHILDE, to mention a few). Of the books, there were translations with commentaries of original sources (e.g., WILLIAM HARVEY), new works by colleagues (e.g., GARCÍA BACCA), and translations from contemporary historians and philosophers of science (e.g., JOHN D. BERNAL, PHILIP FRANK, MAURICE FRECHET, and HANS REICHENBACH, among others). It is very significant that none of these texts was directly associated with the history of mathematics.

had been done. Socially, ideologically, and politically, the university was inevitably different.

Above all, members of the university had lost their political innocence. The university community was politically polarized, and those interested in philosophical or historical questions were largely associated with left wing ideologies. In general, the well-established and more traditional members of the staff regarded the development of mathematics and the sciences as free from an ideological agenda. At one point, they opposed — both secretly and openly — several attempts to popularize the study of the history and philosophy of mathematics. Most of these scholars claimed that students only needed to learn technical mathematics. Some courses (including classical mechanics and electricity) were omitted from the standard curricula. On the other hand, mostly younger members of the community (the majority of them students) became increasingly critical of the uses and roles of science in society. Others, with different goals and interested in the proper study of the history of mathematics, decided to travel abroad to pursue their graduate studies. Among the world's leading historians of mathematics, KENNETH O. MAY (1915–1977, **B**) greatly influenced the professionalization of the history of mathematics in Mexico. He did not visit Mexico, but like the Spaniard CID (RODRIGO DÍAZ DE VIVAR (ca. 1043–1099)) who, even after his death, won military battles, MAY inspired the Mexican community through his example to establish its first historical and philosophical journal (*Mathesis*) in 1985. His administrative and organizational skills also inspired others to establish the *Asociación para la Historia, Filosofía y Pedagogía de las Ciencias Matemáticas* (Association for the History, Philosophy, and Pedagogy of the Mathematical Sciences) and marks the final step in the successful professionalization of the discipline in Mexico. Today, historians represent one of the most productive and dynamic groups within the Mexican mathematical community. Although they remain a small group of scholars, they edit a highly respected research journal, and have organized several meetings of an international character. Undergraduate and graduate students take courses and develop dissertations in this area.

15.3.6 Conclusion

The mathematics department of the Faculty of Sciences, the Institute of Mathematics, and the Mexican Mathematical Society — all physically located at the *Universidad Nacional Autónoma de México* (1551) and main components in the process of professionalization of modern mathematics in Mexico — were founded, almost simultaneosly, between 1939 and 1943. Unfortunately, SOTERO PRIETO, who believed the history of mathematics should be a major component of the working knowledge of any mathematician, who guided the first generation of professional mathematicians, and who even lectured on historical topics, died just before this process of professionalization materialized. PRIETO's immediate followers and students did not share some of his basic principles. Consequently, the history of mathematics lost its most important driving force and initial momentum.

Interest in the history of mathematics, as well in philosophical and pedagogical concerns, fell well behind that of other branches of mathematics. In fact, in the mid 1970s, members of the *UNAM* staff proposed to eliminate these subjects from the curricula offered to students.

Nevertheless, by the mid 1980s, the first graduate students, professionally trained as historians of mathematics, returned to Mexico from abroad. As a consequence, over the next few years, and inspired by similar activities promoted by KENNETH O. MAY, these scholars offered undergraduate and graduate courses, launched a new research journal, and established the first society of its kind in Mexico for the history of mathematics.

15.4 United States of America

JOSEPH W. DAUBEN

15.4.1 Introduction

It will come as no surprise that the development of mathematics in the United States in many respects mirrors the history of the nation itself. As the country became economically sound and politically viable, especially after the Civil War at mid-century, professionalization in American life was reflected in its universities, newly-founded institutions of technology and graduate schools. Late in the 19th century interest in the history of science began to stir, and within the first quarter of the 20th century, some major practitioners had emerged who can truly be said to have been not simply mathematicians with historical interests, but indeed, serious historians of mathematics. The full story of how the history of mathematics came to establish itself as a discipline in the decades following World War II is a long, complex yet very interesting one, to which historians of mathematics have only recently begun to devote serious attention (see, for example, the section on history of mathematics in the 20th century in [DAUBEN 1985, 203–215] and [DAUBEN/LEWIS 2000], as well as [MAY 1973, 648–649], which lists articles on 20th-century mathematics, [MAY/GARDNER 1972] and subsequent editions of the *International Directory of Historians of Mathematics*, for listings of scholars whose research either concentrates on or includes history of 20th-century mathematics, and [MERZBACH 1989]).

By the end of the 19th century, the history of mathematics as a useful adjunct to teaching was also recognized, and across the country courses on the history of mathematics were taught with the instruction of future teachers in mind. Popularizations were also written, primarily for mathematicians but also with an eye on the general, literate population that found mathematics of historical and recreational interest. In what follows, these diverse aspects of the writing of the history of the history of mathematics in the U.S. will be covered, including an account of how the subject has emerged as a strong, academic discipline in the latter half of the 20th century.

15.4.2 History of Mathematics in the United States: Early Efforts Through World War I

Although a number of writers in the 19th century took an interest in history of mathematics, none can be said to have been primarily an historian of mathematics. Nevertheless, there were several important figures who contributed to the subject and who did so at a very high level. The earliest of these was NATHANIEL BOWDITCH (1773–1838, **B**), whose English edition with commentary (1829/1839) of LAPLACE's *Traité de mécanique céleste* (Treatise on Celestial Mechanics) was much more than a translation. BOWDITCH wanted to fill in steps LAPLACE had omitted from his mathematical presentation, add more recent results to the text, and provide credit to authors whose work had not been mentioned by LAPLACE. BOWDITCH did this for the first four volumes, the last of which was published posthumously in 1839 (the fifth volume of the *Mécanique céleste* appeared in 1823–25, but too late for BOWDITCH to translate). Much of the effort he put into the comments and additions to the first four volumes were clearly historical in nature, and represent a striking contribution to the history of astronomy and mathematics by an early American.

More systematic — and serious — about the history of science was another American, the philosopher CHARLES S. PEIRCE (1839–1914, **B**). Much has been written about this eccentric and in his later life reclusive figure, the creator of pragmatism and founder of semiotics. Most of what PEIRCE contributed to the history of mathematics was the result of a deep and abiding interest in the logic of science. Here important parallels, and in some measure influence as well, might be drawn between the English philosopher WILLIAM WHEWELL (1794–1866, **B**) and PEIRCE. Above all, reflecting his historical and methodological interests, in 1898 PEIRCE contracted to write a history of science in which he intended to include a significant amount of material on the history of mathematics.

This was part of a grand scheme PEIRCE had to reveal the logic of science, which was never realized (but see the two volumes of selections from PEIRCE's writings related to this history of science, including history of mathematics, edited by CAROLYN EISELE (1902–2000, **B**) in 1985). Unfortunately, most of what PEIRCE wrote was never published, and just as for BOWDITCH it would be too much to claim that he was a historian of mathematics above all else. If what both these figures contributed to the subject went beyond purely antiquarian or anecdotal interest, the fact remains that they regarded the history of mathematics as an adjunct to the history of science in general, or as primarily of philosophical interest, as an example of method. The question, therefore, remains to be answered: who was the first writer in the United States who should be classified as a dedicated historian of mathematics? The honor should probably go to FLORIAN CAJORI or to DAVID EUGENE SMITH (with several not insignificant rivals).

CAJORI (1859–1930, **B**) was among those who enriched America by emigrating in 1875 by choice, not by political necessity as was the case on a large scale later — much to the benefit of history of mathematics in America. CAJORI was sixteen when he left his native Switzerland for Wisconsin where he studied at the University of Wisconsin, earning a B.Sc. degree in 1883 and an M.Sc. degree in 1886. After further graduate study at The Johns Hopkins University, he taught applied mathematics briefly at Tulane University befor accepting a position in 1889 at Colorado College, where he first taught physics and then mathematics.

CAJORI's first historical work, published by the U.S. Bureau of Education: *The Teaching and History of Mathematics*, appeared in 1890. D. E. SMITH praised this work for setting out "a line of study which was quite new in this country and which served a worthy purpose in creating an interest in the subject," although lacking the scholarship and thoroughness of his later publications [SMITH 1930, 778].

Indeed, CAJORI was the first in a continuing American tradition of mathematicians interested in the history of mathematics due to its perceived value in teaching. But this was only one of several factors that dramatically changed the intellectual climate at the end of the century:

> It is symptomatic of the period that between 1870 and 1890 more than 200 "learned societies" were founded; these included the New York Mathematical Society (1888), the American Historical Association (1884), and the National Education Association (1870). Other factors, of special relevance to history of mathematics, include the rise of graduate education, the establishment of professional schools in engineering and business, and the conversion of the nineteenth century teachers' training institutes and normal schools to graduate "schools of education" [MERZBACH 1989, 641].

CAJORI was also the first in the U.S. to write serious textbooks devoted to the history of mathematics, beginning with *A History of Mathematics*, which first appeared in 1893 and was subsequently reissued in numerous editions and several revisions. It was, as he said himself,

> ... an increased interest in the history of the exact sciences manifested in recent years by teachers everywhere, and the attention given to historical inquiry in the mathematical class-rooms and seminaries of our leading universities [that] cause me to believe that a brief general history of Mathematics will be found acceptable to teachers and students [CAJORI 1893, v].

When it came to antiquity, almost all of what CAJORI had to say concerned Greek geometry; the Middle Ages included both "Hindoos and Arabs"; coming down to recent times, individual subjects were treated separately, including synthetic and analytic geometry, algebra, analysis, theory of functions, theory of numbers, and applied mathematics. Three years later CAJORI published *A History of*

Elementary Mathematics with Hints on Methods of Teaching (1896), which he admitted was taken to a large extent "with only slight alteration" from his history of mathematics. Meanwhile, he was also interested in the history of physics and the roles mathematics and experimentation played, which led to CAJORI's *A History of Physics in its Elementary Branches Including the Evolution of Physical Laboratories* (1899), wherein he sought to show that "some attention to the history of a science helps to make it attractive, and that the general view obtained of the development of the human intellect by reading the history of science, is in itself stimulating and liberalizing" [CAJORI 1899, v]. Presumably a similar philosophy explained his concern for history of mathematics.

Knowing of these works, MORITZ CANTOR (1829–1920, **B**) invited CAJORI to contribute a section to volume four of his monumental *Vorlesungen über Geschichte der Mathematik* (1908) [CANTOR 1880/1908], and this represented a significant turning point in CAJORI's career. According to SMITH it was writing for CANTOR that finally prompted CAJORI to study original sources, and "thus there opened a new era in his work, the era in which he found himself" [SMITH 1930, 778]. Subsequently, CAJORI's *History of the Logarithmic Slide Rule* (1909) led naturally to a biography of WILLIAM OUGHTRED (1916). CAJORI also revised his earlier works on history of mathematics, while all the time working on his *History of the Concepts of Limits and Fluxions in Great Britain from Newton to Woodhouse* (1919). SMITH has called this "a kind of inaugural address at the University of California" [SMITH 1930, 779], for the year before CAJORI had been called to the University of California at Berkeley, as Professor of the History of Mathematics. He was certainly the first to receive such a position and title in the United States.

By then, just after World War I, CAJORI had been contributing to the history of mathematics for nearly thirty years. During the last decade of his life, however, he continued to write, publishing *The Early Mathematical Sciences in North and South America* (1928), which appeared almost simultaneously with his comprehensive *The History of Mathematical Notations* (in two volumes, 1928 and 1929). The idea for such a work had been suggested to CAJORI by ELIAKIM H. MOORE (1862–1932) at the University of Chicago. As CAJORI himself explained its rationale:

> This history constitutes a mirror of past and present conditions in mathematics which can be made to bear on the notational problems now confronting mathematics. The successes and failures of the past will contribute to a more speedy solution of the notational problems of the present time [CAJORI 1928/29, vol. I, 1].

Above all, CAJORI hoped that,

> if the contemplation of mistakes in past procedure will afford a more intense need of some form of organized effort to secure uniformity, then this history will not have been written in vain [CAJORI 1928/29, vol. II, xvii].

For a bicentenary meeting held in New York in 1927 to commemorate the death of ISAAC NEWTON, CAJORI produced a major essay on NEWTON's fluxions. He also prepared critical editions of NEWTON's *Principia* and *Optics*, both of which were nearly finished when pneumonia claimed his life in August 1930 in Berkeley, California.

More typical of historians of mathematics in the United States early in the 20th century was CAJORI's slightly junior contemporary, JULIAN LOWELL COOLIDGE (1873–1954, **B**). COOLIDGE had a great respect for the history of mathematics, especially MICHEL CHASLES's *Aperçu historique sur l'origine et le développement des méthodes en géométrie* (Historical Conspectus of the Origin and Development of Methodes in Geometry) [CHASLES 1837]. But the remarkable advances made in classical scholarship throughout the 19th century had rendered much of CHASLES's (1783–1880, **B**) research obsolete, having "almost completely altered our outlook on the mathematics of ancient times" [COOLIDGE 1940, vii]. The fact that CHASLES himself contributed to the subject of projective geometry was another indication that even as he wrote, a new era in the history of geometry was just beginning. Moreover, confessing that he was unable to read German, CHASLES had failed to include such important figures as STEINER, GAUSS, and PLÜCKER. COOLIDGE, however, fully appreciated the difficulty of writing contemporary history, and had to admit that his own history of geometry, although written a century later than Chasles, omited almost entirely any references to topology, in part because it was a subject with which he was only partially familiar. As COOLIDGE observed, "topology is so new, and is changing so rapidly, that it is hard to say what will be its final place in the geometrical scheme" [COOLIDGE 1940, ix].

Such apologies aside, COOLIDGE set a high standard. He not only explained original works, but commented on proofs, discussed notation, and drew comparisons between older and more recent geometrical methods. Unlike CAJORI, however, COOLIDGE was writing neither with teachers in mind nor for pedagogical reasons. COOLIDGE, it seems, was interested primarily in two questions: "Can we draw any general conclusions from the five-thousand-year study we have now completed? Is there any general tendency discernible throughout the whole?" [COOLIDGE 1940, 422]. In addition to pointing out the trend towards more and ever more generalization, there was also greater abstractness. In response to these tendencies, COOLIDGE raised a caution of his own: "How much farther can it go, how general, how abstract can geometry become without losing all specific content?" [COOLIDGE 1940, 423].

In contrast to COOLIDGE and his relatively late interest in the history of mathematics, his Ivy-League contemporary, RAYMOND CLARE ARCHIBALD (1875–1955, **B**), maintained a life-long interest. He was also an inveterate bibliophile:

> [At Brown] he threw himself into this task [of acquiring books] with a will, and in the course of ten or fifteen years brought the library to a position of excellence [...] During the period 1920–1940 Archibald

> without doubt knew more about mathematical books and their values than anyone else in this country. Frequently he went to Europe for the summer, always provided with funds to spend for mathematics books for Brown. He carried on a flourishing correspondence with scientific booksellers throughout the world, and if a dealer offered an item for sale at less than it was worth he was on occasion quick to recognize the bargain and to cable an order for purchase [ADAMS/NEUGEBAUER 1955a, 743–744].

ARCHIBALD was also responsible for the American Mathematical Society's collection, which as librarian he built up over a span of twenty years (1921–1941), primarily through publication exchanges.

Among ARCHIBALD's best-known contributions to the history of mathematics is his *Outline of the History of Mathematics* (1932, revised through a sixth edition which appeared in 1949). He also wrote the *Semicentennial History of the American Mathematical Society* (1938), and contributed the bibliography on Egyptian and Babylonian mathematics to the CHACE edition of the *Rhind Mathematical Papyrus* (1927–1929). As ADAMS and NEUGEBAUER summed it up:

> Archibald was both a scholar of the old school and a gentleman of the old school as most of us now regard it. He was brought up in the classical tradition with much emphasis on Latin and Greek. He had a very remarkable memory, and he carried with him at all times an enormous store of factual information in the fields of his interest [ADAMS/NEUGEBAUER 1955a, 745].

Antiquarians and Philanthropists

Before drawing comparisons between ARCHIBALD and two other, major contributors to the history of mathematics in the United States, it is appropriate to mention briefly ARNOLD BUFFUM CHACE (1845–1932, **B**), remembered for his handsome and painstaking edition of the *Rhind Papyrus* for the Mathematical Association of America.

CHACE is in fact representative of an important tradition, that of the wealthy amateur who developed a serious interest in an esoteric subject and then devoted much of his own wealth to its pursuit and development. As a mathematical amateur, CHACE studied quaternions with BENJAMIN PEIRCE (1809–1880) and sat in on higher mathematics courses at Brown. In 1910, on a trip to Egypt (twelve years before the spectacular discovery of the tomb of King TUTANKHAMEN in 1922), CHACE and his wife became "much interested in [Egypt's] monuments and literature." The original *Rhind Mathematical Papyrus (RMP)* was found at Thebes in the ruins of a small building near the Ramesseum, and was purchased in 1858 by A. HENRY RHIND, who bequeathed it upon his death to the British Museum. The Museum published a lithographic facsimile in 1898, and it was a copy of this version that CHACE purchased in 1912, two years after the trip he made to Egypt in 1910. He began intensive study of the latter at age 67. When CHACE published his version of the *RMP* in 1927, he included photographs of the original, along with additional fragments of the *RMP* that had been discovered in the collections of the New York Historical Society. For details, see [CHACE 1927/29, Vol. 1, Preface and Introduction, 1–2].

The last twenty years of his life were devoted to working with his wife on the *Rhind Papyrus* (she copied the hieratic from the British Museum's original, and drew the corresponding hieroglyphic transcription). When the edition was finally completed, it was published in two magnificent volumes (1927, 1929), which included 109 two-color plates, all at CHACE's expense.

David Eugene Smith, Libraries and Publishers

Although CHACE cannot by any means be considered a professional historian of mathematics, or as typical of those who contributed to its success, he made one of America's earliest contributions to the history of ancient mathematics. More representative of the increasingly professional level to which the history of mathematics had risen in the U.S., and the only serious rival to CAJORI as the most prominent, was DAVID EUGENE SMITH (1860–1944, **B**). Unlike CAJORI, however, SMITH was a native American.

But like CAJORI, SMITH was a tireless researcher and prolific writer. He compiled bibliographies for teaching mathematics (1912); edited the Indian classic text, the *Gaṇitāsārā-Sangraha* of MAHĀVĪRĀCĀRYA (with M. RAṆGĀCĀRYA, 1912); co-authored *A History of Japanese Mathematics* (with MIKAMI YOSHIO; 1914) produced a facsimile edition with notes of the earliest mathematical work of the New World, *The Sumario Compendioso* of Brother JUAN DÍEZ (1921); and even wrote a guide to *Historical-Mathematical Paris* (1924). Among his best-known works are, in chronological order: *History of Modern Mathematics* (1906), *Rara Arithmetica* (1908), *Hindu-Arabic Numerals* (co-authored with L. C. KARPINSKI, 1919), *Number Stories of Long Ago* (1919), *History of Mathematics* (in two volumes, 1923/1925), and a *History of Mathematics in America Before 1900* (co-authored with J. GINSBURG, 1934).

SMITH's greatest legacy to the history of mathematics, however, was the library which he amassed and which now forms the D. E. Smith Collection at Columbia University, alongside the Plimpton and Dale Collections. Like his colleague at Brown, R. C. ARCHIBALD, SMITH was not only an inveterate book collector, but also served as one of the early AMS librarians (from 1902–20; it was from SMITH that ARCHIBALD inherited the position from 1921–1938).

In fact, libraries and publishing houses also have their part in the history of mathematics in the United States. As G. W. HILL noted in his presidential address to the American Mathematical Society in 1895, "in America we are not well situated for [historical] investigations [...] on account of the meagreness of our libraries" [HILL 1896, 126]. It was due to the efforts of figures like ARCHIBALD, SMITH, and somewhat later, KARPINSKI, that substantial libraries for both mathematics and the history of mathematics were established early in the 20th century. Meanwhile, PAUL CARUS (1852–1919), as editor of the *Monist*, a general philosophical journal, emphasized articles of a historical and philosophical character. CARUS, with the support of the "Zinc Mogul" EDWARD C. HEGELER (1835–1910) and his daughter MARY (whom CARUS married in 1888), founded the Open Court Publishing Company. "Under its imprint appeared numerous monographs in the history of mathematics. Many were translations of classics in modern mathematics" [MERZBACH 1989, 649]. Among the translations were LAGRANGE's *Lectures on Elementary Mathematics* (trans. T. J. McCORMACK, 1898) (1865–1932) and DEDEKIND's *Essays on the Theory of Numbers* (trans.W. W. BEMAN, 1901).

Histories, Bibliographies, Rare Books and the Dedicated Collector

Another of the great bibliophiles and a truly professional historian of mathematics in every sense of the word was LOUIS C. KARPINSKI (1878–1956, **B**). KARPINSKI, who taught mathematics at the University of Michigan, was on leave for a year in 1909–1910 as a Fellow and University Extension Lecturer at Teachers College at Columbia University, New York City. There he met D. E. SMITH, and it was SMITH who sparked KARPINSKI's interest in history of mathematics. Together they produced *The Hindu-Arabic Numerals* (1911), after which KARPINSKI went on to study numerous manuscripts on his own, especially those dealing with algebra and algorithms in various European languages. This eventually resulted in his own *History of Arithmetic* (1925). Among his more significant publications are various studies devoted to JORDANUS DE NEMORE, SACROBOSCO, and ABU KAMIL. KARPINSKI also edited ROBERT OF CHESTER's (ca. 1140) Latin translation of the *Algebra* of AL-KHWĀRIZMĪ, and NICHOMACHUS OF GERASA's *Introduction to Arithmetic.*

Meanwhile, KARPINSKI also compiled *A Bibliography of Mathematical Works Printed in America through 1850* (1940), which may be regarded as a complement of sorts to the SMITH-GINSBURG volume, *A History of Mathematics in America before 1900* (1934). There was, however, a sense of rivalry between KARPINSKI and SMITH as KARPINSKI's student, PHILLIP S. JONES, explained:

> The rivalry with D. E. Smith and Columbia University, at least in the mind of Karpinski, developed after he returned to the University of Michigan and began to build there a collection of source materials in the history of mathematics. He did this by badgering librarians and alumni for funds as he scanned book catalogues and traveled to visit dealers and auctions. In later years he would say, "I always said Michigan had the better collection even when it didn't." This statement was made especially with respect to Columbia's collection which later became the Smith-Plimpton collection, when two bibliophiles gave their private collections to Columbia University. However, he was also aware of the collection at Brown University built up through the efforts of R.

> C. Archibald. It is hard to know which was his more basic urge, that of the collector, or that of the competitor [JONES 1976, 189].

In the course of his scholarly career, KARPINSKI also published bibliographies of algebraic and trigonometric works that he continuously updated. According to JONES:

> After his retirement, Professor Karpinski continued to combine business with pleasure. He bought atlases and books in varied fields from dealers and at auctions, loaded them in his car, and called at many libraries to help them improve their holdings and to introduce famous rarities into collections at small colleges [JONES 1976, 189].

Moreover, in his role as teacher, KARPINSKI excelled in getting the message into every medium:

> He was an expositor and transmitter of new historical developments to those with less time, expertise, or contact with the field. This was done via radio talks, speeches at teacher's conventions, expository articles and reviews. He was, for example, the invited speaker at the first organizational meeting of the Mathematical Association of America at Cleveland, Ohio, in December, 1915 (KARPINSKI gave an illustrated lecture on the history of algebra) [JONES 1976, 189].

In a similar spirit of popularizing the history of mathematics, KARPINSKI designed a series of slides on the histories of arithmetic, algebra, geometry, and higher mathematics which were projected on a central column in the Hall of Science at the "Century of Progress Exposition" organized to celebrate the centenary of the City of Chicago in 1933. Beginning that same year, and in fact until his death, KARPINSKI's professional commitment to the history of mathematics was reflected in his service as an associate editor of *Scripta Mathematica*, founded by JEKUTHIEL GINSBURG (1889–1957, **B**) in 1932.

The appearance of a journal devoted in large measure to the history of mathematics was a sign of the progress a dedicated group of scholars had made towards truly professionalizing the subject:

> [These] individuals and others formed an active group, promoting the history of mathematics as an independent research field, as a motivating subject for teachers of mathematics, as a stimulus for mathematical research, and as a source of general edification and pleasure. Conscious of the limited availability of reference materials and libraries, they collaborated in making requisite primary and secondary source materials more easily available, be it through book purchases, through translations, through bibliographies, through text editions and analyses, or simply through reviews [MERZBACH 1989, 649].

History of Mathematics in Transition

Another immigrant who came to the United States early in the century, in order as he said to escape being shoved into the Royal Arsenal at Woolwich or the India Civil Service [Bell 1955, 70], was Eric Temple Bell (1883–1960, **B**), a truly remarkable individual.

He was a prolific author, writing more than 250 papers, five books, eleven popularizations, and, under his pseudonym John Taine, publishing 17 science fiction novels, many short stories, and some poetry. *Before the Dawn* (1934), the only science fiction he signed as E. T. Bell (and his own admitted favorite), was inspired, he said, by the models of dinosaurs he had seen as a boy in Croydon Park near London.

Among mathematicians, however, Bell's best-known writings were unabashedly popularized histories of mathematics, the first of which, *The Queen of the Sciences* (1931), he was commissioned to write in connection with the Chicago World's Fair of 1933. Bell went on to write, among others, *The Handmaiden of the Sciences* (1937), *Men of Mathematics* (1937), *The Development of Mathematics* (1940), and *Mathematics, Queen and Servant of Science* (1951).

Bell typified an important transitional period in the historiography of mathematics in the U. S. Prior to World War II, most historians of mathematics were trained as mathematicians. If they developed an interest in the history of mathematics early in their careers, as did Cajori and Smith, they were more than likely interested in the value of history for pedagogical purposes. If they were primarily research mathematicians, like Coolidge and Bell (whose first book on the history of mathematics did not appear until 1937, when he was well over 50), they came to their historical interests having first made their mark as theoretical mathematicians. Bell, however, was a mathematician who appreciated the history of mathematics for its own intrinsic, if largely anecdotal, interest, and his work must honestly be categorized more in the tradition of popularization than of serious scholarly history. In fact, Bell often protested that his works should not be considered histories of mathematics, for which he had considerable misgivings:

> Theirs is a difficult and exacting pursuit; and if controversies over the trivia of mathematics, of but slight interest to either students or professionals, absorb a considerable part of their energies, the residue of apparently sound facts no doubt justifies the inordinate expense of obtaining it. Without the devoted labors of these scholars, mathematicians would know next to nothing, and perhaps care less, about the first faltering steps of their science. Indeed, an eminent French analyst of the twentieth century declared that neither he nor any but one or two of his fellow professionals had the slightest interest in the history of mathematics as conceived by historians. He amplified his statement by observing that the only history of mathematics that means anything to a mathematician is the thousands of technical papers cramming the journals devoted exclusively to mathematical research. These, he

> averred, are the true history of mathematics, and the only one either possible or profitable to write. Fortunately, I am not attempting to write a history of mathematics; I hope only to encourage some to go on, and decide for themselves whether the French analyst was right [BELL 1940, quoted from the 2nd rev. ed. of 1945, x].

Nevertheless, BELL also appreciated the lessons the history of mathematics could teach, especially to younger students or mathematicians in search of a career or an area of specialization. He found it: "astonishing how few students entering serious work in mathematics or its applications have even the vaguest idea of the highways, the pitfalls, and the blind alleys ahead of them." Nevertheless, BELL was convinced that "a survey of the main directions along which living mathematics has developed would enable them to decide more intelligently in what particular field of mathematics, if any, they might find a lasting satisfaction" [BELL 1940, quoted from the 2nd rev. ed. of 1945, vi].

BELL also regarded his descriptive accounts of the lives and labors of the great mathematicians as of intrinsic human interest: "With a knowledge of who these men were, [...] and an appreciation of their rich personalities, the magnificent achievements of science fall into a truer perspective and take on a new significance" [BELL 1937, 4]. In this regard, however, BELL was less interested in getting the facts of his historical narratives right than he was in the more colorful details of personal anecdotes and heated academic rivalries. As the historian of mathematics K. O. MAY once remarked in evaluating BELL's work:

> Bell's insights and provocative style continue to influence and intrigue professional mathematicians in spite of their historical inaccuracies and sometimes fanciful interpretations [MAY 1970, 584].

15.4.3 Increasing Professionalization: History of Mathematics and History of Science

Less influential than BELL in popularizing the history of mathematics in the United States but much more significant in its professionalization, was GEORGE SARTON (1884–1956, **B**). In 1912 he founded the first international quarterly journal for the history of science, *Isis*, which included on its editorial board such prominent historians of mathematics as SIR THOMAS HEATH (1861–1940, **B**) and D. E. SMITH, among other notables like ARRHENIUS (1859–1927) and POINCARÉ (1854–1912). While the history of science was still in its infancy professionally, SARTON literally "invented" a research position and a full-time career for himself at a time when "the discipline barely possessed a cognitive, let alone a professional identity" [THACKRAY/MERTON 1972, 476].

SARTON gave strong support to the history of mathematics through his journals, *Isis* (volume I, 1913) and *Osiris* (volume I, 1936), and his constant concern for bibliographies, including *Horus: A Guide to the History of Science* (1952) and annual critical bibliographies accompanying *Isis*. SARTON also produced the widely read *Introduction to the Study of the History of Mathematics* (1936). This was a direct expression of SARTON's own belief that the history of mathematics must be the foundation upon which all of the rest of the history of science should be built. As he put it: "Take the mathematical developments out of the history of science, and you suppress the skeleton which supported and kept together all the rest" [SARTON 1936a, 4].

Closely linked with the success of *Isis* was the founding of the History of Science Society in 1924. This was actually the inspiration of D. E. SMITH, who in 1923 circulated a letter to forty-five individuals, suggesting they meet in Boston to discuss the idea. Early the following year, ARCHIBALD, E. W. BROWN (1866–1938), CAJORI, KARPINSKI, and SMITH were among those who met as founding members of the Society (see the notice of the founding of the Society [as reported in *Isis* **6** (1924), 6–7; see also [MERZBACH 1989, 653]]).

A Journal for History of Mathematics in the U.S.: "Scripta Mathematica"

Equally concerned as was SARTON with the importance of journals to promote the history of science, but with a specific interest in the history of mathematics, JEKUTHIEL GINSBURG was a pioneer as the founding editor of *Scripta Mathematica.* From its inception in 1932, the journal was devoted to the philosophy, history, and expository treatment of mathematics. Just as supportive of the journal's success was its initial board of associate editors: R. C. ARCHIBALD, L. C. KARPINSKI, C. J. KEYSER (1862–1947, **B**), GINO LORIA (1862–1939, **B**), LAO G. SIMONS (1870–1949, **B**), and D. E. SMITH.

As GINSBURG put it in an opening editorial meant to describe the purpose of *Scripta Mathematica*:

> History of Mathematics presents the science in the state of becoming; by means of philosophy the crystalline structure of the finished product is revealed. These two aspects of the subjects are supplementary; one without the other loses the greater part of its value [Ginsburg 1932, 1].

GINSBURG, who had studied with D. E. SMITH at Columbia University's Teachers College, spent his career teaching mathematics at Yeshiva University in New York City. He not only added important historians and mathematicians to the editorial board of *Scripta Mathematica*, including such well-known names as ABRAHAM A. FRAENKEL (1891–1965) and SIR THOMAS LITTLE HEATH, but he also promoted women like LAO SIMONS (Hunter College) and VERA SANFORD (1891–1971, **B**) (of Northwestern University). Unfortunately, his vision of the journal was doomed in the hands of his successor, ABE GELBART (1913–1994). When GINSBURG died in 1957, the issue dedicated to his memory also bore a new title page reflecting the change in editorship and corresponding editorial philosophy. Without explanation, *Scripta Mathematica* dropped its stated front-page interest in history and philosophy of mathematics. Having given up its unique focus on the history of mathematics, however, it was suddenly just another journal of average quality advertising the expository and research aspects of mathematics. *Scripta Mathematica* ceased publication within a few years.

Emigrés and Refugees

Also born in 1878, like KARPINSKI, but unlike him in every other respect as a historian of mathematics was the German mathematician MAX DEHN (1878–1952, **B**). A sense of the history of mathematics was reflected early in his substantial monograph written with POUL HEEGAARD (1871–1948), on "Analysis Situs" for the *Encyklopädie der mathematischen Wissenschaften*, and again later in an article on "Die Grundlegung der Geometrie in historischer Entwicklung," which DEHN contributed to a book he edited with MORITZ PASCH, *Vorlesungen über neuere Geometrie* (Lectures on Recent Geometry) (1926).

DEHN was professor of pure and applied mathematics at the University of Frankfurt until he was forced to give up his position in 1935. For the next few years he lectured wherever he could in Europe, taught for a year at the Technical University in Trondheim (1939–1940), and then emigrated to the United States.

It was in America that DEHN concentrated most of his efforts on the history of mathematics — while moving from one position to another, from Pocatello, Idaho, to Chicago, Illinois; from Annapolis, Maryland, to Madison, Wisconsin; and finally, to Black Mountain College in North Carolina where he taught from 1945–1952.

In addition to publishing major historical studies of his own, for example, on space, time, and number in ARISTOTLE (1936), he produced a whole series of four articles for the *American Mathematical Monthly*, on the history of mathematics from 600 B.C.– 600 A.D. [DEHN 1943/44].

As his student WILHELM MAGNUS (1907–1990) (who later went on to write an historical overview of the history of modern group theory [CHANDLER/MAGNUS 1982]) said of DEHN, his works arose after years of careful study of the sources in their original languages. DEHN also wrote several historical articles with ERNST HELLINGER, especially a joint study of JAMES GREGORY's *Vera Quadratura* (The True Quadrature), produced for the GREGORY tercentenary and published by the Royal Society of Edinburgh [TURNBULL 1939], and another general appreciation of GREGORY's mathematical achievements for the *American Mathematical Monthly.*

ERNST HELLINGER (1883–1950, **B**), born in Breslau, provided another example of how emigration prior to World War II enriched America relative to the history of mathematics. After a brief stint at Marburg, HELLINGER taught with DEHN at the new University of Frankfurt from 1914 until he too was forced to retire by the Nazis in 1936. Two years later he was arrested and placed in a concentration camp until he was allowed to emigrate to the United States in 1939. Subsequently, he taught at Northwestern University in Evanston, Illinois, became a citizen in 1944, and worked with DEHN on their two historical studies of JAMES GREGORY (for DEHN and HELLINGER, cf. Section 5.5.2).

Earlier HELLINGER had also contributed a now-classic article, written over many years with OTTO TOEPLITZ (1881–1940, **B**) for the *Encyklopädie der mathematischen Wissenschaften*, on "Integralgleichungen und Gleichungen mit unendlich vielen Unbekannten" (Integral Equations and Equations with Infinitely many Unknowns) [HELLINGER/TOEPLITZ 1927, reprinted 1928 and 1953].

But among those who emigrated from Europe to the United States as a result of the impending threat of war, was OTTO NEUGEBAUER (1899–1990, **B**; see also Sections 5.4.6 and 6.8). NEUGEBAUER's arrival in the United States prompted the creation of a department specifically with his interests and abilities in mind. The only one of its kind in the United States, the Department for the History of Mathematics was officially founded in 1946 at Brown University. Given his unique and pioneering contributions to the history of Babylonian mathematics and astronomy, as well as his many contributions to the history of the exact sciences in antiquity, NEUGEBAUER had already established his reputation in Germany prior to World War II. Together with JULIUS STENZEL (1883–1935) and OTTO TOEPLITZ, NEUGEBAUER was also a founding editor of the journal, *Quellen und Studien* (1930–1938). At Brown, and later at the Institute for Advanced Study, he contributed significantly to the rising fortunes of the history of mathematics in America. Above all, the small but gifted circle of scholars he trained at Brown in ancient languages and the history of the exact sciences in antiquity proved a unique and lasting legacy.

Joining NEUGEBAUER at Brown was another orientalist, ABRAHAM J. SACHS (1914–1983, **B**). SACHS's major contributions to Assyriology included his publication, in 1945, of *Mathematical Cuneiform Texts*, and in 1955, *Late Babylonian Astronomical and Related Texts.* It was SACHS who found an important archive of Babylonian astronomical diaries in the British Museum, and it was his life's work to publish this collection, although it remained unfinished at the time of his death in 1983.

15.4.4 Recent History of Mathematics in the United States

Following World War II, mathematics in the United States prospered as never before. The amount of mathematics published increased dramatically; the character of the mathematics actually produced was more abstract and sophisticated; and the number of individuals required to teach the subject reached unprecedented levels. With the successful launch of Sputnik, the first artificial Earth satellite in 1957 by the U.S.S.R., the U.S. government increased spending dramatically for scientific research and education at all levels, which in turn had a positive effect on all of the sciences (see [DOW 1991] and [DIVINE 1993]).

Indeed, a number of factors conspired at the same time to increase interest just as dramatically in the history of mathematics. Not only was there a multiplier effect, the ever-growing number of students in the post-war era required a corresponding increase in the number of teachers, and more of these in turn came to appreciate the value of history in both their research and teaching.

The growing discipline of the history of science also served to professionalize the subject in new ways. For the first time in any significant measure, graduate students could actually study the history of mathematics and earn a Ph.D. Concomitantly, mathematicians developed a new appreciation for the inherent interest of the subject. Closely related to this were new publications, including new journals either founded in the U.S. or with strong ties to U.S. scholars, like the *Archive for History of Exact Sciences* (1960), *Historia Mathematica* (1974), and *The Mathematical Intelligencer* (1979), to name three that place great emphasis on the history of mathematics.

Textbooks

The most influential historians of mathematics in the decades immediately following World War II were HOWARD EVES (*1911), RAYMOND WILDER (1896–1982), CARL BOYER, and MORRIS KLINE. EVES's *An Introduction to the History of Mathematics* (1953) has been a highly successful textbook, used in departments of mathematics across the country to introduce students to the history of the subject. When EVES first designed a course to teach the history of mathematics while on the staff at Oregon State College in 1948, he plotted his strategy as follows:

> I conceived the innovation of accompanying Problem Studies. This injection of some genuine mathematics into the course was carried out with much conviction and not a little trepidation. The result was gratifying; the course flourished in an astonishing and exciting way [EVES 1964, ix].

Those who used the book agreed. As EVES later wrote:

> One of the pleasantest experiences connected with the first edition was the extensive correspondence received from high school teachers and from high school students. Many of the teachers have instituted in their schools a course in the history of elementary mathematics based upon this book, and a surprising number of high school students have used the book and its problem studies as a source for project material in various courses and in state mathematics and science fairs. It is hoped that the new amplified bibliographies may better serve these folks [EVES 1964, vii].

Above all, EVES was impressed by the way in which the problem studies served to introduce students to mathematical research. He also generously acknowledged the help of CARL BOYER, "whose scholarly comments have done so much to improve the historical accuracy of the second edition" [EVES 1964, viii]. BOYER, however, was interested in the history of mathematics for reasons that fundamentally went well beyond what EVES had achieved with his elementary textbook.

CARL B. BOYER (1906–1976, **B**) also contributed significantly to the professionalization of the history of mathematics in its most scholarly sense in the United States. MORRIS KLINE, another of the major figures in the recent professional history of the history of mathematics in the United States, describes BOYER's significance as follows:

> A new era in the pursuit of the history of mathematics was inaugurated by the work of Carl Boyer [...]. It was Carl's own insight and courage that led him to become a professional historian. Insight was certainly needed for a young man to perceive the importance of history at a time when it was being derogated by the mathematical community; and only courage could have sustained his long efforts to defy the snobbish mathematicians. Certainly it was not material rewards or the search for prestige that motivated this work for he well knew during the many years in which he pursued scholarly study and writing that mathematicians would hardly grant recognition to an historian. Nevertheless, by persistence, wisdom in the selection of vital themes and sound scholarship Carl succeeded in vitalizing and elevating the status of history in mathematical research and education [KLINE 1976, 391–392].

BOYER's first publication, *The Concepts of the Calculus* (1939), based on his Ph.D. thesis, was a "clarion call" in the words of MORRIS KLINE:

> It is the first American history to pursue a subject in depth and ranged from the work of the Greeks to that of Weierstrass. His second book, *History of Analytic Geometry*, exhibits the same qualities, to trace the developments of that subject in the maze of 19th century creations was an enormous task. His third and fourth books also broke fresh ground. Of these two, the latter, *A History of Mathematics*, is especially commendable because it includes an account of some 20th century developments [KLINE 1976, 392].

Indeed, BOYER regarded his work as of potential use to students and scholars alike, but above all, he regarded it as a contribution to the history of thought. He also hoped his work would help to redress "the lack of mutual understanding too often existing between the humanities and the sciences." BOYER believed firmly that the history of mathematics could give the professional a "sense of proportion":

> No scholar familiar with the historical background of his specialty is likely to succumb to that specious sense of finality which the novitiate all too frequently experiences. For this reason, if for no other, it would be wise for every prospective teacher to know not only the material

> of his field but also the story of its development [Boyer 1939, repr. 1949, v].

RICHARD COURANT (1888–1972), who wrote a "Foreword" to the second printing of [BOYER 1939, repr. 1949], also noted the value of history as an antidote to dogmatism:

> As a matter of fact, reaction against dogmatism in scientific teaching has aroused a growing interest in history of science; during the recent decades a very great progress has been made in tracing the historical roots of science in general and mathematics in particular [BOYER 1939, repr. 1949, i].

In fact, COURANT hoped that BOYER's book would "reach every teacher of mathematics; then it certainly will have a strong influence towards a healthy reform in the teaching of mathematics."

Teaching Mathematics: The Uses of History

In the wake of Sputnik, and as the Cold War heated up, another result was that mathematicians began to question the effectiveness of traditional methods of teaching. Among the critics, none was more outspoken yet more explicit in his suggested remedies, than MORRIS KLINE (1908–1992, **B**). KLINE was committed to the reform of teaching and especially appreciated the role that the history of mathematics could play in this. In the course of his career, he published more than a dozen books, most of which dealt with the history of mathematics or the importance of the subject for the proper teaching of mathematics.

Among KLINE's earliest publications was a book *Introduction to Mathematics* (1937), co-authored with three of his colleagues at New York University, HOLLIS R. COOLEY (1899–1987), DAVID GANS, (1907–1999), and HOWARD E. WAHLERT (1905–1975). The book's subtitle, "A Survey Emphasizing Mathematical Ideas and their Relations to other Fields of Knowledge," revealed its purpose, one that became a philosophical principle KLINE followed for the rest of his life. As the authors lamented in their preface, most undergraduate mathematics courses rarely gave students "any real understanding of the character of the subject or of its relation to the sciences, the arts, philosophy, and to knowledge in general." As a result, "far too many intelligent students are 'soured' for life as far as mathematics is concerned." Instead, *Introduction to Mathematics* sought to remedy this by emphasizing examples from physical experience. Contending that mathematics was not just "a collection of methods, but a vast, unified system of reasoning with many of the characteristics of a fine art," their book intended to show not only that science and philosophy are indebted to mathematics for making precise such concepts as motion, velocity, and the infinite, but also that "mathematics has been an important factor in the development of civilization" in the broadest sense. KLINE adopted virtually the same philosophy in a textbook he wrote on his own some years later: *Calculus: An Intuitive and Physical Approach* (1967).

In a more popular work published in 1953, *Mathematics in Western Culture*, KLINE made a straightforward case for the idea that "mathematics has been a major cultural force in Western civilization," but he railed against teaching and writing that presented mathematics "as a series of apparently meaningless technical procedures" [KLINE 1953, ix]. A decade later he echoed many of the same ideas in *Mathematics: A Cultural Approach* (1962). This book was written as a text for liberal arts students, but KLINE also had in mind those who were training to teach elementary mathematics as well as secondary teachers in non-mathematical subjects. The book ranged over elementary mathematics, and sought to demonstrate "the extent to which mathematics has molded our civilization and our culture." To do so, the book gave numerous examples showing how mathematics was intimately related to physical science, philosophy, logic, religion, literature, the social sciences, music, painting and other arts. In virtually everything he wrote, KLINE was committed to the idea that knowledge is "not additive but an organized whole and that mathematics is an inseparable part of that whole" [KLINE 1962, vi].

The book was also a political statement of KLINE's abiding belief that mathematics was meaningful primarily through its applications, and that concrete problems had stimulated most of its history. He objected strongly to the teaching of mathematics in highly abstract settings, and insisted that "we must overcome the disconnectedness which kills the vitality of our curricula. To help students see the interrelationships of the various branches of knowledge is the greater wisdom" [KLINE 1962, vi].

The greatest and doubtless most lasting contribution Kline made to the history of mathematics was his monumental *Mathematical Thought from Ancient to Modern Times* (1972; subsequently published in a paperback edition, in three volumes, 1990). Although KLINE admittedly ignored non-Western civilizations, including Chinese, Japanese, and Mayan contributions "because their work had no material impact on the main line of mathematical thought," he strongly believed that:

> the roots of the present lie deep in the past and almost nothing in that past is irrelevant to the man who seeks to understand how the present came to be what it is [KLINE 1972, viii].

For students, KLINE hoped his book would:

> ... give perspective on the entire subject and relate the subject matter of the courses not only to each other but also to the main body of mathematical thought [...] Indeed the account of how mathematicians stumbled, groped their way through obscurities, and arrived piecemeal at their results should give heart to any tyro in research [KLINE 1972, ix].

15.4.5 Conclusion

Professionalization of the History of Mathematics

In trying to make some final, if preliminary, assessments of the history of the history of mathematics in the United States, it is necessary to distinguish amateurs from professionals. It is also helpful to delineate those who turned to the history of mathematics because they sought to popularize mathematics, largely through anecdotal accounts, from those who had more serious motives related either to teaching or to scholarly research of a primarily technical or academic nature. In many cases, however, such distinctions blur; in others they may not be of much significance.

Until recently, with the possible exception of CHARLES SANDERS PEIRCE who appreciated its philosophical value, interest in America in the history of mathematics has traditionally stemmed from its usefulness in mathematics teaching, or from its attractiveness as a pastime for mathematicians in old age or retirement. The cases of CHACE and COOLIDGE however, attest to the fact that the contributions of older mathematicians could be of a very high and serious order.

As a subject, the history of mathematics was institutionalized surprisingly early in the position specifically created for the history of mathematics at the University of California for FLORIAN CAJORI. More typical, and certainly more important for professional development of the subject, were the many courses on the history of mathematics taught at a remarkably large number of colleges and universities — primarily for prospective teachers. In one survey conducted by A. W. RICHESON (1897–1966) in the early 1930s of 617 American colleges, more than a quarter, 161 colleges or universities in all, listed one or more courses taught on the history of mathematics. Of the 77 at Teachers Colleges and Normal Schools in the sample, 32 or slightly more than 40% taught the history of mathematics. Only two institutions offered more than one such course — Hunter College in New York City and Dartmouth College in New Hampshire. Ten schools also offered courses on the history of mathematics at the graduate level as well.

While this may give the impression of substantial activity — or at least interest — in the history of mathematics in U.S. institutions of higher education in the 1930s, RICHESON distinguished between two basic approaches:

> The courses offered by the teachers' colleges and normal schools deal almost entirely with the historical development of secondary school subjects and methods in mathematics; while those listed by the liberal arts and technical schools give a broader view of the history of mathematics [RICHESON 1934, 164].

Aside from the one example of Brown University, where the subject was indeed unique with its highly specialized faculty in a Department for History of Mathematics, the history of mathematics elsewhere in the United States depended more on the vision and dedication of unique individuals like CARL BOYER and

MORRIS KLINE than on any organized institutional support of a coordinated or systematic sort.

Indeed, those who have brought the history of mathematics to prominence in the United States include many who have seen it as a passionate mission, as well as a matter of social necessity. Among those who helped to encourage a younger generation of historians of mathematics in the United States, CAROLYN EISELE, (1902–2000, **B**) did much to promote an appreciation for the mathematics of CHARLES S. PEIRCE, while DIRK J. STRUIK, (1894–2000, **B**) wrote the universally appreciated *Concise History of Mathematics* (1948), a landmark for those working in the United States and a guide and example for historians of mathematics everywhere. Perhaps most important, STRUIK's vision of the subject was not limited to internalist perspectives. He viewed the history of mathematics as part of the broader fabric of human history itself, rooted in the cultures and traditions, the values, societies, and times, in which mathematicians, thanks to their mathematics, have flourished. This was also the approach taken by RAYMOND WILDER (1896–1982) [WILDER 1981], and by those like CLAUDIA ZASLAVSKY (*1917) [ZASLAVSKY 1973] and MARCIA ASCHER (*1935) [ASCHER 1981] and [ASCHER 1991], who have contributed to the growing interest in ethnomathematics. ABRAHAM SEIDENBERG also took a wider, anthropological approach to his studies of the diffusion of early number systems and the "ritual origins" of counting and geometry, derived as he believed they were from rituals described in the *Śulba-sūtras* of ancient India.

Meanwhile, serious study of the exact sciences has been promoted to a significant extent by the works of MARSHALL CLAGETT (*1916) and his students who have made substantial contributions to the history of medieval mathematics, and by CLIFFORD TRUESDELL (1919–2000, **B**), whose research on the history of rational mechanics has focused in particular on the works of LEONARD EULER. TRUESDELL's *Archive for History of Exact Sciences*, a journal he founded in 1960, has also contributed greatly to serious research devoted to history of mathematics, and has set a high standard for both technical analysis and historical scholarship.

Among more recent studies, the works of WILBUR KNORR (1945–1997, **B**) represent another important feature of the increasing interest in technical detail and rigorous analysis of primary sources. KNORR's contributions to the history of ancient mathematics also considered the importance of the transmission of knowledge between cultures, especially the traffic of ideas between Greek, Hellenistic, Roman, Islamic, and later medieval scholars.

Globalization of History of Mathematics

In the last quarter of this century, the creation of the International Commission on the History of Mathematics and the Commission's international quarterly journal, *Historia Mathematica* (founded by KENNETH O. MAY in 1972), reflects in a very concrete way the recent fortunes of the history of mathematics, not just in the U.S., but on a truly global scale. Just as plenary lectures devoted to history of mathematics attract large audiences at International Congresses of Mathematicians every four years, so too on the national level, the annual meetings of the American Mathematical Society and the Mathematical Association of America usually include at least one if not multiple sessions, seminars, and invited lectures devoted to the history of mathematics [MERZBACH 1989, 663]. Special workshops devoted specifically to the ways in which history of mathematics can be useful in college teaching have been offered on a yearly basis. The interest of the National Council of Teachers of Mathematics (NCTM) in historical topics, along with special sessions devoted to the history of mathematics at annual meetings of the History of Science Society, the American Association for the Advancement of Science, and other national academies and societies, all reflect a growing interest in history of mathematics throughout the U.S.

The substantial number (411) of Americans listed in the most recent edition of *The World Directory of Historians of Mathematics* [ANDERSEN/DYBDAHL 1995] is but one tangible indication of the extent to which this interest has grown. Another is the special series for works on history of mathematics only recently launched jointly by the American Mathematical Society and the London Mathematical Society, which since 1988 has published more than a dozen volumes.

History of mathematics also has important connections with the broader study of the history of science and to departments for history of science and technology. In one outstanding case, at Brown University, there is the country's only Department for History of Mathematics, founded by OTTO NEUGEBAUER in 1946, which continues to train students at the doctoral level in history of mathematics. At most universities, however, the history of mathematics may be studied most often as a specialization within departments of mathematics or history of science. In rare instances, departments of history or philosophy also permit graduate students to work with faculty whose special interests concern history of mathematics.

Present and Future Prospects

When the American-born KENNETH O. MAY (see Section 15.5.2) launched the first international journal specifically devoted to the history of mathematics in 1974, he greeted new subscribers with an editorial, "Congratulations to the Thousand":

> There are about 10^3 scholars throughout the world teaching or doing research in the history of mathematics. They are the heirs of a long tradition which goes back beyond Eudemus, who flourished in the 4th century B.C., and includes among its practitioners many of the great mathematicians. By the late 19th century, it had a substantial literature and place in mathematical bibliography. Several specialized journals: Boncompagni's *Bulletino* [sic] (1868–1887), Moritz Cantor's *Abhandlungen* (1877–1913), Eneström's *Bibliotheca Mathematica* (1884–1915) and Gino Loria's *Bollettino* (1898–1919) focused an intensive activity and collected rich sources for future workers. But after World War I, with the new growth of general history of science and the

> founding of broad journals such as *Isis*, the historians of mathematics seem to have become less visible. The literature was more than ever scattered in the journals of mathematics, education, popularization, and the general history of science. In the sixties the continuing tradition, which had been kept alive by a number of distinguished workers, enjoyed a revival whose most notable feature was a rapid increase of interest in the mathematics community. At the same time, history of science had developed sufficiently in size and maturity to support specialized journals [*Historia Mathematica*, **1** (1974), 1].

Indeed, history of science, as GEORGE SARTON liked to emphasize, relies upon the history of mathematics as the backbone and foundation for the rest of the history of science. But given the extent to which the history of mathematics is also a technical history, it requires special study and knowledge as prerequisites. Historians of mathematics in the United States may be found teaching in Departments of Mathematics, of History, of Philosophy, of Sociology, and in numerous interdisciplinary faculties and programs across the country. However, despite the substantial number of Departments or Programs for History of Science in many colleges and universities, the history of mathematics is not always well represented. And while many departments of mathematics offer at least one undergraduate level course on history of mathematics, most who teach courses may be considered autodidacts of the subject, and rarely have those who teach such courses in departments of mathematics done any graduate study specifically related to history of mathematics or history of science. Nevertheless, whether through individual courses, public lectures, or professional organizations, history of mathematics is a subject actively studied by more individuals now than ever before. Its future in the United States should be one of continued activity in all areas relevant to the subject, in teaching, lecturing, and publication.

Thus, historians of mathematics in the United States find growing acceptance of their work at all levels, whether in classrooms or at local, regional, national, or international meetings. In addition to the joint meetings of the American Mathematical Society and the Mathematical Association of America, where plenary lectures on history of mathematics always attract substantial audiences, annual meetings of the History of Science Society also include special sessions devoted to history of mathematics on a regular basis, although these generally involve fewer individuals than the meetings of either the AMS or the MAA. As JOHN FAUVEL (1947–2001) observed in commenting on a joint meeting of the American Mathematical Society and the Mathematical Association of America in January of 1997 (held in San Diego, California), where numerous workshops, minicourses, and special sessions were devoted to the history of mathematics: "Clearly there is a considerable interest in this subject among the American mathematical community, and a widespread demand for information about it. The history talks seemed to be among the best-attended of the whole meeting. This augurs well for the future — while placing a keen responsibility on those concerned with the professional development of the subject" [JOHN FAUVEL, *Newsletter of the British Society for the History of Mathematics*, no. 33 (1997), 13].

This is a responsibility shared by serious historians of mathematics across the United States, who collectively have made possible a thriving community in the present, one that shows every indication of continuing to promote actively the history of mathematics in both popular and professional directions in the years ahead.

15.5 Canada

CRAIG G. FRASER

15.5.1 Before 1966

Before the 1960s the history of mathematics was not an active area of study in Canada. Although sometimes included as a topic in the teaching of secondary and college mathematics, particularly in the province of Quebec, it was cultivated primarily as a private pursuit or as a subject of popularization. In Quebec, however, Jesuit teachers of mathematics and astronomy traditionally included some history of mathematics in their teaching. In the 1920s the mathematician ADRIEN POULIOT (1896–1977) used history to help popularize this subject. LÉON LORTIE (1902–1986), a chemist and polymath, published in 1955 an historical account of the teaching of mathematics in Quebec [LORTIE 1955]. From 1959 to 1962, when the historian of mathematics CHRISTOPH J. SCRIBA (*1929) was a member of the mathematics department at the University of Toronto, he taught courses on mathematics with a strong historical emphasis. He also lectured on history of mathematics in courses and at meetings given at the Ontario College of Education, all of which helped to prepare the way for a more permanent presence for history of mathematics in Canada.

Following the arrival of KENNETH O. MAY (1915–1977, **B**) in Toronto in 1966, history of mathematics became established as an organized academic discipline. From the base MAY established the field has grown and developed over the past three decades, reflecting within the Canadian context the larger international growth of interest in the history of science [LEVERE 1988].

It is also of note to mention that the distinguished American historian of mathematics, RAYMOND CLARE ARCHIBALD (1875–1955, **B**), began his career in Canada, as a teacher and professor of mathematics in the Maritime provinces. Although he left Canada in 1908 at age 32 to take up a position at Brown University, Providence, RI (USA), he retained links throughout his life with Mount Allison and Acadia Universities (see Section 15.4.2).

15.5.2 Kenneth O. May

In the decade following his graduation KENNETH O. MAY carried out research in mathematical economics at Carleton College in Northfield, Minnesota. He also became active in the developing field of mathematics education. His *Elements of Mathematics* (1959) was a prominent college textbook in the "New Mathematics" tradition. The emergence of his interest in the history of mathematics coincided with a growing concern for the subject of bibliography and information retrieval. These two subjects would eventually be joined in his 1973 *Bibliography and Research Manual of the History of Mathematics* [MAY 1973].

In 1966 MAY moved from Northfield, Minnesota, to the Department of Mathematics of the University of Toronto. In that year an advisory committee to the president of the University under the directorship of Professor JOHN ABRAMS (1914–1981) recommended the establishment of an institute to promote research and teaching in the history and philosophy of science. Responding to the rise of history of science in the United States and Europe, the committee's report acknowleged the internal resources that already existed in this field at the university and the need for a formal body to coordinate further teaching and research. When the Institute for the History and Philosophy of Science and Technology was established within the School of Graduate Studies in 1967, KENNETH MAY was one of its founding faculty members. He participated actively in the Institute, serving as its director from 1973 to 1975, and ensured that history of mathematics had a prominent place in the first decade of its existence. By the middle seventies Toronto had become an active center for the history of mathematics, attracting numerous doctoral students and sabbatical visitors. MAY died unexpectedly of a heart attack on December 1, 1977, at age 62.

MAY published on a variety of historical topics, and wrote a masterful article on GAUSS for the *Dictionary of Scientific Biography* [MAY 1972]. However his lasting legacy lies in another direction. In a memorial tribute his student CHARLES JONES (*1939) wrote [DRAKE 1978, 5] "his major contributions [to the history of mathematics] were clearly in the yeoman's services he rendered to the discipline. His exceptional ability to organize enabled him to make valuable contributions as an editor and compiler, and it will be those contributions that will mould the conception that the profession will come to have of Ken May as a historian of mathematics." Along with RENÉ TATON (*1915) and A. P. YOUSHKEVICH (1906–1993, **B**), MAY called for the creation of an international journal of the history of mathematics at the Twelfth International Congress of the History of Science in Paris in 1968. In order to obtain financial support from the Canadian government he proposed the formation of a national Canadian association, a body which became established in 1976 as the Canadian Society for the History and Philosophy of Mathematics (CSHPM).

The journal *Historia Mathematica* itself began publication in 1974 under his editorship (cf. [DAUBEN 1999]. During this period he also published his *Research Manual* [MAY 1973], a detailed guide for students explaining every detail about how to organize and undertake scholarly research, including a comprehensive bibliography about mathematicians, mathematical topics, time periods, universities, organizations, and historiography. MAY was an enthusiastic pioneer in the development of computer-aided information retrieval, a project in which he was assisted by a host of student assistants. His *Research Manual* remains today an indispensable reference tool for historical investigation. It provided inspiration for JOSEPH DAUBEN's collaborative *The History of Mathematics from Antiquity to the Present: A Selective Bibliography* [DAUBEN 1985], a work dedicated to the memory of MAY.

In 1969 MAY became chair of the fledgling International Commission on the History of Mathematics, a position he retained until shortly before his death. The Commission was to provide a "common House of Learning for all historians of mathematics in which there is room for everybody doing serious research" (C. J. SCRIBA, as quoted in [DRAKE 1978, 9]). "His dream was that of a world community of historians of mathematics working together unanimously regardless of race, political conviction or other non-scientific barriers" [ibid.]. To advance this goal, MAY prepared his *World Directory of Historians of Mathematics* (1972), a work containing more than 500 names and listing the major interests of each individual [MAY/GARDNER 1972].

15.5.3 The Situation Today

Due in no small part to the legacy of KENNETH MAY, history of mathematics in Canada now has an active national association, the Canadian Society for the History and Philosophy of Mathematics, and a dozen or so researchers who publish work on a regular basis. Canadians participate at international meetings and serve on the editorial boards of leading journals. However, like any new subject seeking recognition, the discipline possesses its own characteristic difficulties. Many historians are located in departments of mathematics unfamiliar with professional work in the history of science. History departments on the other hand tend to see history of mathematics as a very technical subject aimed at a primarily scientific audience. As the discipline continues to become consolidated internationally and a body of historical work accumulates, this situation should gradually improve.

History of mathematics is carried out as an active subject of research at the Universities of Toronto (Toronto, Ontario), Simon Fraser (Burnaby, British Columbia), McMaster (Hamilton, Ontario), Acadia (Wolfville, Nova Scotia), Laval (Québec, Québec), Québec à Montréal (Montréal, Québec), York (Downsview, Ontario), and others. At the annual congress each May of the humanities and social sciences the CSHPM conducts a three-day meeting that attracts scholars from across Canada as well as foreign visitors from the United States and Europe. In 1997 and again in 1999 the CSHPM held a joint meeting with the British Society for the History of Mathematics. The proceedings of the annual meeting have been published since 1989.

Chapter 16

Japan

SASAKI CHIKARA

16.1 The Prewar Period, 1868–1945: The Flowering of the Study of the History of Japanese Mathematics

16.1.1 Endō Toshisada and His Successor Mikami Yoshio

Interest in the history of mathematics in Japan was sparked by ENDŌ TOSHISADA (1843–1915), whose 1896 book on the history of traditional Japanese mathematics, *wasan*, was written just as it was becoming clear that modern Western mathematics was destined to replace *wasan*.[1] His book, *Dai-Nippon Sūgakushi* (A History of Mathematics in Great Japan), thus represented the swan song of premodern mathematics in Japan.

Before the Meiji Restoration (which ended the military Shogun governments with the return to authority of the Japanese emperor in 1868, the year in which the young Emperor MUTSUHITO (1852–1912) declared the Restoration and opened the new Meiji Period), ENDŌ, who was of a samurai family, had been a practitioner of traditional Japanese mathematics. After the Restoration, he began to learn Western mathematics, and found it rather easy to graft it onto *wasan*. Unfortunately for *wasan* practitioners, with the proclamation of a new system of learning in 1872, the new government determined that Western mathematics should be taught in schools instead of traditional mathematics. Moreover, in 1877 the newly founded University of Tokyo, supreme authority for higher education in Japan, began to teach Western mathematics. This coincided with the arrival of its first professor

[1] Japanese names are given in the traditional East Asian format, family name first, given name(s) following; thus ENDŌ is the family name, TOSHISADA is ENDŌ's given name.

of mathematics, KIKUCHI DAIROKU (1855–1917), who had just graduated with an outstanding record from Cambridge University. In the same year, the Tokyo Mathematical Society was established as the very first scholarly society in Japan.

Nevertheless, even after the proclamation of the new system of learning, some *wasan* mathematicians struggled desperately for survival. ENDŌ, one of these frustrated mathematicians, began to study the history of traditional Japanese mathematics in 1878 rather than continue to pursue the old-fashioned *wasan*. The result of his research appeared in 1896 as the *Dai-Nippon Sūgakushi*.

ENDŌ himself seemed dissatisfied with this first edition of his *magnum opus*, and left manuscripts for further revisions. After his death in 1915, MIKAMI YOSHIO (1875–1950, **B**) and others inherited his project of historical study. MIKAMI published the second revised and enlarged edition of ENDŌ's book in 1918.[2] Since then, historians of *wasan* have taken on successive revisions of the work. In 1981, HIRAYAMA AKIRA (1904–1998) published the most recent version under the title *Zoshu Nippon Sūgakushi* (A History of Japanese Mathematics, Enlarged and Revised).[3]

In prewar Japan, MIKAMI YOSHIO was the most internationally renowned historian of mathematics. His fame rests on the publication of two books in English: *The Development of Mathematics in China and Japan* [MIKAMI 1913] (Reprint 1974), and, in collaboration with DAVID E. SMITH (1860–1944, **B**), *A History of Japanese Mathematics* [MIKAMI/SMITH 1914]. MIKAMI's description of Chinese mathematics provided an important foundation for its treatment by JOSEPH NEEDHAM (1900–1995, **B**) in *Mathematics and the Sciences of the Heavens and the Earth*, volume 3 of NEEDHAM's *Science and Civilisation in China* [NEEDHAM 1959].

MIKAMI was trained neither as a mathematician nor as an historian. His knowledge of Japanese mathematics was fundamentally self-taught. As early as 1906, MIKAMI had published his first memoir on *wasan* and had begun his study of the history of Chinese mathematics as background for the flowering of *wasan*. KIKUCHI DAIROKU soon came to appreciate MIKAMI's abilities as an historian of mathematics and helped arrange for his appointment as a part-time employee of the Japan Imperial Academy in 1908. For six years, from 1911 through 1916, MIKAMI studied philosophy at the Imperial University of Tokyo in order to deepen his understanding of epistemological aspects of mathematics. Unlike ENDŌ, he also established through correspondence a close relationship with an historian, D.E. SMITH, who could teach him critical standards of historiography. This contact and his philosophical training seem to have served him well. In 1921, MIKAMI composed a series of significant papers entitled *Bunkashi-jō yori mitaru Nippon no Sūgaku* (Japanese Mathematics from the Viewpoint of Cultural History), published in 1922. This was republished as a single volume in 1947, and again with some important appendices in 1984 [HIRAYAMA/ŌYA/SHIMODAIRA 1984].

[2] MIKAMI YOSHIO reviewed this edition in *Isis* **4** (1921), 70–72.

[3] See the review by SASAKI CHIKARA in *Historia Scientiarum* (Tokyo) **21** (1981), 123–124.

16.1.2 The Tōhoku School

A group of historians at the Mathematical Institute of Tōhoku Imperial University in Sendai, in the northeast district of Japan, was especially committed to the serious study of *wasan*. This group was led by MIKAMI's rival, HAYASHI TSURUICHI (1873–1935, **B**). The two were rivals primarily in their different approaches to history of mathematics, HAYASHI's focusing more on technical mathematical issues, whereas MIKAMI's was more historical. They also differed over whether SEKI TAKAKAZU (†1708) had been the true founder of *enri* (Analytical Calculations on the Circle) as HAYASHI maintained, or whether the honor should go to TAKEBE KATAHIRO (1661–1739), SEKI's best student, as MIKAMI argued.

HAYASHI began studying *wasan* in 1896 at the suggestion of his teacher at the Imperial University of Tokyo, KIKUCHI DAIROKU. In 1911, the year Tōhoku (now Tohoku) University was founded, HAYASHI began to publish his own private journal, *The Tōhoku Mathematical Journal*. This was an epoch-making event, for previously there had been no scholarly journal fully devoted to mathematics and its allied areas in Japan. At first HAYASHI paid the costs of producing the journal himself, but later it was taken over as an official publication of the University beginning in 1916. HAYASHI also had a gift for nurturing young mathematicians and made Tōhoku University a national center in mathematics. A prolific author, HAYASHI wrote influential mathematics textbooks for students in junior and senior high schools, as well as original papers in various fields of mathematics.

Today, however, HAYASHI is remembered above all as an historian of Japanese mathematics [FUJIWARA 1936]. In his 1910 paper, "The *Fukudai* and Determinants in Japanese Mathematics," he claimed for the first time that the mathematician SEKI TAKAKAZU had discovered the concept of determinant even before LEIBNIZ [HAYASHI 1910]. HAYASHI's papers on the history of Japanese mathematics were later collected and published in 1937 [HAYASHI 1937]. His approach to *wasan* had a special character and specific purpose, namely, he sought to clarify the methods and results of *wasan* mathematicians by relying fully upon modern Western mathematics. In this respect, MIKAMI could not rival HAYASHI, who claimed to be sixth in the line of succession from SEKI's school.

HAYASHI's collaborator, the mathematician FUJIWARA MATSUSABURŌ (1881–1946, **B**), had also been lured into the study of *wasan* by 1937, the year in which HAYASHI's two volume *Wasan Kenkyū Shūroku* appeared. FUJIWARA had been supervising the editorial work for these two volumes, and his enthusiasm for *wasan* was further encouraged in 1940 when the Japan Academy suggested a major study, *Meiji-zen Nippon Sūgaku-shi* (The History of Japanese Mathematics before the Meiji Period), intended to commemorate the twenty-sixth centennial of the supposed birth of the Japanese state. FUJIWARA, as a member of the Academy, was entrusted with this editorial work. It is said that from January, 1943, he resolved to study old Japanese mathematics without breaks or holidays. Over the years, he filled numerous notebooks with his *wasan* research, and wrote drafts amounting to some four thousand manuscript pages.[4] When the United States Air Force bombed Sendai City on the night of July 9, 1945, however, FUJIWARA barely escaped from his home. When he later saw the bomb and fire damage, he immediately feared that all his manuscripts had been destroyed. Almost miraculously, the wife of his son had happened to bring FUJIWARA's manuscripts into the air-raid shelter of her house on the day of the bombing [FUJIWARA 1949a].

It is easy to imagine how overjoyed FUJIWARA must have been to see the surviving manuscripts. They were finally published posthumously in five volumes from 1954 to 1960, under the auspices of the Japan Academy. Although the actual author was FUJIWARA, HIRAYAMA AKIRA supervised the editorial work [FUJIWARA 1954/60]. Since their publication, these volumes have been regarded as the most authoritative on the history of *wasan*. FUJIWARA was not only an able mathematician, but he also fully appreciated MIKAMI's views on cultural history. Needless to say, even the excellent work of these two scholars has its shortcomings, but of the historians of Japanese mathematics in prewar Japan, FUJIWARA and MIKAMI are easily the two most significant. Moreover, FUJIWARA also delivered lectures on the history of Western mathematics. His lecture notes were published posthumously in 1956 under the title *Seiyō Sūgaku-shi* (A History of Western Mathematics), which is a very brief account of Western mathematics from antiquity to EULER.

Another notable historian of mathematics who had an affiliation with Tōhoku University, OGURA KINNOSUKE (1885–1962), should also be mentioned. He served as an assistant in the Mathematical Institute during its early period under both HAYASHI and FUJIWARA. OGURA produced a number of influential books and papers, in particular on the history of mathematics and on mathematics education. Although not a strict Marxist historian, OGURA was certainly influenced by Marxist historiography in emphasizing the social aspects of the history of mathematics. OGURA's collected works were published in eight volumes in 1974 and 1975 [OGURA 1974–75].

16.2 The Postwar Period, 1945–1986: Beginnings of the Serious Study of the History of Western Mathematics

Noteworthy for historical study of mathematics in postwar Japan is the fact that serious interest in the history of Western mathematics arose among historians of mathematics, like KONDŌ YŌITSU (1911–1979). While at Kyoto Imperial University from 1930 to 1934, KONDŌ studied philosophy and theoretical physics, as well as the philosophy of mathematics, under TANABE HAJIME (1885–1962). TAN-

[4] For FUJIWARA's account of *wasan* in English, see [FUJIWARA 1942]. FUJIWARA preferred the transliteration "*wazan*" to "*wasan*" for traditional Japanese mathematics. On the works of FUJIWARA, see [FUJIWARA 1949b].

ABE had studied in the early 1920s at the university of Freiburg with EDMUND HUSSERL (1859–1938), the founder of phenomenological philosophy, and became a leading philosopher in prewar Japan, succeeding NISHIDA KITARO (1870–1945) in the chair of philosophy. During his student years, KONDŌ was deeply committed to a radical student movement agitating against the Emperor's totalitarian regime. He was strongly influenced by Marxist thought on science and technology, which enjoyed a certain popularity in Japan around the year 1930. After graduating from Kyoto University, he moved to Tōhoku University in 1935 at the suggestion of TANABE (and perhaps OGURA), in order to study the foundations and history of mathematics under the guidance of FUJIWARA MATSUSABURO. From the time of his move to Sendai, where he stayed until 1938, KONDŌ's interests seem to have shifted from foundations to the history of mathematics. What he studied seriously during the War was the history of non-Euclidean geometry and the mathematical tools of theoretical and experimental physics. The results of these historical studies were published immediately after World War II.

KONDŌ wrote two outstanding books: *Kikagaku Shisō-shi: Hi-yukuriddo Kikagaku to Riman Kikagaku wo Shudai to Site* (The History of Geometrical Thought: with Special Reference to non-Euclidean and Riemannian Geometries), which appeared in Tokyo in 1946, and *Sūgaku Shisōshi Josetsu* (An Introduction to the History of Mathematical Thought), published in Kyoto in 1947. The latter explained how the analytical tools for expressing the propagation of heat were formulated by JEAN BAPTISTE JOSEPH FOURIER (1768–1830), based on physical modeling, and seems to have been written as a critique of the idealist mathematical philosophy especially prominent in German universities after CARL GUSTAV JACOB JACOBI (1804–1851).

KONDŌ's history of geometry was truly a monumental work in the history of mathematics in postwar Japan. Among the most impressive arguments KONDŌ puts forth is the assertion that the abstract concept of manifolds due to BERNHARD RIEMANN (1826–1866) was closely connected to his philosophical consideration of the notion of space. According to KONDŌ, RIEMANN based his new conception of non-Euclidean space on the differential geometry of GAUSS (1777–1855) and on the psychological and epistemological study of space undertaken by HERBART (1776–1841). In the end, KONDŌ's concept of manifolds was not a consequence of static philosophical speculations but was a product of the interrelationship between theory and practice. KONDŌ revealed his approach to the history of geometry in the opening sentences of his preface to the monograph [KONDŌ 1946, 1]:

> Despite its extreme abstractness, mathematics is a cultural form of the products of practical human activities like the natural sciences. Needless to say, it is influenced to a considerable extent by the society in which it is created, technology, the natural sciences, and philosophy, in both its materials and methods.

Even this brief quotation reflects KONDŌ's materialist view of history, but as his later career showed, he was not totally dogmatic. He was, in fact, highly critical of the Stalinist, or pseudo-Marxist, approach to both epistemology and history. His history of geometry was actually completed on October 12, 1944, and represented an expression of his resistance to what he viewed as the imperialist war then underway. Indeed, his monograph on the history of geometrical thought, in particular its preface, symbolized the beginning of a new era for history of mathematics in postwar democratic Japan.

Later, KONDŌ wrote several books devoted to various fields of the history of Western mathematics, for example, one on the mathematics of antiquity and another devoted to the formation of analytic geometry by DESCARTES (1596–1650) and FERMAT (1601–1665). He also published a philosophical-historical monograph entitled *Dekaruto no Shizen-zō* (Descartes's View of Nature), published in 1959. Kondō's collected works appeared in five volumes in 1994 [KONDŌ 1994], edited by SASAKI CHIKARA (*1947). (For details of KONDŌ's life and works see [SASAKI 1994].)

KONDŌ's historical scholarship, which philosophically and philologically set a very high standard, attracted considerable attention from mathematicians and historians of mathematics alike. One of the leading mathematicians in postwar Japan, IYANAGA SHOKICHI (*1906) published a book on the history of Western mathematics from EUCLID to LEBESGUE (1875–1941) in 1944, which reflects his command of both mathematics and philology as well as KONDŌ's influence. Another mathematician deeply impressed by KONDŌ's works, NAKAMURA KOSHIRO (1901–1986), had studied algebraic topology in Berlin toward the end of Weimar Germany, and under HEINZ HOPF (1894–1971) in Zürich. After World War II, NAKAMURA's interests turned to the history of mathematics. In his historical studies, he attached importance to reading original papers and books and interpreting texts in a philologically rigorous manner. For example, in addition to a Japanese translation of EUCLID's *Elements* published with colleagues in 1970, he also wrote his own *Sūgakushi* (A History of Mathematics) devoted to basic concepts of mathematics, of which the first edition was published in 1962, the second in 1981. His main work, which appeared in 1980, was *Kinsei Sūgaku no Rekishi: Bisekibun no Keisei wo megutte* (The History of Mathematics in the Early Modern Period: On the Formation of the Differential and Integral Calculus); see [HARA 1987]. NAKAMURA's successor, KOKITI HARA (*1918), is internationally known for his work on PASCAL (1623–1662) and ROBERVAL (1602–1675).

16.3 Conclusion

To summarize, in the prewar period, study of the history of Japanese mathematics had flourished since the end of the 19th century. On the other hand, the history of Western mathematics was certainly investigated, but publications were generally not of a high standard. In the postwar period, however, important works on the history of Western mathematics of high quality began to appear in increasing numbers. Moreover, the younger generation has once again begun to study seriously the history of Japanese mathematics, inheriting the works of MIKAMI and FUJIWARA, but now informed by a broader perspective on world history.

As for institutions supporting history of mathematics, the History of Mathematics Society of Japan was founded in 1962 as a successor to *Sanyūkai* (The *Wasan* Society), which was active between 1959 and 1962. The Society, whose members are almost exclusively historians of Japanese mathematics, also publishes a journal in Japanese, *Sūgakushi Kenkyū* (The Journal of History of Mathematics, Japan). History of mathematics is taught at the graduate level in Japan in the Department of History and Philosophy of Science at the Graduate School of Arts and Sciences, University of Tokyo. This has become a center for teaching history of mathematics in general, not only for Japanese mathematics but also for Chinese, Sanskrit, Arabic, and European mathematics, among others, which are also studied. The Department has also produced several Ph.D.s in history of mathematics since its founding in 1970. In addition to the graduate program at the University of Tokyo, Kyoto City has an especially active research group that specializes in particular on the history of Sanskrit mathematics. Beginning in 1962, the History of Science Society of Japan (founded in 1940) has published *Japanese Studies in the History of Science*, which in 1980 changed its name to *Historia Scientiarum* and began its second series in 1991. The journal publishes works on the history of mathematics, as well as history of science and technology, in English, French, and German.

Chapter 17

China

LIU DUN and JOSEPH W. DAUBEN

Although limited materials that may be considered "history of mathematics" are to be found in China even as early as pre-QIN times (before 221 B. C.), the subject in any substantial or systematic sense dates from the late 16th century. By then traditional Chinese mathematics had fallen into decay and rudimentary knowledge of Western mathematics had been introduced to China by the Jesuits. Beginning with CHENG DAWEI (1533–1606), the study of the history of mathematics has continued for more than 400 years in China.[1] At first, the major purpose of such research was to preserve traditional mathematics. Through the activities of a group of scholars known as the Qian-Jia School, the most important mathematical classics of antiquity and related historical materials were collected, studied, collated, and edited to a certain extent with limited commentaries of an historical nature. Since the 1920s, mainly thanks to LI YAN (1892–1963, **B**), QIAN BAOCONG (1892–1974, **B**), and their students, considerable effort has been given to collecting historical materials and attempting rational reconstructions. As a result, the history of mathematics has become a highly-developed, specialized part of the modern history of science in China.

17.1 The Decline of Traditional Chinese Mathematics

The term "traditional Chinese mathematics" refers to a long-established research tradition in China, which used counting-rods and a decimal place-value system of numeration as basic calculating tools, emphasized concrete problems and their applications, stressed mechanical algorithms, and, like EUCLID's *Elements* in the

[1] Chinese names are given in the traditional East Asian format, family names first, given name(s) following; thus CHENG is the family name, DAWEI is CHENG's given name.

West, regarded a single work, the *Jiuzhang suanshu* (*JZSS*, Nine chapters on the mathematical arts) (1st century A.D.) as its major paradigmatic work.

Traditional Chinese mathematics reached its summit in the 12th and 13th centuries A.D., after which it began to decline. Except for a few subjects, primarily commercial arithmetic using the abacus and calculations concerning a harmonic theory based upon a Chinese twelve-note scale derived by ZHU ZAIYU (1536–1611), who even calculated the value of π and discussed other mathematical topics in his *Lulu jingyi* (Essence of Musicology) (1596), mathematics was stagnant or even regressed in the course of the MING dynasty (1368–1644). ZHU ZAIYU created his harmonical theory based on an equal twelve-note scale about 1581, some 10 to 20 years earlier than the Dutch mathematician SIMON STEVIN (1548-1620).

Moreover, MING dynasty scholars had only a superficial knowledge of earlier mathematics. Most important mathematical works were virtually unknown, and some had been lost. One leading mathematician of the 16th century even said that the *Tianyuan shu* (Method for Establishing and Solving Higher Equations) employed by mathematicians in the SONG-YUAN dynasties (960–1368) made him feel "so ignorant as to not know where to begin." In a word, traditional Chinese mathematics had indeed been lost, and its rediscovery was one of the main historical preoccupations of a number of mathematicians and scholars beginning in the MING dynasty.

17.2 Early Authors

One important exception to the failure of MING mathematicians to appreciate earlier Chinese achievements is the use they made of practical arithmetic. Thanks to the prosperity of commerce, many books explaining use of the abacus were written in the MING dynasty. Among these so-called "commercial mathematical books," CHENG DAWEI's *Suanfa tongzong* (*SFTZ*, Systematic Treatise on Algorithms) (1592) was the most significant, and not the least because it preserves important historical information on Chinese mathematics.

CHENG DAWEI was born in Xiuning (now Huangshan City) located in Anhui Province. As a trader he went to many places early in his career, and always made an effort to find mathematical books and to visit scholars with mathematical interests. When he finally retired from his business career, he settled at home to write, and eventually finished the distinguished *SFTZ*, a book that was much in demand as soon as it appeared in print. It was not long before this book was also transmitted to Japan, Korea, and other Southeast Asian countries. It is well known that 17th-century Japanese mathematics was particularly influenced by this book. According to YOSIDA MITUYOSI (1598–1672), it was the major reference for his own study, *Jinkōki* (Account of the Fate of the World) (1627), which is regarded as the most important work in *wasan* — traditional Japanese mathematics [YAN/MEI 1990] (see Ch. 16).

From an historiographical point of view, the *SFTZ* closes with an important appendix, *Suanjing yuanliu* (*SJYL*, The Origins and Development of the Mathematical Classics). This is basically a bibliography of mathematical books consisting of a total of 51 editions since 1084 A.D. Most of them, unfortunately, have been lost, and therefore, the *SJYL* provides a rare source of information about mathematical publications between 1084 and 1592 A.D.

In the late MING dynasty, Western mathematics to a limited extent was introduced by the Jesuits. For example, MATTEO RICCI (1552–1610), the earliest of the Jesuits in China, translated the first six volumes of the edition of EUCLID's *Elements* by CHRISTOPH CLAVIUS (1537–1612) into Chinese with the collaboration of XU GUANGQI (1562–1633) in 1607. Nevertheless, the Jesuits' primary goal was to propagate Christianity, and consequently Western mathematics was neither systematically translated nor widely distributed.

By the early QING dynasty (1644–1911), however, Western mathematical knowledge had already displayed its superiority in both logical reasoning and practical applications. This led QING rulers, along with some leading intellectuals, to advocate a theory of so-called *xixue zhongyuan* (Chinese Origins of Western Learning) [LIU 1991], which sought to establish that the newly-imported Western knowledge could also be found even earlier in traditional Chinese mathematics. As the supreme ruler over all the empire, Emperor KANGXI (1654–1722, reigned since 1661) adopted this idea, claiming that he was not introducing anything foreign but rather was restoring the most authentic Chinese tradition in mathematics. This sometimes had a heuristic value as well; interpretation of rediscovered Chinese texts could be facilitated in light of what had been taught by the Jesuits [JAMI 1994]. As a result, some scholars began to pursue historical studies on traditional Chinese mathematics, and of those who did, MEI WENDING (1633–1721) was the most remarkable.

MEI WENDING was born in Xuancheng (now Xuanzhou City) in Anhui Province. Throughout his life he worked hard to advocate traditional Chinese mathematics, while making efforts to link it up with Western mathematics. He wrote more than 80 books and was praised by the Emperor KANGXI, who encouraged and protected mathematics during the course of his reign. In the QING dynasty, MEI WENDING was acknowledged as "the first celebrated mathematician" [Jiang Yong: *Yimei* (The Wing of Mei, 18th century, Preface)], and his works played an important role in reviving interest in traditional mathematics during the middle QING dynasty.

MEI WENDING's first mathematical work was the *Fangcheng lun* (On Simultaneous Linear Equations) (1672). *Fangcheng* is one of the *jiushu* (Nine Subjects Concerning Number) emphasized in Confucian education in the pre-QIN period (before 221 B.C.). When he finished writing this book, MEI WENDING wrote to one of his friends saying that "I am disgusted by those Western missionaries who exclude traditional Chinese mathematics, and therefore I wrote this book about which even MATTEO RICCI could not possibly say a bad word" [MEI 1672]. Indeed, Western missionaries who went to China in the 16th and 17th centuries did not mention simultaneous linear equations because the subject was then only in its infancy in the West. MEI WENDING clearly wished to demonstrate the superiority of early Chinese mathematics over the methods Western scholars had brought to China, and at least in this case, the example of simultaneous linear equations was an excellent one to stress.

In his various works, MEI WENDING compiled ancient mathematical materials and studied a number of almost forgotten topics. For example, the *gougu* theorem (referred to in the West as the Pythagorean theorem) was a well-known and important focus of ancient Chinese geometry, but since the time of LIU HUI and ZHAO SHUANG, two brilliant mathematicians of the 3rd century, no proof of the *gougu* theorem had been given in any mathematical books. MEI WENDING, however, proposed two proofs, along with other applications of the theorem in his *Gougu juyu* (Illustration of the Right-Angled Triangles) (written before 1692). He next introduced the celestial coordinate transformation from the *Shoushi* (Calendar) (1280) of GUO SHOUJING (1231–1316), and interpreted its trigonometric meaning in the *Qiandu celiang* (The Measurement of a Prism with Two Right Triangle Bases) (1700). In his 1704 text, the *Pingliding sancha xiangshuo* (A Detailed Account of 3rd Degree Interpolation), he explained the interpolating method used by ancient Chinese calendar-makers, and in his *Fangyuan miji* (On the Relation Between the Square and the Circle) of 1710, he investigated the formula for the volume of a sphere. The latter result, historically a very interesting topic in traditional Chinese mathematics, inspired later scholars.

MEI WENDING also wrote two essays on the historiography of the history of mathematics which deserve special mention here: the *Jiushu cungu* (Ancient Fragments of the Nine Subjects Concerning Numbers) (ca. 1690), was a compilation of ancient mathematical texts, and the *Gu suanqi kao* (A Study on Ancient Mathematical Tools) (no later than 1693), offered the first textual study of counting-rods and methods of calculation in the history of Chinese mathematics. It is no exaggeration to say that MEI WENDING's works deeply influenced the "Chinese Renaissance" of mathematics during the middle QING period [LIU 1986].

17.3 The "Qian-Jia School" and Its Successors

Generally speaking, the Qian-Jia School refers to a research community — extant during the two successive reigns of the Emperors QIANLONG (1711–1799, emperor since 1735) and JIAQING (1760–1820, emperor since 1796) — whose members worked to rejuvenate Chinese classical studies through textual criticism. During this period, considerable emphasis was placed on the study of the Chinese classics. Subsequently, as one of the *Liuyi* (Six Arts) that ancient Confucians were required to study for their moral education and general knowledge, mathematics acquired a certain importance. Indeed, several masters of the classics became interested in traditional mathematics, and among these, DAI ZHEN (1724–1777) was the most important figure.

DAI ZHEN was also born in Xiuning, Anhui Province. He was a famous philosopher and a universal genius, proficient in paleography, phonology, the collation of editions, and textual criticism, as well as in astronomy, geography, and mathematics. In 1772, he was appointed as one of the official compilers of the *Siku quanshu* (*SKQS*, Complete Library of Four Branches of Literature) (1773–1781) and assumed responsibility for its mathematical sections.

One significant result of the compilation of the *SKQS* was the rediscovery of important works of ancient mathematics, many of which were virtually unknown. From the *Yongle dadian (YLDD)* (The Great Encyclopedia of the Yongle Reign) (1403-1424), compiled during the reign (1402–1424) of the Emperor YONGLE of the MING dynasty, DAI ZHEN drew seven of the so-called "ten" ancient mathematical classics formerly adopted by the faculty of mathematics as textbooks for the TANG Imperial College (656 A.D.): the *Zhoubi suanjing* (Zhou Gnomon Mathematical Classic) (not later than 50 B.C.), *Jiuzhang suanshu* (Nine chapters) (1st century A.D.), *Haidao suanjing (HDSJ)* (Sea Island Mathematical Classic) (3rd century), *Sunzi suanjing* (Master Sun's Mathematical Classic) (4th century), *Wucao suanjing* (Mathematical Classic of the Five Government Departments) (6th century), *Wujing suanshu* (Arithmetic in the Five Classics) (6th century), and the *Xiahou Yang suanjing* (Xiahou Yang's Mathematical Classic) (8th century).[2] In the meantime, other rare copies of ancient mathematics were acquired by private sources. In a famous collection owned by MAO LI (known as the Jiguge Library), in addition to the above-listed works in the *Yongle dadian* (except for the *Haidao suanjing*), there were also copies of the *Zhang Qiujian suanjing* (Zhang Qiujian's Mathematical Manual) (5th century) and the *Qigu suanjing* (Continuation of Ancient Mathematics) (early 7th century). The books compiled by DAI ZHEN, along with his collations and commentaries, were published at the Imperial Printing Office using moveable-type printing, and issued as the *Wuyingdian juzhen ban* (Edition of the Accumulated Treasures of the Wu Ying Palace) (1726).

Not long thereafter, KONG JIHAN (1739–1784), a relative of DAI ZHEN's by marriage, took the nine mathematical texts gathered together in DAI's compilation, and in order to make ten books altogether, he added the *Shushu jiyi* (Memoir on some Mathematical Arts) (6th century). He published this complete collection as the *Suanjing shishu* (Ten Books of Mathematical Classics), which formed part of the *Weiboxie chongshu ban* (Collected Edition of the Ripple Pavilion) (1773). This is a wooden block-print edition. (For details on the sources and editions of the *Ten Classics* of ancient Chinese mathematics, see [LI/DU 1987, 92–104; 225–230], and [MARTZLOFF 1997, 123–141].)

As a result of these efforts, ten mathematical classics, which had been produced in the millennium spanning the HAN through TANG dynasties (206 B.C. – 907 A.D.) and adopted as textbooks by the faculty of mathematics at the TANG

[2] The original *Xiahou Yang suanying* adopted by the TANG Imperial College was a 4th-century version. By the time the NORTHERN SONG government decided to reprint the classics in 1084, this work had been lost, and so an 8th-century "forgery" of a work including quotations from the earlier *Xiahou Yang* was used instead (see [QIAN 1963, vol. 1, Preface]).

Imperial College, were restored as a whole. Thus after 1733, the *Ten Classics* were once again readily available to scholars for serious study. In addition to these, DAI ZHEN also rescued two mathematical books of the SONG-YUAN period which had been lost for several centuries: the *Shushu jiuzhang* (Mathematical Treatise in Nine Sections) (1247) by QIN JIUSHAO (1202–1261), and the *Yigu yanduan* (Illustrations and Computations of Ancient Mathematics) (1259), by LI ZHI (1192–1279).

Another figure who contributed significantly to the preservation and codification of ancient mathematical texts, RUAN YUAN (1764–1849), was born in Yizheng, Jiangsu Province. In the course of his life, he assumed numerous official posts from the local level to the imperial court itself. Wherever he was, he strove to protect and promote knowledge. In mathematics, he enlisted the help of mathematicians like JIAO XUN (1763–1820), LI RUI (1768–1817), and LIN TINGKAN (1757–1809), and collected rare mathematical books including the *Siyuan yujian (SYYJ)* (The Jade Mirror of the Four Elements) (1303) by ZHU SHIJIE (ca. 1300 A.D.).

RUAN YUAN's most outstanding contribution to the historiography of Chinese mathematics centered on the compilation of the *Chouren zhuan* (*CRZ*, Biographies of Astronomers and Mathematicians) [RUAN 1799]. (Here *Chouren* refers to those whom the ancients called professional astronomers or mathematicians.) The *CRZ* consists of 46 chapters and includes more than 300 scholars (of these, 41 are foreigners) from the earliest legendary times to the 1790s. According to JOSEPH NEEDHAM (1900–1995, **B**), "since very often mathematics was only one of the various scientific accomplishments of indiviuals in a non-specialized age, the book serves as the nearest approach to a history of Chinese science ever written in China" [NEEDHAM 1954, 50]. Other scholars, including D. E. SMITH (1860–1944, **B**), LOUIS VAN HEE (1873–1951), and MIKAMI YOSHIO (1875–1950, **B**) have also praised the importance of the *CRZ* for the history of science [FU 1990, 235–237].

In contrast to both DAI ZHEN and RUAN YUAN, LI RUI was a professional mathematician who was devoted to (and who benefited from) traditional Chinese mathematics. LI RUI was born in Yuanhe (now Suzhou City), Jiangsu Province. As a young man he studied with QIAN DAXIN (1728–1804), another master from the Qian-Jia School who knew both the classics and mathematics. LI RUI joined RUAN YUAN's secretarial staff and became the major contributor to the *CRZ*. Later, he served as secretary to ZHANG DUNREN (1754–1834), a senior local official also interested in promoting mathematics.

LI RUI's works primarily concerned traditional topics. Inspired by QIN JIUSHAO (1202–1261) and LI ZHI (1192–1279), two profound scholars of SONG-YUAN mathematics, LI RUI examined the theory of equations following the method of "extraction of higher [roots] of numerical equations" found in mathematical works from the HAN to SONG-YUAN periods in his *Kaifang shuo* (On Equations, 1814, 1819). For example, although only square and cube root extractions are found in the *JZSS*, more general methods applicable to equations as high as degree 10 are found in later works. In particular, by analyzing the process of the *zengchen kaifang fa* (Mechanical Method for Solving Numerical Equations), LI RUI discovered an equivalent version of DESCARTES's "rule of signs" for determining the number of positive roots in an equation with real coefficients.

LI's *Rifa shuoyu qiangruo kao* (An Exploration of Data for Fractions of the Tropical Year) (1799) was a study of the approximation methods of ancient Chinese astronomers. This work dealt with discrepancies between the lunar and solar calendars, and from a mathematical point of view, was concerned with accurate measurement of the tropical year which required fractions more precise than 365 and 1/4 days. In response to the *JZSS*, LI RUI wrote three books: the *Hushi suanshu xicao* (A Detailed Manual on the Arithmetic Concerning Arcs and Vectors) (1798), the *Gougu suanshu xicao* (A Detailed Manual on the Arithmetic of Right Triangles) (1806), and the *Fangchen xinshu cao* (Manual on a New Method for Solving Simultaneous Linear Equations) (1808). In these works he carefully studied traditional topics like *gougu* (right triangle properties) and *fangcheng* (solution of simultaneous linear equations), and interpreted their meanings which had been unknown for a very long time. In addition, LI RUI also studied *HDSJ* and *JJSS*, both belonging to the *Ten Mathematical Classics*, clarifying his explanations with diagrams and explicit procedures. For the SONG-YUAN mathematical books which had been discovered in his time, he collated a number of works by YANG HUI (13th century), including his *Chengchu tongbian benmo* (Alpha and Omega of Variations on Methods of Computation) (1274), *Tianmu bilei chengchu jiefa* (Practical Rrules of Arithmetic for Surveying) (1275), and *Xugu zhaiqi suanfa* (Continuation of Ancient Mathematical Methods for Elucidating the Strange [Properties of Numbers]) (1275). Unfortunately, LI RUI's collation of ZHU SHIJIE's *SYYJ* was not completed due to LI's death in 1817.

At virtually the same time, LI HUANG (†1812) was also held in high esteem for his knowledge of traditional mathematics in Northern China. Although his hometown was Zhongxiang, Hubei Province, he was called "Northern LI" because he secured most of his official positions in Beijing, whereas LI RUI was called "Southern LI." LI HUANG's representative work is the *Jiuzhang suanshu xicao tushuo* (A detailed Account with Illustrations of the Nine Chapters on the Mathematical Art) (1812, 1820), which was taken as the basic text for studying the *JZSS* until the appearance of QIAN BAOCONG's critical edition in 1963. LI HUANG also collated the *HDSJ* and *Qigu suanjing*.

LUO SHILIN (1789–1853) was considered one of the last to uphold the mathematical tradition of the Qian-Jia School. In addition to his study of the *SYYJ*, he contributed ten chapters to the *Xu chouren zhuan* (A Supplement to the Biographies of Astronomers and Mathematicians) (1840). Several decades later, ZHU KEBAO (1845–1903) wrote seven and HUANG ZHONGJUN wrote twelve more chapters constituting new supplements to this on-going biographical work.

The mathematical sections of the *Siku quanshu zongmu tiyao* (A General Bibliography of the Complete Library of the Four Branches of Literature), published in 1782, were written by DAI ZHEN, GUO ZHANGFA, CHEN JIXIN, and NI TINGMEI (all 18th century). At the end of the QING dynasty, several additional bibliographies of mathematical works also appeared, including FENG ZHENG's

Suanxuekao chubian (A Preliminary Exploration of Traditional Mathematics) in two volumes, unpublished, finished in 1898, LIU DUO's *Gujin suanxue shulu* (Bibliography of Mathematical Books from Ancient to Modern Times) (1898), DING FUBAO's *Suanxue shumu tiyao* (Bibliography of Mathematical Books) (1899) in three volumes, and LIANG ZHAOKENG's *Tianwen suanxue kao* (Exploration of Astronomical and Mathematical Books) in 16 volumes, completed in 1901 but unpublished. Based on these works, ZHOU YUNQING (20th century), one of DING FUBAO's students, compiled the *Sibu zonglu suanfa bian* (Bibliography of Mathematical Books in the Four Branches of Literature) (1956), which covers more than 800 mathematical books written up to the end of the QING dynasty. This work includes brief descriptions of the contents of each work, the various editions, and quotes from prefaces, all in a few sentences. Each of these works helped to prepare the foundation for modern studies on the history of Chinese mathematics, which has facilitated the recent compilation in five volumes under the general editorship of GUO SHUCHUN (*1941) of the *Zhongguo gudai kexue jishu dianji tonghui. Shuxue juan* (Complete Edition of Ancient Books on Chinese Science and Technology. Mathematics Section) [GUO 1993].

As for the history of mathematics outside of China, scholars of the MING and QING dynasties had only a fragmented and hazy understanding based upon very limited sources. For example, EUCLID, ARCHIMEDES, and APOLLONIUS were introduced by the first Jesuits through translations of Renaissance editions. Based on this knowledge, RUAN YUAN and his colleages, in their *CRZ* (1799), included 41 Western authors whose names were known in China, but remarkably neither ISAAC NEWTON nor GOTTFRIED W. LEIBNIZ was mentioned. After the Opium War in 1840, the QING rulers were forced to open China's doors to Western powers, and Western mathematics was reintroduced. For instance, *An Essay on Probabilities* (London 1838) by the British mathematician AUGUSTUS DE MORGAN (1806–1871, **B**) was jointly translated by HUA HENGFANG (1833–1902) and JOHN FRYER (1839–1928) in 1880. It opened with a detailed introduction to this subject's history by HUA HENGFANG.

17.4 The Modern Scholars

Among the most significant contemporary studies of the history of Chinese mathematics are those due to LI YAN (1892–1963, **B**) and QIAN BAOCONG (1892–1974, **B**), both prominent pioneers active in the half-century following the 1920s. As the late Chinese mathematician HUA LUOGENG (L. K. HUA, 1910–1985) once said, "It is mainly thanks to the works of both LI YAN and QIAN BAOCONG that we now recognize the major features in the development of ancient Chinese mathematics" [HUA 1983, Preface, ii]. Indeed, their many publications continue to be of considerable value for anyone interested in the history of Chinese mathematics.

YAN DUNJIE (1917–1988), although younger than LI YAN and QIAN BAOCONG, is often considered a member of the first generation of scholars who contributed to modern studies on the history of Chinese mathematics. From 1980 to 1988, YAN DUNJIE served as Vice-Director of the Institute for History of Natural Sciences in Beijing. In the course of his scholarly career, YAN DUNJIE published over 100 papers on the history of science. In particular, he was a well-known expert on Chinese mathematical astronomy. From 1978-1980 he offered a series of lectures on "The History of History of Science" at the Graduate School of the University of Science and Technology in Beijing. He is the first scholar to teach such a course at an advanced level in China.

Compared with the long tradition of studying the history of Chinese mathematics, research on the history of mathematics in other countries is comparatively recent. Only a few works written in English have been translated into Chinese — for example DIRK STRUIK's *A Concise History of Mathematics* (Beijing: Science Press, 1956), and MORRIS KLINE's *Mathematical Thought from Ancient to Modern Times* (Shanghai: Science and Technology Press, 1979–1981). On the other hand, a few scholars have begun to write and publish their own studies of Western mathematics. Among these, LIANG ZONGJU's (1924–1995) *Shijie shuxueshi jianbian* (A Concise History of Mathematics in the World) (Shenyang: Liaoning People's Press 1979), is quite popular with Chinese readers (for a biography of LIANG, see [WANG 1996]).

17.5 Conclusion

After the end of World War II and the Civil War in China which resulted in the creation of the People's Republic of China in 1949, history of science — and with it history of mathematics — entered a new phase. Institutionally, as part of the Chinese Academy of Sciences, the Institute for History of Natural Sciences was founded in Beijing in 1957 under the auspices of ZHU KEZHEN (1890–1974), a well-known naturalist and meteorologist who was also vice-president of the Chinese Academy of Sciences.

LI YAN was appointed the first Director of the Institute. Since its foundation, the research group for history of mathematics has been the most active and fruitful part of the Institute. In addition to LI YAN, QIAN BAOCONG, and YAN DUNJIE, several younger historians of mathematics, including DU SHIRAN, (*1929), MEI RONGZHAO (*1935), HE SHAOGENG (*1939), and GUO SHUCHUN, have joined the Institute which constitutes the largest and most productive research group of its kind in the country.

During China's Cultural Revolution (1966-1976), virtually all research on history of science, including history of mathematics, was suspended. Following the Revolution, the Institute was reconstituted under the auspices of the Chinese Academy of Sciences.

Since 1978, the Institute has trained postgraduate students for both masters and doctoral degrees. The library of the Institute has now collected 12,000 specialized volumes, of which the number of ancient books and records on science and technology, especially original editions and manuscripts on Chinese mathematics, is unsurpassed in China, and even in the world. The two academic quarterlies edited and sponsored by the Institute: *Ziran kexueshi yanjiu* (Studies in the History of Natural Sciences) and *Zhongguo keji shiliao* (Chinese Historical Materials on Science and Technology), are widely read in both Chinese and foreign scientific circles. Among journals in China that regularly publish articles on history of mathematics, in addition to the Institute's two quarterlies, *Ziran bianzhengfa tongxun* (Journal of the Dialectics of Nature), edited by the Chinese Academy of Sciences, should also be mentioned.

Other active centers for history of mathematics have emerged at several universities in the People's Republic of China, including Beijing, Hangzhou, Dalian, Shanghai, Xi'an, Tianjin, Wuhan, Xuzhou, and Qufu. Especially noteworthy is the Institute for History of Science, founded and directed by the historian of mathematics Li Di (*1927) at Inner Mongolia Normal University (Huhhot). Under the editorship of Li Di, the Institute in Huhhot, in conjunction with Jiuzhang Press (Taipei), publishes *Shuxueshi yanjiu wenji* (Collection of Studies on History of Mathematics), which appears annually. Recently the Chinese Society for History of Mathematics founded in Dalian in 1981 approved the adoption of this periodical as its official journal.

The Chinese Society for History of Mathematics is affiliated with both the Chinese Society for the History of Science and Technology, and with the Chinese Mathematical Society. Since 1982, Chinese historians of mathematics have organized more than ten international symposia on the history of mathematics. Although the total number of scholars and teachers working on history of mathematics throughout the People's Republic of China is not exactly known, it is estimated to be considerably more than 350.

Chapter 18

India

RADHA CHARAN GUPTA

18.1 Introduction

Despite a significant record of mathematical achievements in ancient India, writings devoted to describing and analyzing the historical record are rare. Ancient India produced no HERODOTUS, and historical literature in general did not grow in ancient India. This is sometimes attributed to deep-rooted peculiarities of Indian thought, culture, and philosophical notions which rendered the mind of the Indian scholar indifferent to the search for the bare truths of historical facts, and have therefore discouraged serious study of historical development and change.

Acknowledging the absence of a CLIO among the Indian deities, it is not possible to examine here the causes which prevented the appearance of a genuine historical literature in ancient India, a subject which has prompted some controversial differences of opinion. Nevertheless, most would agree that while there was a sense of appreciation for history in ancient India, it was oriented not to objective narrative history but to a different purpose related to spiritual endeavours. Rather than build historical mansions, it can be said that scholars mined the quarries, cut the stone, amassed the materials, and drew the plans, but did not erect the edifice.

One major difficulty with scientific historiography in ancient India reflects the fact that Hindus attach a divine origin to almost every branch of their science. Thus the origin of *Jyotiḥ-śāstra* (Science of Luminaries), a discipline which includes astronomy, mathematics, and astrology, is attributed to BRAHMĀ, who created the world. Magic squares are said to have been taught first by Lord ŚIVA of the Hindu Trinity to MAṆIBHADRA, brother of KUBERA (the god of wealth). Such divine attributions automatically assign a venerable past to specific discoveries but do an injustice in failing to recognize the historical persons who actually made the discoveries (or inventions) themselves.

In many cultures establishing accurate chronologies has been problematic, and agreeing on reliable dates for specific ancient Indian works or discoveries and their authors is no exception. Moreover, certain typical historiographical vices add further complications — as do claims for priority in discoveries and the tendency to see modern scientific theories literally in the ancient sources. Such misreadings — due more often than not to overly zealous pride in one's own nation, culture and homeland — cannot withstand the test of time (nor are such examples confined to ancient India alone).

18.2 Beginnings of Indigenous Historiography

PĀṆINI's *Aṣṭādhyāyī* (ca. 500 B.C. or earlier) has a significant place in the history of ancient Indian sciences. It presents the first sample of an artificial language (created by some sort of symbols) and other aspects of linguistics drawn from the oral culture and traditions of ancient India. The *Arthaśāstra* of KAUṬILYA is another work of similar importance since it records traditional practices related to various aspects of Indian life and society, and includes an historical survey regarding metrology, the ancient system of weights and measures, as well as units and devices for measuring time. Its author was also named VIṢṆUGUPTA, CĀṆAKYA etc.. He was the chief minister of CANDRAGUPTA MAURYA and ruled from about 322 to 298 B.C. (although the surviving version of the *Arthaśāstra* seems to be a compilation dating to the third century A.D.).

The earliest known work that may specifically be called (in a sense) a history of the exact sciences in India is the *Pañca Siddhāntikā* (Five Astronomical Works) (abbreviated hereafter as *PS*) of VARĀHAMIHIRA (6th century A.D.). According to [NEUGEBAUER/PINGREE 1970/71, Part I, p. 3], it is "unquestionably one of the most important sources for the history of Indian astronomy and its relation to its Babylonian and Greek antecedents." His *Bṛhat Saṁhitā* (Great Compendium), encyclopedic in nature, is equally important for the history of astrology in India. Note that in ancient times in India mathematics is found embedded in astronomy (which was mostly developed for astrological purposes); hence any account of historical writings must refer heavily to astronomical sources.

VARĀHAMIHIRA was a native of Avantī or Ujjain in Western India. His father ĀDITYADĀSA was also his teacher. They were worshippers of the Sun (god). VARĀHAMIHIRA was a prolific author and wrote works in all three branches of the traditional *Jyotiḥ-śāstra*. His treatises were generally composed after he had gone through the opinions of earlier authors whom he often quoted. Consequently, his works are of considerable historical interest, all the more because they also served as models for subsequent writers.

Given the multitude of astronomical theories and systems that prevailed in India, VARĀHAMIHIRA was motivated to compare their merits as well as their faults. The result was his *PS* which gives a good summary of the following five astronomical works):

(i) The *Paitāmaha-siddhānta* (or *Brahma-siddhānta*) which took 80 A.D. as a point of reference for astronomical calculations (this reference-point is taken to indicate the approximate date of the work).

(ii) The *Vasiṣṭha-siddhānta* (3rd century A.D. or earlier).

(iii) The *Romaka-siddhānta* (of Western origin).

(iv) The *Paulisá-siddhānta* (of Hellenistic origin).

(v) The (old) *Sūrya-siddhānta* which selected March 505 as its reference point.

Since the original versions of each of these works are lost, the *PS* is all the more important as an historical source. It is also interesting to note that the *PS* was a basic text used by an Indian called CHUTHAN HSITA (GOTAMA SIDDHA) when he wrote a work in China, the *Jiu gi li* (on calendrical calculations involving the nine "planets"), in 718 A.D. [NEUGEBAUER/PINGREE 1970/71, Part I, p. 16].

Another category called *Saṅgraha* (collection) is comprised of compendia of rules and results from previous works. The *Gaṇitasāra-saṅgraha* (Compendium of the Essence of Mathematics) of MAHĀVĪRA (ca. 850 A.D.) is a rich source for the history of ancient Indian mathematics, all the more so since the *Patīgaṅita* (Mathematics of the Board) of ŚRĪDHARA (ca. 750 A.D.) is not extant in full. There are also many *saṅgraha ślokas* (collected verses) which are often found as well in such noted commentaries as the *Kriyākramakarī* (Operational Exposition) on the *Līlāvatī* (The Beautiful) and the *Yuktidīpikā* on the *Tantra-saṅgraha.* These two commentaries are especially important for the historiography of the Late Aryabhaṭa School of South India.

A more critical survey of Indian astronomical theories is found in the *Jyotirmīmāṁsā* (Investigations of Astronomical Theories) of NĪLAKAṆṬHA SOMAYĀJI (16th century), who discusses various hypotheses and methods along with their historical significance. NĪLAKAṆṬHA critically examined various corrections to planetary parameters (such as those proposed by HARIDATTA in 683 A.D. and LALLA in the 8th century A.D.), as well as other novelties suggested by earlier astronomers. In so doing he provided an historical presentation of various Indian schools of mathematical astronomy. His *Jyotirmīmāṁsā* thus reveals a glimpse of rational scientific thinking in India. His assertion that no one who devised a new system would be "ridiculed for doing so in this world, nor punished in the next" is a significant statement in a traditionally bound yet tolerant Indian society. (Compare this attitude with that in Renaissance Italy where GALILEO was prosecuted for propounding new theories.) Such conditions affect free thinking and hence the development of science.

18.3 Modern Historical Studies and Historiography of Indian Mathematical Sciences

Although the first modern studies of Indian exact sciences began to appear by the close of the 17th century, the first noteworthy account of Indian mathematical astronomy in a European language did not appear until the 18th century. In 1772, GUILLAUME LE GENTIL (1725–1792) published a more-or-less comprehensive account of Indian astronomy in his "Mémoire sur l'Inde" which was included in the *Histoire de l'Academie Royale des Sciences* (Paris, 1772). LE GENTIL was a trained astronomer who was sent to India to observe a transit of Venus in 1761. Although he reached India too late to observe the event and missed another transit eight years later, because the sky was not clear, he had the chance to stay in India and study the native astronomy. As a result, he produced his historical account which generated further European interest in the subject. Deeper studies were produced by JEAN SYLVAIN BAILLY (1736–1793), whose *Traité de l'astronomie indienne et orientale* (Paris, 1787) has become a classic work on Indian astronomy. Historiographically BAILLY emphasized the antiquity, originality, and methodology of the Indian exact sciences, and this approach attracted the attention of such Europeans as P. S. LAPLACE of France and JOHN PLAYFAIR of the University of Edinburgh.

Meanwhile, the Asiatic Society was founded in Calcutta by WILLIAM JONES (1746–1794), a famous British orientalist and jurist. The Society aimed to encourage the study of the ancient civilizations of Asia and India, including their history, arts, crafts, sciences, and literature. The long historical paper by SAMUEL DAVIS (†1819), "On the Astronomical Computations of the Hindus," appeared in *Asiatick Researches* (1790). A decade later, at the turn of the century, JOHN BENTLEY carried out his pioneering investigations of the earliest developments of Indian mathematical astronomy, which culminated in BENTLEY's *Historical View of the Hindu Astronomy* (published posthumously in 1825).

A new chapter in the historiography of Indian mathematical sciences began early in the 19th century with the studies of HENRY THOMAS COLEBROOKE (1765–1837, **B**), who was both an able mathematician and an expert orientalist. He collected a large number of Sanskrit mathematical and astronomical texts, and was especially interested in their unpublished commentaries, the study of which resulted in a far better understanding of the historical development of these subjects. His *Algebra with Arithmetic and Mensuration from the Sanscrit of Brahmegupta and Bháscara* (London, 1817) has become a classic used by both modern historians and historiographers of Indian mathematics. His long "Dissertation on the Algebra of the Hindus" prefixed to the above work is of great importance. As COLEBROOKE observed:

> Had an earlier translation of Hindu mathematical treatises been made and given to the public, especially to the early mathematicians in Europe, the progress of mathematics would have been much more rapid, since algebraic symbolism would have reached its perfection long before

the days of Descartes, Pascal, and Newton [*The Mathematics Student* **1** (1933), 11–12].

CHARLES M. WHISH inaugurated another new chapter in the historiography of Indian mathematics with his paper "On the Hindu Quadrature of the Circle and the Infinite Series of the Proportion of the Circumference to the Diameter Exhibited in the Four Śāstras, etc." [WHISH 1835]. This study presented material showing that foundations for analysis had been laid down in India in the early 15th century. The Rev. EBENEZER BURGESS (1805–1870), an American missionary who was in India from 1839 to 1854, was also a devoted scholar of Indian exact sciences. His famous translation of the *Sūrya-siddhānta* (On Mathematical Astronomy) appeared in 1860, with further material by WILLIAM D. WHITNEY (1827–1894) and mathematical notes by HUBERT A. NEWTON (1830–1896), professor of mathematics at Yale University.

Later in the century, AUFRECHT WEBER's "Über den Vedakalender, namens Jyotisham" (*Abhandlungen der (königlichen) Akademie der Wissenschaften zu Berlin, phil.-hist. Kl.* (1862), 1–130) provided an edition and translation of the *Vedāṅga Jyotiṣa* (dated between 1200 and 500 B.C.). In 1864 BHĀU DĀJI (1821–1874) discovered a manuscript of the *Āryabhaṭīya* of ĀRYABHAṬA I (*476 A.D.). Soon more manuscripts were located, and shortly thereafter H. KERN (1833–1917) published an edition of the *Āryabhaṭīya* with commentary of PARAMEŚVARA (Leiden, 1874).

GEORGE FREDERICK WILLIAM THIBAUT (1848–1914) played the most significant role among scholars in India during the last quarter of the 19th century as far as historical writing is concerned. Born in Germany, he worked briefly in England before being appointed professor at Government Sanskrit College (India) in 1875. His detailed essay on the *Śulba-sūtras* (the most ancient mathematical texts of India), and his translation of the *Baudhāyana Śulba-sūtra* (the oldest of them all), were published between 1875 and 1877. Jointly with SUDHAKARA DVIVEDI (1855–1910/11), he edited and translated the *Pañca-siddhāntikā* which, as already mentioned, is a very important source for the history of the ancient period. In fact, THIBAUT correctly pointed out that a sound understanding of the evolution of the various astronomical systems (summarized in the above-mentioned work) is necessary to appreciate the development and transition of Indian astronomy from the elementary empirical stage to the more siddhāntic works in which geometry and trigonometry played their important roles.

18.4 Indian Historians

Ancient India's famous international universities at Takṣilā, Nālandā, and Odantapura were all destroyed by medieval feuds. In the 19th century, new universities were founded in the three most cosmopolitan cities of India: Calcutta, Bombay, and Madras (now Bombay is called Mumbai, and Madras is called Chennai) in 1857 under the British rule. These institutions were set up to provide education (in English) in European sciences and other Western subjects. Prior to these, several colleges were opened, some of which offered instruction using Sanskrit. Thus opportunities existed to learn both ancient Indian as well as modern European subjects in the fields of mathematics and mathematical astronomy, and indeed, study, research, and publishing in the modern sense attracted many Indians.

Among the early native scholars who received a sound education in Indian as well as Western (European) exact sciences, BĀPŪDEVA ŚASTRĪ (1821–1900) was perhaps unique. He was born on November 1, 1821, to a Brahmin family of Maharashtra. In addition to his early education in Western arithmetic and algebra at the Marathi School in Nagpur, he also studied — under ḌHUṆḌHIRĀJA MIŚRA — the *Līlāvatī* (on arithmetic and mensuration) and the *Bījagaṇita* (on algebra), both of which are standard Sanskrit works. Noting the talent of BĀPŪDEVA, the British political agent LANCELOT WILKINSON secured his admission to the Sehore Sanskrit School where he studied *Siddhānta Śiromaṇi* (on mathematical astronomy), Euclidean geometry, and European science in general under PANDIT SEVĀRĀMA (19th century) and with the general guidance of WILKINSON himself. In 1842, BĀPŪDEVA ŚASTRĪ joined the faculty of the Government Sanskrit College where he taught *rekhā-gaṇita* (Euclidean geometry). Through his various works, he enriched the exposition of Indian mathematics and astronomy, especially by using European material and notation. Although he was the first to bring to light India's contribution to calculus, his exposition of "Bhāskara's Knowledge of the Differential Calculus" (see details in [ŚĀSTRĪ 1858]) was not entirely satisfactory. His historical writings primarily aimed to uncover the gems of ancient Indian mathematics and to advance the claim that Indian mathematicians anticipated the invention of calculus.

A better-known scholar, teacher, and writer who produced influential studies of Indian mathematics and astronomy was SUDHAKARA DVIVEDI (1855–1910/11) (see [GUPTA 1990b]). His editions of various Sanskrit works helped other scholars in turn to investigate and write about the history of Indian mathematics. His works also provide a synthesis of sorts of the Eastern and Western mathematical sciences. Moreover, he himself was an historian of Indian mathematics and mathematical astronomy, and contributed as well to the historiography of these disciplines.

DVIVEDI was born in 1855 in Khajuri, a village near Varanasi (=Benares). At an early age he was attracted to mathematics, which he studied under PANDIT DEVAKṚṢṆA. In 1883 he was appointed librarian at the Government Sanskrit College in Benares, where he became Professor of Mathematics and Astronomy when BĀPŪDEVA ŚASTRĪ retired in 1889. To encourage Indian students "towards the cultivation of Western science," DVIVEDI wrote a number of textbooks in native languages on new topics such as the theory of equations, conic sections, calculus, and logarithms, incorporating historical notes as well.

In 1890 he wrote the famous *Gaṇaka Taraṅgiṇī* (in Sanskrit) on the lives and works of Indian astronomers and mathematicians from antiquity to his own time. But he conceived mathematics as a universal subject to which contributions were made by different countries, nations, and cultures. He also realized the value of using the history of mathematics in teaching and learning the subject. In 1910 his *A History of Mathematics, Part I (Arithmetic)* appeared in Hindi. This was a unique work of its kind in India, full of rich material. DVIVEDI wanted to educate the masses in the history of mathematics (not confined to India). Unfortunately, due to his early death (in 1910 or 1911), he was unable to finish the other three parts of the work he had planned to write.

In the meantime, KṚṢṆAŚĀSTRĪ GOḌABOLE (1831–1886) — jointly with GOVINDA VIṬṬHALA (19th century) — translated the first four books of EUCLID into Marathi in 1874. The translation helped in popularizing axiomatic-type geometry which was foreign to traditional Indian mathematics. GOḌABOLE also wrote a short essay on the history of astronomy (about 1885). Some ten years later a voluminous history of Indian astronomy, written by SAṄKARA BĀLAKṚṢṆA DĪKṢITA (1853–1898) in Marathi, was published in Poona in 1896. This work traced the gradual development of mathematical and physical astronomy in India only since Vedic times, because the Indus Valley civilizations had not as yet been discovered.

18.5 The Twentieth Century

ASUTOSH MUKHERJEE (1864–1924) was a brilliant mathematician. Unfortunately, he was not appointed as a research professor at the University of Calcutta because the University was unable to raise the needed funds (9000 rupees annually). Consequently, MUKHERJEE started practicing law in 1888. When he became a judge in the Calcutta High Court in 1904, he was known as an "historical and sociological" jurist.

From 1906 to 1914, MUKHERJEE served as the Vice-Chancellor of the University of Calcutta. Under his administrative guidance, the University was transformed into a modern institution which served as a model for other Indian universities. He firmly believed that teaching and research were essential for discovery, preservation, and application of knowledge, and above all, for the creation of new scholars. He insisted on an all-India or national character of the University, with an emphasis on the international character of science.

Despite attempts on the part of vested (European) interests (e.g. Eurocentric views) "to establish that Indians had no originality in the departments of science, mathematics and astronomy" [SINHA 1974, 90], the University of Calcutta collected relevant manuscripts and promoted the exploration and investigation of such historical materials. Study and research in the history of Indian mathematics and astronomy was also encouraged. Eventually, this led to interest in historiography as well.

Further, the free and open organization of Calcutta and other universities in India provided ample opportunities for faculty, students, and other scholars to study the global history of mathematics (without national or geographical boundaries) through various publications both old and new. Thus most scholars with an interest in the history of mathematics tried to understand the contributions of other countries and cultures, and to place Indian achievements in a context of world historical achievements.

In 1913 GANESH PRASAD (1876–1935, **B**) was appointed the first Rashbehari Ghosh Professor of Applied Mathematics in the University of Calcutta, and in 1923 he became the Hardinge Professor of Pure Mathematics there. Besides doing significant research in mathematics, he had a great interest in the history of mathematics, which he propagated and popularized. PRASAD's *Some Great Mathematicians of the Nineteenth Century: Their Lives and Their Works* (2 vols., 1933/1934), a pioneering work, includes GAUSS, CAUCHY, ABEL, JACOBI, WEIERSTRASS, and RIEMANN in volume one; while the other included CAYLEY, HERMITE, BRIOSCHI, KRONECKER, CREMONA, DARBOUX, CANTOR, MITTAG-LEFFLER, KLEIN, and POINCARÉ. In fact, if "history is nothing but biography of great men" (THOMAS CARLYLE), then PRASAD's classic of more than 700 pages and 16 portraits can be considered to be a marvelous history of 19th-century mathematics. Unfortunately, a planned third volume never appeared, due to PRASAD's death in 1935; the manuscript now seems to be lost.

Meanwhile, GANESH PRASAD endowed a fund to enable the Calcutta Mathematical Society to award a regular prize for a work related to the history of Indian mathematics. This organization, although responsible for doing so, has never made an award. Nevertheless, under the influence of PRASAD, two of his students — AVADHESH NARAYAN SINGH (1901–1954, **B**) and BIBHUTIBHUSAN DATTA (1888–1958, **B**) — proved to be important scholars of the history of mathematics, especially with reference to India, although their basic doctoral research was respectively in pure and applied mathematics.

DATTA made in-depth studies of original Sanskrit and other sources, and also read related papers on the history of Indian as well as other sciences. In 1927 he delivered an address on "Contribution of the Ancient Hindus to Mathematics" at the Allahabad University Mathematical Association, which published the lecture in its *Bulletin.* Later, the address formed the nucleus of DATTA's major work (with A. N. SINGH), the well-known *History of Hindu Mathematics: A Source Book* (2 vols., [DATTA/SINGH 1935/38]).

Another significant work of DATTA is *The Science of the Sulba* (Calcutta, 1932), which is based on a thorough study of the *Śulba-sūtras.* He also published about four dozen papers devoted to history of mathematics. DATTA's work corrected the misinterpretations (historical, chronological, linguistic, mathematical, etc.) of GEORGE RUSBY KAYE (1866–1926) [KAYE 1925] (see below). DATTA retired voluntarily in 1929 and became a *Saṁnyāsi* (a wandering monk) in 1938. He came to be called SWAMI VIDYARANYA, and his interests changed to religion and philosophy [GUPTA 1980].

Although A. N. SINGH was quite enthusiastic about history of Indian mathematics, administrative duties perhaps prevented him from publishing more than he did. Only the first two parts of his *History of Hindu Mathematics. A Source Book* (written jointly with DATTA) were published in his life-time. But his creation of the Section for Hindu Mathematics at Lucknow University was an important event. After his death, the section continued to make contributions under the guidance of KRIPA SHANKAR SHUKLA (*1918), who edited and translated a number of Sanskrit texts.

What DATTA and SINGH did for the history of Indian mathematics, PRABODH CHANDRA SENGUPTA (1876–1962) did for the history of Indian astronomy. He studied original Sanskrit and other sources and published a number of papers [GUPTA 1979]. There were also other historical writers in the neighboring discipline of astronomy and mathematics. These include A. A. KRISHNASWAMI AYYANGAR (1892–1953), LAXMAN VASUDEVA GURJAR (1909–1982) (whose *Ancient Indian Mathematics and Vedha* (Poona, 1947) was quite popular), NARENDRA KUMAR MAJUMDAR (20th century), SARADA KANTA GANGULI (*1881), and SUKUMAR RANJAN DAS (ca. 1930). The monographs of BENOY KUMAR SARKAR [SARKAR 1918], like [KAYE 1925], are not considered reliable because the first makes exaggerated claims and the other denies even genuine ones.

When India became independent from British rule in 1947, the need to boost national sentiment by making such exaggerated historical claims about Indian accomplishments was no longer necessary. This is reflected to some extent in the post-independence works of T. A. SARASVATI AMMA (20 century), AMULYA KUMAR BAG (*1937), K. R. RAJAGOPALAN (20th century), SAMARENDRA NATH SEN (1918–1992), and C. N. SRINIVASIENGAR (1901–1972).

However, some serious problems, including questions of accurate chronology, continue to hamper the historiography of mathematics in India, especially for the ancient period [GUPTA 1990a]. Deliberate attempts to make exaggerated claims for India on the one hand, and to deprive Indians of even their original achievements on the other, still continue.

18.6 Conclusion

At various times the motivation and purpose for studying the history of mathematics in India have differed, sometimes significantly. At first it was studied out of curiosity and to help determine how much mathematics was developed in ancient India *vis à vis* the rest of the world. The purpose of such historical writings was to show that Indian achievements were quite high and original. To counter the culturally demoralizing effect of India's supposed borrowing from the Greeks (and others), chronologies established Indian's priorities in scientific achievements. Later, during the political struggle that eventually ended in independence for India from British rule, the Indian *Vedas* were heralded as the fountainhead of all modern sciences. Indian historians subsequently strove for more balanced historical treatment, based on sounder corroborative evidence. Despite this shift, historians of mathematics in India have yet to settle on a common philosophy to guide their research. They have, however, an important outlet for communicating their evolving ideas in the journal, *Gaṇita Bhāratī*, founded in 1979.

Chapter 19

Arab Countries, Turkey, and Iran

SONJA BRENTJES

19.1 Introduction

In this chapter, some historical conditions are described which shaped the emerging historiography of mathematics in the Arab countries, Turkey, and Iran before 1970.[1] It makes no claim to be a systematic and well-balanced survey of all the relevant books and papers written. One of the purposes of this chapter is to highlight attractive topics for future research in the history of mathematics, science, and medicine, and the history of encounters between the Middle East, Europe, and the Americas in these disciplines since early modern times.

A multifaceted historiography of mathematics within a broader disciplinary and cultural setting has existed in the Muslim world at least since the 9th century. It was partly appropriated from Greek and Sasanian sources in the translation movement of the early Abbasid caliphate (750–ca. 900) and was extended and reshaped through different literary genres such as biographical and bibliographical dictionaries, historical chronicles, encyclopedias, prefaces of and introductions to mathematical works, and in the presentation of particular mathematical problems [BRENTJES 1992]. This production of historiographies continued at least until the 18th century. Most of the later works such as the *Ghunyat al-rāʾid fī ṭabaqāt ahl al-ḥisāb wa'l-farāʾiḍ* (The Searcher's Wealth, on the Classes of the People of Calculation and Inheritance Mathematics) by IBN AL-QĀḌĪ (†1616/17) have never been studied by historians of mathematics.

[1] I thank M. BAGHERI, Tehran, for his help regarding Iran.

Changes in the social and cultural context of mathematics in Muslim societies and their impact on mathematics proper under the Turkish and Mongol speaking dynasties since the late 11th century, as well as the consequences of growing contacts between European and Muslim scholars and craftsmen, particularly from the 16th century onwards, have recently received more attention from historians of science and mathematics. What follows, however, is restricted primarily to the potentially most significant developments in the Muslim world for the history of mathematics, and to the emergence of a new type of historiography of mathematics during the 19th and 20th centuries.

19.2 Exchanges with Western Europe

From the late 12th century, European Christians visited the Middle East with increasing frequency. Their travel accounts, diaries, reports, and letters delivered certain information about either the state of the arts or earlier texts and authors, in particular those dealing with astronomy, geography, medicine, and systems of calculation or calendars (see, for instance, [DU MANS 1890, 163–170], [FERRIER 1996, 129–142]). In the 16th and 17th centuries, European scholars as well as Catholic missionaries came to live for extended periods in the Ottoman, the Safavid, or the Moghul Empires. These travellers produced a number of books — some of which dealt with scientific topics — for either European or Middle Eastern audiences; as yet, these have not received a systematic treatment by historians of science or mathematics.

On the other hand, Muslim travellers, mostly ambassadors with their retinues, and merchants, also came to Europe. A number of them reported on the sciences and technologies they observed. In 1720/21, for instance, an Ottoman ambassador was sent to Paris. His official mission was to convey to the French king the Ottoman consent to repair the Church of the Holy Sepulcher in Jerusalem. The real aim, however, was "to visit fortresses and factories, and to make a thorough study of means of civilization and education, and report on those suitable for application in the Ottoman Empire" [GÖÇEK 1987, 4]. Several accounts of this visit and its results are extant both in French and Turkish. Although they contain long passages about the sciences and scientific instruments in France [GÖÇEK 1987, 17–18, 57–60, 139], these have not been studied by historians of science or mathematics, nor has the embassy's impact on science and technology in the Ottoman Empire been assessed. An important event closely connected with the preparation for this mission, as well as with its subsequent impact, was the foundation, in Constantinople, of the first Muslim printing press in 1726/27 by the ambassador's son (and later Grand Vezir), MEHMED SAID EFENDI (†1761) and by IBRAHIM MÜTEFERRIKA (†1745). IBRAHIM MÜTEFERRIKA, a Hungarian convert to Islam who served in the Ottoman army, printed a map of the Marmara Sea in 1719/20, of the Black Sea in 1724/25, and of Iran in 1729/30. (A map of Egypt made by MÜTEFERRIKA was discovered in 1976.) Moreover, in 1732, MÜTEFERRIKA published a book by

Ḥājjī Khalīfa's (†1657) on mapping the world in addition to his own treatise on magnetism and the compass. Most of the books printed in Müteferrika's lifetime, however, treated themes of political and military history such as the rise of the Safavid Empire, the discovery of the New World, Ottoman naval wars, or military organization. Grammar texts were also prevalent. This trend was largely followed until 1796/97 when the press was closed down due to the death of its owner. Further Ottoman printing presses were established in Constantinople at the beginning of the 19th century, for example, in the School of Engineering and Artillery, where medieval mathematical works and textbooks for the new schools were printed [*EI*, **6** (1991), 800–802].

19.3 A New Start for Historiography of Science and Mathematics

The evolution of a new type of historiography of science and mathematics resulted from reform policies introduced in different regions of the Muslim world in the course of the 19th century. The Ottoman printing presses and the School of Engineering and Artillery were both linked to these institutional reforms. Other efforts were carried out in Egypt, formally a part of the Ottoman Empire, and in Persia, where Muslim printing presses were founded at the beginning of the 19th century (Egypt: Bulaq, ca. 1822; Persia: Tabriz, 1817; Tehran, 1825) [*EI*, **6** (1991), 797–798, 803–804]. Among reforms to affect historiography were the establishment of new state schools (military, medical, engineering, industrial), the sending of students to European universities and companies, and the translation of European (and non-European) books on science, mathematics, medicine, history, philosophy, geography, military organization and engineering, politics, and culture. In some cases, books on the history of philosophy and science were translated, too [Ṭūqān 1963, 494], [Abu-Lughod 1965, 51, 63].

Approximately eleven medieval mathematical texts and encyclopedias with mathematical chapters from authors who flourished between the 13th and the 17th centuries were printed during the 19th century. Most of these texts, however, were not meant as contributions to history, but as textbooks for courses in the traditional schools (see Table 1).

The major biographical work printed in the 19th century is by Ibn Khallikān (†1282), *Kitāb wafayāt al-aʿyān wa-anbāʾ abnāʾ al-zamān* (Book on Deceased Eminent People and News on Contemporary People), Tehran, 1867 and Cairo, 1853, 1882, 1892/93, 1910. Ibn Abī Uṣaybiʿa's (†1270) *ʿUyūn al-anbāʾ fī ṭabaqāt al-aṭibbāʾ* (Sources of Nnews on the Classes of Physicians), printed in Cairo in 1882, represented a new approach. Ibn al-Ṭaḥḥāna (19th century) not only reproduced an ancient manuscript, but edited it as an historical object in its own right. Although the printing of medieval manuscripts continued during the first half of the 20th century, increased attention was directed to preparing scholarly editions of manuscripts. Important medieval mathematical manuscripts

Table 1. 19th-century printed editions of medieval mathematical writings

Author	Title	Place of publication	Date of publication
NAṢĪR AL-DĪN AL-ṬŪSĪ (1200–1274)	*Taḥrīr kitāb al-uṣūl li-Uqlīdis* (Recension of Euclid's *Elements*)	Istanbul Fez Tehran	1801 1876 1881
MUḤAMMAD BARAKA AL-ʿABADĪ (13th century)	*Taḥrīr kitāb al-uṣūl li-Uqlīdis* (Recension of Euclid's *Elements*)	Tehran	1879
MUḤAMMAD B. MAḤMŪD AL-ĀMULĪ († 1352)	*Nafāʾis al-funūn fī ʿarāʾis al-ʿuyūn* (The Charms of the Disciplines on the Brides of the Sources) Encyclopedia	Tehran	1892 1897–1899
GHIYĀTH AL-DĪN JAMSHĪD AL-KĀSHĪ († 1429)	*Miftāḥ al-ḥisāb* (Key of Arithmetic)	Tehran	1889
ʿALĪ AL-QALAṢĀDĪ († 1486)	*Kashf al-asrār ʿan ʿilm ḥurūf al-ghubār* (The Unveiling of the Secrets on the Science of the Dust-Letters) Arithmetic	Fez	1892/93
SIBṬ AL-MĀRIDĪNĪ († ca. 1495)	*Kifāyat al-qanūʿ fi'l-ʿamal bi'l-rubʿ al-maqṭūʿ* (What is sufficient for the modest to work with the divided quadrant) *Luqṭat al-jawāhir fi'l-khuṭūṭ wa'l-dawāʾir* (The finding of jewels, on lines and circles) *Maṭlab fi'l-ʿamal bi'l-rubʿ al-mujayyab* (The search for working with the sine quadrant)	Istanbul Cairo Cairo	1858 1876 1892
ʿABD AL-RAḤMĀN AL-AKHḌARĪ († 1510)	*Al-Sirāj fī ʿilm al-falak* (The lamp on the science of the orbs) Astronomy	Cairo	1896/97
BAHĀʾ AL-DĪN AL-ʿĀMILĪ († 1622)	*Khulāṣat al-ḥisāb* (The Essence of Calculation) *Al-Kashkūl* (The Begging Bowl = The Album) Encyclopedia	Istanbul Tehran Tabriz Tehran Cairo	1851 1859 1860 1850 1879 1871 1885 1888 1900

were also published in cities of the Moghul Empire and later British India during the 19th and early 20th centuries, as well as in India and Pakistan after 1948 [MATVIEVSKAYA/ROZENFELD 1985, vol. 1].

In the second half of the 19th century, partly in order to emulate French and British patterns, national and city libraries and scientific societies were founded either by ruling houses and their administrations or by individual intellectuals. The National Library of Egypt in Cairo was formed in 1870 on the initiative of ʿALĪ PĀSHĀ MUBĀRAK (1823–1893), the Minister of Education. It acquired European books and Arabic, Persian, and Turkish manuscripts, including hundreds of mathematical and astronomical texts. The first catalogues were prepared between 1883 and 1891 by MUḤAMMAD AL-BIBLAWĪ (1863–1954), SHAYKH ʿABD AL-RAḤMĀN AL-SAYYID (1855–1902), and KARL VOLLERS (1857–1909) [MATVIEVSKAYA/ROZENFELD 1985, vol. 1, 220], [NEUMANN 1893, 156], [SUTER 1893, 1]. SUTER's German translation of the part describing the mathematical manuscripts appeared in 1893. Other catalogues of traditional or newly-founded manuscript libraries were prepared in Fez (1887), Istanbul (1875, 1882/83, 1883, 1885/86, 1886/87, 1888/89, 1892/93, 1893/94, 1894/95), Damascus (1881/82), Algiers (1893, 1909), Jerusalem (1900), Tlemcen (1907), and Tunis (1908–1911).

19.4 Diverse Attitudes towards History of Science and Mathematics

Until recently, the origin and cause of the Arabic *nahḍa* (awakening) movement — which shaped the intellectual climate of the Arab world until the end of the First World War — has been associated with the early 19th-century reforms in Egypt [*EI*, **7** (1993), 900–903]. A different interpretation, however, sees the *nahḍa* movement resulting from three factors: the partial failure of these early reforms; the opening of missionary schools and their impact upon the rise of a new, Western-oriented intellectual elite; and a growing dissatisfaction with British influence in Egypt in the later decades of the 19th century [*EI*, **7** (1993), 180–185], [*EI*, **5** (1986), 1090]. There are also diverging views about the attitude of this movement towards the past [*EI*, **7** (1993), 901–902]. Prominent writers, graduates of the missionary schools, and Muslim thinkers hoped that *nahḍa* could yield a modern Arabic or Islamic society that combined the best from its own past with the best from contemporary Europe. The following five examples illustrate the variety of 19th-century attitudes.

RIFĀʿA AL-ṬAHṬĀWĪ (1801–1873), one of the most successful members of the third group of Egyptian students sent to Paris in 1826 and later director of the School of Translation (1836), reputedly recognized the modern European sciences, despite their foreign appearance, as Islamic. He saw them as resulting from European translations of Arabic books, an opinion he justified by referring to one of the scholars of the Azhar mosque [DAMARDĀSH 1965, 6].

His compatriot, MAḤMŪD PĀSHĀ AL-FALAKĪ (1815–1885), was even more successful in mastering the new sciences and embracing the career opportunities they opened. After completing a traditional elementary education in his village, he graduated from the Naval School in Alexandria and then taught mathematics and astronomy at the School of Engineering. In 1850/51 he was sent to Paris where he studied with ARAGO (1786–1853). Later, he returned to Egypt in 1859 and joined the freemasons, the Egyptian Institute in Alexandria (from 1880 on in Cairo), and the Geographical Society, becoming president of the latter. In 1882 he served as Minister of Public Works for two months, and afterwards as Minister of Education. At the beginning of his career, he translated books on mathematics and physics from French into Arabic. After returning from France, he directed a team to map Egypt. AL-FALAKĪ also published several papers on contemporary scientific matters such as magnetism and on historical problems such as the character and age of the pre-Islamic Arabic calendar [DAMARDĀSH 1965, 151, 153, 155], [NEUMANN 1893, 156–158], [*EI*, **2** (1991), 764].

In 1876, YAʿQŪB ṢARRŪF (1839–1912) and FĀRIS NIMR (1859–1951), both graduates of the Syrian Protestant College of Beirut (1866; today: American University of Beirut), founded the journal *al-Muqtaṭaf* to popularize European science and culture. It also published papers on the history of science and mathematics such as ṢARRŪF's paper *Tārīkh al-jabr wa'l-muqābala* (History of Algebra) which appeared in volume 7, 1883, together with two papers on modern algebra. The editors may have been following, at least partially, the example set by the first director of the College, CORNELIUS VAN DYCK (1818–1895), who published a book on the history of astronomy in Arabic in the mid-1870s.

The influential Muslim thinker and spokesman for Pan-Islamism, JAMĀL AL-DĪN AL-AFGHĀNĪ (†1897), drew attention to past achievements, since in his view, earlier Muslim mastery of science augured well for future scientific success in the Muslim world [*EI*, **2** (1991), 417f]. Reformers who propagated a national or regional perspective, such as MUṢṬAFĀ KĀMIL PĀSHĀ (1874–1908) from Egypt, looked at the Muslim heritage of scientific achievements as a source from which modernizing efforts in the Arab world, in particular Egypt, could draw sustenance. The invitation to CARLO A. NALLINO (1872–1938) to teach in Cairo at the newly founded Egyptian University (1909/10), as well as his lecture course on the history of Arab astronomy, illustrate this attitude [DAMARDĀSH 1965].

A further motivation to study the history of science and mathematics arose from efforts to modernize and purify the classical Arabic language. Several private societies or circles devoted to this project were created during the late 19th and early 20th centuries, e.g. the *Nādi Dār al-ʿUlūm* (The Club of the House of Science) or the *Lajnat al-muṣṭalaḥāt al-ʿilmiyya* (Committee on Scientific Technical Terms) (1907). A prominent proponent of this movement was AḤMAD ZĀKĪ PĀSHĀ (1863–1914) who published a study on Arabic encyclopedias in 1891 [*EI*, **5** (1986), 1090], [MATVIEVSKAYA/ROZENFELD 1985, vol. 1, 245]. After the demise of the Ottoman Empire, state-sponsored societies were created in Damascus (1919), Cairo (1932), and Baghdad (1945). These were often modelled on the French

Academy (Académie Française) and simultaneously pursued the goals of protecting and modernizing the classical Arabic language. These academies sponsored historical research published in their journals. AḤMAD S. SAʿĪDĀN (1914–1991, **B**), for instance, published some of his studies on the history of Arabic arithmetic in the *Journal of the Institute of Arabic Manuscripts*, which was affiliated with the Egyptian Academy [MATVIEVSKAYA/ROZENFELD 1985, vol. 1, 130, 256].

In a recent paper, AḤMAD DJEBBAR (*1941) pointed out that Maghrebian scholars started to investigate aspects of history of science and mathematics only in the 1930s. Some of this research focused on the history of Arabic literature in the Maghreb in which scientific and mathematical writings were integrated as minor components. The framework of these early studies is reflected in titles such as *Al-Nubūgh al-maghribī fi'l-adab al-ʿarabī* (The Moroccan Genius in Arabic Literature) by ʿABDALLĀH GHANNŪN (†1989), published in 1938 in Tétouan [DJEBBAR 1998, 17].

19.5 Trends in History of Mathematics in Iran and Turkey

A movement similar to that in the Arab world arose in Persia during the 19th century under the Qajar dynasty (1779–1921), and gained strength with the accession to power of the PAHLAVI dynasty (1921–1978). The goal was to purify the Persian language and to create new words for concepts introduced by the reforms. In 1935, under the patronage of the shah, a Society for the Iranian Language was created after the model of the French Academy *(Académie Française)*. Some of its founding members published books on the history of mathematics in Iran, for instance ʿALĪ AKBAR DIHKHUDĀ (1879–1955) and SAʿĪD NAFĪSĪ (1895–1967) [*EI*, **5** (1986), 1096], [MATVIEVSKAYA/ROZENFELD 1985, vol. 1, 229, 346]. Like their Arabic contemporaries, they saw history as the vehicle for national and cultural identification. Their work is mainly biographical and represents the "nationalization" of medieval scholars. The medieval scholars who attracted the most attention were almost the same as those studied in Egypt, Syria, and Lebanon — MUḤAMMAD B. MŪSĀ AL-KHWĀRIZMĪ (fl. ca. 780–830), IBN SĪNĀ (†1036), ABŪ L-RAYḤĀN AL-BĪRŪNĪ (973–1048), ʿUMAR AL-KHAYYĀM (1048?–1131?) and NAṢĪR AL-DĪN AL-ṬŪSĪ (†1274).

In the Ottoman Empire and later Turkey, the same drive to construct national identity is evident. Thus, in the 19th century, pan-Islamic ideas and sentiments prevailed and were even expressed as pan-Ottomanism. The collapse of the Ottoman Empire during the First World War sparked a movement to base national identity on the concepts of language and ethnicity. This was accompanied by efforts to purify the Turkish language. At the same time, reforms of the 19th and 20th centuries forced Turkish intellectuals either to coin new Turkish words for newly introduced foreign concepts or to assimilate their foreign designations. In 1932, by order of the government, the Turkish Linguistic Society was founded. Its historical approach followed that of its sister organisation, the Committee for Research into Turkish History,[2] founded two years earlier. This so-called maximalist approach to history embraced all Turk or Turko-Mongol languages and their peoples as part of Turkish heritage [*EI*, **5** (1986), 1100]. The two societies inherited the bulk of the estate of MUSTAFA KEMAL ATATÜRK (1881–1938, president of Turkey since 1923) [*EI*, **5** (1986), 1099–1101]. The journals of the historical society published numerous papers on the history of mathematics, astronomy, and related fields, e.g. by AYDIN SAYILI (1913–1993, **B**) and his student SEVIM TEKELI (*1924). The tension between these two cultural approaches towards history of science and mathematics (Islamic versus national or ethnic) continued to influence Arabic, Persian, and Turkish writings until 1970, the period covered here. In the history of mathematics, there was a slight preference for the ethnic/national approach, while in history of science both attitudes were prevalent. This is illustrated by the titles given in Table 2.

19.6 The New Institutions

In the first half of the 20th century, in almost all countries under consideration, the institutional base for the history of science and mathematics grew through the creation of universities, journals, and committees for the publication of cultural heritage. Advanced degrees in history of science and mathematics (master or doctoral thesis) were introduced at several Middle Eastern universities. In Egypt, scholars such as MUṢṬAFĀ NAẒĪF (20th century), ʿALĪ MUṢṬAFĀ MUSHARRAFA (1898–1950), AḤMAD SAʿĪD AL-DAMARDĀSH (20th century) or AḤMAD MUḤAMMAD MURSĪ (20th century) formed an Egyptian Society for the History of Science in 1939. The first issue of the society's journal included papers on MUḤAMMAD B. MŪSĀ AL-KHWĀRIZMĪ and IBN AL-HAYTHAM (965–ca. 1040).

The earliest university journals to publish papers on the history of science and mathematics were those of the Catholic University St. Joseph, Beirut (*al-Mashriq*), the faculty of letters of Ankara University, the faculty of letters of Alexandria University, the faculty of letters of Tehran University and the American University of Beirut (*al-Abḥāth*). In Tunisia, the University of Tunis created an outlet for publishing studies by scholars affiliated with the university. For example, MUḤAMMAD SOUISSI published his edition and translation of IBN AL-BANNĀʾ's *Talkhīṣ aʿmāl al-ḥisāb* (Survey of the Operations of Arithmetic). He included in this edition a commentary based upon another of IBN AL-BANNĀʾ's arithmetical writings and the works of two later Maghrebian and Andalusian scholars, AL-HUWĀRĪ (13th century) and AL-QALAṢĀDĪ [DJEBBAR 1998, 8, 19].

The most concerted efforts to embed the history of mathematics and science in the university curriculum were made at Ankara University through the appointment of A. SAYILI, and at the American University of Beirut through that

[2] Its name was changed twice in the following years, first to Society for Research into Turkish History (1931), and later to Society for Turkish History (1935).

Table 2. Examples of the two cultural approaches to history of science and mathematics in Arab countries, Turkey, and Iran in the 20th century

Title	Approach	Place of publication	Date of publication
La Pensée iranienne à travers l'histoire	ethnic/national	Paris	1930
La science chez les turks ottomans	ethnic/national	Paris	1939
Turāth al-ʿarab al-ʿilmī fi'l-riyāḍiyyat wa'l-falak (The Scientific Heritage of the Arabs in Mathematics and Astronomy)	ethnic	Cairo	1941
Ārāʾ al-falāsifa al-islāmiyyīn fiʿl-ḥaraka (The Opinions of the Islamic Philosophers on Movement)	Islamic	Cairo	1943
Falāsifat al-ʿarab (The Philosophers of the Arabs)	ethnic	Beirut	1950–1954
Ta'rīkh-e ʿulūm-e ʿaqlī dar tamaddon-e eslāmī tā awāsiṭ-e qarn-e panjom (History of the Rational Sciences in Islamic Civilization until the Middle of the Fifth Century)	Islamic	Tehran	1952
Büyük türk alimi Nasireddin Tusi (The Great Turkish Scholar Naṣīr al-Dīn al-Ṭūsī)	national	Istanbul	1953
La pensée de l'Islam	Islamic	Istanbul	1956
Türk matematikcileri (Turkish Mathematicians)	national	Istanbul	1958
Un Mathématicien Tuniso-Andalou: al-Qalaṣādī	ethnic/national	Madrid	1970
A Short History of Mathematics in Iran from the Ninth to the Seventeenth Centuries	national	Tehran	1973

of EDWARD S. KENNEDY (*1912). The former was appointed to the first chair for the history of science in 1952. The latter, along with numerous students and associates, published papers primarily on the history of astronomy and mathematics [MATVIEVSKAYA/ROZENFELD 1985, vol. 1, 165, 220, 265, 287–288, 385].

The Catholic University St. Joseph, founded in Beirut in 1875 to counterbalance the influence of the protestant missions in Syria, encouraged editions and studies of Arabic texts on mathematics, both of which were published in its journals and through its printing house. Among these were THĀBIT B. QURRA's (826–901) translation of the *Arithmetike Eisagoge* (Introduction to Arithmetic) by NIKOMACHOS OF GERASE (1959). The university also collaborated closely with the French Institute of Damascus, which was established in 1922 for archaelogical exploration of the Levant. A similar institution had been created in 1888 in Cairo. PAUL KRAUSS (1904–1944), a Czech immigrant affiliated with the two institutes, published major works on the so-called JABIR corpus, on the 10th-century philosopher and physician MUḤAMMAD B. ZAKARIYĀʾ AL-RĀZĪ, and on other subjects.

These two French institutes highlight a third context within which the emerging history of science and mathematics in the Arab world, Turkey, and Iran must be seen. Primarily, they were established to provide French orientalists the long-sought opportunity to work and live in the cultures they were studying. They were also founded to stabilize and deepen the political and economic influence of the French Republic in the Middle East. Finally, their establishment was intimately linked to the Franco-British struggle over colonies and mandates in the Middle East, as well as to Franco-German battles over similar issues [AVEZ 1993, 9–17]. Writings on the history of science and mathematics by Middle Eastern authors either placed the medieval Arabic, Persian, Turkish, or Islamic achievements in opposition to those of Europe, or they emphasized the universal character of science and mathematics and the illustrious place occupied by leading Muslim scholars in world history.

In the Maghreb, certain institutions founded by the French and Spanish colonial administrations provided scholars with outlets for research on manuscripts including those on science and mathematics. At least one of these institutions, the *Maʿhad Mawlāy al-Ḥasan li'l-abḥāth* (The Mawlāy al-Ḥasan institute for research) in Tétouan also permitted Moroccan scholars to publish biographical studies which included information on scholars such as IBN AL-BANNĀʾ [DJEBBAR 1998, 19, footnote 107].

A fourth incentive to support research on Arabic history, including history of science and mathematics, arose from political considerations with regard to the Palestinian-Israeli conflict. In the early 1950s, Spain and Egypt agreed to establish an institute for Islamic studies, the *Maʿhad al-dirāsat al-islāmiyya*, in Madrid. Later, other Arab countries came to contribute to the funding of this institute. Despite its political background, the institute also published historical studies, such as editions of sources on medieval philosophy and biographical works. In 1958, a Moroccan scholar, MUḤAMMAD AL-FĀSĪ (20th century), published a paper in the institute's journal on the life and works of IBN AL-BANNĀʾ [DJEBBAR 1998, 19, footnote 108].

A fifth major point of orientation was an emphasis on scientific progress. Editing and analyzing texts served to show that medieval Muslim scholars had made major contributions to mathematics and science, and so had participated in the overall progress of science. Other questions — such as the institutional settings of astronomy, medicine, and mathematics (the royal courts, hospitals, and observatories), the relationships between religion and the so-called secular sciences, and the contributions of Ottoman scholars to early modern science — were also raised within this framework. The scholarship of A. SAYILI is exemplary in this regard.

Prior to 1970, the lion's share of studies on the history of mathematics and science published in the Arab countries, Turkey, and Iran were biographies of leading medieval scholars (such as MUḤAMMAD B. MŪSĀ AL-KHWĀRIZMĪ, AL-BĪRŪNĪ, IBN AL-HAYTHAM, KŪSHYĀR B. LABBĀN (fl. ca. 1000), ʿUMAR AL-KHAYYĀM, NAṢĪR AL-DĪN AL-ṬŪSĪ or GHIYĀTH AL-DĪN JAMSHĪD AL-KĀSHĪ), editions of original sources, manuscript catalogues, and analytical studies of particular problems and methods. The most prolific Middle Eastern historians of mathematics and science before 1970 were LOUIS CHEIKHO (1889–1927), ʿALĪ MUṢṬAFĀ MUSHARRAFA, MUṢṬAFĀ NAẒĪF, AḤMAD SAʿĪD AL-DAMARDĀSH, ABDÜLHAK ADNAN (1882–1955), AYDIN SAYILI, AḤMAD SAʿĪDĀN, ʿĀDIL ANBŪBA, JAMĀL AL-DĪN HUMĀʾĪ (1900–1980) and ABŪ L-QĀSIM QURBĀNĪ (*1911). The majority of these, including other scholars not explicitly discussed here, concentrated on the history of science and mathematics of Muslim societies between the 9th and 15th centuries. Some exclusively studied the history of mathematics, while others explored the entire medieval canon of the mathematical and rational sciences. Only a very few, such as AYDĪN SAYILI, AḤMAD SAʿĪD AL-DAMARDĀSH, PARVĪZ SHAHRĪYĀRĪ (*1926), or ʿABD AL-ḤAMID I. ṢABRĀ (*1924), also considered other periods, cultures, or mathematical disciplines (Greek antiquity, medieval Europe, early modern Europe, 19th-century Egypt, analytical geometry, algebra, calculus, etc.).

19.7 Conclusion

History of mathematics in the Muslim world has been a field of literary activity of long standing. Having its origin in ancient Greek and Sasanian traditions, it has encompassed a variety of questions and approaches, although biographical, bibliographical, linguistic, and problem-oriented contributions have dominated. During the 19th century, in almost all regions of the Ottoman Empire, Iran, and India, a new start towards history of mathematics and science engendered new fields of work, such as modern editions of historical writings and comparisons with modern European developments. The motivations for this new start were manifold. Most of them arose from the need to reform the societies of the Muslim world in the face of expanding European colonialism, rapidly developing Western technology, and the impact that repeated wars (the Napoleonic wars at the turn of the 18th century and the two World Wars of the 20th century) had upon these regions. The social and cultural changes produced by these developments have in turn affected three main perspectives from which the histories of science and mathematics have been viewed: an ethnic and national perspective, a pan-Islamic perspective, and a perspective concentrating on progress.

Efforts to institutionalize the history of science and mathematics were first linked to movements to modernize Muslim societies. The founding of colleges, universities, research institutes and their journals, publishing houses by Catholic and Protestant missionaries, national and pan-Islamic movements, and certain activities of the European colonial powers, allowed Middle Eastern, Maghribian, and foreign scholars to publish papers, monographs, catalogues, and editions of manuscripts in local languages.

A second wave of institutionalizing the history of science and mathematics was the outcome of efforts at the beginning of the 20th century to adapt the languages of the region (Arabic, Persian, Turkish) to the new technological and scientific terminology of the West. This led to the creation of private study circles, state academies, scholarly societies, journals, etc., which also encouraged publications, meetings, and debates on history of science and mathematics. After the Second World War, the first professors teaching history of mathematics were appointed in Lebanon, Turkey, and Tunisia, and university curricula in history of science were launched. In the years after 1970, further efforts to stabilize the teaching and research base of history of science and mathematics were made in Syria, Turkey, and Algeria, and most recently in Egypt. This led to the foundation of the Institute for the History of Arabic Science at the University of Aleppo in Syria, the introduction of a history of science program for graduate students at Istanbul University in Turkey, the establishment of the Society for the History of Mathematics in Algeria, and the opening of the Science Heritage Center at Cairo University in Egypt. The focus of history of mathematics in the Arab countries, Turkey, and Iran until 1970 was the Islamic one in Arabic, Persian, and Turkish. This interest served to pave the way for more locally confined studies recognizing the need to differentiate the more than 1000 years of history in the Muslim world in terms of the histories of numerous different dynasties and cultures. It also led to the recognition of various different origins and causes of the Arabic *nahḍa* (awakening) movement which shaped the intellectual climate in the Arab world until the end of the First World War, and which influenced as well approaches Arab scholars took to the history of the mathematical sciences.

Chapter 20

Postscriptum

JOSEPH W. DAUBEN, JEANNE PEIFFER, CHRISTOPH J. SCRIBA,
and HANS WUSSING

20.1 The Character of Historiography

Historiography may be regarded as a kind of meta-history. According to EDMUND B. FRYDE, Professor of History at the University of Wales:

> Modern historians aim mainly at reconstructing an accurate record of human activities and at achieving a more profound understanding of them. This conception of their task is quite recent, dating only from the development in the late 18th and early 19th centuries of scientific history, cultivated largely by professional historians. It springs from an outlook that is very new in human experience: the assumption that the study of history is a natural, inevitable kind of human activity. Before the late 18th century, historiography (the writing of history) did not stand at the center of any civilization. History was almost never an important part of regular education, and it never claimed to provide an interpretation of human life as a whole. This was more appropriately the function of religion, philosophy, even perhaps of poetry and other forms of imaginative literature [FRYDE 1990, 945].

20.2 George Sarton's Views

Historiography of mathematics, too, may not as yet stand at the center of civilization, nor is it the core of any regular educational curriculum, for the sciences or otherwise, although if GEORGE SARTON were to have had his way, this might have been a very different story. For SARTON truly believed that history of mathematics was not only the apex of the history of the sciences, the most perfect and sublime of all as might befit a true Platonist view of the world and the rightful place mathematics should assume therein, but SARTON also took mathematics to be the foundation, the very backbone upon which all of the other sciences must depend. It was the armature that gave coherence and substance to both the physical and life sciences. He was explicit about this in a fundamental essay on the subject:

> [T]he history of mathematics should really be the kernel of the history of culture. Take the mathematical developments out of the history of science, and you suppress the skeleton which supported and kept together all the rest. Mathematics gives to science its innermost unity and cohesion, which can never be entirely replaced with props and buttresses or with roundabout connections, no matter how many of these may be introduced [SARTON 1955, 4].

Having noted the centrality of the history of mathematics not only to intellectual history in general, but also to the entire history of science, or more properly, the histories of the sciences, SARTON then went on to raise several issues which, in closing, *Writing the History of Mathematics* must also consider. Noting that it was impossible to contemplate the "mathematical past" without asking some fundamental questions, these, SARTON added, "are as simple to formulate as they are difficult to answer" — namely:

> To what extent were the filiation and development of ideas determined either by outside circumstances or by a kind of internal necessity? [SARTON 1955, 14–15].

In responding to this question, SARTON offered a brief catalog of external circumstances that may affect mathematics. "There is no doubt," he insisted, "that mathematical discoveries are conditioned by outside events of every kind, political, economic, scientific, military, and by the incessant demands of the arts of peace and war" [SARTON 1955, 15].

20.3 Interrelations

Indeed, in the course of writing *Writing the History of Mathematics*, it became clear that the historiography of mathematics, especially in its early phases, had close interrelations to diverse influences, to the various branches of mathematics itself, of course, but to philosophy, history, and philology as well.[1]

[1] Several efforts to survey the historiography of mathematics are worth noting here, namely those by DIRK J. STRUIK: [STRUIK 1980]; S. S. DEMIDOV and M. FOLKERTS (Eds.): [DEMIDOV/FOLKERTS 1992]; and IVOR GRATTAN-GUINNESS: "Talepiece: The history of mathematics and its own history," in [GRATTAN-GUINNESS 1994, vol. 2, 1665–1675].

Throughout the Renaissance, for example, humanism with its emphasis on recapturing ancient texts played a decisive role in the renaissance of mathematics. This was not only the case in the great centers of translation in Spain and Italy, but in northern Europe as well, for example in Benelux, where humanism did not concentrate solely on literary texts but also included science and mathematics as well. Editorial work played a central role there from the 15th century onwards. Catalogues of authors and titles were produced, mostly in chronological order, often subdivided according to subdisciplines of mathematics. Influential far beyond his home country was G. J. VOSSIUS' *De universae mathesios natura et constitutione liber, cui subiungitur chronologia mathematicorum* (Book on the Nature and Constitution of the Universe of Mathematics, to Which is Attached a Chronology of Mathematicians) (1650).

20.4 On the History of Historiography

During the 17th and 18th centuries, mathematics was considered to be less a special discipline and more a part of general education and scholarship. To write a scientific monograph or survey without an historical introduction was considered an affront against the ideal of all-around scholarship. Similarly, any history of literature had to include the sciences, and here too, mathematics has its own role to play.

Historiographic studies could also take on strategic nationalistic interests. Here Italy was typical in attempts, following the *Risorgimento* and unification of the country, to establish a national Italian school of mathematics. In connection with this, historically interested mathematicians tried to rediscover or to reconstruct an Italian tradition of mathematics. It comes as no surprise that LEONARDO OF PISA, who had worked at the court of FREDERICK II, was a central figure in such attempts.

A similar interest in local mathematicians is also evident in the case of Portugal, and often national landmarks will encourage historical studies, as did the American bicentennial in 1976 which sparked considerable interest in the United States in history of American mathematics, a phenomenon that holds for many other countries included in this book as well.

Philosophical preconceptions have exerted their influences, too. Thus, for example, the development of history of mathematics in France was determined more than elsewhere by the philosophical sentiments of the Enlightenment, while in German-speaking countries the strong philological tradition, especially in the 19th century, was remarkable and predominant.

The example of the Danish mathematician and historian of mathematics HIERONYMOUS GEORG ZEUTHEN indicates that his reputation as a mathematician contributed substantially to the acceptance of the history of mathematics in Denmark, and on the international scene as well. Likewise, the prestige of BOUR-

BAKI has doubtless had a positive effect on legitimizing history of mathematics in France, if of a very particular sort.

The example of Mexico, on the other hand, demonstrates how the political context of the moment may influence if not determine the interrelations of historiography of mathematics with other disciplines. As the political order there changed dramatically between 1920 and 1940, with the process of industrialization and the application of modern technology, the important role of mathematics was increasingly appreciated. In a newly-founded university, opened in 1954 (the *Universidad Nacional Autónoma de México (UNAM)*, a very active seminar arose, oriented to philosophy and history, whose members began to produce a substantial number of publications. Reflecting an intensive collaboration on the borders of history and philosophy of science and mathematics, few such examples are to be found elsewhere. Unfortunately, the student upheaval of 1968 in Mexico effectively brought this activity to an end, from which it has yet to recover. Further examples of how political ideologies may have significant consequences for the history of mathematics — affecting for example the extent of the support it receives and the kinds of research it is expected to carry out — could be cited in the case of the former East Germany or the USSR, among others that might also be mentioned.

20.5 Functions of Historiography

The functions of historiography, as apparent in the narrative accounts of the history of mathematics from country to country as related in Part I of *Writing the History of Mathematics*, are diverse, but may be summarized briefly if incompletely as follows. At its most basic, history of mathematics may be used as LEIBNIZ once said, simply to give mathematicians their due, to establish priorities or to propagandize on the special value of one discovery, theorem, method, or specific result over others. This may be for reasons of chronology, clarity, coherence, perhaps even on aesthetic grounds that one mathematician's work may be better than another's due simply to the elegance of the arguments made, or the solutions derived. Others have used history of mathematics to establish their claims that progress is indeed the inevitable course of history, a view that was dominant for the 18th and much of the 19th centuries. History of mathematics may also serve simply to recount the life of a great individual, or to trace the development and historical significance of a particularly potent idea or useful theorem.

History of mathematics has also been used to serve narrow political or social ends, to express the achievements of a people or nation. Histories of women mathematicians, histories of African mathematics, or works written in China to show that all of Western mathematics had already been contemplated, if forgotten, by early Chinese mathematicians, are all examples of ways of doing history of mathematics that may serve as political or social commentaries while making valid contributions to the history of mathematics with respect to its own historiography. It comes as no surprise that national interests — not to be confused with the sort of nationalistic interests just mentioned — also have a place in the historiography of mathematics. This is often true, for instance, with respect to biographies. It is altogether natural that Scandinavian historians, whether by nationality or birth, should be interested in investigating the life and work of ABEL, LIE, or CASPAR WESSEL. Such a tendency is especially noticeable in the choices made by 18th- and 19th-century authors.

Explicitly political factors related to the rise of liberal states in the 19th century offer another case in point. In Spain, in connection with a movement begun by the aristocracy and clerics of the Church, it was argued that at the time of its peak Spain was leading the world in every respect, including mathematics and the sciences, but that the enemies of Spain had succeeded in suppressing the knowledge of Spanish achievements. Hence it was considered an important task of the history of science to unearth and note these national achievements again. As a consequence, long lists of authors and titles were compiled in order to establish such claims. Similarly, but perhaps with more important results, studies of the history of applied mathematics, including astronomy and navigation in the 16th and 17th centuries, were undertaken to reveal the role of the sciences for the success of Spanish navigators and the expansion of the Spanish Empire.

A similar result may be found in Belgium, where following the dissolution of the United Netherlands and the founding of Belgium in 1830, a movement was launched to exhibit the achievements of the arts and sciences of the Belgian people. GERARD MERCATOR, GEMMA FRISIUS, and SIMON STEVIN were thus rescued from historic oblivion by what can only be described as a romantic, patriotic venture with positive benefits for the history of mathematics. As in Spain, the nationally oriented biographic-bibliographic sides of historiographic interests were also emphasized. The widening of themes beyond what was strictly relevant to Belgian history eventually led to a second phase of broader historical interests towards the end of the 19th century, but even then, interest in Belgium in classical antiquity remained minimal. To a lesser degree this was also true of the Netherlands. BIERENS DE HAAN, for example, relied upon his knowledge of publishing to pursue bibliographical and biographical subjects of purely Dutch interest. Although HUYGENS should have been a logical candidate for serious study in Holland, the magnificent Huygens edition was not undertaken until the next generation of historians of mathematics became interested in doing so.

As for the Americas, both North and South, apart from some isolated examples in the 19th century, interest in history of mathematics is predominantly a 20th-century phenomenon. This is no doubt due to the fact that modern mathematics itself only took hold and began serious development in the Americas in the last century. As a consequence, the pursuit of the history of mathematics is likewise comparatively recent.

Ethnomathematics has also been increasingly a focus of concern over the past few decades in particular. Pre-Columbian mathematics and the mathematical and proto-mathematical concepts of the native Indian populations before the arrival of the Europeans has been a subject of considerable interest. Attention is now being paid not only to the number systems, calendar arithmetics, and astronomical ideas of the Aztecs, Mayans, and Incas, but especially to games, ornaments, architectural decoration, and pottery designs, all of which may implicitly involve mathematical structures and hidden order relations.

20.6 Institutional Factors

Writing the History of Mathematics also reveals how important institutional factors have been for the success of mathematics, especially in the 20th century. Indeed, for reasons that were clearly well established by the 19th century, universities, institutes, academies, journals, publishing houses, and a wide variety of other institutions emerged or were created to sustain mathematics.

Here the case of Switzerland provides a particularly good example, for the role of its many local *Naturforschende Gesellschaften* (societies of natural sciences) were instrumental in promoting studies in the history of mathematics. In many countries, mathematics journals were the first to provide a place for the publication of historical research on a regular basis. Moreover, the first journals devoted specifically to history of mathematics were allied with mathematical journals, like TERQUEM's *Bulletin de bibliographie, d'histoire et de biographie mathématiques*, issued as a supplement to the *Nouvelles annales de mathématiques*; the *Abhandlungen zur Geschichte der mathematischen Wissenschaften mit Einschluss ihrer Anwendungen*, which grew out of the *Zeitschrift für Mathematik und Physik*; and LORIA's *Bollettino di Storia e Bibliografia Matematica*, which was at first associated with the Italian *Giornale di Matematiche*. For a time LORIA's journal was published independently as the *Bollettino di Bibliografia e Storia delle Scienze Matematiche*, and later (from 1922), it was issued as a separately-paginated section of *Il Bollettino di Matematica*. It should be noted, however, that BALDASSARE BONCOMPAGNI's *Bullettino di Bibliografia e di Storia delle Scienze Mathematiche e Fisiche* (1868–1887) was a notable exception to this pattern, in that the *Bullettino* was founded independently of any allied mathematical journal.

In the 20th century the success of journals has prompted several publishers to support book series featuring works on history of mathematics. These include both the *Science Networks · Historical Series* and *Vita Mathematica* from Birkhäuser Verlag (Basel and Boston), the French collection *Un savant, une époque* from Belin, as well as the *History of Mathematics Series* produced jointly by the American Mathematical Society (Providence, Rhode Island) and the London Mathematical Society (London), the *History of Modern Mathematics Series* published on an occasional basis by Academic Press (New York and Orlando), and the *Sources in the History of Mathematics and Physical Sciences* published by Springer-Verlag (Berlin and New York).

20.7 History of Mathematics and Mathematics Education

Recently, the history of mathematics has emerged as especially relevant to the interests of teachers, and here the lead has been taken by the International Commission of Mathematics Instruction (ICMI). In addition to sponsoring, organizing, and encouraging regional as well as international meetings, the Commission has proved instrumental in promoting history of mathematics through a recent collaborative effort, a book edited by JOHN FAUVEL and JAN VAN MAANEN, *History in Mathematics Education* (2000). As the pre-publication information about the book explains its goals:

> This book investigates how the learning and teaching of mathematics can be improved through integrating the history of mathematics into all aspects of mathematics education, lessons, homework, texts, lectures, projects, assessment, and curricula .... Resulting from an international study on behalf of ICMI (International Commission of Mathematics Instruction), the book draws upon evidence from the experience of teachers as well as national curricula, textbooks, teacher education practices, and research perspectives across the world. Together with its 300-item annotated bibliography of recent work in the field in eight languages, the book provides firm foundations for future developments. Focusing on such issues as the many different ways in which the history of mathematics might be useful, on scientific studies of its effectiveness as a classroom resource, and on the political process of spreading awareness of these benefits through curriculum design, the book will be of particular interest to teachers, mathematics educators, decision-makers, and concerned parents across the world [FAUVEL/VAN MAANEN 2000].

As early as 1926 the National Council of Teachers of Mathematics (NCTM) in the United States supported the value of history in the teaching of mathematics. (See, for example, the volume edited by JOHN K. BAUMGART, *et al.*, *Historical Topics for the Mathematics Classroom* (2nd rev. ed. 1989, [BAUMGART 1989]). In part this is because, as many believe, mathematics recapitulates its own history, and historical material can therefore help to illuminate the actual development of complex mathematical ideas. Others value history in the classroom as a means of promoting cultural diversity, as a means of showing that mathematics is universal to all cultures.[2]

Just as there may be striking differences, there may be equally striking similarities, as between the Pythagorean theorem discovered in ancient Greece and its

[2]There has been considerable interest over the past several decades in ethnomathematics, especially as it pertains to teaching mathematics. See, among many others: [ZASLAVSKY 1973], [D'AMBROSIO 1988], [ASCHER 1991], [POWELL/FRANKENSTEIN 1997], and [GERDES 1998].

counterpart, known at virtually the same time, in ancient China ([Li/Du 1987], [Martzloff 1987]). Recently, considerable attention has been paid to developing strategies for incorporating more history in the mathematics classroom, and to the pedagogical value of telling students about historical origins, relating anecdotes, using timelines, and introducing historical problems and exercises. As Frank J. Swetz reported in the *Encyclopedia of Mathematics Education* (2001):[3]

> The history of mathematics is a bountiful resource that can enrich and strengthen mathematics learning. History provides human interest to mathematics; it associates concepts with their situational origins and the people who helped shape and formalize those concepts. Further, it reveals the universal and culturally diverse nature of mathematical involvement. Such revelations have helped to reshape perceptions of mathematics education and opened questions for scholarly debate and research such as, Do different people view mathematics in different ways and thus use it and learn it differently? This question, arising largely from a historical perspective, has spawned a new discipline, ethnomathematics, which rests on the premise that all people possess and articulate a natural mathematics [Grinstein/Lipsey 2001, 322].

20.8 History of Mathematics: Recent Trends

As the sophistication of history has risen noticeably in the post-war period, so has the sophistication of history of science and, with it, history of mathematics. No longer content to write narrow, internalist histories of the various branches and sub-disciplines of mathematics, or straightforward biographies of the great figures, historians of mathematics now view the subject prosopographically, institutionally, and even as a reflection of modernist tendencies in general. There is a noticeable increase in the attention paid to the history of applied mathematics and its interactions with both utilitarian demands of society in general and with pure mathematics in particular.[4] Studies of mathematics from sociological points of view are reflected in a volume published by Henk Bos, Herbert Mehrtens, and Ivo Schneider ([Bos/Mehrtens/Schneider 1981]). Closely related are works that examine mathematics as a community, as its development under Bourbaki in France certainly was, and as it has been studied in a volume on the emergence of the American community of mathematicians recently written by Karen Parshall and David Rowe ([Parshall/Rowe 1994]).

History of mathematics has also been influenced by historians of science interested in studying the history of laboratories and experiments. For example, the role of teamwork for mathematicians working in industry, or cases of multiple

[3] Frank J. Swetz: "History of Mathematics, Overview." In: [Grinstein/Lipsey 2001, 316–323].

[4] As an example of this genre see Ivor Grattan-Guinness's 3-volumes on French mathematics [Grattan-Guinness 1990].

authorship in the production of new mathematics, represent new ways in which mathematics is conceived and organized. Here MORITZ EPPLE's recently-published article, "Genies, Ideen, Institutionen, mathematische Werkstätten: Formen der Mathematikgeschichte" (Geniuses, Ideas, Institutions, Mathematical Workshops: Forms of History of Mathematics) [EPPLE 2000], is a case in point. In such works, the aim is to study mathematics in terms of research practice.

While some historians have focused upon the history of mathematics in specific cultures or regions, others have begun increasingly to study the question of transmission of mathematical knowledge across cultures. This sort of research calls for teamwork between specialists on different cultures, languages, and mathematical traditions. A recent example of such teamwork is reflected in a meeting held at the Rockefeller Foundation's Research and Conference Center in Bellagio, Italy, where a group organized by JOSEPH DAUBEN and YVONNE DOLD-SAMPLONIUS met to discuss "Transmission of Mathematical Ideas: 2000 Years of Exchange and Influence from Late Babylonian Mathematics to Early Renaissance Science."[5] This book, too, on writing of history of mathematics, is yet another example of how the collaboration of many experts can contribute not only to the history of mathematics, but to its own history as well.

20.9 Electronic Resources

Computers have had a special importance for mathematics since the beginning of the 20th century, first as an area in which mathematical ideas and logical principles might be applied in their design, but increasingly because the speed and complexity of computers have made them a valuable instrument in the development of mathematics itself. This in turn has stimulated a particular branch of the history of mathematics, with its own journals and special associations devoted to history of computing.

The computer has also served the community of historians of mathematics through the establishment of electronic bulletin boards and electronic journals for history of mathematics. *Historia Mathematica*, for example, is available in an on-line edition (through the IDEAL network of Academic Press), and recently the American Mathematical Society and London Mathematical Society jointly issued the CD-ROM version of *The History of Mathematics from Antiquity to the Present: A Selective Annotated Bibliography* (1985), revised and updated under the editorship of ALBERT C. LEWIS: [DAUBEN/LEWIS 2000].

As for websites, these come in many varieties, but one of the most comprehensive is maintained by faculty and students at the University of St. Andrews in

[5]The proceedings of this International Conference, sponsored by the U.S. National Science Foundation, the International Union of Mathematicians, and the International Commission on History of Mathematics, is being edited by the organizers and will appear as a volume of *Boethius*, a series published by Steiner Verlag in Stuttgart, Germany, under the general editorship of MENSO FOLKERTS.

Scotland ≪http://www-history.mcs.st-andrews.ac.uk/history/index.html≫. This includes extensive hypercard files covering such areas as biographies of mathematicians and specific periods of the subject, with links to other resources and websites of interest to historians of mathematics. Among these, museum websites often include exhibition catalogues, while other sites put books on-line. Another very useful website is maintained by Jeff Miller of Gulf High School at New Port Richey, Florida: ≪http://members.aol.com/jeff570/mathword.html≫ and ≪http://members.aol.com/jeff570/mathsym.html≫, devoted to the earliest uses of mathematical words and symbols. All of these sites keep up with recent scholarship through international collaborations, and provide up-to-date resources that could not exist without the Web and the cooperative efforts of historians of mathematics in all parts of the world [Barrow-Green 2000]. Additionally, there are several active discussion groups that should be mentioned, some private and accessible by invitation only, and some open to anyone who wishes to subscribe. Of the latter, one maintained by Fred Rickey in the United States for the Mathematical Association of America is accessible by e-mail: "math-history-list@enterprise.maa.org." Further links to sites of interest to historians of mathematics may be found by accessing ≪http://www.maths.tcd.ie/pub/HistMath/Links/≫.

Another innovative use of the computer to further broaden accessibility to history of mathematics is a new program offered by The Open University, which offers a course of university lectures using the resources of the World Wide Web, through which students can avail themselves of "Virtual learning. Virtually anywhere" according to one of the Open University's recent advertisements. The Open University, based in Milton Keynes, UK, but with a recently-opened branch, the United States Open University, currently offers two on-line courses, "History of Mathematics I" (covering topics from the third millennium B.C. through the 17th century), and "History of Mathematics II" (which goes forward to the 20th century). Each course "combines multimedia learning materials, on-line faculty support and student interaction, along with technical assistance." Details are available at ≪http://www.open.edu≫.

20.10 The Humanism of Mathematics

In closing, it may be helpful to recall the words of the historian of science George Sarton, who appreciated in the most eloquent terms the significance of the history of mathematics for the rest of the history of science. Sarton's was a visionary sense of the importance of the enterprise itself, and if his rhetoric may at times seem excessive, it is nevertheless worth keeping in mind his understanding of what the history of mathematics has to offer, and its importance on a truly universal scale:

> The main duty of the historian of mathematics, as well as his fondest privilege, is to explain the humanity of mathematics, to illustrate

> its greatness, beauty, and dignity, and to describe how the incessant efforts and the accumulated genius of many generations have built up that magnificent monument .... The study of the history of mathematics will not make better mathematicians but gentler ones, it will enrich their minds, mellow their hearts, and bring out their finer qualities [SARTON 1955, 28].

Most historians of mathematics would probably claim much less for the significance or ultimate import of the research in which they are engaged, but one impressive conclusion that emerges from *Writing the History of Mathematics* is that time and again, history has played influential roles in the making of mathematicians. As the example of the German mathematician A. G. KÄSTNER makes plain, his historical study of parallel lines directly stimulated interest in the problem which eventually led to the discovery of non-Euclidean geometries. A serious regard for history of mathematics has helped mathematicians in formulating and solving problems, it has enriched the curriculums of those who teach mathematics, and it has contributed in fundamental ways to the work of historians and humanists generally, especially in the century just past. The emphasis the intellectual historian JOHN THEODORE MERZ placed on the history of mathematics — and the history of science generally — in his history of 19th-century European thought is an early case in point [MERZ 1904/1912]. But as the world becomes increasingly dependent upon and characterized by mathematics and the sciences, we are certain that the relevance of mathematics and its history will make historiographic appraisals such as this all the more useful in the decades and centuries to come.

JOSEPH W. DAUBEN, JEANNE PEIFFER, CHRISTOPH J. SCRIBA, and HANS WUSSING

Hamburg, April 2001

Part II

Portraits and Biographies

Portraits

Jean Etienne Montucla (1725–1799)

Moritz Cantor (1829–1920)

Hieronymus Georg Zeuthen (1839–1920)

Paul Tannery (1843–1904)

Zoel García de Galdeano (1846–1924)

Walter William Rouse Ball (1850–1925)

Francisco Gomes Teixeira (1851–1933)

Gustaf Hjalmar Eneström (1852–1923)

Johan Ludvig Heiberg (1854–1928)

Florian Cajori (1859–1930)

David Eugene Smith (1860–1944)

Sir Thomas Little Heath (1861–1940)

Gino Loria (1862–1954)

Raymond Clare Archibald (1875–1955)

Mikami Yoshio (1875–1950)

Julio Rey Pastor (1888–1962)

Kurt Vogel (1888–1985)

Li Yan (1892–1963)

Qian Baocong (1892–1974)

Dirk Jan Struik (1894–2000)

José Babini (1897–1984)

Otto Neugebauer (1899–1990)

JOSEPH EHRENFRIED HOFMANN (1900–1973)

ADOLF PAVLOVICH YOUSHKEVICH (1906–1993)

Biographies

Agostini, Amedeo (* March 6, 1892, Capugnano, Italy; † June 27, 1958, Livorno, Italy). AGOSTINI's mathematical studies were interrupted by World War I, during which he was taken prisoner. After the end of the War he continued his studies at the University of Bologna, where he received his degree in 1919, having studied with ENRIQUES, PINCHERLE, and BORTOLOTTI. After serving as PINCHERLE's assistant from 1919 to 1925, he taught analytic and projective geometry at the Naval Academy in Livorno. But it was as a student at Bologna, under the influence of ENRIQUES and, above all, BORTOLOTTI, that AGOSTINI's interests turned to the history of mathematics. He obtained the *venia legendi* (*libera docenza*) which qualified him to teach history of mathematics at the university level, and from 1927 on he gave courses at the University of Pisa. In 1937 he was elected as a corresponding member of the International Academy of History of Science. In addition to textbooks on calculus and analytic geometry, AGOSTINI published some 75 papers on various historical subjects.

Secondary literature: *May* 49. *P* VI, 28; VIIb(1), 40. — CHISINI, OSCAR: "In memoria di Amedeo Agostini." *Period. Mat.* (4) **37** (1959), 245–251 (with bibliography). U.B.

Aiton, Eric John (* September 8, 1920, Dunfermline, Scotland; † February 22, 1991, Oldham, England). After study at Manchester University in 1943, ERIC AITON went into school-teaching for 20 years. In 1975 he moved to the Didsbury College of Education (later part of Manchester Polytechnic) until retiring in 1985.

AITON's research focused mainly upon history of mathematics, astronomy, mechanics, and optics from COPERNICUS to EULER. His contributions largely comprised conceptual analyses of the original sources, and reconstructions of the processes of discovery. He studied the published (and some unpublished) works of major figures like KEPLER, DESCARTES, LEIBNIZ, and NEWTON, and also brought light to neglected ones.

AITON's M.Sc. thesis (1953) at the University of London dealt with the theory of the tides in the mid-eighteenth century. His Ph.D. (1958) was revised and published as a book: *The Vortex Theory of Planetary Motions* (1975). This was a major scholarly contribution, for the vortex theory had been rather eclipsed by

Newtonian mechanics. Aiton showed it to be a fascinating network of theories of both astronomical and cosmological interest, and integral to the thought of several pre-Newtonian figures. He also studied KEPLER's astronomy in detail.

LEIBNIZ was a major preoccupation of Aiton's, not only in connection with the vortex theory. Starting from his astronomy and mathematics, AITON became familiar with all aspects of his work, and produced *Leibniz: A Biography* (1984). He also examined LEIBNIZ's interest in the *I ching*. Concerning NEWTON himself, AITON's contributions included analysis of supposed blunders concerning the inverse orbit problem (1964, 1989).

AITON also prepared translations/editions of classic texts. He produced PEURBACH's *Theoricae novae planetarum* (New Theory of Planets) (1987) on his own; in collaboration with A. M. DUNCAN and later also with J. V. FIELD he prepared KEPLER's *Mysterium cosmographicum* and *Harmonices mundi libri V* (Five Books on the Harmony of the World) (1981, 1995). He planned to edit the remaining three outstanding volumes in the second series of the *Opera omnia* of EULER, but he did not quite complete the very tricky volume 31, *Kosmische Physik* (Cosmic Physics) (posthumously published in 1996). This contains a melange of papers on tides, geodesy, meteorology, aether modelling, and the passage of comets.

In terms that were once widely discussed by some historians of his generation, AITON was largely an "internalist." He thought that the historian should take seriously the scientific components of the history which he was studying. In addition, in preparing editions and translations, he increased the accessibility to scholars of the primary literature, especially for neglected figures such as PEURBACH, whose textbook on astronomy had been a major source for the late 15th and 16th centuries.

But in additon to the "facts" of the matters that he studied, AITON was also greatly interested in the philosophies of science which his historical figures upheld. In an important article: "Celestial Spheres and Circles" (*Hist. Sci.* **19** (1981), 75–114), he showed that astronomers up to and including KEPLER held both ontological and instrumentalist positions simultaneously: the spheres of the universe could be regarded as real, but theoretical superstructures could be taken as mere apparatus. As a result of his extensive experience with teaching and teaching supervision, AITON was also concerned with the bearing of history upon mathematics education. He was also active in the Manchester branch of the Mathematical Association, and served as its treasurer for a decade from 1962 and then as president for three years from 1973.

From the early 1980s AITON developed good contacts with Japanese historians; his lectures were published in Japanese, and his *Leibniz* was translated. He also became well known in Britain, dying in office as president of The British Society for the History of Mathematics.

Secondary literature: GRATTAN-GUINNESS, IVOR: "Eric Aiton. An Appreciation." *Ann. Sci.* **48** (1991), 305–308. — WILSON, CURTIS: "Eric John Aiton (1920–1991)." *HM* **18** (1991), 390–392.

I.G.G.

Amodeo, Federico (* October 8, 1859, Avellino, Italy; † November 3, 1946, Napels, Italy). AMODEO obtained his degree (*laurea*) in mathematics from the University of Napoli in 1883. There he taught as professor at the Technical Institute from 1890 to 1923, when he retired. As an instructor (*libero docente*) he also taught projective geometry at the university from 1885 to 1923. In addition, from 1905 to 1922 he gave a course on the history of mathematics and collected his lectures in four manuscript volumes: *Compendio di storia della matematica* (Compendium of the History of Mathematics). AMODEO published some 150 papers on projective geometry, algebraic geometry, and history of mathematics as well. In addition to his *Compendio*, his main works include: *Origine e sviluppo della geometria projettiva* (Origin and Development of Projective Geometry) (1939); *Vita matematica napoletana* (Mathematical Life in Naples) Part I (1905), Part II (1924); and *Sintesi storico-critica della geometria delle curve algebriche* (Historical-critical Synthesis of the Geometry of Algebraic Curves) (1945).

Secondary literature: *P* IV, 19; V, 20; VI, 48; VIIb(1), 74. — CAFIERO, FEDERICO: "Federico Amodeo." *Atti Accad. Pont.*, N.S. **11** (1963), 365. — [PALLADINO 1989, (with portrait)]. U.B.

Archibald, Raymond Clare (* October 7, 1875, Stewiacke, Nova Scotia, Canada; † July 26, 1955, Sackville, New Brunswick, Canada). RAYMOND CLARE ARCHIBALD was raised by his mother, a teacher at Mount Allison Ladies College in Sackville, New Brunswick, where ARCHIBALD himself also studied, graduating from Mt. Allison with first class honors in mathematics and a teacher's diploma in violin when he was only 18. Thereafter, he went to the United States where he earned an MA at Harvard University in 1897. He then spent two years in Europe, continuing his studies first at the University of Berlin, then at the University of Strasbourg, where he finished his Ph.D. in 1900. From 1900–1907 he was back in Canada teaching (both mathematics and violin) at Mt. Allison. Among his interests there was the library, which he "developed from nothing to 12,000 volumes" [ADAMS/NEUGEBAUER 1955, 293]. Following a year as professor of mathematics at Acadia University in Nova Scotia, ARCHIBALD accepted a position as instructor at Brown University, Providence, Rhode Island, in 1908.

At Brown, ARCHIBALD was an inspiring and dedicated teacher. He gave up music for intensive concentration on his mathematics, which included historical interests and, as at Mt. Allison, a devotion to the mathematics library. Over the years he was at Brown, the library's holdings increased substantially, often the result of purchases ARCHIBALD would make while traveling abroad. At the same time, ARCHIBALD served as librarian (1921–1941) for the American Mathematical Society, whose collection he built up over a span of twenty years, primarily through publications exchanges. He also acquired works related to English and American literature, with an emphasis on poetry and drama, which he later bequeathed to Mt. Allison University, in memory of his mother, as the Mary Mellish Archibald Memorial Library, including 2,700 recordings and piano sheet music for 70,000 songs.

ARCHIBALD was the recipient of several honorary degrees, and was a member of numerous academies and foreign mathematical societies. He served for two years as editor in chief of the *American Mathematical Monthly* (1919–1921). The following year he was elected president of the Mathematical Association of America (in 1922), and twice later as vice president and chair of Section A for Mathematics of the American Association for the Advancement of Science (AAAS) (in 1928, and again in 1937). He also served as vice president and chair of the AAAS Section L for History and Philosophy of Science.

In addition to literary works which ARCHIBALD wrote (including *Carlyle's First Love, Margaret Gordon, Lady Bannerman* (1910)), he contributed as well to the *Encyclopedia Britannica* and the *Dictionary of American Biography.* ARCHIBALD also prepared a valuable bibliography on Egyptian and Babylonian mathematics for the edition of the *Rhind Mathematical Papyrus* published by ARNOLD B. CHACE (1929).

Among his best-known contributions to the history of mathematics is his *Outline of the History of Mathematics* (1932, revised through a sixth edition which appeared in 1949; repr. 1966). ARCHIBALD also wrote the *Semicentennial History of the American Mathematical Society, 1888–1938* (1938, repr. 1988).

Secondary literature: *May* 56. *P* V, 30–31; VI, 70; VIIb(1), 117–118. — ADAMS, C. R., and OTTO NEUGEBAUER: "Obituary. Raymond Clare Archibald. *In Memoriam.*" *AMM* **62** (1955), 743–745. — ADAMS, C. R., and OTTO NEUGEBAUER: "Raymond Clare Archibald. A Minute." *Scripta Math.* **21** (1955), 293–295. — ADAMS, C. R., and OTTO NEUGEBAUER: "R. C. Archibald and Mathematics Libraries." *Science* **123** (1956), 622–623. — SARTON, GEORGE: "Raymond Clare Archibald." *Osiris* **12** (1956), 5–34 (with bibliography and portrait, facing p. 5). — Portrait in *Scripta Math.* **8** (1941), facing p. 145. J.W.D.

Arneth, Arthur (* September 19, 1802, Heidelberg, Germany; † December 16, 1858, Heidelberg, Germany). ARNETH was a German teacher of mathematics and physics in Hofwil (Kanton Bern, Switzerland) and Heidelberg. In 1828 he became *Privatdozent* at the University of Heidelberg, and in 1838 professor of mathematics and physics at the *Lyceum.* He published numerous mathematical papers and a history of pure mathematics seen in relation to the "development of the human mind" (1852).

Main works: *Systeme der Geometrie* (1840). — [ARNETH 1852].

Secondary literature: *P* I, 63, 1529; III, 42. M.F.

Arvesen, Ole Peder (* March 27, 1895, Frederikstad, Norway; † January 21, 1991, Trondheim, Norway). ARVESEN was educated as an engineer, but later turned to descriptive geometry, which he taught first as reader at the Technical University in Trondheim in 1927, and then as professor beginning in 1939 — after having obtained his doctorate in 1934. Arvesen's historical publications include the following main works: *Mennesker og matematikere* (Human Beings and

Mathematicians) (1940), *Gi meg et fast punkt* (Give Me a Fixed Point) (1950), *Fra åndens verksted* (From the Workshop of the Spirit) (1973).

Secondary literature: *P* VI, 81; VIIb(1), 133. — ARVESEN, OLE PEDER: *Men bare om løst og fast, - Erindringer.* Trondheim: Tapir 1976. — JOHNSON, DAG: "Ole Peder Arvesen: minnetale i fellesmøte 9. desember 1991." *Det Kgl. norske videnskabers forhandlinger* 1991, 85–90 (with portrait)). K.A.

Babini, José (* May 10, 1897, Buenos Aires, Argentina; † May 18, 1984, Buenos Aires, Argentina). JOSÉ BABINI, mathematician and historian of mathematics, occupied major administrative posts, among which he was the dean of the famous *Facultad de Ciencias Exactas y Naturales* of the National University of Buenos Aires during a decisive period in its development (1955–1966). He was in fact a prolific author ([NICOLÁS BABINI (1994)] lists 69 books, 121 booklets, 239 articles, 207 book reviews, and 75 notes, a total of 711 items). These works, some of them translations, constitute a significant contribution to the diffusion of classical sources and of classical historiography. While the majority are works on history of mathematics, the rest of BABINI's publications are devoted to the history of the sciences.

Among his most important studies, BABINI co-authored (with J. REY PASTOR) *Historia de la matemática* (History of Mathematics) (1953). Then BABINI published *Historia sucinta de la matemática* (Concise History of Mathematics) (1953), followed by his *Biografía de los infinitamente pequeños* (Biography of the Infinitely Small) (1957). Of his works on the history of science, notably his *Origen y naturaleza de la ciencia* (Origin and Nature of Science) (1947), *Arquímedes* (Archimedes) (1948), *Historia de la ciencia argentina* (History of Science in Argentina) (1949), *Historia sucinta de la ciencia* (Concise History of Science) (1951), and *La evolución del pensamiento científico en la Argentina* (Evolution of Scientific Thought in Argentina) (1953) deserve to be mentioned. Throughout this period BABINI published nearly one book each year; all included important parts devoted to the history of mathematics. Likewise the prefaces and introductions to his translations of books also deserve mention, especially his translation of FEDERICO AMODEOS's history of projective geometry.

Apart from his own publications, BABINI participated in important national and international enterprises in the history of mathematics. He was involved, for example, with the publication (first in Buenos Aires and then for a short period in Paris) of *Archeion*, he helped to establish *Quipu*, was a member of the editorial committee of *Historical Studies in the Physical Sciences*, and assisted in organizing the Washington meeting of the International Commission on Mathematical Education.

Although BABINI was an historian of mathematics who helped to diffuse the works of others, he was foremost an organizer. He was a leader in the organization of new Argentinian universities, took a hand in the revitalization of established universities, was a member of the *Consejo Nacional de Investigaciones Científicas (CONICET)* (National Research Council), took part in international projects, promoted research and education in history of science in Argentina, and was also involved in an effort to reconstitute ALDO MIELI's library. Although successive military dictatorships impeded the development of a community of historians of mathematics in Argentina, BABINI helped establish the first firm foundations for the emergence of new and interesting efforts to promote the subject that is now being advanced with greater impetus.

Secondary literature: *P* VI, 98; VIIb(1), 168. — ORTIZ, EDUARDO L., and LEWIS PYENSON: "José Babini: Matemático e historiador de la Ciencia." *Lull* **7** (1984), 77–98. — PAPP, DESIDERIO: "José Babini, 1897–1984." *Arch. int. Hist. Sci.* **35** (1985), 414–416. — ORTIZ, EDUARDO L., and LEWIS PYENSON: "José Babini, 11 May 1897 - 18 May 1984." *Isis* **76** (1985), 567–569 (with portrait). — KOHN LONCARICA, ALFREDO G.: "José Babini (1897–1984)." *Quipu* **2** (1985), 129–147 (with bibliography and portrait). — BABINI, NICOLÁS: *José Babini: bio-bibliografía (1897–1984).* Buenos Aires 1994. M.O.

Bachelard, Gaston (* June 27, 1884, Bar-sur-Aube, France; † October 16, 1962, Paris, France). BACHELARD, before becoming a philosopher of science, taught physics and chemistry at a secondary school in Bar-sur-Aube (1919–1930), then philosophy at the University of Dijon (1930–1940). In 1940, he accepted the professorial chair of history and philosophy of science at the *Sorbonne* and became the head of the *Institut de philosophie et d'histoire des sciences* (1940–1955).

Main works: *Etude sur l'évolution d'un problème de physique* (1928). — *La formation de l'esprit scientifique* (1938).

Secondary literature: *DSB* **1**, 365–366. — CANGUILHEM, GEORGES: "Hommage à Gaston Bachelard." *Ann. Univ. Paris* **33** (1963), no. 1, 24–39. — HYPPOLITE, JEAN: "L'épistémologie de G. Bachelard." *Rev. Hist. Sci.* **17** (1964), 1–12. — DAGOGNET, FRANÇOIS *Gaston Bachelard. Sa vie, son œuvre avec un expoé de sa philosophie.* Paris 1965. — VRIN, J. (Ed.): *Problèmes d'histoire et de philosophie des sciences.* Paris 1968. J.P.

Bachmann, Paul (* June 22, 1837, Berlin, Germany; † March 31, 1920, Weimar, Germany). BACHMANN started his academic career as *Privatdozent* at the University of Breslau in 1864, where he was promoted to extraordinary professor in 1867. From 1875–1890 he was ordinary professor in Münster. He published numerous books on the theory of numbers.

Main works: "Über Gauß' zahlentheoretische Arbeiten" (= *Materialien für eine wissenschaftliche Biographie von Gauß*, Heft 1) (1911). — *Das Fermat-Problem in seiner bisherigen Entwicklung* (1919).

Secondary literature: *DSB* I, 370. *May* 64. *P* III, 57; IV, 51; V, 47; VI, 99; VIIa(1), 72. — *NDB* **1**, 497. — HAUSSNER, ROBERT (Ed.): "Vorwort." In: BACHMANN, PAUL: *Grundlehren der neueren Zahlentheorie.* Berlin and Leipzig, 2nd ed. 1921. — HENSEL, KURT: "Paul Bachmann und sein Lebenswerk." *Jber. DMV* **36** (1927), 31–73 (with bibliography and portrait). — Portrait: [BÖLLING 1994, 21.7]. M.F.

Baldi, Bernardino (* June 5, 1553, Urbino, Italy; † October 10, 1617, Urbino, Italy). BERNARDINO BALDI was among the last representatives of the Renaissance in Italy. He was a poet and a man of letters, an erudite scholar, an historian, and a biographer. He also cultivated mathematics and mechanics, and edited scientific texts of Greek antiquity. First educated by private tutors, BALDI then studied mathematics in his home-town under FEDERICO COMMANDINO, and after the latter's death in 1575, under GUIDOBALDO DAL MONTE. At COMMANDINO's suggestion he translated HERO's *Automata* into Italian, which only appeared in print in 1589, prefaced with a history of mechanics by BALDI. He also translated book VIII of PAPPUS's *Collections*, ARATUS's *Phenomena*, and the PSEUDO-ARISTOTELIAN *Questions of Mechanics*. These translations, like the majority of BALDI's scientific works, were left unpublished. (The *Questions* was posthumously printed in 1621, with a commentary by BALDI, which represents his most important contribution to physics.)

In 1573 BALDI moved to Padua where he studied Greek philology and philosophy. But the university was closed in 1576 due to the plague, and BALDI returned to Urbino, where he continued to study mathematics with COMMANDINO's pupil, GUIDOBALDO DAL MONTE. In 1579 BALDI was invited by DUKE FERRANTE II GONZAGA to serve as mathematician to the Gonzaga Court in Mantua. There BALDI began to work on his *Dizionario Vitruviano* (Vitruvian Dictionary), which eventually appeared in several volumes and earned him great fame. After a period spent in Milan with cardinal CARLO BORROMEO, BALDI returned to the service of FERRANTE GONZAGA as Abbot of Guastalla, a small town not far from Mantua. There BALDI completed his first biography, the *Vita di Federigo Commandino* (Life of Federigo Commandino), in 1587. This was based on an account of COMMANDINO's life given to BALDI by COMMANDINO himself in his final years. (This work remained in manuscript until 1714, when it was published in the *Giornale de' Letterati d'Italia*, **19** (1714), 140–185.)

While preparing COMMANDINO's biography, BALDI conceived the vast project of writing accounts of the lives of all the great mathematicians, but it took nearly ten years for him to complete this project. BALDI's *Vite de' matematici* (Lives of Mathematicians) is an impressive work of more than 2000 manuscript pages, which includes the biographies of 201 mathematicians from antiquity up to BALDI's own times. The first part was undertaken in the years 1587–1589, but the manuscript of BALDI's *Vite* was not completed until 1595–1596. BALDI wrote his biographies based on what material and information he was able to collect, all of which he eventually presented in two volumes, following a chronological order. The first volume includes the lives of mathematicians in antiquity, from THALES up to the birth of Jesus Christ; the second volume includes subsequent biographies, up to CLAVIUS. In the words of PAUL ROSE [Rose 1974, 272]: "By combining immense humanistic erudition with mathematical learning [Baldi] accomplished one of the great monuments of the mathematical Renaissance, his *Vite de' matematici* which stands as the first large scale history of mathematics." Even after the completion of his *Vite*, BALDI continued to collect biographical material about mathematicians, some of which he included in his later work, *Cronica de' matematici o vero epitome dell'istoria delle vite loro* (Chronicle of Mathematicians, or in Other Words, Epitomes of the History of Their Lives), in two volumes. In addition to summarizing biographies already included in the *Vite*, the *Cronica* provides information about the lives and works of more than 150 additional mathematicians. It begins with EUPHORBUS and ends with GUIDOBALDO DAL MONTE.

Both the *Vite* and the *Cronica*, however, were left unpublished by BALDI. The *Cronica* was eventually published by ANGELO MONTICELLI in Urbino in 1707. As for the *Vite*, ENRICO NARDUCCI published a volume entitled *Vite de' matematici scritte da Bernardino Baldi* (Lives of Mathematicians written by Bernardino Baldi) in Rome in 1861, which included the lives of some 30 medieval mathematicians. More *Vite*, (including PACIOLI's), were later published by BONCOMPAGNI in his *Bullettino* in 1879. NARDUCCI also edited 28 of BALDI's *Vite Inedite di Matematici Italiani* (Unpublished Lives of Italian Mathematicians) for the same journal, (*Bull. Boncompagni*, **19** (1886), 335–640; **20** (1887), 197–308). This work by NARDUCCI still represents the best source of information about BALDI's *Vite*. NARDUCCI had planned to publish the whole of BALDI's works, but in spite of his efforts, 125 of BALDI's *Vite* (amounting to nearly 1300 manuscript pages) remain unpublished.

In 1601 BALDI entered the service of the DUKE OF URBINO FRANCESCO II, who charged him with writing the biography of FEDERICO DA MONTEFELTRO. As a result, BALDI resigned his position as Abbot of Guastalla and settled in Urbino. Eventually BALDI's *Vita e fatti di Federico di Montefeltro duca di Urbino* (Life and Deeds of Federico di Montrefeltro, Duke of Urbino) was published (in ten books) in Rome (1824), while his *Vita e fatti di Guidobaldo I* (Life and Deeds of Guidobaldo I) was published in Milan in 1821. In the final years of his life, BALDI began research on geography which also is as yet unpublished.

Secondary literature: *DSB* **1**, 419–420. *May* 65. *P* I, 94. — *DBI* **5**, 461–464. — AFFÒ, IRENEO: *Vita di Monsignore Bernardino Baldi da Urbino, primo Abbate di Guastalla.* Parma 1873 (with list of Baldi's publications and manuscripts). — ROSE, PAUL L.: "Rediscovered manuscripts of the 'Vite de' matematici' and mathematical work by Bernardino Baldi (1553–1617)." *Rend. Accad. Naz. Lincei, Classe Sci. Fis., Mat. e Nat.* **56** (1974), 272–279 (with list of Baldi's printed works and scientific manuscripts). — BILINSKI, BRONISLAW: "Prolegomena alle Vite dei matematici di Bernardino Baldi (1587–1596)." *Accademia Polacca delle Scienze. Biblioteca e Centro di Studi a Roma, Conferenze e Studi* **71** (1977), 1–135. — [BALDI 1998]. U.B.

Becker, Oskar (* September 5, 1889, Leipzig, Germany; † November 13, 1964, Bonn, Germany). BECKER was a German philosopher with a strong interest in the history of Greek mathematics. He became *Privatdozent* for philosophy at the University of Freiburg in 1922, was the assistant of EDMUND HUSSERL and MARTIN HEIDEGGER from 1923 to 1931, *ausserplanmässiger Professor* in Freiburg (1927), and *ordentlicher Professor* at the University of Bonn (1931–1946, 1951–1955). His main areas of interest were mathematical ontology and logic, the foundations of mathematics, and the history of Greek mathematics.

Main works: "Eudoxos-Studien I–V." *Q. St. G. Math. Astron. Phys.* **B 2** (1933), 311–333, 369–387; **B 3** (1936), 236–244, 370–388, 389–410. — "Die Lehre vom Geraden und Ungeraden im 9. Buch der euklidischen Elemente." *Q. St. G. Math. Astron. Phys.* **B 3** (1936), 533–553. — *Geschichte der Mathematik* (with JOSEPH E. HOFMANN). Bonn 1951. — *Grundlagen der Mathematik in geschichtlicher Entwicklung.* Freiburg 1954. — *Das mathematische Denken der Antike.* Göttingen 1957, 2nd ed. 1966.

Secondary literature: *P* VIIa(1), 120. — *Rheinische Friedrich-Wilhelms-Universität Bonn. Chronik und Bericht über das akademische Jahr 1964/65.* Jahrgang 80 (= N. F. Jahrgang 69). Bonn 1965, 31–32 (with portrait). — HOFMANN, JOSEPH EHRENFRIED: "Oskar Becker †." *Prax. Math.* **7** (1965), 245. M.F.

Bell, Eric Temple (* February 7, 1883, Peterhead, Scotland; † December 21, 1960, Watsonville, California, USA). BELL, an American mathematician, historian of mathematics, and science fiction novelist, was born in Scotland. But by 1884 he had moved with his family to the United States where he grew up in San José, California. Following the death of his father, BELL returned to England with his mother and older brother, and in 1898 entered the Bedford Modern School, where he studied mathematics with EDWARD M. LANGLEY, who inspired BELL's life-long interest in elliptic functions and number theory. Four years later (in 1902), largely to avoid government service, BELL returned to the United States, where at first he supported himself with a variety of odd jobs ranging from ranch hand and mule skinner to land surveyor.

Returning to California, BELL entered Stanford University. Within two years he had completed all of his undergraduate courses, finishing with honors and election to the academic honor society, Phi Beta Kappa. In 1908 he received his master's degree from the University of Washington, and four years later, in 1912, BELL received his Ph.D. in mathematics from Columbia University. There BELL studied with CASSIUS J. KEYSER, "one of the main reasons," he later admitted, "for my going to Columbia." BELL had read KEYSER's writings and was impressed; he liked both KEYSER's philosophical approach to mathematics and his literary style, which showed that "mathematics was more than an interminable grind of theorems" [Reid 1993, 135].

Having lectured at Harvard, the University of Chicago, and the University of Washington (where he remained as a teacher-scholar from 1912 through 1926), BELL assumed his final position that same year at the California Institute of Technology, where he remained until his retirement in 1953. BELL died in Watsonville, California, in 1960, not far from where he had been raised as a child.

BELL was a prolific author, and in addition to more than 250 mathematical papers on a range of subjects covering arithmetic, number theory, and elliptic functions, he published nine texts devoted either to mathematics or history of mathematics: *Algebraic Arithmetic* (1927), *Numerology* (1933), *The Search for Truth*

(1934), *The Handmaiden of the Sciences* (1937), *Men of Mathematics* (1937), *The Queen of the Sciences* (1938), *The Development of Mathematics* (1940), *Mathematics, Queen and Servant of Science* (1951), and *The Last Problem* (1961). *The Last Problem*, published posthumously, was indeed his last publication and was devoted to a history of FERMAT's so-called last theorem.

BELL was elected president of the Mathematical Association of America for 1931–1932. Earlier, an article he had written on "Arithmetical Paraphrases" in the *Transactions of the American Mathematical Society* won the Society's Bôcher Prize for 1924. But his popularized histories of mathematics were his best-known writings, especially *Men of Mathematics*. The historian of mathematics, KENNETH O. MAY, believed that BELL's insights and provocative style continue to influence and intrigue professional mathematicians in spite of their historical inaccuracies and sometimes fanciful interpretations [MAY 1970, 584]. Others have been more critical, objecting to BELL's usually opinionated and often prejudiced views; his description of the Russian mathematician SOF'YA (SOPHIA) KOVALEVSKAYA, for example, has been denounced as an "infuriatingly patronizing, innuendo-laden mistreatment" [Cooke 1992, 92h:01056].

Whatever may be said of the shortcomings of BELL's research and historical scholarship, however, one cannot help but admire the enthusiasm with which he promoted mathematics. He was active in organizations that supported research, teaching, and writing. While in religion and politics he has been described as an "individualist and uncompromising iconoclast" [DSB 1970/90, **1**, 584], these traits were no less true of his historical works, which ultimately have had the greatest influence and have often served as the first introductions for many mathematicians to the history of mathematics.

BELL claimed that he was prompted to write his *The Development of Mathematics* (1940, 2nd ed. 1945) due to requests from "numerous correspondents, principally students and instructors, for a broad account of the general development of mathematics" [v], and because of his personal association with creative mathematicians. But he cautioned that what he wrote were not histories of the traditional kind, but rather narratives of the decisive epochs in the development of mathematics. He also "believed that a survey of the main directions along which living mathematics has developed would enable [students] to decide more intelligently in what particular field of mathematics, if any, they might find a lasting satisfaction" [vi].

As for the demands of history, BELL appreciated the challenges historians had to face:

> Theirs is a difficult and exacting pursuit; and if controversies over the trivia of mathematics, of but slight interest to either students or professionals, absorb a considerable part of their energies, the residue of apparently sound facts no doubt justifies the inordinate expense of obtaining it. Without the devoted labors of these scholars, mathematicians would know next to nothing, and perhaps care less, about the

> first faltering steps of their science. Indeed, an eminent French analyst of the twentieth century declared that neither he nor any but one or two of his fellow professionals had the slightest interest in the history of mathematics as conceived by historians. He amplified his statement by observing that the only history of mathematics that means anything to a mathematician is the thousands of technical papers cramming the journals devoted exclusively to mathematical research. These, he averred, are the true history of mathematics, and the only one either possible or profitable to write. Fortunately, I am not attempting to write a history of mathematics; I hope only to encourage some to go on, and decide for themselves whether the French analyst was right [*Development of Mathematics*, 1940, x].

BELL's own answer would have been negative, for he believed that "With a knowledge of who these men were, [...] and an appreciation of their rich personalities, the magnificent achievements of science fall into a truer perspective and take on a new significance" [Bell 1937, 18].

Secondary literature: *DSB* **1**, 584–585. *May* 69. *P* VI, 165–166, VIIb(2), 299. — Obituary notice: *New York Times*, December 22, 1960. — COOKE, ROGER L.: Review of "New Material on S. V. Kovalevskaya" by ANNE HIBNER KOBLITZ, in *Math. Rev.* (1992), 92h:01056. — REID, CONSTANCE: *The Search for E. T. Bell, also known as John Taine.* Washington, D. C., 1993 (with photographs). J.W.D

Berr, Henri (* January 31, 1863, Lunéville, France; † November 19, 1954, Paris, France). Professor of rhetoric at the *Lycée Henri IV* in Paris, BERR founded the *Revue de synthèse historique* in 1900 (*Revue de synthèse* from 1931 on). In 1925 he created the *Centre international de synthèse*, located in Paris at the *Hotel de Nevers*, 12 rue Colbert. It was the Center's section for history of science, headed by ALDO MIELI as its first director, that was important for institutionalizing the discipline in France. For BERR, history of science was to play an important role in the explicit synthesis he intended to create from both the natural and the human sciences.

Main work: *La synthèse en histoire. Essai critique et théorique.* Paris 1911.

Secondary literature: DELORME, SUSANNE: "Henri Berr." *Osiris* **10** (1950), 5–9 (with bibliography and portrait). — SARTON, GEORGE: "Henri Berr (1863–1954). La synthèse de l'histore et l'histoire de la science." *Centaurus* **4** (1955/56), 185–197. — Centre international de synthèse: *Henri Berr et la culture du XXe siècle.* BIARD, AGNÈS, *et al.* (Eds.). Paris 1997, 301–338 (with bibliography and portrait). J.P.

Bertrand, Joseph Louis François (* March 11, 1822, Paris, France; † April 5, 1900, Paris, France). A brilliant mathematician and important textbook writer, JOSEPH BERTRAND was editor (beginning in 1865) of the *Journal des savants*, to which he also contributed a number of historical studies. As *secrétaire* *perpétuel* of the *Académie des Sciences*, he wrote biographies of PONCELET, LAMÉ, CHASLES, and CAUCHY, among others. BERTRAND wrote two historical books, on D'ALEMBERT and PASCAL.

Main works: *Rapport sur les progrès les plus récents de l'analyse mathématique.* Paris 1867. — *L'Académie des sciences et les académiciens de 1666 à 1793.* Paris 1869.

Secondary literature: *DSB* **2**, 87–89. *May* 73. *P* I, 171; III, 121; IV, 110. — DARBOUX, GASTON: "Eloge historique de J. L. F. Bertrand." *Eloges académiques et discours.* Paris 1912, 1–60. J.P.

Bessel-Hagen, Erich Paul Werner (* September 12, 1898, Charlottenburg near Berlin, Germany; † March 29, 1946, Bonn, Germany). After having been the private assistant (1921–1923) to FELIX KLEIN, the German mathematician ERICH BESSEL-HAGEN in 1925 became a *Privatdozent* in Göttingen. From 1925 to 1931 he taught at the universities of Göttingen, Halle, and Bonn. In Bonn he was promoted to the position of *Extraordinariat* in 1931, and assisted OTTO TOEPLITZ in building up a historico-didactic division of the *Mathematisches Seminar*. After TOEPLITZ was expelled from Germany by the National Socialists in 1935, BESSEL-HAGEN was left in charge of this division.

Main works: Co-editor (with R. FRICKE and H. VERMEIL) of vol. 3 of FELIX KLEIN's *Gesammelte mathematische Abhandlungen* (1923). — "Das Buch über die Ausmessung der Kreisringe des Aḥmad ibn ʿOmar al-Karābīsī" (with OTTO SPIES). *Q. St. G. Math. Astron. Phys.* **B 1** (1931), 502–540. — "Ṯābit b. Qurra's Abhandlung über einen halbregelmässigen Vierzehnflächner" (with OTTO SPIES). *Q. St. G. Math. Astron. Phys.* **B 2** (1932), 186–198. — "Die Geometrie des Descartes." *Semester-Ber.* **14** (1939), 39–70.

Secondary literature: *P* VI, 203; VIIa(1), 166. — NEUENSCHWANDER, ERWIN: "Der Nachlass von Erich Bessel-Hagen im Archiv der Universität Bonn." *HM* **20** (1993), 382–414 (with bibliography). M.F.

Beth, Hermanus Johannes Elisa (* July 5, 1880, Rozendaal-Nispen, Netherlands; † February 6, 1952, Amersfoort, Netherlands). The mathematician HERMANUS BETH taught at the State Secondary School in Almelo, where he became director of the school in 1912. Later he held corresponding posts at Deventer (1922–35) and at Amersfoort (1935–46). Together with E. J. DIJKSTERHUIS, BETH founded the *Historische Bibliotheek voor de Exacte Wetenschappen* (Historical Library for the Exact Sciences).

Main works: [BETH 1929], [BETH 1932].

Secondary literature: *May* 74. *P* VI, 203; VIIb(1), 361. — BOTTEMA, OENE: "H. J. E. Beth 1880 – 5 Juli – 1950." *Euclides* **25** (1950), 241–252 (with portrait). P.B.

Bierens de Haan, David (* May 3, 1822, Amsterdam, Netherlands; † August 12, 1895, Santpoort, Noord-Holland, Netherlands). BIERENS DE HAAN, who began teaching mathematics at the *Gymnasium* in Deventer in 1848, gave up his post in 1853 to devote himself to mathematical and historical studies. He became an extraordinary professor at the University of Leiden in 1863, and ordinary professor there in 1867.

Secondary literature: *May* 75. *P* III, 128; IV, 120. — *NNBW* **7**, 512–513. — KLUYER, J. C.; D. J. KORTEWEG, and D. H. SCHOUTE: "David Bierens de Haan 1822–1895." *Nieuw Arch. Wiskd.* (2) **2** (1896), i–xxviii (with bibliography). — SCHREK, D. J. E.: "David Bierens de Haan." *Scripta Math.* **21** (1955), 31–41 (with portraits). — BOS, HENK J. M.: "David Bierens de Haan 1822–1895." *Nieuw Arch. Wiskd.* (3) **26** (1978), 65–73. P.B.

Biernatzki, Karl Leonhard (* December 28, 1815, Altona, Germany; † January 23, 1899, Altona, Germany). BIERNATZKI studied theology in Erlangen and Kiel, before he became rector in Friedrichstadt in 1841. After having worked for a while as *Redakteur* (editor) for the newspaper *Altonaer Merkur*, BIERNATZKI served since 1851 as secretary of the *Zentralverein für chinesische Mission* (Central Association for the China Mission) in Kassel. Between 1855 and 1859 he worked in a similar function in Berlin, before he returned to his home town as a protestant minister. BIERNATZKI's interest in the history of mathematics seems to have been limited to the study of Chinese mathematics. His article in *Crelles Journal* was mostly based on a French publication by A. WYLIE (see WYLIE's biography).

Main works: "Die Arithmetik der Chinesen." *Crelles J.* **52** (1855), 59-94; repr. Walluf near Wiesbaden: Sändig 1973.

Secondary literature: *BBKl* **1** (1975), 620. M.F.

Biot, Edouard-Constant (* July 2, 1803, Paris, France; † March 12, 1850, Paris, France). EDOUARD-CONSTANT, the son of JEAN-BAPTISTE BIOT (**B**), studied sinology and published a number of papers in the *Journal asiatique* on mathematics in China. These were devoted to positional arithmetic, the so-called Pythagorean theorem, and PASCAL's arithmetical triangle.

Secondary literature: *P* I, 199. J.P.

Biot, Jean-Baptiste (* April 21, 1774, Paris, France; † February 3, 1862, Paris, France). Educated at the newly-founded *Ecole Polytechnique*, BIOT became professor of mathematical physics at the *Collège de France* in 1800, followed by his appointment in 1808 as professor of astronomy in the science faculty at the University of Paris. Elected a member of the First Class of the *Institute*, he was also elected to the *Académie française* in 1856, largely by virtue of his research on Egyptian, Babylonian, and Chinese astronomy. At the Academy he strongly opposed the ideas of LOUIS AMÉLIE SÉDILLOT (**B**) on the originality of Arabic contributions to the history of mathematics. Late in life BIOT studied the scientific past of India and China, although he did not have the necessary linguistic skills to do so adequately. Unfortunately, his publications on these subjects reflect typically

Eurocentric racist stereotypes. BIOT also authored a biography of ISAAC NEWTON for the *Biographie universelle.*

Main works: *Essai sur l'histoire générale des sciences pendant la Révolution française.* Paris 1803. — *Etudes sur l'astronomie indienne.* Paris 1859. — *Etudes sur l'astronomie indienne et sur l'astronomie chinoise.* Paris 1862.

Secondary literature: *DSB* **2**, 133–140. *May* 76. *P* I, 195; III, 133. J.P.

Birkhoff, Garret D. (* January 10, 1911, Princeton, New Jersey, USA; † November 22, 1996, Water Mill, Long Island (New York), USA). BIRKHOFF (son of the mathematician GEORGE D. BIRKHOFF), received his Bachelor of Arts degree in mathematics from Harvard University in 1932. He never went on to receive the traditional advanced graduate degrees, but began teaching at Harvard in 1936, where he spent the rest of his career. BIRKHOFF was named Putnam Professor of Pure and Applied Mathematics at Harvard in 1969, a position he held until his retirement from teaching in 1981. Although a specialist in modern algebra and lattice theory, BIRKHOFF also contributed important works to fluid mechanics, scientific computing, reactor theory, differential equations, and to the history of mathematics. In 1974 he organized with I. B. COHEN a Workshop on History of Mathematics, the proceedings of which comprise two numbers of *Historia Mathematica* **2** (1975), 425–615. Among his contributions to the history of mathematics is a volume he edited (with UTA C. MERZBACH), a *Source Book in Classical Analysis* (1973).

Main works: *Collected Mathematical Papers* (D. V. WIDDER *et al.*, eds.), 3 vols. (1950, repr. 1968).

Secondary literature: SAXON, WOLFGANG: "Garrett Birkhoff, 85, Mathematical Theorist, Dies." *New York Times*, 28 November 1996, Section D, 22 (with photograph). J.W.D.

Bjerknes, Carl Anton (* October 24, 1825, Christiania, Norway; † March 20, 1903, Christiania, Norway). BJERKNES was educated in mineralogy, but became increasingly interested in mathematics and hydrodynamics. In 1861 he obtained a position at Christiania University (now University of Oslo) in applied mathematics and eight years later, a professorship for pure mathematics. In 1880 he published *Niels Henrik Abel, en Skildring af hans Liv og videnskabelige Virksomhed* (Niels Henrik Abel, a Description of his Life and Scientific Activity), which also appeared in French in 1885.

Secondary literature: *DSB* **2**, 166–167. *May* 77. *P* III, 137; IV, 129; V, 119; VI, 233. — *NBL* **1**, 581–583. — Obituary in *Christiania Videnskabs-Selskabs Forhandlinger* **7** (1903). — [*Acta Math.*, Tab. Gen. **1–35** (1913), 38; portrait 131]. — [STUBHAUG 2000b, portrait: 82]. — Portrait: [BÖLLING 1994, 7.3]. K.A.

Björnbo, Axel Anthon (* April 20, 1874, Copenhagen, Denmark; † October 6, 1911, Hellerup, Denmark). AXEL BJØRNBO, the son of the philologist RICHARD CHRISTENSEN, changed his family name in 1901, so that he would not be mistaken for his namesake who also wrote on the history of mathematics. At the *Borgerdydskolen* (School of Civil Virtue) in Copenhagen, where his father had taught and where JOHANN LUDVIG HEIBERG still taught Greek, BJØRNBO passed the school-leaving examination in 1891. He had already read the Greek mathematicians in the original, and at the University of Copenhagen he studied not so much mathematics as its history, which he did primarily with HIERONYMUS GEORG ZEUTHEN. Before the turn of the century he moved to Munich for further studies, both at the University and the Technical University of Munich. This proved decisive for his future, for not only did he take part in ANTON VON BRAUNMÜHL's seminars on the history of mathematics, but he also attended LUDWIG TRAUBE's lectures on Latin paleography. BJØRNBO was accordingly qualified to make a serious study of the transmission of mathematical writings in the Middle Ages. In 1901 BJØRNBO was awarded his doctorate at the University of Munich for his work on MENELAUS's *Spherics*. He analyzed the structure, mathematical content, and transmission of this important text, and also treated the general development of Greek spherics and spherical trigonometry (published 1902). This work was based on manuscripts that BJØRNBO had found and examined in an extended study-tour of libraries in Germany, Austria, Italy, and France.

Beginning in 1902, BJØRNBO was a librarian at the Royal Library in Copenhagen, a post that he held until his sudden death in 1911. Despite his duties at the library, BJØRNBO continued his intensive study of mathematical medieval manuscripts. It was clear to him that serious statements about medieval mathematics were only possible if all the relevant manuscripts had been scrutinized and catalogued, because the attributions and even the readings of individual manuscripts were not always reliable. Therefore he suggested that all manuscripts of mathematical texts should be listed before authoritative editions could be established. He outlined this program in detail in an article, "Über ein bibliographisches Repertorium der handschriftlichen mathematischen Literatur des Mittelalters" (On a Bibliographical List of Medieval Mathematical Literature in Manuscripts) (*Bibl. Math.* (3) **4** (1903), 226–333).

With immense energy and enthusiasm, BJØRNBO himself described and excerpted all the mathematical and astronomical manuscripts that he could find. Within a few years he was able to collect reliable and comprehensive material for the history of mathematics in the Western Middle Ages. In this context he published descriptions of important individual manuscripts (Basel, F.II.33, and Paris, lat. 9335, both in 1902) and groups of manuscripts (Florence, San Marco, 1903–1912). The principal result of his researches is a card index (containing more than 2000 cards), in alphabetical order, of *incipits* (the first few words of the texts), for all medieval mathematical and astronomical texts known to him. Unfortunately, this card index, which has been preserved as part of ENESTRÖM's papers in the Stockholm Academy of Science, has never been used by historians of science — not even by LYNN THORNDIKE and PEARL KIBRE for their well-known *Catalogue of Incipits of Medieval Scientific Writings in Latin* (2nd ed., 1963).

In the course of his research, BJØRNBO found numerous hitherto unknown texts and was even able to clarify the transmission of well-known texts, particularly of the medieval Latin translations of Arabic works. He discovered, for instance, the translation of EUCLID's *Elements* by GERARD OF CREMONA (1905) and the direct translation of PTOLEMY's *Almagest* from the Greek made in Southern Italy (1909). He also wrote an important article on the transmission of AL-KHWĀRIZMĪ's astronomical tables (1909); the edition itself came out only after BJØRNBO's death (1914). In the Vatican BJØRNBO found a manuscript that contained two very important works by JOHANNES WERNER (1468–1528) for the history of trigonometry: the *De triangulis sphaericis* (On Spherical Triangles) and *De meteoroscopiis* (On the Meteoroscope, *i.e.* an instrument invented by WERNER to solve problems in spherical astronomy). BJØRNBO's editions of these works were published in 1907 and 1913. He also edited optical writings — from AL-KINDĪ and Pseudo-EUCLID, among others (1912). Also posthumous was the publication of his edition of the translation of THĀBIT IBN QURRA's tract on MENELAUS's theorem (1924). In his last years BJØRNBO spent much time on historical questions about geography and cartography, particularly the cartographical description of Scandinavia by the Danish geographer CLAUDIUS CLAVUS (fl. 1426) and the associated cartographic tradition. One result was the *Anecdota cartographica septentrionalia* (Unpublished Northern Cartographical Material) (1908; with CARL PETERSEN (1813–1880)).

There are some editions of medieval texts in BJØRNBO's *Nachlass* in the Royal Library in Copenhagen. Although these are in various states of completion, most by now have been edited by other scholars. But among those that remain unpublished are Latin translations of THEODOSIUS's *Sphaerica* and *De habitationibus*, MENELAUS's *Sphaerica*, and the *algorismus* writings of JORDANUS OF NEMORE.

All of BJØRNBO's work is characterized by thoroughness and exactitude. He always formed a workable and sensible plan, and then kept to it. He never accepted unsupported statements and never published an edition before seeing all of the relevant related manuscripts. This approach sets him apart from most of his contemporaries, in particular from MAXIMILIAN CURTZE, who edited numerous Latin mathematical texts. (BJØRNBO was in close contact with CURTZE and enjoyed the use of his *Nachlass* after CURTZE died in 1903). Even today, BJØRNBO remains a model for the editing of medieval mathematical texts.

Main Publications: "Studien über Menelaos' Sphärik. Beiträge zur Geschichte der Sphärik und Trigonometrie der Griechen." *Abh. Gesch. math. Wiss.* **14** (1902). 154 pp. — "Über zwei mathematische Handschriften aus dem vierzehnten Jahrhundert." *Bibl. Math.* (3) **3** (1902), 63–75. — "Die mathematischen S. Marcohandschriften in Florenz. I – IV." *Bibl. Math.* (3) **4** (1903), 238–245; **6** (1905), 230–238; **12** (1911/1912), 97–132, 193–224. — "Gerhard von Cremonas Übersetzung von Alkwarizmis *Algebra* und von Euklids *Elementen.*" *Bibl. Math.* (3) **6** (1905), 239–248. — "Ioannis Verneri de triangulis sphaericis libri quatuor, de meteoroscopiis libri sex, cum prooemio Georgii Ioachimi Rhetici." *Abh. Gesch. math. Wiss.* **24** (1907 and 1913). 184, 260 pp. — "Al-Chwarizmis trigonometriske Tavler." In:

Festskrift til H. G. Zeuthen. Kopenhagen 1909, 1–17. — "Alkindi, Tideus und Pseudo-Euklid. Drei optische Werke. Herausgegeben und erklärt von † Axel Anthon Björnbo und Seb. Vogl. Mit einem Gedächtniswort auf A. A. Björnbo von H. G. Zeuthen, einem Verzeichnis seiner Schriften und seinem Bildnis." *Abh. Gesch. math. Wiss.* **26**, no. 3 (1912). 176 pp. — "Die astronomischen Tafeln des Muḥammed ibn Mūsā al Khwārizmī in der Bearbeitung des Maslama ibn Aḥmed al-Madjrīṭī und der latein. Übersetzung des Athelhard von Bath auf Grund der Vorarbeiten von A. Bjørnbo † und R. Besthorn in Kopenhagen, herausgegeben und kommentiert von H. Suter in Zürich." *D. Kgl. Danske Vidensk. Selsk. Skrifter.* 7. Række. historisk og filosofisk Afd. III.1. Kopenhagen 1914. — "Thabits Werk über den Transversalensatz (liber de figura sectore). Mit Bemerkungen von H. Suter. Herausgegeben und ergänzt durch Untersuchungen über die Entwicklung der muslimischen sphärischen Trigonometrie von Dr. H. Bürger und Dr. K. Kohl." *Abh. Gesch. Med. Nat.*, Heft 7 (1924). —

Secondary literature: *May* 77. *P* V, 121; VI, 234. — *DBL* **3** (1934), 201–203. — HEIBERG, JOHANN LUDWIG: "Axel Anthon Björnbo (1874–1911)." *Bibl. Math.* (3) **12** (1911/12), 337–344 (with portrait). — GARBOE, AXEL: "Axel Anthon Björnbo." *MGMN* **11** (1912), 132. — ZEUTHEN, H. G.: (Gedächtniswort). As cited above in *Abh. Gesch. math. Wiss.* **26**, no. 3 (1912), 1–6 (with portrait). — VANGENSTEEN, O. C. L.: in *Norsk Geogr. Selsk. Årbog* **22** (1912), 139–146. — FOLKERTS, MENSO: "Der Nachlaß Axel Anthon Björnbos." *HM* **5** (1978), 333–337.

M.F.

Bobynin, Viktor Viktorovich (* November 8, 1849, Shili village, Smolenks Region, Russian Empire; † November 25, 1919, Tula, Russia). VIKTOR BOBYNIN was born into a modest noble family in the Shili village, part of the Roslavl district of the Smolensk Region. From the *Gymnasium* in Tula (1860–1867) he went on to graduate from the University of Moscow in 1872. While teaching at the Military High School in Nizhniĭ Novgorod, he began his research on the history of mathematics. In 1881 BOBYNIN returned to Moscow, and the following year defended his dissertation on *Mathematics in Ancient Egypt*, whereupon he received his master's degree from the University. Thus qualified, he began his academic career in 1882 as a *Privatdozent* at the University of Moscow, where he began to lecture on the history of mathematics, a subject he continued to teach until his death. Each year BOBYNIN changed the content of his lectures. He also published the first journal in Russia devoted to the history of mathematics, *Physical and Mathematical Sciences in the Past and Present*, for which he used his own funds. BOBYNIN also contributed numerous works of his own to the journal, including a great number of biographies, bibliographies, obituary notices, reviews, and reports concerning scientific life at home and abroad. During 1885–1898 he published 13 volumes in all, and in 1899–1904 he issued one more volume entitled *Development of Physical and Mathematical Sciences.* Meanwhile, BOBYNIN actively struggled to promote history of mathematics in secondary education. His fields of research were diverse, and ranged from the mathematical ideas of primitive peoples to arithmetic of the

19th century. His works were distributed not only in Russia, but also in the West. Beginning in 1883, he also published articles regularly in G. ENESTRÖM's journal, *Bibliotheca Mathematica.* BOBYNIN wrote the chapter on elementary geometry in the 18th century for the fourth volume of CANTOR's *Vorlesungen über Geschichte der Mathematik* (1908) ([CANTOR 1880/1908]).

According to BOBYNIN, the object of the history of mathematics should be "the investigation of the gradual development of mathematics and the discovery of the laws that rule this development" [Rybnikov 1950]. He attached great importance to the problem of the origins of mathematical knowledge and to the factors that determine its development. This helps to account for his lively interest in the Egyptian Rhind papyrus as soon as it was published, which was the topic of his dissertation. In what he called "mathematics of the people," he was interested in the relics of ancient mathematical ideas as preserved in folklore and in living peasant practices. He saw this as one possible way to penetrate the ancient origins of the history of mathematics, as well as medieval European science. As a result of his research, BOBYNIN discovered what he took to be the key to understanding the genesis of mathematical knowledge.

If, for the earliest emerging mathematical cultures, it was the "mathematics of the people" that one should examine, for the developed cultures the most important elements, according to BOBYNIN, were biographies and bibliographies, of which he produced many. He published three volumes of *Russian Bibliography in the Physical and Mathematical Sciences* which covered scientific work from the beginning of printing up to 1816 (1886–1900). As a result of his efforts, by the turn of the century BOBYNIN was one of the most important historians of mathematics in Europe, and his work succeeded in uniting a truly historical conception of sources with a philosophical approach to the development of mathematical ideas.

Secondary literature: *May* 79. — RYBNIKOV, KONSTANTIN A.: "Viktor Viktorovich Bobynin." *IMI* **3** (1950), 343–357 (in Russian), with portrait, and bibliography by A. M. LYKOMSKAYA, pp. 358–396. — ZUBOV, VASILIĬ PAVLOVICH: "V. V. Bobynin and his Works on the History of Mathematics." *TIIET* **15** (1956), 277–322 (in Russian). S.S.D.

Boncompagni Ludovisi, Baldassarre (* May 10, 1821, Rome, Italy; † April 13, 1894, Rome, Italy). BALDASSARRE BONCOMPAGNI may be considered the first truly professional historian of mathematics in Italy, and one of the most prominent figures in this field in the second half of the 19th century. Through his works, history of mathematics acquired the standards of a rigorous discipline. He contributed substantially to the promotion of historical research, and he may well be credited as the founder of the Italian school for history of mathematics even though, as a noble and wealthy man, he had no teaching position and produced all of his work as a dedicated amateur.

Prince BALDASSARRE BONCOMPAGNI was born into a family which also boasted among its members UGO BONCOMPAGNI, who in 1572 was elected as POPE GREGORY XIII and is universally known for the reform of the (Gregorian) calendar (1585) named after him. The young BALDASSARRE was first educated at the *Archiginnasio* in Rome, then at the *Collegio Romano* where he attended lectures in mathematics and physics offered by BARNABA TORTOLINI, the future editor of the *Annali di scienze matematiche e fisiche* (beginning in 1850). Although his first scientific publication was a paper on definite integrals published in *Crelle's Journal* in 1843, BONCOMPAGNI's interests very soon turned towards the history of science. In 1846 he published his first paper in this field, which was devoted to the development of physics in Italy during the 16th and 17th centuries. As a result of these works, Prince BONCOMPAGNI was appointed as one of 30 members of the *Accademia Pontificia dei Nuovi Lincei*, newly established by POPE PIUS IX in 1847 after the model of the celebrated Lincean Academy founded by Prince FEDERICO CESI in 1602. By the end of the 1840s BONCOMPAGNI had become interested in medieval mathematics, the rediscovery and appreciation of which was one of his major contributions to the history of mathematics.

BONCOMPAGNI's first relevant publications were *Della vita e delle opere di Platone Tiburtino, traduttore del secolo XII* (On the Life and Works of PLATO OF TIVOLI, Translator of the 12th Century) (1851); and *Della vita e delle opere di Gherardo Cremonese, traduttore del secolo XII, e di Gherardo da Sabbioneta, astronomo del secolo XIII* (On the Life and Works of GHERARD OF CREMONA, Translator of the 12th Century; and of GHERARD OF SABBIONETA, Astronomer of the 13th Century) (1851), both of which appeared with facsimiles. GUGLIELMO LIBRI (1803–1869, **B**) had already called attention to both of these figures in his *Histoire*, and BONCOMPAGNI listed all of the translations they made, describing as well the existing codices and their editions. Because the edition of ancient texts required a number of uncommon typographical characters, in order to improve the quality of his publications, BONCOMPAGNI established his own printing plant. For almost forty years it served not only for BONCOMPAGNI's own publications but also for publications by his friends and colleagues, including in particular the printing of the transactions of the *Accademia Pontificia dei Nuovi Lincei*, which he published at his own expense beginning in 1871. Indeed, when Rome became the capital of unified Italy in 1870, BONCOMPAGNI refused to join the *Accademia dei Lincei*, re-founded on the initiative of the scientist and politician QUINTINO SELLA in opposition to the papal academy. Although he did not share the strong opposition to the new state articulated by the Pope and the Roman Catholic Church, BONCOMPAGNI preferred to maintain his membership in the *Accademia Pontificia*. He also refused a seat in the Italian Senate offered to him by SELLA.

After his early works on medieval mathematics, which also included a paper "Della vita e delle opere di GUIDO BONATTI, astronomo e astrologo del secolo XIII" (On the Life and Works of GUIDO BONATTI, Astronomer and Astrologer of the 13th Century) (1851), BONCOMPAGNI focused his interests on LEONARDO PISANO (FIBONACCI), whose life and works only came to be appreciated by historians of mathematics after COSSALI's and LIBRI's historical studies. BONCOMPAGNI contributed to a better evaluation of FIBONACCI's contributions to mathematics, and in fact established FIBONACCI as the greatest figure of medieval mathematics.

BONCOMPAGNI's first publications about FIBONACCI were *Intorno ad alcune opere di Leonardo Pisano matematico del secolo XIII* (On Some Works of LEONARDO OF PISA, Mathematician of the 13th Century) (1854, with facsimiles), and *Tre scritti inediti di Leonardo Pisano, secondo la lezione di un codice della Biblioteca Ambrosiana di Milano* (Three Unpublished Writings of Leonardo of Pisa, According to the Reading of a Codex of the Ambrosiana Library in Milan) (1854). Among previously unpublished writings by FIBONACCI were the *Liber Quadratorum*, which according to COSSALI, LIBRI, and CHASLES had been lost, and the *Flos*, a collection of problems by FIBONACCI which until then was completely unknown. BONCOMPAGNI discovered both of these works in a codex kept in the Ambrosiana library in Milan. From a mathematical point of view, these writings represent FIBONACCI's more important works and their publication excited considerable interest among historians of mathematics, including mathematicians like GIOVANNI CODAZZA and ANGELO GENOCCHI. GENOCCHI, who at the time was a lawyer in Turin without any academic position, was particularly interested in number theory. He pointed out some inaccuracies in BONCOMPAGNI's edition, which prompted the latter to edit a revised version of the book, *Opuscula of Leonardo of Pisa*, published by BALDASSARRE BONCOMPAGNI as *Opuscoli di Leonardo Pisano pubblicati da Baldassarre Boncompagni* (2nd ed., 1856). In addition, GENOCCHI sent BONCOMPAGNI a detailed, critical study of FIBONACCI's papers, which BONCOMPAGNI presented for publication in TORTOLINI's *Annali*. BONCOMPAGNI himself edited the *Scritti di Leonardo Pisano, matematico del secolo decimoterzo, pubblicati da Baldassarre Boncompagni. Vol. I, Liber Abaci* (Writings of LEONARDO OF PISA, Mathematician of the 13th Century, Published by Baldassarre Boncompagni. Vol. I: *Liber Abaci*) (1857). This was the first printed edition of FIBONACCI's treatise. BONCOMPAGNI asked GENOCCHI to provide the publication with mathematical and critical commentaries. When GENOCCHI declined to do so, BONCOMPAGNI published the work as *Scritti inediti di Pietro Cossali chierico regolare teatino, pubblicati da B. Boncompagni* (Unpublished Writings by the Regular Theatine Clerk PIETRO COSSALI, Published by B. BONCOMPAGNI) (1857), which included notes on the *Liber Abaci* left unpublished by PIETRO COSSALI. This was followed by the publication of both LEONARDO's *Practica Geometriae*, and a 3rd revised edition of the *Opuscoli* under the title: *Scritti di Leonardo Pisano matematico del secolo decimoterzo pubblicati da Baldassarre Boncompagni. Vol. II, Practica Geometriae ed Opuscoli* (Writings of LEONARDO OF PISA, Mathematician of the 13th Century, Published by BALDASSARRE BONCOMPAGNI. Vol. II, *Practica Geometriae ed Opuscula* (1862). BONCOMPAGNI's edition of FIBONACCI's collected works is a unique source for medieval mathematics. Along with information about FIBONACCI's life and work, it includes an impressive amount of historical material on medieval mathematics and on mathematicians like PAOLO DELL'ABACO and ANTONIO DE' MAZZINGHI, as well as bibliographical information on rare books and manuscripts.

In addition to the edition of FIBONACCI's writings, BONCOMPAGNI's contributions to the history of medieval mathematics also include *Trattati d'aritmetica*

— *pubblicati da Baldassarre Boncompagni. vol. 1, Algorismi de numero Indorum; vol. 2, Joannis Hispalensis, liber Algorismi de pratica Arismetricae* (Treatises of Arithmetic Published by BALDASSARRE BONCOMPAGNI. vol 1, *Algorithms of Indian Numbers*; vol. 2, *Joannis Hispalensis, Book of Algorithms of the Practice of Arithmetic* (1857). The first of these is a Latin translation of a treatise by AL-KHWĀRIZMĪ, the second is a work by the 13th-century mathematician, JOHN OF SEVILLE. Moreover, in 1862 in the *Atti* of the *Accademia dei Nuovi Lincei*, BONCOMPAGNI reprinted (with commentaries) the first printed textbook on arithmetic, generally known as the *Aritmetica di Treviso* (Arithmetic of Treviso). This is a booklet consisting of 62 charts, which BONCOMPAGNI edited with more than 700 pages of historical and critical commentaries!

However impressive all of this editorial work may be, there is no doubt that BONCOMPAGNI's fame is above all due to the *Bullettino di bibliografia e di storia delle scienze matematiche e fisiche*, which he founded in 1868 and edited for twenty years, printing it at his own expense, until 1887. This is a journal of extraordinary scientific value, one which contributed greatly to establishing standards of rigor and precision in writing the history of mathematics. As LORIA [LORIA 1916, 2nd ed., 1946, 70] aptly wrote, BONCOMPAGNI's *Bullettino* "must be kept on the bookshelves of every scholar of the history of mathematics." Reprinted in 1968, the journal remains a valuable source of information and an extremely useful research tool. The volumes of the *Bullettino* cover virtually every period of mathematics, from Greek geometry through Arabic and medieval mathematics to Renaissance and modern mathematics, including biographies and editions of correspondence. BONCOMPAGNI's *Bullettino* also served to promote the history of mathematics both in Italy and abroad. In this respect, it is worth mentioning that BONCOMPAGNI sent free copies of the journal to academies, libraries, and to anyone interested in the subject. Moreover, the *Bullettino* was a truly international journal. In addition to papers by Italian historians, it published papers by French scholars, as well as Italian (and French) translations of papers originally published in German (e.g. by CANTOR, CURTZE, S. GÜNTHER, and H. HANKEL).

BONCOMPAGNI himself contributed some 130 of his more than 200 publications to the journal. In editing the *Bullettino*, he not only displayed his immense historical erudition, but he stressed the need for precision in quoting original sources and demanded rigorous accuracy in bibliographical references. BONCOMPAGNI also provided each issue of the journal with an updated list of recent mathematical and historical publications. When he decided to give up the editorship of the *Bullettino* due to illness, he asked ANTONIO FAVARO (1847–1922, **B**) to succeed him, but FAVARO, already responsible for producing a National Edition of GALILEO's works, was forced to decline the offer, and unable to find a suitable successor, the journal ceased publication in 1887.

In the course of his life, BONCOMPAGNI established a network of correspondents and secretaries throughout Italy and abroad. Scholars with whom he was in contact provided him with transcriptions of codices kept in private and public libraries, and occasionally BONCOMPAGNI would ask them to buy a rare book or manuscript on his behalf. In this way he was able to amass a celebrated library amounting to some 600 manuscripts and 20 000 volumes. (A catalogue of the entire collection was compiled by his private secretary; see ENRICO NARDUCCI, *Catalogo dei manoscritti ora posseduti da Baldassarre Boncompagni* (Catalogue of the Manuscripts Now in Possession of BALDASSARRE BONCOMPAGNI) (1862). A second edition of this work, published by NARDUCCI in 1892, amounted to some 520 pages). Although BONCOMPAGNI's own library was of exceptional scientific value, and according to FAVARO, "unique in the world," after his death it was unfortunately dispersed and partly sold at auction.

Secondary literature: *DSB* **2**, 283–284. *May* 83–84. *P* III, 157; IV, 156. — CODAZZA, GIOVANNI: "Il principe Boncompagni e la storia delle scienze matematiche in Italia." *Il Politecnico* **20** (1864), 5–27. — [ARRIGHI 1989]. U.B.

Bonola, Roberto (* November 14, 1874, Bologna, Italy; † May 16, 1911, Bologna, Italy). ROBERTO BONOLA obtained his degree (*laurea*) in mathematics from the university of Bologna in 1898 with a thesis on non-Euclidean geometry supervised by FEDERIGO ENRIQUES. He was also ENRIQUES's assistant for descriptive and projective geometry until 1900, when he was named professor at a secondary school *(Scuola Normale)*, first in Petralia Sottana (Palermo), and later, in 1901, in Pavia. There he was also an assistant for the chair of calculus at the university, where he became an instructor (*libero docente*) for descriptive and projective geometry in 1908. Eventually, in 1911, just shortly before his untimely death, BONOLA was named professor of mathematics at the *Istituto Superiore di Magistero Femminile* in Rome.

From the time he was a student, the history of non-Euclidean geometry was BONOLA's favorite research subject. After the completion of his thesis in 1899, he published an extensive "Bibliografia sui fondamenti della geometria in relazione alla geometria non euclidea" (Bibliography on the Foundations of Geometry in Relation to Non-Euclidean Geometry), various parts of which appeared in LORIA's *Bollettino di Bibliografia e Storia delle Scienze Matematiche* between 1899 and 1902. At the same time BONOLA contributed a long essay to ENRIQUES's *Questioni riguardanti la geometria elementare* (Questions Concerning Elementary Geometry) (1900), "Sulla teoria delle parallele e sulla geometria non euclidea" (On the Theory of Parallels and Non-Euclidean Geometry). After BONOLA's death this article was reprinted in ENRIQUES's *Questioni riguardanti le matematiche elementari* (Questions Concerning Elementary Mathematics), (Part I, Vol. II, 1912, 311–428). An extended version of this essay was published by BONOLA as *La Geometria non euclidea. Esposizione storico critica del suo sviluppo* (Non-Euclidean Geometry. A Critical Historical Exposition of its Development) (1906; repr. 1975). This was also translated into English by HORATIO S. CARSLAW as *Non-Euclidean Geometry* (1912), with a preface by ENRIQUES. The Dover reprint of 1955 includes as appendices English translations by G. B. HALSTED of papers by BÓLYAI and LOBACHEVSKIĬ, and remains a standard reference on the subject of the history of non-Euclidean geometry.

Secondary literature: *DSB* **2**, 283–284. *May* 84. *P* V, 142. — VENERONI, EMILIO: "Roberto Bonola." *Period. Mat.* (3) **9** (1911), 319–320. U.B.

Bopp, Karl (* March 28, 1877, Rastatt, Germany; † December 5, 1934, Heidelberg, Germany). As *Privatdozent* in Heidelberg for mathematics, BOPP received a *Lehrauftrag* in 1906 for history of mathematics (as the successor of M. CANTOR), as well as for political arithmetic and insurance; he was promoted to the position of extraordinary professor in 1915. His main areas of interest were history of mathematics in the 17th and 18th centuries. In addition to a study of GREGORIUS A SANCTO VINCENTIO (1907), BOPP published the correspondence of J. H. LAMBERT (1914–1916) with EULER (1925) and with A. G. KÄSTNER (1928).

Secondary literature: *May* 85. *P* V, 142; VI, 278. — LOREY, WILHELM: "Karl Bopp † (28. 3. 1877 – 5. 12. 1934)." *Jber. DMV* **45** (1935), 116–119. M.F.

Bortolotti, Ettore (* March 6, 1866, Bologna, Italy; † February 17, 1947, Bologna, Italy). As a student in Bologna, after graduating from the Technical Institute in 1884, BORTOLOTTI enrolled in the School of Engineering of the University of Bologna, where he studied algebra and both analytic and advanced geometry with SALVATORE PINCHERLE. At the same time, because of the poor economic conditions of his family, he also worked as an administrator for the Italian telegraph company. At PINCHERLE's suggestion, BORTOLOTTI then moved to the faculty of mathematics and in 1889 obtained his degree (*laurea*) in mathematics, *summa cum laude.* After serving as PINCHERLE's assistant until 1891, BORTOLOTTI obtained a professorship at the *Liceo* in Modica (Sicily).

BORTOLOTTI spent the academic year 1892/1893 in Paris, where he attended lectures by POINCARÉ, PICARD, DARBOUX, and JORDAN. Beginning in 1894, upon his return to Italy, BORTOLOTTI taught mathematics at the *Liceo* in Rome. Two years later, he obtained his *venia legendi* (*libera docenza*) in algebra (in 1896), and in 1900 he began his academic career as professor of calculus at the University of Modena. Over the next twenty years BORTOLOTTI taught algebraic analysis and calculus, and also served as dean of the faculty from 1913 to 1919. Then he moved to the University of Bologna, where he taught analytic geometry until his retirement in 1936, whereupon he was named professor *emeritus.* From the year of the founding of the Italian Mathematical Union by PINCHERLE in 1923, BORTOLOTTI served as its secretary until 1945. He also assisted PINCHERLE as secretary of the International Congress of Mathematicians, which was held in Bologna in 1928.

As a student of PINCHERLE, BORTOLOTTI first devoted his attention to mathematical analysis, and published papers on the calculus of finite differences, the theory of distributive operations, continuous fractions, and infinite algorithms. He also published papers related to his teaching (including papers on the asymptotic behaviour of series, the orders of infinity, and the theory of improper integrals). However, the majority of his publications (some 180 of his 220 publications) were devoted to the history of mathematics. Along with GINO LORIA (1862–1939 **B**),

BORTOLOTTI may indeed be considered the most prominent historian of mathematics working in Italy in the first half of the 20th century.

BORTOLOTTI's interests turned to the history of mathematics when he was in Rome, under the influence of the mathematician VALENTINO CERRUTI, who had promoted the National Edition of GALILEO's works as well as the edition of FAGNANO's collected works. After BORTOLOTTI's move to Modena, his historical research focused on the Modenese mathematician PAOLO RUFFINI. RUFFINI's contributions to the theory of equations and modern algebra were the subjects of BORTOLOTTI's first paper (1902) devoted to the history of mathematics. BORTOLOTTI claimed priority for RUFFINI over both CAUCHY and GALOIS with respect to the foundations of group theory and the use of GALOIS's resolvent. This study reveals some of the main themes of BORTOLOTTI's historical work. In particular, he felt it was his primary duty to emphasize the merits and priority of the accomplishments of Italians with respect to foreign mathematicians. Indeed, his aim was to "vindicate to Italy the glory of its mathematicians, not because of shabby national spirit but for pure love of truth and justice," as BORTOLOTTI himself wrote in 1931 when introducing a list of his own publications. Accordingly, his work was characterized by a search for primary sources combined with a strong patriotic sentiment which often turned into blatant nationalism — even before the Fascist regime was established in Italy, which made nationalism an essential ingredient of Italian politics and culture. BORTOLOTTI's historical work on RUFFINI culminated in an edition of RUFFINI's *Opere matematiche* (Mathematical Works) in three volumes (1914–1954), including a volume of correspondence.

In addition to RUFFINI's life and work, BORTOLOTTI's main research subject was the history of mathematics in the Middle Ages and the Renaissance. Contrasting MORITZ CANTOR's views on the priority of JORDANUS NEMORARIUS (JORDANUS DE NEMORE) in establishing algebra in western Europe, BORTOLOTTI championed LEONARDO PISANO's pioniering role. Of particular interest are historical papers BORTOLOTTI devoted to the history of Renaissance algebra and the solution of the cubic equation by SCIPIONE DAL FERRO, TARTAGLIA, and CARDANO. BORTOLOTTI claimed SCIPIONE DAL FERRO's priority in solving the cubic equation, thereby opposing the views of LORIA and other foreign historians; BORTOLOTTI also re-evaluated CARDANO's contributions to equation theory with respect to those of TARTAGLIA.

BORTOLOTTI's tireless search for Renaissance sources of mathematics (especially material related to Bologna) led him to discover manuscripts of books IV and V of BOMBELLI's *Algebra*, which he published in 1929. Along with the first three volumes of BOMBELLI's work, they were all posthumously reprinted in a single volume: *L'algebra, opera di Rafael Bombelli* (*Algebra*, a Work of Rafael Bombelli), edited by UMBERTO FORTI (1966). In this same vein, BORTOLOTTI also called attention to the pioneering discovery of continuous fractions made by the 17th-century Bolognese mathematician PIETRO CATALDI.

In fact, after moving to Bologna, the history of mathematics at the University of Bologna during the 16th and 17th centuries became BORTOLOTTI's favorite subject. His relevant papers were collected together as a singe work: *Studi e ricerche sulla storia della matematica in Italia nei secoli XVI e XVII* (Studies and Research on the History of Mathematics in Italy in the 16th and 17th Centuries) (1928). In addition to writings devoted to the history of the solution of the cubic equation, the book included papers on the contributions of Bolognese mathematicians (including CATALDI, CAVALIERI, TORRICELLI, MENGOLI, and G. MANFREDI) to the early development of calculus. This was followed by a *Seconda serie di studi e ricerche sulla storia della matematica in Italia* (Second Series of Studies and Research on the History of Mathematics in Italy) (1944) — although the title is rather misleading because this collection also includes papers on the history of ancient Egyptian and Babylonian mathematics as well. A few years later, BORTOLOTTI completed another book, *La storia della matematica nell'università di Bologna* (The History of Mathematics at the University of Bologna) (1947). This was a revised and extended version of an earlier booklet, *L'ecole mathématique de Bologna* (The Mathematical School of Bologna), published by BORTOLOTTI on the occasion of the International Congress of Mathematicians held in Bologna in 1928. In addition to the above-mentioned works, BORTOLOTTI's major publications include a long (posthumous) essay: "Storia della matematica elementare" (History of Elementary Mathematics) (in: *Enciclopedia delle matematiche elementari*, LUIGI BERZOLARI (Ed.), vol. III, Part II, 1949, pp. 540–750), and an edition of the book *I cartelli di matematica disfida* (The Mathematical Challenges) (1933) concerning the challenge issued by TARTAGLIA to CARDANO (and FERRARI), which were subsequent to the publication of CARDANO's *Ars Magna*.

Along with CARDANO and BOMBELLI, EVANGELISTA TORRICELLI was another of BORTOLOTTI's heroes. The publication of TORRICELLI's *Opere* (Works) (1919–1944) — edited by LORIA and GIUSEPPE VASSURA — met with BORTOLOTTI's sharp criticism and prompted a number of subsequent papers in which he sought to point out LORIA's misunderstandings and mistakes. Although LORIA honestly recognized the flaws in the edition of TORRICELLI's *Opere*, this gave rise to a lasting argument between them. "As for [TORRICELLI's] unpublished works," BORTOLOTTI eventually observed, "their characteristic disorder was aggravated by the work of the editor [LORIA], who was unable to penetrate into the author's thought and, in spite of this, wanted to have a hand in it" [BORTOLOTTI 1947, 133]. Although perhaps less internationally recognized than LORIA, BORTOLOTTI nonetheless exerted a remarkable influence on the Italian community of historians of mathematics. Among his students there were in particular AMEDEO AGOSTINI and ETTORE CARRUCCIO, who was also his son-in-law.

Secondary literature: DSB **2**, 319–320. May 86. *P* IV, 160; V, 146; VI, 284; VIIb(1), 481. — BOMPIANI, ENRICO: "In ricordo di Ettore Bortolotti." *Atti Mem. Accad. Sci. Modena* (5) **7** (1947), 185–202 (with bibliography). — CARRUCCIO, ETTORE: "Ettore Bortolotti." *Period. Mat.* (4) **26** (1948), 1–13. — SEGRE, BENIAMINO: "Ettore Bortolotti – Commemorazione." *Rend. Accad. Sci. Ist. Bologna, Classe Sci. Fis.*, N.S. **52** (1949), 47–86 (with bibliography). — PANZA, MARIO: "Ettore Bortolotti, storico della matematica." In: *La matematica italiana tra le* *due guerre mondiali*, a cura di ANGELO GUERRAGGIO. Bologna 1987, 279–306 (with bibliography). U.B.

Bosmans, Henri (* April 7, 1852, Mechelen, Belgium; † February 3, 1928, Brussels, Belgium). As a Jesuit, BOSMANS taught mathematics at the Collège Saint-Michel in Brussels. He published numerous studies on the history of mathematics in the 16th and 17th centuries, and on mathematicians mainly from the Southern Netherlands. Among these were SIMON STEVIN, MICHEL COIGNET, ADRIANUS ROMANUS, GODFRIED WENDELEN, and the Jesuit mathematicians and astronomers J. C. DELLA FAILLE, GREGORIUS A SANCTO VINCENTIO, ANDREAS TACQUET, and FERDINAND VERBIEST.

Secondary literature: *May* 86. *P* VI, 287. *BNB* **30**, 182–183. — PEETERS, PAUL: "Le R. P. Henri Bosmans, S. J." *Rev. Quest. Sci.* (4) **13** (1928), 201–214. — PELLIOT, PAUL: "Henri Bosmans." *T'ouang Pao* **26** (1928), 190–199. — ROME, ADOLPHE: "Le R.P. Henri Bosmans, S.J. (1852–1928)." *Isis* **12** (1929), 88–112. — BERNARD-MAÎTRE, HENRI: "Un historien des mathématiques en Europe et en Chine: Le Père Henri Bosmans S.J. (1852–1928)." *Arch. int. Hist. Sci.* **3** (1950), 619–656. P.B.

Bosscha, Johannes (* November 18, 1831, Breda, Netherlands; † April 15, 1911, Heemstede, Netherlands). Trained as a physicist, BOSSCHA was professor at the Royal Military Academy in Breda (1860–63), and inspector of secondary education (1863–73). In 1873 he was appointed professor of physics at the Delft Polytechnical School. BOSSCHA succeeded DAVID BIERENS DE HAAN as editor of the *Œuvres* of CHRISTIAAN HUYGENS.

Secondary literature: *P* I, 247; III, 166; IV, 162; V, 148. — *BWN* **1**, 79. — LORENTZ, HENDRIK ANTOON: "Obituary Johannes Bosscha." *Programma van de Hollandsche Maatschappij der Wetenschappen te Haarlem vor het jaar 1911*, 2–8. P.B.

Bossut, Charles (* August 11, 1730, Tartaras, France; † January 14, 1814, Paris, France). Educated by Jesuits (like MONTUCLA and LALANDE) at Lyon, BOSSUT taught mathematics at the *Ecole du Génie* at Mézières, then hydrodynamics at the *Louvre*. He also contributed much to European scientific education by writing a series of textbooks. He edited the works of PASCAL, wrote for the *Encyclopédie méthodique*, and produced a general history of mathematics whose style is not technical but largely biographical. A close friend of D'ALEMBERT, he was a member of the *Académie des Sciences* in Paris.

Main works: *Essai sur l'histoire générale des mathématiques.* 2 vols., Paris 1802; Engl. trans., by JOHN BONNYCASTLE, Clerkenwell 1803; German trans., by N. TH. REIMER, Hamburg 1804.

Secondary literature: *DSB* **2**, 334–335. *May* 87. *P* I, 249. J.P.

Bourbaki, N(icolas) (Twentieth century). "N(ICOLAS) BOURBAKI" is the pseudonym of a 20th-century group of mathematicians, mainly French, which has published a series of treatises under the general title, *Éléments de mathématique* (Elements of Mathematics). The works of BOURBAKI provide overviews of different areas of modern mathematics, primarily in terms of the structures they have in common, and combine axiomatics with a meticulously abstract method of presentation. BOURBAKI's publications also contain historical notes which were collected as a separate volume and published under the title *Éléments d'histoire des mathématiques* (Elements of History of Mathematics).

The Bourbaki team first came together in the academic year 1934–1935, and continues to operate to this day (2001). The *Éléments de mathématique* were published under the pen name "N. BOURBAKI" as a series of booklets, starting with a digest of results on Set Theory (1939). The group reached its publication peak in the 1950s, but persists and will soon publish a new work on algebra which is presently nearing completion.

In 1948 the group established a mathematics seminar in Paris known as the "Séminaire Bourbaki," which has since become the most notable in the Western world, a position as yet unchallenged. Although the members of BOURBAKI selected the topics as well as the seminar participants and edited the talks presented in the seminar for publication under the same title, the seminar activities remained quite separate from the behind-the-scenes discussion and revision of the various parts of the *Éléments de mathématique.*

Indeed, this enterprise was a real and unusual collective effort: BOURBAKI co-opted its collaborators, kept its membership secret, and did not acknowledge individual contributions. At any point in time there were, on average, ten active members in the group. The way they organized their work sessions has varied over the years, but usually, especially after World War II, they met three to four times a year in general meetings, called "congresses," where members submitted their contributions to the critical eyes of their peers. Thus each part of the *Éléments de mathématique* went through many successive versions, first written by individuals and then reread, discussed, and reworked by the whole team before being handed over to a new writer who would have to incorporate the group's recommendations into a revised text. Sometimes, as many as ten or more revisions would be made, with the result that such on-going amendments have blurred the record, making it difficult to find traces of particular authors in the final version.

An expository work, the *Éléments de mathématique* — BOURBAKI deliberately uses the unusual singular rather than the common plural of the French word "mathématique(s)" in order to demonstrate a faith in the unity of mathematics — is nevertheless an original piece of craftsmanship. The authors' ground-breaking approach was to choose particular conceptual settings into which they cast whole mathematical theories. They also strove to give their presentations a clear but strict organic unity, and achieved a particularly homogenous exposition by adopting the same approach throughout the treatise and by stressing ties between the different theories. In this way they wrote the following volumes that make up the *Éléments*: Set Theory, Algebra, General Topology, Functions of a Real Variable, Topological Vector Spaces, Integration Theory, Lie Groups and Lie Algebras, Commutative Algebra, Spectral Theories, and Differentiable and Analytic Manifolds. The first six volumes form a tightly knit set subtitled "Les Structures Fondamentales de l'Analyse" (The Fundamental Structures of Analysis), while the others are more akin to monographs. All but Spectral Theories and Differentiable Manifolds, which are mere digests of results, contain at least one companion historical note.

BOURBAKI's first explicit project about historical notes dates back to 1936; it is mentioned in a list of suggestions concerning the writing style the group should adopt and sundry items to include in the treatise. According to this plan, historical notes would appear at the end of chapters or volumes. The BOURBAKI meeting of September 1938 summoned ANDRÉ WEIL to write a history of topology for the volume on General Topology, then in the making; this first historical note appeared with the first chapters on General Topology in 1940. Over the years, many historical notes were written by different members and published within BOURBAKI's *Éléments de mathématique.* They often contain historical findings and insights that were novel at the time, and each article refers amply to primary sources. Some notes had but one single author and, although members of BOURBAKI usually discussed historical notes, they did not subject them to the many thorough revisions undergone by the strictly mathematical sections of the *Éléments de mathématique.* At the suggestion of Hermann, BOURBAKI's publisher in Paris, the historical notes were assembled in one volume, the *Éléments d'histoire des mathématiques* (Elements of History of Mathematics), and published under the pseudonym "NICOLAS BOURBAKI." The first edition (1960) included the twenty-one historical articles which had appeared up to then in the *Éléments de mathématique*, with some revisions and a short preface as well as a title for each section (all provided by JEAN DIEUDONNÉ) (1906–1992, **B**). Meanwhile, old, new, and modified historical notes continued to feature within the *Éléments de mathématique* as volumes were reedited and fresh chapters completed. In time, the *Éléments d'histoire* were reprinted, revised, corrected, and extended through consecutive editions (1964; 1969 — which included a name index for the first time; 1974; and 1984 — by a new publisher, Masson in Paris).

Closely related to specific chapters of the *Éléments de mathématique*, the collected notes do not constitute a comprehensive history of mathematics. Like BOURBAKI's monument itself, many areas of mathematics are not included at all. The chapters in the historical volume appear in the same order as the individual volumes of the Bourbaki series (beginnig with Set Theory and ending with Commutative Algebra), and although cross-referencing is common, each essay is relatively independent of the others. Yet the fragments all display the same historiography, one that is stringently internal, focused on pure mathematics, and strongly driven by concepts and texts. BOURBAKI's historical essays do not set mathematics in the foreground of a history of ideas, they scarcely discuss applications of mathematics or relations between mathematics and other sciences, and deliberately ignore biographical as well as general historical aspects. The more substantial notes trace ideas and questions back to ancient times and follow their development sometimes up to the twentieth century. All the accounts are teleological in that the mathematical achievements of recent years mould them; they are retrospective as well (i.e. "Whiggish") inasmuch as posterior knowledge bears heavily on historical facts. Not only does BOURBAKI deploy modernized terminology and modernist interpretations in its historical reports, it also subordinates them to the conceptual frames of the *Éléments de mathématique*: concepts, questions, texts — and hence mathematicians — are all deemed historically worthy if in line with Bourbakian choices. BOURBAKI's historical essays were written mainly for mathematicians by world-class mathematicians who pursued their interest for the history of their field and found — intentionally or not — a support for their views of pure mathematics among works of the past. Thus, BOURBAKI's historical notes bring some indirect explanations to Bourbakian options.

The *Éléments de mathématique* have had considerable influence on the development of mathematics and its teaching after World War II, especially in France. BOURBAKI's historical notes made a particular mark as there had been few notable French histories of mathematics devoid of explicit epistemological considerations since the historical notices of the French edition of the *Encyklopädie der mathematischen Wissenschaften* (Encyclopedia of Exact Sciences) at the beginning of the 20th century (see Section 1.9). BOURBAKI's essays were also innovative in emphasising the mathematics of the 19th and 20th centuries, and thus attracted readers who were particularly interested in recent history. Whereas most of the *Éléments de mathématique* were translated into Russian (beginning in 1958), English (beginning in 1966), and Japanese (beginning in the late 1960s), translations of the *Éléments d'histoire* into Italian (1963), Russian (1963), German (1971), Spanish (1972), Polish (1980), and English (1994), have brought BOURBAKI's historical notes to an even more diversified readership. When, especially in the 1980s, BOURBAKI's influence and views on mathematics came under severe criticism, the group's historical writings were challenged as well, often by historians of mathematics who took issue with particular points that they strove to redress in their own studies. In this respect, the fate of the *Éléments d'histoire* followed that of the *Éléments de mathématique* to which they remained intimately connected.

The BOURBAKI archive is preserved at the *Ecole normale supérieure* in Paris where the group, a corporation by the name of *Association des Collaborateurs de Nicolas Bourbaki*, keeps its headquarters and secretariat.

Secondary literature: *DSB* **2**, 351–353. *May* 87. — CORRY, LEO: "Nicolas Bourbaki and the Concept of Mathematical Structure." *Synthèse* **92** (1992), 315–348. — ZADDACH, ARNO: "Regressive Products and Bourbaki." In: SCHUBRING, GERT (Ed.): *Hermann Günther Grassmann (1809–1877): Visionary Mathematician, Scientist and Neohumanist Scholar.* Dordrecht 1996, 285–295). — BOREL, ARMAND: "Twenty-Five Years with Nicolas Bourbaki, 1949–1973." *Mitt. DMV* 1998, no. 1, 8–15). — BEAULIEU, LILIANE: *La Tribu N. Bourbaki* (Springer: forthcoming, with portraits). L.B.

Boutroux, Pierre (* December 6, 1880, Paris, France; † August, 1922, France). Son of the philosopher EMILE BOUTROUX and cousin of the mathematician HENRI POINCARÉ, BOUTROUX studied mathematics and lectured on integral calculus at the University of Poitiers (1908–1920). He then accepted a chair for the general history of science at the *Collège de France* in Paris. In writing the history of mathematics, BOUTROUX refused to limit himself only to technicalities. The underlying ideas, major lines of development, and scientific context were of the utmost importance for BOUTROUX's historical method. Studying analytic functions and the singularities of differential equations, he contributed "Fonctions analytiques" (largely following W. F. OSGOOD's contributions to the subject) to the *Encyclopédie des sciences mathématiques* (1904–1913) **2**/II.

Main works: *Les principes de l'analyse mathématique. Exposé historique et critique.* 2 vols., Paris 1914 and 1919. — *L'idéal scientifique des mathématiciens dans l'antiquité et dans les temps modernes.* Paris 1920; German trans. Leipzig and Berlin 1927 (repr. Wiesbaden 1968).

Secondary literature: *DSB* **2**, 358. *May* 88. *P* V, 154; VI, 299. — [Acta Math., Tab. Gen. **1–35** (1913), 40; portrait 133]. — BRUNSCHVICG, LÉON: "L'œuvre de Pierre Boutroux." *Revue de métaphysique et de morale* **29–30** (1922), 285–289.
J.P.

Bowditch, Nathaniel (* March 26, 1773, Salem, Massachusetts, USA; † March 16, 1838, Boston, Massachusetts, USA). BOWDITCH, an American astronomer, corrected and extended JOHN HAMILTON MOORE's *New American Practical Navigator* through numerous editions, and from the third edition on, beginning in 1802, the work bore BOWDITCH's name. Best known for his translation of LAPLACE's *Mécanique céleste*, BOWDITCH's aim was historical in his efforts to incorporate later results into the text and to provide credit for predecessors whom LAPLACE had failed to mention.

Secondary literature: *DSB* **2**, 368–369. *May* 88. *P* I, 266. — PICKERING, JOHN: *Eulogy on Nathaniel Bowditch, LL.D.* Boston 1838 (with portrait). — BERRY, ROBERT E.: *Yankee Stargazer. The Life of Nathaniel Bowditch.* New York 1941 (with portrait).
J.W.D.

Boyer, Carl Benjamin (* November 3, 1906, Hellertown, Pennsylvania, USA; † April 26, 1976, New York, New York, USA). CARL BOYER, American historian of mathematics, studied mathematics at Columbia University, from which he graduated in 1928 with honors in mathematics. He was also elected a member of two prestigious undergraduate academic honor societies, Phi Beta Kappa and Sigma Xi. In 1928 he began teaching as a tutor in mathematics at Brooklyn College, while continuing his study of mathematics, for which he received his master's degree from Columbia University in 1929. A decade later BOYER completed his Ph.D. in intellectual history at Columbia, and this subsequently became his first book, *Concepts of the Calculus* (1939), reprinted as *History of the Calculus and its Conceptual Development* (1959).

From 1934–1940 BOYER taught as an instructor of mathematics at Brooklyn College (and from 1935–1941 was also lecturer in science at University College of Rutgers University, New Jersey). In 1941 he was promoted to the position of assistant professor at Brooklyn, in 1948 to associate, and in 1953 he was promoted to the rank of full professor. Meanwhile, BOYER had been elected to the Council of the History of Science Society, on which he later served a second term, 1950–1953. He was also the book review editor for *Scripta Mathematica* from 1947 until 1970. In 1954–55 BOYER's scholarship was recognized with a prestigious Guggenheim Fellowship. In 1957 he was elected a member of the International Academy of the History of Science, and the following year (1958–59) he served as vice-president of the American Association for the Advancement of Science. Among the editorial boards on which BOYER served were those of the *Archive for History of Exact Sciences* (1960–1976), *Dictionary of Scientific Biography* (for which he was responsible for mathematics, 1960–1976), and *Historia Mathematica* (1973–1976).

In addition to his position as professor of mathematics at Brooklyn College of the City University of New York, BOYER also taught as visiting professor at Yeshiva University, New York, and at summer institutes at the University of California (Berkeley), University of Northern Iowa, University of Michigan, University of Kansas, and at the Catholic University of America. A special session honoring BOYER, "On the Tercentenary of Leibniz's Invention of the Calculus: a Tribute to Carl Boyer," was held at the annual meeting of the History of Science Society in December of 1976 (in Philadelphia).

BOYER's research interests included history of the calculus and analytic geometry. Among his most important historical works are *History of Analytic Geometry* (1956), *The Rainbow: From Myth to Mathematics* (1959), and *A History of Mathematics* (1968; Portuguese translation, 1974). BOYER was primarily an internalist historian of mathematics. His focus was usually on mathematics rather than mathematicians. As he noted in his *History of Analytic Geometry* (1956), especially for 20th-century mathematics, his publications "largely omitted" biographical details "because often they have little bearing upon the growth of concepts" [vii–viii]. On the other hand, in evaluating the importance of history of mathematics for individual mathematicians and their work, BOYER emphasized (in the same volume) the value of knowing the past: "Perhaps nowhere does one find a better example of the value of historical knowledge for mathematicians than in the case of FERMAT, for it is safe to say that, had he not been intimately acquainted with the geometry of Apollonius and Viète, he would not have invented analytic geometry" [viii].

Secondary literature: GILLISPIE, CHARLES C.: "Carl B. Boyer. November 3, 1906 – April 26, 1976." *Isis* **67** (1976), 610–614 (with portrait). — KLINE, MORRIS: "Carl B. Boyer – *In Memoriam.*" *HM* **3** (1976), 387–401 (with bibliography and portrait). — KLINE, MORRIS: "Carl B. Boyer (1906–1976)." *Arch. int. Hist. Sci.* **27** (1977), 275–276. J.W.D.

Braunmühl, Anton (von) (* December 22, 1853, Tiflis (today: Tbilisi), Grusinia, Russia; † March 7, 1908, Munich, Germany). BRAUNMÜHL, who was German and taught mathematics at secondary schools in Munich from 1877 to 1888, was a *Privatdozent* at the *Technische Hochschule* Munich 1884–1888, where he subsequently became *ausserordentlicher Professor* in 1888, and then *ordentlicher Professor* in 1892. He concentrated his research on the history of trigonometry.

Main works: Biographical sketches of SCHEINER (1891), GALILEI (1892), COPERNICUS (1896), WEIERSTRASS (1898), SEIDEL (1898), LIE (1900). — "Historische Studie über die organische Erzeugung ebener Curven von den ältesten Zeiten bis zum Ende des achtzehnten Jahrhunderts." In: DYCK, WALTHER (Ed.): *Katalog mathematischer und mathematisch-physikalischer Modelle ....* München 1892, 54–88. — "Beiträge zur Geschichte der Trigonometrie." *Abhandlungen der Kaiserlichen Leopoldinisch-Carolinischen Deutschen Akademie der Naturforscher* **71** (1898), 1–30. — Nassîr Eddîn Tûsi und Regiomontan." *Abhandlungen der Kaiserlichen Leopoldinisch-Carolinischen Deutschen Akademie der Naturforscher* **71** (1898), 31–67. — *Vorlesungen über Geschichte der Trigonometrie.* 2 vols. Leipzig 1900/1903.

Secondary literature: *DSB* **2**, 430. *May* 89. *P* IV, 176; V, 161. — *NDB* **2**, 560. — BURKHARDT, HEINRICH; SEBASTIAN FINSTERWALDER, and SIEGMUND GÜNTHER: "Anton Edler von Braunmühl." *Berichte der TH München* 1907/08 (15 pp.). — GÜNTHER, SIEGMUND: "Zum Gedächtnis A. von Braunmühls." *MGMN* **7** (1908), 362–367. — WIELEITNER, HEINRICH: "Anton von Braunmühl." *Bibl. Math.* (3) **11** (1910/11), 316–330. M.F.

Bretschneider, Carl Anton (* May 27, 1808, Schneeberg (Saxony), Germany; † November 6, 1878, Gotha, Germany). BRETSCHNEIDER started his career as *Privatdozent* for law at the University of Leipzig (1830–1832), and then served for a time in the court system of Gotha (1832–1835). From 1836 to 1878 he taught mathematics and geography at the *Real-Gymnasium* (from 1859 at the *Gymnasium Ernestinum* in Gotha. His areas of interest were the pedagogy of mathematics and Greek mathematics.

Main works: "Beiträge zur Geschichte der griechischen Geometrie." In: *Programm des Gymnasiums Ernestinum zu Gotha*, 1869. 12 pp. — [BRETSCHNEIDER 1870].

Secondary literature: *May* 89. *P* I, 294; III, 189. — BRETSCHNEIDER, ALFRED: "Carl Anton Bretschneider. Ein Gedenkblatt für seine Freunde und Schüler." *Z. Math. Phys.* **24** (1879), *Hist.-.lit. Abt.*, 73–91. M.F.

Broch, Ole Jacob (* January 14, 1818, Frederikstad, Norway; † February 5, 1889, Paris, France). BROCH, a Norwegian mathematician and administrator, obtained his doctorate in 1847. A year later he was offered a position in applied mathematics at Christiania University, where he became professor of pure mathematics in 1858. BROCH worked intensively at improving the teaching of mathematics at all levels in the Norwegian education system, and in his own teaching he included history of mathematics. He served as minister of defense in the Norwegian government from 1869–1872. Afterwards, he became increasingly involved in administrative matters and stopped lecturing in 1880.

Secondary literature: May 91. *P* I, 203; III, 198; IV, 187. — *NBL* **2**, 184–195. — [STUBHAUG 2000b, portrait: 82]. K.A.

Bruins, Evert Marie (* January 4, 1909, Woudrichem near Dordrecht, Netherlands; † November 20, 1990, Amsterdam, Netherlands). EVERT BRUINS learned Latin and Greek at a *Gymnasium* in Amsterdam, from which time his interests in cuneiform and hierogpyphic writing also date. He studied mathematics, physics, and chemistry at the University of Amsterdam, and in 1932 became an assistant in experimental physics, working under Dr. J. CLAY on earth magnetism. In the mid-thirties he joined a scientific expedition lasting several months on the Dutch ship *Boskoop* meant to explore the North Sea, South Atlantic, and Pacific Ocean. Continuing the research of two Norwegian scientists, C. STÖRMER and K. BIRKELAND on the phenomena of the radiation of the sky, BRUINS discovered the cosmic ray belt later known as the Van Allen belt, but his doctoral dissertation of 1938 which contained this discovery was largely ignored due to the outbreak of World War II. Thus the American JAMES A. VAN ALLEN rediscovered the phenomenon 16 years later.

In 1935 BRUINS, who was vice-director for the Institute of Physics at the University of Amsterdam, also lectured on mathematics and began to do work on the history of mathematics. During the Second World War he became lecturer in mathematical analysis, and his serious interest in the history of mathematics dates from this time. As a professor of mathematics at the University of Bagdad from 1952 to 1954 and again in 1955/56, he established the Institute of Mathematics there. Upon his return to Holland, BRUINS lectured on mathematics and history of mathematics, and in 1969 was promoted to professor of history of mathematics at the University of Amsterdam. He retired from this position in 1979.

After writing a series of papers about viscosity, BRUINS discovered the lyotropic numbers in colloid chemistry and cosmic rays. He also began, in 1945, to publish papers on mathematics and the history of mathematics. In mathematics, his interests concentrated on invariant theory, non-Euclidean geometry, the theory of equations, and mathematical aspects of quantum theory. In the history of mathematics, BRUINS devoted himself mostly to Egyptian, Babylonian, and Greek mathematics. In addition to these languages he also acquired some knowledge of Arabic, Russian, and a variety of European languages.

Among his publications, two works deserve to be singled out. In 1961 he edited, together with MARGUERITE RUTTEN, *Textes mathématiques de Suse* (Mathematical Texts of Susa), and in 1964 he published the *Codex Constantinopolitanus Palatii Veteris No. 1*, in three parts, which contains the sole surviving text of HERON's *Metrica* (BRUINS had been allowed to make photographs of the manuscript in Istanbul on his way back from Bagdad in 1956).

In his historical scholarship, BRUINS worked more as a mathematician than as a philologist or historian. A great mental calculator and familiar with methods in numerical analysis, he enjoyed showing how the construction of tables or the computation of seemingly difficult procedures in texts from antiquity could be explained, thereby contradicting explanations given by OTTO NEUGEBAUER and other scholars. His different, occasionally controversial interpretations of ancient texts sometimes provoked scientific disputes in which he defended his convictions emphatically. It is characteristic of BRUINS that he entitled one of his articles "Printing and Reprinting of Theories Contrary to Facts and Texts" (*Janus* **57**, 134–149; with a reply by HANS FREUDENTHAL and a further retort from BRUINS in the following volume).

Not the least of his services to the history of mathematics (apart from publishing more than 100 papers) was his long-time editorship of the journal *Janus*. BRUINS became co-editor in 1957 with volume 46, and acted as its sole editor from 1963 until his death; at times he even supported its publication by his own private means. Many of his articles, some of which have more the character of editorials and express his personal views about how research in the history of mathematics ought to be performed, may be found in the pages of *Janus*. BRUINS kept an enormous correspondence with colleagues from numerous countries. He was an enthusiastic traveller, usually accompanied by his wife, CORNELIA OOSTERKAMP, and was often invited to give lectures in various parts of the world.

Further publications: *Fontes matheseos: Hoofdpunten van het prae-griekse en griekse wiskundig denken.* Leiden 1953.

Secondary literature: KNOBLOCH, EBERHARD: "Evert Marie Bruins (1909–1990)." *HM* **18** (1991), 381–382 (with portrait and bibliography by JAN P. HOGENDIJK: 383–389). — HOGENDIJK, JAN P., and EBERHARD KNOBLOCH: "Evert Marie Bruins, 1909–1990." *Arch. int. Hist. Sci.* **42** (1992), 317–319. C.J.S.

Brun, Viggo (* October 13, 1885, Lier, Norway; † August 15, 1978, Drøbak, Norway). BRUN graduated from the University of Oslo in mathematics in 1909, went to Göttingen in 1910, became professor of mathematics at the *Norges Tekniske Högskole* in Trondheim in 1924, and at the University of Oslo in 1946. In number theory he is known for "BRUN's sieve," which is a further development of the sieve of ERATOSTHENES. BRUN's contributions to the history of mathematics include *Regnekunsten i det gamle Norge* (The Art of Reckoning in Norway in Former Times) (1962) and *Alt er tall. Matematikkens Historie fra Oldtid til Renessanse* (Everything is Number. The History of Mathematics from Antiquity to the Renaissance) (1964). He also contributed to research on the Norwegian mathematician ABEL. While in Florence, Italy, he found the important manuscript known as the "Paris Memoir," a work by ABEL that had been lost after being submitted to the *L'Académie des Sciences* in Paris in 1826.

Secondary literature: *P* V, 176; VI, 354; VIIb(1), 607. — *NBL* **2**, 244–245. — [SCRIBA 1980, with portrait]. K.A.

Brunschvicg, Léon (* November 10, 1869, Paris, France; † January 18, 1944, Aix-les-Bains, France). A French philosopher and pupil of EMILE BOUTROUX, BRUNSCHVICG became interested in mathematics along with the latter's son PIERRE (**B**). BRUNSCHVICG went on to teach at the *Sorbonne* and at the *Ecole*

Normale Supérieure for thirty years, and exerted a lasting influence on such epistemologists as GASTON BACHELARD (**B**), JEAN CAVAILLÈS (**B**), and ALEXANDRE KOYRÉ (**B**).

Main work: *Les étapes de la philosophie mathématique.* Paris 1912.

Secondary literature: *DSB* **2**, 544–546. *P* VI, 357; VIIb(1), 611. — Special issues of the *Revue de métaphysique et de morale* **55**, nos. 1–2 (1945), and of the *Bulletin de la Société française de philosophie* **57**, no. 2 (1963). J.P.

Bubnov, Nikolai Mikhailovich (* February 2, 1858, Kiev, Ukraina; † after July 17, 1939, in Yugoslavia?). While studying history at the University of St. Petersburg, Bubnov decided to edit the correspondence of GERBERT, later Pope SYLVESTER II (999–1003). Having collected the relevant manuscript material during a long research trip through France, England, Italy, Germany, Switzerland, and Austria (1882–1884), he was awarded his doctorate at St. Petersburg in 1891 for his study on GERBERT's letters. In 1890–1891 he taught at the *Ecole supérieure féminine* (Girls' High School) in St. Petersburg, and from 1891 to 1919 he was professor of ancient and medieval history at the University of Kiev.

Later BUBNOV turned to GERBERT's mathematical writings, especially his work on calculation with "Gerbert's" abacus, *i.e.* a special form of counting-board on which the operations were carried out with marked counters. After further travels to libraries (1894–1898) BUBNOV presented the fundamental edition of GERBERT's mathematical works in 1899, which includes not only his treatise on the abacus, but also many commentaries on it, as well as related abacus texts from the 11th and 12th centuries, critical editions of GERBERT's other mathematical and astronomical writings, and a large appendix on the transmission of the texts of the *Corpus agrimensorum* (texts of the Roman land surveyors). This collection of editions of Latin texts is based upon thorough studies of the manuscripts.

After the October Revolution BUBNOV emigrated in 1919 from Soviet Russia to Yugoslavia, where he taught ancient history at the University of Ljubljana (1920–1928). He planned to make a French translation of GERBERT's letters, but died before publishing it, at an unknown date after July 17, 1939.

GERBERT apparently used counters on his counting-board marked with the Hindu-Arabic numerals. This led BUBNOV to study the origin of our numeral system. In several lengthy articles, most of which were written in Russian (see the list in [Lattin 1933, 181–182, note (1)]), he tried to prove that this system did not come from India, but that the fundamental elements were known in classical antiquity, and that our debt is to the Greeks, not to the Hindus. While this idea is outdated today, BUBNOV's edition of GERBERT's works is still valuable.

Main Works: (Ed.) *Gerberti postea Silvestri II papae Opera mathematica (972–1003).* Berlin 1899; repr. Hildesheim 1963. — *Arithmetische Selbstständigkeit der europäischen Kultur. Ein Beitrag zur Kulturgeschichte.* Berlin 1914.

Secondary literature: *May* 92. — LATTIN, HARRIET PRATT: "The Origin of Our Present System of Notation According to the Theories of Nicholas Bubnov." *Isis* **19** (1933), 181–194. — MATVIICHINE, IAROSLAV: "Nicolas Boubnov commentateur de Gerbert: une édition non réalisé." In: *Gerbert l'Européen. Actes du colloque d'Aurillac 4–7 juin 1996* (= Société des lettres, sciences et arts "La Haute-Auvergne", Mémoires **3** (1997)), 347–361. Aurillac 1997. M.F.

Cajori, Florian (* February 28, 1859, St. Aignan, Switzerland; † August 14, 1930, Berkeley, California, USA). FLORIAN CAJORI was the first professional historian of mathematics in the United States. Although born in Switzerland, he was sixteen when he left in 1875 to join his older brother who had also emigrated to Wisconsin in the U.S. CAJORI went on to graduate from the University of Wisconsin in 1883 with a B.Sc., earning his M.Sc. in 1886. He then studied for a year and a half at The Johns Hopkins University before accepting a position teaching mathematics and applied mathematics at Tulane University (from which he earned a Ph.D. in 1894). Six years earlier, he had already accepted a position as professor of physics at Colorado College (in 1889). Beginning in 1898, he also taught as professor of mathematics, and somewhat later served as dean of the Department of Engineering from 1903 until 1918. And then, remarkably, he was called to the University of California at Berkeley, where he was named "Professor of the History of Mathematics," the first to receive such a position and title in the United States.

CAJORI's earliest historical book appeared in 1890 — *Teaching and History of Mathematics* — which D.E. SMITH has kindly described as "although quite naturally lacking in the scholarship and thoroughness displayed in the publications of the last years of his life, it laid out a line of study which was quite new in this country and which served a worthy purpose in creating an interest in the subject" [Smith 1930, 778]. This was followed in 1894 by CAJORI's *History of Mathematics*, and two years later by his *History of the Concepts of Limits and Fluxions in Great Britain from Newton to Woodhouse* (1896; repr. 1919; Japanese trans. 1929). He also published a *History of Physics* in 1899. Meanwhile, MORITZ CANTOR invited CAJORI to contribute a section (§20) on arithmetic to volume four of his monumental *Vorlesungen über Geschichte der Mathematik* (Lectures on History of Mathematics) (1909), and this represented another significant turning point in CAJORI's career.

According to SMITH, it was writing for CANTOR that finally prompted CAJORI to consult original sources, and "thus there opened a new era in his work, the era in which he found himself" [Smith 1930, 778]. Subsequently, CAJORI's *History of the Logarithmic Slide Rule* (1909) led naturally to a biography of WILLIAM OUGHTRED (1916). Meanwhile, CAJORI revised his earlier works on history of mathematics, while all the time working on his *History of the Concepts of Limits and Fluxions in Great Britain from Newton to Woodhouse* (1919) (which SMITH has called "a kind of inaugural address at the University of California" [Smith 1930, 779]). Within the next decade, CAJORI also published *The Early Mathematical Sciences in North and South America* [Cajori 1928a], and his *The History of Mathematical Notations* [CAJORI 1928/29, in two volumes]. For a bicentenary meeting in honor of ISAAC NEWTON held in New York City in 1929, CAJORI produced a major essay on NEWTON's fluxions and began to work on critical editions of NEWTON's *Principia* and *Optics*, both of which were nearly finished at the time of his death, which came suddenly due to pneumonia in August of 1930.

Not only did CAJORI occupy the first chair for history of mathematics in the United States, he served as president of the Mathematical Association of America in 1917, and as vice president of Section L for History and Philosophy of Science of the American Association for the Advancement of Science in 1923. He also served as vice president of the History of Science Society in its first year, 1924–1925 (with D. E. SMITH as president), and in 1929 was elected president of the *Comité internationale d'histoire des sciences.*

Main works: *The Teaching of the History of Mathematics in the United States.* Washington, D.C. 1890.

Secondary literature: *May* 95. *P* IV(1), 212–213; V, 195–196; VI 387. — SIMONS, LAO: "Florian Cajori." *AMM* **37** (1930), 460–462. — [SMITH 1930, with a portrait facing p. 777].— ARCHIBALD, R. C.: "Florian Cajori, 1859–1930." *Isis* **17** (1932), 384–407. J.W.D.

Cantor, Moritz (* August 23, 1829, Mannheim, Germany; † April 10, 1920, Heidelberg, Germany). CANTOR, a leading German historian of mathematics, came from Baden in south-west Germany. In Mannheim he passed the school-leaving examination at the *Gymnasium* in 1848. CANTOR studied from 1848 to 1851 in Heidelberg and Göttingen (under CARL FRIEDRICH GAUSS, WILHELM WEBER, and MORITZ A. STERN). He was especially influenced by GAUSS's lecture on the method of least squares (Winter-Semester 1850–51). Having received his Ph. D. in Heidelberg in 1851 with the dissertation *Ueber ein weniger gebräuchliches Coordinaten-System* (On a Coordinate System, Little in Use) which was based upon CARNOT's *Géométrie de position* (Geometry of Position) (1803), he continued his studies in Berlin under J. STEINER and P. G. LEJEUNE DIRICHLET (1852). In 1853 CANTOR became *Privatdozent* at the University of Heidelberg. He then gave lectures on elementary mathematics including "political arithmetic." In 1863 he became *außerordentlicher* and in 1875 *Honorar-Professor*. He retired in 1913 and died in Heidelberg on April 10, 1920.

CANTOR's interest and historical importance lay in history of mathematics rather than mathematics itself. The young CANTOR seems to have been led by ARTHUR ARNETH (1802–1858, **B**) — *Gymnasiallehrer* in Heidelberg and *Privatdozent* at the university — to the history of mathematics. In the late 1850s CANTOR was in Paris, staying with relatives. There he met the *doyen* of the history of mathematics, MICHEL CHASLES, who was impressed by the young man and published a letter that CANTOR had written to him, on ZENODORUS and his times, in the *Comptes Rendus* of the Paris *Académie des Sciences.* CHASLES also encouraged CANTOR to continue his historical work.

In the summer semester of 1860, CANTOR started to give lectures on the history of mathematics at the University of Heidelberg (Cantor was the first to begin teaching this subject at Heidelberg). In 1863–64, 1873–74, and 1874–75, CANTOR gave three-hour lectures on history of mathematics. One of his students in these early days was SIEGMUND GÜNTHER, himself later a noted historian of science. Beginning with the winter semester of 1875, CANTOR lectured regularly on the development of mathematics from the beginnings up to the 19th century, the whole cycle taking three semesters; there was no comparable course in other German universities, and indeed, Cantor's course seems to have been unique in the world. He also gave lectures on mathematics itself, e.g. on the theory of numbers, determinants, and analysis. Furthermore, he was responsible for the students of the "cameralistic disciplines" (including the study of insurance). One result of all these teaching activities was CANTOR's work on "political arithmetic" (*Politische Arithmetik oder die Arithmetik des täglichen Lebens* (Political Arithmetic or the Arithmetic of Daily Life) (1898; 2nd ed. 1903).

CANTOR's first publication on the history of mathematics was devoted to the introduction of Hindu-Arabic numerals into Europe ("Ueber die Einführung unserer gegenwärtigen Ziffern in Europa" (On the Introduction of our Present Numerals into Europe) *Z. Math. Phys.* **1** (1856), 65–74). He also contributed historical articles to the second and third volumes of the *Zeitschrift* (on RAMUS, STIFEL, CARDANO, and EUCLID's *Porisms*). From the fourth volume (1859) on, he was co-editor of the *Zeitschrift für Mathematik und Physik*, at first with EMIL KAHL and O. SCHLÖMILCH (1861–93), then with SCHLÖMILCH alone (1893–96), and finally with RUDOLF MEHMKE (1896–1900). CANTOR was responsible for the "historical-critical" part (called *Historisch-kritische Abteilung* from the second year of its existence), and himself contributed both articles and numerous reviews to the journal. Indeed, CANTOR was the moving force behind the increasing tendency of the *Zeitschrift für Mathematik und Physik* to be a journal for the history of mathematics: in 1875 he started the *Historisch-literarische Abtheilung*, and in 1877 the *Abhandlungen zur Geschichte der Mathematik*, which began as a supplement of the *Zeitschrift für Mathematik und Physik*, but became an independent publication in 1901 beginning with volume 11. Up to the end of World War I, there were only three serious periodicals for the history of mathematics anywhere in the world: the *Zeitschrift für Mathematik und Physik* and its supplements, BALDASSARRE BONCOMPAGNI's *Bullettino*, and ENESTRÖM's *Bibliotheca mathematica.*

CANTOR's first extended account of the history of mathematics was his *Mathematische Beiträge zum Kulturleben der Völker* (Mathematical Contributions to the Cultural Life of the Peoples) (1863). In this book he conjectured that the Pythagoreans had invented the Hindu-Arabic numerals, which had then been transmitted to the Middle Ages by BOETHIUS. His main argument was based on the appearance of these numerals in the so-called "Geometry II" attributed to BOETHIUS, which G. FRIEDLEIN was to publish. But in fact, this work is an 11th-century compilation. CANTOR, however, to his dying day believed that it was genuine, although FRIEDLEIN himself and later, above all, N. BUBNOV (1899), brought forward decisive arguments against its authenticity. It was through the *Mathematische Beiträge* that CANTOR came to the notice of M. CURTZE, who had found a manuscript of works from the 14th century in his school library in Thorn. Thus began a lively correspondence between CURTZE and CANTOR, which led to a life-long friendship. Another of CANTOR's close friends was BONCOMPAGNI, in whose *Bullettino* numerous articles by CANTOR (mostly translations from the German) also appeared.

In the following years, CANTOR concentrated on medieval Western mathematics. In 1865 he published a text on Hindu-Arabic arithmetic from a manuscript in Heidelberg, "Ueber einen Codex des Klosters Salem" (On a Codex of the Cloister in Salem), (*Z. Math. Phys.* **10** (1865), 1–16). A lengthy publication, *Euclid und sein Jahrhundert* (Euclid and His Century) (1867), treated the works and influence of the three great Greek mathematicians: EUCLID, ARCHIMEDES, and APOLLONIUS, although basically this book is a rather unoriginal collection of facts. CANTOR's *Die römischen Agrimensoren und ihre Stellung in der Geschichte der Feldmesskunst* (The Roman Surveyors and Their Position in the History of Surveying) (1875) was based on the fundamental edition by KARL LACHMANN *Gromatici Veteres. Die Schriften der römischen Feldmesser* (The Ancient Surveyors. Writings of the Roman Surveyors) (F. BLUME and A. RUDORFF, eds., in 2 vols., 1848–1852), which is still in use today. To this end CANTOR used not only the oldest known Latin mathematical manuscript (the so-called "Arcerianus," from the 5th or 6th century, now in Wolfenbüttel), but also a manuscript from the Benedictine monastery of St. Peter in Salzburg, which contains GERBERT's geometry and other geometrical writings as well. CANTOR investigated both the sources and influence of these treatises, especially because they played an important part in transmitting Egyptian and Greek practical geometry to the Latin Middle Ages. He also published the text of EPAPHRODITUS and VITRUVIUS RUFUS (which he misattributed to NIPSUS) which was not included in LACHMANN's collection. In the late 1870s CANTOR published further works on Hindu-Arabic arithmetic ("Gräko-indische Studien" (Greco-Indian Studies), *Z. Math. Phys., Hist.-lit. Abt.* **22**, 1877, 1–23), on GALILEO, on LEONARDO DA VINCI, and on COPERNICUS.

These may all be considered as preparation for CANTOR's monumental *Vorlesungen über Geschichte der Mathematik* (Lectures on History of Mathematics), of which the first volume (covering mathematics from its origins to 1200 A.D.), appeared in 1880 [CANTOR 1880/1908]. The work as a whole, which is partly based on CANTOR's own lectures and the results of his contemporaries' research, was very well received. Even today it is the most comprehensive history of mathematics from the beginnings to the modern period. The second volume, for the period 1200 to 1668, appeared in 1892. During the preparation of this volume it became clear that a second edition of volume I was already necessary. Eighty pages longer than the first edition, this appeared in 1894. The fascicles of the third volume, for 1668 to 1758, were published between 1894 and 1898. CANTOR was of the opinion that a fourth volume, treating the period 1758 to 1799, the year of GAUSS's dissertation, could only be written by a team of scholars. This he explained in a lecture at the Second International Mathematicians' Congress in Paris (1900). The plan took concrete form in 1904: the work was to be written by nine authors (SIEGMUND GÜNTHER, VIKTOR BOBYNIN, ANTON BRAUNMÜHL, FLORIAN CAJORI, EUGEN NETTO, GINO LORIA, VIKTOR KOMMERELL, GIUSEPPE VIVANTI, and CARL RAIMUND WALLNER). Four years later the final volume appeared under CANTOR's editorship in 1908. CANTOR also wrote the final and crowning contribution, a summary-article, "Entwicklung der Mathematik zwischen 1759 und 1799, Geschichte der Ideen in diesem Zeitraume" (Development of Mathematics between 1759 and 1799, History of Ideas in this Period). Here CANTOR put forward his own approach to the history of mathematical ideas: that they showed the essential unity of the human intellect. There were further editions of the first three volumes (vol. I, third edition 1907, unaltered reprint 1922; second edition of vol. II in 1900, and of vol. III in 1901). The four volumes (in their latest editions) total more than 3900 pages.

CANTOR tried to put all of the information available to him into his *Vorlesungen.* He not only collected masses of new material which had been discovered in the preceding decades, which he put into order, but he added his own results and fused the numerous items into a whole, according to his own plan. In the process he included many apparently unimportant details — in fact, his love of detail is apparent on every page.

Such a large and comprehensive work could not help but contain mistakes. Also, in his subjective assessments of some matters, for instance his ideas about the transmission of Hindu-Arabic numerals to the West, he tended to repeat his earliest opinions too uncritically. It is therefore easy to understand why, despite the generally enthusiastic reception of CANTOR's *Vorlesungen,* there were many who complained of mistakes. Among critics of the book was GUSTAF ENESTRÖM: having reviewed the first volume positively, he later established a special section in the journal *Bibliotheca mathematica,* which he edited, for reporting inaccuracies and outright errors. Many well-known historians of mathematics contributed to this section, which was called "Kleine Bemerkungen" (Small Remarks or Annotations). Among them were BRAUNMÜHL, CURTZE, RUDIO, SUTER, TANNERY, and WERTHEIM. As the years went by, the tone of ENESTRÖM's remarks became more hostile: his two recuring complaints were that CANTOR was not a good enough mathematician for such a large undertaking, and that he had not worked carefully enough. On the other hand, as ENESTRÖM's criticism became sharper, the number of contributors to the "Kleine Bemerkungen" correspondingly shrank. For the last volume (1913–14) before the demise of the journal at the beginning of World War I, ENESTRÖM was left to write most of the section himself. Many of the criticisms were justified, but they serve to show primarily that no comprehensive and detailed history of mathematics could be written by a single person. Moreover, ENESTRÖM's collection of criticisms was itself criticized. As the Belgian historian of mathematics HENRI BOSMANS wrote in 1914: "Comment s'y prendre pour éviter d'accréditer l'erreur dans l'histoires des mathématiques, par M. G. Eneström," ("How can one avoid acknowledging error in the history of mathematics, by M. G. Eneström") [*Revue Quest. sci.* **76**, Oct. 1914]. "Pourquoi l'érudit critique tarde-t-il tant à faire paraître son volume? C'est, il es vrai, infiniment plus difficile que de se borner à annoter des 'Vorlesungen' de Cantor" ("Why has the learned critic delayed so long in bringing out his volume? It is true that that is infinitely more difficult than merely annotating Cantor's *Vorlesungen*.")

In addition to his *Vorlesungen*, CANTOR also published several works on individual mathematicians and particular problems. The most important are: on a mathematical fragment originating in the Bobbio monastery (1881), ALBRECHT DÜRER (1888), AḤMAD IBN YŪSUF's treatise on proportions (1888), NICHOLAS OF CUSA (1889), LEONARDO DA VINCI (1890), COPERNICUS (1899), GAUSS (1899), and CARDANO (1903).

From 1875 on CANTOR was engaged in contributing to the *Allgemeine deutsche Biographie* (General German Biographical Dictionary), for which he wrote several hundred biographies of mathematicians and scientists. He also contributed to various literary journals, including the *Jenaer Literaturzeitung*, the *Münchner allgemeine Zeitung*, and the *Literarisches Zentralblatt*.

CANTOR received many honors for his academic achievements. On the occasions of both his seventieth and eightieth birthdays there appeared compendious *Festschriften*. He was a member of the academies of St. Petersburg, Turin, Vienna, and Heidelberg. In his time CANTOR was, without question, one of the leading historians of mathematics in the world. His importance lies mostly in his "Vorlesungen," which even today remain unrivalled for their comprehensiveness. MORITZ CANTOR tried to demonstrate the connections he discerned between different cultures, which he took to reflect the universality of thought which operates in the human mind. Although he was interested in everything great that he believed defined the human spirit, he was a realist and always disliked speculative philosophy.

Main works: *Mathematische Beiträge zum Kulturleben der Völker* (1863). — *Die römischen Agrimensoren und ihre Stellung in der Geschichte der Feldmesskunst* (1875). — [CANTOR 1880/1908]. — *Politische Arithmetik oder die Arithmetik des täglichen Lebens* (1898, 2nd. ed. 1903). —

Secondary literature: *DSB* **3**, 58–59. *May* 97. *P* III, 232; IV, 218; V, 201; VI, 397. — *DBJ* (1917–1920), 509–512. — *NDB* **3**, 129. — CURTZE, MAXIMILIAN: "Verzeichnis der mathematischen Werke, Abhandlungen und Recensionen des Hofrat Professor Dr. Moritz Cantor." *Abh. Gesch. Math.* **9** (1899) (= Supplement zum 44. Jahrgang der *Z. Math. Phys.*), 625–650 (with portrait). — BOPP, KARL: "Moritz Cantor †. Gedächtnisrede gehalten im Mathematischen Verein zu Heidelberg am 19. Juni 1920." *Sb. Akad. Heid., Math.-nat. Kl.*, Abt. A, Jahrgang 1920, 14. Abhandlung (16 pp.). — [CAJORI 1920/21]. — JUNGE, GUSTAV: "Zum 100. Geburtstage von Moritz Cantor (geb. 23. August 1829)." *Unterrichtsbl. Math. Nat.* **35** (1929), 239–241 (with portrait). — SMITH, DAVID EUGENE: "Moritz Cantor." *Scripta Math.* **1** (1932/33), 204–207. — [SARTON 1951]. — [LÜTZEN/PURKERT 1994, with portrait] — Portrait: [BÖLLING 1994, 18.2]. M.F.

Carra de Vaux, Bernard (* February 3, 1867, Bar-sur-Aube, France; † December 28, 1953, Nice, France). A French Arabist and professor at the *Institut catholique de Paris*, CARRA DE VAUX, in his *Les penseurs de l'Islam*, vol. 2 on geography, mathematics, and the sciences, dealt with the great astronomers and mathematicians. He published a major study of HERON OF ALEXANDRIA as known from an Arabic edition of his work by QUSṬĀ IBN LŪQĀ and one on ABŪ 'L-WAFĀ' (*Journal asiatique* **19** (1892), 408–471).

Main works: *Les penseurs de l'Islam.* 5 vols. Paris 1921–1926. — "Les mécaniques ou l'élévateur de Héron d'Alexandrie, publ. pour la première fois sur la version arabe de Qostá ibn Lûqá et traduites en français." *Journal asiatique* (9) **1** (1893), 386–472; **2** (1893), 152–269, 420–514.

Secondary literature: MILLAS, J. M.: "Carra de Vaux." *Arch. int. Hist. Sci.* **7** (1954), 61–62. J.P.

Carruccio, Ettore (* June 3, 1908, Velletri, Italy; † July 5, 1980, Bologna, Italy). After completing his secondary, classical school education, CARRUCCIO attended the University of Rome where he studied with CASTELNUOVO, SEVERI, LEVI-CIVITA, VACCA, VOLTERRA, and ENRIQUES, who were his professors in the faculty of mathematics. Under the influence of ENRIQUES and VACCA, CARRUCCIO's interests soon turned to the history and philosophy of mathematics. While still a student, he edited and translated LEIBNIZ's *Nova methodus* (New Method) into Italian in 1927. This was published, along with his subsequent translations and editions of both NEWTON's *De quadratura curvarum* (On the Quadrature of Curves) and TORRICELLI's *De infinitis spiralibus* (On Infinite Spirals), as an appendix to a new edition of GUIDO CASTELNUOVO's booklet, *Le origini del calcolo infinitesimale nell'era moderna* (The Origins of the Infinitesimal Calculus in the Modern Era) (1938; repr. 1962).

In 1930 CARRUCCIO received his degree (*laurea*) in mathematics and was appointed assistant lecturer for history of science at the University of Rome. Due to his domestic circumstances, however, he was forced to give up this (unpaid) position, and instead taught mathematics and physics in a secondary school. In 1940, after he obtained his *venia docendi* (*libera docenza*), CARRUCCIO began lecturing on descriptive geometry at the University of Modena. Following World War II, during the 1950s and 1960s, CARRUCCIO offered courses on "elementary mathematics from an advanced standpoint," as well as courses on history of mathematics, at the Universities of Modena, Turin, and Bologna. In 1975, his pioneering work in promoting the history of mathematics was officially recognized with a professorship for CARRUCCIO in the history of mathematics at the University of Bologna. This was actually the first full professorship for the subject at any Italian university.

CARRUCCIO's scientific writings are mainly devoted to the history of mathematics, the history of mathematical logic, and the foundations of mathematics. CARRUCCIO described himself as a faithful disciple of ENRIQUES, and he combined his historical research with a deep interest in the philosophy of mathematics and logic. His more than 90 publications include: *Matematica e logica nella storia del pensiero contemporaneo* (1958; Engl. trans.: *Mathematics and Logic in History and in Contemporary Thought* (1964); *Mondi della logica* (Worlds of Logic) (1971); and *Appunti di storia delle matematiche, della logica, della metamatematica* (Notes on the History of Mathematics, Logic, and Metamathematics) (1977).

Secondary literature: D'AMORE, BRUNO: "Ettore Carruccio — the Man and the Scientist." *HM* **8** (1981), 235–242 (with bibliography and portrait). U.B.

Caspar, Max (* May 7, 1880, Friedrichshafen, Germany; † September 1, 1956, Munich, Germany). CASPAR, a leading German specialist on JOHANNES KEPLER, taught mathematics in Ravensburg, Rottweil, and Cannstatt from 1909 to 1934. Beginning in 1934, he served as editor of the *Gesammelte Werke* (Collected works) of KEPLER.

Works: Editor of vols. 1, 3, 4, 5, 7, 13–16 of KEPLER's *Werke.* — German translations of KEPLER's *Mysterium cosmographicum* (1923), *Astronomia nova* (1929), and *Harmonice mundi* (1939). — *Johannes Kepler.* Stuttgart 1948; 4th ed. Stuttgart 1995; English trans.: New York 1959.

Secondary literature: *May* 101. *P* VI, 410; VIIa(1), 336. — ROSSMANN, FRITZ: "Zum 70. Geburtstag von Prof. Dr. Max Caspar." *Naturw. Rdsch.* **3** (1950), 237–238. — Anonymous obituary: *Das Antiquariat* **12** (1956), 225. — LEIBBRAND, WERNER: "Max Caspar († 1. 9. 1956, München)." *Arch. int. Hist. Sci.* **10** (1957), 89. — VOLK, OTTO: "Max Caspar (1880–1956)." *Jber. DMV* **62** (1960), 93–98 (with bibliography). M.F.

Cassina, Ugo (* April 1, 1897, Polesine Parmense, Italy; † October 5, 1964, Milan, Italy). Forced to interrupt his studies because of World War I (during which he distinguished himself by obtaining two medals for military valor and two war-crosses), CASSINA graduated (*laurea*) in mathematics from the University of Turin in 1921, where he studied with GIUSEPPE PEANO. PEANO exerted a tremendous influence on CASSINA. According to HUBERT KENNEDY (1980, 172), CASSINA was the "most knowledgeable and the most devoted" pupil, whose "admiration and affection for Peano lasted a lifetime." In 1922 CASSINA became TOMMASO BOGGIO's assistant for analytic geometry, and PEANO's for algebraic analysis. The following year CASSINA became director of the School of Algebraic Analysis at the University of Turin (see [KENNEDY 1980, 156]), and in 1925 he was named professor at the Naval Academy in Livorno. After becoming an instructor (*libero docente*) for algebra and analytic geometry, he was charged with teaching higher geometry, elementary mathematics from an advanced standpoint, and history of mathematics at the University of Milan.

In 1941 CASSINA obtained a professorship in analytic geometry at the Military Academy in Caserta, and in 1948 was named professor for the same subject at the University of Pavia. Eventually, in 1951, he obtained a chair for "complementary mathematics" at the University of Milan. He was also elected a member of the *Académie Internationale d'Histoire des Sciences.* His more than one hundred publications include works on geometry, analysis, mathematical logic, and history of mathematics. Among his major publications are several books: *Calcolo numerico* (Numerical Calculus) (1928) and, of historical interest, *Critica dei principi della matematica e questioni di logica* (Critique of the Principles of Mathematics and Questions of Logic) (1961), and *Dalla matematica egiziana alla matematica moderna* (From Egyptian Mathematics to Modern Mathematics) (1961). Both of these include papers devoted to the life and work of PEANO. As a faithful, if at times over-zealous student of PEANO, CASSINA also devoted himself to PEANO's *Academia pro Interlingua*, for which he served as editor of the Academy's journal, *Schola et vita*, founded in 1926. In 1956 the Italian Mathematical Union (UMI) appointed CASSINA as editor of PEANO's *Opere scelte* (Selected Papers), which appeared in three volumes between 1957 and 1959. CASSINA also edited, with introduction and notes, a facsimile edition of the 1908 version of PEANO's *Formulario mathematico* (Mathematical Formulary), which — as CASSINA said repeatedly — is "an inexhaustible mine of science" (quoted in [KENNEDY 1980, 121]).

Secondary literature: *P* VI, 411; VIIb(2), 738. — MANARA, CARLO FELICE: "Ugo Cassina." *Rend. Ist. Lomb. Sci. Lett. Parte Generale* (3), **98** (1964), 108–111 (with portrait). — [KENNEDY 1980]. U.B.

Castro Freire, Francisco de (* September 23, 1809, S. Silvestre do Campo, Coimbra, Portugal; † March 10, 1884, Nisa, Portugal). CASTRO FREIRE joined the Faculty of Mathematics in Coimbra after his graduation in 1830. In 1838/39 he issued a translation of a course in pure mathematics by LOUIS BENJAMIN FRANCOEUR, and in later years he published works on rational mechanics and geometry. Beginning in 1863, he collaborated in the compilation of the *Portuguese Astronomical Almanachs.* Apart from his important *Historical Memoir* of 1872 (see Section 14.3), CASTRO FREIRE published an article in *O Instituto* on "Mathematics in the First Two Dynasties" [FREIRE 1884]. He became a sub-dean of the University of Coimbra in 1876, and president of the scientific and literary academy, *O Instituto*, in 1877.

Secondary literature: *GEPB* **6**, 259. L.S.

Cavaillès, Jean (* May 15, 1903, Saint-Maixent, France; † 1944, Arras, France). JEAN CAVAILLÈS was born into a Protestant family in St. Maixent. His father, an officer, taught geography at the *École militaire.* Young JEAN's early education took place more at home than at school; his father's frequent changes of orders effectively prevented any regular schooling. CAVAILLÈS learned early to work hard and in isolation.

After receiving a double *baccalaureat* in mathematics and philosophy from the *lycée* in Bordeaux in 1920 with an honorable mention, CAVAILLÈS entered the *Lycée Louis-le-Grand* in Paris to prepare for the entrance examination for the *Ecole Normale Supérieure (ENS)*, in the course of which he received his *licence* in philosophy. In the summer of 1921, his father sent him to Germany for several weeks to improve his knowledge of the language. This first contact with the homeland of IMMANUEL KANT, "one of mankind's greatest thinkers," as the young man of eighteen wrote to his parents, occurred after his vocation for philosophy had already been affirmed.

In 1923 he entered the *Ecole Normale Supérieure*, where he studied Greek, began his study of mathematics, and took courses in philosophy under EMILE

Bréhier and Léon Brunschvicg. Under the influence of the latter, he became interested in history and philosophy of mathematics, and after four years of study, Cavaillès received the *agrégation* in philosophy and the *licence* in mathematics. In October of 1927, while visiting Germany, at the *Staatsbibliothek* in Berlin he read Felix Klein's work on the history of mathematics, which made a great impression. As a result, Cavaillès was determined to write a thesis on the history of set theory. Created by Georg Cantor, the theory had been attacked by Leopold Kronecker, but defended by Richard Dedekind and David Hilbert. The paradoxes of set theory which Cantor himself had discovered, along with related results due to Ernst Zermelo and Bertrand Russell, shook set theory at its very foundations and raised questions as to its legitimacy.

At about this same time, Cavaillès read the works of the German philosopher Edmund Husserl, who had developed, during the years from 1901 to 1916 when he held a Göttingen professorship, close ties with Hilbert. In public philosophical discussions that took place during a university summer course in Davos (Switzerland) in 1929, Cavaillès encountered the thought of Ernst Cassirer, the noted representative of neo-Kantianism, and Martin Heidegger, the rising star of phenomenology. In the autumn of 1929, Cavaillès was in Tübingen to study the manuscripts of the mathematician Paul Du Bois Reymond, who was also interested in the mathematical infinite. Cavaillès's investigations in the history of mathematics and his philosophical interests were thus tightly intertwined; for both, he oriented himself toward Germany.

In 1929–30, Cavaillès received a Rockefeller grant to pursue research for his thesis in Germany. After spending some time in Berlin and Hamburg, he went to Göttingen. Abraham Fraenkel had told him of the existence there of a part of the unedited correspondence between Cantor and Dedekind, which Cavaillès eventually edited and published with the assistance of the German mathematician Emmy Noether. Meanwhile, Cavaillès did not forget his interests in philosophy, and also attended courses of Husserl and Heidegger at Freiburg.

Upon returning to Paris, he taught (as an *agrégé répétiteur*) at the *Ecole Normale*, where he offered courses on philosophy to science students, and courses on mathematics to philosophy students. Given a place in 1936 at the *lycée* in Amiens, he there met Lucie Aubrac, professor of history, who later would recruit him for the Resistance in France. After further study at the Sorbonne and having defended his theses in 1938 (these were later published by Hermann), he went to Strassbourg as Professor, where he got to know Henri Cartan, André Weil, and Charles Ehresmann, founding members of the Bourbaki group.

Having become decoding officer at Army Headquarters early in World War II, Cavaillès was captured by the Germans in 1940, but later escaped and fled to the southern zone. In October, he returned to his position at the University of Strasbourg, which was moved back to Clermont-Ferrand. There, in 1941, he founded, together with Emmanuel d'Astier de la Vigerie, the resistance movement, *Libération Sud*, which Lucie Aubrac joined. Called to the *Sorbonne* as an acting lecturer in logic and the methodology of science, he helped direct the *Libération*

Nord. After he had created an intelligence network in 1942, he was arrested and interned at the camp at Saint-Paul d'Eyjaux. CAVAILLÈS escaped again at the end of 1942. In 1943, when he was relieved of his duties as professor at the *Sorbonne* by the Vichy government, he left France for London, where he met RAYMOND ARON, creator of the review *La France libre.* Returning to France, he was arrested in August of 1943 and taken to Fresnes. In January 1944, he was condemned to death and executed at an unknown date in the prison at Arras.

CAVAILLÈS's last manuscript was later published by GEORGES CANGUILHEM and CHARLES EHRESMANN under the title *Sur la logique et la théorie de la science.* His theses reflected a considerable understanding of the history of mathematics after the beginning of the 19th century, and a true appreciation of logic, including the results of GÖDEL of 1931 and his work on the continuum hypothesis.

Secondary literature: RAMÍREZ, SANTIAGO: "La obra de Jean Cavaillès." *Mathesis* (Mexico) **1**, no. 1, (1985), 149–158 (bibliographic essay). — SINACEUR, HOURYA: "L'épistémologie de Jean Cavaillès." *Critique* **461** (1985), 964–988. — SINACEUR, HOURYA: "Structure et concept dans l'épistémologie mathématique de Jean Cavaillès." *Rev. Hist. Sci.* **40** (1987), 5–30. — SINACEUR, HOURYA: *Jean Cavaillès. Philosophe mathématique.* Paris 1994. — GRANGER, GILLES GASTON: "Jean Cavaillès et l'histoire." *Rev. Hist. Sci.* **49** (1996), 569–582. — Portrait in CAVAILLÈS, JEAN: *Sur la logique et la théorie de la science.* Paris 1947, 4th ed. 1987. H.S.

Chace, Arnold Buffum (* November 10, 1845, Valley Falls, Rhode Island, USA; † February 28, 1932, Providence, Rhode Island, USA). CHACE was a self-made businessman who graduated from Brown University with a B.A. degree in 1866. Later, he served as chancellor of Brown University for 28 years. CHACE maintained a life-long interest in mathematics, which was reflected in the office he held, as treasurer of the Mathematical Association of America (MAA), for 18 years. CHACE became fascinated with Egyptian mathematics following a trip to Egypt in 1910. Subsequently, he undertook an intensive study of the Rhind mathematical papyrus and financed a scholarly edition, to which he himself contributed, published by the MAA in two volumes [CHACE 1927/29].

Secondary literature: *May* 106. — ARCHIBALD, RAYMOND C.: "Arnold Buffum Chace." *AMM* **40** (1933), 139–142. J.W.D.

Chasles, Michel (* November 15, 1793, Épernon, France; † December 18, 1880, Paris, France). Following his studies at the *Ecole Polytechnique* from 1812 to 1815, CHASLES experienced wartime service and tried his hand at business before returning to the scientific studies that would earn him the professorship of geodesy and theory of machines that he held at his *alma mater* from 1841 to 1851. In 1846, he also accepted the chair of higher geometry created for him in the Faculty of Science at the University of Paris, which he held until his death. If most of his principal publications — works on geometry — reflect teaching duties which required him to edit a new style of text, they attested equally to the various facets of his interest in the history of mathematics, especially following the publication in 1837 of his *Aperçu historique* [CHASLES 1837]. Four such facets are considered here.

First, his taste for collecting historical documents betrayed a flawed historian, a victim targeted by forgers who even succeeded in selling him letters in French reputedly written by CLEOPATRA [Koppelman 1971].

Second, he was interested in the mathematics "of the Indians [and] of the Arabs," which translations by FREDERIC ROSEN, HENRY THOMAS COLEBROOKE, and others from the early decades of the 19th century had opened up more fully to European audiences. CHASLES's main interest in this subject lay in questions of priority as evidenced by the history of arithmetic that he published in 1843. There, he argued for Pythagorean, as opposed to Indian, origins of the decimal place-value number system [Koppelman 1971], [RAINA 1999, chapter 6].

Third, CHASLES focused on the restoration of lost texts. Such efforts had been increasingly undertaken since the 16th century, and CHASLES traced the lineage of his most complete work in this vein, the *Porisms* of EUCLID, in note XII in his *Aperçu historique.* There, he explicitly enunciated the principles of the genre as he understood them (pp. 14–15) and presented EUCLID as a precursor of projective geometry. There are reasons to believe that he was better inspired when he attempted to reconstruct, on the basis of secondary sources, the contents of DESARGUES's writings; more recently recovered documents allow us to verify the soundness of his intuitions here [BKOUCHE 1994].

Fourth, although of unequal value, both of the latter efforts reflect the main features of CHASLES's studies in the history of mathematics: their intimate connection with his own researches in geometry. The continuity is evident in the facts themselves. The *Aperçu historique* represented the publication of the writings that won for CHASLES on May 16, 1830, a mathematical competition instigated by QUETELET and sponsored by the Royal Belgian Academy.

As announced, the competition called for "a philosophical examination of the different methods used in recent geometry and particularly of the method of polar reciprocals." On the manuscript he submitted (which awaits study in the archives of the Academy), additions and erasures reflect the process that led to the introduction, originally restricted in length, to the two prize-winning memoirs; it was this introduction that ultimately gave the book its title (see CHASLES's preface to the second edition, 1875, pp. i–ii). Moreover, the celebrated notes mixed contemporary geometrical developments and historical results. The continuity between CHASLES's mathematical and his historical interests was especially evident, however, in his seeming subjection of history to the ends of contemporary reflection on mathematics. His work anchored him in the tradition, inspired by GASPARD MONGE and LAZARE CARNOT, that aimed to transform geometry in such a way as to allow it to become the equal, the mirror, of analysis, indeed, to supplant analysis in the treatment of geometrical problems. This program, which JEAN VICTOR PONCELET had already explicitly enunciated from an historical perspective at the beginning of his *Traité de propriétés projectives des figures* (Treatise on the Projective Properties of Figures) (1822), hinged on identifying exactly what it was that conferred this privileged status on analysis and to find the means to transfer those same properties to geometry. Generality proved key, and this oriented the historical presentation. If generality was what had distinguished analysis, then what did geometry since classical antiquity share of this property? What form does generality take in the latter discipline, and what advantages can be expected from it? What methods reveal it? Animated by these questions, CHASLES cut, and he selected. For him, history was at the same time a reservoir of ideas and an instrument for reflection on the actual state of things. History went hand in hand with a mathematical practice, peculiar to the early 19th century, that oriented itself based on philosophical questions. In this sense, one might say that the predominant role of philosophical questioning in CHASLES's approach to history was merely a reflection of the practice of mathematics in his day, a practice for which the Royal Belgian Academy served as a mouthpiece in formulating, as it did, the prize competition announcement.

Main works: [CHASLES 1837] — *Les trois livres de Porismes d'Euclide, rétablis pour la première fois, d'après la notice et les lemmes de Pappus, et conformément au sentiment de R. Simson sur la forme des énoncés de ces propositions.* Paris 1860. — *Rapport sur les progrès de la géométrie.* Publication faite sour les auspices du Ministère de l'Instruction Publique. Paris 1870.

Secondary literature: *DSB* **3**, 212–215. *May* 106. *P* I, 423; III, 263. — LORIA, GINO: "Michel Chasles et la teoria delle sezioni coniche." *Osiris* **1** (1936), 421–450 (with bibliography and portrait). — [RAINA 1999]. K.C.

Christensen, Sophus Andreas (* September 18, 1861, Holbæk, Denmark; † January 3, 1943, Copenhagen, Denmark). CHRISTENSEN graduated in mathematics from the University of Copenhagen in 1884, received his Ph.D. in 1895, was a teacher at *Odense Katedralskole* 1890–1909, and head of *Nykøbing Katedralskole* 1909–1930. In 1888 he won a prize from the University of Copenhagen for an essay on the tenth book of EUCLID's *Elements.* His doctoral dissertation was devoted to *Matematikens Udvikling i Danmark og Norgei det XVIII Aarhundrede* (The Development of Mathematics in Denmark and Norway in the 18th Century) (1895). In 1919 he published a Danish translation of *The Sand-Reckoner* by ARCHIMEDES and *On Sizes and Distances* by ARISTARCHUS.

Secondary literature: *DBL* **3**, 285. K.A.

Colebrooke, Henry Thomas (* June 15, 1765, London, England; † March 10, 1837, London, England). Son of SIR GEORGE COLEBROOKE — the head of a wealthy banking firm, Member of Parliament, and (from 1769) chairman of the Board of Directors of the East India Company — HENRY T. COLEBROOKE was tutored at home in classical languages, French, German, and mathematics, which became a "ruling passion" [DNB **11** (1887), 282]. In 1782 he began thirty-two years of service to the East India Company. In his role as a provincial magistrate, he undertook serious study of Sanskrit in order to read Hindu law books. This led to his *Digest of Hindu Law* (1798, in 4 vols.). Subsequently appointed to the Court of Appeal at Calcutta (1801), he was also appointed professor of Hindu Law and Sanskrit at the College of Fort Williams (1805), where he did not teach but undertook his *Sanskrit Grammar* (1805). He also wrote many essays on the Sanskrit and Pracrit languages, poetry, Indian religious ceremonies, and an important "Essay on the Vedas." In October of 1814 COLEBROOKE returned to England, where he completed a volume on mathematics in India. He helped found the Royal Asiatic Society in 1823, and in 1824 served as president of the Royal Astronomical Society. He was also a member of the Geological and Linnean Societies, and a foreign member of the French Institute and the Imperial Academy of St. Petersburg.

Main works: *A Grammar of the Sanskrit Language.* Calcutta 1805. — *Dictionary of the Sanskrit Language.* Serampore 1808. — *Algebra, with Arithmetic and Mensuration, from the Sanskrit of Brahmegupta and Bhāscara.* London 1817. — *Miscellaneous Essays.* 2 vols., London 1857; rev. ed., 1873.

Secondary literature: *P* I, 462. — COLEBROOKE, T. E.: *The Life of H. T. Colebrooke.* London 1873, vol. 1 of 3, including the revised edition of COLEBROOKE's *Miscellaneous Essays.* 2 vols., London 1857. The edition of 1873 also includes COLEBROOKE's essay, "Indian Weights and Measures," which originally appeared in *Asiatic Researches* **5** (1798), 91–108. — Portrait: An engraving of a portrait bust of Colebrook serves as a frontispiece to [T. E. Colebrooke 1873, vol. 1]. J.W.D.

Coolidge, Julian Lowell (* September 28, 1873, Brookline, Massachusetts, USA; † March 5, 1954, Cambridge, Massachusetts, USA). COOLIDGE, an American mathematician, taught at Harvard University, having studied abroad (from 1902–04) with CORRADO SEGRÈ at Turin and EDUARD STUDY at Bern. Although an expert on geometry, especially non-Euclidean and complex geometries, he also wrote important works on the circle and the sphere, on algebraic plane curves, and an introduction to mathematical probability (1925). Upon his retirement from Harvard University, his interests turned to history, and he published three influential works: *A History of Geometrical Methods* (1940), *A History of the Conic Sections and Quadric Surfaces* (1943), and *The Mathematics of Great Amateurs* (1949). Whereas the first two of these works are basically internalist accounts of the development of technical mathematics, COLLIDGE's study of "great amateurs" is of a more expository character.

Secondary literature: *DSB* **3**, 399. *May* 113. *P* V, 240; VI, 475; VIIb(2), 894. — *DAB, Supp.* **5** (1951–1955), 129–131. — STRUIK, DIRK J.: "J. L. Coolidge (1873–1954)." *AMM* **62** (1955), 669–682 (with bibliography). J.W.D.

Cossali, Pietro (* June 29, 1748, Verona, Italy; † December 20, 1815, Padua, Italy). Son of a noble family, COSSALI first studied at a Jesuit college and then privately at home with a Theatine tutor as his personal instructor. In 1768 COSSALI moved to Milan where he also became a member of the Theatine order. Attracted by philosophy, mathematics, and physics, COSSALI declined to teach canon law (for which he had been offered a position by the University of Padua), and instead returned to Verona where he founded a private, scientific academy in 1778 for the promotion, in particular, of experimental physics.

As a rather polemical figure, COSSALI had numerous disagreements with various Italian scientists during this period, including ANTONIO MARIA LORGNA, who was the director of the Military School in Verona. In turn, LORGNA opposed COSSALI's admission to the *Società Italiana di Scienze*, founded by LORGNA himself in 1782. Eventually, however, COSSALI was admitted as a member of this learned society (in 1793).

Meanwhile, COSSALI had moved to Parma, where he had been named professor of theoretical physics at the local university in 1787. When this professorship was changed into a chair for astronomy, COSSALI's scientific interests apparently shifted to astronomy as well, as reflected in the seven volumes of ephemerides he published, and for which he provided introductory essays, beginning in 1791.

COSSALI's first contribution to the history of mathematics was an article of 1779 on CARDANO's method for solving the cubic equation in the irreducible case. In the early 1790s, COSSALI began his investigation of the history of algebra, which culminated in his celebrated *Origine, trasporto in Italia, primi progressi in essa dell'algebra. Storia critica di nuove disquisitioni analitiche e metafisiche arricchita* (Origins, Transmission to Italy, and Early Development of Algebra There), (in 2 vols., 1797–1799). According to UGO BALDINI (DBI **30** (1985), 107), this work is "a masterpiece of the history of science in the seventeenth century." Although COSSALI was very much influenced by MONTUCLA's *Histoire des mathématiques*, he did not hesitate to criticize the *Histoire* by pointing out various inaccuracies and mistakes. For example, COSSALI established exactly the date of 1202 for the *Liber Abaci* by LEONARDO OF PISA (FIBONACCI), and he emphasized LEONARDO's role in "transporting" algebra to Italy. COSSALI also tried to establish the innovativeness of the Italian development of algebra in comparison to the Greek and Arabic traditions. Although outdated, COSSALI's work is still a valuable source, particularly for the medieval and Renaissance periods.

In 1806 COSSALI was named professor of higher calculus at the University of Padova by NAPOLEON. Over the following years, he continued his historical research, as reflected in his manuscripts (which are preserved in the public library of Verona), some of which were published by BONCOMPAGNI in 1857. In his introduction, BONCOMPAGNI lists some 30 sources containing biographical information about COSSALI. The book itself includes manuscripts by COSSALI on LEONARDO OF PISA and LUCA PACIOLI, along with excerpts from their works; COSSALI's lectures on arithmetic and papers on NICCOLÒ TARTAGLIA are also included. COSSALI's manuscript devoted to *La storia del caso irriducibile* (The History of the Irreducible Case), which apparently was ready for publication when COSSALI died, has been edited by ROMANO GATTO (1996)

Secondary literature: May 115. *P* I, 482. — *DBI* **30** (1985), 104–109. — BONCOMPAGNI, BALDASSARRE: *Scritti inediti del P. D. Pietro Cossali chierico regolare teatino*. Roma 1857. — FRANCI, RAFFAELA: "Pietro Cossali, storico dell'algebra." In: [BARBIERI/DEGANI 1989, 199–217]. U.B.

Costabel, Pierre (* October 24, 1912, Draguignan, France; † November 20, 1989, La Varenne Saint-Hilaire, France). Trained in physics at the *Ecole Normale Supérieure*, PIERRE COSTABEL was not only a French historian of the mathematical sciences, but was an ordained priest. He was responsible, along with RENÉ TATON, for developing the history of exact sciences at the *Centre Alexandre Koyré*, founded in 1958 in Paris. His main field of research was mechanics in the 17th and 18th centuries. He also did remarkable editorial work, and produced editions of works by DESCARTES, FLORIMOND DE BEAUNE, LEIBNIZ, JOHANN I BERNOULLI, MALEBRANCHE, and VARIGNON.

On June 20, 1983, COSTABEL was elected a corresponding member of the Parisian Academy of Sciences, one of the few historians of science to be accorded this honour since 1900. At an international level, he was *secrétaire perpétuel* of the *Académie internationale d'histoire des sciences* (1965–1983).

Main works: *Leibniz et la dynamique, les textes de 1692.* Paris 1960. — *Nicolas Malebranche, Mathematica (= Œuvres complètes de Malebranche* XVII/2). Paris 1979. — *Démarches originales de Descartes savant.* Paris 1982.

Secondary literature: TATON, RENÉ: "Pierre Costabel (1912–1989)." *Rev. Hist. Sci.* **43** (1990), 297–311. — TATON, RENÉ, and B. BILODEAU: "Liste des publications de Pierre Costabel." *Rev. Hist. Sci.* **43** (1990), 313–324. — PEIFFER, JEANNE: "*Les leçons de l'histoire.* Hommage à Pierre Costabel, * 24 octobre 1912, † 20 novembre 1989." *Stud. Leibnitiana* **23** (1991), 127–132 (with portrait facing p. 124, and additions to the bibliography). — HALL, RUPERT A.: "Pierre Costabel, 1912–1989." *Arch. int. Hist. Sci.* **41** (1991), 105–108. — TATON, RENÉ: "Pierre Costabel (1912–1989)." *HM* **18** (1992), 292–295. J.P.

Couturat, Louis (* January 17, 1868, Paris, France; † August 3, 1914, between Ris-Orangis and Melun, France). COUTURAT was trained in philosophy at the *Ecole Normale Supérieure (ENS)* in Paris, where he improved his knowledge of mathematics by following the courses of JULES TANNERY, who was then a member of the science faculty of the university, as well as courses taught by EMILE PICARD and CAMILLE JORDAN. His thesis (1896), devoted to a study of the mathematical infinite, contributed to the spread of transfinite set theory created by GEORG CANTOR.

COUTURAT lectured on mathematical philosophy at the University of Caen and taught a history of logic course at the *Collège de France* (1905–1906). Earlier (1900–1901) he undertook important archival research in the famous Leibniz archives in Hannover and published new material on LEIBNIZ's logic.

Main works: *De l'infini mathématique.* Paris 1896. — *La logique de Leibniz.* Paris 1901. — *Opuscules et fragments inédits de Leibniz.* Paris 1903. *L'Œuvres de Louis Couturat.* Paris 1983.

Secondary literature: *DSB* **3**, 455. *May* 118. — LALANDE, ANDRÉ: "L'œuvre de Louis Couturat." *Revue de métaphysique et de morale* **22**, suppl. (Sept. 1914), 644–688 (with bibliography and portrait). — SCHMID, ANNE-FRANÇOISE: "La correspondance inédite entre Bertrand Russell et Louis Couturat." *Dialectica* **37** (1983), 75–109. *Louis Couturat – de Leibniz à Russell* .... Paris 1983. J.P.

Curtze, Maximilian (* August 4, 1837, Ballenstedt near Halle, Germany; † January 3, 1903, Thorn, West Prussia, Germany). MAXIMILIAN CURTZE was the fourth son of a physician who lived in Ballenstedt near the Harz mountains. After having graduated from the *Gymnasium Carolinum* in Bernburg, he studied in Greifswald from 1857 to 1860. There he was influenced by the professor of mathematics JOHANN AUGUST GRUNERT, who took a fatherly interest in him. GRUNERT is known as the founder of the journal *Archiv der Mathematik und Physik* (generally known as *Grunert's Archiv*), which published historical articles at an early date.

After teaching at the *Höhere Bürgerschule* in Lennep, Rhine province (1861–1864), CURTZE began to teach in 1864 at the *Gymnasium* in Thorn in West Prussia, where he lived up to his death. In 1879 he became *Oberlehrer* and in 1887 professor. After teaching for 30 years, he retired in 1894.

CURTZE was largely responsible for founding in Thorn (the birth-place of COPERNICUS) the *Coppernicus-Verein für Wissenschaft und Kunst* (Copernicus Association for Science and Art). In its *Mitteilungen*, he published the *Inedita Copernicana*, which were based upon manuscripts he had studied in Berlin, Frauenburg, Uppsala, and Vienna (1878, 1882). CURTZE was the one primarily responsible for the edition of *De revolutionibus* (On the Revolutions) which was published in Thorn in 1873 on the occasion of the 400th anniversary of the birth of COPERNICUS. CURTZE's edition was based upon COPERNICUS' autograph. Other publications by CURTZE on COPERNICUS include *Reliquiae Copernicanae* (Copernican Remains) (1875) and a biography of COPERNICUS (1899).

Even more important were CURTZE's contributions to medieval Western mathematics. Given his unusual competence not only in classical languages, but also in French and Italian, he was well equipped to understand the mathematical writings of classical antiquity, the Middle Ages, and modern times. At GRUNERT's suggestion he translated several modern mathematical textbooks from Italian into German, for example works by BELTRAMI, BRIOSCHI, and CREMONA. But the Latin Middle Ages, which had been little investigated at that time, became CURTZE's main interest. A mathematical manuscript in the school library in Thorn which contained writings of NICOLE ORESME, BRADWARDINE, and the BANŪ MŪSĀ, induced CURTZE to study these texts. In 1864 he came into contact with MORITZ CANTOR, and this was the beginning of a friendship which lasted up to CURTZE's death. Over the next four decades CURTZE edited a large number of medieval mathematical texts, mostly for the first time and from manuscripts up to then unknown, among the most important of which were: NICOLE ORESME, *Algorismus proportionum* (1868); BANŪ MŪSĀ, *Liber trium fratrum* (1885); (Ps.) JORDANUS NEMORARIUS, *De triangulis* (1887); JOHANNES DE SACROBOSCO, *Algorismus*, with the commentary of PETRUS DE DACIA (1897); AL-NAYRĪZĪ, commentary on EUCLID I–X (1899); medieval texts on trigonometry (1900); SAVASORDA,

Liber embadorum (1902); REGIOMONTANUS' correspondence (1902); LEONARDO MAINARDI, *Practica geometriae* (1902); the *Algebra* attributed to "Initius Algebras" (1902).

In addition to such editions, CURTZE also wrote numerous articles, mostly on mathematics in the Western Middle Ages. *Inter alia* he studied the approximations to square and cube-roots, the history of the Chinese *Ta Yen* rule for simultaneous indeterminate equations, the theory of numbers, and trigonometry.

Through CURTZE's works knowledge of mathematics in the West was enormously extended: a series of authors (e.g. JOHANNES DE MURIS, DOMINICUS DE CLAVASIO, ROBERTUS ANGLICUS, JOHANNES DE LINERIIS) and their works were first evaluated by CURTZE, in some cases even rediscovered. In his time he was certainly the most important researcher in the field of Western mathematics in the Middle Ages. There are, however, aspects of his works which may be criticized: in contrast to his younger contemporary A. A. BJØRNBO, CURTZE did not trouble himself with questions related to the transmission of the texts that he edited; nor did he try to find the best manuscripts for his editions. In fact, almost all of his editions relied on only one manuscript, the one which happened to be at hand. This is easily understandable since prior to 1896, CURTZE had visited only a very few libraries. After his study-tour of libraries in middle and southern Germany, Austria, and Poland, the manuscripts which he needed were for the most part sent to him at Thorn. Since CURTZE prepared his editions with enormous speed, and in most cases without any commentary, it is clear that many of his editions contain faulty readings and misprints, Moreover, CURTZE knew no Arabic and was not in contact with an Arabist who could have helped with his editions of translations from the Arabic. In spite of these weaknesses, however, CURTZE's output represents a great advance over the work of most of his predecessors.

Further publications: "Ueber die Handschrift R. 4°. 2, Problematum Euclidis explicatio der Königl. Gymnasialbibliothek zu Thorn." *Z. Math. Phys.*, Supplement, **13** (1868), 45–104. — "Ein Beitrag zur Geschichte der Algebra in Deutschland im fünfzehnten Jahrhundert." *Abh. Gesch. Math.* **7** (= *Z. Math. Phys.*, Suppl. zum 40. Jahrgang) (1895), 31–74. — "Ueber eine Algorismus-Schrift des XII. Jahrhunderts." *Abh. Gesch. Math.* **8** (1898), 1–27. — "De inquisicione capacitatis figurarum. Anonyme Abhandlung aus dem fünfzehnten Jahrhundert." *Abh. Gesch. Math.* **8** (1898), 29–68. — "Practica Geometriae. Ein anonymer Tractat aus dem Ende des zwölften Jahrhunderts." *Mh. Math. Phys.* **8** (1897), 193–224; **9** (1898), 266–268. — "Die Abhandlung des Levi ben Gerson über Trigonometrie und den Jacobstab." *Bibl. Math.* (2) **12** (1898), 97–112. — "Eine Studienreise." *Centralbl. Bibl.* **16** (1899), 257–306. — "Urkunden zur Geschichte der Trigonometrie im christlichen Mittelalter." *Bibl. Math.* (3) **1** (1900), 321–416.

Secondary literature: *DSB* **3**, 512. *May* 118. *P* III, 317; IV, 288; V, 256. — CANTOR, MORITZ: "Maximilian Curtze †." *Jber. DMV* **12** (1903), 357–368 (with portrait). — GÜNTHER, SIEGMUND: "Maximilian Curtze." *Bibl. Math.* (3) **4** (1903/04), 65–81 (with portrait). — FAVARO, ANTONIO, in *Atti Adunanze Ist. Veneto Sci. Lett. Arti* **63** (1904), 377–395. M.F.

Czwalina, Arthur Gottlieb (* May 5, 1884, Posen, Prussia, Germany; † April 28, 1964, Berlin, Germany). ARTHUR CZWALINA studied mathematics, physics and geography at the universities of Berlin and Marburg. He was a teacher of mathematics in Berlin (1910), Allenstein (East Prussia) (1919), Gumbinnen (East Prussia) (1923–1937), and Niesky (Oberlausitz) (1939–1945, and again in the early 1950s). Interested mainly in Greek mathematics, CZWALINA translated into German, with commentaries, the writings of ARCHIMEDES (1922–1925) (published in the series *Ostwalds Klassiker*), APOLLONIUS (1926), CLEOMEDES (1927), AUTOLYCUS (1931), THEODOSIUS (1931), and DIOPHANTUS (1952).

Secondary literature: *P* VI, 507; VIIa(1), 371. — HOFMANN, JOSEPH EHRENFRIED: "Arthur Czwalina †." *Prax. Math.* **6** (1964), 214. — KNABE, PAUL: "Dr. Arthur Czwalina †." *Sudh. Arch.* **49** (1965), 435–437. M.F.

da Cunha, Pedro José (* June 8, 1867, Lisbon, Portugal; † February 4, 1945, Lisbon, Portugal). PEDRO JOSÉ DA CUNHA, after graduating from the Polytechnic School in Lisbon, continued his studies at the Army School (also in Lisbon), from which he graduated with a degree in military engineering in 1891. Upon joining the faculty at the Polytechnic School in 1896, he not only taught astronomy, but revised the courses that were offered, developed the teaching of the solar system, and introduced new courses such as the theory of errors and astrophysics. Beginning in 1914, he lectured on the infinitesimal calculus, and in mathematics published several papers on the theory of series, curves, and surfaces. Among his historical works are a book on the Polytechnic School of Lisbon [CUNHA 1937], two papers on the introduction of the concepts of limit and of the infinitely small [CUNHA 1940a], and on the evolution of the teaching of differential calculus at the Polytechnic School of Lisbon [CUNHA 1940b].

Main works: [CUNHA 1929a], [CUNHA 1929b], [CUNHA 1940a], [CUNHA 1940c].

Secondary literature: *GEPB* **8**, 269 (with portrait). — SANTOS ANDREIA, E. I.: "A IV^a Cadeira e os seus Professores," in: *Faculdade de Ciências de Lisboa. Primeiro Centenário da Escola Politécnia de Lisboa 1837–1937*, (portrait after p. 24). Lisboa 1937. — [GONÇALVES 1966]. L.S.

Dahlin, Ernst Mauritz (* August 4, 1843, Handog, Sweden; † February 24, 1929, Hälsingborg, Sweden). DAHLIN graduated in mathematics from Uppsala University in 1875 with the thesis *Bidrag till de matematiska vetenskapernes historia i Sverige före 1679* (Contributions to the History of Mathematics in Sweden Before 1679). When it was suggested he be offered a position for history of mathematics at Uppsala University, this was considered too specialized, and so he was offered one for mathematics instead. Dahlin continued his historical research only for a brief period, and then left the university for financial reasons.

Secondary literature: *SBL* **9**, 724–726 (with portrait). K.A.

Datta, Bibhutibhusan (* June 28, 1888, Kanungopara (Chittagong), now in Bangladesh; † October 6, 1958, Pushkar (Rajasthan), India). DATTA was a science graduate of Calcutta University from which he obtained his master's degree in mathematics in 1914, and the D.Sc. in 1921 for work in applied mathematics (especially hydrodynamics). He served as a lecturer in the University Science College and then as the Rashbehari Ghosh Professor of Applied Mathematics for three years (from 1924 to 1929). But Datta was disinterested in a professorship and indifferent to worldly life. He finally retired in 1933, and became a *Sannyasi* (named SWAMI VIDYARANYA) in 1938. His historical writings include about 50 research papers and the following books: *The Science of the Sulba* (1932, repr. 1991) and *History of Hindu Mathematics. A Source Book* (with A. N. SINGH, in two parts: [DATTA/SINGH 1935/38]).

Secondary literature: *P* VI, 522. — JONES, PHILLIP S.: "Bibhutibhusan Datta. A Note and a Question." *HM* **3** (1976), 77–78. — [GUPTA 1980]. — DUTT, SUKOMAI: "Bibhuti Bhusan Datta (1888–1958) or Swami Vidyaranya." *Gaṇita Bhāratī* **10** (1988), 3–15 (with bibliography and portrait). R.C.G.

Dehn, Max (* November 13, 1878, Hamburg, Germany; † June 27, 1952, Black Mountain, North Carolina, USA). DEHN was a German mathematician who studied with DAVID HILBERT at Göttingen. After World War I, he taught pure and applied mathematics at the University of Frankfurt from 1921 to 1935. Because he was Jewish, Dehn was forced to resign from his position by the National Socialist regime, and in 1935 he emigrated to the U.S. where he taught at various universities before settling at Black Mountain College in North Carolina, where he died in 1952. Primarily a geometer, DEHN found solutions for two of HILBERT's famous unsolved problems (numbers 3 and 23). In 1907 he published (with P. HEEGAARD) a report for the *Encyklopädie der mathematischen Wissenschaften* on topology (analysis situs) which offers a synoptic historical overview of the subject. He also wrote, together with ERNST HELLINGER, two articles on the mathematical achievements of JAMES GREGORY ([TURNBULL 1939, 468–478]; and "Certain Mathematical Achievements of James Gregory," *AMM* **50** (1943), 149–163).

Secondary literature: *DSB* **4**, 9. *May* 121. *P* V, 272; VI, 540; VIIa(3), 391. — MAGNUS, WILHELM, and RUTH MOUFANG: "Max Dehn zum Gedächtnis." *Math. Ann.* **127** (1954), 215–227 (with portrait). — [SIEGEL 1965]. — [PINL 1969, Teil I: 211–213]. J.W.D.

Delambre, Jean-Baptiste Joseph (* September 19, 1749, Amiens, France; † August 19, 1822, Paris, France). A French astronomer best remembered for contributing to the measurement of the earth (in order to establish a base for the metric system), DELAMBRE studied Latin, Greek literature, and history at the *Collège du Plessis* in Paris, as well as mathematics and astronomy which he studied privately. Not only was DELAMBRE LALANDE's scientific collaborator, but he was also a member of the *Academie des Sciences* and the *Institut de France.* From 1812 on, DELAMBRE devoted his time to compiling a technical history of astronomy written mainly for astronomers and mathematicians in general.

Main works: *Rapport historique sur les progrès des sciences mathématiques depuis 1789 ....* Paris 1810. — *Histoire de l'astronomie.* 6 vols., Paris 1817–1827.

Secondary literature: *DSB* **4**, 14–18. *May* 121. *P* I, 539; III, 344. — *MARS* **1** (1816), xci–cii. — COHEN, I. BERNARD: Introduction to the reprint of *Histoire de l'astronomie moderne.* New York and London 1965–1969. J.P.

De Morgan, Augustus (* June 27, 1806, Madurai, India; † March 18, 1871, London, England). Though today better remembered for his work on symbolic algebra and logic (which resulted in the famous De Morgan Laws), AUGUSTUS DE MORGAN was, throughout his career, an enthusiastic historian of mathematics. He regarded the subject as being of great value in mathematical investigation and learning, as well as intrinsically interesting. DE MORGAN wrote no major historical monograph, but in an immense variety of shorter contributions to publications such as the *Penny Cyclopaedia* (1833–46), the *Companion to the Almanac* (1836–55), *Charles Knight's Gallery of Portraits* (1833–36), and WILLIAM SMITH's *Dictionary of Greek and Roman Biography* (1844–49), he contributed greatly to ensuring that history was presented in a style at once lively and scholarly.

Having studied at Cambridge under the tutorial guidance of, *inter alia*, WILLIAM WHEWELL and GEORGE PEACOCK, themselves both keen historians of the mathematical sciences, DE MORGAN left Trinity College with an encyclopedic knowledge of the history and philosophy of the subject, having virtually exhausted the college library's supply of relevant material. Shortly after his appointment to the professorship of mathematics at the newly-founded University College London, at the age of twenty-one, he began a copious output of mathematical publications. Historical writings occupy one-sixth of this work and are generally of a remarkably high quality. They can be roughly divided into five main topics:

- astronomy
- the calculus controversy: vindication of Leibniz
- Newton and his niece
- arithmetic
- bibliography

The bulk of his work on the history of astronomy focused on the 16th and 17th centuries and, in particular, on the debate caused by the emergence of the Copernican system in that period. Of particular value are his accounts of the new system's reception during the decades before the discoveries of GALILEO — an area strangly disregarded by previous historians. In two complementary articles, written nearly two decades apart, he exploited his wide reading to excellent effect by outlining the views of over twenty-five contemporaneous scientists on the new astronomy. His disclosure of the viewpoints of obscure philosophers such as THOMAS FIENUS and FRANCIS PATRICIUS, in addition to the little-known opinions of ROBERT RECORDE, SIMON STEVIN, and FRANÇOIS VIÈTE on the matter, renders his work in this area particularly interesting to the historian of science.

His desire to set the historical record straight, coupled with a fascination for anything Newtonian, led him to investigate the infamous calculus priority dispute of the early 18th century. After examination of various Royal Society documents, including two editions of its report on the affair, the *Commercium Epistolicum*, he came to the conclusion that NEWTON and his supporters had been decidedly underhanded in their treatment of LEIBNIZ; and, being "unfortunate enough to differ from the general opinion in England as to the manner in which LEIBNIZ was treated," ["On a Point Connected with the Dispute between KEILL and LEIBNITZ about the Invention of Fluxions." *Philosophical Transactions of the Royal Society* **136** (1846), 107–109; here: 108] felt it his duty to publish his findings. It was the start of the rehabilitation of LEIBNIZ among British historians of mathematics.

For DE MORGAN, the study of the history of mathematics was crucial to understanding mathematicians as human beings as well as cerebral forebears. Consequently, his research on the question of the marital status of NEWTON's niece, while it may seem inconsequential today, was, to him and his contemporaries, of vital importance in forming a complete picture of NEWTON's psychological make-up. Where he differed from historians such as WILLIAM WHEWELL and DAVID BREWSTER was that for DE MORGAN, the question of NEWTON's private morals and religious beliefs had no bearing on his intellectual greatness.

His research on the history of arithmetic was wide-ranging and varied. Of particular interest are his discussions of ancient and modern reckoning and the origin of the symbols for addition and subtraction. On this latter question, the perusal of an incunabula in his possession resulted in the discovery that the employment of + and − signs dated from at least thirty years earlier than was previously believed. His best known work in this area, *Arithmetical Books* [DE MORGAN 1847], is a chronological catalogue of nearly 400 publications on the subject and was the most comprehensive work of its kind undertaken up to that time. Its subject area also coincides with another of DE MORGAN's favorite topics: mathematical bibliography.

DE MORGAN was one of the 19th century's most notable bibliophiles; his own library was one of the finest accumulations of books on the history of mathematics in the country. (The library is now housed within the University of London Library. It was the model for LORD CRAWFORD's library in the Royal Observatory at Edinburgh, which was founded on the purchase of BABBAGE's library, and recently received volumes from the estate of E. J. AITON.) At his death, the collection stood at well over three thousand items. Not surprisingly, much of DE MORGAN's work on history is devoted to mathematical bibliography since, as he said himself, "the history of science is almost entirely the history of books and manuscripts" ["On the earliest printed almanacs." *Companion to the Almanac for 1846*, 1–31; here: 1]. *A Budget of Paradoxes* (1872), probably his most famous work, is a good example of his research in this area. Grounded entirely on the vast array of vintage mathematical oddities in his possession, the *Budget* cheerfully sends up scientific ignorance in all its many forms, from circle-squaring to perpetual motion, remaining testimony to the fact that DE MORGAN's historical knowledge was matched by a keen sense of humour.

De Morgan's historical works are characterised by two related features. Firstly, a great emphasis on primary and archival sources, and secondly, a desire to construct, from the complex mass of evidence available, an accurate historical picture of events as they really occurred. Nowhere is his historiography more clearly exhibited than in his *Arithmetical Books*, where he emphasized the importance of the historian's direct personal experience of the sources:

> The most worthless book of a bygone day is a record worthy of preservation. Like a telescopic star, its obscurity may render it unavailable for most purposes; but it serves, in the hands which know how to use it, to determine the places of more important bodies [De Morgan 1847, ii].

Secondary literature: *DSB* **4**, 35–37. *May* 122. *P* II, 203. — De Morgan, Sophia Elisabeth: *Memoir of Augustus de Morgan.* London: Longmans 1882 (with bibliography, including a list of his many *Penny Cyclopaedia* articles). — Richards, Joan L.: "Augustus De Morgan, the History of Mathematics, and the Foundations of Algebra." *Isis* **78** (1987), 7–30 (with portrait). — Rice, Adrian: "Augustus De Morgan: Historian of Science." *Hist. Sci.* **34** (1996), 201–240. A.R.

Depman, Ivan Yakovlevich (* July 17, 1885, Tarvast village (Estonia), Russian Empire; † March 26, 1970, Leningrad, USSR). A professor in the Vyatka Pedagogical Institute (1918–1925), and the Leningrad Pedagogical Institute (1925–1970), Ivan Depman was interested in mathematics in Russia and in the history of arithmetic.

Main Works: "Karl Mikhailovich Peterson and his Dissertation." *IMI* **5** (1952), 134–164 (in Russian). — *History of Arithmetic.* Moscow 1959 (in Russian).

Secondary literature: [Kurosh 1959]. — [Fomin/Shilov 1969]. — [Borodin/Bugai 1979]. — Portrait in [Shtokalo 1966/70, vol. 4, book 2, p. 486]. S.S.D.

De Vries, Hendrik (* August 25, 1867, Amsterdam, Netherlands; † March 3, 1954, in Palestine). De Vries, a Dutch mathematician, studied at the *Eidgenössische Polytechnicum* in Zürich, where he was an assistant in descriptive and projective geometry (1890–1894). From positions as an assistant in physics (1896) and then *privaatdocent* (1901) at the University of Amsterdam, he went on to become professor at the *Technische Hogeschool* in Delft (1905), and two years later professor at the University of Amsterdam (1907). After his retirement, he went to Palestine, where he died at the age of 86. Among his areas of special interest, De Vries contributed to the development of analytic and projective geometry, as well as to the history of the theorems of Pascal, Brianchon, Desargues, Möbius, Plücker, Steiner, Mascheroni, and Monge.

Secondary literature: *P* V, 1420; VI, 2780; VIIb(9), 5824. — Van der Waerden, Bartel L.: "Levensbericht van Hendrik de Vries (25 Augustus 1867 –

3 Maart 1954)." *Jaarboek der Koninklijke Nederlandse Akademie van Wetenschappen* 1953–1954 (1954), 275–277 (with portrait). P.B.

De Waard, Cornelis (* August 19, 1879, Bergen op Zoom, Netherlands; † May 6, 1963, Vlissingen, Netherlands). DE WAARD studied mathematics and physics at Amsterdam, after which he went on to teach at schools first in The Hague and then Winschoten. From 1909 until his retirement in 1944, he taught physics at Vlissingen. De Waard was the first editor of the correspondence of MARIN MERSENNE, of which he published eight volumes. He was mainly interested in the first half of the 17th century, and studied primarily figures around DESCARTES, including MERSENNE, ROBERVAL, FERMAT, DE BEAUNE, DESARGUES, BEAUGRAND, FRÉNICLE, and PASCAL. DE WAARD's main contribution to the history of science, in addition to his annotations to the *Correspondance* of MERSENNE, are the edition of ISAAC BEEKMAN's *Journal* (1947) and publication of documents concerning PASCAL.

Secondary literature: *May* 374. — *BWN* **2**, 608. — VAN PROOSDIJ, B. A.: "Liste des travaux de Cornelis de Waard." *Janus* **47** (1958), 128–131. — HOOYKAAS, R.: "A l'occasion du 80e anniversaire de Cornelis de Waard." *Arch. int. Hist. Sci.* **12** (1959), 173–175. — ROCHOT, B.: "Cornelis de Waard 1879–1963." *Rev. Hist. Sci.* **16** (1963), 253–256. P.B.

Dickstein, Samuel (* May 12, 1851, Warsaw, Poland; † September 28, 1939, Warsaw, Poland). DICKSTEIN studied mathematics at the University of Warsaw from 1866 to 1870 and graduated with the degree of *kandydat* of the mathematical and physical sciences (similar to a doctor's degree). He worked at a grammar school as the head-master and lecturer for mathematics, and from 1915 to 1919 as professor of mathematics and history of mathematics at the Polish University of Warsaw. DICKSTEIN organized mathematical life in Poland, and created and edited the periodicals *Prace Matematyczno-Fizyczne* (Mathematical and Physical Works) (1888) and *Wiadomości Matematyczne* (Mathematical News) (1897). His scientific interests centered on algebra and the history of mathematics. He edited letters of G. W. LEIBNIZ and ADAM KOCHAŃSKI ([DICKSTEIN 1901/02]), a volume on the life and works of JÓZEF HOENE-WROŃSKI ([DICKSTEIN 1896]), and numerous other publications on history of mathematics. He was a member of the *Akademia Umiejętności* (Academy of Art) (1893) and of the *Polska Akademia Umiejętności* (Polish Academy of Art), and in 1907 founded and was a member of the *Towarzystwo Naukowe Warzawskie* (Warsaw Society of Sciences).

Secondary literature: *DSB* **4**, 83–84. *May* 127. *P* IV, 327; VI, 563; VIIb(2), 1057. — BIRKENMAJER, ALEXANDRE: "Samuel Dickstein." *Arch. int. Hist. Sci.* **1** (1948), 328–329. — Portrait in [IWIŃSKI 1975], 21. Z.P.-B.

Dieudonné, Jean Alexandre (* July 1, 1906, Lille, France; † November 11, 1992, Paris, France). JEAN DIEUDONNÉ graduated from the *Ecole Normale Superieure (ENS)* and then taught at the universities of Rennes, Nancy, and at the

Institut des hautes études scientifiques (Bures-sur-Yvette). He was also dean of the Faculty of Sciences at the University of Nice. DIEUDONNÉ helped to author the historical notes to the different volumes of BOURBAKI's *Eléments de mathématiques.* Teaching algebraic geometry, he introduced his *Cours* (1974) with an historical outline of the discipline's development, which filled an entire volume. Subsequently, he continued to pursue highly technical historical research. He oversaw, in particular, the edition of a collective work entitled *Abrégé d'histoire des mathématiques, 1700–1900* (Abridged History of Mathematics) (1978). A founding member of the BOURBAKI group (**B**), DIEUDONNÉ was elected a member of the French Academy of Sciences in 1968.

Main works: *Cours de géométrie algébrique*, vol.1: *Aperçu historique sur le développement de la géométrie algébrique.* Paris 1974. — *Panorama des mathématiques pures: le choix bourbachique.* Paris 1977. — *Abrégé d'histoire des mathématiques, 1700–1900.* 2 vols., Paris 1978; 2nd ed. in 1 vol., Paris 1986; German trans. Berlin 1985.

Secondary literature: *P* VI, 568; VIIb(2), 1060. — DUGAC, PIERRE: *Jean Dieudonné, mathématicien complet.* Paris 1995 (with portraits). J.P.

Dijksterhuis, Eduard Jan (* October 28, 1892, Tilburg, Netherlands; † May 18, 1965, Bilthoven, Netherlands). EDUARD DIJKSTERHUIS was born in Tilburg, an industrial city in the province of Brabant, in 1892. His father taught history at a small secondary school. DIJKSTERHUIS studied mathematics at the University of Groningen, although he had contemplated studying classical languages or music. In June of 1918, under the supervision of J. A. BARRAU, he defended a thesis in mathematics devoted to the theory of screws. After spending another year studying the history of science, DIJKSTERHUIS taught physics and mathematics at the secondary school in Tilburg where his father was director. From 1919 until 1953, DIJKSTERHUIS taught elementary mathematics and physics to children between ages 12 and 18. In 1920, he married J. C. E. NIEMEIJER, the daughter of a wealthy Groningen industrialist.

The first major contribution DIJKSTERHUIS made to the history of exact sciences was his *Val en worp. Een bijdrage tot geschiedenis der mechanica van Aristoteles tot Newton* (A History of the Mechanics of Free Fall from Aristotle to Newton) (1924). This book was written in the tradition of PIERRE DUHEM, whom DIJKSTERHUIS admired greatly. In his book, DIJKSTERHUIS severely criticized historians like ERNST MACH or those whose books were based on secondary sources using modern science as a yard-stick to evaluate earlier contributions to science. Like DUHEM, DIJKSTERHUIS stressed the gradual development of science and the continuity between the "science" of the Middle Ages and the early modern period (primarly, the age of GALILEO).

In the 1920s, DIJKSTERHUIS also became very active in the movement for the reform of secondary education in mathematics. He favored a method emphasizing rigor in teaching mathematics, more or less inspired by the axiomatic style of EUCLID and other Greek mathematicians. He underscored the formal value of mathematical thinking and was opposed to stressing its practical value. With respect to mechanics, Dijksterhuis was impressed by its mathematical character and denied that it was a part of physics.

Included in his reform program was the historical training of secondary school teachers. Because their pupils would experience the same problems great mathematicians had considered and resolved in the past, it was necessary for school teachers to have a sound knowledge of the history of mathematics in order to help their pupils successfully overcome these same problems. Several articles in journals for secondary education and a (partial) Dutch translation, with commentary, of *De Elementen van Euclides* (The Elements of Euclid) (1929–1930) served as a significant contribution towards improving the historical knowledge of his colleagues. In this same spirit, he also wrote a book on Archimedes (1938, translated into English with additional material in 1956).

In 1930 Dijksterhuis became an unsalaried lecturer in the history of mathematics at the University of Amsterdam and two years later he was also offered a similar position at the University of Leiden. Unfortunately, his courses on Greek and early modern mathematics were not very successful, and so he resigned his positions in 1936. Two years earlier, however, he had joined the editorial board of an influential cultural magazine, *De Gids*, of which he became the secretary in 1940. Given the fact that he had also been serving as a member of an influential government committee on secondary education since 1934, Dijksterhuis was no longer just an historian of mathematics. Not only did he publish on the history of science and mathematics, but also on the history of culture in general. Crucial for understanding the development of his ideas is an article he published in *De Gids* on the early modern introduction of mathematics in science, *De intrede der wiskunde in de natuurwetenshap* (The Advent of Mathematics to Science) (1934). Characteristic of his attention to philological detail is a brief work he published on the sources of mathematical terminology in 1939.

During the German occupation of Holland from 1940–1945, Dijksterhuis tried to continue with his activities as best as he could. He published a biography of Simon Stevin (*Simon Stevin*) (1943, abbreviated English translation 1970), whom he admired both as a mathematician and as a reformer of the Dutch (mathematical) language, and he helped to continue publication of *De Gids* until the last year of the War. In 1944, he accepted a part-time lecturership (this time salaried) in the history of the exact sciences at the University of Amsterdam. This position had been promised to him by the university in 1941, but accepting this lectureship in 1944 smacked of collaboration with the Germans. This appointment also failed, this time because there were hardly any students left at the university due to the War. Dijksterhuis was consequently unable to give the courses he had planned; after the War when the position was discontinued, he had to fall back completely upon his position as a secondary school teacher in Tilburg.

Fortunately, the year 1950 was a turning point of sorts in his career, for in that year Dijksterhuis published his magnum opus, *De mechanisering van het wereldbeeld* (The Mechanization of the World Picture). Here he once again defended his thesis concerning the gradual development of science (as opposed to the idea of a scientific revolution as popularized by ALEXANDRE KOYRÉ). According to DIJKSTERHUIS, the development of science can be characterized in terms of its gradual mathematization: from the 16th century on, mechanics, which he took to be a part of mathematics, became central to all branches of science. According to DIJKSTERHUIS, this ideal was essentially Greek in origin, so he started with a discussion of the Greek heritage.

Mechanization of the World Picture was not just a major contribution to the history of science, it was also highly valued for its literary style. In 1952, DIJKSTERHUIS was awarded the most important literary award in Holland for this book. Meanwhile, he also, at last, had become a member of the Royal Dutch Academy of Arts and Sciences. The crowning achievement of his career came in 1953 when he was appointed professor of the history of exact sciences at the University of Utrecht. From the small town of Oisterwijk (near Tilburg), where he had lived since getting married and where his three children had grown up, he moved to Bilthoven, a village near Utrecht. There he adopted the typical life style of a traditional professor. For five years, from 1955 to 1960, he was also professor of the history of exact sciences at the University of Leiden. In the years that followed, apart from his university courses, he gave many lectures throughout the country in which he discussed the cultural and social meaning of mathematics, science, and technology. In these lectures, he often stressed the role of the history of science as a bridge between "the two cultures."

On January 1, 1959, Dijksterhuis suffered a stroke from which he never fully recovered. Although in 1960 he resumed his university lectures and was able to publish a popular history of science and technology with R. J. FORBES, *Overwinning door gehoorzaamheid. Geschiedenis van natuurwetenschap en techniek* (1961, English translation in 2 vols. under the title *A History of Science and Technology*, 1963), his career came to a premature, untimely end. After resigning his university position in 1963, his health gradually deteriorated and DIJKSTERHUIS died on 18 May 1965.

Further publications: *Vreemde woorden in de wiskunde* (1939, repr. 1948).

Secondary literature: *P* VI, 569; VIIb(1), 1063. — *BWN* **1** (1979), 169–171. — COHEN, H. FLORIS: *The Scientific Revolution. A Historiographical Inquiry* (1994), esp. 59–73. — DIJKSTERHUIS, EDUARD JAN: *Clio's stiefkind*, KLAAS VAN BERKEL (Ed.) (1990). — BERKEL, KLAAS VAN: *Dijksterhuis. Een biografie* (with portraits) (1996).

K.v.B.

Dodt van Flensburg, Johannes Jacobus (* 1800, Flensburg, Germany; † August 25, 1847, Utrecht, Netherlands). To his surname DODT he added the name of his birthplace Flensburg in Germany. As a private teacher he went to Holland, where he settled in Utrecht. There he became a teacher at the Municipal Gymnasium, and in 1828 an assistant at the University Library. DODT published numerous studies on the history of The Netherlands, including articles on mathematicians such as LUDOLF VAN CEULEN, SIMON STEVIN, WILLEBRORD SNELLIUS, and others.

Secondary literature: *NNBW* **4**, 509–510. P.B.

Droysen, Johann Friedrich (* July 19, 1770, Greifswald, Germany; † October 10, 1814, Greifswald, Germany). DROYSEN's academic career is closely connected to Greifswald University, where he became extraordinary professor of mathematics in 1806, and ordinary professor in 1812. In 1800 he published an historical work on the contributions of Swedish scholars to mathematics and physics [DROYSEN 1800].

Secondary literature: *P* I, 603. — *SBL* **11**, 457–459. K.A.

Dugac, Pierre (* July 12, 1926, Bosanska Dubica, Yugoslavia; † March 7, 2000, Paris, France). Leaving communist Yugoslavia at the age of 19, PIERRE DUGAC found shelter in a refugee camp in Italy before moving to Paris in 1946, where he received French citizenship twenty years later (in 1966). First trained in literature, in Paris he turned to mathematics and its history. As an associate professor of mathematics at the University Pierre et Marie Curie in Paris (1964–1991), he also lectured on historical topics beginning in 1975. His dissertation (1978) dealt with the foundations of analysis from CAUCHY to BAIRE. DUGAC played a pioneering role in the professional development of history of mathematics as a discipline in French universities. As director of the history of mathematics seminar at the *Institut Henri Poincaré*, in 1980 he launched the series of *Cahiers* published by this seminar, which included numerous archival documents and correspondence (for example, editions of the letters of LEBESGUE and POINCARÉ). In 1990, DUGAC was elected a corresponding member of the Paris Academy of Sciences, where he was in charge of the history of science. His research focussed primarily on 19th-century analysis and its foundational problems, especially with respect to the French school of function theorists (BOREL, BAIRE, LEBESGUE, DENJOY, etc.). DUGAC also published on the history of the concepts of real number, limit and continuity, on the rise of set theory, and on the changing demands of rigor in mathematics.

Main works: *Richard Dedekind et les fondements des mathématiques* (with numerous unpublished texts). Paris 1976. — "Notes et documents sur la vie et l'œuvre de René Baire." *Arch. Hist. ex. Sci.* **15** (1976), 297–383. — "Histoire du théorème des accroissements finis." *Arch. Hist. ex. Sci.* **30** (1980), 86–101. — *Jean Dieudonné, mathématicien complet.* Paris 1995.

Secondary literature: [La rédaction]: "Pierre Dugac. Bosanska Dubica, le 12 juillet 1926 – Paris, le 7 mars 2000." *Rev. Hist. Math.* **5** (1999), 302–315 (with bibliography and portrait). J.P.

Dugas, René (* August 11, 1897, Caen, France; † June 15, 1957, Paris, France). Trained at the *Ecole Polytechnique*, DUGAS was an engineer who worked for the French Railway Company, taught mechanics at the *Ecole Polytechnique*

(1942–1955), and devoted himself, during his spare time, to the history of mechanics. He completed the history of mathematics written by D'OCAGNE (1955) by contributing chapters on the second half of the 19th and the beginning of the 20th century.

Main works: *Histoire de la mécanique.* Paris 1950. — *La mécanique au XVIIe siècle.* Paris 1954.

Secondary literature: COSTABEL, PIERRE: "René Dugas." *Arch. int. Hist. Sci.* **10** (1957), 305–307. — TATON, RENÉ: "René Dugas (1897–1957)." *Rev. Hist. Sci.* **10** (1957), 263–264. J.P.

Duhem, Pierre (* June 10, 1861, Paris, France; † September 14, 1916, Cabrespine, France). A French physicist who graduated from the *Ecole Normale Supérieure*, DUHEM taught at the universities of Lille (1887–1893), Rennes (1893–1894), and Bordeaux (1894–1916). He did creative work in thermodynamics, hydrodynamics, and elasticity. In the history of exact sciences, he made a substantial contribution to the knowledge of medieval science. In particular, he discovered an important group of medieval Parisian schoolmasters (notably JORDANUS), and credited them with results later discovered by Renaissance and 17th-century scientists, and which played a seminal role in the so-called "Scientific Revolution," a concept which DUHEM bitterly opposed. His thesis, supported by extremely conservative religious and political views, argued for a continuous unbroken tradition of natural philosophy in the Latin West. DUHEM was elected a corresponding member of the French Academy of Sciences in Paris in 1913.

Main works: *L'évolution de la mécanique.* Paris 1902. — *Les orgines de la statique.* 2 vols., Paris 1905–1906. — *Etudes sur Léonard de Vinci, ceux qu'il a lus et ceux qui l'ont lu.* 3 vols., Paris 1906–1913. — *Le système du monde. Histoire des doctrines cosmologiques de Platon à Copernic.* 10 vols., Paris 1913–1959.

Secondary literature: *DSB* **4**, 225–233. *May* 132. *P* IV, 354; V, 310; VI, 612. — PICARD, EMILE: *La vie et l'œuvre de Pierre Duhem.* Paris 1921 (with portrait). — PIERRE-DUHEM, HÉLÈNE: *Un savant français: Pierre Duhem.* Paris 1936. — [SARTON 1951]. — BROUZENG, PAUL: *Duhem. Science et providence.* Paris 1987. — BRENNER, ANASTASIOS: *Duhem, science, réalité et apparence: la relation entre philosophie et histoire dans l'œuvre de Pierre Duhem.* Paris 1990. J.P.

Dunnington, Guy Waldo (* January 15, 1906, Bowling Green, Missouri, USA; † April 10, 1974, Natchitoches (Louisiana), USA). WALDO DUNNINGTON, after having lectured on German literature and history of mathematics in St Louis, Kansas City, and La Crosse (Wisconsin), in 1946 became a professor of the German language and literature at Northwestern State University of Natchitoches (Louisiana). An enthusiastic admirer of C. F. GAUSS since the age of twenty, he built up a large Gauss archive (later given to his university). DUNNINGTON is the author of the standard biography of GAUSS: *Carl Friedrich Gauss — Titan of Science. A Study of his Life and Work* (1955). In 1985, the *Gauss-Gesellschaft* in Göttingen published posthumously his article "Der Unsterblichkeits-Gedanke bei Gauß und Jean Paul" (The Idea of Immortality according to Gauss and Jean Paul) in its *Mitteilungen* no. 22, 3–7 (with portrait of DUNNINGTON).

Secondary literature: DOHSE, FRITZ-EGBERT: "G. Waldo Dunnington †." *Gauss-Gesellschaft e.V. Göttingen: Mitteilungen* no. 12 (1975), 3 (with portrait). — DOHSE, FRITZ-EGBERT: "Das Gauß-Dunnington-Archiv in Natchitoches (USA)." *Gauss-Gesellschaft e.V. Göttingen: Mitteilungen* no. 34 (1997), 63–66 (with photographs). C.J.S.

Echegaray y Eizaguirre, José (* March 1833, Madrid, Spain; † September 14, 1916, Madrid, Spain). This dramatist (1904 Nobel Prize for Literature) was a civil engineer who was also interested in mathematics as a professor at the Civil Engineering School, where he taught until 1868, and as a scientific publicist. A free-trader politician, he was a government minister on several occasions — which at one point included responsibilities he assumed for public education. As regards mathematical policy, he presided over the Mathematics Section of the *Junta para Applicación de Estudios e Investigaciones Científicas* (Council for Scientific Research) and the Spanish Mathematical Society, serving both from their inception (1907 and 1911, respectively). With a well-known speech on the history of pure mathematics in Spain on the occasion of the incorporation of the Royal Academy of Exact, Physical, and Natural Sciences in Madrid in 1866, he appears as one of the most outstanding figures of the so-called Polemics of Spanish Science (see Sections 13.1 and 13.2).

Secondary literature: *May* 133. *P* VI, 631. — OLMET, LUIS ANTÓN DEL and ARTURO GARCÍA CARRAFFA: *Echegaray.* Madrid 1912. E.A.; M.H.

Eibe, Thyra (* November 3, 1866, Copenhagen, Denmark; † January 1, 1955, Copenhagen, Denmark). THYRA EIBE graduated in mathematics from the University of Copenhagen in 1895, and was a teacher at HANNA ADLER's *Fællesskole* (later *Sortedams Gymnasium*) in the period 1899–1934. She published a Danish translation of HEIBERG's Greek-Latin edition of EUCLID's *Elements* (Copenhagen, 1897–1912). Together with KIRSTINE MEYER, EIBE edited OLE RÖMER's *Adversaria* in 1910.

Secondary literature: *DBL, Supplement*, 277. — HØYRUP, ELSE: "Thyra Eibe – Danmarks først kvindelige matematiker." *Normat* **2** (1993), 41–44 (English summary). K.A.

Eisele, Carolyn (* June 13, 1902, Bronx, New York, USA; † January 15, 2000, New York, New York, USA). EISELE, an American historian of mathematics, was a graduate of Hunter College in New York City, from which she received a B.A. degree in 1923. She did graduate work in mathematics at Columbia University and the University of Chicago, and earned her M.A. degree from Columbia University in 1925. Her teaching career as a member of the mathematics department at her *alma mater*, Hunter College, spanned almost fifty years, commencing with her graduation in 1923 until her retirement in 1972. It was at Hunter that her interest in history of mathematics was sparked when she was called upon to teach an undergraduate course on history of mathematics. Later, following World War II, together with C. DORIS HELLMAN and CARL BOYER, she helped to found The Metropolitan New York Section of the History of Science Society in 1953, which was consolidated in the early 1990's with the History and Philosophy of Science Section of The New York Academy of Sciences. As a scholar, EISELE was one of the leading experts on the life and works of the American founder of Pragmatism, CHARLES S. PEIRCE. Many of her publications dealing with history of mathematics are collected in *Studies in the Scientific and Mathematical Philosophy of Charles S. Peirce: Essays by Carolyn Eisele* (1979). Among numerous articles in professional journals, one of her most significant early contributions was "C. S. Peirce: A Nineteenth Century Man of Science" (*Scripta Mathematica* **24** (1959), 305–324). In 1982, EISELE became an advisor to the Peirce Edition Project at Indiana University-Purdue University at Indianapolis, which now houses her extensive library. She was also actively involved in the Charles S. Peirce Society and the preservation and restoration of the PEIRCE homestead in Milford, Pennsylvania. EISELE was awarded an honorary Doctor of Humanities degree in 1980 by Texas Tech University (Lubbock), where she was also a member of the Institute for Studies in Pragmatism, and an honorary Doctor of Science degree in 1982 by Lehigh University (Bethlehem, PA).

Major works: *The New Elements of Mathematics by Charles S. Peirce.* 4 vols., The Hague 1976. — *Historical Perspectives on Peirce's Logic of Science: A History of Science.* 2 vols., Berlin 1985.

Secondary literature: Special issue in honor of Carolyn Eisele: *HM* **9**, no. 2 (1982) (with portrait, p. 262; additional photographs, pp. 288–289). J.W.D.

Eisenlohr, August (* October 6, 1832, Mannheim, Germany; † February 24, 1902, Heidelberg, Germany). EISENLOHR became *Privatdozent* for Egyptology in Heidelberg in 1869, in 1872 extraordinary professor, and in 1885 honorary professor, all in Heidelberg. As an Egyptologist, he published an edition, with commentary, of the *Rhind Papyrus*: *Ein mathematisches Handbuch der alten Ägypter* (The Rhind Papyrus: A Mathematical Handbook of the Ancient Egyptians) (1877).

Secondary literature: *NDB* **4**, 417. — Chronik der Stadt Heidelberg für das Jahr 1902 (1904), 132–133. — *Biographisches Jahrbuch und Deutscher Nekrolog* **7** (1905), 26*. — RANKE, H.: "August Eisenlohr." *Bad. Biogr.* **6** (1935), 64–67. — Portrait: *DGB* **101**, 163. — M.F.

Ekama, Cornelis (* March 31, 1773, Paesens, Friesland, Netherlands; † February 24, 1826, Leiden, Netherlands). EKAMA studied theology, mathematics, and physics at Franeker, after which in 1803 he was appointed lecturer in mathematics, physics, and the art of navigation at Zierikzee. In 1809 he became professor at Franeker, where he taught rhetoric, metaphysics, and astronomy. After the suppression of Franeker University in 1811 (in 1810 the Northern Netherlands had been annexed to France, and in 1811 Franeker University had been closed by

French imperial decree), he was appointed professor of philosophy, mathematics, physics, and astronomy at the University of Leiden. EKAMA is the author of the first serious study of the life and work of GEMMA FRISIUS [EKAMA 1823].

Secondary literature: *P* I, 655. — *VdAA* **5**, 24. — *NNBW* **6**, 472–474. P.B.

Eneström, Gustaf Hjalmar (* September 5, 1852, Nora, Sweden; † June 10, 1923, Stockholm, Sweden). GUSTAF ENESTRÖM was the second and last child of JAKOB FILIP ENESTRÖM, a coal mine owner, and ANNA AUGUSTA LUNDQUIST, his second wife (from the father's first marriage, there were seven children). Neither ENESTRÖM nor his elder sister EMMA LOVISA ever married. ENESTRÖM finished grammar school in Stockholm in 1870, where he distinguished himself in mathematics by answering a mathematical examination problem so remarkably that his solution was published. He went on to study in Uppsala, and after much less time than usual, passed the *artes liberales* examination in 1871, for which he was awarded the title *kand. fil.* Rather than continuing in mathematics, ENESTRÖM began working in 1874 for the library of the observatory at Uppsala University, and the following year accepted a position at the library of the university. In 1879 he moved to the Royal Library in Stockholm, and from 1901 on he worked in various governmental and court libraries, his last position being at the *Svea Hovrätt* (the Swedish court of Appeal), beginning in 1919. ENESTRÖM was elected a member of three learned societies, the *Reale accademia di scienze, lettere ed arte* in Padova (1886), the *Real academie de ciencias exactas, fisicas y naturales* in Madrid (1894), and the *Schweizerische Naturforschende Gesellschaft* (1909). Moreover, he was designated *Doctor honoris causa* at Lund University in 1919.

Parallel with his work as a librarian, ENESTRÖM was engaged in statistical work concerning superannuation schemes, some of which he published. In connection with his investigations, he was led to consider the roots of a polynomial $a_nx^n + a_{n-1}x^{n-1} + \cdots + a_0$ with positive coefficients. His result, published in Swedish in 1893, states that the absolute values of the roots lie between the minimum and the maximum of the numbers $a_{n-1} : a_n$, $a_{n-2} : a_{n-1}$, $\ldots$, $a_0 : a_1$. Later this was called KAKEYA's theorem after the Japanese mathematician who published it in 1912. A few years before his death, ENESTRÖM published a note in French in *The Tôhoku Mathematical Paper* (1920), which presented his earlier proof of the theorem. The result is now known as the KAKEYA-ENESTRÖM theorem. A list of his works on actuarial sciences and statistics can be found in *Nordisk Försäkringstidskrift* (1950).

Nearer to ENESTRÖM's heart was the history of mathematics. His first publication, an article on the history of the isoperimetric problem, appeared in the *Yearbook of Uppsala University* for 1876, followed three years later by another paper on the history of calculus. From then on, he published more than 150 historical papers and notes, many of which appeared in the *Jahrbuch über die Fortschritte der Mathematik*, or in his own journal, *Bibliotheca Mathematica*. ENESTRÖM covered many fields and periods, but his strongest interests were the Middle Ages, including Islamic science, the 16th and 17th centuries, and EULER. He also contributed to such major projects as the edition of EULER's works and the French edition of the *Encyklopädie der mathematischen Wissenschaften.* ENESTRÖM was also helpful to a number of his colleagues, and often commented on manuscripts sent to him by SMITH, TROPFKE, and WIELEITNER before they were published.

A few years after ENESTRÖM had settled in Stockholm, GÖSTA MITTAG-LEFFLER moved there as well, and soon took charge of the organization of mathematical research in Sweden. Among MITTAG-LEFFLER's many initiatives was the creation of a new journal, *Acta Mathematica*, for which he agreed that ENESTRÖM should be responsible for an appendix, *Bibliotheca Mathematica*, surveying the mathematical literature. This was the beginning of ENESTRÖM's thirty-years career as editor of *Bibliotheca Mathematica*, which appeared in three series. The first consists of only three volumes (1884–1887), and contains, in addition to bibliographies, a few historical notes written by ENESTRÖM and various colleagues.

In 1887 *Bibliotheca Mathematica* separated from *Acta mathematica*, and thereafter appeared as an independent journal specifically devoted to history of mathematics. This second series consists of 13 volumes (each containing 124 pages) and was published until 1899. ENESTRÖM himself financed the journal, but received intellectual support from colleagues abroad; in fact, his list of editors is quite impressive (Table 1). Among the still useful materials in this series are reports on the history of mathematics in various countries. Moreover, the series contains a general index for the period 1887–1896, with short biographical notes on the contributors, along with their portraits.

G. J. ALLMAN (Galway)	K. HUNRATH (Rendsburg)
D. BIERENS DE HAAN (Leiden)	P. MANSION (Gand)
C. A. BJERKNES (Christiania)	A. MARRE (Paris)
V. BOBYNIN (Moskwa)	E. NARDUCCI (Roma)
H. BROCARD (Montpellier)	E. NETTO (Berlin)
G. CANTOR (Halle a/S.)	P. RICCARDI (Modena)
M. CANTOR (Heidelberg)	M. STEINSCHNEIDER (Berlin)
L. DE MARCHI (Milano)	H. SUTER (Zürich)
A. FAVARO (Padova)	P. TANNERY (Tonneins)
E. GELCICH (Lussinpiccolo)	G. VALENTIN (Berlin)
G. GOVI (Napoli)	E. WOHLWILL (Hamburg)
S. GÜNTHER (München)	R. WOLF (Zürich)
J. L. HEIBERG (Köbenhavn)	H. G. ZEUTHEN (Köbenhavn)

Table 1. The list of the editorial board published in the first volume of the second series of *Bibliotheca Mathematica*

As soon as he started the second series, ENESTRÖM approached B. G. TEUBNER about publishing *Bibliotheca Mathematica*, but an agreement was not reached until 1899. This resulted in the third series, of which 14 volumes were issued between 1900 and 1914. The First World War put an end to the journal, and despite various later initiatives, it was never revived.

As recounted in Section 6.5, ENESTRÖM was very concerned about the state of the art of history of mathematics. Not much work found favour in his critical eyes, but nothing upset him as much as MORITZ CANTOR's *Vorlesungen über Geschichte der Mathematik* (Lectures on History of Mathematics). ENESTRÖM's reaction to this work illustrates his eagerness and willingness to fight for what he considered important, namely a scientific history of mathematics; it also shows that in the process he lost his sense of proportion. At first, in volume one of the third series of *Bibliotheca Mathematica*, ENESTRÖM characterized CANTOR's *Vorlesungen* as an excellent accomplishment. One almost gets the impression that he thought CANTOR's work was a first approximation to the kind of history of mathematics that ENESTRÖM thought should be written. But gradually ENESTRÖM became more and more critical. He created a special section in *Bibliotheca Mathematica* for notes devoted to CANTOR's mistakes. This part of the journal grew substantially over the years, and altogether comprises several hundred pages, of which ENESTRÖM himself wrote the largest part.

In volume thirteen of *Bibliotheca Mathematica* ENESTRÖM explained that at first he was positive about CANTOR's *Vorlesungen* because he had read an approving review by STÄCKEL, but at the time had been too busy to study it carefully himself. When he finally found time to do so, he discovered numerous errors. ENESTRÖM's reason for spending so much time pursuing mistakes in the *Vorlesungen* was that he wanted to warn the many readers who already considered it a standard work. His efforts prompted one mathematician to comment: "Sie haben uns unseres mathematisch-historischen Gottes beraubt" (you have deprived us of our god of the history of mathematics). In the end, identifying CANTOR's mistakes became something of an obsession for ENESTRÖM, to which he devoted much time which could have been used more constructively to promote the discipline to which he was so devoted.

Occasionally CANTOR responded to ENESTRÖM's criticisms, but considering the extent to which he was attacked, his reaction was not as strong as it might have been. Indeed, according to K. G. HAGSTROEM, the two scholars maintained a friendly private correspondence despite their disagreements [HAGSTROEM 1953, 154].

Undoubtedly ENESTRÖM's colleagues found that his attacks on CANTOR went too far, and they may also have felt that his views on how history of mathematics should be written were rather eccentric (see for example the views expressed by HENRI BOSMANS, as quoted above in the biography of MORITZ CANTOR). Apparently his colleagues for the most part valued ENESTRÖM for his editorial work, his historical research, and his help, and they chose therefore to overlook his idiosyncracies and peculiarities.

ENESTRÖM was respected internationally, but did not receive much acknowledgement at home. Although MITTAG-LEFFLER collaborated with him for a time, he does not seem to have accepted ENESTRÖM as a member of his circle of mathematicians who dominated Swedish mathematics. Whether this was due to a lack of interest in ENESTRÖM's work, or to a lack of respect for him as a mathematician, is difficult to say. In either case, ENESTRÖM was left isolated in Sweden because his work was so clearly addressed to mathematicians that they were his only natural audience. In general ENESTRÖM did not express his feelings, but the fact that he was hurt by MITTAG-LEFFLER can be inferred from his decision not to leave his rich collection of books to the library at the Mittag-Leffler Institute. Instead, the *Bibliotheca Enestroem* came to be housed at the *Stockholms Högskola*, where ENESTRÖM's papers, including a substantial correspondence, are also kept.

ENESTRÖM once wrote:

> When in the beginning of the seventies I began to deal with mathematical-historical studies, most mathematicians with whom I came in contact believed that research in this direction was only a useless waste of time, so that such work should be regarded not as meritorious but rather, derogatory. If an old mathematical work — that is what they say — contains really lasting conclusions they would be found in an improved form in the modern systematic descriptions and therefore, studying an old work seems to be useless; if, however, a work represents an obsolete point of view, studying it is useless or even pernicious [LOREY 1926, 314].

Although he never received the approval of contemporary Swedish mathematicians, ENESTRÖM contributed much to change the attitudes described above towards the history of mathematics.

Main works: *Bibliotheca Mathematica.*

Secondary literature: *May* 135. *P* IV, 382; V, 338; VI, 665. — *SBL* **13**, 538–543 (with portrait). — LOREY, WILHELM: "Gustav Eneström. *In memoriam.*" *Isis* **8** (1926), 313–320 (with portrait). — HAGSTROEM, K. G.: "Gustaf Eneström. Till hunddraårsminnet av hans födelse." *Normat* **1** (1953), 145–155, 182 (with portrait).

K.A.

Engel, Friedrich (* December 26, 1861, Lugau near Chemnitz, Germany; † September 29, 1941, Giessen, Germany). ENGEL, a German mathematician, went to Christiania in 1884 at the behest of FELIX KLEIN in order to assist SOPHUS LIE in the publication of his theories. In 1885 he became *Privatdozent* and in 1889 extraordinary professor at the University of Leipzig. He was called to an ordinary professorship in Greifswald in 1904, and then held an ordinary professorship in Giessen (1913–1931). Apart from the service he rendered to LIE's geometrical ideas by giving them an exact analytical form, he published several books on the history of non-Euclidean geometry, e.g. on N. I. LOBACHEVSKIĬ (1895, 1899) and (with P. STÄCKEL) *Die Theorie der Parallellinien von Euklid bis auf Gauss* (The Theory of Parallel Lines from Euclid to Gauss) (1895). He also edited the *Gesammelte mathematische und physikalische Werke* (Collected Mathematical and Physical Works) of HERMANN GRASSMANN (1894–1911, with biography), and the *Gesammelte Abhandlungen* (Collected Papers) of SOPHUS LIE (1922–1937).

Secondary literature: *DSB* **4**, 370–371. *May* 135. *P* IV, 384; V, 339; VI, 665; VIIa(1), 508. — *NDB* **4**, 501. — *Sächsische Akademie der Wissenschaften zu Leipzig, Jahrbuch 1949–1953.* Berlin: Akademie-Verlag 1954, 50. — PURKERT, WALTER: "Zum Verhältnis von Sophus Lie und Friedrich Engel." *Wiss. Z. Univ. Greifswald, Math.-Nat. Reihe* **23** (1984), no. 1/2, 29–34. — SCRIBA, CHRISTOPH J.: "Friedrich Engel (1861–1941) / Mathematiker." In: GUNDEL, HANS GEORG; PETER MORAW, and VOLKER PRESS (Eds.): *Gießener Gelehrte in der ersten Hälfte des 20. Jahrhunderts* (= *Lebensbilder aus Hessen* **2**). Marburg 1982, 212–223 (with portrait). M.F.

Enriques, Federigo (* January 5, 1871, Livorno, Italy; † June 14, 1946, Roma, Italy). FEDERIGO ENRIQUES was a leading figure not only in the recent history of Italian mathematics, but also in Italian culture. Along with CASTELNUOVO and SEVERI, ENRIQUES led the school which dominated algebraic geometry in the early decades of the 20th century. He also gained an international reputation for his works in both the philosophy and the history of science, and played a major role in promoting the history of mathematics in Italy. The young ENRIQUES was educated at the University of Pisa and the *Scuola Normale Superiore*, where ENRICO BETTI, ULISSE DINI, LUIGI BIANCHI, and RICCARDO DE PAOLIS were among his professors. At the age of 20 ENRIQUES obtained his degree (*laurea*) in mathematics, and in 1891–1892 a post-graduate grant enabled him to perfect his studies first at the *Scuola Normale Superiore* and then in Rome where he had planned to work with LUIGI CREMONA, the founder of the Italian school of geometry. In Rome, however, he became acquainted with GUIDO CASTELNUOVO, a former student of GIUSEPPE VERONESE in Padua and of CORRADO SEGRE in Turin. CASTELNUOVO had been appointed to a professorship of analytic and projective geometry at the University of Rome in 1891, and his friendship with ENRIQUES proved to be of decisive importance for the direction of ENRIQUES's work. CASTELNUOVO introduced ENRIQUES to the more recent developments of algebraic geometry and the theory of algebraic surfaces in particular, which was the subject of the first important paper published by ENRIQUES in 1893.

After several months working in Turin with CORRADO SEGRE, ENRIQUES accepted a position in January 1894 teaching projective and descriptive geometry at the University of Bologna, where three years later he was promoted to full professor of projective geometry. That same year ENRIQUES published his lectures in a celebrated volume, *Lezioni di geometria proiettiva* (Lectures on Projective Geometry) (1st ed. 1897, 4th ed. 1920, repr. 1996; German translation 1903, French translation 1930). The Bolognese period was undoubtedly the most fruitful in ENRIQUES's mathematical career. As is clear from his recently published correspondence with CASTELNUOVO (see [BOTTAZZINI, GARIO, CONTE 1996]) working in close collaboration with his friend — who in 1896 also became his brother-in-law — ENRIQUES laid the foundations and then developed the theory of algebraic surfaces. Together ENRIQUES and CASTELNUOVO first summarized the results of their joint work in a paper published in 1897 in *Mathematische Annalen*, and then in an appendix to the second volume of *Théorie des fonctions algébriques de deux variables independantes* (Theory of Algebraic Functions of Two Independent Variables) (1906) by Charles Emile Picard and Georges Simart, where (at Picard's request) they presented the methods and results of the Italian school.

In 1907 Enriques and Severi were awarded the Bordin Prize of the Paris Academy of Science for their joint memoir on the classification of hyperelliptic surfaces. (A comprehensive account of the theory of algebraic surfaces and their classification was given by Enriques and Castelnuovo in *Die algebraischen Flächen vom Gesichtspunkt der birationalen Transformationen* (Algebraic Surfaces from the Viewpoint of Birational Transformations) (1914), published as part of the *Encyklopädie der mathematischen Wissenschaften.* Also in 1907, Enriques received the Royal Prize of the Accademia dei Lincei for his contributions to algebraic geometry.

By that time, however, his interests had partly turned to philosophy. Indeed, in 1906 Enriques published his first, influential volume in the philosophy of science, *Problemi della scienza* (Problems of Science, 2nd ed. 1909; repr. 1985), which was translated very soon into French (1909), German (1910), Russian (1911), and English (1914). Enriques's early philosophical interests dated back to the mid-1890s when, due to his teaching, he tackled the problem of the foundations of geometry and, consequently, the philosophical problem of the nature of space. (Enriques presented his views on this subject in an article, *Prinzipien der Geometrie* (Principles of Geometry, 1907), which was also included as part of the *Encyklopädie der mathematischen Wissenschaften.*) Although he had begun to publish philosophical papers as early as 1901, the appearance of the *Problems of Science* marked a turning point in Enriques's career. Indeed, from that time on Enriques became more and more involved in philosophical questions. In 1907, in collaboration with Eugenio Rignano, he founded the *Rivista di scienza* (generally known as *Scientia* as it was called after 1910), which Enriques edited until 1914. From 1906 to 1913 he served as chair of the *Società Filosofica Italiana,* in which capacity he organized the Fourth International Congress of Philosophy held in Bologna in 1911. Enriques's philosophical thought, with its emphasis on rationalism and the value of science as expounded in the *Problems,* was opposed by the Italian representatives of idealism headed by Benedetto Croce and Giovanni Gentile.

The congress gave them a public opportunity to launch a strong critique of Enriques's arguments, to which he reacted by collecting and reworking some of his philosophical papers in a new work, *Scienza e razionalismo* (Science and Rationalism) (1912, repr. 1990), wherein he contrasted positivism with the idealistic philosophy of Croce and Gentile. However, his isolated efforts to renew Italian philosophy were doomed to fail. Meanwhile, Enriques's concerns for philosophy intertwined with his increasing interest in the history and teaching of mathematics. In 1900 he edited a book of *Questioni riguardanti la geometria elementare* (Questions Concerning Elementary Geometry), (German edition in two volumes 1907, 1910), which is really a collection of essays written by a number of his pupils and colleagues. Years later this work was extended into *Questioni riguardanti le matematiche elementari* (Questions Concerning Elementary Mathematics) (two volumes, 1912–1914). The second edition of this work almost doubled in size to 4 volumes, which appeared between 1924 and 1927). This work was reprinted in 1983 and is still regarded, even today, as a standard reference for problems concerning elementary mathematics. In addition, in collaboration with UGO AMALDI, ENRIQUES published a textbook, *Elementi di Geometria* (Elements of Geometry). From the first edition of 1903, and through many subsequent editions and reprints, this textbook provided the first introdution to geometry for generations of Italian students. In 1918 ENRIQUES was elected president of *Mathesis*, the Italian association of teachers of secondary schools, a position which also reflects his interest in teaching and mathematical education. ENRIQUES's lectures in Bologna attracted a number of students, including historians like ROBERTO BONOLA and AMEDEO AGOSTINI, and mathematicians like GIUSEPPE VITALI and OSCAR CHISINI.

Together with CHISINI, ENRIQUES published his lectures on algebraic geometry in a monumental work, *Lezioni sulla teoria geometrica delle equazioni e delle funzioni algebriche* (Lectures on the Geometric Theory of Algebraic Equations and Functions) (four volumes, 1915–1934; repr. 1985), which represents an overall summary of results achieved by the Italian school of algebraic geometry. In 1922 ENRIQUES was appointed to a chair at the University of Rome. There he taught "complementary mathematics" as well as higher geometry from 1923 to 1938, when he was forced to leave his position at the university because of the racial laws decreed by the Fascist Regime. In his place FRANCESCO SEVERI, his former student who over the years had become one of ENRIQUES's strongest opponents and the "chief" of Italian mathematics under Fascism, assumed the chair of higher geometry. Earlier, ENRIQUES's lectures on geometry given at the University of Rome were edited by his student, LUIGI CAMPEDELLI, in a volume of *Lezioni sulla teoria delle superficie algebriche* (Lectures on the Theory of Algebraic Surfaces) (1932). ENRIQUES's interests in the history of mathematics (and in the history of science) increased substantially while he was in Rome. From 1921 until 1938 he edited the *Periodico di Matematiche*, a journal mainly addressed to secondary school teachers in which he reserved special room for the publication of historical papers (by BORTOLOTTI and AGOSTINI in particular).

ENRIQUES himself contributed to the history of logic with an influential book, *Per la storia della logica* (On the History of Logic) (1922, repr. 1987; French translation 1925, German translation 1927, English edition 1929). With the help of students and various collaborators he edited (with historical notes and commentaries) *Gli Elementi di Euclide e la critica antica e moderna* (Euclid's *Elements* and its Ancient and Modern Critics) (4 volumes, 1924–1935). It is worth mentioning that vol. 3, which included book X of the *Elements*, was edited by RUTH STRUIK, and vol. 4 (books XI–XIII) was edited by AGOSTINI. These volumes appeared in the series *Per la storia e la filosofia della matematica* (On the History and Philosophy of Mathematics), founded and edited by ENRIQUES himself. In addition to EUCLID's *Elements*, this series included an Italian edition of works by

ARCHIMEDES, BOMBELLI, GALILEO, NEWTON, and DEDEKIND (the latter with historical and critical notes by OSCAR ZARISKI). ENRIQUES was also involved in the realization of the *Enciclopedia Italiana Treccani*, a monumental work in several volumes directed by his former rival in philosophy, GIOVANNI GENTILE, who in the meantime had became minister of Public Education under the Fascist government. The *Enciclopedia Italiana* was intended to present an overview of Italian culture, and at GENTILE's request ENRIQUES agreed to be responsible for the coverage of mathematics. ENRIQUES himself contributed to the *Enciclopedia* by writing a number of entries (many of them historical). At the University of Rome ENRIQUES also founded and directed the *Istituto per la storia della scienza* (Institute for the History of Science) were GIOVANNI VACCA was appointed to teach history of mathematics. Together with his former student GIORGIO DE SANTILLANA, ENRIQUES planned to write a comprehensive history of science from antiquity to modern times. However, they were only able to publish the first volume of the planned work, *Storia del pensiero scientifico. Il mondo antico* (History of Scientific Thought. The Ancient World) (1932). Some years later they published an abridged version of the entire project, a *Compendio del pensiero scientifico dall'antichità fino ai tempi moderni* (Compendium of Scientific Thought from Antiquity to Modern Times) (1937). In 1938 ENRIQUES published *Le matematiche nella storia e nella cultura* (Mathematics in History and Culture) (repr. 1971), which was based on his lectures and seminars, and edited by his student ATTILIO FRAJESE. Although the Fascist racial laws prevented him from publishing papers and books in Italy after 1938, he still managed to do so by resorting to different tricks. For example, he published a number of essays in France, and some papers in Italy as well under a pseudonym, ADRIANO GIOVANNINI (a combination of the names of his dauther ADRIANA and his son GIOVANNI). His volume on *Le superficie razionali* (Rational Surfaces) (1939), based on his university lectures, simply appeared under the name of ENRIQUES's student, FABIO CONFORTO.

After the fall of the Fascist regime and the subsequent liberation of Rome from the Nazis, ENRIQUES reclaimed his position at the university in 1944. After the War he also resumed his editorial activities, and together with CASTELNUOVO he founded a new series devoted to "classic" texts, the first of which was GALILEO's *Dialogo … sopra i due massimi sistemi del mondo, Tolemaico, e Copernicano* (Dialogue Concerning the Two Chief World Systems) prefaced by ENRIQUES himself. ENRIQUES had also planned to publish GALILEO's *Discorsi e dimostrazioni matematiche intorno à due nuove scienze* (Dialogues Concerning Two New Sciences) and GAUSS's *Disquisitiones generales circa superficies curvas* (General Investigations of Curved Surfaces), but these projects were never completed due to ENRIQUES's unexpected death. His last mathematical work, *Le superficie algebriche* (Algebraic Surfaces), was published posthumously by CASTELNUOVO in 1949. In addition to his tremendous influence on the development of algebraic geometry, ENRIQUES influenced an entire generation of historians of mathematics in Italy, in particular his students ATTILIO FRAJESE and ETTORE CARRUCCIO. His philosophical thought has also been reevaluated and appreciated by philosophers and historians of philosophy, especially by LUDOVICO GEYMONAT in his last writings.

Secondary literature: *DSB* **4**, 373–375. *May* 135. *P* IV, 388; V, 340; VI, 667. — [*Acta Math.*, Tab. Gen. **1–35** (1913), 51; portrait 138]. — CHISINI, OSCAR: "Federigo Enriques." *Boll. UMI* (3) **1** (1946), 70–72. — A special issue of the *Periodico di Matematiche* (4) **25** (1947) is devoted to ENRIQUES's life and work. It includes papers by GUIDO CASTENUOVO, LUIGI CAMPEDELLI, FABIO CONFORTO, OSCAR CHISINI, and biographical recollections contributed by ENRIQUES's daugther, ADRIANA. — See also: *Riposte armonie. Lettere di Federigo Enriques a Guido Castelnuovo*, BOTTAZZINI, UMBERTO, PAOLA GARIO, and ALBERTO CONTE (Eds.). Torino 1996 (with portrait). U.B.

Favaro, Antonio (* May 21, 1847, Padua, Italy; † September 29, 1922, Padua, Italy). Born to a noble and educated family (his father was also a mathematician), ANTONIO FAVARO graduated in 1866 from the University of Padua, where he studied mathematics with GIUSTO BELLAVITIS, SERAFINO MINICH, and DOMENICO TURAZZA. Three years later he obtained a degree in engineering from the University of Turin, whereupon he enrolled at the Polytechnicum in Zurich. However, he only studied there for a short time because, in 1870, he returned to Padua to accept an assistantship under his former teacher TURAZZA. TURAZZA taught rational mechanics at the university, and later became FAVARO's father-in-law. FAVARO spent the rest of his academic career at the University of Padua; in 1872 he was promoted to extraordinary professor for graphical statics, and ten years later, he was made full professor in 1882. In all, FAVARO taught graphical statics for some fifty years, and also gave "free" courses on the history of mathematics beginning in 1878. Although his early papers were devoted to engineering, his scientific interests soon turned towards history. As FAVARO himself recognized, BALDASSARRE BONCOMPAGNI and his *Bullettino* proved to be decisive influences on his career. It was in BONCOMPAGNI's *Bullettino*, that FAVARO published his early papers on history of mathematics, including his long essay, "Notizie storiche sulle frazioni continue dal secolo XIII al secolo XVII" (Historical Notices on Continued Fractions from the 13th to the 17th Century) (1875), as well as papers on GALILEO's life and work.

Among FAVARO's favorite subjects was the history of the University of Padua, to which he devoted a number of publications, including: *Lo Studio di Padova e la Compagnia di Gesù sul finire del secolo decimosesto* (The University of Padua and the Company of Jesus at the End of the 16th Century) (1878), and *Le matematiche nello Studio di Padova dal principio del secolo XIV alla fine del XVI* (Mathematics at the University of Padua from the Beginning of the 14th Century to the End of the 16th) (1880). These studies intertwined with his research on GALILEO. Indeed, nearly 300 of FAVARO's more than 500 publications are concerned with the figure of GALILEO. In 1883 FAVARO published *Galileo Galilei e lo Studio di Padova* (Galileo Galilei and the University of Padua) (in 2 vols., repr. 1966). As an appendix to the second volume of this work, FAVARO reprinted the proposal for a national edition of GALILEO's works, which he had already presented to the *Istituto Veneto* in 1881. In addition, FAVARO printed a number of GALILEO's manuscripts and related documents in BONCOMPAGNI's *Bullettino* in 1884 and 1885. Subsequently, he published several collections of essays, including *Miscellanea galileiana inedita* (Miscellaneous Unpublished Galilean Papers) (1887, running to more than 300 printed pages!), and some 400 pages of *Nuovi studi galileiani* (New Galilean Studies) (1891). FAVARO's project to produce a national edition of GALILEO's works was eventually approved by the Minister of Public Education, M. COPPINO, in 1887, and three years later the first volume appeared; the twentieth and last was published by FAVARO in 1909.

Incredibly, this excellent work went almost unnoticed at the time. In fact, only 500 complimentary copies of it were originally printed, although after FAVARO's death, the entire work of 20 volumes was reprinted between 1929 and 1939 by ANTONIO GARBASSO and, after his death, by GIORGIO ABETTI with the collaboration of ANGELO BRUSCHI and ENRICO FERMI, who also published a number of additions and corrections, often based on FAVARO's manuscripts.

The national edition of GALILEO's works served as FAVARO's starting point for a number of parallel and subsequent studies, all related to GALILEO. In addition to 41 papers, "Amici e corrispondenti di Galileo" (Friends and Correspondents of Galileo), published between 1894 and 1919 (and reprinted in 3 volumes by PAOLO GALLUZZI (1983), FAVARO's publications include *Oppositori di Galileo* (Opposers of Galileo; six monographs, 1892–1921), *Scampoli galileiani* (Galilean Remnants) (1892–1914), and *Adversaria galileiana* (Galilean Adversaries; 7 contributions, 1916–23). The latter two works have been reprinted by LUCIA ROSSETTI and MARIA LAURA SOPPELSA (1992). FAVARO's research on the history of the University of Padua culminated in his *Saggio di bibliografia dello studio di Padova, 1500–1920* (Essay on Works Devoted to the University of Padua, 1500–1920) (in 2 vols., 1922).

Moreover, from the beginning of the century FAVARO became increasingly involved in the project of a national edition of the works of LEONARDO DA VINCI. As a member of the *Commissione Vinciana*, he worked with ENRICO CARUSI in editing LEONARDO's *Del moto e misura dell'acqua* (On the Motion and Measure of Water), which appeared in 1923, one year after FAVARO's death.

Secondary literature: *May* 149. *P* III (1), 429; IV (1), 406; V, 357; VI (2), 711. — *DBI* **45** (1995), 441–444.— FAVARO, GIUSEPPE: "Antonio Favaro. Biobibliografia." *Atti Ist. Veneto Sci. Lett. Arti* **82** (1922–23), 221–303 (with bibliography and portrait). — BORTOLOTTI, ETTORE: "Antonio Favaro storico delle scienze matematiche." *Atti e memorie della R. Deputazione di storia patria per le provincie di Romagna*, (4) **14** (1924), 1–26. — Many documents related to FAVARO's life and work, including his teaching, are kept in the archives of the University of Padua. The correspondence between FAVARO and PAUL TANNERY has been published in TANNERY, PAUL: *Memoires scientifiques* **14** (Correspondence) (1937), 439–517.

U.B.

Fernández de Navarrete, Martín (* November 9, 1765, Abalos (La Rioja), Spain; † Ocober 8, 1844, Madrid, Spain). FERNÁNDEZ DE NAVARRETE studied at the Seminar of Vergara and, at the age of fifteen, he joined the navy, where he studied with CIPRIANO VIMERCATI, an outstanding 18th–century Spanish mathematician. During his service in the navy, FERNÁNDEZ DE NAVARETTE participated in military activities, and at the same time performed astronomical duties under the direction of JOSÉ DE MAZARREDO (1745–1812). In 1786 he was sent to Cartagena, where he continued to study mathematics with GABRIEL CISCAR (1759–1829). FERNÁNDEZ DE NAVARETTE's *History of the Spanish Navy and its Connection with Mathematics* was published posthumously in 1846 ([FERNÁNDEZ DE NAVARRETE 1846]).

Secondary literature: *Primer Centenario de Don Martín Fernández de Navarrete conmemorado por el Instituto de España en los Salones del Museo Naval el día 27 de enero de 1945.* Madrid 1945. E.A.; M.H.

Fettweis, Ewald (* July 23, 1881, Eupen near Aachen, Germany; † July 24, 1967, Aachen, Germany). FETTWEIS taught mathematics in Düsseldorf and Koblenz, 1911–1945. From 1945 to 1954 he held a professorship at the Pedagogical Academy in Aachen. His main areas of interest were related to the teaching of mathematics and to ethnomathematics.

Main works: *Wie man einstens rechnete* (1923). — *Das Rechnen der Naturvölker* (1927).

Secondary literature: *P* VI, 731; VIIa(2), 30. — SCRIBA, CHRISTOPH J.: "Ewald Fettweis † ." *Nbl. DGGMNT* **19** (1969), 17–18. — REICH, KARIN; MENSO FOLKERTS, and CHRISTOPH J. SCRIBA: "Das Schriftenverzeichnis von Ewald Fettweis (1881–1967) samt einer Würdigung von OLINDO FALSIROL." *HM* **16** (1989), 360–372 (with bibliography and portrait). M.F.

Fleckenstein, Joachim Otto (* July 7, 1914, Düsseldorf, Germany; † February 21, 1980, München, Germany). After graduation from the *Gymnasium*, FLEKKENSTEIN emigrated in 1933 to Switzerland and studied mathematics, physics, astronomy and philosophy at the University of Basel, where he received his Ph.D. in astronomy in 1939 and his *Habilitation* for history of the exact sciences and astronomy in 1947. From 1938 to 1951 he was assistant in astronomy in Basel. During the following years he lectured at various universities (apart from Basel, where he remained active until the end of his life) in Mainz, Darmstadt, Free University of Berlin, and Izmir. He also did observational work at the observatory of the University of Milan from 1954 to 1963. In 1963 he was appointed to the newly created chair for the history of the exact sciences and technology at the *Technische Hochschule* in Munich. In addition, he continued to lecture regularly at the University of Basel and, since 1978, at the University of Turin. Having become a Swiss citicen in 1953, he also was elected in 1968 a member of the *Weiterer Bürgerrat* (extended city council) in Basel.

FLECKENSTEIN, a fascinating, multilingual personality with widespread interests and a strong philosophical bent, also contributed to several editorial projects: he edited the volumes II, 5 ("Principia mechanica," 1957) and II, 23 ("Sol et luna I," 1969) of the second series of LEONHARD EULER's *Opera omnia* and collaborated in the production of vol. IV A,1 (1975), the inventory of EULER's correspondence. As the successor of OTTO SPIESS, he headed the Bernoulli-Commission and edited vol. 1 of *Die Werke von Jakob Bernoulli* ("Astronomie, Philosophia naturalis," 1969). Beginning in 1977 he also was in charge of the Copernicus edition. Further publications include biographical studies in the series of *Beihefte* to the journal *Elemente der Mathematik* on NEWTON and LEIBNIZ, the brothers JAKOB and JOHANN BERNOULLI, and the book *Leibniz. Barock und Universalismus* (1958).

Secondary literature: *P* VIIa (2), 69. — Scientific estate in University Library of Basel. — HOSKIN, MICHAEL: "Joachim Otto Fleckenstein (1914–1980)." *JHA* **11** (1980), 144–145. — FIGALA, KARIN: "Joachim Otto Fleckenstein, 1914–1980." *Arch. int. Hist. Sci.* **33** (1983), 126–127. — FOLKERTS, MENSO (Ed.): *Gemeinschaft der Forschungsinstitute für Naturwissenschafts- und Technikgeschichte am Deutschen Museum, 1963–1988.* München: Deutsches Museum 1988 (esp. pp. 78–82, with portraits). C.J.S.

Frajese, Attilio (* November 11, 1902, Rome, Italy; † August 5, 1986, Rome, Italy). After attending a classical, secondary school (*liceo*), ATTILIO FRAJESE studied at the University of Rome where he received his degree (*laurea*) in engineering at the age of 21 with a thesis on hydrodynamics and the theory of turbines. Instead of looking for a job in industry, FRAJESE devoted himself to teaching in secondary schools. This was the first step in his career as a government official in the Ministry of Public Education. In 1942 FRAJESE became a school inspector, and in 1949 was promoted to director of the Ministry of Education. His first interest in the history of mathematics began in 1934, when GIORGIO DE SANTILLANA, who had been FRAJESE's classmate in secondary school, introduced him to FEDERIGO ENRIQUES. With the encouragement of ENRIQUES, FRAJESE undertook his first historical research and audited the lectures ENRIQUES was giving on history of mathematics at the university. In 1938 Frajese edited the text of these lectures which was published as a book: *Le matematiche nella storia e nella cultura* (Mathematics in History and Culture) (1938). The following year FRAJESE was invited to teach the history of mathematics in place of ENRIQUES, who had been forced to give up his teaching due to the Fascist racial laws. However, this did not break off their collaboration, and during the Nazi occupation of Rome in 1943, FRAJESE helped ENRIQUES and his family to escape the Nazi deportation of Jews by hiding them in his own house. FRAJESE continued to teach history of mathematics at the University of Rome for some 33 years, until he retired. In the course of his career, and largely under the influence of ENRIQUES, FRAJESE devoted himself mainly to the history of Greek mathematics, and became one of the most authoritative experts in this field. In addition to nearly 70 papers, FRAJESE published several books, including: *La matematica nel mondo antico* (Mathematics in the Ancient World) (1951), which he later included in another work: *Attraverso la storia della matematica* ([A Journey] Through the History of Mathematics) (1962, 2nd ed. 1969); *Platone e la matematica nel mondo antico* (Plato and Mathematics in the Ancient World) (1963); and *Galileo matematico* (Galileo the Mathematician) (1964). In addition, FRAJESE edited EUCLID's *Elements* with notes and historical commentaries (1970), and published an Italian edition of the collected works of ARCHIMEDES (1974).

Secondary literature: MARACCHIA, SILVIO: "Attilio Frajese, storico della matematica." In: [BARBIERI/DEGANI 1989, 161–169]. U.B.

Freudenthal, Hans (* September 17, 1905, Luckenwalde (near Potsdam), Germany; † October 13, 1990, Utrecht, Netherlands). FREUDENTHAL studied at Berlin and Paris. In 1930 he became L. E. J. BROUWER's assistant and in 1931 *privaatdocent* at the University of Amsterdam, but he was dismissed by the Nazis in 1940. After the War he returned to his position in Amsterdam, and in 1946 was appointed professor in Utrecht. FREUDENTHAL held this position until his retirement in 1975. The main themes of his numerous historical publications were the "foundational crisis" in Greek mathematics, LEIBNIZ, KANT, QUETELET, and the history of probability and statistics, the history of geometry around 1900, BROUWER and the history of the foundations of mathematics, and the history of topology.

Further works: *Schrijf dat op, Hans. Knipsels uit een leven* [personal memoirs]. Amsterdam 1987. — *Revisiting Mathematics Education. China Lectures* (1991, with bibliography).

Secondary literature: *P* VIIa (2), 111. — [PINL 1969, 178]. — HOGENDIJK, JAN P.: "*In memoriam* Hans Freudenthal 1905–1990." *Arch. int. Hist. Sci.* **41** (1991), 353–354. — VAN EST, W. T.: "Hans Freudenthal: 17 September 1905 – 13 Oktober 1990." *Nieuw Arch. Wiskd.* (4) **9** (1991), 131–138. — BOS, HENK J. M.: "*In memoriam* Hans Freudenthal (1905–1990)." *HM* **19** (1992), 106–108. — STREEFLAND, L.: *The Legacy of Hans Freudenthal.* Dordrecht 1993. — Portrait in: *Mathematisches Forschungsinstitut Oberwolfach. Festveranstaltung, 5. Oktober 1984.* Freiburg: Gesellschaft für Mathematische Forschung e.V. 1984, 27. P.B.

Friedlein, Johann Gottfried (* January 5, 1828, Regensburg, Germany; † May 31, 1875, Hof, Germany). As a teacher of mathematics in Regensburg (1851–1853), Erlangen (1853–1863), Ansbach (1863–1868), and Hof (1868–1875), FRIEDLEIN was interested in Greek and Roman mathematics, as well as the history of counting.

Main works: *Gerbert, die Geometrie des Boethius und die indischen Ziffern* (1861). — *Zur Geschichte unserer Zahlzeichen und unseres Ziffernsystems* (1864/65). — *Die Zahlzeichen und das elementare Rechnen der Griechen und Römer und des christlichen Abendlandes vom 7. bis 13. Jahrhundert* (1869). — Editions of PEDIASIMUS on geometry (1866); BOETHIUS on arithmetic, music and geometry (1867); VICTORIUS, *Calculus* (1871); PROCLUS, *Commentary on Euclid* (1873).

Secondary literature: *May* 159. *P* III, 477. — CANTOR, MORITZ: "Gottfried Friedlein †." *Z. Math. Phys., Hist.-.lit. Abt.* **20** (1875), 109–113. — CANTOR, MORITZ: "Goffredo Friedlein. *Necrologia.*" *Bull. Bonc.* **9** (1876), 531–553 (translation of the preceding obituary, with bibliography). M.F.

Frobesius (Frobes), Johann Nikolaus (* January 7, 1701, Goslar, Germany; † September 11, 1756, Helmstedt, Germany). FROBESIUS studied mathematics in Goslar, Helmstedt, and Halle, was professor of philosophy, of logic and metaphysics, and beginning in 1741, of mathematics and physics at the University of Helmstedt. Strongly influenced by CHRISTIAN WOLFF's philosophy of the Enlightenment, FROBESIUS published several introductory works, among them *Historica et dogmatica ad mathesin introductio* (Introduction to the History and Theory of Mathematics) (1750) and *Rudimenta biographiae mathematicae* (Rudiments of Mathematical Biographies) (1751/54/55).

Secondary literature: *P* I, 810. — *ADB* **8**, 129. C.J.S.

FUJIWARA Matsusaburō (* February 14, 1881, Tsu, Japan; † October 12, 1946, Fukushima, Japan). One of the most erudite mathematicians in modern Japan, FUJIWARA MATSUSABURŌ was born in Tsu city of Mie prefecture in central Japan. He learned mathematics at the Third Higher School in Kyoto and at the Department of Mathematics of the Imperial University in Tokyo, where the mathematician who influenced him most was FUJISAWA RIKITARO (1861–1933). After having graduated from the Imperial University in 1905, FUJIWARA continued to study mathematics, especially analysis, in its graduate school.

In 1906 he became a lecturer of mathematics at the First Higher School in Tokyo and, in 1907, was promoted to professor of mathematics there. In the same year, along with HAYASHI TSURUICHI (1873–1935), he was also named professor of mathematics at the newly established Tōhoku Imperial University in Sendai. As a prerequisite before accepting a position at an imperial university, FUJIWARA first went abroad for further study in Göttingen, Berlin, and Paris. While in Europe he simultaneously made an effort to buy books on mathematics for his new institute. Upon his return to Japan in 1911, he assumed the professorship at Tōhoku.

On the recommendation of the president of the University, he was awarded a doctorate in 1914; barely a decade later, he and TAKAGI TEIJI (1875–1960) of the Imperial University in Tokyo were both elected to the Imperial Academy of Japan in 1925. FUJIWARA was known to have been an extremely erudite mathematician in comparison to TAKAGI, who was the more original researcher. Soon after the death of his colleague HAYASHI in 1935, FUJIWARA made up his mind to succeed HAYASHI by devoting himself to the study of history of traditional Japanese mathematics, and began giving lectures on the history of both Japanese and European mathematics for undergraduate students of the Tōhoku Mathematical Institute. His well-written monographs on algebra (2 vols., 1928–29) and the differential and integral calculus (2 vols., 1934–39) are regarded as examples of the most outstanding books on mathematics written in pre-war Japan. FUJIWARA's manuscripts on the history of mathematics in Japan survived the bombing of Sendai by the U. S. Air Force on July 10, 1945, and were published posthumously as *Meiji-zen Nippon Sugakushi* (The History of Mathematics of Pre-Meiji Japan) in 5 volumes between 1954 and 1960. His lecture notes on the history of mathematics were also published under the editorship of his disciples as *Nippon Sugakushi-yo* (A Concise History of Japanese Mathematics) in 1952, and as *Seiyo Sugakushi* (A History of Western Mathematics) in 1956. FUJIWARA is remembered among mathematicians as a rival of TAKAGI and as equally qualified an historian of mathematics as MIKAMI YOSHIO.

Secondary literature: *May* 160. *P* V, 404; VI, 835; VIIb(3), 1518. — [SASAKI 1996, with portrait on p. 73]. S.C.

García de Galdeano y Yanguas, Zoel (* July 5, 1846, Pamplona, Spain; † March 28, 1924, Zaragoza, Spain). The most cosmopolitan among Spanish mathematicians of his day, GARCÍA DE GALDEANO was professor at the University of Zaragoza from 1891 onwards. His work includes over two hundred titles devoted to the modernization of Spanish mathematics through the introduction of the most relevant foreign works, and by using history of mathematics as a tool for this purpose, as well as for a deeper understanding of current mathematical developments.

Secondary literature: *P* III, 491; IV, 478; VI, 851. — HORMIGÓN, MARIANO: "Una aproximación a la biografía científica de García de Galdeano." *El Basilisco* **16** (1984), 38–47. E.A.; M.H.

García de Zuñiga, Eduardo (* September 30, 1867, Montevideo, Uruguay; † April 2, 1951, Progreso, Uruguay). GARCÍA DE ZUÑIGA, an engineer of major public works as well as a mathematician and historian of mathematics, created one of the world's most important specialized libraries for the history of mathematics (the library is now located at the *Facultad de Ingeniera, Universidad de la Republica*, Montevideo). GARCÍA DE ZUÑIGA studied mathematics in Montevideo and Berlin (1903–1905), and later played an important role in the foundation of the Uruguayan school of mathematics. GARCÍA DE ZUÑIGA himself published a number of mathematical works, and devoted considerable attention to acquiring a superb mathematical library, beginning in 1905, to which he added a large number of books in 1912 and 1923, and continued to expand the collection into the 1940s. He also collected important periodicals for his library, including the *Acta Eruditorum, Journal des Savants*, the *Annales de Gergonne*, and *Crelle's Journal*, as well as many specialized 20th-century journals. Today the collection of periodicals is up to date (except for the period of dictatorship in Uruguay, 1973–1984). The library also contains works devoted to mathematics and mathematicians of all periods and mathematical subjects. As a historian of mathematics, GARCÍA DE ZUÑIGA gave detailed lectures that were later published in periodicals, especially on Greek mathematics (he mastered Latin and Greek).

Main works: A short but important collection of works has recently been gathered together as *Lecciones de historia de las matemáticas* (Lectures on the History of Mathematics) (1992).

Secondary literature: The most important holdings of the mathematical library founded by GARCÍA DE ZUÑIGA are described in [MARTÍNEZ 1994]. M.O.

Genocchi, Angelo (* March 5, 1817, Piacenza, Italy; † March 7, 1889, Turin, Italy). As early as his student days in secondary school (*liceo*) the young GENOCCHI revealed a special gift for mathematics. In his home-town of Piacenza, however, the local university offered no courses in advanced mathematics; there was only a Law Faculty, and so Genocchi enrolled there as a student, although he continued to study mathematics on his own. In 1838 he obtained his degree (*laurea*) in jurisprudence, and two years later he began his practice as an attorney. In 1845 he was also named professor of Roman Law in the Law Faculty at Piacenza. At the time Piacenza was under Austrian rule, which explains why GENOCCHI enthusiastically welcomed the beginning of the war for independence from Austria in 1848. One year later, following the defeat of Custoza that ended the war, and hopes of independence, GENOCCHI left Piacenza in a state of self-imposed exile. Having decided to give up the practice of law, he settled in Turin where he devoted himself completely to mathematics, and to number theory in particular. In Turin GENOCCHI also became acquainted with the mathematicians FELICE CHIÒ and GIOVANNI PLANA. GENOCCHI audited their lectures at the university, and at CHIÒ's suggestion, he began to teach algebra and complementary geometry in November of 1857. GENOCCHI spent the rest of his academic career in Turin. In 1860 he was named professor of advanced analysis, and from 1865 to his death he taught infinitesimal analysis, with GIUSEPPE PEANO among his students. GENOCCHI's lectures were edited by PEANO as: *Calcolo differenziale e principii di calcolo integrale pubblicato con aggiunte dal Dr. Peano* (Differential Calculus and the Principles of Integral Calculus, Published with Additions of Dr. Peano) (1884). This was to become one of the most influential treatises on analysis in its day, to the point that in his article for the German *Encyklopädie*, ALFRED PRINGSHEIM listed it among the most important treatises on analysis ever published.

GENOCCHI's interest in the history of mathematics was aroused in the mid-1850s, when BALDASSARRE BONCOMPAGNI published LEONARDO PISANO's *Liber Quadratorum* (Book of Squares). GENOCCHI pointed out some inaccuracies in the edition, which prompted BONCOMPAGNI to edit a revised version of the book. This led to an intensive correspondence between them — about 1500 letters from BONCOMPAGNI to GENOCCHI survive and are kept in the *Genocchi Archive* in the public library in Piacenza. Some 220 more letters between the two have recently been discovered in Rome (see [Israel 1990]). Most of GENOCCHI's contributions to the history of mathematics were devoted to medieval mathematics, and to the work of LEONARDO PISANO in particular. Most were published in TORTOLINI's *Annali di scienze matematiche e fisiche* in the 1850s. In addition, some 15 papers of historical content were later published in BONCONPAGNI's *Bullettino*. GENOCCHI's last historical papers, concerning the foundations of geometry, were devoted to the history of non-Euclidean geometry. GENOCCHI was skeptical about the consistency of non-Euclidean geometry and combined his critical remarks about BELTRAMI's interpretation of LOBACHEVSKIĬ's geometry — which were later to be confirmed by HILBERT — with a penetrating historical analysis of a forgotten paper by FRANÇOIS DAVIET DE FONCENEX on the theory of the lever, where the latter pointed out the relationship between the principles of mechanics and EUCLID's fifth postulate.

Secondary literature: *May* 169. *P* III, 505; IV, 1708; V, 419. — PEANO, GIUSEPPE: "Angelo Genocchi." *Annuario dell'Universita' di Torino* (1889/90), 195–202. — SIACCI, FRANCESCO: "Cenni necrologici di Angelo Genocchi." *Memorie della Regia Accademia delle Scienze di Torino. Classe di Scienze Fisiche, Matematiche e Naturali* (2) **39** (1889), 463–495 (with bibliography). — ISRAEL, GIORGIO: "On Correspondence between B. Boncompagni and A. Genocchi." *HM* **17** (1990), 48–54. — CONTE, ALBERTO, and LIVIA GIACARDI (eds.): *Angelo Genocchi e i suoi interlocutori scientifici.* Torino 1991 (with portrait). — Portrait: [BÖLLING 1994, 3.6]. U.B.

Gerhardt, Carl Immanuel (* December 2, 1816, Herzberg (Saxony), Germany; † May 5, 1899, Halle, Germany). GERHARDT, a teacher of mathematics in Salzwedel (1839–1853), Berlin (1853–1855), and Eisleben (1856–1891), was interested in Greek mathematics, but above all in the history of the calculus and especially in LEIBNIZ.

Main works: *Leibnizens mathematische Schriften* (7 vols., 1849–1863). — *Briefwechsel zwischen Leibniz und Christian Wolf* (1860). — *Die philosophischen Schriften von Gottfried Wilhelm Leibniz* (7 vols., 1875–1890). — *Der Briefwechsel von Gottfried Wilhelm Leibniz mit Mathematikern*, vol. 1 (1899). — Editions of MAXIMUS PLANUDES on arithmetic (1865); PAPPUS, *Collections*, books 7 and 8 (1871). — *Geschichte der Mathematik in Deutschland* (1877).

Secondary literature: *May* 170. *P* I, 882; III, 507; IV, 492. — *NDB* **6**, 285–286. — [CANTOR 1900, with portrait]. — MÜLLER, FELIX: "Carl Immanuel Gerhardt." *Bibl. Math.* (3) **1** (1900), 205–216 (with portrait). — HESS, HEINZ-JÜRGEN: "Karl Immanuel Gerhardt. Ein großer Leibniz-Editor." In: *Stud. Leibnitiana, Supplementa* **26** (1986), 29–64 (with portrait). — Portrait: [BÖLLING 1994, 32.4]. M.F.

Geymonat, Ludovico (* May 11, 1908, Barge, Cuneo, Italy; † November 29, 1991, Rho, Milan, Italy). LUDOVICO GEYMONAT was one of the more influential philosophers and a leading intellectual figure in Italy in the second half of the twentieth century. Opposed to the idealistic philosophy still dominant in Italy at that time, he helped to create a new dimension in Italian culture by promoting the development of modern logic, philosophy and history of science (and history of mathematics in particular). Educated at a Jesuit college, GEYMONAT enrolled at the University of Turin, from which he graduated (*laurea*) in philosophy in 1930. Two years later he completed his degree in mathematics, having studied with

Giuseppe Peano and Guido Fubini, who supervised his thesis. In addition to Peano, neopositivism exerted a major influence on Geymonat's early philosophical research. He spent some time in Vienna where he heard Moritz Schlick's lectures and became acquainted with members of the Vienna Circle. After returning to Italy, Geymonat published *La nuova filosofia della natura in Germania* (The New Philosophy of Nature in Germany) (1934) which, according to Schlick, was "the best presentation of our ideas hitherto published by a neutral observer."

Over the course of the next decade Geymonat taught philosophy and mathematics in private secondary schools in Turin, since his refusal to join or endorse the Fascist party excluded him from any public university career. During World War II he participated in the Resistance movement agains the Nazis. In 1945 Geymonat published a collection of his philosophical writings as *Studi per un nuovo razionalismo* (Studies for a New Rationalism), which was symbolically dated April 25, 1945, the day of Italy's liberation from the Nazis which marked the end of the War in Italy. Then, for some 20 years Geymonat combined his philosophical and scientific work with intense political activity as a member of the Italian Communist Party.

Beginning in 1946, he also taught a course on history of mathematics at the University of Turin, and soon thereafter published his lectures in a pioneering book, *Storia e filosofia delli analisi infinitesimale* (History and Philosophy of Infinitesimal Analysis) (1947). In this work, which is still worth studying, Geymonat combined the "historical problems" of the development of analysis from antiquity to the beginning of the 20th century with his favorite subjects, logic and philosophy of science. The core of Geymonat's book (indeed two-thirds of it) is devoted to "the success of the demand for rigor" in the 19th century. In particular he focused on both the emergence of set theory and the problem of the foundations of mathematics with a thorough discussion of Cantor's theory of transfinite numbers and the axiomatization of set theory. In the final chapter Geymonat provided a survey of Lebesgue integration theory and modern measure theory. In 1949 Geymonat was appointed extraordinary professor of theoretical philosophy at the University of Cagliari, where he also taught advanced mathematics. Three years later he moved to the University of Pavia, after which he was appointed to a newly-created chair for philosophy of science at the University of Milan in 1955. This was the first full professorship for philosophy of science at any Italian university. (For a number of years Geymonat also taught logic at Milan.) After his retirement in 1978 Geymonat was named an *emeritus* professor of the University of Milan.

The years Geymonat spent in Milan were the most fruitful and influential of his entire academic career. There he created a philosophical and scientific "school" which attracted an impressive number of young students interested in logic, philosophy, and history of mathematics (and of science in general). He promoted the diffusion of these disciplines in Italian universities and combined his academic work with intense editorial activity, founding and editing new series of books devoted to the history and philosophy of science. Geymonat himself did original research in various fields, including history of science and of mathematics as well. In addition to works of philosophical interest like his *Filosofia e filosofia della scienza* (Philosophy and Philosophy of Science) (1960), GEYMONAT published *Galileo Galilei* (1957), and some years later, *Storia della matematica* (History of Mathematics), which was included in the *Storia della scienza* (History of Science) (in 3 vols., 1963) edited by NICOLA ABBAGNANO. By the end of the 1960s GEYMONAT's approach to the philosophy of science turned away from his former neorationalism to advocate Marxist dialectical materialism and, somewhat later, a kind of dialectical realism as explained in his book *Scienza e realismo* (Science and Realism) (1977). GEYMONAT's activity in the field of the history of science culminated in the monumental work, *Storia del pensiero filosofico e scientifico* (History of Philosophical and Scientific Thought) (in 9 vols., 1970–1977). This very influential work, largely written by GEYMONAT himself, covers the subject from antiquity to the second half of the 20th century, and includes substantial chapters on the history of logic and mathematics written by CORRADO MANGIONE. (Two additional volumes of the *Storia*, edited by ENRICO BELLONE and CORRADO MANGIONE, were published in 1995). After the completion of his *Storia*, in the last decade of his life GEYMONAT was mainly interested in general philosophical questions. However, the history of science still influenced his epistemological views which, in his final years, turned towards what he called "scientific historicism," a philosophical view inspired by ENRIQUES's philosophy of science and the role of history in it.

Secondary literature: MANGIONE, CORRADO (Ed.): *Omaggio a Ludovico Geymonat. Saggi e testimonianze.* Padova 1992. — Obituaries of LUDOVICO GEYMONAT by FRANCESCO BARONE, NORBERTO BOBBIO, PIETRO BUZANO, and GABRIELE LOLLI are collected in: *Atti Acc. Sci. Torino, Classe Sci. Morali Stor. Fil.* **127** (1993), 73–96. — ROERO, CLARA SILVIA: "Ludovico Geymonat." In: ROERO, CLARA SILVIA (Ed.): *La Facoltà di Scienze matematiche fisiche naturali di Torino (1848–1998)*, vol. 2: *I docenti.* Torino 1999, 625–628 (with portrait).

U.B.

Gherardi, Silvestro (* December, 17, 1802, Lugo di Romagna, Ravenna, Italy; † July 29, 1879, Florence, Italy). After completing his studies at the University of Bologna, GHERARDI joined the faculty as a professor at the age of 25. He taught mechanics, hydrology and, later, physics from 1827 until 1849. GHERARDI combined his scientific work with political activity — participating in the Bologna uprisings of 1831 and 1848 (at the time, Bologna was part of the Papal States). In 1849 he served the Roman Republic as minister of Public Education. After the fall of the Republic, GHERARDI was also forced to leave his position at the university. From 1857 to 1861 he was professor of physics at the University of Turin and from 1862, after the political unification of Italy, he was named a professor *emeritus* of the University of Bologna.

GHERARDI's scientific work was mainly devoted to experimental physics, to electricity and magnetism in particular. However, he was also interested in the history of science and of mathematics as well. GHERARDI devoted his early historical papers to the lives and works of LUIGI GALVANI and ALESSANDRO VOLTA. Then his interests turned to the history of mathematics at the University of Bologna. Related to this was his work, *Di alcuni materiali per la storia della Facoltà Matematica nell'antica Università di Bologna* (Some Materials for the History of the Mathematical Faculty of the Ancient University of Bologna) (1846), in which GHERARDI published part of the results of his studies. This widely-read book was eventually translated into German by M. CURTZE in 1871. It is mainly concerned with the early periods of the Bologna mathematical faculty, beginning with CECCO D'ASCOLI who lectured on astrology around 1320, and who wrote the encyclopedic work *Acerba* (from the Latin "coacervus" (pile)). This incomplete poem in rhyme was first appreciated by LIBRI in his *Histoire*. However, the main part of GHERARDI's book is devoted to the role played by mathematicians at Bologna, and by SCIPIONE DAL FERRO in particular, in discovering the solution of the cubic equation and in contributing generally to the emergence of modern algebra. GHERARDI was the first to collect and study the *cartelli di matematica disfida* (the mathematical challenges) exchanged between TARTAGLIA and FERRARI in 1547–48. Abstracts and excerps from these documents, which are of exceptional value for the history of Renaissance mathematics, had been published earlier by GIOVANNI FANTUZZI in his *Notizie degli scrittori bolognesi* (Notes on Bologna Writers) (1794). Eventually they were published in extenso by ENRICO GIORDANI as *I sei cartelli di matematica disfida primamente intorno alla generale risoluzione delle equazioni cubiche di Ludovico Ferrari coi sei contro-cartelli in risposta di Niccolò Tartaglia* (Six Mathematical Challenges Primarily Related to the General Solution of Cubic Equations by Ludovico Ferrari and Six Challenges in Response by Niccolò Tartaglia) (1878). These studies on Renaissance mathematics in Bologna made GHERARDI one of the most authoritative Italian historians of his day.

Another favorite subject of GHERARDI's historical research was the life and work of GALILEO. In particular, in 1869–70 GHERARDI published some important papers devoted to GALILEO's trial, including a number of relevant documents he was able to find in the archives of *S. Uffizio* during his stay in Rome in 1849 at the time of the Roman Republic.

Secondary literature: *May* 172. *P* III, 511. — PROCISSI, ANGIOLO: "Silvestro Gherardi scienziato e storico della scienza." *Studi Romagnoli* **4** (1953), 87–101 (with bibliography). — FIOCCA, ALESSANDRA: "Scritti inediti di Silvestro Gherardi sulla storia delle matematiche." In: [BARBIERI/DEGANI 1989, 181–198]. — BRIATORE, LUIGI, and FRANCESCO GHEPARDI: " Silvestro Gherardi." In: ROERO, CLARA SILVIA (Ed.): *La Facoltà di Scienze matematiche fisiche naturali di Torino (1848–1998)*, vol. 2: *I docenti.* Torino 1999, 243–245 (with portrait). U.B.

Gilbert, (Louis) Philippe (* February 7, 1832, Beauraing, Belgium; † February 4, 1892, Leuven, Belgium). GILBERT studied mathematics at Leuven and Paris, whereupon he became professor at the University of Leuven in 1855. Among his historical writings are a study of the problem of the rotation of a rigid body about a point, and articles on ADRIANUS ROMANUS, RENÉ FRANÇOIS DE SLUSE,

AMPÈRE, MICHEL CHASLES, LÉON FOUCAULT, VICTOR PUISEUX, and GASPAR PAGANI.

Secondary literature: *May* 173. *P* III, 515; IV, 498. — *BNB* **38**, 250–252. — MANSION, PAUL: "Liste des publications de Louis-Philippe Gilbert." *Rev. Quest. sci.* **31** (1892), 628–641. — MANSION, PAUL: *Notice sur les travaux scientifiques de Louis-Philippe Gilbert* (1893, with portrait). — MAWHIN, JEAN: "Louis Philippe Gilbert, De l'analyse mathématique aux sources du Nil, en passant par la rotation de la Terre et le proces de Galilée." *Rev. Quest. sci.* **160** (1989), 385–396. — MAWHIN, JEAN: "Louis-Philippe Gilbert et le proces de Galilée." In OPSOMER, CARMELIA (Ed.): *Copernic, Galilée et la Belgique. Leur réception et leurs historiens* (1995), 47–61. P.B.

Ginsburg, Jekuthiel (* August 15, 1889, Lipniki, Wolhynia, Russia; † October 7, 1957, New York, New York, USA). GINSBURG, who immigrated to the United States in 1912, studied mathematics at Columbia University, where he received his master's degree in 1916. He taught mathematics at Yeshiva University in New York City, and was the founding editor of *Scripta Mathematica* (see [GINSBURG 1932]). His publications include *Ketavim mivharim* (in Hebrew), a study of Jewish mathematics (1960). In collaboration with D. E. SMITH, GINSBURG also wrote *History of Mathematics in America Before 1900* (1934).

Secondary literature: *May* 173. *P* III, 1234; IV, 1432; V, 1233. — BELKIN, SAMUEL: "Professor Jekuthiel Ginsburg." *Scripta Math.* **23** (1957), 7–9 (with portrait, facing p. 7). — BOYER, CARL B.: "Jekuthiel Ginsburg (1889–1957)." *Isis* **49** (1958), 335–336. — Portrait: *Scripta Math.* **6** (1939), facing p. 69. J.W.D.

Gloden, Albert (* March 5, 1901, Luxembourg; † March 2, 1966 Luxembourg). After his studies in Paris, GLODEN took a teaching post at the Lycée in Esch (Grand Duchy of Luxembourg). In 1935 he became a teacher of mathematics and physics at the *Athénée* in Luxembourg. Among his historical writings are a sketch of the history of number theory in Belgium, and articles on Belgian mathematicians born in the Grand Duchy of Luxembourg.

Secondary literature: TATON, RENÉ: "Albert Gloden (1901–1966)." *Rev. Hist. Sci.* **19** (1966), 274. — ITARD, JEAN: "Albert Gloden (1901–1966)." *Arch. int. Hist. Sci.* **20** (1967), 103–105. — AREND, SYLVAIN: "*In memoriam* Albert Gloden." *Ciel et Terre* **84** (1968), 453–454 (with portrait). P.B.

Gnedenko, Boris Vladimirovich (* January 1, 1912, Simbirsk, Russian Empire; † December 27, 1995, Moscow, Russia). BORIS GNEDENKO taught at the Ivanovo Textile Institute (1930–1934) and at Moscow University (1938–1945, 1960–1995). From 1945 until 1960 he was a researcher in Kiev at the Institut of Mathematics of the Ukrainian Academy of Sciences, which he also served as director (1956–1958). In 1960 he returned to Moscow as professor in the faculty of mathematics and mechanics at Moscow University. A well-known mathematician, he was a specialist in probability theory and its applications. His interests in the history of mathematics included the history of probability theory and mathematical statistics, as well as the history of mathematics in Russia.

Main works: *A Concise History of Mathematics in Russia* (1946, in Russian). — *Probability Theory* (1948, in Russian). — In collaboration with O. SHEYNIN: "The Theory of Probability," Chapter 4 of A. N. KOLMOGOROV and A. P. YUSHKEVICH (Eds.), *Mathematics of the 19th Century. Mathematical Logic. Algebra. Number Theory. Probability Theory* (1992).

Secondary literature: *May* 174. — [KUROSH 1959] — [FOMIN/SHILOV 1969] — [BORODIN/BUGAI 1979] — [BOGOLYUBOV 1983, with portrait on p. 137]. S.S.D.

Godeaux, Lucien (* October 11, 1887, Morlanwelz, Belgium; † April 21, 1975, Liège, Belgium). GODEAUX studied mathematics at the University of Liège, as well as in Bologna, Göttingen, and Paris. He was lecturer in mathematics at the Royal Military Academy in Brussels (1920–1925), and in 1925 became professor at the University of Liège. He published numerous biographical notes on Belgian and foreign mathematicians, among them FRANCESCO SEVERI, JACQUES HADAMARD, ELIE CARTAN, and FEDERIGO ENRIQUES. He is the author of chapter VII on geometry in TATON's *Histoire générale des sciences*, vol. III, part 2 (1964).

Secondary literature: *P* V, 432; VI, 911; VIIb, 1663. — SEGRE, BENIAMINO: "Lucien Godeaux." *Boll. UMI* **11** (4) (1975), 639–644. — GODEAUX, JEAN and PAUL GODEAUX: "Lucien Godeaux (1887–1975). Sa vie – son œuvre." *Bull. Soc. R. Sci. Liège* **64** (1995), 3–77 (with bibliography and portrait). P.B.

Golius, Jacobus (* 1596, The Hague, Netherlands; † September 28, 1667, Leiden, Netherlands). GOLIUS, who studied medicine, mathematics, and astronomy at Leiden, went on to become professor of Arabic (1625) and mathematics (1629) at the University of Leiden. He was the founder of the University Observatory. GOLIUS is known as an early collector and editor of Arabic mathematical and astronomical manuscripts ([GOLIUS 1669]).

Secondary literature: *P* I, 927. — *NNBW* **10**, 287–289. P.B.

Gravelaar, Nicolaas Lambertus Willem Anthonie (* November 29, 1851, Groningen, Netherlands; † February 18, 1913, Deventer, Netherlands). NICOLAAS GRAVELAAR, a primary school teacher at Groningen, headmaster in Zwolle, and (from 1880) lecturer at the teacher training college in Deventer, published studies on PITISCUS's *Trigonometria*, the works of JOHN NEPER, STEVIN's *Problemata geometrica*, the contributions of CARDANO and FERRARI to the solution of algebraic equations, and the origin of the multiplication sign ×.

Secondary literature: *May* 178. *P* IV, 529; V, 446. — STRUIK, DIRK J.: "N. L. W. H. Gravelaar (1851–1913)." *Euclides* **6** (1929), 204–207. P.B.

Günther, Adam Wilhelm Siegmund (* February 6, 1848, Nürnberg, Germany; † February 3, 1923, Munich, Germany). After having been *Privatdozent* at Erlangen since 1872 and Munich (*Polytechnikum*) since 1874, SIEGMUND GÜNTHER taught mathematics in Ansbach from 1876 to 1886, when he became ordinary professor of geography at the *Technische Hochschule* in Munich (1886–1919). Among his numerous publications on the history of mathematics and geography are: *Vermischte Untersuchungen zur Geschichte der mathematischen Wissenschaften* (Various Studies on History of the Mathematical Sciences) (1876), *Geschichte der Mathematik. I. Teil: Von den ältesten Zeiten bis Cartesius* (History of Mathematics, Part I. From the Earliest Times to Descartes) (1908), and *Geschichte des mathematischen Unterrichts im deutschen Mittelalter bis zum Jahre 1525* (History of Mathematics Instruction in the German Middle Ages to the Year 1525) (1887).

Secondary literature: *DSB* **5**, 573–574. *May* 181. *P* III, 560; IV, 546; V, 460; VI, 971; VIIa(2), 312. — *NDB* **7** (1966), 266. — FAVARO, ANTONIO, in *Atti Adunanze Ist. Veneto Sci. Lett. Arti* (5) **3** (1877), 913–958. — HAAS, ARTHUR: "Siegmund Günther. Zum 70. Geburtstage." *Arch. Gesch. Nat. Techn.* **8** (1917), 129–133 (with portrait). — KISTNER, A.: "Zu S. Günthers siebenzigstem Geburtstag." *MGMN* **17**, 1–4 (with portrait). — WIELEITNER, HEINRICH: "Siegmund Günther †." *MGMN* **22** (1923), 1–2. — SCHÜLLER, WERNER: "Siegmund Günther †." *ZMNU* , 109–113 (with portrait). — GÜNTHER, LUDWIG: "Günther, Siegmund. Mathematiker und Geograph. 1848–1923." In: *Lebensläufe aus Franken* (= Reihe 7 der *Veröffentlichungen der Gesellschaft für fränkische Geschichte*). Vol. 4. Würzburg 1930, 204–219. — FOLKERTS, MENSO: "Erinnerung an Siegmund Günther (1848–1923)." *Nbl. DGGMNT* **34** (1984), 78–81 (with portrait). — Portrait: [BÖLLING 1994, 39.1]. M.F.

Guimarães, Rodolfo Ferreira Dias (* January 4, 1866, Oporto, Portugal; † July 9, 1918, Lisbon, Portugal). GUIMARÃES started his advanced studies at the Oporto Polytechnic Academy, followed by the Army School in Lisbon, where he studied military engineering and from which he graduated in 1891. In mathematics he published mainly on topics related to geometry. From 1900 onwards his main research field was the history of mathematics. Apart from his *Aperçu* (see Section 14.4), he published papers on PEDRO NUNES: [GUIMARÃES 1914/15], [GUIMARÃES 1915c], [GUIMARÃES 1916a], and [GUIMARÃES 1916b]; and on other historical subjects (see [GUIMARÃES 1911], [GUIMARÃES 1917], and [GUIMARÃES 1921]).

Secondary literature: Anonymous obituary in *Revista Militar* **70** (1918), 1–15 (with portrait). — SARAIVA, LUIS: "Historiography of Mathematics in the Works of Rodolfo Guimarães." *HM* **24** (1997), 86–97. L.S.

Hankel, Hermann (* February 14, 1839, Halle, Germany; † August 29, 1873, Schramberg (Black Forest), Germany). HANKEL's short career as a mathematician began in 1863 as *Privatdozent* at Leipzig, where he was promoted to the rank of extraordinary professor in 1867. There followed ordinary professorships in Erlangen (1867) and Tübingen (1869). Apart from his interest in various branches of mathematics, he published two original works on the history of mathematics:

Die Entwicklung der Mathematik in den letzten Jahrhunderten (The Development of Mathematics During the Past Few Centuries) (1871), and *Zur Geschichte der Mathematik in Alterthum und Mittelalter* (On the History of Mathematics in Antiquity and the Middle Ages) (1874).

Secondary literature: *DSB* **6**, 95–96. *May* 185. *P* III, 582. — *ADB* **10**, 516–519. — *NDB* **7**, 618–619. — ZAHN, W. VON: "Einige Worte zum Andenken an Hermann Hankel." *Math. Ann.* **7** (1874), 583–590. (Italian translation with bibliography in *Bull. Bonc.* **9** (1876), 290–308). — HOFMANN, JOSEPH EHRENFRIED: Introduction to reprint of [HANKEL 1874, ix–xiv] (with portrait). M.F.

HAYASHI Tsuruichi (* June 13, 1873, Tokushima (Shikoku Island), Japan; † October 4, 1935, Matsue City, Shimane Prefecture, Japan). HAYASHI TSURUICHI, after having been educated at the Third Higher School in Kyoto, studied mathematics at the Imperial University in Tokyo under professors KIKUCHI DAIROKU (1855–1917) and FUJISAWA RIKITARO (1861–1933). He graduated from the University in 1897, along with TAKAGI TEIJI (1875–1960). HAYASHI's interest in the history of Japanese mathematics is said to have been stimulated by KIKUCHI. HAYASHI taught mathematics successively at the Higher Normal School in Tokyo and at the newly-founded Kyoto Imperial University. When plans were made in 1907 to establish Tōhoku Imperial University, he was named professor of mathematics, as was FUJIWARA MATSUSABURŌ (1881–1946). When the university was officially inaugurated in 1911, HAYASHI not only served as director of its Mathematical Institute, but launched a new journal, *The Tōhoku Mathematical Journal*, the first international journal devoted to mathematics in Japan. HAYASHI and his younger colleague, FUJIWARA, aspired to model their institution after the Mathematical Institute at Göttingen.

HAYASHI wrote a number of mathematics textbooks for secondary schools. His rivalry with MIKAMI YOSHIO over the history of Japanese mathematics began in 1906, and ended only with HAYASHI's sudden death on October 4, 1935. His papers on the history of mathematics were collected and published posthumously in two volumes in Tokyo in 1937 under the title *Wasan Kenkyu Shuroku* (Collected Papers on Japanese Mathematics).

Secondary literature: *May* 187. *P* V, 509; VI, 1055. — [FUJIWARA 1936, with portrait]. S.C.

Heath, Thomas Little (* October 5, 1861, Barnetby-le-Wold, Lincolnshire, England; † March 16, 1940, Ashtead, Surrey, England). Sir THOMAS HEATH established his reputation as an historian of Greek mathematics and the author of many specialist works on individual mathematicians while holding one of the most exacting as well as important posts in British public life, Joint Permanent Secretary of the Treasury. He was one of six children of SAMUEL HEATH, a farmer of Thornton Abbey, Barnetby-le-Wold, Lincolnshire. All six were mathematically and musically gifted. Thomas was educated at Caistor Grammar School and Clifton College before going to Trinity College, Cambridge, as a Foundation Scholar. There he obtained a First Class in the Classical Tripos, Part I (the first part of the undergraduate degree) in 1881, took a First Class in the Mathematical Tripos, Part I, being Twelfth Wrangler (twelfth in order of merit) in 1882, and a First Class in the Classical Tripos, Part II, in 1883. He obtained the first place in the examination for the Home Civil Service (Class I) and was appointed a clerk in the Treasury in 1884. He was made a fellow of Trinity in 1885 while pursuing his career at the Treasury, where by rapid steps he rose to be Joint Permanent Secretary in 1913, sharing the post with Sir JOHN (afterwards Lord) BRADBURY.

HEATH's responsibility was preparation of the public service for the condition of war. In 1919, when the Treasury was reorganized after World War I, HEATH became Comptroller General and Secretary to the Commissioners for the Reduction of the National Debt. His special task was the financing of the Irish Land Acts. By the testimony of his colleagues he was an excellent civil servant in the old tradition of written minutes rather than oral discussion. He retired from public service in 1926.

By that time he had gained another reputation on which his fame mainly rests. It had begun with a dissertation on *Diophantos of Alexandria* in 1885, a remarkable work devoted to the only extant Greek writer on algebra, especially considering that HEATH's book was written before the appearance of HEIBERG's text. Later, HEATH brought out in 1910 a thoroughly revised second edition in which the problems suggested to FERMAT and EULER by DIOPHANTUS are examined. In 1886 HEATH produced *Appolonius of Perga: a Treatise on the Conic Sections*, the first easily intelligible translation of that seminal work in any language, with a valuable historical introduction (reprinted in 1961). The *Works of Archimedes* followed in 1897, with translations of all the books of that genius then known, for which HEATH provided scholarly essays in which he showed *inter alia* how the Syracusan had anticipated the integral calculus. A gap in HEATH's writing was explained in 1908 when he produced a massive work, *The Thirteen Books of Euclid's Elements*, in three volumes. Despite the popularity of EUCLID, there had not previously been a reliable English translation (a second edition followed in 1925). In 1920 Heath published *Euclid in Greek, Book I* in the vain hope that it might lead to a revival of the teaching of EUCLID in the schools. In the meantime, after HEIBERG discovered a manuscript version of the *Method* in which ARCHIMEDES revealed his manner of working, HEATH brought out a supplement with translation in 1912. In 1913 HEATH turned his attention to astronomy with *Aristarchus of Samos: The Ancient Copernicus*, in which a translation of *On the Sizes and Distances of the Sun and Moon* is preceded by a complete history of Greek astronomical science up to ARISTARCHUS. HEATH devoted an entire book to the subject in his *Greek Astronomy* (1932); although it makes some mention of HIPPARCHUS and PTOLEMY, it is a pity that HEATH was never able to give his attention to the serious mathematical writings of those authors.

HEATH's specialist works culminated in 1921 in *A History of Greek Mathematics* in two volumes. Arranged partly by date and partly by subject, these volumes gave a magnificent conspectus of the whole of Greek mathematics from

THALES to Byzantine commentators. There had been a previous history, by GINO LORIA in Italian, which can still be read with profit. Although HEATH's work would need some revision if a new edition were brought out today, by and large it still remains the best history in any language.

HEATH showed his awareness of the direction revision might take when he brought out in 1931 *A Manual of Greek Mathematics*, a simplified and abbreviated version of the 1921 treatise. In this he paid some attention to OTTO NEUGEBAUER's interpretation of Babylonian mathematics as algebraic, a feature which was then thought to have been just below the surface of the more geometrical Greek mathematics and to have influenced DIOPHANTUS in particular.

When he died on March 16, 1940, HEATH left behind a manuscript on *Mathematics in Aristotle* which was almost ready for the printer and which Sir DAVID ROSS saw through the press in 1949; it went into a second edition in 1970. HEATH had hoped to produce an edition of PAPPUS, but it was not to be.

In 1914 HEATH married ADA MARY THOMAS. The marriage produced a son and a daughter, and some credit for HEATH's amazing work must be given to his family as most of his work had to be done at home in the evenings when rigid silence had to be observed. ADA was an accomplished musician, and HEATH himself was no mean pianist. When he had an intractable problem he would move over to the piano, and while playing would often find that the solution would come to him. He was a great admirer of BRAHMS, and early in his Treasury career made a trip to Vienna merely to see him at a distance.

HEATH was made a K.C.B. (Knight Commander (of the Order) of the Bath) in 1909 and K.C.V.O. (Knight Commander of the Royal Victorian Order) in 1916 for his work in the Treasury. He was made a Fellow of the Royal Society in 1912 and served for two periods on its Council. He was President of the Mathematical Association in 1922–25, and was elected a fellow of the British Academy in 1932.

My personal acquaintance with HEATH began in 1928 when I elected to read Greek mathematics as a special subject in "Greats" in Oxford and he was called upon to examine me. My *Loeb* volumes of *Selections* illustrating the history of Greek mathematics were in effect a handbook to his *History*. I was later asked to read the proofs of his *Mathematics in Aristotle* before its posthumous publication.

Secondary literature: *DSB* **6**, 210–211. *May* 188. *P* VI, 1057; VIIb(3), 1913. — *DNB*, Supplement 1931–1940 (1949), 416. — HEADLAM, M. F., with a mathematical contribution by J. GILBART SMYLY: "Sir Thomas Little Heath." *Proc. BA* **26** (1940), 1–16. — SMITH, DAVID EUGENE: "Sir Thomas Little Heath." *Osiris* **2** (1936), v–xxvii (with bibliography and portrait). — THOMPSON, D'ARCY W.: "Sir Thomas Little Heath. 1861–1940." *Obituary Notices Fellows RS* **3** (1939–1941), 409–426 (with portrait, 408).

I.B-T. †

Heiberg, Johan Ludvig (* November 27, 1854, Aalborg, Denmark; † January 4, 1928, Copenhagen, Denmark). HEIBERG, from a wealthy medical doctor's family, was appointed professor of classical philology at the University of Copenhagen in 1896, but his academic career had already begun some twenty years earlier. In 1883 he was elected a member of the Royal Danish Academy of Arts and Sciences; in addition to his membership in a number of learned societies abroad, he also received honory doctorates from Oxford, Leipzig, and Berlin.

In 1879 HEIBERG earned his doctorate with the epoch-making dissertation, *Quaestiones Archimedeae* (Archimedean Questions), devoted to ARCHIMEDES's life, works, and, especially, the transmission of his texts. HEIBERG's dissertation initiated a new concept of textual criticism, pinning down every detail with a complete explanation of every step in the transmission of a given text. From 1883 on he was almost always occupied with editing Greek texts on mathematics: six volumes of EUCLID's *Opera* (1883–95), the extant Greek parts of APOLLONIUS's *Conics* (1891–93), and SIMPLICIUS's commentary on ARISTOTLE's *De Caelo* (On the Heaven) (1894). HEIBERG was also co-director of a *Gymnasium* in Copenhagen and, until his death, taught high school Greek every morning from 8 to 10.

With his editions HEIBERG cleared an immense and confused territory in history of science, and was considered the obvious successor for a vacant chair in philology at the University of Copenhagen in 1896. Over the next thirty years he continued to contribute to the history of numerous branches of science: geometry, mechanics, astronomy, medicine, and alchemy, and even paved the way to Byzantine studies (the latter has only been acknowledged recently ([Alpers 1988]).

To the Danish public he offered scores of popular yet authentic and original papers on ancient Greek culture, dominated by his special interest in the history of religion. Beginning in 1901, he helped to edit the complete works of the Danish philosopher SØREN KIERKEGAARD.

HEIBERG's finest achievement, however, was the identification and reconstruction of the text of ARCHIMEDES's *Ephodos* (The Method) discovered in Constantinople in 1906. Owing to his familiarity with ARCHIMEDES's idioms and way of thinking, he was able to decipher almost immediately the barely-legible palimpsest, in which he was greatly helped by his friend, professor of mathematics H. G. ZEUTHEN, whose deep understanding of Greek mathematical problems was also a great asset. So much new material was "excavated" by this discovery and by HEIBERG's thorough work in editing Greek mathematical texts that his edition of ARCHIMEDES of 1880–81 was outdated, and therefore HEIBERG himself undertook a revision which appeared 1910–15. (The palimpsest disappeared in the 1920s but emerged in France in the 1990s and was purchased at an auction by an anonymous buyer for $ 2 million. It is now (Dec. 2000) being deciphered by two competing teams of scientists using different modern techniques to recover text and images.)

By the time of his death in 1928, almost all Greek texts on mathematical sciences had passed through HEIBERG's editorial hands. After assuming his professorship in 1896, he continued with SERENUS (1898), PTOLEMY (1898–1907), HERO (1912–14), and as a finale in 1927, THEODOSIUS and the anthology *Mathematici graeci minores* (Minor Greek Mathematicians). The magnitude of HEIBERG's achievement may be measured by the fact that when he started his career as an editor of ancient Greek mathematics, only the works of one Greek mathematician, PAPPUS, were available in a proper critical edition (F. Hultsch, 1876–79).

In 1907 HEIBERG's interest in Greek medicine led him to the idea of creating a great international project, *Corpus medicorum graecorum* (Collection of Greek Medical Authors), to be based in Berlin. Of more importance to the history of mathematics was his contribution on "Exakte Wissenschaften und Medizin" (Exact Sciences and Medicine) for *Einleitung in die Altertumswissenschaften* (Introduction to Studies of Antiquity) edited by ALFRED GERCKE and EDUARD NORDEN in 1910. HEIBERG also collaborated with ZEUTHEN in editing the *Mémoires scientifiques* of PAUL TANNERY, their common friend over many years. That monument, in eight volumes (1912–1927; nine more were published after HEIBERG's death), stands as a memorial to all three of these great contributors to the history of mathematics (see Section 1.9).

Although deeply engaged in Greek mathematics, HEIBERG wrote little on the subject from a purely mathematical point of view. His aim was nearly always to elucidate the transmission of ancient mathematical texts. On many occasions he must have discussed mathematical problems with ZEUTHEN, but he probably never thought of himself as a mathematician, but rather as a mediator and editor. Throughout his work the history of texts and the history of culture were conmingled with the history of the human mind.

Secondary literature: *May* 188. *P* III, 604; IV, 605; V, 512; VI, 1063. — RAEDER, HANS: "Johan Ludvig Heiberg. 27/11 1854 – 4/1 1928." *Isis* **11** (1928), 367–374 (with portrait). — HØEG, CARSTEN: "Johan Ludvig Heiberg. Geboren 27. November 1854, gestorben 4. Januar 1928." *Jahresbericht über die Fortschritte der klassischen Altertumswissenschaft* **233** (1931) [1933–34], 38–77. — SPANG-HANSEN, ESBERN: *Filologen J. L. Heiberg.* Copenhagen: Det Kongelige Biblioteks Nationalbibliografiske Afdelings Publikationer, 1929; 2nd ed. 1969. — ALPERS, KLAUS: "Klassische Philologie in Byzanz." *Class. Philol.* **83** (1988), 342–360. C.M.T.

Heilbronner, Johann Christoph (baptized March 13, 1706, Ulm, Germany; † January 17, 1745, Leipzig, Germany). HEILBRONNER, about whom little is known, studied theology and mathematics and gave private mathematical lectures in Leipzig with emphasis on practical aspects, such as surveying. (For some years at about the same time (1739–1756), A. G. KÄSTNER was teaching mathematics at the University of Leipzig as extraordinary professor.) HEILBRONNER is mainly remembered for his *Historia matheseos universae a mundo condito ad saeculum p[ost] C[hristum] n[atum] XVI* (History of Universal Mathematics from the Origin of the World to the 16th Century) (1742). MONTUCLA acknowledged that some of the material HEILBRONNER collected (among others, compilations of mathematical passages from ARISTOTLE and PSELLOS, and reports about manuscripts in European libraries) was of value to him when he was writing his *Histoire.*

Further works: [HEILBRONNER 1739] — *Geometrische Probleme mit ihren Auflösungen.* Leipzig 1745.

Secondary literature: *P* I, 1046; III, 605. — *NDB* **8**, 259. C.J.S

Heller, Siegfried (* December 1, 1876, Rohrbach, Lorraine, Germany; † June 9, 1970, Schleswig, Germany). HELLER studied mathematics, physics, and chemistry at the university of Kiel (with interruptions in Berlin, Munich, and Göttingen). In Kiel he attended PAUL STÄCKEL's seminar on the history of non-Euclidean geometry, under whose supervision he wrote his dissertation. Later, after World War I, HELLER was a regular participant in the seminar on history of mathematics offered jointly in Kiel by HEINRICH SCHOLZ, JULIUS STENZEL, and OTTO TOEPLITZ. Having taught mathematics at *Gymnasia* in Kiel (1907–1926) and Schleswig (1926–1935), HELLER became *Oberschulrat* in Kiel (1935–1947). After his retirement, a close collaboration with J. E. HOFMANN developed. HELLER's historical interests centered on Greek mathematics and the history of number theory.

Main Works: "Ein Fehler in einer Archimedes-Ausgabe, seine Entstehung und seine Folgen." *Abh. Bayer. Akad. der Wiss., math.-nat. Kl.*, Neue Folge **63**. München 1954, 40 pp. — "Ein Beitrag zur Deutung der Theodoros-Stelle in Platons Dialog *Theaetet*." *Centaurus* **5** (1956), 1–58. — "Die Entdeckung der stetigen Teilung durch die Pythagoreer." *Abh. dt. Akad. Wiss., Kl. Math., Phys. u. Techn.* Jahrgang 1958, Nr. 6. Berlin 1958, 28 pp. — "Lösung einer Fermatschen Dreiecksaufgabe für Torricelli." *Arch. int. Hist. Sci.* **23** (1970), 67–79. — "Untersuchungen über die Lösungen der Fermatschen Dreiecksaufgabe." *Revista Matemática Hispano-Americana* (4) **29**, 195–210; **30** (1970), 69–98.

Secondary literature: [HOFMANN 1970] — HOFMANN, JOSEPH EHRENFRIED: "Siegfried Heller † ." *Nbl. DGGMNT* **20** (1970), 93–95. — SCRIBA, CHRISTOPH J.: "Siegfried Heller (1. Dezember 1876 bis 9. Juni 1970)." *Jber. DMV* **73** (1971), 1–5. M.F.

Hellinger, Ernst (* September 30, 1883, Striegau (Silesia), Germany; † March 28, 1950, Chicago, Illinois, USA). HELLINGER was a Jewish mathematician who studied in Heidelberg, Breslau, and Göttingen, where he received his Ph.D. in 1907 as a student of DAVID HILBERT. After teaching at the universities of Marburg (1909–1914) and Frankfurt (1914–1936), HELLINGER was forced to give up his teaching. After he had been held for some weeks in a concentration camp, he emigrated to the U.S. in March 1939, where he taught at Northwestern University in Evanston, Illinois. HELLINGER wrote several historical articles with MAX DEHN.

Main works: "On Gregory's *Vera Quadratura*" (with MAX DEHN). In: [TURNBULL 1939, 468–478]. — "Certain Mathematical Achievements of James Gregory" (with MAX DEHN). *AMM* **50** (1943), 149–163.

Secondary literature: *DSB* **6**, 235–236. *P* V, 515; VI, 1075; VIIa(2), 436. — [SIEGEL 1965]. — [PINL 1969, Teil I: 211, 214–215]. J.W.D.

Hermelink, Heinrich (* December 11, 1920, Marburg, Germany; † August 31, 1978, Lahr (Baden), Germany). HERMELINK wrote a Ph.D. dissertation in physics (1947) and pursued the study of Oriental languages. Beginning in 1957, he earned his living as a patent lawyer in Munich, but continued to do research on Arabic and Persian mathematics and astronomy and the history of magic squares.

Main works: *Archimedes: Über einander berührende Kreise. Aus dem Arabischen übersetzt und mit Anmerkungen versehen* (with YVONNE DOLD-SAMPLONIUS and MATTHIAS SCHRAMM) (1975).

Secondary literature: FOLKERTS, MENSO, and PAUL KUNITZSCH: "Heinrich Hermelink 1920–1978." *Janus* **66** (1979), 209–215 (with bibliography). — SCRIBA, CHRISTOPH J.: "Heinrich Hermelink †." *Nbl. DGGMNT* **29** (1979), 15–16. — SCRIBA, CHRISTOPH J.: "Heinrich Hermelink *in Memoriam*." *HM* **6** (1979), 233–235 (with portrait). M.F.

Hjelmslev, Johannes Trolle (* April 7, 1873, Hørning, Denmark; † February 16, 1950, Copenhagen, Denmark). JOHANNES HJELMSLEV graduated in mathematics from the University of Copenhagen in 1894, received his Ph.D. in 1897, began teaching as professor of mathematics at the *Polyteknisk Læreanstalt* in 1905, and at the University of Copenhagen in 1917. It was thanks to HJELMSLEV that the work of GEORG MOHR became known, which (among other places) HJELMSLEV described in "Beiträge zur Lebensbeschreibung von Georg Mohr (1640–1697)," *Det Kongelige Danske Videnskabernes Selskab, Mathematisk-fysiske Meddelelser*, **11**, no.4 (1931). HJELMSLEV also published several other papers on history of mathematics, the most notable being "Eudoxus's Axiom and Archimedes's Lemma," *Centaurus* **1** (1950), 2–11.

Secondary literature: *May* 195. *P* V, 545; VI, 1129; VIIb(4), 2032. — *DBL 6*, 379. K.A.

Hofmann, Joseph Ehrenfried (* March 7, 1900, Munich, Germany; † May 7, 1973, Günzburg, Germany). JOSEPH HOFMANN studied mathematics and physics at the University and *Technische Hochschule* in Munich from 1919 to 1924, and received his doctorate in 1927 at the *Technische Hochschule* in Munich with a dissertation, *Über die gestaltliche Diskussion des durch eine gewisse Differentialgleichung 1. Ordnung 2. Grades definierten Kurvensystems* (On the Discussion of the Shape of a System of Curves, Defined by a Certain Differential Equation of 1st Order and 2nd Degree). Several of his university professors had strong historical interests themselves: FERDINAND VON LINDEMANN had published papers to confute the one-sided account of J. M. CHILD about the NEWTON–LEIBNIZ priority dispute (see [CHILD 1920]), WALTER DYCK was engaged in the preparation of a new KEPLER edition (later continued by MAX CASPAR), and GEORG FABER asked HOFMANN for assistance when proof-reading EULER's articles on infinite series for the edition of EULER's *Opera omnia*. Thus as a graduate student and assistant (in Munich, 1925–1927, and Darmstadt, 1927–1928) HOFMANN had very strong encouragement to occupy himself with history of mathematics. Perhaps the most important influence on him was that of HEINRICH WIELEITNER, who quickly won him wholly for the history of mathematics. Together they made several studies in the history of the calculus.

From 1928 to 1938 HOFMANN taught in schools in Bavaria (1928–1929 Günzburg, 1929–1938 Nördlingen), yet at the same time he was deeply engaged in historical research. This resulted in his *Habilitation* in the history of mathematics at the university in Berlin (1939). Shortly afterwards (1940) he became director of the *Leibniz-Arbeitsstelle* of the *Preußische Akademie der Wissenschaften* in Berlin for the edition of the collected works of LEIBNIZ. As editor, he inherited from DIETRICH MAHNKE the task of preparing the first volume of LEIBNIZ's mathematical correspondence for publication. Towards the end of World War II, the destruction of Berlin forced HOFMANN to return to Southern Germany. He was among the first mathematicians after the War to spend several months doing research at the *Mathematisches Forschungsinstitut Oberwolfach* (Mathematical Research Institute Oberwolfach) in the Black Forest. It was here that he wrote his report about German publications in history of mathematics that had appeared between 1939 and 1946 ([HOFMANN 1948a]; for the *FIAT* report series see Section 5.5.3, footnote 2).

From 1947 until his retirement in 1963 HOFMANN again taught mathematics and physics at the *Gymnasium* in Günzburg. In addition, he gave lectures on history of mathematics at the *Technische Hochschule* in Karlsruhe and at the University of Freiburg, and became honorary professor in Tübingen (1950). During those years he wrote *Die Entwicklungsgeschichte der Leibnizschen Mathematik während des Aufenthaltes in Paris (1672–1676)* (published in 1949), (English edition: *Leibniz in Paris (1672–1676): His Growth to Mathematical Maturity*, 1974).

In 1954 he organized for the first time an international colloquium at Oberwolfach — the beginning of the still regularly-held week-long meetings on the history of mathematics. He died on May 7, 1973, in Günzburg, having been hit by a car on his morning walk.

Two features might be singled out as especially characteristic for HOFMANN's conception of the historiography of mathematics. He strongly emphasized what he called *Problemgeschichte der Mathematik*, which may be translated roughly as "internal history" of mathematics, yet at the same time seen in cultural perspective. (Somewhat surprisingly, he paid relatively little attention to the development of applied mathematics in his research.) HOFMANN was also very much interested in the creative process of "doing" mathematics; indicative are titles of such articles as "Von der Feinfühligkeit des mathematischen Genies" (Of the Sensitivity of a Mathematical Genius). *Arch. int. Hist. Sci.* **8** (1955), 339–350, or "Vom Einfall zur Methode" (From Sudden Inspiration to Method). *Sb. Berliner Math. Ges.*, Jahrgänge 1969, 1970 und 1971 (1.4.1969 – 30.9.1971), 30–31.

LEIBNIZ occupied a central place in HOFMANN's research throughout his life (see Section 5.5.3). Of even greater impact, perhaps, is the fact that his long-standing aim to create positions for full-time editors of the vast amount of LEIBNIZ's unpublished mathematical and scientific papers and letters was realized with the support of the director WILHELM TOTOK of the *Niedersächsische Landesbibliothek* in Hannover (where the bulk of LEIBNIZ's manuscripts is kept). Since HOFMANN's death, several volumes of mathematical papers and letters have been published in *Gottfried Wilhelm Leibniz: Sämtliche Schriften und Briefe*, (3. Reihe and 7. Reihe).

The wide scope of HOFMANN's historical research may be indicated by a list of the more important mathematicians who occur in titles of his papers: ARCHIMEDES, ALHAZEN, ALBERTUS MAGNUS, NICOLAUS CUSANUS, REGIOMONTANUS, BRADWARDINE, DÜRER, STIFEL, BOMBELLI, VIÈTE, KEPLER, GREGORIUS A S. VINCENTIO, DESCARTES, FERMAT, ROBERVAL, VAN SCHOOTEN, WALLIS, MICHELANGELO RICCI, N. MERCATOR, BROUNCKER, PASCAL, HUDDE, HUYGENS, J. GREGORY, OZANAM, NEWTON, LEIBNIZ, TSCHIRNHAUS, JAKOB and JOHANN BERNOULLI, COTES, EULER, A. VON HUMBOLDT, GAUSS, VON STAUDT, and M. CANTOR. HOFMANN's concentration on the period from early modern times to the 18th century is obvious. As to subject areas, the history of the calculus (including logarithms, infinite series, differential equations, and other related areas) clearly stood at the center of his interests. Next in number are publications about problems on the history of geometry (including trigonometry).

Apart from the history of mathematical analysis, the history of number theory was probably the area of research to which HOFMANN felt most strongly attracted. His discovery of FERMAT's *Supplement* to the *Solutio duorum problematum* (Solution of Two Problems Posed by the French Mathematician FRÉNICLE DE BESSY) (see *Neues über Fermats zahlentheoretische Herausforderungen von 1657*, (1943)) prompted HOFMANN over a span of many years to investigate FERMAT's methods in number theory and, in several publications, to provide a re-evaluation of this outstanding number theorist of the 17th century. In the course of his investigations HOFMANN studied in detail several special problems that had fascinated FERMAT and a few of his contemporaries, such as the six square problem (1958) or the problem of the four cubes (1961). Often large computations became necessary, for which in some cases SIEGFRIED HELLER assisted him. Over the years HOFMANN extended his research in number theory to the 18th century. For example, in a fundamental paper on the methods of FERMAT and EULER in number theory (*Arch. Hist. ex. Sci.* **1** (1961), 122–159) he compared their methods and indicated how they were related to methods developed during the 19th century.

In 1951 HOFMANN published, together with OSKAR BECKER, a history of mathematics, *Geschichte der Mathematik* ([BECKER/HOFMANN 1951]). His contribution consisted of Parts II and III, the history of Oriental and Western mathematics, while BECKER wrote Part I on mathematics in antiquity. Between 1953 and 1957, HOFMANN's own *Geschichte der Mathematik* appeared in three small pocket books ([HOFMANN 1953/57/63]), crammed with facts and very rich bibliographical information. An English translation was published in 1957/59 (without the bibliography), a Spanish one in 1960.

Beginning in the late 1960s, HOFMANN was active as an advisor for reprint editions to the publishing house *Georg Olms Verlag*, Hildesheim. He always provided an historical introduction, and in several cases he also added an index to works that originally had been published without one. Reprints to which he contributed include JOHANN BERNOULLI: *Opera Omnia* (1968), CHRISTIAN WOLFF:

Elementa Matheseos Universae (1968/71), François Viète: *Opera Mathematica* (1970), Abraham Gotthelf Kästner: *Geschichte der Mathematik* (1970), Christian Wolff: *Anfangsgründe aller Mathematischen Wissenschaften* (1973), and Christian Wolff: *Kurtzer Unterricht von den vornehmsten Mathematischen Schriften* (1973). Finally, the posthumously published *Register zu Gottfried Wilhelm Leibniz: Mathematische Schriften* und *Der Briefwechsel mit Mathematikern (herausgegeben von C. I. Gerhardt)* (Olms Paperbacks, vol. 49) (1977), is an excellent example of Hofmann's concern for careful, detailed documentation of both source material and secondary literature.

Last but not least, a considerable number of papers were written by Hofmann to provide historical material for the teacher of mathematics. As a teacher himself, he was greatly interested in interactions between history and pedagogy of mathematics, and as a result, he published numerous articles and two textbooks for the *Gymnasium* with this aspect of historical writing in mind.

Main works: *Studien zur Vorgeschichte des Prioritätstreites zwischen Leibniz und Newton...* (1943). — *Nicolaus Mercator (Kauffman), sein Leben und Wirken, vorzugsweise als Mathematiker* (1950). — Introduction to *Nikolaus von Cues, Die mathematischen Schriften* (1951). — *Über Jakob Bernoullis Beiträge zur Infinitesimalmathematik* (1956). — *Frans van Schooten derJüngere* (1962). — *Michael Stifel, Leben, Wirken und Bedeutung für die Mathematik seiner Zeit* (1968). — *Gottfried Wilhelm Leibniz: Sämtliche Schriften und Briefe.* 3. Reihe, 1. Band (1976). —

Secondary literature: *P* VI, 1141; VIIa(2), 525. — Joseph Ehrenfried Hofmann: "Vier Jahrzehnte im Ringen um mathematikgeschichtliche Zusammenhänge." In: *Wege zur Wissenschaftsgeschichte*, Sticker, Bernhard, and Friedrich Klemm (Eds.). Wiesbaden 1969, 25–37 (with selected bibliography and portrait). — Folkerts, Menso: "Joseph Ehrenfried Hofmann †." *Sudh. Archiv* **57** (1973), 227–230 (with portrait). — Scriba, Christoph J.: *Joseph Ehrenfried Hofmann, Ausgewählte Schriften*, 2 vols., (1990); vol. 1, 5–40 (Bibliography and éloges, with portrait). Hildesheim and New York 1990. M.F.; C.J.S.

Holmboe, Bernt Michael (* March 23, 1795, Vang, Norway; † March 28, 1850, Christiania, Norway). Holmboe, who was a teacher at *Christiania Katedralskole*, is primarily known for having identified Abel's great gifts for mathematics. It was with Holmboe that Abel first read works by Euler, Lagrange, and Gauss. In 1826 Holmboe became an associate professor and in 1834 professor of mathematics at the university in Christiania. Five years later he edited the first collection of Abel's works.

Secondary literature: *DSB* **6**, 474. *P* I, 1134. — *NBL* **6**, 246–247. K.A.

Holst, Elling Bolt (* July 19, 1849, Drammen, Norway; † September 2, 1915, Holtet, Norway). Elling Holst studied mathematics in Germany, Denmark, England and France, and in 1883 he obtained a doctorate at Kristiania University. He served as a teacher from 1886 to 1912. In 1878 he published a book in Norwegian on PONCELET's influence on geometry, and later wrote some biographies of Norwegian mathematicians, including ABEL. He was also one of the editors of ABEL's letters. His interest in history of mathematics led him to lecture on the subject. Besides his mathematical activities, like his English colleague CHARLES L. DODGSON, he also wrote books for children.

Secondary literature: *May* 196. *P* III, 652; IV, 660. — *NBL* **6**, 282–284. — [*Acta Math.*, Tab. Gen. **1–35** (1913), 66; portrait 147]. — [STUBHAUG 2000b, portrait 195]. K.A.

Hoppe, Edmund (* February 25, 1854, Burgdorf near Hannover, Germany; † August 12, 1928, Göttingen, Germany). HOPPE, who taught mathematics and physics in Hamburg from 1877 until 1919, then went on to teach history of exact sciences as a *Dozent* at the University of Göttingen from 1919 until 1928. His areas of interest were the history of mathematics and physics.

Main Works: *Mathematik und Astronomie im klassischen Altertum* (1911). — "Bedeutung der *νεύσεις* in der griechischen Mathematik." *Mitt. Math. Ges. Hamburg* **5** (1911–1920), 289–304. — "Ist Heron der Verfasser der unter seinem Namen herausgegebenen Definitionen und der Geometrie?" *Philologus* **29** (= old series **75**) 1919, 202–226. — "Heron von Alexandrien." *Hermes* **62** (1927), 79–105. — "Geschichte der Infinitesimalrechnung bis Leibniz und Newton." *Jber. DMV* **37** (1928), 148–187.

Secondary literature: *P* III, 657; IV, 664; V, 557; VI, 1160. — *NDB* **9**, 617–618. — ROSENFELD, LEON: "Edmund Hoppe (geb. 1854, 02, 25, gest. 1928, 08, 12)." *Isis* **13** (1929/30), 45–50 (with portrait). — SCHIMANK, HANS: "Edmund Hoppe oder über Inhalt, Sinn und Verfahren einer Geschichtsschreibung der Physik." *Arch. Gesch. Math. Nat. Techn.* **11** (1928/29), 345–351 (with portrait). M.F.

Hoüel, Jules (* April 7, 1823, Thaon, France; † June 14, 1886, Périers, France). Trained at the *Ecole Normale Supérieure*, HOÜEL received his doctorate in 1855 from the *Sorbonne* and in 1859 was appointed to a chair of pure mathematics at the University of Bordeaux. He contributed to the spread of non-Euclidean geometry in France. As editor of the *Bulletin des sciences mathématiques*, he used his knowledge of foreign languages to introduce many historical works to France.

Secondary literature: *DSB* **6**, 522. *May* 197. *P* IV, 667. — BRUNEL, G.: "Notice sur l'influence scientifique de Guillaume-Jules Hoüel." *Mémoires de la Société des sciences physiques et naturelles de Bordeaux*, (3) **4** (1888), 1–78 (with bibliography and portrait). J.P.

Hultsch, Friedrich (* July 22, 1833, Dresden, Germany; † April 6, 1906, Dresden, Germany). HULTSCH, who taught Latin and Greek in Leipzig and Dresden (1857–1889), was interested in metrology and mathematics in classical antiquity.

Main works: editions of HERON, geometry and stereometry (1864); writings on metrology (1864–1866); PAPPUS ([HULTSCH 1876/78]); and AUTOLYCUS (1885). — *Griechische und römische Metrologie* (1862). — Many articles in PAULY-WISSOWA's *Real-Encyclopädie*, e.g. "Abacus," "Apollonios v. Perge," "Archimedes," "Arithmetica," "Diophantos," "Dioptra," "Eudoxos."

Secondary literature: *May* 198. *P* III, 667; IV, 673; V, 563. — *NDB* **10**, 30–31. — RUDIO, FERDINAND: "Friedrich Hultsch." *Bibl. Math.* (3) **8** (1907/08), 325–402 (with bibliography). M.F.

Itard, Jean (* June 16, 1902, Serrières (Ardèche), France; † May 8, 1979, Paris, France). After completing his studies at the *Ecole Normale des Instituteurs* in 1921, ITARD graduated with a degree in mathematics (*agrégation*) from the University of Marseilles in 1924. Upon obtaining his *agrégation* in June of 1925, he taught first at the *lycée* in Alençon until 1928, when he began teaching at the *Lycée Saint Charles* in Marseilles. Several years later he went to Paris, where he taught advanced mathematics at high schools named after BUFFON, MICHELET, and finally HENRI IV, until he retired in 1962. He also lectured on Greek mathematics at the *Ecole Pratique des Hautes Etudes* (6th section).

ITARD was not simply a professor or a historian of science. Indeed, his career as a trade union supporter and political activist helps to clarify aspects of his scholarly career, especially his interest in making the history of mathematics accessible to a wider public, but without sacrificing rigor or exactitude. He was also involved in the founding of the *Institut supérieur ouvrier* of the Socialist trade union, and the *Confédération Générale du Travail (CGT)* (General Labor Union), where for many years he taught differential calculus and probability theory to trade union members. Although he was often a candidate for the French Parliament, with no realistic hope of ever being elected, Itard did contribute for ten years (1926–1936) to *Etudes socialistes*, the journal of the CGT. ITARD also contributed, with IGNACE KOHEN, GEORGES LEFRANC, JEAN BOIVIN, and others, to producing a bulletin whose title was in itself a program for action: *Pour la révolution constructive* (For the Constructive Revolution), from 1931 onwards.

Beginning in 1936, pedagogy and the history of science began to take precedence over ITARD's political activities. In 1938 he started to prepare a collection of mathematical textbooks in which, contrary to the official curriculum, he tried to integrate elements of the history of mathematics. These historical texts were intended to be didactic, but they were not strictly scholarly, moreover they were never a commercial success. In 1961 ITARD undertook a new series of textbooks with ANDRÉ HUISMAN, and in 1972 he wrote another textbook in collaboration with his son GILLES ITARD. Apart from his textbooks, JEAN ITARD was also involved in works which popularized scientific and interdisciplinary culture. For example, between 1929 and World War II, he collaborated with E. WEILL (editorial secretary and professor of mathematics at the *Lycée Saint-Louis*) in editing *L'Enseignement scientifique*, and from 1959 until 1961 he edited another journal (with GILBERT WALUSINSKI), *L'Enseignement des sciences.*

ITARD's works on history of mathematics are considerable. He was among the first in France to take advantage of the new light shed on ancient mathematics by the publication of Egyptian and Babylonian mathematical texts, and his judicious commentaries contributed to a more profound understanding of these difficult documents. He never failed to include new findings in the historical research he published on methods and problems going back to antiquity, which comprise the second part of his book, *Mathématiques et mathématiciens* (Mathematics and Mathematicians) (1959). Likewise, ITARD produced a French edition of the arithmetical books of EUCLID in hopes of overcoming the widespread idea of a simply geometric EUCLID. Due to the precise questions ITARD poses, his concern for historical context, and the analysis he offers of the mathematical content of EUCLID's theorems and their connection to calculation techniques, this book constitutes a resource of the first order, whose suggestions are always stimulating. ITARD never lost sight of the fact that the scarcity of ancient texts made it inevitable that the historian must often present conjectures, and he knew how to link fruitful hypotheses of the mathematician with the careful prudence of the historian.

In the history of modern mathematics, ITARD was interested (as the list of his contributions to the *Dictionary of Scientific Biography* [DSB 1970/90] testifies) in every period from the end of the 15th down to the beginning of the 20th century, but the majority of his most original studies were related to the 17th century, from KEPLER to NEWTON, and especially to the great figures of French mathematics from the middle of the century, including FERMAT, DESCARTES, ROBERVAL, and PASCAL, among others. Working for the most part by himself in his office on the avenue Paul Appell where his library included most of his favorite authors and their contemporaries, ITARD took a particular pleasure in reading, rereading, and confronting these texts, whereby he acquired a true familiarity with them. It was only thereafter, following long periods of patience and reflection, that he began the preparation of an article or more ambitious study. Many of his publications deal with particular points of the history of mathematics which bring new insights thanks to the presentation of new documents, or to a more attentive rereading of texts he already knew. His most synthetic studies, like his research on the origins of the infinitesimal calculus, his work on FERMAT, and the various biographical notes which he wrote, reveal at the same time the extent of his historical knowledge and his profound understanding of the main lines of the evolution of mathematical thought and techniques, as well as the diverse aspects of the personalities of the mathematicians about whom he wrote.

Main works: *Mathématiques et mathématiciens* (in collaboration with PIERRE DEDRON), Paris 1959; repr. Paris 1982; Eng. trans. by J. V. FIELD in two vols., London 1973, 1975. — *Les livres arithmétiques d'Euclide* (in the series *Histoire de la pensée*), Paris 1962. — *Arithmétique et théorie des nombres* (in the series *Que sais-je?*), Paris 1963; 2nd ed. 1967; 3rd ed. 1973; Japanese trans. 1971. — *Les nombres premiers* (in the series *Que sais-je?*), Paris 1969; 2nd ed. 1976.

Contributions to encyclopedic works: In R. TATON (Ed.): *Histoire générale des sciences*, "Les mathématiques (Science hellène)" (vol. 1, Paris, 1957, 226–245; 2nd ed. 1966, 226–242); "Mathématiques pures et appliquées (Science hellénistique et romaine)" (vol. 1, Paris 1958, 321–354); "De l'algèbre symbolique au calcul infinitésimal," (vol. 2, 1958, 207–241; 2nd ed. 1969, 217–251);

"Analyse mathématique et théorie des nombres" (vol. 3, Paris 1961, 49–76. — In [DSB 1970/90], articles on ARBOGAST, BILLY, BOBILLIER, BOUQUET, BRET, CHUQUET, CLAIRAUT, A. GIRARD, HENRION, KRAMP, LACROIX, LAGRANGE, LAURENT, LEGENDRE, OCAGNE, J. RICHARD, L. RICHARD, ROLLE, and SAINT-VENANT. — The principal papers of JEAN ITARD have been published with an introduction by ROSHDI RASHED: *Essais d'histoire des mathématiques*, Paris 1984 (with bibliography and portrait).

Secondary literature: CAVEING, MAURICE; ROSHDI RASHED, and RENÉ TATON: "Nécrologie Jean Itard (1902–1979)." *Rev. Hist. Sci.* **32** (1979), 345–350. — TATON, RENÉ: "Jean Itard (1902–1979)." *Arch. int. Hist. Sci.* **31** (1981), 195–197. — CAVEING, MAURICE; ROSHDI RASHED, and RENÉ TATON: "Jean Itard (1902–1979)." *HM* **12** (1985), 1–5 (with bibliography and portrait). R.R.

Jourdain, Phillip Edward Bertrand (* October 16, 1879, Ashbourne (Derbyshire), England; † October 1, 1919, Fleet (Hampshire), England). PHILLIP JOURDAIN studied at Trinity College Cambridge, became a freelance scholar, and in 1913 European editor of *Open Court Publishing Company.* Areas of interest: history of mathematical logic and set theory, and of mechanics.

Main works: *The Nature of Mathematics* (1912). — See *Selected Essays on the History of Set Theory and Logics* (1909–1918), I. GRATTAN-GUINNESS (Ed.). Bologna 1991.

Secondary literature: *May* 207. *P* V, 598; VI, 1258. — JOURDAIN, L., and GEORGE SARTON: "Phillip E. B. Jourdain." *Isis* **5** (1923), 126–136 (with bibliography and portrait). I.G.G.

Jørgensen, Jens Jørgen Frederik Theodor (* April 1, 1894, Haderup, Denmark; † July 30, 1969, Copenhagen, Denmark). JØRGEN JØRGENSEN graduated in philosophy from the University of Copenhagen in 1918, where he accepted a chair for philosophy in 1926. In 1931 he published *A Treatise of Formal Logic. Its Evolution and Main Branches with its Relation to Mathematics and Philosophy*, a work in three volumes which he had written some years before and which at the time was considered a major contribution to the history of logic. In his later career he mainly concentrated on contemporary philosophy and logic.

Secondary literature: *DBL* **7**, 547–549. — CHRISTENSEN, NIELS EGMONT: "Jørgen Jørgensen as a Philosopher of Logic." *Danish Yearbook of Philosophy* **13** (1976), 242–248 (with portrait). K.A.

Junge, Gustav (* April 29, 1879, Ludwigslust near Schwerin, Germany; † December 17, 1959, Bad Lauterberg (Harz), Germany). A teacher of mathematics in Berlin from 1902 to 1924, JUNGE's main interest was Greek mathematics.

Main works: "Wann haben die Griechen das Irrationale entdeckt?" *Novae Symbolae Joachimicae* 1907, 221–264. — "Besonderheiten der griechischen Mathematik." *Jber. DMV 35* (1926), 66–80, 150–172, 251–268. — "Flächenanlegung und Pentagramm." *Osiris* **8** (1948), 316–345. — Edition of the Arabic translation of

EUCLID's *Elements* by AL-HAJJĀJ (1932, with W. THOMSON and H. H. RAEDER), and of the Arabic text of PAPPUS's commentary on Euclid X (1930, with W. THOMSON).

Secondary literature: *P* VI, 1264; VIIa(2), 661. — *NDB* **10**, 680–681. M.F.

Kästner, Abraham Gotthelf (* September 27, 1717, Leipzig, Germany; † June 20, 1800, Göttingen, Germany). As extraordinary professor of mathematics in Leipzig (1739–1756) and ordinary professor in Göttingen (1756–1800), KÄSTNER not only published widely-read general textbooks on mathematics, but was also known in German literature, notably for his epigrams. He had a general interest in the history of mathematics and through his lectures stimulated some historical research.

Main work: [KÄSTNER 1796/1800].

Secondary literature: *DSB* **7**, 206–207. *May* 208. *P* I, 1217. — HOFMANN, JOSEPH EHRENFRIED: Introduction to reprint of [KÄSTNER 1796/1800, vol. 1, vii–xvii] (with portrait). M.F.

Kagan, Veniamin Fiodorovich (* March 9, 1869, Shavli in Lithouania, Russian Empire; † May 8, 1953, Moscow, USSR). VENIAMIN KAGAN taught as *Privatdozent* and later as professor at the University of Odessa from 1904 until 1923, when he accepted a position as professor at the University of Moscow. KAGAN, a geometer, contributed primarily as a historian of mathematics to the history of geometry.

Main works: *Foundations of Geometry.* 2 vols. Odessa 1905–07 (in Russian). — *N. I. Lobachevskiĭ.* Moscow-Leingrad 1948 (in Russian; French trans. Moscow 1974).

Secondary literature: *DSB* **7**, 207–208. *May* 208. *P* V, 604; VI, 1268; VIIb(4), 2341. — [KUROSH 1959] — [FOMIN/SHILOV 1969] — [BORODIN/BUGAI 1979] — [BOGOLYUBOV 1983, (with portrait on p. 200)]. S.S.D.

Karpinski, Louis Charles (* August 5, 1878, Rochester, New York, USA; † January 25, 1956, Winter Haven, Florida, USA). LOUIS KARPINSKI, an American mathematician and educator, was a student of HEINRICH WEBER in Strasbourg, where he received his Ph.D. in 1903. Having studied at Cornell University, KARPINSKI went on to accept a position at Teachers College, Columbia University, until he left in 1918 to assume a position as professor of mathematics at the University of Michigan. KARPINSKI was a charter member of the Mathematical Association of America, president of the History of Science Society (1943), and was elected a *membre effectif* (# 14), of the *Comité internationale d'histoire des sciences.* His publications include a study of Hindu Arabic Numerals (with D. E. SMITH, 1911), and ROBERT OF CHESTER's Latin translation of AL-KHOWARIZMI's (1915) *Algebra* (1911).

Main works: *Robert of Chester's Translation of al-Khowarizmi.* Leipzig 1915. — *Bibliography of Mathematical Works Printed in America Through 1850.* Ann Arbor, MI, 1940.

Secondary literature: *DSB* **15**, 255–257. *May* 210. *P* V, 612; VI, 1282; VIIb(4), 2384. — JONES, PHILLIP S.: "Louis Charles Karpinski, Historian of Mathematics and Astronomy." *HM* **3** (1976), 185–202 (with bibliography). — MOULTON, F. R.: "American Association in Action." *Scientific Monthly* **50** (1940), 85–91 (portrait of KARPINSKI on p. 90). J.W.D.

Kary-Niyazov, Tashmukhammed Niyazovich (* September 2, 1896, Khodzhent, Russian Empire; † March 17, 1970, Tashkent, USSR). TASHMUKHAMMED KARY-NIYAZOV taught at the University of Tashkent, where he was professor from 1931, and served the University as rector from 1931–1933. From 1943 until 1946 he was president of the Academy of Sciences of Uzbek SSR. His primary historical interests were related to Arabic mathematics.

Main work: *Ulug Bek's Astronomical School.* Moscow 1950 (in Russian).

Secondary literature: [KUROSH 1959] — [FOMIN/SHILOV 1969] — [BORODIN/BUGAI 1979] — [BOGOLYUBOV 1983, (with portrait on p. 211)]. S.S.D.

Klein, Christian Felix (* April 25, 1849, Düsseldorf, Germany; † June 22, 1925, Göttingen, Germany). FELIX KLEIN was one of the last mathematicians who was able to survey virtually all the many branches of mathematics around 1900. In an attempt to retain the unity of mathematics, the natural and human sciences, as well as technology, while trying to improve the general mathematical education of students, he launched several historiographical projects (as an editor), gave lectures on history of mathematics, and encouraged young scholars interested in the subject. KLEIN, however, though highly influential as a mathematician, cannot be called a professional historian of mathematics. He was mainly interested in using history to support his large universal mathematical interests.

KLEIN was educated at the *Humanistisches Gymnasium* (high school) in Düsseldorf before attending the University of Bonn in 1865. He studied mathematics and science with JULIUS PLÜCKER and ALFRED CLEBSCH and obtained his doctorate at the age of nineteen in 1868. After a few months of study in Paris along with his friend SOPHUS LIE, he obtained his postdoctoral qualification as *Privatdozent* in 1871, and in October of 1872 received his first chair of mathematics at the University of Erlangen. In 1875 KLEIN was made a professor at the *Technische Hochschule* (Technical College) in Munich, and this brought him into close contact with the applied side of mathematics. In 1880 he accepted a position at the University of Leipzig, and in 1886 went on to the University of Göttingen, a position he held until his retirement in 1913.

KLEIN's tendency to unite different directions of mathematical research under a common point of view revealed itself early in his well-known *Erlangen Program.* For almost half a century, he was editor of the *Mathematische Annalen* (1875–1924). KLEIN arranged under a new system of German mathematical journals for the publishing house of Teubner to take over the historical journal *Bibliotheca mathematica* in 1900, which was edited by GUSTAF ENESTRÖM. KLEIN was the supervisor of more than fifty doctoral candidates. He also undertook a serious reform of teacher training and mathematics instruction.

After having published papers and books about higher geometry, the theory of algebraic equations, and the theory of functions, in particular his pathbreaking research on automorphic functions carried out in competition with HENRI POINCARÉ, KLEIN devoted himself primarily to the management of science after 1892. He supported the project of the *Deutsche Mathematiker-Vereinigung* (German Mathematicians Society, *DMV*, founded in 1890), to obtain comprehensive reports on the development of different branches of mathematics and their applications. These historical reports were published in the *Jahresbericht der DMV*, or incorporated into the great collaborative undertaking, the *Encyklopädie der mathematischen Wissenschaften mit Einschluss ihrer Anwendungen* (Encyclopedia of the Mathematical Sciences, Including Their Applications, published from 1898 to 1935), which was initiated by KLEIN, HEINRICH WEBER, and FRANZ MEYER in 1894, and which was chiefly managed by KLEIN (see Sections 5.4.4 and 1.9).

Another of the major book projects in which KLEIN was active is the series *Kultur der Gegenwart* (Today's Culture) (the original plan comprised 40 volumes). The part devoted to "Mathematics" contained articles on history of mathematics by H. ZEUTHEN, A. VOSS, and H. E. TIMERDING.

Yet another major project of KLEIN's was the publication of the *Abhandlungen über den mathematischen Unterricht in Deutschland* (Treatises on Mathematical Education in Germany) which appeared in five volumes. Articles in the *Abhandlungen* series dealt with the historical development of mathematical instruction in different types of schools and educational institutions. A comprehensive report on the teaching of mathematics in German universities in the 19th century was written for this series by WILHELM LOREY [LOREY 1916]. This undertaking was initiated by the International Commission on Mathematical Education *(ICME)* (*Internationale Mathematische Unterrichtskommission, IMUK*) that had been founded under the chairmanship of KLEIN at a meeting in Rome in 1908.

KLEIN's most famous work on history of mathematics is his collection of *Vorlesungen über die Entwicklung der Mathematik im 19. Jahrhundert* (Lectures on the Development of Mathematics in the 19th Century) (2 vols., 1926, 1927). In preparation he had offered student seminars and held a colloquium on the specific mathematical literature, attended among others by C. CARATHÉODORY, P. DEBYE, and R. COURANT. KLEIN himself lectured on historical topics for three semesters, beginning in the winter term, 1914/15, and made use of research assistance from women during the First World War. His youngest daughther ELISABETH STAIGER, who had studied mathematics, physics, and English, prepared two parts of the written version of KLEIN's lectures, while a third part was completed by two other women. R. COURANT, O. NEUGEBAUER, and S. COHN-VOSSEN — with the help of D. J. STRUIK and others — edited KLEIN's lectures posthumously.

KLEIN was sensitive to many of the prerequisites for working on history of mathematics professionally, and he contributed significantly to the editions of the collected papers of PLÜCKER, F. MÖBIUS, and C. F. GAUSS. The collected papers of H. GRASSMANN and the volume of B. RIEMANN's papers (edited by M. NOETHER and W. WIRTINGER) appeared through KLEIN's initatives. In addition, KLEIN wrote a number of obituaries of contemporary mathematicians and initiated the posting of commemorative plaques on the former residences of important Göttingen mathematicians. Thus, in all of these different ways KLEIN sought to generate a stimulating climate which helped to give history of mathematics an accepted place in both research and education.

Main works: *Gesammelte mathematische Abhandlungen*, 3 vols. Berlin 1921, 1922, 1923; repr. 1973. — [KLEIN 1926/27]. — "Gauss' wissenschaftliches Tagebuch 1796–1818," in: *Festschrift zur Feier des 150jährigen Bestehens der Königlichen Gesellschaft der Wissenschaften zu Göttingen (Beiträge zur Gelehrtengeschichte)*, Berlin 1901, 1–44, repr. in: *Math. Ann.* **57**, 1–26 and in: BIERMANN, KURT-R.; HANS WUSSING, and OLAF NEUMANN (Eds.), *Ostw. Klass.* **256**, Leipzig 1979. — *Materialien für eine wissenschaftliche Biographie von Gauß*, collected by FELIX KLEIN and MARTIN BRENDEL. Leipzig 1911–1919. — *Encyklopädie der mathematischen Wissenschaften mit Einschluß ihrer Anwendungen.* 6 vols., Leipzig 1898–1935. — HINNEBERG, PAUL (Ed.): *Die Kultur der Gegenwart. Ihre Entwicklung und ihre Ziele.* Part Three, First Section: Die mathematischen Wissenschaften. FELIX KLEIN (Ed.) Leipzig and Berlin 1912–1914. — *Abhandlungen über den mathematischen Unterricht in Deutschland, veranlaßt durch die IMUK.* FELIX KLEIN (Ed.), 5 vols. Leipzig and Berlin 1909–1916.

Secondary literature: *DSB* **7**, 396–400. *May* 214. *P* III, 724; IV, 756; V 636; VI, 1329. — TOBIES, RENATE, in collaboration with FRITZ KÖNIG: *Felix Klein* (= *Biogr. hervorr. Nat. Techn. Med.* **50**). Leipzig 1981. — ROWE, DAVID E.: *Felix Klein, David Hilbert, and the Göttingen Mathematical Tradition.* 2 vols. (Dissertation, City University of New York, 1992). — TOBIES, RENATE: "Mathematik als Bestandteil der Kultur. Zur Geschichte des Unternehmens *Encyklopädie der mathematischen Wissenschaften mit Einschluss ihrer Anwendungen.*" *Mitt. Österr. Ges. Wiss.Gesch.* **14** (1994), 1–90. — Portrait: [BÖLLING 1994, 20.3]. R.T.

Kline, Morris (* May 1, 1908, New York, New York, USA; † June 10, 1992, Brooklyn, New York, USA). KLINE, an American applied mathematician and historian of mathematics, received his Ph.D. in mathematics from New York University (NYU) in 1936, whereupon he spent two years as an assistant at the Institute for Advanced Study in Princeton, New Jersey. Thereafter, he taught at the Courant Institute, New York University, where he was professor of mathematics and director of the Division of Electromagnetic Research.

Among KLINE's earliest publications was a book he co-authored with three of his colleagues at NYU, HOLLIS COOLEY, DAVID GANS, and HOWARD WAHLERT, *Introduction to Mathematics* (1937). The book's subtitle revealed its purpose, one that was to serve KLINE as a philosophical principle for the rest of his life: "A Survey Emphasizing Mathematical Ideas and their Relations to other Fields of Knowledge." As the authors lamented in their preface, students were rarely given "any real understanding of the character of the subject or of its relation to the sciences, the arts, philosophy, and to knowledge in general." As a result, "far too many intelligent students are 'soured' for life as far as mathematics is concerned." This book sought to remedy this problem by emphasizing examples from physical experience. The authors believed that mathematics was not "just a collection of methods, but a vast, unified system of reasoning which possesses many of the characteristics of a fine art."

KLINE focused on just these ideas in a more popular work he published on his own some fifteen years later: *Mathematics in Western Culture* (1953). Early on KLINE was railing against teaching and writing that presented mathematics "as a series of apparently meaningless technical procedures." Here his point was simple: "mathematics has been a major cultural force in Western civilization" [ix].

Similar historical themes were struck in KLINE's *Mathematics and the Physical World* (1959), and were even more emphatically stressed in his *Mathematics. A Cultural Approach* (1962). This book attempts to show "the extent to which mathematics has molded our civilization and our culture. [Mathematics] is shown to be intimately related to physical science, philosophy, logic, religion, literature, the social sciences, music, painting and other arts" [v].

KLINE's most important contribution to the history of mathematics, however, is his monumental study, *Mathematical Thought from Ancient to Modern Times* (1972). Due to the fact that this work unfortunately ignores several civilizations (as KLINE readily admits), including Chinese, Japanese, and Mayan contributions, "because their work had no material impact on the main line of mathematical thought," this is an entirely Western-centric treatment. KLINE's conviction is that "the roots of the present lie deep in the past and almost nothing in that past is irrelevant to the man who seeks to understand how the present came to be what it is" [viii].

For students, KLINE hoped his study would serve several purposes. "[History] may give perspective on the entire subject and relate the subject matter of the courses not only to each other but also to the main body of mathematical thought" [ix]. KLINE also believed there was an object lesson to be learned from the history of mathematics as well: "Indeed the account of how mathematicians stumbled, groped their way through obscurities, and arrived piecemeal at their results should give heart to any tyro in research" [ix].

As a corollary to KLINE's strong criticism of abstract mathematics, all too often taught as if it had no history, no roots in concrete, physical experience, KLINE was severely critical of the so-called "New Math," and in 1973 wrote a provocative book, *Why Johnny Can't Add. The Failure of the New Math.* This was followed a few years later with an equally critical indictment, *The Professor Can't Teach. Mathematics and the Dilemma of University Education* (1977). Here KLINE criticized the over-emphasis on the part of universities on research, which meant that teachers often gave their students and classroom time short shrift in preference for working on publications. By the late seventies KLINE was thoroughly disillusioned with academic mathematics, and his discontent was reflected in a widely-read book, *Mathematics. The Loss of Certainty* (1980). As KLINE explained, "Mathematics was regarded as the acme of exact reasoning, a body of truths in itself, and the truth about the design of nature. How man came to the realization that these values are false and just what our present understanding is constitute the major theme." Again, KLINE's major dissatisfaction was directed against overly-theoretical, abstract approaches to mathematics. For the past 100 years, he complained, "most mathematicians have withdrawn from the world to concentrate on problems generated within mathematics. They have abandoned science" [278].

KLINE's last book, written only seven years before his death in 1992, returned guardedly to the optimism reflected in his earlier works, and again to themes familiar from his textbooks and historical writings. In *Mathematics and the Search for Knowledge* (1985), the emphasis was upon what can be known about the reality of the physical world around us, not by experiment, observation, or the senses, but only through mathematics. As KLINE stressed, "for many vital phenomena, mathematics provides the only knowledge we have" [vi].

KLINE retired from New York University in 1975, and taught for another year as distinguished professor of mathematics at Brooklyn College of the City University of New York. In addition to his writing and teaching, he was also an associate editor of the *Archive for History of Exact Sciences.*

Main works: [KLINE 1953]. — [KLINE 1972].

Secondary literature: *P* V, 638; VI, 1333; VIIb(4), 2495. — PACE, ERIC: "Morris Kline, 84, Math Professor and Critic of Math Teaching, Dies," *The New York Times* (Thursday, June 11, 1992), D-23, cols. 1–3 (with portrait of 1964).
J.W.D.

Knorr, Wilbur Richard (* August 29, 1945, Brooklyn, New York, USA; † March 18, 1997, Stanford, California, USA). WILBUR KNORR was an American historian of mathematics who received his B.A. degree from Harvard University in History and Science in 1966, and went on to complete his Ph.D. at Harvard in History of Science in 1973. At Harvard he studied ancient science and mathematics with JOHN MURDOCH, G. E. L. OWEN, and A. I. SABRA. He read Greek, Arabic, and Hebrew, all of which contributed to his special interest: history of mathematics in antiquity, and the transmission of mathematics between cultures. He served on the editorial boards of *Historia Mathematica*, the *Archive for History of Exact Sciences*, and *Isis*, and is best known for his detailed, technical studies of EUCLID, APOLLONIUS, and ARCHIMEDES. In addition to teaching positions he held at Brooklyn College of the City University of New York and Stanford University, he was also a member of the Institute for Advanced Study, Princeton (1978–79).

Main works: *The Evolution of the Euclidean Elements.* Dordrecht 1975. — *Ancient Sources of the Medieval Tradition of Mechanics.* Florence 1982. — *The Ancient Tradition of Geometric Problems.* Boston 1986, repr. New York 1993. — *Textual Studies in Ancient and Medieval Geometry.* Boston 1989.

Seconday literature: FOWLER, DAVID: "*In Memoriam.* Wilbur Richard Knorr (1945–1997): An Appreciation." *HM* **25**, 123–132 (with bibliography and portrait). J.W.D.

Kolmogorov, Andrei Nikolaevich (* April 25, 1903, Tambov, Russian Empire; † October 20, 1987, Moscow, USSR). ANDREI KOLMOGOROV was a famous mathematician who spent his entire career at Moscow State University, where he was a professor of the Faculty of Mathematics and Mechanics from 1931 until 1987. He was also a member of the Academy of Sciences of the USSR. His major areas of research interests in history of mathematics included history of analysis and of probability theory, as well as general questions.

Main works: "Mathematics." In: *Great Soviet Encyclopedia* **38** (in Russian, with a periodization of the history of mathematics). Moscow 1938. — "I. Newton and Modern Mathematical Mentality." In: *Moscow University to the Memory of I. Newton* (in Russian). Moscow 1946.

Secondary literature: *May* 216–217. *P* VI, 1368; VIIb(4), 2522. — [KUROSH 1959] — [FOMIN/SHILOV 1969] — [BORODIN/BUGAI 1979] — [BOGOLYUBOV 1983, (with portrait on p. 232)] — SHIRYAEV, A. N. (Ed.): *Memories about Kolmogorov* (in Russian). Moscow 1993 (with portraits). S.S.D.

Korteweg, Diederik Johannes (* March 31, 1848, s-Hertogenbosch, Netherlands; † May 10, 1941, Amsterdam, Netherlands). KORTEWEG was professor of mathematics, rational mechanics, and astronomy at the University of Amsterdam (1881–1918). Through his work on HUYGENS he contributed greatly to the history of 17th-century mathematics.

Secondary literature: *DSB* **7**, 465–466. *P* III, 744; IV, 794; V, 671; VI, 1385; VIIb, 2568. — *BWN* **4**, 266–267. — BETH, H. J. E., and W. VAN DER WOUDE: "Levensbericht van D. J. Korteweg." *Jaarboek Koninklijke Nederlandsche Akademie van Wetenschappen 1945–46* (1946), 194–208 (with portrait). P.B.

Koyré, Alexandre (* August 29, 1892, Taganrog, Russia; † April 28, 1964, Paris, France). The son of a prosperous Russian importer and oil investor, KOYRÉ studied at Tiflis and Rostov-on-Don before going to Göttingen, where he studied with HUSSERL and attended HILBERT's lectures on mathematics. In 1911 KOYRÉ went to Paris for further study with BERGSON, PIERRE-ANDRÉ LALANDE, and BRUNSCHVICG (**B**), among others. Although his early works were devoted to theology, especially the philosophies of ANSELME and BOEHME (for which he received his diploma from the *Ecole Pratique des Haute Etudes* (Practical School of Advanced Studies) and the doctorate from the *Sorbonne*, he went on to more profound studies of DESCARTES, whose mathematics he took to be essential for the appreciation of the rest of DESCARTES's philosophy, especially with respect to notions of continuity and the infinite.

KOYRÉ later devoted studies to COPERNICUS, KEPLER, GALILEO, and NEWTON, among others. Less well-known to historians of science are KOYRÉ's studies of various German mystics, of HEGEL and HUSSERL, and of Russian intellectual history in the 19th century.

After the German invasion of France in 1940, KOYRÉ joined the *Free French* and worked with a group of scholars in creating, in New York, the *Ecole Libre des Hautes Etudes* (Free School of Advanced Studies). After World War II KOYRÉ taught at various American universities, including Harvard, Yale, and Chicago, while continuing to teach at the *Ecole Pratique des Haute Etudes.* In 1955 he visited the Institute for Advanced Study in Princeton, New Jersey (USA), and was made a permanent member the following year, after which he divided his time between Princeton and Paris. In 1958, he succeeded in creating a *Centre de recherches d'histoire des sciences et des techniques* (Center for Research on the History of Science and Technology) in Paris, later renamed the *Centre A. Koyré.*

His main areas of interest in history of science concerned the Scientific Revolution. He collaborated with I. B. COHEN on the preparation of a critical edition of NEWTON's *Principia* ([KOYRÉ/COHEN 1972]). His essential contributions to the history of mathematics are embodied in such works as *Etudes galiléenes* (1939), *From the Closed World to the Infinite Universe* (1957; French trans. 1961); and *Newtonian Studies* (published posthumously in 1965; French trans. 1966).

Further works: *Etudes d'histoire de la pensée philosophique.* Paris 1961. — *Etudes d'histoire de la pensée scientifique.* Paris 1961.

Secondary literature: *DSB* **7**, 482–490. — BELAVAL, YVON: "Les recherches philosophiques d'Alexandre Koyré." *Critique* **20** (1964), no. 207/208, 675–704. — COSTABEL, PIERRE, and CHARLES COULSTON GILLISPIE: "*In memoriam* Alexandre Koyré (1892–1964)." *Arch. int. Hist. Sci.* **17** (1964), 149–156. — DELORME, SUZANNE; PAUL VIGNAUX, RENÉ TATON, and PIERRE COSTABEL: "Hommage à Alexandre Koyré." *Rev. Hist. Sci.* **18** (1965), 129–159 (with portrait). — KUHN, THOMAS S.: "Alexander Koyré and the History of Science." *Encounter* **34** (1970), 67–69. — JORLAND, GÉRARD: *La science dans la philosophie. Les recherches épistémologiques d'A. Koyré.* Paris 1981. — REDONDI, PIETRO (Ed.): *Science: The Renaissance of a History. Proceedings of the International Conference Alexandre Koyré, Paris, Collège de France, 10–14 June 1986* (= *Hist. Techn.* **4** (1987) [numerous articles about KOYRÉ, his work and influence]. — STOFFEL, JEAN-FRANÇOIS: *Alexandre Koyré: Bibliographie de la littérature primaire et secondaire* (= *Reminisciences* **4**). Louvain-la-Neuve 1996. — REDONDI, PIETRO: "Koyré, Alexandre (1892–1964)." In: CRAIG, E. (Ed.): *Routledge Encyclopedia of Philosophy.* London and New York 1998, vol. 5, 296–298. — STOFFEL, JEAN-FRANÇOIS: *Bibliographie d'Alexandre Koyré.* Florence 2000. J.P.

Kragemo, Helge Bergh (* March 18, 1897, Norway; † August 12, 1968, Norway). Although KRAGEMO completed his university education in 1926, having studied the sciences, he had already started working as a librarian at the University of Oslo in 1919, a position he maintained until his retirement in 1967. In the 1930's he published three biographies of the Norwegian mathematician LUDVIG SYLOW. K.A.

Krause, Max (* April 20, 1909, Darmstadt, Germany; † about February 18, 1944, near Winniza, Russia). KRAUSE studied mathematics and oriental languages and had a position as an assistant in the department of oriental languages of the University of Hamburg, before he was drafted to the army. Interested in Arabic mathematics and astronomy, he published his dissertation in 1936 on *Die Sphärik von Menelaos aus Alexandrien in der Verbesserung von Abū Naṣr Manṣūr b. ʿAlī b. ʿIrāq* (The Spherics of Menelaus of Alexandria in the Revised Version by Abū Naṣr Manṣūr b. ʿAlī b. ʿIrāq), and in the same year "Stambuler Handschriften islamischer Mathematiker" (Istanbul Manuscripts of Islamic Mathematicians). *Q. St. Gesch. Math. Astr. Phys.*, Abt. **B 3** (1936), 437–532.

Secondary literature: *P* VIIa(2), 904. — DIETRICH, ALBERT: "Max Krause *in Memoriam*." *Islam* **29** (1949), 104–108 (with portrait). M.F.

Krylov, Aleksei Nikolaevich (* August 15, 1863, village Vissyaga, Simbirs region, Russian Empire; † October 26, 1945, Leningrad, USSR). Beginning in 1890 ALEKSEI KRYLOV worked at the St. Petersburg Naval Academy, at the St. Petersburg Technological Institute, and then, from 1900, as director of the Experimental Section of the Naval Departement. In 1908–1910 he was the principal inspector of shipbuilding and chief of the Naval Technical Committee of the Russian Empire.

From 1927 until 1934 he served as director of the Physico-Mathematical Institute of the Academy of Sciences of the USSR. KRYLOV was a specialist in the mathematics and mechanics of shipbuilding. His interests in the history of mathematics included history of mathematical analysis.

Main work: Russian translation of ISAAC NEWTON's "Principia." St. Petersburg 1915–1916.

Secondary literature: *DSB* **7**, 513–514. *May* 221. — [KUROSH 1959] — KHANOVICH, I. G.: *Academician Aleksei Nikolaevich Krylov* (in Russian). Leningrad: Nauka 1967 (with portrait). — [BORODIN/BUGAI 1979] — [BOGOLYUBOV 1983, with portrait on p. 250]. S.S.D.

La Cour, Poul (* April 13, 1846, Ebeltoft, Denmark; † April 24, 1908, Askov, Denmark). LA COUR studied physics and meteorology at the University of Copenhagen, from which he graduated in 1869. Beginning in 1878, he taught mathematics at the folk high school in Askov. The first edition of his textbook, *Historisk Matematik* (Historical Mathematics), appeared in 1881 and was reedited regularly until 1942.

Secondary literature: *May* 222. *P* III, 305; IV, 279; V, 698. — *DBL* **8**, 447–449. — HANSEN, H. C.: *Poul La Cour. Grundtvigianer, opfinder og folkeoplyser.* Vejen: Askov Højskole 1985 (with portrait). K.A.

Lacroix, Sylvestre François (* April 28, 1765, Paris, France; † May 24, 1843, Paris, France). The author of famous mathematical textbooks, LACROIX included descriptions of historical developments in his works. He also contributed a chapter on partial differential equations to volume III of the second edition of MONTUCLA's

Histoire des mathématiques. In 1799 he was elected a member of the *Institut de France.*

Main work: "Preface" to the first volume of *Traité du calcul différentiel et du calcul intégral.* 2nd ed., 3 vols., Paris 1810–1819.

Secondary literature: *DSB* **7**, 549–551. *May* 223. *P* I, 1340. J.P.

Lalande, Jérôme de (* July 11, 1732, Bourg-en-Bresse, France; † April 4, 1807, Paris, France). Educated by the Jesuits at the *Collège de Lyon*, LALANDE studied astronomy in Paris and became, in 1760, professor at the *Collège Royal.* He contributed to the volumes on mathematics of the *Encyclopédie méthodique* (1784–1789) and edited the last two volumes of the second expanded edition of MONTUCLA's *Histoire des mathématiques* [MONTUCLA 1799/1802].

Secondary literature: *DSB* **7**, 579–582. *May* 225. *P* I, 1349. J.P.

Lejeune, Albert (* January 2, 1916, Liège, Belgium; † March 3, 1988, Visé, Belgium). LEJEUNE, a teacher at the Royal Athenaeum in Visé, produced studies on the history of optics in antiquity and the middle ages, in particular on PTOLEMY's *Optics.*

Main works: [LEJEUNE 1956]. — [LEJEUNE 1957]. P.B.

Lenoble, Robert (* March 1, 1900, Orléans, France; † January 4, 1959, Orléans, France). LENOBLE was a priest who taught philosophy at colleges of the Oratoire in Paris, but following an accident in 1949, he devoted the rest of his life to research. His main area of interest was 17th-century science, especially MERSENNE, on whom he wrote a thesis.

Main work: *Mersenne ou la naissance du Mécanisme.* Paris 1943.

Secondary literature: COSTABEL, PIERRE: "Robert Lenoble (1902 [misprint for 1900] – 1959)." *Rev. Hist. Sci.* **12** (1959), 167–169. — COSTABEL, PIERRE: "Robert Lenoble (1902 [misprint for 1900] – 1959)." *Arch. int. Hist. Sci.* **11** (1958), 385–386. — COSTABEL, PIERRE: "Robert Lenoble (1900–1959)." *Physis* **2** (1960), 92–94. — POPKIN, RICHARD H.: "Robert Lenoble (1900–1959)." *Isis* **51** (1960), 200–202. J.P.

Le Paige, Constantin Marie Michel Hubert Jérôme (* March 9, 1852, Liège, Belgium; † January 27, 1929, Liège, Belgium). LE PAIGE was professor of mathematics and astronomy at the University of Liège from 1882 until his retirement in 1922. In addition to research on the origins of the symbols of operation, his historical publications include studies of mathematics in the Prince-bishopric of Liège [LE PAIGE 1888], in particular the works of RENÉ FRANÇOIS DE SLUSE [LE PAIGE 1884], and of the astronomer GODFRIED WENDELEN.

Secondary literature: *DSB* **8**, 250. *May* 237. *P* III, 797; IV, 268; V, 731. — *BNB* 30, 653–655. — [*Acta Math.*, Tab. Gen. **1–35** (1913), 79; portrait 155]. — GODEAUX, LUCIEN: "Notice sur Constantin le Paige." *Annuaire Académie royale* *de Belgique* **105** (1939), *Notes biographiques*, 239–269 (with bibliography and portrait). — Portrait: [Bölling 1994, 3.1]. P.B.

LI Yan (* August 22, 1892, Fuzhou City, Fujian Province, China; † January 14, 1963, Beijing, China). It is well known that Li Yan and Qian Baocong, two prominent Chinese scholars, were active for half a century in studying the history of mathematics in ancient and medieval China. Li Yan was the son of an intellectual but poor family in Fuzhou City, Fujian Province. He passed the college entrance examination and went on to study civil engineering at the Tangshan Railway and Mining College in 1912. Unfortunally, he had to discontinue his study in 1913 to work on the Longhai Railway. Thereafter, as one of the contemporary engineers of China, he devoted his life to the Long Hai Railway for 42 years.

At virtually the same time, he began his academic career as an historian of Chinese mathematics. Between 1915 and 1917, he corresponded with the American historian of mathematics D. E. Smith on the history of Chinese mathematics. In the course of his scholarly carreer, Li published more than twenty books and sligthly more than one hundred treatises. He was an historian of Chinese mathematics who exerted an international influence.

During the period of the 1930s and the 1940s, he collected a number of his articles together and published them as books (four in all, in 1933, two in 1935, and 1947, respectively), under the title *Zhongsuanshi luncong* (Collected Essays on the History of Chinese Mathematics) (4 vols.). In the 1950s, a revised edition was published (5 vols.).

Volume One was devoted to achievements of ancient Chinese mathematicians, including the theory of fractions, research on the *gougu* (right-angled triangle) theorem (the so-called Pythagorean theorem), the *pingfang lingyue* (approximate expression of irrational roots) method, the method of *dayan qiuyi shu* (the so-called Chinese remainder theorem by Western scholars), Jia Xian's triangle (the Chinese "Pascal triangle"), the *zongheng tu* (the magic square), the *fangcheng* (a method for solving simultaneous linear equations, later also applied to higher equations), and the *duoji zhaocha* (the methods of series and interpolation).

Volume Two studied mathematical books in every dynasty. Here Li Yan's bibliography of mathematical works of the Ming and Qing dynasties proved especially valuable.

Volume Three discussed the transmission of Western mathematics into China during the Ming and Qing periods, and the study of Western mathematics including logarithms, conic sections, trigonometry, and infinite power series by Chinese scholars at that time. In particular, Li Yan studied the chronicle of MEI Wending, the famous early Qing dynasty mathematician.

Volume Four dealt with the counting-rod arithmetical operations, the application of the abacus, the history of mathematical education systems, the *Ceyuan haijing* (Sea Mirror of Circle Measurement), and the chronicles of LI Shanlan and HUA Hengfang, two famous mathematicians in the late Qing dynasty.

Volume Five contained the investigation and source materials for the history of mathematical interchange and the historiography of the history of mathematics in China.

Among his many books, the most representative are his *Zhongguo suanxue shi* (A History of Chinese Mathematics) (1937; Japanese trans. by YABUUTI KIYOSHI and SHIMAMOTO KAZUO, 1940); the *Zhongguo shuxue dagang* (An Outline of the History of Chinese Mathematics) (2 vols., 1958); the *Zhongguo gudai shuxue jianshi* (A Concise History of Ancient Chinese Mathematics) (2 vols., written in collaboration with DU SHIRAN, Engl. trans. by J. N. CROSSLEY and A. W.-C. LUN, 1986).

LI YAN was also a famous collector of ancient and medieval Chinese mathematical books. After his death, his extensive collection was donated to the Institute for History of Natural Sciences, Academia Sinica, Beijing, where it is now used by students and scholars from all parts of the world. From 1957 to 1963, LI YAN was the Institute's first director. In the course of his career he made many valuable contributions to the cause of the history of modern science in China, and facilitated cultural exchanges with foreign scholars. His works are still quoted by historians of science and mathematics alike, and remain useful resources for Chinese and foreign scholars even today.

Secondary Literature: WONG MING: "Le professeur Li Yen, 1892–1963." *Rev. Hist. Sci.* **16** (1963), 256–257. — ZHANG DIANZHOU: "Correspondence between Li Yan and D. E. Smith" (in Chinese, with English summary). *Zhongguo keji shiliao* (Chinese Historical Materials on Science and Technology) **12**, no. 1 (1991), 75–83. — DU SHIRAN: "In Memory of Prof. Li Yan" (in Chinese, with English summary). *Zhongguo keji shiliao* (Chinese Historical Materials on Science and Technology) **13**, no. 4 (1992), 31–36 (with portrait on inside front cover). — LI DI: "Yan LI as a Founder of the Science of the History of Chinese Mathematics" (in Chinese; English and Chinese summaries). *Neimenggu shida xuebao (ziran kexue Hanwen ban)* 1994, no. 2, 73–80. — [LIU 1994, **4**, 108–109]. L.D.

Libri, Guglielmo Bruto Icilio Timoleone Conte Carrucci della Somaia (* January 2, 1803, Florence, Italy; † September 28, 1869, Fiesole, Florence, Italy). GUGLIELMO LIBRI is one of the more controversial figures among historians of mathematics. The son of a noble family, he attended secondary school in Florence and then studied at the University of Pisa. At the age of 17 he published his first scientific paper, *Memoire sur la theórie des nombres* (Memoir on the Theory of Numbers) (1820), which caught CAUCHY's attention. In 1823 (at the age of 20!) LIBRI was appointed professor of mathematical physics at the University of Pisa, but the following year, due to a serious illness which prevented him from teaching, he was forced to give up this position. However, because his scientific abilities were highly appreciated by the Grand Duke of Tuscany, LIBRI was named professor *emeritus* of the University of Pisa. Between 1824 and 1830 he published several papers on analysis, on the theory of equations, and on mathematical physics (which were praised by FOURIER, who by then was secretary of the French Academy of Sciences).

In 1825 LIBRI left Florence for Paris, where he was welcomed into the circle of French mathematicians. In 1830 he participated in the July Revolution in Paris, which overthrew the BOURBONS, giving power to LOUIS PHILIPPE. LIBRI then returned to Tuscany where he intended to take part in the Italian independence movement. After failure of the uprisings in 1831, LIBRI was banished from Florence for his political activities, and was exiled to Paris. In 1833 he became a French citizen and was elected to succeed LEGENDRE as a member of the French Academy of Sciences. Three years later LIBRI was made a professor at the Sorbonne. An enthusiastic supporter of the new regíme in France, LIBRI very soon became a prominent and very powerful figure in French academic and political life. However, his excessively great academic power in contrast to his actual mathematical ability, along with his arrogant and distasteful character, prompted envy and rivalry. Indeed, LIBRI was a mediocre mathematician, apparently overestimated in his day, but one who eventually revealed himself to be a first-rate historian of mathematics. He possessed an extraordinary erudition and easily mastered classic and modern languages. Moreover, LIBRI was a passionate bibliophile and spent a fortune collecting old book and manuscripts. His library amounted to some 35,000 books and 2000 manuscripts, some of which he apparently presented to French libraries. Indeed, some rare books and codices in the Mazarine Library in Paris come from LIBRI's own collection. Eventually, this true bibliomania combined with the downfall of his political supporters proved the main cause of his personal ruin.

Beginning in 1829, LIBRI was increasingly interested in the history of medieval mathematics, and he studied in particular the manuscripts of LEONARDO DA VINCI. LIBRI's major historical work, however, was the *Histoire des sciences mathématiques en Italie depuis la Renaissance des lettres jusqu'à la fin du XVIIIe siècle* (History of the Mathematical Sciences in Italy since the Renaissance of Learning until the End of the 18th Century) (in four volumes, 1838–1841, repr. 1989; [LIBRI 1838/41]).

According to LORIA, although flawed by "some inaccuracies," LIBRI's *Histoire* is a "classic work" and "still one of the more important among those devoted to the history of positive sciences" (quoted in [PROCISSI 1989, 179]). Even today it remains a valuable source for the history of Italian mathematics, particularly because of the appendices which include a number of manuscripts and excerpts from rare texts first published by LIBRI. In the *Histoire* LIBRI strongly emphasized the importance of the contributions of Italians to Renaissance mathematics, and the book culminates with the vindication of GALILEO's role as the founder of modern science.

Although his academic power was beginning to decline, in 1843 LIBRI was elected a professor at the *Collège de France.* After teaching for two years, he left this position to become secretary of a Royal Commission in charge of compiling a catalog of all manuscripts kept in French libraries. In performing this task, LIBRI succeeded in estranging all of the members of the *Ecole des Chartes.* In secret, his enemies sent a report to F. P. G. GUIZOT, a powerful Minister of Foreign Affairs, serving in fact as the Prime Minister from 1840 to 1848, and accused LIBRI of having stolen rare books and manuscripts from various French libraries. The scandal only became public in 1848 following the revolution which overthrew the king LOUIS PHILIPPE D'ORLEANS, whereupon LIBRI fled to England. Although LIBRI strongly maintained his innocence, in 1850 as the result of a rather unreliable trial which denied him any possibility of defence, he was sentenced *in absentia* to ten years in prison, and was banished from all of his official positions in France, including his membership in the Academy of Sciences.

A request for LIBRI's rehabilitation, made by the French Minister of Justice in 1861, was eventually rejected by the Senate. The Libri affair divided the opinions of his contemporaries. Among his enemies there were the members of the *Ecole des Chartes* and mathematicians like his former friends (but subsequently political adversaries) ARAGO and LIOUVILLE. On the other hand, in addition to most Italian scientists and patriots, his supporters included GUIZOT, MERIMÉE, DE MORGAN, and others. (A selection of writings and documents for and against LIBRI is appended to BORTOLOTTI's biographical notice of LIBRI published in the *Enciclopedia Italiana Treccani*). LIBRI spent the rest of his life in exile in London, and was only able to return to Italy in 1869, a few months before his death. His celebrated library was dispersed, and now, most of his papers (including letters and personal documents) are kept in the National Library in Paris and in the Moreniana Library in Florence (see [Candido 1942]).

Secondary literature: *May* 240. *P* I, 1450. — *EIT* **21**, 67–68. — STIATTESI, ANDREA: *Commentario storico-scientifico sulla vita e le opere del Conte Guglielmo Libri.* 2nd ed. Florence 1879. — CANDIDO, GIACOMO: "Il 'Fondo Palagi-Libri' della Biblioteca Moreniana di Firenze." *Atti del II Congresso dell'Unione Matematica Italiana.* Roma 1942, 841–885. — FUMAGALLI, GIUSEPPE: *Guglielmo Libri,* a cura di B. MARACCHI BIGIARELLI. Firenze 1963. U.B.

Lie, Marius Sophus (* December 17, 1842, Nordfjordeid, Norway; † February 18, 1899, Kristiania, Norway). After having graduated from the university at Kristiania in 1865, SOPHUS LIE was uncertain about his future. Two years later, however, he found himself drawn to geometry, to which he later made substantial contributions. He obtained his doctorate in 1871, and was offered an extraordinary professorship in 1872. The following year he and LUDVIG SYLOW began to make a new edition of ABEL's works. In the period 1886–1898 LIE was professor at Leipzig, a position he left when colleagues in Norway persuaded him to return home, where he served only a brief time as professor before he died. A hundred years after WESSEL's paper on complex numbers had been published, LIE had it reprinted in the *Archiv for Mathematik og Naturvidenskab*, a journal of which he had been co-founder.

Secondary literature: *May* 240. *P* III, 808; IV, 882; V, 742. — *NBL* **8**, 353–357. — [STUBHAUG 2000b, numerous portraits]. — Portrait: [BÖLLING 1994, 7.5]. K.A.

Liebmann, Heinrich (* October 22, 1874, Strassburg, Germany; † June 12, 1939, Solln near Munich, Germany). LIEBMANN, a German mathematician, became *Privatdozent* in Leipzig in 1899, was promoted to extraordinary professor in 1905, and as such went to the *Technische Hochschule* in Munich in 1910. From 1920 he was ordinary professor at the University of Heidelberg. In 1935, when he was attacked by the Nazis because he was Jewish, he went into retirement. His areas of interest included differential equations, geometry, and the history of non-Euclidean geometry.

Main works: *N. J. Lobatschefskijs imaginäre Geometrie und Anwendung der imaginären Geometrie auf einige Integrale* (1904). — German edition of R. BONOLA: *Die nichteuklidische Geometrie. Historisch-kritische Darstellung ihrer Entwicklung* (1908).

Secondary literature: *P* IV, 887; V, 744; VI, 1525; VIIa(3), 93. — *NDB* **14**, 508. — "Mathematische Abhandlungen. Heinrich Liebmann zum 60. Geburtstag am 22. Oktober 1934 gewidmet von Freunden und Schülern. Vorgelegt vom Klassensekretär Paul Ernst." In: *Sb. Akad. Heid., Math.-nat. Kl.*, Jahrgang 1934, 8.–17. Abhandlung (with portrait, following p. vi). — [PINL 1972, 162–167]. M.F.

Lietzmann, Walter (* August 7, 1880, Drossen, Neumark, Germany; † July 12, 1959, Göttingen, Germany). LIETZMANN, a German mathematician and educator, began his career as a teacher of mathematics in Barmen (1906), Jena (1914), and Göttingen (1919), where in 1920 he became *Dozent* and in 1934 *Honorarprofessor* of pedagogy for the exact sciences at the university, from which he retired in 1946. LIETZMANN contributed much to the realization of FELIX KLEIN's program to reform the teaching of mathematics in secondary schools. Although mainly remembered for his numerous publications on the pedagogy of mathematics, LIETZMANN also published a number of historical works, including *Überblick über die Geschichte der Elementarmathematik* (Survey of the History of Elementary Mathematics) (1926; 2nd ed. 1928); *Aus der Mathematik der Alten* (Selections from the Mathematics of the Ancients) (1928); *Altes und Neues vom Kreis* (The Old and New About Circles) (1935); and *Frühgeschichte der Geometrie auf germanischem Boden* (Early History of Geometry on Germanic Soil) (1940).

Secondary literature: *May* 241. *P* V, 745; VI, 1530; VIIa(3), 99. — ZÜHLKE, PAUL: "Walter Lietzmann zum 75. Geburtstage." *Math.-phys. Semesterber.* **4** (1955), 161–164 (with portrait). — PROKSCH, RUTH: "Walter Lietzmann †." *MNU* **12** (1959), 227–228 (with portrait). — WANSING, J. H., in *Euclides* **35** (1959), 81–82. — WOLFF, GEORG [?]: "Walter Lietzmann † (1880–1959)." *Prax. Math.* **1** (1959), 127–129. — SCHOEN, R.: "Walther Lietzmann †." *Prax. Math.* **2** (1960), 19 (with portrait). — STENDER, RICHARD: "Walther Lietzmann zum Gedächtnis." *Math.-phys. Semesterber.* **7** (1960), 1. — STRUBECKER, KARL: "Walter Lietzmann †." *Physik. Bl.* **16** (1960), 30–31. M.F.

Lohne, Johannes August (* January 21, 1908, Flekkefjord, Norway; † January 29, 1993, Flekkefjord, Norway). JOHANNES LOHNE studied mathematics, geography, and physics at the University of Oslo (1927–1932) and obtained an M.A. with a thesis in optics. He taught mathematics at various places, and since 1950 at Flekkefjord *Gymnasium*. He also studied history of mathematics, physics, and particularly optics. Of the considerable number of papers that he wrote (in particular on IBN AL-HAYTHAM, HARRIOT, JOHANNES KEPLER, and ISAAC NEWTON, LOHNE himself regarded his "Essays on Thomas Harriot," *Arch. Hist. ex. Sci.* **20** (1979), 189–312, to be the most important. K.A.

Lorey, Wilhelm (* January 23, 1873, Frankfurt (Main), Germany; † July 3, 1955, Königstein near Frankfurt, Germany). LOREY taught mathematics in Leer, Quakenbrück, Remscheid, Görlitz, and Minden (1896–1912), became director of the *Öffentliche Handelslehranstalt* (Public Commercial School) in Leipzig in 1912, and *Dozent* for actuarial mathematics at the University of Leipzig (1920–1933). After World War II he taught at the University of Frankfurt (beginning in 1946), where he founded and directed a regular colloquium on the history of mathematics; from 1953 onwards he held an honorary professorship. Apart from actuarial mathematics, he was also interested in didactics and the history of mathematics.

Main works: *Das Studium der Mathematik an den deutschen Universitäten seit Anfang des 19. Jahrhunderts* (1916). — *Der Deutsche Verein zur Förderung des mathematischen und naturwissenschaftlichen Unterrichts 1891–1938* (1938).

Secondary literature: *May* 247. *P* V, 765; VI, 1565; VIIa(3), 136. — BEHNKE, HEINRICH: "Wilhelm Lorey zum Gedächtnis." *Math.-phys. Semesterber.* **5** (1956), 1–3 (with portrait). — SCHUBRING, GERT: "Wilhelm Lorey (1873–1955) und die Methoden mathematikgeschichtlicher Forschung." *Math. Didact.* **9** (1986), 75–87. M.F.

Loria, Gino (* May 19, 1862, Mantua, Italy; † January 30, 1954, Genoa, Italy). Together with ETTORE BORTOLOTTI, GINO LORIA may be considered the leading historian of mathematics in Italy in the first half of the 20th century, and a prominent figure in the international community of historians of science in his day. Coming from a cultivated family — his father SALOMON was a lawyer and his brother ACHILLE was a celebrated economist — LORIA attended the University of Turin and at the age of 21 graduated (*laurea*) in mathematics with a thesis on higher geometry under the supervision of ENRICO D'OVIDIO. By the end of 1883 LORIA obtained a grant for a postgraduate course at the University of Pavia, where he studied with EUGENIO BELTRAMI, EUGENIO BERTINI, and FELICE CASORATI. One year later, in November of 1884, he returned to the University of Turin as D'OVIDIO's assistant for algebra and analytic geometry. LORIA kept this position until November of 1886, when he was appointed extraordinary professor of higher geometry at the University of Genoa. There he spent the rest of his academic career, becoming full professor in 1891 and teaching higher geometry for 49 years until 1935, when he retired. For many years he also taught descriptive geometry and history of mathematics. After his retirement, LORIA was named professor *emeritus* of the University of Genoa in 1936.

LORIA's first publication was a paper on the geometry of straight lines jointly written with his classmate, CORRADO SEGRE, and published in the *Mathematische Annalen* (1883). However, in a short time their research interests went in divergent directions. While SEGRE was interested in the more recent developments of algebraic geometry, and very soon became the leading figure of the new school of algebraic geometry in Italy, LORIA was mainly concerned with "classical" subjects such as the geometry of straight lines and spheres and the algebraic correspondences between fundamental forms, even though he did some work in the new fields of hyperspatial projective geometry and Cremona transformations in space. However, LORIA took no part in the modern development of geometry, and his interests turned rather soon to history of mathematics. Nonetheless, even after he left Turin for Genoa, his close friendship with SEGRE lasted until SEGRE's untimely death in 1924. According to TERRACINI [1954, 407], his teachers at the University of Turin may have played a role in stimulating LORIA's interest in history. For example, D'OVIDIO used to provide historical notes for his lectures on analytic geometry, and ANGELO GENOCCHI also raised historical issues. "Times were favorable for the history of mathematics" in Italy, as TERRACINI observes in referring to the lasting influence of PRINCE BONCOMPAGNI and his *Bullettino*.

The first hint of LORIA's historical interests dates back to 1886, when he published a short note in *Acta mathematica* in which he called attention to C. V. MOUREY's proof of the fundamental theorem of algebra. The history of this theorem became a main subject of LORIA's historical research, which culminated five years later in the publication of a comprehensive essay of more than 60 pages on the subject in PEANO's *Rivista di matematica* (1891, with additions in 1892 and 1893). In this paper LORIA gave a fairly complete account of the different proofs of this theorem, from D'ALEMBERT's and EULER's to more recent ones. At ENESTRÖM's suggestion, LORIA published a short version of his essay in *Bibliotheca mathematica* (1891), which included a list of some 80 publications concerning the proof of the fundamental theorem of algebra.

In the meantime, in 1887 LORIA had published a lecture, "Il passato e il presente delle principali teorie geometriche" (Past and Present of the Main Geometric Theories) in the *Memorie dell'Accademia delle Scienze di Torino.* Over the years LORIA revised and expanded this paper into a book with the same title, which in its final edition (1931) consisted of two volumes. The first one included the development of geometry from antiquity to the end of the 19th century, with special emphasis on modern subjects like the theory of algebraic curves and algebraic surfaces, analysis situs, geometry of straight lines, enumerative geometry, non-Euclidean geometry, and the geometry of hyperspaces. In the second volume LORIA discussed the developments of these topics in the first 30 years of the 20th century.

As a working geometer LORIA was mostly interested in the history of geometry, which represents the main subject of his more than 380 publications. His first relevant work was the book, *Nicola Fergola e la scuola di geometri napoletani che lo ebbe a duce* (Nicola Fergola and the School of Neapolitan Geometers which he led) (1892). This was devoted to the history of a school of "synthetic" geometry founded by NICOLÒ FERGOLA (1753–1824) and which flourished in Naples in the early decades of the 19th century. However, by the end of the century this school had been almost forgotten in Italy, to the point that LORIA himself was surprised to learn about its existence when he read CHASLES's *Aperçu historique sur l'origine et le developpement des méthodes en géométrie* (Historical Overview of the Origins and Development of Methods in Geometry).

In 1889 BELTRAMI called LORIA's attention to a prize for a textbook on history of mathematics to be awarded by the *Istituto Veneto di Scienze.* At BELTRAMI's suggestion, LORIA undertook a thorough study of the works of the ancient Greek geometers. Fascinated by the arithmetic and geometric productions of the "Greek genius," LORIA was "unable to give up" [LORIA 1914, v] his study even after the deadline for the prize was over. Subsequently, he wrote a series of long papers which PIETRO RICCARDI urged him to publish, and these appeared between 1893 and 1902 in the *Memorie dell'Accademia di Scienze, Lettere e Arti di Modena*, and then later as separate booklets under the same title, *Le scienze esatte nell'antica Grecia* (The Exact Sciences in Ancient Greece). LORIA also issued a revised edition of these papers as a book (again with the same title), which came to almost one thousand (!) pages (1914, reprinted in 1987). As LORIA noted, this book was written "by a mathematician for persons who, although with a modest scientific education, are interested in mathematics." As for the historical material, LORIA chose "a compromise between arrangement by subjects and strict adherence to chronological order, each of which has advantages and disadvantages of its own." Accordingly, his work was divided into five parts: 1) The Greek Geometers, Forerunners of Euclid; 2) The Golden Age of Greek Geometry; 3) The Mathematical Substratum of Greek Natural Philosophy; 4) The Silver Age of Greek Geometry; 5) The Arithmetic of the Greeks. Within each part, subjects are arranged in chronological order under the names of persons or schools. At the time, this book was highly praised; according to THOMAS HEATH (1921, v) it was "undoubtedly the best history of Greek mathematics which exists at present." Even today LORIA's work still represents a basic reference. An abridged French version was published as *Histoire des sciences mathématiques dans l'antiquité ellénique* (History of the Mathematical Sciences in Greek Antiquity, 1929).

By 1897 LORIA, who had contributed to ENESTRÖM's *Bibliotheca mathematica* since 1891, resurrected the heritage of BONCOMPAGNI's *Bullettino* by launching a new *Bollettino di bibliografia e storia delle scienze matematiche*, first as a supplement to BATTAGLINI's *Giornale di matematiche* and then, from 1898 until 1922, as a journal on its own. In addition to LORIA himself, among those who contributed papers and reviews to the 21 volumes of the *Bollettino* were (among others) GIOVANNI VACCA, ROBERTO BONOLA, ETTORE BORTOLOTTI, GIOVANNI VAILATI, and ROBERTO MARCOLONGO. Beginning in 1922 (until 1938), and reduced in size, LORIA published the *Bollettino* as the "historical and bibliographical section" of the *Bollettino di matematica* edited by ALBERTO CONTI.

At the first International Congress of Mathematicians held in Zurich in 1897, LORIA presented a paper on the development of the theory of plane curves. Listing all of the known plane curves, including their history and properties, was an impressive task that LORIA was able to realize five years later in his *Spezielle algebraische und transzendente ebene Kurven. Theorie und Geschichte* (Special Algebraic and Transcendental Plane Curves. Theory and History) (1902; second German edition in two volumes, 1910–1911; Italian revised and extended edition, 1930, in two volumes amounting to more than one thousand pages). This extraordinarily rich catalogue of plane curves is an extremely useful tool for both mathematicians and historians. For each curve LORIA gave not only the equation(s) and its geometrical properties, but also the relevant historical information. This work is divided into seven parts: the first four are devoted to algebraic curves of determinate orders, part V includes algebraic curves of any order, part VI deals with transcendental curves, and part VII discusses the laws for deducing curves from other curves. As skew curves were explicitly excluded by LORIA, this subject was continued in a sequel: *Curve sghembe speciali, algebriche e trascendenti* (Special Algebraic and Transcendental Skew Curves) (two volumes, 1925). An abridged version of his treatise on plane curves was published in 1915 as "Spezielle ebene algebraische Kurven von einer Ordnung höher als der vierten" (Special Algebraic Plane Curves of Order Greater than Four) in the *Encyklopädie der mathematischen Wissenschaften.* Also related to this was a long essay of LORIA's, "Curve e superficie speciali" (Special Curves and Surfaces), which he wrote for the *Enciclopedia delle matematiche elementari* (Encyclopedia of Elementary Mathematics) edited by LUIGI BERZOLARI.

Descriptive geometry was a favorite subject of LORIA's for many years, one he liked to teach and to which he also devoted a considerable amount of historical research. In addition to textbooks like his *Vorlesungen über darstellende Geometrie* (Lectures on Descriptive Geometry) (two parts, 1907–1913), *Metodi di Geometria Descrittiva* (Methods of Descriptive Geometry) (1909; second edition, 1919; third edition, 1925), *Poliedri, curve e superficie secondo i metodi della geometria descrittiva* (Polyhedrons, Curves and Surfaces According to the Methods of Descriptive Geometry) (1912), and *Complementi di geometria descrittiva. Visibilità, ombre, chiaroscuro, prospettiva lineare* (Complements of Descriptive Geometry. Visibility, Shadows, Chiaroscuro, Linear Perspective) (1924), LORIA published many papers devoted to the history of descriptive geometry, including in particular a long article on "Perspective und darstellende Geometrie" (Perspective and Descriptive Geometry) published in the fourth volume of CANTOR's *Vorlesungen* (1908).

LORIA's historical work on descriptive geometry culminated in the book *Storia della geometria descrittiva dalle origini ai giorni nostri* (History of Descriptive Geometry from its Origins to the Present Day) (1921). This book of some 580 pages provides a comphehensive overview of the development of the subject from antiquity to the end of the 19th century. In addition to emphasizing the fundamental role played by MONGE and his school in France, LORIA also devoted the concluding chapters of the book to the development of descriptive geometry in various countries, including Italy and Germany in particular.

Along with descriptive geometry, LORIA also considered writing a comprehensive history of analytic geometry, a project he outlined at the International Congress of Mathematicians at Heidelberg in 1904. However, he was never able to finish this, even though he published a number of related papers, including a long memoire on the development of analytic geometry from DESCARTES and FERMAT to MONGE and LAGRANGE, which appeared in 1924 in the *Memorie dell'Accademia dei Lincei*.

LORIA was also invited to give a lecture on mathematical education in Italy at the Heidelberg Congress. Indeed, for many years he was involved in questions concerning secondary teaching and the reform of the educational system in Italy, and he devoted many papers to this subject. On the occasion of the International Congress of Mathematicians at Zurich in 1932, he presented a comparative view of the mathematical training of secondary school teachers in various countries.

LORIA was also interested in training professional historians of mathematics. At the fourth International Congress of Mathematicians held in Rome in 1908, he suggested a collaborative project to write a textbook for training in history of mathematics, which he tried to produce with the cooperation of many historians from different countries. The sort of international cooperation this required proved impossible with the outbreak of World War I, and so he published what material he was able to gather himself in a *Guida allo studio della storia della matematica* (Handbook for the Study of the History of Mathematics) (1916; second expanded edition, 1946) ([LORIA 1916]). The handbook is divided into two parts. The first lists major works devoted to the history of mathematics (arranged according to countries and subjects) and related professional journals. The second part provides an annotated list of bibliographies, biographies, and other critical tools, including encyclopedias, catalogues, and reviews. In particular LORIA listed bibliographical material about Greek and Latin mathematics and the mathematics of other ancient non-European peoples, along with biographies and biographical resources concerning modern mathematics arranged according to countries, collected works, and the scientific correspondence of mathematicians. Although outdated, LORIA's handbook still represents a valuable reference tool.

Contrary to many great historians of his generation, LORIA was not very much interested in editing primary sources and collected works. He limited himself to working with VITO VOLTERRA and DIONISIO GAMBIOLI in the edition of FAGNANO's collected works, *Opere matematiche del marchese G. C. de' Toschi di Fagnano* (Mathematical Works of the Marquis G. C. de' Toschi di Fagnano) (3 volumes, 1912). Then he continued work on the edition of TORRICELLI's collected works begun by GIUSEPPE VASSURA, which he considered "a national task of universal interest." However, his edition of the *Opere di Evangelista Torricelli* (Works of Evangelista Torricelli, in 4 volumes, 1919–1944), which included the publication of many manuscripts, was sharply criticized, mainly by BORTOLOTTI, who pointed out a number of misunderstandings and mistakes made by the editor.

LORIA's historical work culminated in his *Storia delle matematiche dall'alba della civiltà al tramonto del secolo XIX* (History of Mathematics from the Beginning of Civilization to the End of the 19th Century) (three volumes, 1929–1933; the second, revised edition appeared in one volume of more than 970 pages, 1950; reprinted 1982). As LORIA himself pointed out in his introduction, this book is a general history of mathematics written "by a mathematician for mathematicians." He was proud to state that "this is the first general history of mathematics which includes the whole nineteenth century" ([LORIA 1916, 2nd edition, 33]. In addition, it is also worth mentioning that the final chapter is devoted to historians of mathematics.

After his retirement, LORIA collected many of his papers and lectures, and published them together as a book: *Scritti, conferenze, discorsi sulla storia delle matematiche* (Writings, Lectures, Adresses on the History of Mathematics) (1937). In addition to papers of a general and methodological character, this book contains many of LORIA's biographies, including those for LAGRANGE, BELTRAMI, DE JONQUIÈRES, CREMONA, MANNHEIM, CORRADO SEGRE, and TANNERY. Indeed, writing biographies was one of LORIA's favorite tasks. In addition to many biographical papers and obituaries, he also published biographies of NEWTON (1920), ARCHIMEDES (1928), and GALILEO (1938).

After the Fascist regime decreeded the racial laws in 1938, which barred anyone of Jewish descent from publishing papers and books in Italy, LORIA continued to work, but published his papers in foreign journals, something he continued to do even during World War II. After his death in 1954, his library, which is a very valuable resource for the history of mathematics, especially for 19th-century mathematics, was given to the University of Genoa.

Secondary literature: *DSB* **8**, 504–505. *May* 247. *P* IV, 914; V, 766; VI, 1565; VIIb(5), 2950. — [*Acta Math.*, Tab. Gen. **1–35** (1913), 84; portrait 157]. — TERRACINI, ALESSANDRO: "Commemorazione del Socio Gino Loria." *Rend. Accad. Naz. Lincei, Classe Sci. Fis. Mat. Nat.* (8) **17** (1954), 402–421 (with bibliography). — HEATH, THOMAS L.: "A History of Greek Mathematics." Oxford 1921, vol. I, introduction. U.B.

Luckey, Paul (* December 26, 1884, Elberfeld, Germany; † July 21, 1949, in the Lake of Constance). LUCKEY was a high school teacher in Elberfeld from 1912 to 1924. As a grammar school teacher of mathematics in Marburg (1924–1932), he also taught applied mathematics and history of mathematics at the University of Marburg in 1924 and 1928. After his early retirement, LUCKEY began to study Arabic in Heidelberg, Berlin, Bonn, and Tübingen. LUCKEY was especially interested in the history of Arabic mathematics and astronomy.

Main works: "Ṯābit ben Qurra's Buch über die ebenen Sonnenuhren" (Thābit ibn Qurra's Book on Plane Sundials), *Q. St. G. Math. Astron. Phys.* **B 4** (1938), 95–148. — "Zur Entstehung der Kugeldreiecksrechnung" (On the Origins of Calculating Spherical Triangles), *Dt. Math.* **5** (1940), 405–446. — "Die Ausziehung der n-ten Wurzel und der binomische Lehrsatz in der islamischen Mathematik"

(Extraction of the n-th Root and the Binomial Theorem in Islamic Mathematics), *Math. Annalen* **120** (1948), 217–274. — "Die Rechenkunst bei Ǧamšīd b. Māsūd al-Kāšī" (The Art of Calculating by Jamshīd b. Māsūd al-Kashī), *Abhandlungen für die Kunde des Morgenlandes* **31**, no. 1. Wiesbaden 1951. — [LUCKEY 1999].

Secondary literature: *P* VI, 1573; VIIa(3), 144. — HOGENDIJK, JAN P.: "Paul Luckey (1884–1949)." In: [LUCKEY 1999, vii–xv] (with bibliography). M.F.

Lurie (Luria), Solomon Yakovlevich (* January 6, 1891, Mogiliov, Russian Empire; † October 30, 1964, Lvov, USSR). SOLOMON LURIE taught at Samara University (1919–1920), at the Leningrad Herzen Pedagogical Institute (1921–1923), and at the University of Leningrad (1922–1930, 1935–1942, 1943–1949). He was also a member of the Institute for the History of Science of the USSR Academy of Sciences in Leningrad (1932–1935). In 1950–1953, as a prominent historian and philologist, he was a professor at the Lvov Institute of Foreign Languages, and then professor at Lvov University from 1953 until 1964. He was particularly interested in history of mathematics in antiquity and in medieval Europe. His publications include "Die Infinitesimaltheorie der antiken Atomisten" (Infinitesimal Theory of the Ancient Atomists), *Q. St. G. Math. Astron. Phys.* (Abt.B) **2** (1933), 106–185; and *Archimedes* (1945) (in Russian).

Secondary literature: [KUROSH 1959]. — Portrait in [SHTOKALO 1966/70, vol. 4, book 2, p. 489]. S.S.D.

Lyusternik, Lazar Aronovich (* January 31, 1899, Zdunska-Wola, Poland, Russian Empire; † July 25, 1981, Moscow, USSR). LAZAR LYUSTERNIK was a prominent professor of mathematics at the University of Nishniĭ Novgorod (1928–1930) and at the University of Moscow (1930–1981). He was a corresponding member of the USSR Academy of Sciences. Among his areas of research interest in the history of mathematics were history of mathematical analysis and history of the Moscow mathematical school. Among his publications are "The Early Years of the Moscow Mathematical School," *Russian Mathematical Surveys* **22** (no. 1) (1967), 133–157; (no. 2), 171–211; (no. 4), 55–91. — "From the History of Symbolic Calculus," *IMI* **23** (1977), 85–101 (in collaboration with S. PETROVA, in Russian).

Secondary literature: *May* 249. *P* VI, 1593; VIIb(5), 2999. — [KUROSH 1959] — [FOMIN/SHILOV 1969] — [BORODIN/BUGAI 1979] — [BOGOLYUBOV 1983, with portrait on p. 298]. S.S.D.

Mahnke, Dietrich (* October 17, 1884, Verden, Germany; † July 25, 1939, near Fürth, Germany). MAHNKE, who was a teacher in Stade and Greifswald from 1911 to 1927, received his Ph.D. from the University of Freiburg in 1925. His thesis, "Leibnizens Synthese von Universalmathematik und Individualmetaphysik" (Leibniz's Synthesis of Universal Mathematics and Individual Metaphysics) was published by the philosopher EDMUND HUSSERL in his influential *Jahrbuch für Philosophie und phänomenologische Forschung* **7** (1925), 305–612. In 1926 MAHNKE became *Privatdozent* for philosophy and history of exact sciences in Greifswald, and the following year he received an ordinary professorship in philosophy at Marburg. From 1936 until his death he worked at the first volume of the mathematical correspondence of LEIBNIZ which was later completed by J. E. HOFMANN.

Interested in development of the calculus, and notably in LEIBNIZ, MAHNKE published *Neue Einblicke in die Entdeckungsgeschichte der höheren Analysis* (New Insights in the History of the Discovery of Higher Analysis) (1925); "Zur Keimesgeschichte der Leibnizschen Differentialrechnung" (On the Early History of the Leibnizian Differential Calculus), *Sitzungsberichte der Gesellschaft zur Beförderung der gesamten Naturwissenschaften Marburg* **67** (1932) [1933], 31–69; and *Unendliche Sphäre und Allmittelpunkt* (Infinite Spheres and the Centerpoint of the Universe) (1937).

Secondary literature: *P* VI, 1623; VIIa(3), 183. — *NDB* **15**, 691–692. — ZAUNICK, RUDOLPH: "Zum Gedächtnis von Dietrich Mahnke (1884–1939)." *MGMNT* **38** (1939), 353–356 (with bibliography). — WOHLTMANN, HANS, in *Stader Archiv* N.S. **30** (1940), 135–144. — WOHLTMANN, HANS: "Dietrich Mahnke 1884–1939." *Niedersächsische Lebensbilder* **3** (1957), 157–166 (with bibliography and portrait). M.F.

Maĭstrov, Leonid Efimovich (* January 22, 1920, Dnepropetrovsk, Ukraine, USSR; † August 1, 1982, near Moscow, USSR). Beginning in 1954, LEONID MAĬSTROV taught at the Moscow Financial Institute, and was also a member of the Institute for the History of Science and Technology of the USSR Academy of Sciences in Moscow until his death in 1982. His areas of special interest, the history of probability theory and the history of computers, were reflected in his publications, which included *Probability Theory. Historical Sketch* (1974), and *The History of Computers* (1976, in collaboration with I. APOKIN, in Russian).

Secondary literature: [KUROSH 1959]. — [FOMIN/SHILOV 1969]. S.S.D.

Mansion, Paul (* June 3, 1844, Marchin near Huy, Belgium; † April 16, 1919, Ghent, Belgium). By the age of twenty-three, MANSION was a lecturer in mathematics at the University of Ghent. In 1870 he was made an extraordinary professor, and in 1874 ordinary professor of mathematics. But he was more interested than most mathematicians in the history of the subject. As historian and biographer, he wrote on Greek mathematics, non-Euclidian geometry, mathematics in Belgium, and on the lives and works of (among others) COPERNICUS, GALILEO, KEPLER, FERMAT, SYLVESTER, CHEBYSHEV, CLEBSCH, and HERMITE.

Secondary literature: *DSB* **9**, 80–81. *May* 253. *P* III, 866; IV, 953; V, 802; VI, 1640. — *BNB* **30**, 540–542. — [*Acta Math.*, Tab. Gen. **1–35** (1913), 86; portrait 158] — DEMOULIN, ALPHONSE: "La vie et l'œuvre de Paul Mansion." *Annuaire Académie royale de Belgique* **95** (1929), 77–147 (with bibliography and portrait). — GILLIS, J.: "Paul Mansion en George Sarton." *Mededelingen Koninklijke Academie voor Wetenschappen, Letteren en Schone Kunsten van België, Klasse der Wetenschappen* **35**, no. 2 (1973) . — Portrait: [BÖLLING 1994, 3.3]. P.B.

Marcolongo, Roberto (* August 28, 1862, Rome, Italy; † May 16, 1943, Rome, Italy). The son of a working-class family, ROBERTO MARCOLONGO attended the *Istituto Tecnico* (Technical Institute) and then University of Rome, where he studied mathematics with BATTAGLINI, CERRUTI, CREMONA, and BELTRAMI. He graduated in 1886, but as a student he had already been a temporary assistant to the chair of mechanics, a position he held for ten years. In addition, from 1888 to 1895, he was also an assistant for algebra and calculus. In 1890 he became a *libero docente* teaching mechanics, and five years later he was named extraordinary professor of rational mechanics at the University of Messina, where he was promoted to full professor in 1900. In 1908 he moved to Naples where he taught mechanics until he retired.

Although most of MARCOLONGO's more than 200 scientific publications are devoted to analysis, mechanics, the theory of elasticity, and mathematical physics, they also include a number of important works devoted to the history of mechanics. His first important article in this field was "Progressi e sviluppo della teoria matematica dell' elasticità in Italia (1870–1907)" (Progress and Development of the Mathematical Theory of Elasticity in Italy (1870–1907)), which appeared in *Nuovo Cimento* in 1907. This fifty-page long paper describes the important contributions made by Italian mathematicians — beginning with ENRICO BETTI — to the modern development of elasticity theory. The list of 238 items by 72 authors appended to the paper is still the most complete bibliography on the subject. In 1915 MARCOLONGO published an essay on "Il problema dei tre corpi da Newton (1686) ai nostri giorni" (The Three-Body Problem from Newton (1686) to Our Own Time), which is almost one-hundred pages long. This too was published in the *Nuovo Cimento*, and covers the early work of NEWTON down to contributions to the problem by KARL SUNDMAN in 1909. As MARCOLONGO shows, the history of this "classical problem" is closely related not only to the development of the methods of mechanics and astronomy, but also to the development of mathematical analysis from NEWTON to the beginning of the 19th century.

A more comprehensive work on the history of mechanics from its origins to GALILEO was MARCOLONGO's monograph, *Lo sviluppo della meccanica sino ai discepoli di Galileo* (The Development of Mechanics Down to the Disciples of Galileo) (1919). Contrasting his views with those expressed by DUHEM in the latter's history of mechanics, MARCOLONGO vindicated GALILEO's pioniering role in establishing modern dynamics through a detailed discussion of both the development of pre-Galileian mechanics and GALILEO's own contributions.

In 1924 MARCOLONGO was appointed to succeed ANTONIO FAVARO (see biography) as a member of the National Commission in charge of editing the works of LEONARDO DA VINCI. This began a period of historical research for MARCOLONGO that lasted until his death. His research focused primarily on LEONARDO's *Codex Arundel* and *Codex Forster*, in which LEONARDO showed his mastery of geometry and mechanics. In particular this involved such matters as the quadrature of lunulae, research on centers of gravity, and the construction of mathematical instruments. Once more in opposition to DUHEM's opinions, which underestimated LEONARDO's mathematical abilities, MARCOLONGO undertook to evaluate LEONARDO's knowledge of mathematics by comparing it with the general development of mathematics at that time. LEONARDO, of course, was not a professional mathematician in a 20th-century sense, but MARCOLONGO stressed that he nevertheless attributed a primary role to mathematics in his mechanical and physical research. "There can be no certainty without application of one of the mathematical sciences" — a sentence of LEONARDO's that MARCOLONGO liked to quote. Among MARCOLONGO's many publications (some 30 in all) devoted to LEONARDO, the most significant include his *Memorie sulla geometria e la meccanica di Leonardo da Vinci* (Memoire on Leonardo da Vinci's Geometry and Mechanics) (1937), and *Leonardo da Vinci artista-scienziato* (Leonardo da Vinci: Artist-Scientist) (1939; 2nd ed. 1943), a book that contributes primarily to explaining the value of LEONARDO's scientific knowledge.

Secondary literature: *May* 254. *P* IV, 957; V, 805; VI, 1646; VIIb(5), 3119. — MARCOLONGO, ROBERTO: *Quaranta anni di insegnamento* (an autobiographical account with a list of MARCOLONGO's publications). Naples 1935. — GATTO, ROMANO: "Roberto Marcolongo e le sue ricerche sulla storia della Meccanica." In: [BARBIERI/DEGANI 1989, 147–159]. U.B.

Marie, Charles François Maximilien (* January 1, 1819, Paris, France; † May 8, 1891, Paris, France). MAXIMILIEN MARIE studied at the *Ecole Polytechnique* in Paris (1838) and the *Ecole d'Application d'Artillerie* in Metz (1840). He taught mathematics in Auteuil, became *Répétiteur* (1862) and *Examinateur d'admissions* (1879) at the *Ecole Polytechnique* in Paris. A prolific writer, MARIE published numerous books and articles on mathematical topics as well as on educational and social questions. His *Histoire des sciences mathématiques et physiques* (History of the Mathematical and Physical Sciences) appeared in 12 volumes (Paris 1883–1888), and covered the subject in 17 sections as far as "Arago et Abel," mostly following biographical lines.

Secondary literature: *May* 254. *P* III, 871; IV, 960. C.J.S.

Markushevich, Aleksei Ivanovich (* April 2, 1908, Petrozavodsk, Russian Empire; † June 7, 1979, Moscow, USSR). Beginning in 1935, ALEKSEI MARKUSHEVICH taught at the University of Moscow, where he was made a professor in 1946; he continued to teach there until his death in 1979. From 1958 until 1963 he served as vice-minister of Education of the Russian Soviet Federate Socialist Republic (RSFSR), having previously been vice-president of the Academy of Pedagogical Sciences of the RSFSR from 1950 to 1958. Later he was also vice-president of the USSR Academy of Pedagogical Sciences (1964–1976). In the history of mathematics he was primarily interested in the theory of functions of complex variables.

Main works: *Skizzen zur Geschichte der analytischen Funktionen* (Outline of the History of Analytic Functions) (Russian edition 1951, German trans. 1955). — "Analytic Function Theory," in *Mathematics of the 19th Century*, KOLMOGOROV, A. N. and A. P. YUSHKEVICH (Eds.), Basel 1996, pp. 119–272.

Secondary literature: *May* 255. — [KUROSH 1959] — [FOMIN/SHILOV 1969] — [BORODIN/BUGAI 1979] — [BOGOLYUBOV 1983, (with portrait on p. 312)].
S.S.D.

May, Kenneth Ownsworth (* July 8, 1915, Portland, Oregon, USA; † December 1, 1977, Toronto, Canada). After early training at the Music Education School in Portland, Oregon, MAY moved with his family to Berkeley, California. There he finished his primary and secondary education, having already shown a strong interest in mathematics, especially as a result of his reading WHITEHEAD and RUSSELL's *Principia Mathematica.* At the University of California at Berkeley he majored in mathematics, and was an excellent student, winning several academic honors, including election to Phi Beta Kappa, the national honor society for scholarly distinction. In the Department of Mathematics at Berkeley, MAY revealed an ability for statistics, especially as applicable to national planning. This was an area in which his father, SAMUEL CHESTER MAY (1887–1955), was an expert. As a professor in the political science department at Berkeley, the elder MAY established the university's Bureau of Public Administration in 1930, which he directed until his retirement.

As a student at Berkeley, KEN MAY not only studied mathematics, but played on the varsity soccer team. He became involved as well in the Institute of Pacific Relations, devoted to promoting peaceful relations among countries in the Pacific area. And before graduating from Berkeley, he jointed the Communist Party. When, in 1936, he received his B.A. degree in mathematics, it was awarded with highest honors. A year later he had earned his M. A. degree, whereupon he became a research fellow of the Institute of Current World Affairs, based in New York City, which gave him a stipend to study the state of science and technology in the Soviet Union. Not only did he begin to study Russian, but at the same time he managed to prepare for his preliminary examination for his Ph.D. in mathematics. This was an oral examination which in MAY's case was based upon a presentation he made on Galois theory, for which he provided some historical perspective that was apparently his first sign of serious interest in history of mathematics.

MAY did not go to the Soviet Union directly, but first went to the University of London where he continued to study Russian (at the School of Slavonic Studies), pursued his interest in statistics at University College, and informally followed courses on general planning and Marxism through lectures at the London School of Economics. In October of 1937 he went to Russia, but was not permitted to meet with anyone at the Russian National Planning Commission. Instead, he used his free time to visit book stores, and sent back a substantial quantity of books for later study in England.

Upon his return to London, he became increasingly involved in statistics and with the group working with RONALD A. FISHER in particular at the London School of Economics. He also became interested in the ideas of J. D. BERNAL. In July of 1938 MAY married a fellow American, RUTH MCGOVNEY, in London, and they spent the following academic year together in Paris, studying at the

Sorbonne by day and at the Worker's University at night. Meanwhile, MAY's research in statistics and planning began to focus increasingly on issues of science, society, and the organization of science. The following spring the MAYs visited the Soviet Union. This time KEN MAY was allowed to meet with theorists at the Kharkov Engineering-Economic Institute, where he was interested in the training of "engineer-economists." The following year, back at Berkeley, MAY taught finance and calculus as a teaching assistant, and began working on the completion of his Ph.D. He also returned as an active member to the campus branch of the Communist party, which led to an irreconcilable break with his father, who publicly disowned and privately disinherited his son as a result of his Communist sympathies. The Board of Regents of the University of California followed suit, and voted to revoke MAY's teaching assistantship, whereupon he withdrew from the university. For a time MAY worked actively in support of the Communist party, but with the outbreak of World War II, and recently divorced, he joined the U.S. army and was made a paratrooper squad leader in the 87th Mountain Infantry, no doubt a reflection of his experience as a skier and mountain climber. As a result, he saw a good deal of the world, and served in both the Aleutian and Italian campaigns.

Following the War, MAY not only reconciled with his father, but also remarried and completed his Ph.D., which he received from the University of California at Berkeley in 1946 with a thesis "On the Mathematical Theory of Employment." Subsequently, he was offered a position as assistant professor at Carleton College in Northfield, Minnesota, where he taught for twenty years. It was at Carleton that in the late 1950's MAY's interests began to turn towards mathematics education and the improvement of teaching at both the high school and college levels. In the 1960s he became increasingly concerned with history of mathematics and information retrieval. In 1964 he spent a year back at the University of California, Berkeley, as a visiting scholar, and in 1966, was offered a joint appointment for history of mathematics in both the Department of Mathematics and the College of Education at the University of Toronto in Ontario, Canada.

It was during the eleven years MAY spent at Toronto that the history of mathematics became the main focus of his energies, which culminated several years later in his *Bibliography and Research Manual of the History of Mathematics* (1973). This was in part the result of a major project to amass a card index of titles on the history of mathematics. He also made a series of ten historical 30-minute films for television, produced by the Ontario Education Communication Authority. These were broadcast in the fall of 1970 and were devoted to such figures in history of mathematics as GALOIS, CANTOR, BÓLYAI, HAMILTON, CHARLES S. PEIRCE, GALILEO, GAUSS, RAMANUJAN, and RUSSELL. At about this same time he was commissioned to write a number of entries for the encyclopedic *Dictionary of Scientific Biography* ([DSB 1970/90]), of which his contribution on GAUSS represents a major piece of historical research and writing.

The idea of founding an International Commission on History of Mathematics and starting a new journal was the result of efforts made by RENÉ TATON and A. P. YOUSHKEVICH, along with MAY, at the XIIth International Congress for History of Science (Paris, 1968) to advance the position of history of mathematics internationally. This led to the founding of the Commission at the next ICHS held in Moscow in 1971, where KEN MAY was elected Chair of the Executive Committee. Immediately he made preparations to issue a series of newsletters, and that same year he launched *Notae de Historia Mathematica*, a *Newsletter* of the Commission on History of Mathematics. MAY not only undertook an international correspondence to lay a foundation for the new journal, but also compiled the first *World Directory of Historians of Mathematics*, which was published by the University of Toronto Press in 1972 ([MAY/GARDNER 1972]). Two years later, with a subscription base of nearly 700 subscribers from 39 countries and considerable institutional support, including endorsements from the leading societies for mathematics and history of science throughout the world, the first volume of the journal, *Historia Mathematica*, made its appearance. Within three years, the journal's subscription base had nearly doubled.

In the meantime, MAY had been serving as director (1973–1975) of the University of Toronto's graduate Institute for the History and Philosophy of Science and Technology. Following a heart attack in 1975, he relinquished this position, and began to make preparations to reduce the extent of his own responsibility for editing and producing *Historia Mathematica*. But he continued to direct the Institute's weekly seminar, called the "History of Math Conspiracy," and remained the editor responsible for overseeing production of the journal. Despite careful attention to his diet and regular exercise, KEN MAY died unexpectedly of another heart attack in December of 1977.

Only months before, at the XVth International Congress for History of Science held in Edinburgh in August of 1977, he passed on the chairmanship of the International Commission on the History of Mathematics to CHRISTOPH J. SCRIBA, and had taken important steps to secure the financial security of the journal by negotiating with a commercial firm, Academic Press, to take over the role of publisher from the University of Toronto. Systematically, MAY had also been increasingly delegating editorial responsibility for the journal to younger colleagues. As C. J. SCRIBA wrote in reflecting on MAY's contributions to the professionalization of the history of mathematics, the journal will "continue as a living tribute to its unpretentious and yet energetic founder and first editor" [Memorial Tribute 1978, 9].

Major works: [MAY 1968]; [MAY 1972]; [MAY 1973] — "Should We Be Mathematicians, Historians of Science, Historians or Generalists?" *HM* **1** (1974), 127–128. — "What is Good History and Who Should Do It?" *HM* **2** (1975), 449–455.

Secondary literature: Reminiscences [Memorial Tribute 1978] by a number of MAY's colleagues, including STILLMAN DRAKE, H. S. M. COXETER, CHARLES V. JONES, HENRY S. TROPP, CHRISTOPH J. SCRIBA, BRUCE SINCLAIR, MICHAEL S. MAHONEY, and DIRK J. STRUIK may be found in "A Memorial Tribute to Kenneth O. May." *HM* **5** (1978), 3–12 (with portrait, p. 2). — TROPP, HENRY S.: "Eloge." *Isis* **70** (1979), 419–422 (with portrait, p. 419). — [JONES/ENROS 1984,

with portrait, p. 358]. — ENROS, PHILIP C.: "Kenneth O. May – Bibliography." *HM* **11** (1984), 380–393.
J.W.D.

Medvedev, Fëdor Andreevich (* February 3, 1923, in a little village near Kozelsk, Kaluga Region, USSR; † February 5, 1993, Moscow, Russia). A serious childhood illness (tuberculosis of the legs) contributed to the formation of FËDOR MEDVEDEV's strong and independent character. After graduating in 1952 from the Kaluga Pedagogical Institute, he began to work in a local school as a teacher. In 1955 he entered the Institute for the History of Science and Technology of the USSR Academy of Sciences in Moscow as a post-graduate student. His supervisor, A. P. YOUSHKEVICH, proposed the early history of set theory in Russia as the subject of MEDVEDEV's thesis, but the education he had received at the Pedagogical Institute was not sufficient for serious research on such a difficult topic. Thanks to his determination and persistence, MEDVEDEV successfully defended his dissertation in 1963 and went on to become a distinguished specialist on the history of set theory and functions of a real variable. From 1958 until his sudden death in February 1993, he worked at the Institute for the History of Science and Technology in Moscow.

In 1965 MEDVEDEV published a monograph, *On the Development of Set Theory in the XIXth Century*, in which he investigated the geometric and algebraic origin of the subject, and presented R. DEDEKIND as one of its founders. For the next several years MEDVEDEV worked on the history of the theory of functions of a real variable. In 1974 he published *On the Development of the Notion of Integral*, and in 1975 his well-known work, *Scenes from the History of Real Functions* appeared (trans. into English in 1991). These were followed by his monograph on *The French School of Set Theory and of the Theory of Real Functions at the End of the 19th Century and the Beginning of the 20th Century*, published in 1976. These books, based on extensive references, offered a detailed analysis of the main concepts, theorems, and proofs, many of which were evaluated from a new point of view. For example, Medvedev emphasized the notion of function in the works of DIRICHLET and LOBATCHEVSKIĬ, and AMPÈRE's theorem about the differentiability of continuous functions. Taken as a whole, MEDVEDEV's publications of the mid-1970s constitute a veritable encyclopedia on the theory of real functions, written from an historical point of view. At the end of the 1970s, the focus of MEDVEDEV's interests shifted to the foundations of set theory and mathematics in general. In 1982 he published a book entitled *On the Early History of the Axiom of Choice*, and several years later, thanks in part to his efforts, the Russian edition of GEORG CANTOR's works on set theory appeared (in 1985). In his last years MEDVEDEV turned his attention to the history of the idea of infinity and the history of nonstandard analysis.

MEDVEDEV's research on the history of mathematics began in the 1950s. In this period the Soviet school on the history of mathematics approached historical analysis, including the sources and the entire process of the development of mathematics, from the point of view of its present state. Consequently, the history of mathematics was regarded as cumulative, progressive, and linear, and as leading exclusively to the "truths" of modern mathematics. This perspective was encouraged at the time by the official doctrine of dialectical materialism in the USSR, and the positivism of which many scientists approved. MEDVEDEV, who from his first papers was among the severest critics of such philosophical prejudices, contributed substantially to eliminating such assumptions in writing the history of mathematics.

Secondary literature: YOUSHKEVITCH, ADOLF; SERGEI DEMIDOV, and PIERRE DUGAC: "F. A. Medvedev et son apport à l'histoire de la théorie des fonctions." *HM* **10** (1983), 396–398 (with portrait). — ZAITZEV, EVGENY: "*In Memoriam* Fyodor Andreevich Medvedev." *Modern Logic* **4** (1994), 283–285. — ZAITSEV, EVGENY A.: "*In Memoriam* Fedor Andreevich Medvedev." *HM* **22** (1995), 88–92 (with bibliography). S.S.D.

Menninger, Karl (* October 6, 1898, Frankfurt, Germany; † October 2, 1963, Heppenheim near Darmstadt, Germany). A teacher of mathematics in Heppenheim 1923–1963, MENNINGER published numerous popular writings on mathematics as well as a history of number words and number systems: *Zahlwort und Ziffer* (Number Words and Number Symbols) (1934; 3rd ed. 1979; Eng. trans. 1969).

Secondary literature: *NDB* **17**, 85. — Anonymous: "Karl Menninger †." *Prax. Math.* **5** (1963), 299. — SIEBER, H.: "Karl Menninger †." *MNU* **16** (1964), 368. M.F.

Michel, Paul-Henri (* April 25, 1894, Amiens, France; † April 20, 1964, Paris (?), France). MICHEL studied German and Italian at the *Sorbonne*, and wrote a thesis on L. B. ALBERTI (1930) before becoming a librarian in Paris. He is best known for studies of Pythagorean arithmetic and of Renaissance science, especially ALBERTI, LEONARDO DA VINCI, GIORDANO BRUNO, and GALILEO.

Main works: *De Pythagore à Euclide – Contribution à l'histoire des mathématiques préeuclidiennes.* Paris 1950. — *La Cosmologie de Giordano Bruno.* Paris, coll. *Histoire de la Pensée*, 1962.

Secondary literature: COSTABEL, PIERRE: "Paul-Henri Michel (1894–1964)." *Rev. Hist. Sci.* **19** (1966), 267–269. — HAHN, ROGER: "Paul-Henri Michel (1894–1964)." *Isis* **57** (1966), 260–261. J.P.

Michel, Henri (* June 17, 1885, Liège, Belgium; † January 18, 1981, Elsene near Brussels). While MICHEL's professional career as a mining-engineer was entirely devoted to industry, his interest in history of science was directed to the history of astronomy and to ancient mathematical and astronomical instruments. His publications include [MICHEL 1939] and [MICHEL 1947].

Secondary literature: HOGE, EDMOND: "*In memoriam:* Henri Michel (1885–1981)." *Ciel et Terre* **97** (1981), 349–351. — HOGE, EDMOND, and JAN DE GRAEVE: "Ciel et Terre rend hommage à Henri Michel. Biographie. Bibliographie." *Ciel et Terre* **98** (1982), Supplément (with portrait). P.B.

Mieli, Aldo (* December 4, 1879, Livorno, Italy; † February 16, 1950, Florida, Argentina). MIELI originally studied chemistry at the University of Pisa, after which he went to Leipzig for further study with the physical chemist WILHELM OSTWALD. He also studied mathematics with the Italian mathematician ULISSE DINI, and served as EMANUELE PATERNÒ's assistant at the University of Rome from 1905 until 1912, having become a *docent* there in his own right in 1908. Beginning in 1912, MIELI's interests increasingly turned to history of science, and he began significant collaborations with several important journals on the subject, including *Isis, Scientia*, and the *Rivista di storia delle scienze mediche e naturali.* In 1921 he became the founding editor of *Archivio di storia della scienza*, which later became *Archeion.*

Political circumstances caused MIELI to leave Italy for Paris in 1928, where he headed the *Centre International de Synthèse*, to which he gave his vast library of works on history of science. It was MIELI who not only helped to found the International Committee for the History of Science, but he served as the organizer for the First International Congress of the History of Science held in Paris in 1929. It was at this meeting that the International Committee became the International Academy for the History of Science, with *Archeion* adopted as the Academy's official journal.

In 1939 MIELI left France for Argentina (again as a political refugee), where he taught at the *Universidad Nacional del Litoral* in Santa Fé and Rosario where he created an Institute for the History and Philosophy of Science. He also edited the first four volumes (1919–1923) of *Archeion* and continued to serve as perpetual secretary of the International Academy of the History of Science. He published two important collections of classics in the history of science, *Biblioteca del Instituto* (Library of the Institute, in Rosario, Argentina), and *Panorama general de historia de la ciencia* (General Panorama of the History of Science) (12 vols. in all, beginning in 1945; the series was completed by DESIDERIO PAPP and JOSÉ BABINI). It contains chapters on the history of mathematics. MIELI also published several other works devoted to the history of science (one of which was about the life and work of GEORGE SARTON). The collaboration between MIELI and BABINI was extremely close.

Secondary literature: *DSB* **9**, 377–379. *P* VI, 1729; VIIb(5), 3270. — CORSINI, ANDREA: "Aldo Mieli." *Rivista di storia delle scienze mediche e naturali* **41** (no. 1) (1950), 111–113. — SERGESCU, PIERRE: "Aldo Mieli." *Actes du VIème Congrès international d'histoire des sciences* (Amsterdam, 1951), 79–95. — [SARTON 1951]. — BABINI, JOSÉ: "Aldo Mieli y la Historia de la Ciencia en la Argentina." *Physis* **4** (1962), 74–84. — BABINI, JOSÉ: *Ciencia, historia e historia de la ciencia* Chapter III. Buenos Aires: CEDAL 1967. — POGLIANO, CLAUDIO: "Aldo Mieli, storico della scienza (1879–1950)." *Belfagor* **38**, no. 1 (1983), 537–557. M.O.

MIKAMI Yoshio (* February 16, 1875, Kōtachi (now part of Kōda) in Hiroshima Prefecture, Japan; † December 31, 1950, Hiroshima, Japan). MIKAMI YOSHIO was born as the son of a land owner in Hiroshima Prefecture. He entered the Second Normal School in Sendai, but had to drop out because of an eye disease. Consequently, he was largely self-taught. Before studying the history of mathematics, he was interested in the philosophy of science, especially that of HENRI POINCARÉ. It was in 1905 at the age of 30 that MIKAMI began to study the history of mathematics, thanks to the encouragement of the American mathematician GEORGE B. HALSTED. MIKAMI's enthusiasm for the history of traditional Japanese mathematics was acknowledged by the leading mathematician at that time, KIKUCHI DAIROKU. Thanks to KIKUCHI's efforts, MIKAMI became a researcher at the Japan Imperial Academy in 1908. As a historian of mathematics in Japan, he succeeded ENDŌ TOSHISADA, the author of *Dai-Nippon Sūgakushi* (A History of Mathematics in Great Japan), the first comprehensive monograph on the history of Japanese mathematics, published in 1895. The first fruits of MIKAMI's own studies were *The Development of Mathematics in China and Japan* (1913), and, in collaboration with D. E. SMITH of Columbia University, *A History of Japanese Mathematics* (1914). From 1911 to 1916, MIKAMI studied philosophy part-time at the University of Tokyo, while working at the Imperial Academy. He edited the second edition of ENDŌ's *magnum opus*, published in 1918 under the title *Zōshū Nippon Sūgakushi* (A History of Japanese Mathematics, Enlarged and Revised). In 1923 he published a monumental paper, *Bunkashi-jō yori mitaru Nippon no Sūgaku* (Japanese Mathematics from the Viewpoint of Cultural History), in which MIKAMI characterized Japanese mathematics more as an art form such as the tea ceremony, not an exact science. This point of view, which was critical of Japanese culture, caused a conflict between MIKAMI and FUJISAWA RIKITARO, a nationalist and an influential mathematician at the University of Tokyo. The disagreement with FUJISAWA caused MIKAMI to resign his position at the Academy.

This unfortunate episode did not weaken MIKAMI's passion for the history of mathematics. In 1926 he published *Shina Sūgaku no Tokushoku* (Main Characteristics of Chinese Mathematics), which was also translated into Chinese, and in 1928 he completed a comparative study on the history of mathematics under the title *Tōzai Sūgakushi* (History of Mathematics East and West). In 1929 he was elected a corresponding member of the *Comité International d'Histoire des Sciences*, thanks to the efforts of its secretary, ALDO MIELI. Between the years 1933 and 1943, MIKAMI delivered lectures on the history of mathematics at Tokyo School of Physics, a private college of mathematics and physical sciences. He also drafted several thousands of manuscript pages on these subjects, which have never been published.

During World War II MIKAMI left Tokyo for his family home in Hiroshima. From that time on his health began to deteriorate. He was in a desperate condition soon after the War. His land was confiscated due to agricultural reforms, but material support from GEORGE SARTON in the United States comforted him very much. Remarkably, in 1949 MIKAMI took his doctorate from *Tōhoku University* in Sendai for his work on SEKI TAKAKAZU, the *de facto* founder of a school of Japanese mathematics which flourished during the Edo Period. MIKAMI is said to have worn traditional Japanese *kimono* without exception, and to have looked as if he were a Confucian scholar. He died on December 31, 1950, in Hiroshima at the age of 75.

Main works: [MIKAMI 1913]; [MIKAMI/SMITH 1914].

Secondary literature: *May* 261. *P* V, 853; VI, 1732; VIIb(5), 3277. — OGURA KINNOSUKE and ŌYA SHIN-ICHI: "Life and Works of Dr. Yoshio Mikami" (in Japanese). *Kagakushi Kenkyū* (Journal of History of Science) **18** (1951), 1–11 and (1952), 45–47. — [KAWAHARA/YANO 1996, 123–125]. — [SASAKI 1996, 69–71, with portrait]. — SASAKI CHIKARA: "The Formation of Cultural History of Mathematics: The Life and Work of MIKAMI Yoshio (1875–1950)" (in Japanese). In: SASAKI CHIKARA (Ed.), *Bunkashijo-yori mitaru Nippon no Su-gaku* (Japanese Mathematics from the Viewpoint of Cultural History). Tokyo 1999, pp. 285–341. S.C.

Milhaud, Gaston (* August 10, 1858, Nîmes, France; † October 1, 1918, Paris, France). MILHAUD studied mathematics at the *Ecole Normale Supérieur* in Paris with GASTON DARBOUX, after which MILHAUD taught at the University of Le Havre beginning in 1881. Over the next ten years he developed strong interests in the philosophy of mathematics. In 1891 he became professor of mathematics at the University of Montpellier, where he gave a series of lectures on the origins of Greek science. In 1894 he defended a Ph.D. dissertation at the University of Paris on the conditions and limits of logical certainty, which after it was published ran through four editions in thirty years. MILHAUD was appointed to the chair of philosophy at Montpellier in 1895, and in 1909, a chair for "history of philosophy in relation to science" was created for him at the *Sorbonne*, where he continued to probe the significance of the sciences for the history of philosophy. In addition to his study of ZENO, he also contributed stimulating ideas to questions related to logical contradictions and the role of demonstration in both mathematics and physics. Although most of MILHAUD's research was devoted to Greek science, he is also responsible for rekindling interest in DESCARTES as a scientist.

Main works: *Les philosophes-géomètres de la Grèce: Platon et ses prédécesseurs.* Paris 1902. — *Etudes sur la pensée scientifique chez les Grecs et chez les modernes.* Paris 1906. — *Descartes savant.* Paris 1921.

Secondary literature: *DSB* **9**, 382–383. *May* 261. *P* IV, 1011; V, 855. — JANET, PIERRE: "Notice sur Gaston Milhaud." *Annuaire de l'association des anciens élèves de l'Ecole normale supérieur*, Paris 1919, 56–60. — GOBLOT, EDMOND: "Gaston Milhaud." *Isis* **3** (1921), 391–395 (with portrait). — POIRIER, RENÉ: *Philosophes et savants français du XXe siècle. Extraits et notices,* vol. II: *La philosophie de la science.* Paris 1926, 55–80. — NADAL, ANDRÉ: "Gaston Milhaud (1858–1918)." *Rev. Hist. Sci.* **12** (1959), 97–110 (with bibliography). J.P.

Mogenet, Joseph (* February 26, 1913, Melreux, Belgium; † February 18, 1980, Soignies, Belgium). In 1951 MOGENET became lecturer at the University of Leuven, and in 1957 professor of classical philology. He was editorial secretary (1950), and later co-director (with ADOLPHE ROME) of the history of science journal *Osiris*. Most of MOGENET's writings deal with the history of Greek mathematics and astronomy, in particular with the work of AUTOLICUS OF PITANE [MOGENET 1950] and PTOLEMY. MOGENET died before he could achieve his main project, namely publication of an edition of the *Great Commentary* of THEON OF ALEXANDRIA on PTOLEMY's *Handy Tables* [MOGENET 1985].

Secondary literature: TIHON, ANNE: "In memoriam Joseph Mogenet (1913–1980)." *Byzantium* **50** (1980), 636–641. — TIHON, ANNE: "Éloge Joseph Mogenet, 26 February 1913 – 18 February 1980." *Isis* **72** (1981), 265–266 (with portrait). P.B.

Molk, Jules (* December 8, 1857, Strasbourg, France; † May 7, 1914, Nancy, France). After graduating from the Zürich polytechnic (1877), MOLK continued his studies in Berlin and Paris, where he obtained a doctorate in mathematics. From 1890 on, he taught applied mathematics and rational mechanics at the University of Nancy. In 1902 he was entrusted with the editorship of the French version of the *Encyclopédie des sciences mathématiques* (1904–1916), which he also regarded as a means of strengthening interest for historical studies in France.

Secondary literature: *May* 264. *P* IV, 1025; V, 870. — [*Acta Math.*, Tab. Gen. **1–35** (1913), 90; portrait 160]. — ENESTRÖM, GUSTAF: "Jules Molk (1857–1914) als Förderer der exakten mathematisch-historischen Forschung." *Bibl. Math.* (3) **14** (1913/14), 336–340. — VOGT, H.: "Jules Molk, 8 décembre 1857 – 7 mai 1914." *Enseign. Math.* (1) **16** (1914), 380–383, 387. — Portrait: [BÖLLING 1994, 3.2]. J.P.

Moll, Gerard (* January 18, 1785, Amsterdam, Netherlands; † January 17, 1838, Amsterdam, Netherlands). MOLL studied mathematics, physics, and astronomy at the *Athenaeum* of Amsterdam and then, during the years 1810–1811, continued his studies in Paris. In 1812 he was appointed director of the Observatory of the University of Utrecht, and in 1815 he became professor of physics at the University as well. He published biographical essays on JAN HENDRIK VAN SWINDEN, PIERRE SIMON DE LAPLACE, and JEAN BAPTISTE DELAMBRE, as well as a study on the inventors of the telescope.

Secondary literature: *DSB* **9**, 459–460. *May* 264. *P* II, 179. — *VdAA* **12**, 291–294. — VAN REES, RICHARD: "Berigt aangaande het leven en de wetenschappelijke verdinsten van wijlen den Hoogleeraar Gerrit Moll." *Algemeene Konst- en Letterbode* (1838 II), 83–91, 98–107. — QUETELET, ADOLPHE: "Notice sur G. Moll." *Annuaire Académie royale de Belgique* **5** (1839), 63–79. P.B.

Monna, Antonie (Anton) Frans (* March 10, 1909, La Haye, The Netherlands; † October 7, 1995, De Bilt (near Utrecht), The Netherlands). After graduating from the University of Leiden in 1933, ANTON MONNA went on to receive his doctorate in 1935 for a thesis written under the direction of JOHANNES DROSTE (1886–1963), "Het probleem van Dirichlet" (On Dirichlet's Problem). After a brief stint teaching in secondary schools in La Haye and Middelburg, MONNA spent five years working in the life insurance business. Beginning in 1942, he served in the

National Department of Education. His relative isolation from the mathematical community in Holland ended in 1961 with his appointment at the University of Utrecht. In 1965 he was promoted to a professorship in the faculty of sciences; his inaugural lecture was devoted to research and teaching in mathematics. When he retired from teaching, he was made Professor Emeritus in 1979.

MONNA was a member and for a time secretary of the governmental Committee for the Modernization of the Mathematics Curriculum (1961–1972), the tasks of which were later taken over by the Institute for the Development of Mathematics Education (IOWO), the present Freudenthal Institute at the University of Utrecht. Most of MONNA's career was devoted to work on Dirichlet's Problem and non-archimedean analysis, but from about 1972 on, his efforts were directed primarily to history and pedagogy of mathematics.

Main works: *Het probleem van Dirichlet* (The Dirichlet Problem). The Hague 1935. — *Beschouwingen over onderzoek en onderwijs in de wiskunde*, (Inaugural lecture). Utrecht 1965. — "The Concept of Function in the 19th and 20th Century." *Arch. Hist. ex. Sci.* **9** (1972), 57–84. — *Functional analysis in historical perspective*. Utrecht 1973. — "Gauss and the Physical Sciences." In: MONNA, A. F. (Ed.): *Carl Friedrich Gauss, 1777–1855: Four Lectures on his Life and Work* (= Communications of the Mathematical Institute, vol 7). Utrecht: Rijksuniversiteit 1978.

Secondary literature: BERTIN, E.; F. VAN DER BLIJ, and E. HORNIX: "Curriculum vitae de A. F. Monna." In *Symposion dédié à A. F. Monna. Utrecht, le 18 décembre 1979* (= Communications of the Mathematical Institute 12-1980, with bibliography). Utrecht: Rijksuniversiteit 1980, 4–5 (with portrait on p. 2). J.W.D.

Montucla, Jean-Etienne (* September 5, 1725, Lyon, France; † December 19, 1799, Versailles, France). JEAN-ETIENNE MONTUCLA is an emblematic figure for the historiography of mathematics, not only because of the remarkable influence his history of mathematics has had, epistemologically, on theories of knowledge, but also because his work appeared at a significant turning point. His *Histoire des mathématiques* (History of Mathematics) appeared in 1758 [MONTUCLA 1758], at the same time the seventh volume of the *Encyclopédie* was published, and just before a condemnation of the already published volumes by the Parliament of Paris, leading to D'ALEMBERT's defection from the editorial enterprise. Intending to cover the entire 18th century, a second edition of the *Histoire* appeared in four volumes between 1799 and 1802. A history of mathematics during the Enlightenment, and thus history up to MONTUCLA's own time, was to close the *Histoire*. And perhaps the idea was to signal the close of mathematics as well, at least what was classical mathematics and its meaning for natural philosophy.

Born at Lyon, it was at the famous Jesuit college of the city that MONTUCLA received a solid education and learned to master ancient languages. His knowledge of Latin and Greek aided him in his historical work, and enabled him to translate directly from sources in the original languages. But he also mastered modern languages, including English, which was not so common at this time in France, along with Italian which was normal, as well as German and Dutch. As was typical of his social class, his scientific knowledge did not come from university courses, and must have come from personal reading. After studying law in Toulouse in a more or less serious way, he went to Paris where at the age of thirty MONTUCLA met CHARLES-ANTOINE JOMBERT (1712–1784), a clever man in the book trade who had already published numerous mathematical books, in particular books for navigation with an emphasis on mathematical methods.

This JOMBERT was in his own way part of the *gens de lettres* (men of letters), a milieu which historians like ROBERT DARNTON have studied in its intellecutal and social diversity [DARNTON 1979]. JOMBERT was in constant contact with many *savants* and *literati.* All were competing to acquire some part of glory, as recently described by ELISABETH BADINTER [BADINTER 1999]. Among these, MONTUCLA met JEAN PAUL DE GUA DE MALVES, a man who had frequent disputes and worked on algebraical curves, but in the old fashionded style of DESCARTES; the astronomer LALANDE, who was an active writer against all sorts of superstitions like astrology, still not professor of astronomy at the *Collège royal*, and certainly not an admirer of the strongly mathematized form that celestial mechanics was taking; and D'ALEMBERT, who was a major enthusiast for this kind of mechanics, and wrote in such terms in the *Encyclopédie*, thus changing the face of natural philosophy. The man responsible for the *Encyclopédie* was DIDEROT, not so excited about mathematics which appeared not sufficiently materialistic for him, even if he privately worked to square the circle. DIDEROT too visited JOMBERT, who was impressed by the success of the editorial enterprise his publisher colleague ANDRÉ FRANÇOIS LE BRETON (1708–1779) had had with the *Encyclopédie.* Thus JOMBERT commissioned MONTUCLA to write books, first on the quadrature of the circle (1754), then on the statistics of inoculation (1756), and eventually what came to be MONTUCLA's best-known work, his volumes on the history of mathematics (1758). This really became a research program for MONTUCLA. He was to express militant modernity — of the sort historians have in mind when they follow ERNST CASSIRER and DANIEL MORNET in speaking of a *clan philosophique.* The purpose of this active group was to prove that through reason, and without any help of religious faith but with precise criticism, the human mind could bring progress to mankind, a sort of replacement for Providence. History of mathematics was thus a perfect theme for MONTUCLA. It was less exposed to Church reaction than was explicit materialism, and in the *Encyclopédie méthodique* (1785), which was a more technical and politically quieter remake of the *Encyclopédie*, the kind of history MONTUCLA had provided was still judged as "the best thing we may have wished on the subject."

Most of what we now know about MONTUCLA's life comes from an obituary written for the Free Society of Agriculture of Seine-et-Oise of which MONTUCLA had surprisingly been a member. In fact, during the 1760s, when money for writing books was scarce, he was offered an official but minor position with the grandiloquent title: Superintendant of Royal Buildings and Gardens in Versailles, a position he was to hold until the outburst of the French Revolution in 1789. After the revolution he naturally regained the post, in 1795, which explains the place where his obituary appeared. Professionals of the book trade — like CHARLES-JOSEPH PANCKOUCKE (1736–1798), the active editor of the *Encyclopédie méthodique* — tried to convince MONTUCLA to bring his successful *Histoire des mathématiques* up to the present. His only response, however, was a new, extensively revised edition of *Récréations mathématiques et physiques* (Mathematical and Physical Recreations) by JACQUES OZANAM (1640–1714), a book originally published as early as 1696. This is a genre far from encyclopedic ambitions. Such books belong to an old tradition, one not directly affected by changes in the sciences, but in any case, the *Récréations* met with considerable success! At least outside of scientific circles, MONTUCLA was viewed as a scientist by 1788, as a letter sent to him by the young JOSEPH FOURIER, then in a Benedictine monastery, testifies. This letter contained proofs of DESCARTES's rule for counting the real roots of a polynomial. FOURIER did not know that such proofs had already been published by DE GUA DE MALVES, but he was invited to read something at the Academy at the end of 1789. MONTUCLA seems to have played no other role in FOURIER's career, and FOURIER was not that satisfied by MONTUCLA's *History.*

Some years later, the printer LOUIS AGASSE succeeded in convincing the now aged MONTUCLA to complete his *Histoire*, in the same spirit as nearly forty years earlier. But this time MONTUCLA could no longer rely on a very small number of figures like D'ALEMBERT, and the group of JOMBERT's visitors. But instead, he could now draw upon all the members of the First Class of the *Institut de France* (which at the end of 1795 served to replace the Academy of sciences (which had disappeared in August of 1793). MONTUCLA was elected to the *Institut* on February 28, 1796. As a result, the *Histoire des mathématiques* was to become a sort of official history, acknowledged as such by the astronomer DELAMBRE, then perpetual secretary of the First Class, in his *Rapport sur le progrès des sciences mathématiques depuis 1789* (Report on the Progress of the Mathematical Sciences since 1789), a book which was published in 1810.

This sort of collective enterprise could not help but modify the approach MONTUCLA had originally taken. Even if the basic ideas of the encyclopedic coverage of all that pertained to the mathematical sciences was his, the detailed minutia of areas like optics, astronomy, and geodesy, with rather little mathematics, represented the collective spirit of common achievements in the sciences. It was very similar to the propaganda for the new Republican metrology that tried to show how all the sciences jointly promoted the new units, meter and kilogram, and so contributed to progress for all mankind. Standardization using mathematics also became a goal, as reflected in the use of decimals. The transition to a collective enterprise is dramatically emphasized by a sentence written on page 336 of the third volume of the *Histoire*: "the printing of this folio was on the verge of being completed when the author died, on December 19, 1799."

The task of completing the *Histoire* subsequently became the responsibility of the rather excentric astronomer JÉRÔME DE LALANDE. However, it would be quite wrong to attribute the innovations in the remaining parts of the *Histoire* to

LALANDE and his numerous collaborators. MONTUCLA's new work really has to be understood as a general European enterprise. The history of science was conceived didacticly as being part of contemporary science; history could serve as the active memory of science for science itself. The first *Histoire*, published in 1758, might better be described as a rational history of discoveries, and a glorification of the human mind. But this was immediately problematic because it made the *Histoire* into a sort of *Bible*, presenting itself as the final word in the sense that MONTUCLA created references, periodizations, and questions. It is a history which not only established an historical tradition, but judged scientific traditions as well. Thus, to give one example, MICHEL CHASLES was not contradicting but writing in the spirit of the *Histoire* when he attributed the best "seeds" for the birth of the calculus to FERMAT rather than to DESCARTES, as had MONTUCLA. Indeed, MONTUCLA said that "if Descartes had not been given to mankind, Fermat might have replaced him in geometry." There could be but only one way to progress! MORITZ CANTOR, using new facts and a philological approach, tried to extend the historical questions raised by MONTUCLA, and thus completely changed the approach to the history of mathematics during the Middle Ages. But, like MONTUCLA, he judged matters according to his own idea of what progress meant. Even if GINO LORIA used a better knowledge of scientific milieux and communities, his ways of judging remained those of MONTUCLA. And among those who have challenged MONTUCLA, like CARL BOYER who questioned the emphasis he gave to DESCARTES in the matter of negative quantities, for example, the question remained one of priority.

MONTUCLA based his historical judgments on a specific model of science which gave his *History* a kind of objectivity, emulating how the academies of the 18th century worked — the Academy of Sciences in Paris, the Royal Society in London, as well as the academies of Berlin or Saint Petersburg — rather than how the universities worked. A specific set of rules was followed concerning publications in order to settle priority matters and other quarrels. These rules only permitted different interpretations if they followed the same uniform standards of argumentation and presentation. Like the *Encyclopédie*, MONTUCLA's *Histoire* operated to a large extent on the model of the academies, and may thus be taken to represent a public sense of truth. It was in this spirit that MONTUCLA's history of mathematics was written.

MONTUCLA was not reluctant to report controversies, but he tried to rank mathematicians according to a hierarchy of merit in favor of the public life of science; his history refused to present different conceptions of mathematics on the same level. This is particularly evident in his presentation of the most famous quarrel of the 18th century, between LEIBNIZ and NEWTON over the calculus. This quarrel has prompted historians to serve, like academies, as mediators in the name of good science. MONTUCLA was fighting errors, and perhaps even more ignorance, having to contend in the same way as academies with a badly justified science, in order to promote true science. By describing the "main characteristics of the most famous mathematicians," and by judging the merits of their individual accomplishments for posterity, he certainly extended the academic genre of the *éloges*, through which collectively he tried to assemble a portrait of what the perfect mathematician should be. This explains why he claimed, without hesitation, to have charted the progress of the mathematical sciences from their origins down to his own time. The model for this kind of history was MONTUCLA's *Histoire des recherches sur la quadrature du cercle* (History of Research About the Quadrature of the Circle), a pocket-sized monograph of 304 pages, which is less a compilation of mathematical endeavors than criticism of a state of mind, directed as a sign of progress against those who undertook to square the circle. The booklet subsequently won MONTUCLA a corresponding membership in the Berlin Academy.

This was a militant history of knowledge, so that in the aftermath of the French Revolution, it prompted a new history, at the very moment when French society began to put so much emphasis on scientific education, in the fresh spirit of the new Republic. It should come as no surprise that an historian came to be so highly appreciated at the same moment a new curriculum in mathematics was developed; for MONTUCLA justified which parts of past mathematics could be forgotten. MONTUCLA created a literary genre, and he may indeed be regarded as the first professional historian of mathematics, the "greatest pioneer of the 18th century specialized history of science" as the late ALISTAIR CROMBIE described him in *Styles of Scientific Thinking in the European Tradition* [CROMBIE 1994, vol. 2, 1599].

Main works: *Histoire des recherches sur la quadrature du cercle ...* Paris 1754; rev. ed. Paris 1831, with a preface by SYLVESTRE FRANÇOIS LACROIX. — *Histoire des mathématiques, dans laquelle on rend compte de leurs progrès depuis l'origine jusqu'à nos jours ....* 2 vols., Paris 1758; 2nd ed. 4 vols., Paris 1799–1802 (completed by JÉRÔME DE LALANDE; repr. Paris 1968 (with a preface by CH. NAUX). — *Récréations mathématiques et physiques d'Ozanam ...* nouvelle édition. Paris 1778.

Secondary literature: *DSB* **9**, 500–501. *May* 265. *P* II, 198. — LE BLOND, A. S.: "Sur la vie et les ouvrages de Montucla." Extrait de la notice historique lue à la Société de Versailles, le 15 janvier 1800. Avec des additions par Jérôme Lalande. In: MONTUCLA, J. E.: *Histoire des Mathématiques*, vol. 4 (1802), 662–672. — J. B. DELAMBRE: *Rapport historique sur les progrès des sciences mathématiques depuis 1789 et sur leur état actuel.* Paris 1810; new annotated edition by J. DHOMBRES, Paris 1985. — [SARTON 1936c, with bibliography and portrait]. — DARNTON, R.: *The Business of Enlightenment. A Publishing History of the Encyclopédie 1775–1800.* Cambridge, MA. 1979. — [CROMBIE 1994]. — BADINTER, E.: *Les passions intellectuelles; I, Désirs de gloire, 1735–1751.* Paris 1999. J.D.

Mordukhaĭ-Boltovskoĭ, Dmitriĭ Dmitrievich (* July 27, 1876, Pavlovsk, near St. Petersburg, Russian Empire; † February 7, 1952, Rostov-on-Don, USSR). DMITRIĬ MORDUKHAĬ-BOLTOVSKOĬ was a professor at the Warsaw Polytechnic Institute (1898–1914), at the University of Warsaw (1909–1914), at the Don Polytechnic Institute (1914–1915), at Rostov University (1915–1949), and at the end of his career, at the Pyatigork Pedagogical Institut (1949–1952). A well-known mathematician, he was interested in such areas of the history of mathematics as the history of ancient mathematics and the evolution of the principal notions of modern mathematics.

Main works: "Studies on the Genesis of Some Principal Ideas of Modern Mathematics." *Izvestiya Severo-Kavkazskogo Universiteta* **3** (15) (1928), 35–129 (in Russian). — Russian trans. of EUCLID's *Elements*, Moscow, 1948–1950.

Secondary literature: *May* 266. *P* VI, 1777; VIIb(5), 3393. — [KUROSH 1959] — [BORODIN/BUGAI 1979] — [BOGOLYUBOV 1983, with portrait on p. 335]. S.S.D.

Mouy, Paul (* 1888, Lille, France; † October 8, 1946, Paris, France). A French philosopher influenced largely by BRUNSCHVICG, MOUY taught philosophy in secondary schools and was an active member of the history of science section at the *Centre international de synthèse* in Paris. He was mainly interested in Cartesian physics.

Main works: *Les lois du choc des corps d'après Malebranche.* Paris 1927. — *Le développement de la physique cartésienne (1646–1712).* Paris 1934.

Secondary literature: DELORME, SUZANNE: "Paul Mouy." *Rev. Hist. Sci.* **1** (1947), 70–72. J.P.

Müller, Conrad Heinrich (* December 12, 1878, Bremen, Germany; † January 9, 1953, Hannover, Germany). CONRAD MÜLLER, *Privatdozent* in Göttingen (1908–1910), then became ordinary professor of advanced mathematics at the *Technische Hochschule* in Hannover (1910–1948). His areas of interest related to history of mathematics were didactics of mathematics, Indian mathematics, and LEIBNIZ.

Main works: *Studien zur Geschichte der Mathematik, insbesondere des mathematischen Unterrichts an der Universität Göttingen im 18. Jahrhundert* (1904). — "Die Mathematik der Sulvasûtra. Eine Studie zur Geschichte der indischen Mathematik." *Abh. math. Sem. Univ. Hamburg* **7** (1929), 173–204. — "Descartes' Geometrie und die Begründung der höheren Analysis." *Sudh. Arch.* **40** (1956), 240–258.

Secondary literature: *May* 267. *P* VI, 1795; VIIa(3), 360. — QUADE, W.: "Conrad Müller." *Jber. DMV* **57** (1954), 1–5. — Portrait: *Catalogus Professorum TH Hannover 1831–1956* (1956), 9. M.F.

Müller, Felix (* April 27, 1843, Berlin, Germany; † 1928, Dresden, Germany). MÜLLER taught mathematics in Berlin (1870–1897). He was mainly interested in the history of mathematical terminology, bibliography, and documentation.

Main works: *Historisch-etymologische Studien über mathematische Terminologie* (1887). — *Zeittafeln zur Geschichte der Mathematik, Physik und Astronomie bis 1500* (1892). — "Zur Terminologie der ältesten mathematischen Schriften in deutscher Sprache." *Abh. Gesch. Math.* **9** (1899), 301–333. — "Über bahnbrechende Arbeiten Leonhard Eulers aus der reinen Mathematik." *Abh. Gesch. math. Wiss.* **25** (1907), 61–116. — *Führer durch die mathematische Literatur. Mit besonderer Berücksichtigung der historisch wichtigen Schriften* (1909). — *Gedenktagebuch für Mathematiker* (3rd ed. 1912).

Secondary literature: *May* 267. *P* III, 946; IV, 1039; V, 884; VI, 1797. — MANSION, PAUL: *Mémorial mathématique d'après le Professeur Dr. Félix Mueller.* Gand 1905 (16 pp.). M.F.

Muir, Sir Thomas (* August 25, 1844, Biggar, Scotland; † March 21, 1934, Cape Town, South Africa). MUIR acted as Administrator for Education in the Union of South Africa. He left some manuscripts to the Royal Society of Edinburgh (he was elected FRS in 1900 on the nomination of KELVIN [Royal Society Archives]).

Main works: *The Theory of Determinants in the Historical Order of Development.* 4 vols. (1906–1923), and related papers. — His manuscripts are preserved in the National Library of Scotland.

Secondary literature: *May* 268. *P* III, 947; IV, 1046; V, 886; VI, 1803; VIIb(5), 3438. — TURNBULL, HERBERT WESTREN: "T. Muir." *Obituary Notices Fellows RS* **1** (1932–33), 179–184; also *J. Lond. Math. Soc.* (2) **10** (1932), 76–80. I.G.G.

Needham, Noël Joseph Terrence Montgomery (* December 9, 1900, London, England; † March 24, 1995, Cambridge, England). JOSEPH NEEDHAM, who first studied medicine and later specialized in biochemistry, was a student, fellow, and later master of Gonville and Caius College, Cambridge. NEEDHAM was the author of *Chemical Embryology* (1931), which he followed by an historical study, *A History of Embryology* (1934). In 1941 he was elected a Fellow of the Royal Society, and during World War II he was sent by the Royal Society to China, where he set up an office for the Sino-British Science Cooperation Office in Chongqing. After extensive travels throughout the country, he published *Chinese Science* in 1945. Following the War, he headed the Science Division of UNESCO.

NEEDHAM's greatest contribution to the history of science has been the monumental series of volumes begun in 1954, devoted to *Science and Civilisation in China.* Needham adopted a Marxist approach to the history of science, but as a devout Christian, brought his own special interpretation to appreciating the historical and social contexts within which science developed on a world scale.

As for his contributions to the history of mathematics, his most visible effort was produced in collaboration with WANG LING, notably volume III of *Science and Civilisation in China*, which was devoted to *Mathematics and the Sciences of the Heavens and Earth* ([NEEDHAM 1959]). CATHERINE JAMI offers a specific assessment of NEEDHAM's contributions to the history of mathematics in her [JAMI 1996, 3–5]. As she notes, NEEDHAM believed that mathematics "provided a language in which to express and verify hypotheses and, in so doing, granted modern science its universal character." It was NEEDHAM's belief that one of the major reasons why modern science failed to develop in China was due to the fact that there was no "vivifying demand from the side of the natural sciences" for mathematics, and that it was a mercantile rather than an agrarian and bureaucratic civilization that was the necessary prerequisite for a mathematicized, experimental science.

Secondary literature: *P* VI, 1831; VIIb(6), 3538. — GAZAGNADOU, DIDIER: *Un taoïste d'honneur: autobiographie. Joseph Needham: De l'embryologie à la civilisation chinoise: entretiens avec Didier Gazagnadou.* Paris 1991. — GOLDSMITH, MAURICE: *Joseph Needham: 20th-Century Renaissance Man.* Paris 1995 (with numerous photographs). — JAMI, CATHERINE: "*In Memoriam.* Joseph Needham (December 9, 1900 – March 24, 1995)." *HM* **24** (1996), 1–5. — MULTHAUF, ROBERT P.: "Joseph Needham (1900–1995)." *Techn. Culture* **37** (1996), 880–891. — BLUE, GREGORY: "Joseph Needham. A Publication History." *Chinese Science* **14** (1997), 90–132. — DAVIES, MANSEL: "Joseph Needham (1900–95)." *Brit. J. Hist. Sci.* **30** (1997), 95–100. — HABIB, S. IRFAN, and DHRUV RAINA (Eds.): *Situating the History of Science. Dialogues with Joseph Needham.* New Delhi 1999.
J.W.D.

Nesselmann, Georg Heinrich Ferdinand (* February 14, 1811, Fürstenau near Elbing, East Prussia, Germany; † January 7, 1881 Königsberg, East Prussia, Germany). NESSELMANN studied mathematics and oriental languages at the University of Königsberg (1831–1837). He received his Ph.D. in 1837, became *Privatdozent* for oriental languages in 1838, extraordinary professor in 1843 and ordinary professor of Sanskrit and Arabic in 1859 at the same university. NESSELMANN acquired a special interest in the history of algebra as a student of C. G. J. JACOBI and F. J. RICHELOT.

Main works: [NESSELMANN 1842]. — *Essenz der Rechenkunst von Mohammed Beha-eddin ben 'Alhossain aus Amul, arabisch und deutsch herausgegeben.* Berlin 1843.

Secondary literature: *P* II, 270; III, 961. — *ADB* **23** (1886), 445–446. — Obituary in *Altpreußische Monatsschrift* **18** (1881), 324–331. M.F.

Neugebauer, Otto (* May 26, 1899, Innsbruck, Austria; † February 19, 1990, Lawrenceville, New Jersey, USA). NEUGEBAUER's father was a construction engineer for the railroad. After graduating from the *Akademisches Gymnasium* in Graz (March, 1917), OTTO NEUGEBAUER spent the next two years in the army as an officer, serving primarily in Italy with an Austrian mountain battery until November of 1918, when he was taken prisoner and held for nearly a year by the Italians at a camp near Monte Casino. After the War he returned to Graz where he studied mathematical physics with MICHAEL RADAKOVIČ and ROLAND WEITZENBÖCK.

In 1921 NEUGEBAUER went to Germany for a year to work with ARTUR ROSENTHAL and ARNOLD SOMMERFELD in Munich. As his interests began to move away from physics to mathematics, NEUGEBAUER decided to transfer to Göttingen in 1922. There he studied mathematics primarily with RICHARD COURANT, EDMUND LANDAU, and EMMY NOETHER, as well as Egyptology with HERMANN KEES and KURT SETHE. Beginning in 1923, he was an assistant in the mathematics department, and the following year was made special assistant to COURANT, who was then head of the department. NEUGEBAUER also spent the spring of 1924 in Copenhagen, where he worked with the mathematician HARALD BOHR. He also studied Sumerian in Rome, and several years later, continued his study of ancient languages with VASILIĬ V. STRUVE and BORIS A. TURAEV in Leningrad.

Meanwhile, in 1926 he completed his thesis for the Ph.D., which was written under the guidance of COURANT and SETHE, on *Die Grundlagen der ägyptischen Bruchrechnung* (The Fundamentals of Egyptian Calculation with Fractions), after which he received the *venia legendi* for history of mathematics on December 17, 1927. Early the following year he began teaching at Göttingen, and by January of 1932 he had been promoted to the rank of Associate Professor. As early as 1928, NEUGEBAUER taught a course on the history of mathematics. As his colleague DAVID PINGREE at Brown University has so aptly summarized NEUGEBAUER's contributions to the subject:

> His interest in the history of mathematics centered initially on two considerations: that non-Greek mathematics can teach us much about the foundation of mathematics itself, and that the history of mathematics, properly investigated, can provide a unifying force in a field that had been torn apart by excessive specialization [Pingree 1991, 87–88].

NEUGEBAUER's own interests focused primarily on Egyptian and Babylonian mathematics. He learned both ancient Egyptian and Akadian, and among his most impressive contributions have been careful editions of ancient texts, which provide the foundation for his own extensive and important historical writings. In 1934 he published a survey of his research as *Vorgriechische Mathematik* (Pre-Greek Mathematics), the only volume of several which NEUGEBAUER planned that was ever published.

At the same time as NEUGEBAUER was pursuing his own research devoted to the history of the exact sciences in antiquity, he was instrumental in establishing several important journals for mathematicians and historians of mathematics alike. To encourage both the production of reliable texts of the highest scholarly quality, as well as expository monographs to accompany and elucidate them, he founded *Quellen und Studien zur Geschichte der Mathematik, Astronomie und Physik* in 1929, along with JULIUS STENZEL and OTTO TOEPLITZ. The journal was actually divided into two sequences, Series A for primary sources, and Series B for scholarly studies. The first to appear was a volume in Series A, which provided a translation of the Moscow mathematical papyrus (by STRUVE), along with a transcription of its Egyptian hieroglyphs (by TURAJEFF), which appeared in 1930. This was followed in 1931 by the first volume in Series B, which was accompanied by an explanatory preface by the editors.

In their foreword, NEUGEBAUER and his colleagues noted, as had their predecessors in the previous century, the continued growth of substantial interest in history of mathematics. But they linked this interest not only to philosophical studies and the pedagogical usefulness of history in teaching, but to the demands of mathematics itself. Recognizing the emphasis early 20th-century mathematicians placed on foundations, they argued that greater appreciation for history of mathematics was a natural consequence. Thus a new venture promoting history of mathematics was timely. Moreover, by choosing for their title *Quellen und Studien*, the editors wished to emphasize that the retrieval of accurate versions of primary sources was of crucial importance. Accurate history of mathematics required, as a prerequisite, reliable and readily-available transcriptions and translations of original sources. To join the talents of the philologist and the mathematician was the most important task facing *Quellen und Studien*. For those who regarded mathematics as more than an esoteric specialty, but as an essential part of human culture, it was hoped that by considering the historical development of mathematical thought, a bridge might be found between the world's intellectual progress and the seemingly a-historical exact sciences. The ultimate goal, in fact, of *Quellen und Studien*, was to help build just such a bridge.

Among the important primary sources to appear in Series A of *Quellen und Studien* were editions of the Moscow mathematical papyrus (STRUVE and TURAJEFF, 1930), the *Mishna ha-Middot* (SOLOMON GANDZ, 1932), and NEUGEBAUER'S own path-breaking study of Babylonian mathematical cuneiform texts, Parts I and II (1935 and 1937). At the beginning of volume 3 of series A, NEUGEBAUER thanked in particular the Rask Ørsted Fond and the Rockefeller Foundation for supporting his continuing work in Copenhagen.

By then, NEUGEBAUER was living in Denmark, where he preferred to work after the National Socialists came decisively to power in Germany in 1933 (officially, NEUGEBAUER was given a leave of absence from Göttingen in June of 1934). In fact, when RICHARD COURANT was dismissed from his position, whereupon he left Germany for the United States, NEUGEBAUER succeeded him as director of the Institute in Göttingen, but only for one day: "He refused to take the oath of fealty to the new state and hence was suspected as *untragbar* (unacceptable)," [Segal 1980, 48]. Rather than remain in Germany, NEUGEBAUER soon left to accept a position at the University of Copenhagen.

Despite the extraordinary list of original sources and path-breaking studies *Quellen und Studien* had put into print since 1930, it all too soon ran afoul of politics and the accelerating fragmentation of Europe through the stormy and difficult decade just prior to Word War II. For example, OTTO TOEPLITZ who was Jewish, was fortunate in being allowed to emigrate, and he left Germany in 1938 to live out the rest of his years in Palestine, where he taught briefly but died in 1940 just at the beginning of World War II [DAUBEN 1995, 49–50]. In fact, 1938 (with the appearance of volume 4 in both Series A and B) was the last year in which *Quellen und Studien* managed to publish. NEUGEBAUER left Copenhagen in 1939, to take up a position in the Department of Mathematics at Brown University in Providence, Rhode Island. Not long thereafter, the only Department for History of Mathematics in the United States was created for him, and Brown soon became an important center for the study of the exact sciences in antiquity.

At Brown, NEUGEBAUER used his experience in founding the reviewing journal *Zentralblatt für Mathematik und ihre Grenzgebiete* to help establish another abstracting journal, *Mathematical Reviews*. The *Zentralblatt* had also run afoul of politics in Germany, and when TULLIO LEVI-CIVITA (who was Jewish) was arbitrarily dropped from its advisory board in 1938, many of the foreign members of the *Zentralblatt*'s advisory board resigned in protest, a list that in addition to NEUGEBAUER included G. H. HARDY, OSWALD VEBLEN, and HARALD BOHR. Not long thereafter, the Council of the American Mathematical Society voted officially to launch and supervise *Mathematical Reviews*, appointing OTTO NEUGEBAUER and JACOB DAVID TAMARKIN (by then also at Brown) as editors [PITCHER 1988, 69–89]. The first volume, which appeared in 1940, acknowledged support from the Carnegie Corporation, along with contributions from the Rockefeller Foundation and the American Philosophical Society.

In helping to establish the history of ancient mathematics on solid foundations, NEUGEBAUER wrote a widely-read textbook, *The Exact Sciences in Antiquity* (1951, 2nd ed. 1957), which includes helpful and detailed bibliographies at the end of each chapter. More technical, and drawing upon all of his earlier publications, is his comprehensive three-volume study, *A History of Ancient Mathematical Astronomy* (1975). NEUGEBAUER, however, did not limit his studies to the ancient Egyptians, Babylonians, and Greeks. He also investigated Ethiopian mathematics, as well as Byzantine and Latin texts drawing on Arabic works in the medieval and Renaissance periods. With his student, NOEL SWERDLOW, these efforts resulted in a careful piece of erudite scholarship, *Mathematical Astronomy in Copernicus's De revolutionibus* (1984).

NEUGEBAUER was a member of numerous scientific societies and international academies, notably the Academies of Copenhagen, Brussels, Paris, London, and Vienna. He was the recipient of the Lewis Prize of the American Philosophical Society, of the Pfizer Prize of the U.S. History of Science Society, and of the prestigious Balzan Prize (1986). He was also awarded several honorary degrees, notably from St. Andrews, Brown, and Princeton Universities. He was a member of the American Philosophical Society, and the U.S. National Academy of Sciences.

Secondary literature: *May* 272. *P* VI, 1842; VIIa(3), 414. — [PINL 1971, 180] — SEGAL, SANFORD: "Helmut Hasse in 1934." *HM* **7** (1980), 46–56. — PINGREE, DAVID: "Otto Neugebauer, 1899–1990. *In Memoriam*." *Arch. int. Hist. Sci.* **40** (1990), 82–84. — PINGREE, DAVID: "Eloge, Otto Neugebauer, 26 May 1899 – 19 February 1990." *Isis* **82** (1991), 87–88. — SWERDLOW, NOEL M.: "Otto E. Neugebauer (26 May 1899–19 February 1900)." *JHA* **29** (1993), 289–299 (with portrait). — SWERDLOW, NOEL M.: "Otto E. Neugebauer." *Proc. APS* **137** (1993), 137–165. — DAVIS, PHILIP J.: "Otto Neugebauer: Reminiscences and Appreciation." *AMM* **101** (1994), 129–131. — PEYENSON, LEWIS: "Inventory as a Route to Understanding: Sarton, Neugebauer, and Sources." *Hist. Sci.* **33** (1995), 253–

282. — DAUBEN, JOSEPH W.: *Abraham Robinson. The Creation of Nonstandard Analysis. A Personal and Mathematical Odyssey.* Princeton 1995; paperback ed. 1998. J.W.D.

Nevanlinna, Rolf Herman (* October 22, 1895, Joensuu, Finland; † May 28, 1980, Helsinki, Finland). ROLF NEVANLINNA received his doctorate in 1919 from the University of Helsinki, where he became docent in 1922 and professor of mathematics in 1926. He made several outstanding contributions to analysis, became interested in how mathematics develops, and encouraged historical studies.

Secondary literature: *May* 273. *P* VI, 1846. — HAYMAN, WALTER K.: "Rolf Nevanlinna (1895–1980)." *Bull. LMS* **14** (1982), 419–436 (with bibliography and portrait). K.A.

Nielsen, Niels (* December 2, 1865, Ørslev, Denmark; † September 16, 1931, Copenhagen, Denmark). NIELSEN graduated from the University of Copenhagen in mathematics in 1891, received his Ph.D. in 1895, and by 1909 was professor of mathematics at the same university. His contributions to the history of mathematics include the following works: *Matematiken i Danmark 1801–1908* (Mathematics in Denmark 1801–1908) (1910); *Matematiken i Danmark 1528–1800* (Mathematics in Denmark 1528–1800) (1912); *Franske Matematikere under Revolutionen* (French Mathematicians During the Revolution) (1927), which also appeared in French; and *Géomètres français du dix-huitième siècle* (French Geometers of the 18th Century, ed. N. E. NØRLUND) (1935).

Secondary literature: *DSB* **10**, 117–118. *May* 279. *P* IV, 1073; V, 905; VI, 1855. — *DBL* **10**, 486. K.A.

d'Ocagne, Maurice (* March 25, 1862, Paris, France; † September 23, 1938, Le Havre, France). D'OCAGNE was a student and (from 1912 on) a professor of geometry at the *Ecole Polytechnique.* His main mathematical work was in nomography. His approach to history of science was strongly biographical. He was elected a member of the French *Académie des Sciences* in 1922.

Main works: *Histoire abrégée des sciences mathématiques,* ed. by RENÉ DUGAS. Paris 1955.

Secondary literature: *DSB* **10**, 170. *May* 281. *P* III, 982; IV, 1086; V, 298; VI, 1892; VIIb(6), 3690. — [*Acta Math.*, Tab. Gen. **1–35** (1913), 92; portrait 161]. J.P.

Ore, Øystein (* October 7, 1899, Oslo, Norway; † August 13, 1968, Oslo, Norway). ORE, Norwegian/American mathematician and historian of mathematics, received his Ph.D. from the University of Oslo in 1924 (thesis: "Zur Theorie der algebraischen Koerper" (On the Theory of Algebraic Fields), *Acta Mathematica* (1923)). He then taught briefly at the University of Oslo until he was recruited to teach mathematics at Yale University, where he began as an Assistant Professor in

1927. Within two years he rose in rank from Associate Professor (1928) to Professor (1929), and in 1931 was named Sterling Professor (1931–1968). ORE's research interests included algebraic number theory, noncommutative arithmetic, lattices, graph theory, and the four-color problem, as well as history of mathematics. His major historical works include biographies and studies of ABEL and CARDANO, along with a mathematical textbook on number theory which also includes historical material.

Main work: *Number Theory and its History.* New York 1948, repr. 1988.

Secondary literature: *May* 283. *P* VI, 1915; VIIb(6), 3761. — ANONYMOUS: "Øystein Ore, 1899–1968." *J. Comb. Theory* **8** (1970), i–iii (with bibliography). — Portrait: *Scripta Math.* **4** (1936), facing page 189. J.W.D.

Ozhigova, Elena Petrovna (* March 1, 1923, Petrograd, USSR; † July 6, 1994, St. Petersburg, Russia). ELENA OZHIGOVA taught at the Leningrad Higher Artillery School beginning in 1956. She was also a researcher in the Leningrad branch of the Institute for the History of Science and Technology of the USSR Academy of Sciences (now the Russian Academy of Sciences) until 1994. Her areas of special interest included the history of mathematics (in particular the history of number theory) in Russia.

Main works: *Egor Ivanovich Zolotariov, 1847–1878.* Moscow-Leningrad 1976 (in Russian). — *The Developement of Number Theory in Russia.* Leningrad 1972 (in Russian).

Secondary literature: [KUROSH 1959] — [FOMIN/SHILOV 1969]. S.S.D.

Paplauskas, Algirdas Boleslavovich (* June 9, 1931, Klaipeda, Lithuania; † September 20, 1984, Moscow, Russia, USSR). As a member of the Institute for the History of Science and Technology of the USSR Academy of Sciences in Moscow (1955–1984), ALGIRDAS PAPLAUSKAS was particularly interested in the history of trigonometric and infinite series in general.

Works: *History of Trigonometric Series from Euler to Lebesgue.* Moscow 1966 (in Russian).

Secondary literature: [FOMIN/SHILOV 1969]. S.S.D.

Papp, Desiderio (* 21 May, 1895, Sopron, Hungary; † 31 January, 1993, Buenos Aires, Argentina). DESIDERIO PAPP received his Ph.D. from the University of Budapest in 1917, having studied philosophy. After teaching as a *Privatdozent* at both the University of Budapest and Vienna, he emigrated to France in 1938, where he wrote extensively as a popularizer, as well as a serious historian of science. In 1942 he left Paris for Buenos Aires, and spent the rest of his life promoting history of science in Latin America. There he worked with ALDO MIELI and JOSÉ BABINI in editing twelve volumes of *Panorama general de la historia de la ciencia* (General Panorama of the History of Science), for which PAPP and BABINI were responsible for the last seven volumes. This has been described as the most important textbook on the history of science written in Spanish.

PAPP taught at universities in Argentina, Chile, Uruguay, Mexico, and Venezuela, as well as at the University of Paris. He was founder and president of the Chilean Group of History of Science (Santiago de Chile) and president of the Argentine Group of History of Science (Buenos Aires). In 1948 he became a corresponding member, in 1961 an effective member, and in 1985 an honorary member of the International Academy of History of Science in Paris.

His writings include more than thirty-six books and numerous articles and reviews. His contributions to history of mathematics were primarily made in works devoted to history of the exact sciences, specifically in an introduction PAPP wrote for a book of selections from the writings of HENRI POINCARÉ: *El legado de Henri Poincaré al siglo XX* (The Legacy of Henri Poincaré to the 20th Century) (Buenos Aires 1944), and in PAPP's contribution, written with JOSÉ BABINI, *Las ciencias exactas en el siglo XIX* (The Exact Sciences in the 19th Century), volume 10 of *Panorama general de historia de la ciencia* (Buenos Aires 1958). A bibliography listing PAPP's most important works may be found in the *éloge* published in *Quipu* in 1993.

Secondary literature: KOHN LONCARICA, ALFREDO G.: "Desiderio Papp (1895–1993). Del Danubio al Plata y los Andes. Nota necrologica, cronología y breve bibliographía." *Quipu* **10** (1993), 199–222. — KOHN LONCARICA, ALFREDO G.: "Desiderio Papp, 21 May 1895 – 31 January 1993." *Isis* **85** (1994), 666–667 (with portrait). — KOHN LONCARICA, ALFREDO G.: "Desiderio Papp, 1895–1993." *Arch. int. Hist. Sci.* **45** (1995), 129–130. M.O.

Pearson, Karl (* March 27, 1857, London, England; † April 22, 1936, Coldharbour (Surrey), England). PEARSON pursued his career at University College London. He formed an important school of statisticians with strong historical interests, and published frequently in *Biometrika* (which he founded in 1901). PEARSON's lectures on *The History of Statistics in the 17th and 18th Centuries* did not appear until 1978 in an edition by his son, but they were of considerable influence much earlier, due to the fact that PEARSON's department was the principal center in Britain for teaching statistics. His work in the history of mathematics is not to be confused with his contemporary contributions (following GALTON) to mathematical (specifically statistical) history, also published in *Biometrika* (for example, the use of regression analysis to infer the authenticity of the embalmed skull of OLIVER CROMWELL).

Manuscripts: University College London.

Secondary literature: *DSB* **10**, 447–473. *May* 293. *P* IV, 1126; V, 950; VI, 1970; VIIb(6), 3916. — PEARSON, EGON SHARPE (Ed.): *The History of Statistics in the 17th and 18th Centuries* (1967; lecture courses, 1921–1933; with portrait). I.G.G.

Peirce, Charles Santiago Sanders (* September 10, 1839, Cambridge, MA, USA; † April 19, 1914, Milford, PA, USA). CHARLES S. PEIRCE, American mathematician, logician, and philosopher, received his AM degree from Harvard University in 1862 and a year later earned his B.Sc. in chemistry, *summa cum laude*, from the Lawrence Scientific School, in 1863. Apart from a brief stint teaching logic at Johns Hopkins University (1874–1884), he spent most of his career working for the U. S. Coast and Geodetic Survey. PEIRCE devoted the greater part of his life to questions of philosophy, scientific method, and logic. He was especially interested in the history of science, including mathematics, for the examples they provided of scientific method at work, and he wrote numerous articles about historical figures and topics in the history of mathematics, including a study of FIBONACCI's *Liber Abaci.*

Secondary literature: *DSB* **10**, 482–488. *May* 293. *P* III, 1013; IV, 1128; V, 953; VI, 1973. — For a biography replete with portraits, see JOSEPH BRENT: *Charles Sanders Peirce. A Life.* Bloomington 1993. J.W.D.

Pelseneer, Jean (* January 7, 1903, Brussels, Belgium; † July 1, 1985, Brussels, Belgium). In 1931, PELSENEER occupied a position as assistant to professor THÉOPHILE DE DONDER, who was in charge of a course on the history of physical and mathematical sciences at the Faculty of Science, University of Brussels. In 1945, after World War II, PELSENEER became a lecturer on "Elements of the History of Physical and Mathematical Sciences." In 1946, a new course on the "History of Scientific Thought" was added to his teaching responsabilities.

Secondary literature: ELKHADEM, HOSAM: "In memoriam Jean Pelseneer 1903–1985." *Arch. int. Hist. Sci.* **38** (1988), 289–290. — STOICA, MARIOARA, and MATEI RADU MARINESCU: "Paul Pelseneer (1863–1945), Savant Belge, Collaborateur et Ami d'Émile Racovitza." *Noesis* **20** (1994), 129–134 (with portrait). P.B.

Peyrard, François (* 1760, Vial, France; † October 3, 1822, Paris, France). PEYRARD was the first librarian of the newly-founded *Ecole Polytechnique* in Paris. From 1807 on he taught mathematics at the *Lycée Bonaparte.* He is mainly remembered for his French translations of ARCHIMEDES and EUCLID.

Main works: *Euclidis quae supersunt. Les Œuvres d'Euclide, en grec, en latin, et en français, d'après un manuscrit très ancien, qui était resté inconnu jusqu'à nos jours.* 3 vols., Paris 1814–1818.

Secondary literature: *P* II, 422. — LACOARRET, MARIE: [Short biographical note.] *Rev. Hist. Sci.* **10** (1957), 56. J.P.

Picatoste Rodríguez, Felipe (* 1834, Madrid, Spain; † 1892, Madrid, Spain). A liberal politician and journalist, PICATOSTE RODRÍGUEZ worked as a substitute professor of mathematics at San Isidro High School in Madrid between 1852 and 1857, and as second chief in the Corps of Archivists and Librarians from 1890 onwards. Among his intellectual interests, mathematics and its history played an important role. His publications include studies on the teaching of elementary mathematics and on mathematical vocabulary, as well as indexes and editions of well-known (mainly Spanish) mathematicians and various mathematical works.

Secondary literature: *Enciclopedia Universal Ilustrada Europeo-Americana* **44** (1921), 518–519. E.A.; M.H.

Pogrebysskiĭ, Iosif Benediktovich (* February 23, 1906, Uman, Ukraine, Russian Empire; † May 20, 1972, Leningrad, USSR). As a researcher in the Institute of Mathematics of the Academy of Sciences of the Ukrainian SSR in Kiev (1935–1941, 1946–1962), IOSIF POGREBYSSKIĬ was later active as a member of the Institute for the History of Science and Technology of the USSR Academy of Sciences in Moscow (1962–1972). His major areas of interest included history of mechanics and applied mathematics from the 17th to the 20th century.

Main work: *From Lagrange to Einstein* (in Russian). Moscow 1996 (with portrait).

Secondary literature: Anonymous, in *Vopr. Ist. Estest. Techn.* 1996, no. 4, 124–132 (in Russian). — [KUROSH 1959] — [FOMIN/SHILOV 1969] — [BOGOLYUBOV 1983]. S.S.D.

Poudra, Noël-Germinal (* April 19, 1794, Paris, France; † 1894, France) POUDRA was a student at the *Ecole Polytechnique* at the same time as CHASLES. His career was that of a military engineer, teaching perspective and descriptive geometry at the *Ecole d'état-major.* He brought his courses to an end with a survey of relevant historical developments. He also edited the works of DESARGUES.

Main works: *Histoire de la perspective ancienne et moderne.* Paris 1864. — *Les Œuvres de Desargues, précédées d'une nouvelle biographie de l'auteur.* 2 vols., Paris 1864.

Secondary literature: *May* 306. *P* II, 512; III, 1063. — ANON.: "Catalogue des travaux de Mr Noel Germinal Poudra." *Bull. Boncompagni* **1** (1868), 302–308. J.P.

Prasad, Ganesh (* November 15, 1876, Ballia, India; † March 9, 1935, Agra, India). PRASAD studied at Muir Central College in Allahabad, where he took his B.A. degree in 1895. After obtaining the D.Sc. from Allahabad University in 1898, he proceeded to Cambridge, England, for further studies. He returned to India in 1904 and became a professor first at Muir College, then at Queen's College in Banaras (now Varanasi). He also taught on two occasions at Calcutta University, first as the *Rashbehari Ghosh Professor* (1914–1917), and later as the *Hardinge Professor* (1923–1935). In the interim, he was principal of Central Hindu College, Benare, and Professor at Banaras Hindu University. PRASAD was founder and lifelong president of the Banaras Mathematical Society. In addition to several original research papers and a number of historical articles, he published the well-known book, *Some Great Mathematicians of the Nineteenth Century: Their Lives and Their Works* (2 vols., Banaras 1933, 1934).

Secondary literature: *May* 307. *P* V, 1003; VI, 2071; VIIb(6), 4125. — SEN, R. N.: "Ganesh Prasad (1876–1935)." *Biogr. Mem. INSA* **5** (1979), 40–45 (with bibliography and portrait). — SINGAL, M. K.: "Ganesh Prasad." *Bulletin of the Mathematical Association of India* **6** (1974), 6–8. R.C.G.

Quetelet, Lambert Adolphe Jacques (* February 22, 1796, Ghent, Belgium; † February 17, 1874, Brussels, Belgium). ADOLPHE QUETELET was a well known mathematician, astronomer, and statistician who studied mathematics at the University of Ghent and astronomy at the Observatory in Paris. He lectured at the *Athenaeum* and Military Academy in Brussels. He was also the founder of the Royal Observatory, which he headed as Director in 1828.

Secondary literature: *DSB* **11**, 236–238. *May* 311. *P* II, 552; III, 1080; VI, 2101. — COLLARD, AUGUSTE: "Adolphe Quetelet historien des sciences et biographe." *Ciel et Terre*, **45** (1929), 89–92, 127–145. — SARTON, GEORGE: "Preface to Volume XXIII of *Isis* (Quetelet)." *Isis* **23**, 6–24 (with portrait). P.B.

QIAN Baocong (* May 29, 1892, Jiaxing City, Zhejiang Province, China; † January 5, 1974, Suzhou City, Jiangsu Province, China). QIAN BAOCONG, prominent historian of Chinese mathematics and mathematics educator, was one of the pioneers who inaugurated serious study of the histories of ancient and medieval Chinese mathematics and astronomy.

Both QIAN BAOCONG and LI YAN were early leaders in studying historical achievements in terms of modern mathematical knowledge. LI YAN emphasized sorting out source materials for the history of mathematics, and was well known for his detailed and reliable materials. While QIAN BAOCONG emphasized textual research into the origin and development of ancient mathematical classics and algorithms, he was also well known for his critical views and profound analyses. JOSEPH NEEDHAM has praised both LI and QIAN as follows: "Among Chinese historians of mathematics, two have been particularly outsstanding, LI YAN and QIAN BAOCONG. The work of the latter, though less in bulk than the former's, is of equally high quality" [NEEDHAM 1959, vol. **3**, 2].

Thanks to a scholarship from the Chinese goverment, QIAN went from studying architecture at Suzhou Railway School to the University of Birmingham, England, where he also studied architecture, obtaining his degree in 1910. After Birmingham, QIAN went on to continue his studies in architecture at Manchester Polytechnic, but only for a year. In 1912 he returned to China, where he taught successively at Suzhou Industrial College, Nankai University, Central University, and Zhejiang University, where he eventually advanced to professor and director of the Mathematics Department. Over the next few decades, QIAN became a member of such important learned organizations in China as the *Zhonghua xueyi she* (Chinese Society for Academic Studies and the Arts) (1921), the Chinese Science Society (1923), the Chinese Astronomical Society (1927), and the Chinese Mathematical Society (1936).

In 1956, QIAN was made a research fellow (at the highest rank) of a special section for the History of Natural Sciences of the Chinese Academy of Sciences. He also became a member of the Research Committee for History of Natural Science, and was named editor in chief of *Kexueshi jikan* (Collection of the History of Science).

QIAN BAOCONG published his first paper in 1921, shortly after deciding upon a research career following the May Fourth New Culture Movement in 1919. In addition to his lasting contribution to texual studies in collating the *Ten Mathematical Manuals* (1963), a series of QIAN's monographic studies covered such important subjects as the ancient Chinese determination of the ratio of the circumference of a circle to its diameter and *geyuan shu* (the method of dividing a circle to calculate the approximate value of its area); the right-angled triangles with integer-valued sides; the *zengcheng kaifang fa* (the method of extracting roots by successive additions and multiplications); QIN JIUSHAO and his *Shushu jiuzhang* (Mathematical Treatise in Nine Sections); WANG LAI and his works.

QIAN placed considerable emphasis upon interconnections between mathematics and other ancient subjects, especially astronomy and the determination of calendars. His monographic studies of the source and development of the Star Charts of GAN DE and SHI SHEN, the origin of the twenty-eight constellations, the development of the *Gaitian* School, and the achievement of the *Shoushi* calendar, laid new foundations for these subjects. QIAN also made a thorough study of ancient Chinese musical harmonics and of the mechanics to be found in the *Mojing* (Mohist Canon).

QIAN also served as editor in chief of *Zhongguo shuxueshi* (A History of Chinese Mathematics) (1964), which included the newest results of Chinese scholars at the time. He also edited *Song Yuan shuxueshi lunwenji* (Collected Papers on the History of Mathematics in the Song and Yuan Periods) (1966), which was the first portion of a dynastic research project which he hoped to pursue. Following QIAN's death in 1974, thirty-three of his more than sixty research papers on the history of science were published in a posthumous collection of his works, *Qian Baocong kexueshi lunwen suanji* (Selected Essays of Qian Baocong on the History of Science) (1983).

Main works: *Gusuan kaoyuan* (Inquiry into the Sources of Ancient Mathematics) (1930). — *Zhongguo suanxue shi* (A History of Chinese Mathematics) (vol. **1**, 1930). — *Suanjing shishu* (Ten Mathematical Manuals) (1963).

Secondary literature: HE SHAOGENG: "The Immortal Achievements of Historian of Mathematics in Commemoration of the Centenary of the Birth of Prof. Qian Baocong" (in Chinese, with English summary). *Zhongguo keji shiliao* (Chinese Historical Materials on Science and Technology) **13** no. 4 (1992), 40–52 (with portraits on inside back cover). — [LIU 1994, **4**, 109–110]. L.D.

Reidemeister, Kurt Werner Friedrich (* October 13, 1893, Braunschweig, Germany; † July 8, 1971, Göttingen, Germany). KURT REIDEMEISTER was an assistant for mathematics in Hamburg (1921), extraordinary professor in Vienna (1922), and ordinary professor in Königsberg (1925), Marburg (1933), and Göttingen (1955). In mathematics he is best known for his results in group and knot theory, and more generally in topology. His historical interests were mainly directed towards the history of Greek mathematics.

Works on history and philosophy: *Die Arithmetik der Griechen* (1940). — *Mathematik und Logik bei Plato* (1942). — *Das System des Aristoteles* (1943). — *Das exakte Denken der Griechen* (1949).

Secondary literature: *DSB* **11**, 362–363. *P* VI, 2144; VIIa(3), 714. — [PINL 1972, 185] — ARTZY, RAFAEL: "Kurt Reidemeister 13. 10. 1893 – 8. 7. 1971." *Jber. DMV* **74** (1972/73), 96–104 (with bibliography). — BACHMANN, FRIEDRICH; HEINRICH BEHNKE, and WOLFGANG FRANZ: "*In memoriam* Kurt Reidemeister." *Math. Ann.* **199** (1972), 1–11. M.F.

Rey, Abel (* December 29, 1873, Châlon-sur-Saône, France; † January 13, 1940, Paris, France). An *agrégé* in philosophy and *licentiate ès sciences*, REY was a student of EMILE BOUTROUX, EMILE PICARD, and PAUL TANNERY. After World War I, he was named professor of the history and philosophy of science at the *Sorbonne*, and in 1932 became the first director of the newly-founded *Institut d'histoire des sciences et des techniques* at the University of Paris. In addition to editing the journal *Thalès*, he collaborated with LUCIEN FEBVRE on the *Encyclopédie française*, and with HENRI BERR on the series *Evolution de l'humanité*, of which he devoted four volumes to the role of Greece in the origins of scientific thought, placing considerable emphasis on history of mathematics.

Main works: *La science dans l'antiquité.* 5 vols., Paris 1930, 1933, 1939, 1946–1948.

Secondary literature: *DSB* **11**, 388–389. *P* VI, 2158; VIIb(7), 4325. — BRUNSCHVICG, LÉON: "Abel Rey." *Thalès* **4** for 1937–1939, (1940), 7–8. J.P.

Rey Pastor, Julio (* August 14, 1888, Logroño, Spain; † February 21, 1962, Buenos Aires, Argentina). REY PASTOR studied mathematics with GARCÍA DE GALDEANO in Zaragoza. Upon his graduation in 1908, he received his Ph.D. a year later in Madrid (1909). After teaching as professor of mathematical analysis first in Oviedo (1911) and then, a year later, in Madrid, he was awarded several grants for postgraduate study in Germany. In 1917 he was invited to lecture at the University of Buenos Aires (Argentina), an opportunity that proved to be decisive for the rest of his life, both personally and professionally. He not only married in Argentina, but he also accepted a number of teaching positions (in addition to the one he had in Madrid). Overwork forced him to neglect many of his most elementary duties on both sides of the Atlantic, which resulted in many conflicts that were only manageable when the dictatorships in one country or the other consented to such an arbitrary situation.

Nevertheless, REY PASTOR was a well known and influential mathematician in both Argentina and Spain. He founded (and directed when present) the Mathematical Laboratory and Seminar of the *Junta para Ampliación de Estudios e Investigaciones Científicas (JAE)* (Council for Scientific Research, Madrid), and edited the *Revista Matemática Hispano-Americana* of the Spanish Mathematical

Society, which appointed him chairman in 1934 — during the Spanish Second Republic (1931–1936). REY PASTOR only accepted this office in 1941, and again in 1955 during the Franco dictatorship. As a result, honors were showered upon him, and these in turn provided him with the means to develop his work further.

As an historian of science and mathematics, some of REY PASTOR's most relevant works include *La ciencia y la técnica en el descubrimiento de América* (Science and Technology in the Discovery of America) (1942), *Historia de la Matemática* (History of Mathematics) (together with J. BABINI) (1951), and *La cartografía mallorquina* (Mallorquian Cartography) (together with E. GARCÍA CAMARERO) (1960).

Further work: *Selecta*. Madrid 1988.

Secondary literature: *DSB* **11**, 394–395. *May* 315. *P* VI, 2159; VIIb(7), 4327. — BABINI, JOSÉ: "Julio Rey Pastor, 1888–1962." *Arch. int. Hist. Sci.* **15** (1962), 361–364. — MILLÁN GASCA, ANA: *El matemático Julio Rey Pastor*. Logroño 1988.
E.A.; M.H.

Reyes Prósper, Ventura (* May 31, 1863, Castuera (Badajoz), Spain; † November 27, 1922, Toledo, Spain). As both a doctor and professor of natural history, mathematics, and physics at the secondary level, REYES PRÓSPER published original contributions to both mathematics and its history in such international and national journals as *Mathematische Annalen, El Progreso Matemático*, and the *Archivo de Matemáticas.*

Secondary literature: *P* IV, 1238. — COBO, JESÚS: *Ventura Reyes Prósper.* Badajoz 1991. E.A.; M.H.

Ribeiro dos Santos, António (* March 30, 1745, Massareios, Oporto, Portugal; † January 16, 1818, Lisbon, Portugal). RIBEIRO DOS SANTOS completed his study of the law at Coimbra University, and became a Doctor of Jurisprudence in 1771, having written a thesis in Canon Law that was published in Holland. He was a supporter of the theories of the sovereignty of the people as propagated by the French Revolution, and was aware of the scientific changes occuring in Europe. RIBEIRO DOS SANTOS published several valuable studies on Portuguese mathematicians. He became head of the library of Coimbra University in 1777, and a member of the Academy of Sciences of Lisbon in 1779.

Main works: [SANTOS 1806a], [SANTOS 1806b], [SANTOS 1812].

Secondary literature: *GEPB* **25**, 620–623 (with portrait). L.S.

Riccardi, Pietro Francesco (* May 4, 1828, Modena, Italy; † September 30, 1898, Modena, Italy). RICCARDI's was an old and prominent family in Modena. His uncle FRANCESCO was secretary of the Ministry of Public Economy of the Duke of ESTE, and his father GEMINIANO (1794–1857) was a mathematician and engineer who taught mathematics at the local military school. He was also a member of the Academy of Sciences of Modena and served as its secretary from 1840 until his death. PIETRO RICCARDI first studied in a Jesuit college, and then in 1845 enrolled in the same institute where his father was a teacher. In January 1848, however, the young PIETRO was expelled from the institute because of his anti-Austrian behavior. Some time later, in March, the Duke FRANCESCO V was forced to leave Modena because of uprisings against the Austrians, and a provisional government was formed. RICCARDI completed his studies and obtained a degree in engineering. He then enrolled in the army but in August of 1848 the uprising was suppressed and the Duke returned to Modena. RICCARDI's degree was cancelled and he was forced to repeat his final examination. In 1849 he obtained a degree in mathematics and, two years later, he passed the qualifying examination in engineering, a profession RICCARDI pursued until 1859 when, as a consequence of the *Risorgimento*, Modena became part of a unified Italy. The teaching system was renewed and in November of 1859, RICCARDI obtained a chair for theoretic geodesy at the University of Modena. He was also involved in the local, political life being a member of the town council since 1861, and of the district council beginning in 1864. RICCARDI taught at the University of Modena until 1877, when he went to teach practical geometry at the (newly-founded) *Scuola di applicazioni per l'ingegneria* (School for Engineering) in Bologna, a position he maintained until his retirement in 1888.

About one half of RICCARDI's more than one-hundert publications are devoted to the history of mathematics. But his name is essentially linked to the *Biblioteca matematica italiana dall'origine della stampa ai primi anni del secolo XIX* (Italian Mathematical Library from the Origin of Printing until the First Years of the 19th Century), published between 1870 and 1893 in two parts with several appendices and additions. The first part (in two volumes) includes a list of works (more than 8000) of 2310 mathematicians born in Italy from ARCHIMEDES until LAGRANGE. The second part is a classification by subjects of the works listed in the first part. The seventh (and last) addition to the *Biblioteca* was posthumously published by BORTOLOTTI in 1928. RICCARDI planned to continue the *Biblioteca* to include the 19th century, and in 1890 published a *Saggio di una Biblioteca Matematica italiana del secolo XIX* (Essay About an Italian Mathematical Library of the 19th Century) which included a list of works on projective and descriptive geometry and their applications to drawing. At the same time he worked at a *Saggio di una bibliografia euclidea* (Essay About a Bibliography on Euclid) which appeared in five parts between 1887 and 1893. This *Saggio* included not only a chronological list of editions and translations of the *Elements* and other minor works of EUCLID, but also a classification of the Euclidean bibliography and a list of monographs on EUCLID's fifth postulate.

Secondary literature: *May* 316. *P* III, 1117; IV, 1240. — CAVANI, FRANCESCO: *Della vita e delle opere del prof. Ing. Pietro Riccardi.* Bologna 1899. — BARBIERI, FRANCESCO: "Il contributo di Pietro Riccardi alla storiografia matematica," in: [BARBIERI/DEGANI 1989, 161–169, with bibliography and portrait].

U.B.

Richard, Claude (* 1588/89, Ornans, France; † October 20, 1664, Madrid, Spain). RICHARD, a Jesuit, taught Hebrew and mathematics at the Jesuit College in Lyon for seven years, after which he then became professor of mathematics at the *Colegio Imperial* in Madrid. RICHARD produced several Latin editions of ancient Greek mathematicians, among them EUCLID, APOLLONIUS, HYPSICLES, and PROKLUS.

Secondary literature: *P* II, 629. — *BUAM* **35**, 584. — *NBG* **42**, 180. — *BCJ* **6**, 1808–1809. P.B.

Rochot, Bernard (* November 2, 1900, Annecy, France; † 1971, France). Trained in philosophy, ROCHOT defended a thesis on GASSENDI at the *Sorbonne* in 1940 before accepting a position at the *Centre National de la Recherche Scientifique (CNRS)*. He is best remembered as an editor of MERSENNE's correspondence.

Main works: *La correspondance de Mersenne*, vols. 6–12. Paris 1960–1972.

Secondary literature: COSTABEL, PIERRE: "Bernard Rochot (1900–1971)." *Rev. Hist. Sci.* **25** (1972), 275–277. J.P.

Romanus, Adrianus [van Roomen, Adriaan] (* September 29, 1561, Antwerp, Belgium; † May 4, 1615, Mainz, Germany). VAN ROOMEN was professor of medicine and mathematics at Leuven University from 1586 to 1592, and then professor of medicine in Würzburg. He resigned from his professorship in 1607. During the years 1610–1612 he stayed in Zamość (Poland), teaching mathematics to a young nobleman. ROMANUS defended ARCHIMEDES against the attacks of JOSEPH SCALIGER ([ROMANUS 1597]) and wrote an introduction to AL-KHWĀRIZMĪ's *Algebra*, which contains a short history of the methods for solving equations of the first to the fourth degree ([ROMANUS ca. 1602]).

Secondary literature: *DSB* **11**, 532–534. *May* 49, 321. *P* VI, 2210. — *NBW* **2**, 751–765. — BOSMANS, HENRI: "Le fragment du commentaire d'Adrien Romain sur l'Algèbre de Mahumed ben Musa al-Chowrizm." *Annales de la Société scientifique de Bruxelles* **30**, 2e partie, (1906), 267–287. — BOCKSTAELE, PAUL: "The Correspondence of Adriaan van Roomen." *Lias* **3** (1976), 85–129, 249–299. — BOCKSTAELE, PAUL: "The Correspondence of Adriaan van Roomen: Corrections and Additions." *Lias* **19** (1992), 3–20. P.B.

Rome, Adolphe (* July 12, 1889, Stavelot, Belgium; † April 9, 1971, Korbeek-Lo near Leuven, Belgium). ROME studied classical philology at Leuven University. There he became lecturer in Greek literature in 1927 and ordinary professor in 1929. His scientific work is mainly devoted to the history of ancient Greek mathematics and astronomy, to ARCHIMEDES, PTOLEMY, HERO OF ALEXANDRIA, and especially to PAPPUS and THEON OF ALEXANDRIA. He took an active part in the publication of *Osiris* as director from 1950 until the publication was discontinued with volume 14 (dedicated to OTTO NEUGEBAUER) in 1962.

Secondary literature: DE RUYT, FRANZ: "Notice sur le chanoine Adolphe Rome." *Annuaire Académie royale de Belgique* **138**, Notes biographiques (1972), 87–99 (with portrait). P.B.

Rouse Ball, Walter William (* August 18, 1850, London, England; † April 4, 1925, Cambridge, England). ROUSE BALL was educated and taught mathematics at Cambridge University. His areas of interest comprised general history, NEWTON, and history of Cambridge mathematics.

Further work: *Mathematical Recreations and Problems* (1892, many later editions); of some historical use. — Correspondence with publisher Macmillan: British Library, Add. Ms. 55208. — Collection of over 600 portraits of mathematicians: library of Trinity College Cambridge.

Secondary literature: *May* 65. *P* IV, 59. — WHITTAKER, EDMUND T.: "W. W. Rouse Ball." *Math. Gaz.* **12** (1925), 449–454 (with portrait). — CAJORI, FLORIAN: "Walter William Rouse Ball (1850–1925)." *Isis* **8** (1926), 321–324 (with portrait). I.G.G.

Rudio, Ferdinand (* August 2, 1856, Wiesbaden, Germany; † June 21, 1929, Zürich, Switzerland). As a professor of mathematics at the *Eidgenössische Technische Hochschule (ETH)* (Swiss Federal Institute of Technology) in Zürich, head of the library, a leading member of the *GeP Society* (an *alma mater* organisation of former "Polytechnicians," cf. [JEGHER/PAUR 1894]), and a member of the lecturers' association of both the University and the Polytechnic (the representative body of the popular "Town Hall Lectures" in Zürich), as well as editor of the *Zürcher Naturforschende Gesellschaft* (Zürich Natural Sciences Society), FERDINAND RUDIO greatly influenced the scientific culture of his chosen country of adoption. He achieved lasting fame, however, as an historian of mathematics, above all through his persistent efforts which led to the publication of LEONHARD EULER's *Opera omnia* (Complete Edition). RUDIO's wide knowledge of mathematics and a sound linguistic-historical education well qualified him for what was to be "essentially his life-work" [FUETER 1926, 125]. RUDIO's official estate (as far as extant) lies scattered around the numerous institutions in which he worked.

RUDIO was born and grew up in solid, middle-class sourroundings in Wiesbaden, Germany. His father, HEINRICH, was a public officer of Nassau and his mother, LOUISE neé KLEIN, was the daughter of a renowned forestry official. He studied at the Polytechnic in Zürich from 1874 to 1877, initially civil engineering and afterwards mathematics. He then moved to Berlin where he participated in the mathematical seminars of KUMMER and WEIERTRASS, and in 1880 obtained his Ph.D. Following the suggestion of his former teacher CARL FRIEDRICH GEISER, he returned to Zürich where, in 1881, he obtained his final qualifications as an academic lecturer with his *Habilitation*, and he subsequently became an honorary professor in 1885 and a full professor in 1889. His textbooks on analytic geometry ensued from his lectures. In 1888 RUDIO married MARIA EMMA MÜLLER from Rheinfeld and they had three daughters.

After the death of RUDOLF WOLF, RUDIO became his successor at the *ETH* library (1894–1920). It was RUDIO who brought this somewhat outmoded institution up to date and also played an active role in founding the Central Library in Zürich in 1914, thereby gaining a reputation as an "authority on libraries"

[FUETER/SCHRÖTER 1929, 34]. With the assistance of the geologist ALBERT HEIM and the zoologist ARNOLD LANG, RUDIO also took over from WOLF the editing of the *Vierteljahrsschrift der Naturforschenden Gesellschaft in Zürich* (The Quarterly Journal of the Natural Sciences Society in Zürich) (vols. **39** (1894) to **56** (1911)), and simultaneously the role of local historian of science (cf. [RUDIO 1896]). At the first International Congress of Mathematicians, which took place in Zürich in 1897, he held the post of general secretary and editor of the proceedings (cf. [RUDIO 1898]).

RUDIO's historical interests and activities are "surrounded by the name of Leonhard Euler" [FUETER/SCHRÖTER 1926, 158]. They began in 1883 with a talk RUDIO gave at the Town Hall in Zürich on the occasion of a memorial service for the 100th anniversary of EULER's death [RUDIO 1884]. This lecture was printed and rendered "outstanding service" in soliciting financial contributions for the publication of EULER's *Opera omnia* [FUETER/SCHRÖTER 1926, 158]. No less famous (and still very readable) is his Town Hall lecture on mathematics and culture in the Renaissance [RUDIO 1892a]. Philological abilities and meticulous working methods provided him with access to Greek sources and to specialists such as HERMANN DIELS or WILHELM SCHMIDT, who published the works of HERON.

RUDIO himself edited fundamental discourses on the measurement of the circle [RUDIO 1892b] as well as commentaries of SIMPLICIUS on the squaring of the circle [RUDIO 1907]. In 1895, together with his colleague ADOLF HURWITZ, he edited letters which the brilliant mathematician GOTTHOLD EISENSTEIN had written to his teacher, MORITZ ABRAHAM STERN [RUDIO 1895], [RUDIO/HURWITZ 1895]. In 1901, RUDIO and a botanist friend, CARL SCHRÖTER, decided to resume the discontinued section "Notes on Swiss Cultural History" in the *Vierteljahrsschrift*, which was devoted in the main part to estimable biographies and obituaries and also, after 1907, to reports on the EULER publication.

RUDIO had initiated the latter at a ceremony in Basel to mark the 200th anniversary of EULER's birth — a celebration attended by many national and international scientists. RUDIO formally proposed, and was supported in this by CARL F. GEISER *(ETH)*, ALFRED KLEINER (University of Zürich) and CHRISTIAN MOSER (University of Berne), that the *Schweizerische Naturforschende Gesellschaft (SNG)* (The Swiss Academy of Sciences) take over the task of publishing EULER's complete works. A corresponding motion was put forward and the way paved for international support. The *SNG* appointed an Euler Commission of eleven members, with RUDIO as president, and within two years "the great work of preparation and financing of the huge project" was achieved [FUETER/SCHRÖTER 1926, 161]. In due course the annual assembly of the *SNG* in Lausanne agreed on "the publication of the complete works of Leonhard Euler in the original language" [RUDIO 1911, XXIX]. In the same year, the Euler Commission formed a financial board, selected a publisher (Teubner in Leipzig), and formed an editorial committee, with RUDIO as chairman and managing editor and the Germans ADOLF KRAZER and PAUL STÄCKEL as co-editors. The editorial plan was completed in a final form by 1910, and the first volume, the *Vollständige Anleitung zur Algebra* (Complete Introduction to Algebra) appeared a year later in 1911 (ed. HEINRICH WEBER). RUDIO himself published EULER's *Commentationes arithmeticae* (Arithmetical Commentaries) in 1915 and 1917, and worked as co-editor on three other volumes. As managing editor he was also responsible for all of the correspondence about the edition, as well as for "contracts with the editorial staff." He had been able to acquire all of EULER's published writings and each editor received the "manuscript of his volumes from Rudio already put together, so that the editor's work was only a matter of looking through the whole and perhaps making remarks in the margins." The final responsibility lay with the managing editor [FUETER/SCHRÖTER 1926, 162].

RUDIO retired from his teaching position in 1928 and, at the same time, relinquished his beloved occupation with EULER. In poor health, he died a year after his retirement in 1929 (cf. [JEGHER 1929]).

The EULER undertaking suffered somewhat during the First World War and in the years following, but twenty volumes had been published by 1928. Many more were in the making and eventually flourished, even though progress was not as rapid as had been hoped. By the end of 2000 the total number of volumes had grown to 73 (with 69 volumes of books, articles and manuscripts, and four volumes of correspondence). At least fourteen more volumes are anticipated before the Euler edition is complete.

Main Works: "Leonhard Euler (1707–1783). Vortrag, gehalten ... in Zürich am 6. Dezember 1883." *Öffentliche Vorträge gehalten in der Schweiz* **8** (1884), no. 3. Zürich 1884. — "Über den Anteil der mathematischen Wissenschaften an der Kultur der Renaissance." *Sammlung gemeinverständlicher wissenschaftlicher Vorträge* ... (Hamburg), N.F. **VI** (1892), 769–802 (= Heft 142) [1892a]. — *Archimedes, Huygens, Lambert, Legendre; vier Abhandlungen über die Kreismessung.* Leipzig 1892 [1892b]. — "Eine Autobiographie von Gotthold Eisenstein." *Abh. Gesch. Math.* **7** (1895), 145–168. — RUDIO, FERDINAND, and ADOLF HURWITZ: "Briefe von G. Eisenstein an M. A. Stern." *Abh. Gesch. Math.* **7** (1895), 169–203. — [RUDIO 1896]. — "Über die Aufgaben und die Organisation internationaler mathematischer Kongresse." *Verhandlungen des ersten internationalen Mathematikerkongresses in Zürich.* Leipzig 1898, 31–37. — *Der Bericht des Simplicius über die Quadraturen des Antiphon und des Hippokrates; griechisch und deutsch* (= *Urkunden zur Geschichte der Mathematik im Altertume*, vol. 1. [No more published]). Leipzig 1907. — [RUDIO 1911].

Secondary literature: *DSB* **11**, 589. *May* 323. *P* IV, 1280; V, 1075; VI, 2236. — *HBLS* **5**, 732–733. — *SL* **5**, 437. — JEGHER, AUGUST, and HANS PAUR (Eds.): *Festschrift der Gesellschaft ehemaliger Studierender der eidgenössischen polytechnischen Hochschule in Zürich.* Zürich 1894. — FUETER, RUDOLF: "Die wissenschaftliche Tätigkeit Ferdinand Rudios." *Vj. Nat. Ges. Zür.* **71** (1926), 124–131. — FUETER, RUDOLF, and CARL SCHRÖTER: "Ferdinand Rudio. Zum 70. Geburtstag." *Vj. Nat. Ges. Zür.* **71** (1926), 147–167 (with bibliography). — FUETER, RUDOLF, and CARL SCHRÖTER: "Professor Dr. Ferdinand Rudio,

1856–1929." *Vh. Schw. Nat. Ges.* **110** (1929), 2. Teil (Anhang: Nekrologe), 33–42. — JEGHER, CARL: "† Ferdinand Rudio." *Schweizerische Bauzeitung* **94** (1929), 231. — [FREI/STAMMBACH 1994]. — Portrait: [BÖLLING 1994, 9.2].
B.G. (Translation: Angela Rast)

Ruska, Julius (* February 9, 1867, Bühl near Baden-Baden, Germany; † February 12, 1949, Schramberg (Black Forest), Germany). RUSKA began his career as a teacher of mathematics in Heidelberg (1889–1910). After studying Semitic languages, he became *Dozent* (1911) and extraordinary professor (1915) of Semitic philology in Heidelberg. This was changed into an honorary professorship in 1927, when RUSKA accepted the post of director of a Research Institute for History of Science in Berlin (1927–1931). When this was enlarged to an Institute for History of Medicine and Science, he was made *Abteilungsvorstand*, a position he held in Berlin until 1937. His areas of primary interest were Arabic chemistry and alchemy, although he also contributed to the history of arithmetic and algebra.

Main works: *Zur ältesten arabischen Algebra und Rechenkunst.* (1917). — "Ursprung und Geschichte eines merkwürdigen Systems von Zahlzeichen." *Abh. Gesch. Nat. Techn.* **9** (1922), 112–126.

Secondary literature: *P* VI, 2247; VIIa(3), 860. — KRAUS, PAUL: "Julius Ruska." *Osiris* **5** (1938), 5–40 (with portrait). — SIGGEL, ALFRED: "Prof. Dr. Julius Ruska." *Arch. int. Hist. Sci.* **3** (1950), 912–915. M.F.

Sachs, Abraham Joseph (* December 11, 1914, Baltimore, Maryland, USA; † April 22, 1983, Providence, Rhode Island, USA). ABRAHAM SACHS, an American Assyriologist, taught at Brown University, where he joined OTTO NEUGEBAUER in 1946 to become one of the founding members of the Department of History of Mathematics. Later he served as chairman until his retirement in 1980. SACHS received his Ph. D. in philology from Johns Hopkins University in 1939, whereupon he spent several years at the University of Chicago's Oriental Institute as a research assistant working on the *Assyrian Dictionary.* In 1941 he accepted a position as a Rockefeller Foundation Fellow at Brown University, and between 1941 and 1948 he advanced from the position of assistant to associate professor. He also spent the year 1945–1946 as a member of the Institute for Advanced Study in Princeton. From 1954 on SACHS was professor of history of mathematics at Brown.

SACHS worked with OTTO NEUGEBAUER on *Mathematical Cuneiform Texts* (1955), in addition to numerous contributions of his own, many of which appeared in the *Journal of Cuneiform Studies.* SACHS discovered an important collection of astronomical tablets in the British Museum, and devoted much of the rest of his life to their publication.

Secondary literature: BERGGREN, JOHN L.: "Abraham J. Sachs (1914–1983): *In Memoriam.*" *HM* **11** (1984), 124–125 (with portrait). — TOOMER, G. J.: "*Obituary.* A. J. Sachs (1914–1983)." *JHA* **15** (1984), 146–149 (with portrait). J.W.D.

Saʿīdān, Aḥmad Salīm (* 1914, Safad, Palestine; † January 23, 1991, Amman, Jordan). AḤMAD SAʿĪDĀN, a Palestinian scholar, studied at the Arabic College in

Jerusalem and the American University in Beirut (AUB), where he obtained the degree of bachelor of science in 1934. In 1966, he submitted his Ph.D. thesis at Khartoum University on "The Development of Hindu-Arabic Arithmetic." This thesis laid the foundation for many of SAʿĪDĀN's best-known contributions to the history of Arabic mathematics, such as his three volumes of *Ta'rīkh ʿilm al-ḥisāb al-ʿarabī* (History of Arabic Arithmetic), published in 1971, 1973, and 1984 in Amman, and the *Dār al-Farqān*, which encompasses material on finger reckoning, the oldest extant Arabic textbook on Indian arithmetic, and works by authors from the Maghrib and al-Andalus. AL-UQLĪDISĪ's Indian arithmetic also appeared in an English translation with commentary in 1978.

SAʿĪDĀN was appointed professor of mathematics at several Arabic colleges and universities in Palestine, Sudan, and Jordan. His activities as an historian of mathematics were directed towards the study of Arabic texts on arithmetic, number theory, algebra, geometry, trigonometry, and astronomy. He edited forty Arabic texts, most of which he also analyzed and commented upon either in Arabic or in English. He also translated some of the major texts into English. Additionally, he published entries on Arabic mathematicians in the *Dictionary of Scientific Biography (DSB)* ([DSB 1970/90]) and wrote surveys on the history of science and mathematics in Islam for a wider audience. In 1979, he retired from active teaching and became president of the University of Abū Dīs in Jerusalem. In 1981, after occupying the city, Israel expelled him. Ten years later, AḤMAD S. SAʿĪDĀN died in Amman.

Secondary literature: ANONYMOUS: "Demise of Ahmad Salim Saidan, eminent historian of science." *IRCICA Newsletter* no. 25 (April 1991), 13. — HOGENDIJK, JAN P., and BORIS A. ROSENFELD: "*In Memoriam* Aḥmad Salīm Saʿīdān (1914–1991)." *HM* **19** (1992), 438–443 (with bibliography by JAN P. HOGENDIJK 439–443). S.B.

Sánchez Pérez, José Augusto (* November 30, 1882, Madrid, Spain; † November 13, 1958, Madrid, Spain). JOSÉ SÁNCHEZ PÉREZ held a doctorate in mathematics and was a professor of mathematics in secondary teaching. He was especially interested in the history of mathematics in Muslim Spain, and published a number of studies in the *Revista de la Sociedad Matemática Española*, in the *Revista Matemática Hispano-Americana*, and monographs such as *Compendio de Algebra de Abenbeder* (A Compendium of the Algebra of Abenbeder) (1916), *Biografías de matemáticos árabes que florecieron en España* (Biographies of Arab Mathematicians who Flourished in Spain) (1921), and *Las Matemáticas en la Biblioteca del Escorial* (Mathematics in the Library of the Escorial) (1929).

Secondary literature: *May* 326. — GARCÍA RÚA, JOAQUÍN: "José Sánchez Pérez." *Gaceta Matemática* **9** (1959), 3–5. E.A.; M.H.

Sanford, Vera (* October 1, 1891, Douglaston, New York, USA; † December 28, 1971, Oneonta, New York, USA). SANFORD, an American mathematician, received her AB degree from Radcliffe College in 1915, and her Ph.D. in mathematics from Columbia University in 1927. She taught mathematics in various schools, most notably at the Lincoln School, Teachers College, Columbia University (1920–29). She then accepted a position as assistant professor at Western Reserve (1929–33), and later assumed the chairmanship of the Department of Mathematics at *SUNY* (State University of New York), Oneonta (1933–43). SANFORD retired as professor *emerita* in 1959. Among her publications are *The History and Significance of Certain Standard Problems in Algebra* (1927), which was the author's thesis at Teachers College, Columbia University, and *A Short History of Mathematics* (1930). J.W.D.

Sarton, George Alfred Leon (* August 31, 1884, Ghent, Belgium; † March 22, 1956, Cambridge, Massachusetts, USA). GEORGE SARTON, sometimes referred to as the "father of the history of science," was a significant catalyst for serious interest in history of science both in the United States and internationally. The only son of a chief engineer and editor of the *Belgian State Railroad*, SARTON's early academic interests focused on philosophy and science, including mathematics, which he studied at the University of Ghent (1902–1911). SARTON's interest in history of science did not begin to develop until 1908–1910, when he studied with professor PAUL MANSION, whose course on "Histoire des sciences mathématiques et physiques" (History of the Mathematical and Physical Sciences) made a strong impression. It was also a course required for the doctorate, which SARTON received in 1911 [De Mey 1984, 44].

SARTON soon found himself committed to the history of science, and in 1912 he founded an international journal for history of science, *Isis*, the first issue of which was published a year later. World War I, however, interrupted academic ventures internationally, and in November of 1914, as German troops occupied SARTON's home in Wondelgem, he sailed with his family for England where he worked briefly for the War Office. The following year he emigrated to the United States, by which time SARTON was already 30 years old.

After brief periods teaching at the University of Illinois and at George Washington University, he received a research stipend from the Carnegie Institutution in Washington, D. C., and not long thereafter began teaching at Harvard University in exchange for a set of offices in Widener Library, where he also edited *Isis*. SARTON regarded the new journal "as the future focus of a new intellectual movement that would bring scientific studies and humanities together and help them become one single and coherent system of knowledge; this is what he later called the doctrine of the 'New Humanism'" [Elkhadem 1984, p. 36]. In addition to the two journals SARTON founded for the history of science, *Isis* (1913) and *Osiris* (1936), the latter intended for longer monographic studies, he also devoted considerable efforts to bibliographies. These resulted in annual critical bibliographies issued as supplements to *Isis*, as well as a separate publication: *Horus: A Guide to the History of Science* (1952).

SARTON's major contribution to history of mathematics — apart from the numerous articles he wrote himself for *Isis*, and the extent to which the subject was prominently reflected in his *Introduction to the History of Science* — was a book SARTON wrote in 1936 devoted specifically to *The Study of the History of Mathematics*. In this work, SARTON discussed various kinds of mathematical history, the "Greek miracle," different forms of discovery in mathematics, including the phenomena of mathematical geniuses, and the danger facing the academic historian of mathematics, "when investigations become very technical." Under such circumstances, he warned, "there is always a danger that the subject be sacrificed to the technique." With this in mind, SARTON appended a short discussion at the end of his monograph, a "Note on the Study of the History of Modern Mathematics," which focused specifically on the special character of the history of mathematics in the 19th and 20th centuries (pp. 29–38).

Secondary literature: *DSB* **12**, 107–114. *May* 326. *P* VI, 2286; VIIb(7), 4663. — An entire number of *Isis*, the journal he founded, is a memorial volume dedicated to SARTON: **48** (3) (1957), 283–350 (with bibliography and photographs). — SARTON, MAY: *I Knew a Phoenix: Sketches for an Autobiography.* New York 1959. — VAN OYE, PAUL: "George Sarton: De mens en zijn werk uit brieven aan vrienden en kennissen." *Brussels: Verhandelingen van de Koninklijke Vlaamse Academie voor Wetenschappen, Letteren en Schone Kunsten van België* (1964). — THACKRAY, ARNOLD: "On discipline-Building: The Paradoxes of George Sarton." *Isis* **63** (1972), 673–695. — FRÄNGSMYR, TORE: "Science or History: George Sarton and the Positivist Tradition in the History of Science." *Lychnos* 1972/73, 104–144. — [GILLIS 1973]. — On the occasion of the 60th anniversary of the founding of the History of Science Society, and the centenary of the birth of GEORGE SARTON, an entire issue of *Isis* was devoted to *Sarton, Science, and History. The Sarton Centennial Issue. Isis* **75**, no. 276 (March, 1984). — GARFIELD, EUGENE: "The Life and Career of George Sarton: The Father of the History of Science." *JHBS* **21** (1985), 107–117. — MERTON, ROBERT K.: "George Sarton: Episodic Recollections by an Unruly Apprentice." *Isis* **76** (1985), 470–486. — PEYENSON, LEWIS: "Inventory as a Route to Understanding: Sarton, Neugebauer, and Sources." *Hist. Sci.* **33** (1995), 253–282. — Portraits: *Osiris* **12** (1956), frontispiece; *Isis* **75** (1984), 6, 10, 18 and 21. J.W.D.

Sayılı, Aydın (* May, 3, 1913, Istanbul, Turkey; † October 15, 1993, Ankara, Turkey). In 1933, SAYILI graduated from the Ankara School for Boys (now: Ankara Atatürk School). Due to the influence of president Atatürk, present in the examination, A. SAYILI decided to study history of science. He applied for a scholarship and was sent to Harvard University where he earned his Ph.D. degree in the history of science under GEORGE SARTON in 1942. His graduation in history of science is said to have been the first such degree granted worldwide. In 1942 he became a member of the Faculty of Letters of the institution which was transformed into Ankara University in 1946. In 1952 he was appointed to the first chair for history of science in Turkey. In 1974, a department for history of science was established at Ankara University and SAYILI became its first director. He was elected as a full or honorary member of the Turkish Historical Society (in 1947), the Society of Turkish Librarians (in 1989), and the *Deutsche Morgenländische Gesellschaft* (in 1989). He received several national and international awards and medals, such as the "service award" of the Turkish Society for Scientific and Technological Research (in 1977), the Polish Copernicus medal (in 1973), and the Nehru medal of UNESCO (in 1992).

SAYILI's first book, "Science is the Truest Guide in Life," 1949, was devoted to popularizing science and history of science. He wrote this book in the spirit of Kemalist ideology and understanding of man's place in society. His adherence to the maximalist approach to Turkish history also guided one of his earliest papers published in *Isis* **31** (1940), 8–24: "Was Ibn Sina an Iranian or a Turk?" SAYILI defended the feasibility of attaching "nationality" to medieval scholars and argued in favor of a Turkish heritage for IBN SĪNĀ. SAYILI's most basic argument was that attributing "nationality" to scholars was a general practice in Western history of science. The notion of nationality applied in this study conspicuously followed 20th-century political divisions in the Middle East. SAYILI spoke of one "Turkish" entity both in terms of language and ethnicity, which he contrasted with one "Iranian" entity. SAYILI's interest in the roles Turkish people have played in Islamic civilization was also manifest in his papers written together with RICHARD N. FRYE, such as "Turks in the Middle East Before the Seljuqs," *Journal of the American Oriental Society* **63** (1943), 194–207, or "The Turks in Khurasan and Transoxania at the Time of the Arab Conquests," *Muslim World* **35** (1954), 308–315. SAYILI also contributed to the study of poetry.

SAYILI's Ph.D. thesis, "Institutions of Science and Learning in Medieval Islam," studied the development of observatories and hospitals in Muslim societies. The part dealing with the history of observatories was published in 1960 under the title: *The Observatory in Islam.* In this book, SAYILI investigated the extant historical sources about major observatories built for rulers and scholars in order to determine whether or not they were actually built, and if so, where, by whom, and with what instruments they were endowed. It is still the only major work on this important subject in the history of science in Islam. A second part of his Ph.D. thesis was published in a paper called "Higher Education in Medieval Islam," *Annales de l'Université d'Ankara* **2** (1948), 30–71. In it, SAYILI discussed the evolution, structure, and profile of Muslim education at the madrasa. He also published another part of his thesis dealing with hospitals in Islam in several extracts in 1947–48.

In his later research, SAYILI was concerned among other topics with the relation between the development of the sciences in Muslim societies and in early modern Western Europe during the 16th and 17th centuries. He published his results in papers such as "Islam and the Rise of the Seventeenth Century Science" or "Murad III's Istanbul Observatory Terrestial Globe and Cultural Contact with Europe," as well as in his book *Copernicus and His Monumental Work* (in English and in Turkish, 1975). In history of mathematics, SAYILI edited, translated, and commented upon writings on quadratic equations, a generalization of the Pythagorean theorem, and the trisection of an angle by ʿAbd al-Ḥamīd b. Turk (early 9th century), Thābit b. Qurra († 901), and Abū Sahl Wayjan b. Rustam al-Kūhī (10th century). Sayili edited several Persian and Arabic poems referring to astronomy, observatories, and the disciplines taught at a madrasa. Almost from the beginning of his scholarly life, he also published on history of astrology, optics, and natural philosophy (the rainbow, theory of motion, the "horror vacui"-problem, etc.), both in Islam and in medieval and early modern Western Europe. In Turkish, he wrote — besides various papers on the subjects mentioned — a book on the history of mathematics, astronomy, and medicine in ancient Egypt and Mesopotamia (1966).

Main work: *The Observatory in Islam And Its Place in the General History of the Observatory* (= *Publications of the Turkish Historical Societies*, Series VII, no. 38.) Ankara 1960; repr. in New York in 1981.

Secondary literature: Ihsanoġlu, Ekmeleddin: "*In memoriam.* Aydīn Sayili 1913–1993." *Arch. int. Hist. Sci.* **45** (1995), 135–148. — [Matvievskaya/Rozenfeld 1985, Vol. 1, 383–385]. S.B.

Schmidt, Olaf Henrik (* December 12, 1913, Sommersted, Denmark; † June 7, 1996, Copenhagen, Denmark). Olaf Schmidt graduated in mathematics from the University of Copenhagen in 1938. The next summer he followed his teacher Otto Neugebauer to the United States for what should have been a shorter stay, but which was prolonged until 1945 because of the Second World War. In 1943 Schmidt obtained a Ph.D. degree at Brown University — his thesis being on ancient spherical astronomy; shortly before his death he agreed to have it published. In 1953 he obtained a position, which in 1965 became a professorship, in the history of exact sciences at the Mathematical Institute at the University of Copenhagen. A great number of students have profited from his clear and extremely well-prepared lectures. His colleagues had less occasion to benefit from Schmidt's insights, primarily because he was always very reluctant to publish, believing that he might be able to express his opinions even more clearly in the future. It is hoped that his lectures and some of his research results will be published in the near future. Schmidt's approach to history of mathematics was similar to Hieronymus Zeuthen's. Schmidt was first of all interested in unravelling the mathematical ideas in earlier texts, keeping to the techniques of the time. In fact, he often introduced an *ad hoc* notation to cover former concepts in order to ensure that the arguments did not stray from then-contemporary insights and rules.

Secondary literature: Brack-Bernsen, Lis: "In Memoriam Olaf Schmidt." *HM* **24** (1997), 131–134. K.A.

Scholz, Heinrich (* December 17, 1884, Berlin, Germany; † December 30, 1956, Münster, Germany). Scholz was a *Privatdozent* for philosophy of religion and systematic theology at the University of Berlin (1910), became an ordinary professor for philosophy of religion at Breslau in 1917, then ordinary professor of philosophy at Kiel (1919), and finally ordinary professor of philosophy, including mathematical logic and foundations, at Münster (1928–1953). His major areas of research interest in history of mathematics were related to philosophy, logic, and Greek mathematics. His historical works include: *Die Grundlagenkrisis der griechischen Mathematik* (The Foundation Crisis in Greek Mathematics) (written with HELMUT HASSE, 1928); "Warum haben die Griechen die Irrationalzahlen nicht aufgebaut" (Why Didn't the Greeks Develop Irrational Numbers?), *Kantstudien* **3** (1928), 35–72; *Geschichte der Logik* (History of Logic) (1931); "Leibniz und der gegenwärtige Stand der mathematischen Grundlagenforschung" (Leibniz and the Present State of Research on the Foundations of Mathematics), *Jber. DMV* **52** (1943), 217–244.

Secondary literature: *May* 330. *P* VI, 2359; VIIa(4), 233. — HERMES, HEINRICH: "Heinrich Scholz zum 70. Geburtstage." *Math.-phys. Semesterber.* **4** (1955), 165–170 (with portrait). — MESCHKOWSKI, HERBERT: "Heinrich Scholz. Zum 100. Geburtstag des Grundlagenforschers." *Hum. Techn.* **27** (1984), 28–52. M.F.

Schoy, Carl (* April 7, 1877, Bittelschies, Hohenzollern, Germany; † December 6, 1925, Frankfurt (Main), Germany). SCHOY taught mathematics in Mülheim (Ruhr) (1908) and Essen (1909), from 1919 to 1921 he was *Privatdozent* in history of mathematics and astronomy at the University of Bonn, and shortly before his death he became *Dozent* of history of the exact sciences in the Orient at the University of Frankfurt (1925). His areas of special interest included: Arabic mathematics (trigonometry in particular) and astronomy.

Main works: *Arabische Gnomonik* (1913). — "Beiträge zur arabischen Trigonometrie." *Isis* **5** (1923), 364–399. — *Die trigonometrischen Lehren des persischen Astronomen Abu'l Raiḥân Muḥ. ibn Aḥmad al-Bîrûnî, dargestellt nach al-Qânûn al-Masʿûdî* (1927). — *Beiträge zur arabisch-islamischen Mathematik und Astronomie* (Reprint of articles). 2 vols. Frankfurt am Main: Institute of the History of Islamic Sciences 1988.

Secondary literature: *May* 331. *P* VI, 2366. — ANDING, ERNST: "Karl Schoy." *Astronomische Nachrichten* **227** (1926), 62–63. — ANDING, ERNST: "Karl Schoy" (Nekrolog). *PGM* **72** (1926), 124. — SMITH, DAVID EUGENE: "The Early Contributions of Carl Schoy." *AMM* **33** (1926), 28–31. — SPIES, OTTO, in *Z. dt. MG.* **5** (1926), 319–327. — RUSKA, JULIUS: "Carl Schoy (Geb. den 7. April 1877, gest. den 6. Dezember 1925.)." *Isis* **9** (1927), 83–95 (with portrait). — WIELEITNER, HEINRICH: "Karl Schoy †." *Jber. DMV* **36** (1927), 163–167 (with portrait). M.F.

Sédillot, Jean-Jacques Emmanuel (* April 26, 1777, Enghien-Montmorency, France; † August 9, 1832, Paris, France). After graduating from the *Ecole Polytechnique*, JEAN-JACQUES SÉDILLOT studied oriental languages with SILVESTRE DE SACY at the newly founded *Ecole des langues orientales vivantes* in Paris, an institution to which he was linked in various ways. Engaged in showing the brilliant progress achieved in the mathematical sciences by the Arabs from the 9th to the 15th century, as an historian of Arabic astronomy, he was given a position as adjoint astronomer at the *Bureau des longitudes* (1814–1832). He published mainly in the *Comptes rendus* of the French Academy of Sciences, but most of his work was left unpublished. DELAMBRE included part of SÉDILLOT's results in his *L'histoire de l'astronomie au moyen âge* (History of Astronomy in the Middle Ages).

Secondary literature: *P* II, 887. J.P.

Sédillot, Louis Pierre-Eugène Amélie (* June 23, 1808, Paris, France; † December 2, 1875, Paris, France). Son of JEAN-JACQUES, LOUIS studied oriental languages and mathematics with his father before teaching history at various Parisian colleges beginning in 1836, notably at the *Collège Henri IV* and the *Collège Saint-Louis*. SÉDILLOT was determined to complete and develop the research begun by his father on the history of the Arabic exact sciences, and to defend the originality of Arabic contributions to mathematics (as opposed to the Indians and Chinese). At the *Académie des Sciences*, he and CHASLES, who supported him, were involved in major controversies with LIBRI, JEAN BAPTISTE BIOT, and BERTRAND, who opposed SÉDILLOT's ideas. Among these, his claim that the Arabs had solved third degree equations was largely confirmed by the research of WOEPCKE. SÉDILLOT published mainly in the *Journal asiatique*, the *Comptes rendus* of the French Academy of Sciences, and in *Notices et extraits des manuscrits de la Bibliothèque royale.*

Main works: *Traité d'astronomie d'Abou'l-Hassan.* 2 vols., Paris 1834–1835. — *Matériaux pour servir à l'histoire comparée des sciences mathématiques chez les Grecs et les Orientaux.* Paris 1845. — *Prolégomènes des tables astronomiques d'Oloug-Beg.* 2 vols., Paris 1846–1853.

Secondary literature: *May* 333. *P* II, 888; III, 1231. — Biographical notices in *L'Orient, l'Algérie et les colonies françaises et étrangères, revue bi-mensuelle suivie de la Biographie des Orientalistes français et étrangers. Histoire, voyages, géographie, sciences, littérature, commerce, industrie, production, colonisation, religion, mours, etc.*, in DUGAT, GUSTAVE: *Histoire des orientalistes de l'Europe du XII*^{e} *au XIX*^{e} *s.*, Paris 1868, 121–142. — SÉDILLOT, C. E.: "Lettre à D. B. Boncompagni sur la vie et les travaux de M. Louis Amélie Sédillot." *Bull. Boncompagni* **9** (1876), 649–655. J.P.

Seidenberg, Abraham (* June 2, 1916, Washington, D.C., USA; † May 3, 1988, Milan, Italy). ABRAHAM SEIDENBERG earned his Ph.D. in mathematics at the Johns Hopkins University in 1943. He spent the rest of his career at the University of California, Berkeley, where he began teaching as an instructor of mathematics in 1945. He was promoted to the rank of professor in 1958, and retired from teaching in 1987. His most important research in mathematics was related to commutative algebra and algebraic geometry. In addition to his work as a pure mathematician, he also published several articles on the history of ancient mathematics. SEIDENBERG was particularly interested in the diffusion of early number systems and the "ritual origins" of counting, geometry, and mathematics generally. It was SEIDENBERG's belief that these were all closely related to number-mysticism and rituals; for example, he argued that counting arose as a means of calling participants into rituals. He also held that the ancient geometries of the Greeks, Babylonians, Egyptians, Indians, and Chinese were derived from ritual practices revealed in the *Śulvasūtras*. SEIDENBERG's papers are preserved in the Bancroft Library at the University of California, Berkeley.

Main works: "The Diffusion of Counting Practices." *University of California Publications in Mathematics* **3**, Berkeley and Los Angeles, CA: University of California Press 1960, 215–299. — "The Ritual Origin of Geometry." *Arch. Hist. ex. Sci.* **1** (1962), 488–527. — "The Ritual Origin of Counting." *Arch. Hist. ex. Sci.* **2** (1962), 1–40. — "The Ritual Origin of Mathematics." *Arch. Hist. ex. Sci.* **18** (1978), 301–342.

Secondary literature: HAHN, ROGER: "Abraham Seidenberg. 1916–1988." *Arch. int. Hist. Sci.* **39** (1989), 146–147. — GANITANAND = GUPTA, R. C.: "Abraham Seidenberg." *Gaṇita-Bhāratī* **11** (1989), 57–59 (with portrait). J.W.D.

Selenius, Clas-Olof (* September 28, 1922, Helsinki, Finland; † September 10, 1991, Uppsala, Sweden). SELENIUS graduated in science from the Swedish university in Helsinki and obtained his doctorate in mathematics at Åbo Academy in 1961. He was active in teaching mathematics, including history of mathematics, from 1945 until 1990, among other places at Åbo Academy and at Uppsala University. Applying the modern theory of specialized continued fractions, he illuminated the cyclic method of Indian mathematics for solving the so-called Bhaskara-Pell equation.
Secondary literature: SCRIBA, CHRISTOPH J.: "*In Memoriam* Clas-Olof Selenius (1922–1991)." *HM* **19** (1992), 325–327 (with bibliography and portrait). K.A.

Sempilius, Hugo [Semple or Sempill, Hugh] (* 1596, Craigevar, Scotland; † September 29, 1654, Madrid, Spain). SEMPILIUS was a Scottish Jesuit who taught mathematics in Madrid. In his *De mathematicis disciplinis libri duodecim* (Twelve Books [= Chapters] on the Mathematical Disciplines) ([SEMPILIUS 1635]) he reported about the work of a great number of mathematicians from antiquity to his own time.

Secondary literature: *P* II, 902. — *DNB* 17, 1173. — *BCJ* **7**, 1117. P.B.

Sergescu, Pierre (* December 17, 1893, Turnu Severin, Romania; † December 21, 1954, Paris, France). SERGESCU simultaneously prepared a *licentiate* in philosophy and in mathematics at the University of Bucharest. After completing his thesis in mathematics (1924), he taught analytical geometry at Cluj University (1926–1943). Under the influence of ALDO MIELI, whom he met in Paris, SERGESCU began to undertake research of his own on history of mathematics. In 1946, as an exile in Paris, he participated actively in the creation of the *International Union for the History of Science.* His main interest was mathematics in France at the turn of the 19th century, and although his style was rather technical, he tried to imbed history of mathematics in a broader scientific context.

Main works: *Coups d'œil sur les origines de la science exacte moderne.* Paris 1951.

Secondary literature: *May* 335. *P* VI, 2417; VIIb(7), 4796. — BODENHEIMER, FREDERICK SIMON: "Petre Sergescu (1893–1954)." *Arch. int. Hist. Sci.* **8** (1955), 3–4. — MONTEL, PAUL: "Discours prononcé aux funérailles de Pierre Sergescu." *Arch. int. Hist. Sci.* **8** (1955), 5–6. — SARTON, GEORGE, in *JHM* **10** (1955), 421–425 (with portrait). — TATON, RENÉ: "Pierre Sergescu (1893–1954)." *Rev. Hist. Sci.* **8** (1955), 77–80. — TATON, RENÉ: "Pierre Sergescu (1893–1954): son œuvre en histoire des sciences et son action pour la renaissance des *Archives internationales d'histoire des sciences.*" *Arch. int. Hist. Sci.* **37** (1987), 104–119. J.P.

Simon, Max (* June 8, 1844, Kolberg, Germany; † January 15, 1918, Straßburg, Germany). SIMON taught mathematics in Berlin (1868–1871) and Straßburg (1871–1912), and in 1903 was made an honorary ordinary professor at the University of Straßburg. His research dealt mainly with the history of the theory of functions, analytical geometry, and Greek mathematics. His historical works include: *Euclid und die sechs planimetrischen Bücher* (Euclid and the Six Planimetric Books) (1901); *Über die Entwicklung der Elementar-Geometrie im 19. Jahrhundert* (On the Development of Elementary Geometry in the 19th Century) (1906); and *Geschichte der Mathematik im Altertum in Verbindung mit antiker Kulturgeschichte* (History of Mathematics in Antiquity in Connection with Ancient Cultural History) (1909).

Secondary literature: *May* 338. *P* III, 1250; IV, 1398; V, 1167; VI, 2455. — FLADT, KUNO: "Max Simon 1844–1918." In: MAX SIMON: *Nichteuklidische Geometrie in elementarer Behandlung* (= Beihefte zur *ZMNU*, **10**). Leipzig, Berlin 1925, ix–xv (with bibliography and portrait). — VOLKERT, KLAUS: "Max Simon als Historiker und Didaktiker der Mathematik." In: [SCHÖNBECK 1994, vol. 1, 73–88]. — Portrait: [BÖLLING 1994, 38.4]. M.F.

Simonov, Nikolai Ivanovich (* February 14, 1910, Moscow, Russian Empire; † January 27, 1979, Moscow, USSR). NIKOLAI SIMONOV taught at Saratov University (1935–1945), Chernovtsy University (1946–1958), the Kiev Institute of Aviation Engineers (1958–1965), and at the Moscow Technological Institute (1965–1979). He was made a professor in 1959. In the history of mathematics, his research interests included the theory of differential and difference equations, and EULER's works.

Main works: *Applied Methods of Analysis in Euler's Works* (Moscow 1957, in Russian). — "Ordinary Differential Equations." In: *History of Mathematics from Ancient Times up to the Beginning of the 19th Century*, A. P. YOUSHKEVICH (Ed.), **3** (1972), 369–408 (in Russian).

Secondary literature: [KUROSH 1959]. — [FOMIN/SHILOV 1969]. — DEMIDOV, S. S., and A. P. YOUSHKEVICH: "Nikolaj Ivanovic Simonov." In: NTM **18**, no. 1, (1981), 92–93 (with portrait). — Portrait in [SHTOKALO 1966/70, vol. 4, book 2, p. 483]. S.S.D.

Simons, Lao G. (* March 29, 1870, San José, California; † November 25, 1949, Greenwich, Connecticut, USA). SIMONS, an American mathematician and historian of mathematics, studied mathematics at Vassar College in Poughkeepsie, N.Y., and received her M.A. and Ph.D. degrees at Columbia University before World War I, after which she taught mathematics at Hunter College until her retirement in 1940. SIMONS served as a review editor for *Scripta Mathematica*, and was a member of the American Association for the Advancement of Science. Her publications include a collection of essays, *Fabre and Mathematics: and Other Essays* (1939).

Secondary literature: *May* 338. — EISELE, CAROLYN: "Lao G. Simons." *Scripta Math.* **16** (1950), 1–3. — *New York Times*, 26 November 1949, 15:5. — Portraits: *Scripta Math.* **7** (1940), facing page 7; *Scripta Math.* **16** (1950), 3. J.W.D.

Singh, Avadhesh Narayan (* 1901, Banaras (now: Varanasi), India; † July 10, 1954, Varanasi, India). SINGH obtained his M.Sc. degree from Banaras Hindu University in 1924 and was awarded the D.Sc. degree in 1929 by the University of Calcutta for his thesis entitled *Derivation and Non-Differentiable Functions.* He served in the Department of Mathematics at Lucknow University, where he became a reader in 1940 and professor in 1943. He was responsible for creating the Hindu Mathematics section there and revived the almost defunct Banaras Mathematical Society under the new name *Bharata Ganita Parisad.* SINGH published about a dozen papers related to the history of Indian mathematics, three dozen papers related to the non-differentiability of functions, and two books: *The Theory and Construction of Non-differentiable Functions* (1935), and *History of Hindu Mathematics. A Source Book* (with B. DATTA, in two parts: [DATTA/SINGH 1935/38]).

Secondary literature: *May* 338. *P* VI, 2457; VIIb(7), 4300. — SINVHAL, S. D.: "Dr. Avadhesh Singh." *Gaṇita* **5**, no. 2 (Dec. 1954), i–vii (with bibliography and portrait). R.C.G.

Smirnov, Vladimir Ivanovich (* June 10, 1887, St. Petersburg, Russian Empire; † February 11, 1974, Leningrad, USSR). VLADIMIR SMIRNOV was a well known mathematician who in 1915 began his career as a professor at Petrograd-Leningrad University. He was a member of the USSR Academy of Sciences. He was primarily interested in the history of mathematics in Russia, especially in the works of EULER and LYAPUNOV.

Main works: "Mathematics" (in collaboration with A. P. YOUSHKEVICH). In: *History of the Academy of Sciences of the USSR*, vol. **2**, Moscow 1964, 34–51, 286–306, 473–482 (in Russian). — (Ed.) *Leonhardi Euleri opera omnia.* Ser. IV A **1**. Basel 1975 (in collaboration with A. P. YOUSHKEVICH *et al.*).

Secondary literature: *May* 340. *P* VI, 2474; VIIb(7), 4953. — [KUROSH 1959]. — [FOMIN/SHILOV 1969]. — [BORODIN/BUGAI 1979]. — [BOGOLYUBOV 1983]. — LADYZHENSKAYA, O. A. (Ed.): *Vladimir Ivanovich Smirnov.* St. Petersbourg 1994 (in Russian, with portrait). S.S.D.

Smith, David Eugene (* January 21, 1860, Cortland, New York, USA; † July 29, 1944, New York City, New York, USA). SMITH was the second of four children. His father, ABRAM P. SMITH, was a lawyer and country judge; his mother, MARY ELIZABETH BRONSON, was the daughter of a physician. Thanks to his mother's self-instruction, SMITH is said to have learned Greek and Latin at home. After attending the State Normal School in Cortland, New York, SMITH studied art, classical languages, and Hebrew at Syracuse University. In 1881, having received his doctorate, he followed his father's wishes, and at age 21 entered law school, also at Syracuse, after which he was admitted to the bar in 1884. But SMITH soon found that he preferred academe to the law, and returned to Cortland where he taught mathematics at the Normal School there, his *alma mater*. In 1884 he received a Ph.M. from Syracuse, followed by another degree only a few years later, a Ph.D. in art history in 1887.

In 1891 SMITH accepted a position as professor of mathematics at the State Normal College in Ypsilanti, Michigan. It was not long before he had finished his first textbook, on *Plane and Solid Geometry* (co-authored with W. BEMAN, 1895), followed a year later by his *History of Modern Mathematics* (1896). In 1898 SMITH was named principal of the State Normal School at Brockport, New York, and three years later (1901), he moved to New York City to take up a new position as professor of mathematics at Teachers College, Columbia University, where he remained until his retirement in 1926. While at Teachers College, SMITH wrote (sometimes in collaboration with others) at least 150 texts for elementary and secondary schools.

SMITH's success was due in part to a combination of interests in teaching and his contacts throughout the world: "Smith's proficiency in languages, combined with his love of travel, early brought him in touch with mathematicians abroad. An appointment to the International Commission on the Teaching of Mathematics (ICTM), headed by Professor Felix Klein of Göttingen , placed him in a position of international influence" [EISELE/BOYD 1973, 721]. Indeed, the commission was founded upon SMITH's initiative, and he served as vice president of the ICTM from 1908 through 1920, and as its president from 1928–1932. He was also an early member of the New York Mathematical Society, established in 1893. When it evolved into the American Mathematical Society (AMS), he was not only appointed its librarian (1902–1920), but for most of this same period served as an associate editor of the Society's *Bulletin*. Later, in 1932, he was elected AMS vice-president.

When the Mathematical Association of America (MAA) was founded in 1916, largely to meet the needs of teachers (in contrast to the AMS, which emphasized research), SMITH was both a charter member and also served on the Association's board of trustees. In addition, he was made an associate editor of the *American Mathematical Monthly*, a position he held for 18 years. In addition to serving two terms as MAA president (1920–1921, and 1928–1932), he also served as vice president of the Association in 1932.

SMITH also played a major role in helping to establish the History of Science Society in 1924, and served as its first president in 1927. He was on the editorial board of *Scripta Mathematica*, was involved in several encyclopedia projects and other major reference works, and even produced a metrical version of the *Rubaiyat* of OMAR KHAYYAM (1933), for which he was decorated by the Shah of Iran, REZA KHAN PAHLEVI.

Among SMITH's best-known works are *Rara Arithmetica* (1908), *Number Stories of Long Ago* (1919), *History of Mathematics* (in two volumes, 1923–1925), *Hindu-Arabic Numerals* (co-authored with L. C. KARPINSKI, 1919), *A History of Japanese Mathematics* (co-authored with Y. MIKAMI, 1914), and a *History of Mathematics in America before 1900* (co-authored with J. GINSBURG, 1934).

Apart from his publications, SMITH's greatest legacy to the history of mathematics was his extraordinary collection of books. An avid collector, he traveled the world buying rare volumes, manuscripts, and mathematical and astronomical instruments. He worked closely with a fellow collector, the publisher GEORGE A. PLIMPTON (1855–1936), and carefully advised PLIMPTON on many purchases [EISELE/BOYD 1973, 722]. Over the course of his lifetime, SMITH amassed nearly 11,000 books, of which he took a particular interest in those related to mathematics in the Far East, India, and Islamic countries, as well as medieval European materials. His impressive collection was donated to Columbia University in 1931.

SMITH, who was married twice, died at his home at the age of 84, in New York City.

Secondary literature: *May* 340. *P* V, 1178; VI, 2475; VIIb(7), 4960. — *DAB, Supplement* **3**, 1941–45, (1973), 721–722. — SARTON, GEORGE: "Dedication to David Eugene Smith." *Osiris* **1** (1936), 5–8 (with portrait facing p. 5). — FRICK, BERTHA MARGARET: "Bibliography of the Historical Writings [of D. E. Smith]." *Osiris* **1** (1936), 13–78. — FRICK, BERTHA MARGARET: "The David Eugene Smith Mathematical Library of Columbia University." *Osiris* **1** (1936), 79–84. — FITE, W. BENJAMIN: "David Eugene Smith." *AMM* **52** (1945), 237–238. — REEVE, WILLIAM DAVID: "David Eugene Smith." *Scripta Math.* **11** (1945), 209–212 (with portrait). — SIMONS, LAO G.: "David Eugene Smith – *In Memoriam*." *Bull. AMS* **51** (1945), 40–50. — DONOGHUE, EILEEN F.: "In Search of Mathematical Treasures: David Eugene Smith and George Arthur Plimpton." *HM* **25** (1998), 359–365. — Portraits: *Scripta Math.* **11** (1945), facing p. 209 and between pp. 368–369. J.W.D.

Snellius, Willebrord (* 1580, Leiden, Netherlands; † October 30, 1626, Leiden, Netherlands). SNELLIUS was a professor at Leiden University (1603), where he taught mathematics, astronomy, and optics. He prepared a Latin translation of SIMON STEVIN's *Wisconstighe Ghedachtenissen* (Thoughts on Mathematics), and a Latin edition of LUDOLF VAN CEULEN's *Van den Circkel* (On the Circle). He also busied himself with the restoration of APOLLONIUS's *On Plane Loci.*

Secondary literature: *May* 341. *DSB* **12**, 499–502. *P* II, 947. P.B.

Sologub, Vladimir Stepanovich (* July 30, 1927, village Polova in Lviv region, Poland; † April 29, 1982, Lvov, Ukraine, USSR). VLADIMIR SOLOGUB was a research member of the Institut for History of the Ukrainian Academy of Sciences. Later, until his retirement, he worked in the Lvov Section of the Ukranian SSR Academy of Sciences. He was primarily interested in the history of partial differential equations in the 18th and 19th centuries.

Main work: *Development of the Theory of Elliptic Differential Equations in the XVIII–XIX Centuries.* Kiev 1975 (in Russian).

Secondary literature: [FOMIN/SHILOV 1969]. S.S.D.

Spiess, Ludwig Otto (* March 1, 1878, Basel, Switzerland; † February 14, 1966, Basel, Switzerland). OTTO SPIESS was the first son of KARL OTTO SPIESS, a Swiss engineer for railway and water-pipeline construction. His mother, MARIE LOUISE FAESCH, was a descendant from a very old and influential Basel dynasty. SPIESS studied mathematics at the University of Basel, a subject for which, from a very early age, he had already shown considerable enthusiasm. For example, he vigorously defended mathematics against the heavy attacks of his co-student CARL GUSTAV JUNG, later famous as a psychologist, who had denied its importance for education and culture. His most prominent professor at Basel was HERMANN KINKELIN, who at the time was the leading expert for insurance problems in Switzerland. He implanted in SPIESS a certain predilection for probability theory and statistics, which later qualified him to act for a long period as a member of the board of the renowned Swiss insurance company *Patria.*

Having obtained his Ph.D. from the University of Basel with a dissertation on "Die Grundbegriffe der Iterationsrechung" (The Basic Concepts of Computation by Iteration), SPIESS continued his studies at the University of Berlin, where from 1902 to 1904 he attended the lectures of H. A. SCHWARZ, F. SCHOTTKY, G. FROBENIUS, and E. LANDAU. Upon his return to Basel he qualified himself as a lecturer at the university. In 1907 SPIESS was made an extraordinary professor, and in 1938 an ordinary professor of mathematics at the University of Basel. Until 1915 he also taught mathematics at the *Gymnasium* in Basel.

SPIESS published only a few of his mathematical results, most of which were related to the theory of functions and certain curves. After 1924 he spent more and more time on historical research work and his lectures on history of mathematics.

The starting point for SPIESS's activities as an historian of mathematics was his book on *Leonhard Euler* (1929). While preparing this book he was also increasingly attracted to study the life as well as the works of the BERNOULLI family. SPIESS discovered the largest and most remarkable parts of the archives of JOHANN I, JOHANN II, and JOHANN III BERNOULLI, which had been hidden and were largely forgotten for more than a century in the library of the Royal Swedish Academy in Stockholm, and in the Ducal Library at Gotha. To his own great astonishment, SPIESS succeeded in his first attempt in the winter of 1935 to buy the Gotha manuscripts. At the same time — also on his demand — the Stockholm manuscripts were given to the Library of the University of Basel to be used in anticipation of the planned edition SPIESS had in mind. And in 1964, two years before his death, SPIESS had the satisfaction of knowing that Basel was also able to buy the second half of the BERNOULLI correspondence as well.

From the beginning SPIESS had decided that his BERNOULLI edition had to start not with the previously published works but with the correspondence. He was convinced that it was a main task of any edition to present texts which had survived as unpublished manuscripts in order to preserve their content and to record the material for posterity in a printed version. Thus he concentrated his activities for more than twenty years on transcribing the manuscripts of the BERNOULLI correspondence. This amounts to nearly 4000 letters which were transcribed before, during, and after the Second World War. The typewritten transcriptions then were compared (often more than twice) with the originals, and an index of names for all of the transcribed letters was also prepared.

Envisaging the prospects of editorial ruin, SPIESS rejected the idea of a comprehensive edition which would present all of the BERNOULLI material in chronological order. On the contrary, he favored an edition in different series which grouped the material together according to its different authors and scientific content. For more than half a century this entire editorial project so clearly bore the stamp of SPIESS that he could abstain from printing his name on the title page of the first volume of the BERNOULLI edition. *Der Briefwechsel von Johann Bernoulli. Band 1* (The Correspondence of Johann Bernoulli. Volume 1) finally appeared in 1955, published by Birkhäuser in Basel. It contains mainly the letters exchanged with the MARQUIS DE L'HÔPITAL. By publishing a carefully screened text of all letters both sent and received, by documenting the lost letters and trying to reconstruct their content, by providing local and global commentaries in separate introductions and footnotes, and by explaining his editorial principles, SPIESS provided the reader with a reliable and useful tool for understanding the nature and development of the works of JOHANN BERNOULLI and his correspondents in their contemporary setting. SPIESS's first volume of the *Bernoulli Edition* was soon recognized by the scientific community as an exemplary model not only for the succeeding volumes of the BERNOULLI series, but also for any modern and pragmatic edition of the works and correspondence of any scientific figure.

The editorial work of SPIESS was accompanied by a series of books and papers on the BERNOULLI family and its circle, for example: "Johann Bernoulli" and "Daniel Bernoulli" in *Grosse Schweizer Forscher* (Great Swiss Scientists, E. FUETER, ed., 1939); "Die Mathematikerfamilie Bernoulli" (The Bernoulli Mathematical Family), in *Grosse Schweizer* (1938); "Über einige neuaufgefundene Schriften der alten Basler Mathematiker" (On Some Newly Found Writings of Basel Mathematicians of Former Times) [*Vh. Nat. Ges. Basel* **56**, no. 1 (1945)]; "Die Summe der reziproken Quadratzahlen" (The Sum of Reciprocal Squares) in *Festschrift für A. Speiser* (1945); "Die Mathematiker Bernoulli (Zum 200. Todestag von Johann I Bernoulli)" (The Bernoulli Mathematicians. On the 200th Anniversary of the Death of Johann I Bernoulli)), in *Basler Universitätsreden* (University of Basel Lectures) (1948); "Bernoulli, Basler Gelehrtenfamilie" (Bernoulli, Basel Family of Scholars), in *Neue Deutsche Biographie* (1955). There are also two other publications by SPIESS worth mentioning: *Basel anno 1760 nach den Tagebüchern der ungarischen Grafen Joseph und Samuel Teleki* (Basel in the Year 1760 According to the Diaries of the Hungarian Counts Joseph and Samuel Teleki) (1936); "Eine akademische Festrede von Daniel Bernoulli über das Leben" (A Formal Academic Lecture by Daniel Bernoulli on Life) [*Vh. Nat. Ges. Basel* **52** (1941)], and "Une édition de l'œuvre des mathématiciens Bernoulli" [*Arch. int. Hist. Sci.* **1** (1948), 356–362]. Before SPIESS died as a bachelor on February 14, 1966, he bequeathed his private fortune to an *Otto Spiess Foundation* in order to support the preparation of future volumes of the *Bernoulli Edition.*

The history of science in Switzerland owes three things to OTTO SPIESS. Firstly, SPIESS made the national and international scientific community aware of the importance of the immortal work of the BERNOULLI mathematicians and their Basel school of mathematics. Secondly, thanks to SPIESS, most of the BERNOULLI manuscripts (especially the letters received and the drafts of letters sent) are today collected together in the library of the University of Basel, where along with the old stock of manuscripts and books from the private library of the Bernoulli family, they form an ideal base for the study of the history of science in the 17th and 18th centuries. And last but not least, SPIESS initiated, prepared, established, and financially supported the editorial research project, *Die gesammelten Werke der Mathematiker und Physiker der Familie Bernoulli*, which has survived him. Having successfully published 14 volumes until now, the Bernoulli Edition should continue with the same purpose and objectives that SPIESS had originated from the beginning.

Secondary literature: *P* V, 1193; VI, 2510. — STRAUB, HANS: "Prof. Dr. Otto Spiess (1878–1966)." *Vh. Nat. Ges. Basel* **77** (1966), 172–180 (with portrait and bibliography). F.N.

Stäckel, Paul (* August 20, 1862, Berlin, Germany; † December 12, 1919, Heidelberg, Germany). STÄCKEL began his peripatetic teaching career at the University of Halle in 1891, and subsequently became extraordinary professor at Königsberg (1895–1897) and then Kiel (1897–1899). He advanced to a position as ordinary professor first at Kiel (1899–1905), then at the *Technische Hochschule* (Technical College) in Hannover (1905–1908), and the Technical College in Karlsruhe (1908–1913), until finally becoming an ordinary professor at the University of Heidelberg (1913–1919). Among his areas of interest were non-Euclidean geometry and the works of GAUSS and EULER.

Historical works: Edition of the correspondence of GAUSS – WOLFGANG BOLYAI (1899). — Edition of *Gauss, Werke*, vol. 8 (1900). — *C. F. Gauss als Geometer* (1917). — *Der Briefwechsel zwischen C. G. J. Jacobi und P. H. von Fuss über die Herausgabe der Werke Leonhard Eulers* (1908, with W. AHRENS). — *Die Theorie der Parallellinien von Euklid bis auf Gauss* (1895, with F. ENGEL). — Edition of WOLFGANG and JOHANN BOLYAI's *Geometrische Untersuchungen* (1913). — German translations of treatises by JACOB BERNOULLI, JOHANN BERNOULLI,

CAUCHY, EULER, LAGRANGE, LEGENDRE, and JACOBI in the series *Ostwalds Klassiker* (see [*Ostwalds Klassiker* 1889]).

Secondary literature: *DSB* **12**, 599. *May* 344. *P* IV, 1427; V, 1194. — [*Acta Math.*, Tab. Gen. **1–35** (1913), 120; portrait 158]. — PERRON, OSKAR: "Paul Stäckel †." *Sb. Akad. Heid., Math.-nat. Kl.*, Abt. A, Jahrgang 1920, 7. Abhandlung (20 pp.). — LOREY, WILHELM: "Paul Stäckel zum Gedächtnis." *ZMNU* **52** (1921), 85–88 (with portrait). — RUDIO, FERDINAND: "Paul Stäckels Verdienste um die Gesamtausgabe der Werke von Leonhard Euler." *Jber. DMV* **32** (1923), 13–32 (with portrait). — VON RENTELN, MICHAEL: "Paul Stäckel (1862–1919) – Mathematiker und Mathematikhistoriker." *Überblicke Mathematik* 1996/97, 151–160 (with portrait). — Portrait: [BÖLLING 1994, 29.1]. M.F.

Stamatis, Evangelos (* September 13, 1898, Thebes, Greece; † March 1, 1990, Athens, Greece). STAMATIS graduated from the newly established National Academy of Gymnastics in 1917, and began his brief career as a gymnast at the Varvakion High School (a distinguished school until 1981) in Athens. Although he had considerable interest in athletics, his interest in the sciences was even greater, so side by side with his professional obligations, he began to study physics at the University of Athens in 1919. In 1923 he graduated from the faculty of physics and mathematics, having already fulfilled his military duty by participating in campaigns in Asia Minor during the war of 1921–22. After 1923, STAMATIS started his long career as a professor in secondary schools. He taught physics, chemistry, geography, and mathematics until his retirement. Beginning in 1959, he also taught professors of secondary schools.

With a fellowship awarded in 1931, STAMATIS went on to study at the University of Berlin, to which he returned from 1936 until 1940. There he attended the lectures of many distinguished professors, including WALTER NERNST, MAX VON LAUE, ERWIN SCHRÖDINGER, and PETER DEBYE. In this period research on the history of mathematics flourished in Germany, and this greatly influenced STAMATIS, who along with his studies in physics began to study the history of mathematics as well. When STAMATIS returned to Athens, he continued his career as a professor but at the same time he entered upon a systematic study analyzing the ancient Greek mathematical heritage. His activities as an historian of ancient Greek mathematics are diverse. He made numerous translations of classic Greek texts, wrote many original papers and books, and published a wide variety of articles in encyclopedias and various journals, most of which were popularizations. STAMATIS also published EUCLID's *Elements* (in 4 vols.), and ARCHIMEDES's *Opera Omnia* (in 3 vols.), the latter of which is still considered to be the most complete edition. From the Arabic version STAMATIS restored the non-extant Greek text, in Dorian dialect, of APOLLONIUS's *Conica* (4 vols.), DIOPHANTUS's *Arithmetica*, as well as ARISTARCHUS's treatise *On the Sizes and Distances of the Sun and Moon.* All these publications are comprised of the ancient text, the translation in modern Greek, along with constructive commentaries and extended ancient and modern bibliographies. STAMATIS's long collaboration with the publishing house of

B. G. TEUBNER also deserves mention, since it produced his editions of EUCLID's *Elements* (1969, 1970, 1972, 1973, 1977) and ARCHIMEDES's *Opera Omnia* (vols. I–II, 1972). In recognition of their collaboration, TEUBNER honored STAMATIS with a special award.

STAMATIS wrote more than ten books and numerous papers. He was a member of the International Academy of the History of Science (1966), and he was also a permanent collaborator of the well-known reviewing journal *Zentralblatt für Mathematik und ihre Grenzgebiete*. STAMATIS devoted his entire life to the ancient Greek contributions to mathematics. He also had to confront many difficulties arising from the fact that he was the only person who worked at this time on the history of mathematics in Greece, and was completely isolated from the university community which basically ignored him, and was sometimes even hostile. This explains why STAMATIS was never made a professor at the University of Athens. Because he was unable to reinvigorate the chair MICHAEL STEPHANIDES had held there, he consequently was unable to reach the next generation of historians of mathematics, and thus they were never exposed to his vast erudition. As a result, STAMATIS was never able to establish a school for the history of mathematics in Greece, or to train any to follow his example. Nevertheless, his name remains an important one for historians of mathematics, and his wide-ranging works constitute his legacy for those scholars who are seriously interested in the history of ancient Greek mathematics.

Secondary literature: CHRISTIANIDIS, JEAN P.: "Evangelos Stamatis (1898–1990)." *Euclide* (3) **7** (1990), fasc. 25, 9–18. — CHRISTIANIDIS, JEAN P., and NIKOS KASTANIS: "*In Memoriam* Evangelos S. Stamatis (1898–1990)." *HM* **19** (1992), 99–105 (with bibliography and portrait). — OREOPOULOS, GEORGE: "The Life and the Work of Evangelos Stamatis." *First National Conference of History and Philosophy of Mathematics: Ancient Greek Mathematics, 1989. Proceedings.* Athens 1993, 15–26. C.P.

Steenstra, Pibo († July 21, 1788, Amsterdam, Netherlands). When, in 1759 STEENSTRA became lecturer in mathematics at the University of Leiden, he gave an inaugural lecture on the rise and development of geometry ([STEENSTRA 1759]). In 1763 he went on to the *Athenaeum Illustre* in Amsterdam, where he taught mathematics, astronomy, and the art of navigation. His schoolbook, *Grondbeginselen der Meetkunst* (Elements of Geometry) (1763), contained a description of the development of geometry since antiquity.

Secondary literature: *P* II, 988. — *VdAA* **17**, 306. P.B.

Steinschneider, Moritz (* March 30, 1816, Prossnitz, Moravia; † January 24, 1907, Berlin, Germany). STEINSCHNEIDER studied at the universities of Prague, Vienna, Leipzig, and Berlin, and received his Ph.D. in Semitic studies in Leipzig (1851). He was a teacher in Prague and Berlin, and acted as director of the Jewish *Töchterschule* (Girls School) in Berlin (1869–1890). He also held a position as an assistant at the oriental department of the *Königliche Bibliothek Berlin* (Berlin

Royal Library). His areas of interest were Jewish and Arabic mathematics and bibliography.

Main works: *Die arabischen Übersetzungen aus dem Griechischen* (1889–1896, repr. 1960). — *Die hebræischen Übersetzungen des Mittelalters und die Juden als Dolmetscher* (1893, repr. 1956). — [STEINSCHNEIDER 1893/1901]. — *Die europäischen Übersetzungen aus dem Arabischen bis Mitte des 17. Jahrhunderts* (1904–1905, repr. 1956).

Secondary literature: *P* III, 1289; IV, 1436; V, 1204. — GÜNTHER, SIEGMUND: "Moritz Steinschneider." *MGMN* **6** (1907), 391–399. M.F.

Stockler, Francisco de Borja Garção (* September 25, 1759, Lisbon, Portugal; † March 6, 1829, Tavira, Portugal). STOCKLER graduated in mathematics from the University of Coimbra after which he followed a military career, attaining the post of general. He was also elected a member of the Academy of Sciences in 1787, and became its secretary. Most of his mathematical works are concerned with analysis and were published either in the Academy's *Memoirs* or by the Academy's press. Among his historical papers (published in the first volume of his *Works*, edited in 1805) are (in Portuguese) *Eulogy of D'Alembert* [STOCKLER 1805a], *Eulogy of José Joaquim Soares de Barros e Vasconcellos* (1721–1793, a Portuguese astronomer and member of the Berlin Academy of Sciences) [STOCKLER 1805b], and a *Memoir on the Originality of Portuguese Naval Discoveries During the XVth Century* [STOCKLER 1805c]. In 1819 STOCKLER published in Paris his *Historical Essay* [STOCKLER 1805d], the earliest-known history of mathematics devoted to a single country in Europe.

Secondary literature: *P* I, 843. — SARAIVA, LUIS MANUEL RIBEIRO: "Garção Stockler and the First History of Mathematics in Portugal." *Arch. int. Hist. Sci.* **42** (1992), 76–86. — SARAIVA, LUIS MANUEL RIBEIRO: "On the First History of Portuguese Mathematics." *HM* **20** (1993), 415–427. — Portraits: There are two portraits of Stockler in the Lisbon Arquivo Histórico-Militar: Pasta 47A, Retratos 17069 and 17070. L.S.

Størmer, Frederik Carl Mülertz (* September 3, 1874, Skien, Norway; † August 13, 1957, Drøbak, Norway). CARL STØRMER graduated in mathematics from Christiania University in 1898 and then was supported on a stipend until 1903, when he became professor of mathematics. The stipend made it possible for him to collaborate with ELLING HOLST and SYLOW in preparing the *Festskrift ved hundredårsjubilet for Abels fødsel* (Festschrift for the One-Hundredth Jubilee of Abel's Birth) (1902), which contained ABEL's correspondence. A large and widely-acknowledged part of STØRMER's research concerned northern lights, but he also had an interest in number theory and published a paper in Norwegian on the life and work of RAMANUJAN in 1934.

Secondary literature: *DSB* **13**, 82–83. *May* 348. *P* IV, 1444; V, 121; VI, 2552; VIIb(8), 5151. — [*Acta Math.*, Tab. Gen. **1–35** (1913), 113; portrait 158]. K.A.

Strabbe, Arnoldus Bastiaan (* June 20, 1741, Zwolle, Netherlands; † March 26, 1805, Amsterdam, Netherlands). STRABBE settled in Amsterdam around 1760, where at first he was active as a bookkeeper, but later he also became the municipal "gauger" responsible for supervising the correctness of weights and measures used in trade and local marketplaces. In his spare time he taught arithmetic, bookkeeping, and commercial arithmetic, and translated the first edition of MONTUCLA's *Histoire* into the Dutch language ([MONTUCLA 1782/1804]). STRABBE was the first secretary of the *Genootschap der Mathematische Weetenschappen* (Society for Mathematical Sciences), founded at Amsterdam in 1778.

Secondary literature: *NNBW* **9**, 1385. — VAN HAAFTEN, MARIUS: *Het Wiskundig Genootschap. Zijn oudste geschiedenis, zijn werkzaamheden en zijn beteekenis voor het verzekeringswezen.* Groningen 1923. — WANSINK, JOH. H.: "Strabbe." *Euclides* **55** (1979/1980), 341–348. P.B.

Struik, Dirk Jan (* September 30, 1894, Rotterdam, Netherlands; † October 21, 2000, Belmont, Massachusetts, USA). DIRK STRUIK, an American mathematician and historian of mathematics, began teaching at the Massachusetts Institute of Technology in 1927. STRUIK published numerous works on history of mathematics, most notably *A Concise History of Mathematics*, which gives particular emphasis to social aspects of mathematics from a generally Marxist perspective. This work, first published in 1948, has been translated into many Western and non-Western languages. STRUIK was also the editor of *A Source Book in Mathematics, 1200–1800* (1969).

Secondary literature: *P* VI, 2569; VIIb(8), 5185. — ROWE, DAVID E.: "An Interview with Dirk Jan Struik." *Math. Intell.* **11**, no. 1, (1989), 14–26 (with portraits; based on a longer version published in *NTM* **25**, no. 2 (1988), 5–23, with portraits). — *Special Issue in Honor of Dirk J. Struik*: *HM* **21**, no. 3 (1994) (with several photographs and bibliography, pp. 268–272). Portraits: A variety of photographs of Struik may be found in COHEN, R. S., and M. W. WARTOFSKY (Eds.): *For Dirk Struik. Scientific, Historical and Political Essays in Honor of Dirk J. Struik* (= *Boston Studies in the Philosophy of Science* **15** (1974)) (with bibliography). J.W.D.

Struve, Vasiliĭ Vasil'evich (* February 3, 1889, St. Petersburg, Russian Empire; † September 15, 1965, Leningrad, USSR). Beginning in 1916, VASILIĬ STRUVE was a professor at Petrograd-Leningrad University, and was elected a member of the USSR Academy of Sciences in 1935. As a well-known Egyptologist, his major interests in the history of mathematics were related to ancient Egypt.

Main work: "Mathematischer Papyrus der Staatlichen Museums der Schönen Künste in Moskau." In: *Q. St. G. Math. Astron. Phys.*, Abt. A., Bd. **1** (1930).

Secondary literature: DIAKONOV, I. M., and A. P. YOUSCHKEVITCH: "Vasily Vasilievich Struve." *Arch. int. Hist. Sci.* **19** (1969), 141–142. — [BOGOLYUBOV 1983]. S.S.D.

Suter, Heinrich (* January 4, 1848, Hedingen (Canton Zürich), Switzerland; † March 17, 1922, Dornach (Canton Solothurn), Switzerland). SUTER was an historian of the exact sciences. He obtained his doctorate from the University of Zürich (1871), which also awarded him an honorary doctorate in 1921. He was elected a corresponding member of the *Physikalisch-medizinische Gesellschaft zu Erlangen* (Erlangen Physico-Medical Society).

SUTER had learned Latin and Greek and later was trained in mathematics in Zürich — both at the *Eidgenössische Technische Hochschule* and at the University of Zürich. Later he studied at the University of Berlin (1869–1870). His first printed work was a two-volume *Geschichte der mathematischen Wissenschaften* (History of the Mathematical Sciences) (1872, 1875). The first volume, SUTER's doctoral thesis, covered the period from antiquity up to the end of the 16th century, and was reprinted in 1973. Volume I was widely recognized, and was translated into Hungarian (1874) and Russian (1876). The second volume (1875) extended the history to the end of the 18th century. The recognition of SUTER's *Geschichte* was, however, shortlived, since the work fell into oblivion after publication of MORITZ CANTOR's extensive *Vorlesungen über Geschichte der Mathematik* (Lectures on History of Mathematics), the first volume of which appeared in 1880.

After having taught in various high schools in the German part of Switzerland, SUTER was appointed as a professor at the *Gymnasium* in Zürich (1886), a position he held until his retirement in 1918. It was only at the time of his nomination as a professor that SUTER, already known and some forty years old, entered the field of Islamic science, of which he was soon to become one of the leading figures. Noting the *lacunae* in our knowledge of Moslem mathematicians, after studying Arabic he then went into an examination of the main bibliographical sources and of various catalogues of Arabic manuscripts. The result was *Die Mathematiker und Astronomen der Araber und ihre Werke* (The Mathematicians and Astronomers of the Arabs and Their Works) (1900). Even today this remains a basic reference work (it has since been reprinted twice, in 1972 and 1986). SUTER's thorough knowledge of the collections of Arabic manuscripts opened a royal road to research for SUTER. The approximately fifty studies he published from 1892 until his death deal almost exclusively with Islamic science (or Greek science in Arabic translations). Choosing his subjects with much intuition and treating them with an exceptional mathematical and philological competence, SUTER has left studies which are neither forgotten nor outdated, but remain essential. Indeed, there is no researcher or student of the exact sciences in the medieval Islamic period who has not come across studies made by SUTER. The recent reprinting of his works is a further testimony of the permanence of his many contributions.

Main works: SUTER's studies published from 1892 to 1922 are reprinted in two volumes, issued by the *Institut für Geschichte der arabisch-islamischen Wissenschaften* in Frankfurt, unter the title *Beiträge zur Geschichte der Mathematik und Astronomie im Islam* (1986); there is a short autobiography in vol. II, 537.

Secondary literature: *DSB* **13**, 155. *May* 350. *P* III, 1315; IV, 1466; V, 1231; VI, 2587. — *HBLS* **6**, 619. — On the life and works of SUTER, there is an obituary by CARL SCHOY in the *Vj. Nat. Ges. Zür.* **67** (1922), 407–413 (reprinted from the *Neue Zürcher Zeitung* of April 8, 1922), and another by JULIUS RUSKA: "Heinrich Suter (Geb. 4. I. 1848, gest. 17. III. 1922)." *Isis* **5** (1923), 409–417 (with portrait). Of these two, the first gives a list of SUTER's publications according to journals, the second according to years of publication. J.S.

Sylow, Peter Ludvig Mejdell (* December 12, 1832, Christiania, Norway; † September 7, 1918, Christiania, Norway). LUDVIG SYLOW studied the sciences at Christiania University. When he graduated in 1856, he was considered a promising mathematician, but there was no vacancy for him at the university. His main occupation in the period 1858–1898 was teaching mathematics at the high school in Frederikshald (now Halden). He became extraordinary professor at the university in 1898. CARL ANTON BJERKNES and OLE JACOB BROCH inspired SYLOW to study ABEL, and SYLOW's mathematical career consisted to a large degree in further developing ABEL's ideas. Moreover, on LIE's initiative and in collaboration with him, SYLOW was engaged to prepare a second edition of ABEL's works. He also collaborated with ELLING HOLST and CARL STØRMER in editing ABEL's letters.

Secondary literature: *DSB* **13**, 215–216. *May* 351. *P* III, 1316; IV, 1468; V, 1235; VI, 2594. — [*Acta Math.*, Tab. Gen. **1–35** (1913), 115; portrait 174]. — Portraits [BÖLLING 1994, 7.1] and [STUBHAUG 2000b, 83]. K.A.

Tacquet, Andreas (* June 23, 1612, Antwerp, Belgium; † December 22, 1660, Antwerp, Belgium). TACQUET was a Jesuit who taught mathematics at the Jesuit Colleges of Leuven (1649–1655) and Antwerp (1645–1649, 1655–1660). He included an historical report, beginning with ADAM and his son SETH, and almost exclusively limited to the Greeks and their predecessors, on the origin and progress of mathematics in his influential textbook for elementary geometry ([TACQUET 1654]).

Secondary literature: *DSB* **13**, 235–236. *May* 352. *P* II, 1064. — *BCJ* **7**, 1806–1811. — *NBW* **11**, 755–759. — BOSMANS, HENRI: "Le jésuite mathématicien anversois André Tacquet (1612–1660)." *De Gulden Passer – Le Compas d'or* **3** (1925), 63–87. P.B.

Tannery, Paul (* December 20, 1843, Mantes-la-Jolie, Yvelines, France; † November 27, 1904, Pantin, Seine-Saint-Denis, France). PAUL TANNERY embodied a type of *polytechnicien* who came to the history of science by combining a solid grounding in the classics with the scientific training characteristic of that prestigious *Ecole*. The breadth of his knowledge, the vast expanse of his research, and the penetration and sureness of his historical judgment made him one of the most influential personalities in a discipline then still in its infancy.

Owing to the fact that his father was a senior railway engineer, TANNERY belonged to a social sphere very much linked to the technological and scientific progress of the second half of the 19th century. He did his secondary studies at the *lycée* in Le Mans and Caen, studying philosophy under JULES LACHELIER. At the age of seventeen, TANNERY entered the *Ecole Polytechnique* in the highest ranks. Not limiting himself to the usual curriculum, he began to study Hebrew and thought about the teaching of mathematics. On leaving the *Polytechnique*, he entered the *Ecole d'application des manufactures de l'état* undoubtedly in response to his family's wishes; they were aware of the advantages presented by this kind of career (in fact, in the service of the state tobacco-monopoly). This type of work, which he carried out very conscientiously, did not satisfy TANNERY's intellectual interests. Throughout his life, he devoted his free time and late nights to these higher pursuits. His career trail took him to a half-dozen major cities — among them Bordeaux and Paris — first as an engineer and then, after 1886, as director and finally as head of the Manufactury of Pantin on the doorsteps of the capital. He occupied the latter post from 1893 until his death. At the time of the siege of Paris, he served as captain in the artillery. For a dozen years, he devoted himself to an intense study of mathematics and philosophy. He read AUGUSTE COMTE; he discussed his evolving thoughts with his brother, the mathematician JULES TANNERY, and with ÉMILE BOUTROUX, the translator of EDUARD ZELLER's *Die Philosophie der Griechen* (The Philosophy of the Greeks) (1844–1852); he perfected his command of ancient languages; and he participated in the activities of the *Société des sciences physiques et naturelles* of Bordeaux, where he published his first works in 1876. In 1881, he married MARIE PRISSET, daughter of a notary from near Poitiers. She supported him in his research and expended great energy in seeing his posthumous publications into print. She worked on the latter until her own death in 1945.

TANNERY met a number of the scholars with whom he corresponded during the course of his travels abroad. On several occasions, he had the opportunity to teach: a private course on the history of mathematics at the Faculty of Science in Paris in 1884–1885 and as a substitute in the chair of Greek and Roman philosophy at the *Collège de France* from 1892 to 1897. In 1903, when the chair in the history of science at the *Collège de France* became vacant, TANNERY was the natural choice of the faculty as well as of the *Académie des Sciences*, but the minister charged with making the actual nomination opted for the second choice, whose name would remain in obscurity. This affair sparked vigorous protests in France and abroad. TANNERY succumbed rapidly to pancreatic cancer the following year.

In the introduction to *La géométrie grecque* (Greek Geometry) (1887), TANNERY recalled the state of the historiography of mathematics during the course of the 19th century, the opening decades of which had remained indebted to the work of MONTUCLA [MONTUCLA 1758]. TANNERY declared that the works of MICHEL CHASLES [CHASLES 1837], [CHASLES 1870], marked a new era. The same can be said for the work of NESSELMANN [NESSELMANN 1842], who argued for a critical history based on the philological study of sources; he also attacked problems of periodization. TANNERY parted with the ambitious syntheses of ARNETH [ARNETH 1852] and of HANKEL [HANKEL 1874]. He mentioned BRETSCHNEIDER [BRETSCHNEIDER 1870], and praised ALLMAN [ALLMAN 1877/85] as well as the edition of PAPPUS by HULTSCH [HULTSCH 1876/78]. He held that the part of MORITZ CANTOR's *Vorlesungen* ([CANTOR 1880/1908]) that had thus far been published (1880–1898) marked the transition to a new current of ideas, according to which one may strictly speak of "science" only with respect to the Greeks. Finally, TANNERY mentioned the new edition of EUCLID that HEIBERG had begun in 1883 ([HEIBERG 1883/86]).

In his work, TANNERY drew inspiration from the methodology of philology, which had come primarily from Germany and which had taken hold in France, supplanting the era of grand historical syntheses that had followed in the wake of the Hegelianism of bygone days. TANNERY separated himself completely from positivism, however, by placing the birth of science in classical Greek antiquity. In his view, the problem of this sudden appearance of science was the first point to be dealt with in a history of mathematics, and so the evaluation of the Greek heritage remained a strong motivation for TANNERY. In order to take up this challenge, he proceeded to produce specialized studies that he published in over a dozen journals, and which concered a range of topics from mathematics to philosophy, from philology to Greek studies, as well as history.

All of these works were collected in the seventeen volumes of his *Mémoires scientifiques*, the first three volumes of which were devoted to the exact sciences in antiquity. Apart from the variety of the questions treated, each of Tannery's contributions proved noteworthy: the quadrature of lunes of HIPPOCRATES OF CHIOS and the measurement of the circle by ARCHIMEDES, the duplication of the cube by ARCHYTAS and EUDOXUS, the language and mathematical passages of PLATO, the geometrical solution of problems of the second degree, questions raised by the corpus of HERON, the history of curved lines and surfaces, Pythagorean number theory, the authenticity of EUCLID's axioms, fractions, the structure of Greek musical scales and the role of music in the development of mathematics, and the origin of algebra. In addition to these, volume seventeen contains work treating the concept of continuity from ZENO OF ELEA to GEORG CANTOR, while volume seven (on ancient philosophy) and volume nine (on philology) include several additional mathematical studies. There were also certain questions to which TANNERY returned, completing, revising, and bringing up to date earlier investigations. TANNERY thus provided a model for modern research in the history of science.

His books do not fail to follow the same method. In *La géométrie grecque* (1887), papers published during eleven years of scientific research were revised and the book itself was conceived only as a starting point towards a broader one, which, however, remained a project. In fact, *La géométrie grecque* is a "critical essay," a book of historical method and philosophy, discussing the key texts that transmit information on geometry before EUCLID: that is to say PROCLUS, HERO, IAMBLICHUS, and SIMPLICIUS, a masterly guide-book, on the highest level of scholarship, forming the basis for later development of research in the field.

Also in 1887, *Pour l'histoire de la science hellène, de Thalès à Empédocle* (On the History of Greek Science from Thales to Empedocles) appeared. Although this work did not concern the history of mathematics *per se*, it nevertheless marked the first time that the pre-Socratic thinkers were treated not from the point of view of the history of philosophy but as scientists. At the same time, TANNERY provided the first French translation of the pre-Socratic fragments, since DIELS's edition had not yet appeared and he had to make do with MULLACH's older edition.[6]

In *Recherches sur l'histoire de l'astronomie ancienne* (Research on the History of Ancient Astronomy) (1893), an area that was still indebted to the work of DELAMBRE [DELAMBRE 1817], TANNERY had the important insight, as for example in the case of EUCLID, to investigate the precursors of the theories treated by CLAUDIUS PTOLEMY. The debate hinged on those parts of the *Almagest* that went back to HIPPARCHUS. TANNERY went even further and highlighted the contributions of earlier astronomers, notably APOLLONIUS OF PERGA.

During the same period, the two-volume edition of the then-known works of DIOPHANTUS was published by TEUBNER. This included Greek commentaries and a complete Latin version.[7] This marked the achievement of many years of profound study of the *Arithmetica*, the ground for which had originally been prepared by NESSELMANN.

TANNERY's philological studies, reaching up to both the Byzantine and the Latin medieval periods, uncovered previously unknown medieval scientific texts, the understanding of which had been distorted by the Romantics. Beginning in 1884, TANNERY became interested in the continuity of ancient thought and did pioneering work on this (*cf.* volumes four and five of the *Mémoires*). Of particular interest are the editions and translations of NICOLAS RHABDAS and MANUEL MOSCHOPOULOS, and TANNERY's study on the history of geomancy, entitled *Rabolion*. The latter remained incomplete owing to a lack of documentation on Arabic science, but TANNERY foresaw its importance. He also studied and compiled editions of PSELLOS, THEODORUS PRODROMUS, and JOHN PHILOPONUS; discovered and published the correspondence, exchanged around 1025, between two schoolmen, one from Cologne and one from Liège; untangled the complex question of the geometrical manuscripts attributed to BOETHIUS and to GERBERT; and brought to light a geometrical manuscript of FRANCO OF LIÈGE (ca. 1050). TANNERY showed that it was not before the 12th century, the century of the Latin translations of EUCLID, that medieval scholars actually had access to geometry. He extended his research in this vein to texts of the 15th and 16th centuries.

TANNERY's works on modern science, particularly those on the 17th century, appear in volume six of his *Mémoires*, while those on epistemology, notably works on set theory and on the concepts of the transfinite and continuity, are collected in volume eight. TANNERY's main body of work on modern science, however, is contained in his great editions: the *Œuvres* of FERMAT (in collaboration with CHARLES HENRY), which includes the "Observations on Diophantus" as well as several texts translated into French; and the *Œuvres* of DESCARTES, to which he

[6] MULLACH, F. W. A.: *Fragmenta philosophorum graecorum.* 3 vols. Paris 1860/1867/1881. "*Bibliothèque Grecque.*" FIRMIN DIDOT, (Ed.).

[7] The French version, which TANNERY also prepared, served as the basis of the edition published by VER EECKE in 1926.

was led by his study, *La correspondance de Descartes dans les inédits du Fonds Libri*, itself instigated by the FERMAT edition.

Undertaken with CHARLES ADAM, the DESCARTES edition ultimately replaced the earlier edition by VICTOR COUSIN (1824–1826); TANNERY worked on volumes one through seven and on volume eleven, and also provided notes in the other volumes. This publication renewed Cartesian studies and particularly the understanding of DESCARTES the scientist. It stands out as defining the modern methodology of critical editions of scientific texts.

The work of TANNERY was not limited, however, to what has been said so far. The *Mémoires* also contain reflections on the history of science and on the relationships between general and topical history (volume ten); notes and analyses of books that reveal the breadth of their author's interests (volumes eleven and twelve; side by side with works on the history of science are those on DU BOIS-REYMOND, FREGE, PEIRCE, HUSSERL, and RUSSELL); and an extensive correspondence (volumes thirteen to sixteen) that is a reflection of TANNERY's place in the international scientific community. Among the names of TANNERY's correspondents are some forty of the foremost philosophers, mathematicians, philologists, and historians of science of the day. Finally, the *Mémoires* contain, thanks to the tireless efforts of MARIE TANNERY, unedited posthumous publications as well as works that were in preparation at the time of TANNERY's death, such as the *Quadrivium* of GEORGIUS PACHYMERES and the *Correspondance* of MARIN MERSENNE.

PAUL TANNERY's correspondence is a valuable source of information on the development of research in the history of science from 1875 to 1904. At the same time, this epistolary activity placed him in a position to play a decisive role in the organization of the discipline at the international level. He was instrumental in organizing the first international meetings for history of science (Paris 1900, Rome 1903, and Geneva 1904). Although these meetings were sections of International Congresses of Philosophy (1900, 1904) or History (1903), they proved very important. After his death, it was not until the founding of the International Academy of the History of Science in 1928 that a new congress, dedicated to his memory, was held in Paris (1929). This marked the beginning of a still ongoing series.

TANNERY also engaged in teaching. In 1892, responding to an official request, he conceived and drafted an interdisciplinary course of studies on the history of science that united philosophers and historians as well as scientists.

Through a body of work the breadth and variety of which never cease to amaze, TANNERY set the course for the history of science in the 20th century by animating, through his example, most of its essential principles: systematic recourse to a philological study of the sources; exact reconstruction of the scientific techniques of the past, notably those of mathematics; methodical production of a number of monographs and specialized studies — he left behind some 250 articles or notes — before any attempt at synthesis; the production of critical editions of fundamental texts; and the organization of international scientific exchanges. The works of TANNERY, through their rigor and precision, often met with the enthusiastic agreement of scholars, a fact reflected in their immediate translation into other languages. In his eyes, though, his work was nothing more than a step toward understanding the complex evolution of science. Even if recent scholarship no longer subscribes to some of his conclusions, his preliminary spadework, his elimination of errors and false trails, and his definition of fruitful research directions were all done with such mastery that his work continues, in most cases, to mark a necessary starting point and an indispensable methodological basis for further research.

Secondary literature: *DSB* **13**, 251–256. *May* 353. *P* III, 1323; IV, 1476. — [MARIE TANNERY]: "Paul Tannery." *Osiris* **4** (1938), 633–689. — BOUTROUX, PIERRE: "L'Œuvre de Paul Tannery." *Osiris* **4** (1938), 690–705 (with bibliography and portrait as frontispiece). — [SARTON 1947]. — [SARTON 1951]. — BERR, HENRY; SUSANNE DELORME, RENÉ TATON, GEORGE SARTON, JEAN ITARD, PAUL-HENRI MICHEL, PIERRE SERGESCU, and ROBERT LENOBLE: "Paul Tannery et l'Histoire générale des Sciences" [several articles]. *Rev. Hist. Sci.* **7** (1954), 297–368 (with portrait of PAUL TANNERY on p. 315 and of MARIE TANNERY on p. 317). M.C.

Teixeira, Francisco Gomes (* January 28, 1851, S. Cosmado, near Viseu, Northern Portugal; † February 8, 1933, Oporto, Portugal). FRANCISCO GOMES TEIXEIRA completed his course of study in the Faculty of Mathematics at the University of Coimbra in 1874, and was given the degree of doctor the following year, with a thesis on integration of second order partial differential equations. In 1876 he was accepted as a member of the faculty, after presenting a paper on the use of non-orthogonal systems of axes in analytic mechanics. Having observed that papers by Portuguese mathematicians were often ignored (and sometimes the results were independently rediscovered abroad) because there was no adequate means to make them known, GOMES TEIXEIRA founded the *Jornal de Sciencias Mathematicas e Astronomicas* (Journal of Mathematical Sciences and Astronomy) in 1877, which turned out to be the most important Portuguese mathematics journal of the 19th century.

GOMES TEIXEIRA represented a new attitude in Portuguese mathematics: he had a large correspondence with many of the most celebrated mathematicians of his time, frequently participated in international meetings, and published over 140 papers in some of the most prestigious jounals of his time. Up to the 1890s he published mainly on analysis, but from that year onwards geometry was his main interest. In 1899 the Royal Academy of Exact, Physical, and Natural Sciences in Madrid awarded him a prize for his *Treatise on Special Curves.* In a revised and considerably augmented version, this text was translated into French as *Traité des Courbes Speciales Remarquables, Planes et Gauches* (1908–1909).

TEIXEIRA also did some research on history of mathematics. His final and most important book, *History of Mathematics in Portugal* (1934), which appeared the year following his death, is still considered to be the best history of Portuguese mathematics. Besides this, GOMES TEIXEIRA wrote other important works on history of mathematics, including a volume *Panegyrics and Con-*

ferences (Coimbra 1925). This volume contains essays on the four mathematicians whom GOMES TEIXEIRA considered the cream of Portuguese mathematics: PEDRO NUNES, ANASTÁCIO DA CUNHA, MONTEIRO DA ROCHA, and DANIEL DA SILVA.

Other works: "Sur les écrits d'histoire des mathématiques publiés en Portugal." *Bibl. Math* (2) **4** (1890), 91–92. — "On an Argument between MONTEIRO DA ROCHA and ANASTÁCIO DA CUNHA." *Annaes Scientificos da Academia Polytechnica do Porto* **1**, (1905–1906), 7–15.

Secondary literature: *May* 355. *P* III, 1328; IV, 1483; V, 1245; VI, 2627; VIIb(8), 5366. — [*Acta Math.*, Tab. Gen. **1–35** (1913), 117; portrait 174]. — HENRIQUE DE VILHENA: *O Professor Doctor Francisco Gomes Teixeira.* Lisbon 1936 (with bibliography and portraits). L.S.

Tennulius, Samuel [ten Nuyl] (* June 23, 1635, Deventer, Netherlands; † Deventer, Netherlands). TENNULIUS was a professor at Steinfurt (1661), and beginning in 1666, he was a lecturer on history, eloquence, and mathematics at the short-lived (1655–1679) University of Nijmegen. He published a Latin translation (with commentary) of NICOMACHUS OF GERASE's *Introduction to Arithmetic.*

Secondary literature: *VdAA* **18**, 20. P.B.

Terquem, Olry (* June 16, 1782, Metz, France; † May 6, 1862, Paris, France). Educated in a Hebrew school and later a student at the *Ecole Polytechnique*, TERQUEM taught mathematics at the *Lycée de Mayence* until 1814. He then became the librarian of the *Dépôt d'artillerie* in Paris. As editor (with CAMILLE CHRISTOPHE GERONO) of the *Nouvelles annales de mathématiques*, in 1855 he launched the first periodical to include history of mathematics, namely the *Bulletin de bibliographie, d'histoire et de biographie mathématiques* (1855–1862), which appeared as a supplement to the *Nouvelles annales.*

Secondary literature: *P* II, 1081: III, 1330. — PROUHET, EUGÈNE: "Notice sur la vie et les travaux d'Olry Terquem." *Bull. Math.* **8** (1862), 81–90. J.P.

Thaer, Clemens Adolf (* December 8, 1883, Berlin, Germany; † January 2, 1974, Detmold, Germany). CLEMENS THAER was made a *Privatdozent* in 1909, and then extraordinary professor in 1921 at the University of Greifswald. Opposing the National Socialist government, he had to give up his university position and from 1935 to 1939 taught mathematics in Cammin (Pomerania). His area of special research interest was Greek mathematics.

Works: German translations of EUCLID's *Elements* (1933–1937) and *Data* (1962). — "Die Euklid-Überlieferung durch Aṭ-Ṭûsî." *Q. St. G. Math. Astron. Phys.* **B 3** (1936), 116–121. — "Antike Mathematik: Bericht über das Schrifttum der Jahre 1906–1930." *Jber. Fortschr. klass. Altertumswiss.* **283** (1943).

Secondary literature: *P* VI, 2637; VIIa(4), 641. — SCHREIBER, PETER: "Clemens Thaer (1883–1974) — ein Mathematikhistoriker im Widerstand gegen den Nationalsozialismus." *Sudh. Archiv* **80** (1996), 78–85 (with portrait). M.F.

Thureau-Dangin, François (* January 3, 1872, Paris, France; † January 24, 1944, Paris, France). THUREAU-DANGIN, a French assyrologist and member of the *Institut de France*, was chief conservator at the *Louvre* in Paris (1895–1928). A leading specialist for cuneiform texts, especially for royal Sumerian inscriptions, he also drew attention to the rich Babylonian sources for the history of mathematics by publishing *Textes Mathématiques Babyloniens* (Babylonian Mathematical Texts) (1938).

Secondary literature: *May* 358. — DORME, E. (Ed.): *Hommage à la mémoire de l'éminent assyriologue François Thureau-Dangin (1872–1944)*. Leiden 1946 (with bibliography and portrait). C.J.S.

Timchenko, Ivan Yurievitch (* February 10 (22), 1863, Odessa, Russian Empire; † August 30, 1939, Odessa, USSR). IVAN TIMCHENKO graduated in 1885 from the Faculty of Physics and Mathematics of the University of Odessa (University of New Russia), where three years later, in 1888, he began to lecture on algebra, analytic geometry, and the theory of analytic functions. Being interested in the theory of analytic functions, TIMCHENKO decided to write a comprehensive work embracing the entire history of the subject, including its connections with different branches of mathematical analysis. He began this work in 1892 with an article in the 12th volume of the *Transactions of the Mathematical Section of the New Russian Society of Naturalists*, followed by subsequent contributions to the 16th and 19th volumes. This part of the work was entitled *Foundation of the Theory of Analytic Functions. Part I. Historical Survey of the Development of the Notions and Methods of the Theory of Analytic Functions. Volume I*, and covered the period up to 1825 — i.e. to CAUCHY's paper "Mémoire sur les intégrales définies prises entre des limites imaginaires" (Memoir on Definite Integrals between Imaginary Limits). TIMCHENKO also published this material as a separate book in 1899, and simultaneously presented it as his master's thesis, which he defended that same year at the University of Odessa. The sequel to this volume (i.e. Volume II, intended to be a volume devoted to pure mathematics) never appeared.

Nevertheless, because of the profundity of its mathematical analysis of primary sources and the wide scope of its historical information, TIMCHENKO's book (Volume I) proved exceptionally popular and is widely utilized by historians of mathematics even today. Throughout the principal text with its numerous notes there is a considerable amount of little-known information, some of it unknown to modern specialists who have rediscovered it again through TIMCHENKO's book. Much of this information was transcribed by TIMCHENKO from archival sources, for example an extract from J. M. C. CONDORCET's manuscript *Integral Calculus*, or citations from the unpublished 8th volume of D'ALEMBERT's *Opuscules*.

TIMCHENKO's history of analytic functions soon became well known in Russia and abroad (cf. for example STÄCKEL's remarks in *Bibl. Math.* (3) **1** (1900), 128) and it is owing to this book that TIMCHENKO soon became one of the best-known European historians of mathematics. Together with other established specialists, he contributed notes for the second edition of M. CANTOR's *Vorlesungen über*

Geschichte der Mathematik (Lectures on History of Mathematics). TIMCHENKO also provided excellent additions for the Russian translation of F. CAJORI's *History of Elementary Mathematics* (Russian trans. 1917).

After defending his thesis, TIMCHENKO started to lecture at the University of Odessa where he offered courses on the history of mathematics, and in 1914 he was promoted to the rank of professor. TIMCHENKO also directed the activities of the mathematics department of the New Russian Society of Naturalists. Following the October Revolution of 1917, he served as rector of the Odessa Polytechnic Institute until 1921. In 1930 he was made head of the Odessa branch of the Kharkov Mathematical Institute, and in 1933 he was dean of the Faculty of Physics and Mathematics at the University of Odessa. TIMCHENKO's scientific productivity, however, decreased sharply in the Soviet period. TIMCHENKO only wrote one article about ARCHIMEDES, which appeared in the first edition of the *Great Soviet Encyclopedia*, and two articles published in the Odessa editions — one on the theory of analytic functions and the other on the history of logarithms.

Secondary literature: KIRO, SERGEI N.: "I. Yu. Timchenko." *Vopr. Ist. Estest. Tekhn.* **17** (1964), 123–126 (in Russian). — Portrait in [SHTOKALO 1966/70, vol. 4, book 2, p. 457]. S.S.D.

Todhunter, Isaac (* November 23, 1820, Rye (Sussex), England; † March 1, 1884, Cambridge, England). TODHUNTER was educated and taught at Cambridge University. His books include histories of probability, calculus of variations, elasticity, and potential theory. In addition to his textbooks, he also wrote a biography of WHEWELL. TODHUNTER was proposed for membership in the Royal Society by JAMES JOSEPH SYLVESTER (1814–1897) and seconded by ARTHUR CAYLEY (1821–1895), and was supported by many leading British mathematicians.

Secondary literature: *DSB* **13**, 426–428. *May* 360. *P* III, 1354. — ROUTH, E. R.: "Isaac Todhunter." *Proceedings of the Royal Society* **37** (1884), 27–32. — R. T., in *Proc. London Math. Soc.* **15** (1884), 284–287 (with bibliography). — [MACFARLANE 1916, 134–146]. I.G.G.

Toeplitz, Otto (* August 1, 1881, Breslau, Germany; † February 15, 1940, Jerusalem, Palestine). TOEPLITZ began teaching as a *Privatdozent* in Göttingen in 1907. He was extraordinary professor at Göttingen (1912–1913) and Kiel (1913–1920), where he was promoted to ordinary professor (1920–1928), after which he was ordinary professor at Bonn (1928–1935). As a Jewish mathematician who lost his university position under the National Socialists in 1935, TOEPLITZ emigrated to Palestine in 1938. His areas of interest included Greek mathematics and the genetic method of teaching mathematics.

Main works: "Mathematik und Antike." *Die Antike* **1** (1925), 175–203. — "Das Verhältnis von Mathematik und Ideenlehre bei Plato." *Q. St. G. Math. Astron. Phys.* **B 1** (1931), 3–34. — *Die Entwicklung der Infinitesimalrechnung. Eine Einleitung in die Infinitesimalrechnung nach der genetischen Methode* (1949).

— Editor (with J. STENZEL and O. NEUGEBAUER) of *Quellen und Studien zur Geschichte der Mathematik, Astronomie und Physik* (1929–1938).

Secondary literature: *DSB* **13**, 428. *May* 360. *P* V, 1261; VI, 2672; VIIa(4), 695. — [PINL 1969, 200–201]. — BEHNKE, HEINRICH, und GOTTFRIED KÖTHE: "Otto Toeplitz zum Gedächtnis." *Jber. DMV* **66** (1963), 1–16. — BEHNKE, HEINRICH: "Otto Toeplitz 1881–1940." In: W. KRULL (Ed.): *Bonner Gelehrte. Beiträge zur Geschichte der Wissenschaften in Bonn. Mathematik und Naturwissenschaften.* Bonn 1970, 49–53. — KLINGENBERG, W.: *Otto Toeplitz 1881–1940.* Bonn 1981. — GOHBERG, I. (Ed.): *Toeplitz Centennial/Toeplitz Memorial Conference in Operator Theory, Dedicated to the 100th Anniversary of the Birth of Otto Toeplitz, Tel Aviv, May 11–15, 1981.* Basel 1982 (with portrait). — SCHUBRING, GERT: "Die Entwicklung des Mathematischen Seminars der Universität Bonn 1864–1929." *Jber. DMV* **87** (1985), 139–163. M.F.

Treutlein, Josef Peter (* January 26, 1845, Wieblingen near Heidelberg, Germany; † July 26, 1912, Karlsruhe, Germany). PETER TREUTLEIN taught mathematics in Karlsruhe from 1866 on. He was especially interested in medieval and early modern mathematics.

Main works: *Geschichte unserer Zahlzeichen und Entwicklung der Ansichten über dieselben* (1875). — Editions of abacus treatises (in *Bull. Boncompagni* **10** (1877), 589–647), and of JORDANUS NEMORARIUS's *De numeris datis* (*Abh. Gesch. Math.* **2** (1879), 1–124. — "Das Rechnen im 16. Jahrhundert." *Abh. Gesch. Math.* **1** (1877), 1–100. — "Die deutsche Coss." *Abh. Gesch. Math.* **2** (1879), 1–124.

Secondary literature: *May* 363. *P* III, 1365; IV, 1523; V, 1269. — BEHM, HANS WOLFGANG: "Peter Treutlein." *ZMNU* **43** (1912), 521–530 (with portrait). — CRAMER, H.: "Joseph Peter Treutlein." *Unterrichtsbl. Math. Nat.* **18** (1912), 121–123. — SCHÖNBECK, JÜRGEN: "Peter Treutlein (1845–1912) und die Entwicklung der geometrischen Propädeutik." In: PETER TREUTLEIN: *Der geometrische Anschauungsunterricht.* Paderborn 1985, pp. E5–E15. — SCHÖNBECK, JÜRGEN: "Der Mathematikdidaktiker Peter Treutlein." In: [SCHÖNBECK 1994, 50–72, with two portraits]. M.F.

Tropfke, Johannes (* October 14, 1866, Berlin, Germany; † November 10, 1939, Berlin, Germany). From 1889 on TROPFKE taught mathematics in Berlin. From 1912 to 1932 he was director of the *Kirschner Oberrealschule* in Berlin. His main field of interest was the history of elementary mathematics.

Main Works: "Erstmaliges Auftreten der einzelnen Bestandtheile unserer Schulmathematik." *Programm Friedrich Realgymnasium Berlin* 1899, 27 pp. — *Geschichte der Elementar-Mathematik in systematischer Darstellung,* 2 vols. (1902–1903); 2nd ed.: 7 vols. (1921–1924); 3rd. ed.: vols. 1–4 (1930–1940).

Secondary literature: *DSB* **13**, 469–470. *May* 363. *P* IV, 1525; VI, 2693; VIIa(5), 726. — LOREY, WILHELM: "Johannes Tropfke 70 Jahre." *Unterrichtsbl. Math. Nat.* **42** (1936), 300. — BRUNS, WALTER J.: "Johannes Tropfke." *Scripta Math.* **7** (1940), 261–262. — HOFMANN, JOSEPH EHRENFRIED: "Johannes Tropfke

(14. X. 1866 bis 10. XI. 1939)." *Dt. Math.* **6** (1941), 114–118 (with bibliography and portrait). M.F.

Truesdell, Clifford Ambrose III (* February 18, 1919, Los Angeles, California, USA; † January 14, 2000, Baltimore, Maryland, USA). CLIFFORD TRUESDELL graduated from the California Institute of Technology, received a certificate in mechanics from Brown University in 1942, and obtained his Ph.D. in mathematics from Princeton in 1943. From 1944 to 1946 he was a staff member of the Radiation Laboratory of the Massachusetts Institute of Technology, from 1946 to 1950 he headed theoretical mechanics laboratories of the US Navy. In 1950 he became professor of mechanics at Indiana University, and in 1961 professor of rational mechanics at Johns Hopkins University in Baltimore.

For his research in mechanics, TRUESDELL received numerous awards and honors. A many-sided scholar with unusually broad interests and linguistic competence, he was also an authority on the literature, arts, crafts, and music of the 16th through 18th centuries.

Apart from numerous books and articles on topics from mechanics, he also published several books and papers on the history of science and edited four volumes of the *Opera Omnia* of LEONHARD EULER. TRUESDELL founded and edited four journals, among them, since 1960, the *Archive for History of Exact Sciences.* In his inaugural editorial, TRUESDELL stated: "The *Archive for History of Exact Sciences* nourishes historical research meeting the standards of the mathematical sciences." The journal soon became one of the leading periodicals for historians of mathematics.

Main works: *The Rational Mechanics of Flexible or Elastic Bodies, 1638–1788* (= *L. Euleri Opera Omnia*, ser. II, vol. 11, part 2). Zürich 1960. — *Essays in the History of Mechanics.* New York 1968. — "The Scholar: A Species Threatened by Professions." *Critical Enquiry*, Summer 1976, 631–648; German trans. in *Hum. Techn.* **17** (1973), 113–127. — *The Tragicomical History of Thermodynamics, 1822–1854.* Berlin and New York 1980.

Secondary literature: ANONYMOUS: "Obituary for Clifford A. Truesdell III." *Caltech News. Newsletter of the Caltech Alumni Association* **34**, no. 2 (2000), 22. — ROBINSON, SARA: "Clifford Truesdell, 80. Master of Two Disciplines of Mechanics. Obituary." *New York Times* (January 22, 2000), A13. — VAKULENKO, A. A., and G. K. MIKHAILOV: "Klifford Trusdell i sovremennaya istoriya mekhaniki." *Vopr. Ist. Estest. Techn.* 2000, no. 3, 59–66. — SPEISER, DAVID: "*In Memoriam:* Clifford A. Truesdell." *Arch. int. Hist. Sci.* **50** (2000), 178–180. J.W.D./C.J.S.

Turnbull, Herbert Westren (* August 31, 1885, Tettenhall, Wolverhampton, England; † May 4, 1961, Grasmere, Westmoreland, England). TURNBULL taught at Hong Kong University from 1911 to 1915 and, after some years of school teaching, was appointed Regius Professor of Mathematics at the University of St. Andrews (Scotland) in 1921. In 1932 he became a fellow of the Royal Society of London. By editing the *James Gregory Tercentenary Memorial Volume* in 1939, TURNBULL called attention to the achievements of GREGORY as an independent discoverer of important methods of the infinitesimal calculus, in particular of infinite series and the Taylor expansion of functions. TURNBULL also edited vol. 1 of *The Correspondence of Isaac Newton* (1959).

Secondary literature: *May* 364. *P* V, 1276; VI, 2703; VIIb(8), 5571. — LEDERMANN, W.: "Herbert Westren Turnbull." *J. Lond. Math. Soc.* **38** (1963), 123–128 (with bibliography). — RUTHERFORD, D. E.: "H. W. Turnbull, F.R.S." *Proc. Edinburgh Math. Soc.* **13** (1963), 273–276. C.J.S.

Vacca, Giovanni Enrico (* November 18, 1872, Genoa, Italy; † January 6, 1953, Rome, Italy). GIOVANNI VACCA studied in the faculty of mathematics at the University of Genoa, where he obtained his degree (*laurea*) in 1897. While still a student, he not only published his first paper in 1894 on the history of mathematics in ENESTRÖM's *Bibliotheca mathematica*, but he was also involved in politics and took part in the foundation and early activities of the Italian socialist party. In the reactionary period that followed, he was exiled from Genoa for one year. Consequently, VACCA moved to Turin where he served as an assistant to GIUSEPPE PEANO from 1897 to 1902. Under PEANO's influence, VACCA devoted himself in particular to the history of logic, and became one of the more active members of PEANO's "school." VACCA contributed a number of bibliographical, biographical, and historical notes and remarks to each of the various (five) editions of PEANO's *Formulario matematico* (Mathematical Formulary). PEANO himself recognized that VACCA "powerfully aided [him] in the whole work" [Preface to the *Formulario matematico*, (quoted in [KENNEDY 1980, 83])].

VACCA spent the summer of 1899 in Hannover studying LEIBNIZ's unpublished manuscripts on logic. His note on this subject prompted COUTURAT, whom VACCA met in Paris during the International Congress of Philosophy in August of 1900, to edit a volume of *Opuscules et fragments inedits de Leibniz* (Works and Unpublished Fragments of Leibniz) (1903).

After several years in Genoa, during which time VACCA was a socialist member of the town council, he resumed his position as PEANO's assistant, from the autumn of 1904 until the spring of 1905, when he moved to Florence and devoted himself to study of the Chinese language. VACCA became increasingly attracted to Chinese culture, and decided to spend two years in China from 1907 to 1909, traveling throughout the country and studying its history and its science. Upon returning to Italy, VACCA taught the Chinese language as well as the history and geography of East Asia at the University of Rome, beginning in 1910 as a *libero docente* (university lecturer), and from 1921 on as a full professor. Nonetheless, VACCA continued to be interested in the history of mathematics, and published papers on a variety of subjects. In 1916 he also edited the first book of EUCLID's *Elements* (along with the Greek text, notes, and historical commentaries). Beginning in 1924 he was responsible for teaching the history of mathematics at the Institute for the History of Science founded by ENRIQUES at the University of Rome

VACCA's various scientific and cultural interests are reflected in his 130 publications. Indeed, these include 38 papers on mathematics, 45 essays on the history of mathematics and science, and 47 publications on the culture of East Asia, China in particular.

Secondary literature: *May* 365. *P* V, 1284; VI, 2727; VIIb(8), 5629. — CARRUCCIO, ETTORE: "Giovanni Vacca, matematico, storico e filosofo della scienza." *Boll. UMI* (3) **8** (1953), 448–456. — CASSINA, UGO: "Giovanni Vacca. La vita e le opere." *Rend. Ist. Lomb. Sci. Lett. Parte Generale* (3) **86** (= 3rd series, vol. 17) (1953), 185–200 (with bibliography and portrait). — See also: OSIMO, G.: "Lettere di Giuseppe Peano a Giovanni Vacca." *Quaderni P.RI.ST.EM* **3**, 1992. — NASTASI, PIETRO, and ALDO SCIMONE (Eds.): "Lettere a Giovanni Vacca." *Quaderni P.RI.ST.EM* **5**, 1995. U.B.

Vailati, Giovanni (* April 24, 1863, Crema, Italy; † May 14, 1909, Rome, Italy). Born of a noble family, GIOVANNI VAILATI studied in Barnabite colleges at Monza and Lodi. In 1880 he enrolled at the University of Turin, where he obtained a degree (*laurea*) in engineering (1884) and another in mathematics (1886). Thereafter he devoted himself to the independent study of modern languages. In 1891 VAILATI published his first paper in PEANO's *Rivista di matematica* and a year later PEANO offered him a position as an assistant teaching calculus at the University of Turin. VAILATI kept this post until 1895, when he was invited to teach projective geometry at Turin and was also appointed as honorary assistant to VOLTERRA, who was then professor of rational mechanics there. This also gave VAILATI the opportunity to teach courses in the history of mechanics from 1896 until 1899. It was the publication of his introductory lectures to these courses, along with papers on the history of mechanics in antiquity, that made him well-known internationally.

In 1899 VAILATI decided to give up his university career in order to teach in secondary schools. At first he taught in Syracuse, then in Bari, Como, and eventually in Florence beginning in 1904. There he was asked by the *Accademia dei Lincei* to edit the official edition of TORRICELLI's works, but VAILATI was unable to complete this project, and the edition was later entrusted to LORIA and VASSURA. At the time, VAILATI was also appointed by the Minister of Public Education to serve on a commission to study the reform of secondary schools.

In the first decade of the 20th century, VAILATI's primary interest turned to philosophy, and he soon gained international recognition as the leading Italian exponent of Pragmatism. During this same period he kept in touch with members of PEANO's "school" (with GIOVANNI VACCA in particular), and, although to a lesser extent, he continued to do research in the history of mathematics. Among his most significant publications were a paper on SACCHERI's *Logica demostrativa* (1903), in which he called attention to a forgotten book by SACCHERI, and a paper on the theory of proportions which was (posthumously) included in ENRIQUES's *Questioni riguardanti le matematiche elementari* (Questions Concerning Elementary Mathematics) (Part I, Tome I, 1912, 143–192).

Secondary literature: *May* 365. *P* IV, 1542; V, 1285. — PEANO, GIUSEPPE: "In memoria di Giovanni Vailati." *Gazzetta del Popolo* (Turin, May 17, 1909), repr. in *Boll. Mat.* **8** (1909), 206–207. — ANONYMOUS [ALESSANDRO PADOA]: "Necrologia di Giovanni Vailati." *Period. Mat.* **24** (1909), 289–292. — See also: LANARO, GIORGIO (Ed.): *G. Vailati, Epistolario.* Torino 1971. U.B.

van der Waerden, Bartel Leendert (* February 2, 1903, Amsterdam, Netherlands; † January 12, 1996, Zürich, Switzerland). VAN DER WAERDEN was one of the most influential mathematicians of the twentieth century. For decades his textbook *Moderne Algebra* (Modern Algebra) (2 vols., 1930/31) was considered a standard work that introduced many generations of students to modern algebra as it had arisen from the works of R. DEDEKIND, H. WEBER, D. HILBERT, E. STEINITZ, E. ARTIN, and EMMY NOETHER. Moreover, VAN DER WAERDEN was an encyclopedic scholar, something that is becoming increasingly rare. He made essential contributions to practically all parts of mathematics ranging from algebraic geometry and abstract algebra to number theory, topology, axiomatic geometry, analysis, and probability theory, including applied mathematics (spinor analysis and quantum mechanics). Equally impressive is his work on the history of science, to which he contributed seven books and nearly 200 articles over a period of fifty years.

VAN DER WAERDEN was the son of THEODORUS VAN DER WAERDEN and his wife DOROTHEA née ENDT. His father, who had trained as an engineer, earned his living as a teacher of mathematics at the *Hogere Burger School* in Amsterdam. But he was also active in politics, and was a member of the left wing of the Social Democrats, a party he represented with a seat in the second chamber of the Dutch Parliament.

From 1919 onwards VAN DER WAERDEN studied mathematics, physics, chemistry, and philosophy at the University of Amsterdam. There he came into contact with L. E. J. BROUWER and R. WEITZENBÖCK, but was influenced more by GERRIT MANNOURY and HENDRIK DE VRIES. It was under the direction of DE VRIES that VAN DER WAERDEN received his doctorate on March 24, 1926, with a thesis on algebraic geometry. In 1924, partly supported by a Rockefeller scholarship, he went for one year to the University of Göttingen where, as he later stated, a new world opened up to him. After completing his military service, he spent another year at the University of Hamburg after which he returned to Göttingen at the beginning of 1927; there he took the *Habilitation* for mathematics on February 26, 1927. In Göttingen VAN DER WAERDEN established especially close contacts with EMMY NOETHER and HELLMUTH KNESER. In Hamburg, EMIL ARTIN, ERICH HECKE, and OTTO SCHREIER were important influences. It was in the summer term of 1926 that VAN DER WAERDEN attended ARTIN's lectures on algebra, and these formed the basis for his best-known book, *Moderne Algebra*, through which he soon became world-famous in the field.

In 1928, when he was only 25 years old, VAN DER WAERDEN was made full professor at the University of Groningen in the Netherlands. Back in Göttingen for the summer of 1929, he met CAMILLA RELLICH, sister of the mathematician FRANZ RELLICH; they were married in September. Two years later VAN DER WAERDEN was appointed to a professorship at the University of Leipzig, where he remained until shortly before the end of the War in 1945. In Leipzig VAN DER WAERDEN made close ties to the physicists WERNER HEISENBERG and FRIEDRICH HUND. This led to his well-known book, *Die gruppentheoretische Methode in der Quantenmechanik* (The Group-Theoretical Method in Quantum Mechanics) (1932), and stimulated his life-long interest in physics. At the same time he began work on his series of articles, "Zur algebraischen Geometrie" (On Algebraic Geometry) (1933–1971), while his friendship with the Leipzig philosopher HANS-GEORG GADAMER awakened his interest in Greek mathematics and in PLATO.

After World War II VAN DER WAERDEN worked briefly for the Royal Dutch Oil Company Shell until he left Europe in 1947 to spend a year as a visiting professor of mathematics at Johns Hopkins University in Baltimore, Maryland (USA). In 1948 he returned to Amsterdam, and three years later he moved again, to the University of Zürich where he succeeded RUDOLF FUETER and where, from 1951 until 1970, he served as director of the Mathematical Institute. During his Swiss period VAN DER WAERDEN directed nearly forty Ph.D. theses. He was a member of the editorial boards of several well-known mathematical series and journals (among them the *Archive for History of Exact Sciences*, 1960–1993). In all, he published more than 20 books and about 300 articles (see the bibliographies [Gross 1972], [Eisenreich 1981], [VAN DER WAERDEN 1983, 469–479], and [Top/Walling 1994]; none of these, it should be noted, is entirely complete).

VAN DER WAERDEN's interest in the history of mathematics was apparently stimulated first by a lecture series given by HENDRIK DE VRIES in Amsterdam. In Göttingen he attended a lecture course by OTTO NEUGEBAUER on Greek mathematics, and it was also through NEUGEBAUER, both in Göttingen and later during a visit to Copenhagen, that VAN DER WAERDEN was introduced to Egyptian and Babylonian mathematics. In NEUGEBAUER's series *Quellen und Studien zur Geschichte der Mathematik, Astronomie und Physik*, VAN DER WAERDEN published his first extensive historical research on the development of Egyptian computations with fractions. Contacts with J. STENZEL, O. TOEPLITZ, O. BECKER, and E. BESSEL-HAGEN — all of whom also contributed to this series — led to VAN DER WAERDEN's publication of "Zenon und die Grundlagenkrise der griechischen Mathematik" (Zeno and the Foundational Crisis of Greek Mathematics) (*Math. Annalen* **117** (1940), 141–161). Within the next few years several more papers followed that marked VAN DER WAERDEN's future areas of research in the history of science: Babylonian and Pythagorean astronomy as well as Pythagorean arithmetic, algebra, and theory of harmony. These studies, which he extended through further investigations over subsequent years, were later collected and presented to a wider public in two very successful volumes, also translated into various other languages, entitled *Erwachende Wissenschaft I* and *II* (Science Awakening) (1950/56 and 1965/68, respectively), as well as in his book *Die Pythagoreer* (The Pythagoreans) (1979). Meanwhile VAN DER WAERDEN's interest in the history of science had broadened once more. He began to study the transmission of Babylonian planetary computations to Egypt and India (1956, 1960), and astronomy in ancient Egypt, India, and Persia generally. Moreover, beginning in the 1960s, he also published a number of papers related to the history of modern algebra. All of this historical research was later summarized in another set of books, including his *Geometry and Algebra in Ancient Civilizations* (1983), *A History of Algebra* (1985), and *Die Astronomie der Griechen* (The Astronomy of the Greeks) (1988).

VAN DER WAERDEN also contributed to the history of modern physics with his volume on *Sources of Quantum Mechanics* (1968). While in Zürich, he supervised seven dissertations on the history of the exact sciences. He was one of the founders of the *Wissenschaftshistorisches Kolloquium* (Colloquium on History of Science) in Zürich, which continues to the present day. Unfortunately, his attempts to establish an *Abteilung Geschichte der Wissenschaft* (History of Science Section) as part of the Mathematical Institute of the University of Zürich had no lasting success, although he did present a large part of his own history of science library to the Institute to support such an *Abteilung*. (VAN DER WAERDEN's books are now part of the library of the Mathematical Institute.)

A basic characteristic of VAN DER WAERDEN's exceptional scientific achievements was his rare gift for being able to analyze difficult technical material quickly and thoroughly, and to organize and describe it in simple and clear language. Often his first drafts were nearly perfect in form and virtually ready for printing. He already had shown such ability at the beginning of his career, when his *Moderne Algebra* introduced the difficult lectures of EMMY NOETHER, and the new ideas of EMIL ARTIN, to a wide mathematical public. VAN DER WAERDEN proceeded in a similar manner in his reconstruction of the four mathematical sciences of the Pythagoreans. Starting out from only a few propositions explicitly attributed to them, mainly by authors from late antiquity, and a detailed mathematical analysis, he was finally able to reconstruct a complete deductive system which in his view the Pythagoreans must have known in its entirety, since all its propositions in the handed-down version of EUCLID's *Elements* "are linked to each other by logical connections and literal quotations." With respect to such reconstructions, however, it must be noted that EUCLID's *Elements* has been preserved primarily in medieval manuscripts, and that these themselves represent the end of a long chain beginning with the now lost pre-Euclidean elements of geometry, until one eventually arrives back at the Pythagorean treatises which VAN DER WAERDEN reconstructed out of these. It is therefore not surprising that many classical scholars viewed his reconstructions with considerable scepticism, because they hold several of these literal quotations and explicit attributions to be later interpolations in the handed-down texts and are therefore not convinced that they are of truely Pythagorean origin. On the other hand it should be realized that philologists, when interpreting the mathematical passages in the *Epinomis* (a late work of PLATO the authenticity of which is somewhat doubtful), "have vainly broken their teeth on this hard nut — it was only in modern times that mathematicians [O. TOEPLITZ and O. BECKER] have been able to throw some light on this matter," as VAN DER

WAERDEN emphasized with obvious satisfaction [*Science Awakening* **1**, New York 1963, 155].

VAN DER WAERDEN received numerous honors for his scientific work: memberships in a great number of scientific academies as well as medals and decorations, including the German *Orden pour le mérite für Wissenschaften und Künste* (Order of Merit for Sciences and Arts) (1973). The *Académie internationale d'histoire des sciences* elected him a corresponding member in 1963, and an effective member in 1967.

VAN DER WAERDEN's scientific estate is preserved in the library of the *Eidgenössische Technische Hochschule (ETH)* in Zürich (cf. [Jakob 1985]). His early papers, however, were almost completely destroyed during an air raid on Leipzig of December 4, 1943.

Secondary literature: a) Bibliographies: *P* VI, 2787; VIIa (4,II), 807. — GROSS, HERBERT: "Herr Professor B. L. van der Waerden feierte seinen siebzigsten Geburtstag." *Elem. Math.* **28** (1973), 25–32 (with bibliography to 1971). — EISENREICH, GÜNTHER: "B. L. van der Waerdens Wirken von 1931 bis 1945 in Leipzig," in: BECKERT, HERBERT, and HORST SCHUMANN (Eds.): *100 Jahre Mathematisches Seminar der Karl-Marx-Universität Leipzig.* Berlin 1981, 218–244 (with bibliography to 1946). — Bibliography to the end of 1972 in [VAN DER WAERDEN 1983, 469–479]. — TOP, JAAP, and LYNNE WALLING: "Bibliography of B. L. van der Waerden." *Nieuw Arch. Wiskd.* (4) **12** (1994), 179–193. — b) Further articles: JAKOB, BARBARA: *Die Briefsammlung B. L. van der Waerden.* Diplomarbeit der Vereinigung Schweizerischer Bibliothekare. Wissenschaftshistorische Sammlungen der ETH-Bibliothek. Zürich 1985. — KNORR, WILBUR R.: "The Geometer and the Archaeoastronomers: On the Prehistoric Origins of Mathematics." *Brit. J. Hist. Sci.* **18** (1985), 197–212. — FREI, GÜNTHER: "Bartel Leendert van der Waerden. Zum 90. Geburtstag." *HM* **20** (1993), 5–11. Revised English trans.: *Nieuw Arch. Wiskd.* (4) **12** (1994), 137–144 (including several additional articles presented in Groningen for "Van der Waerden Day" on May 28, 1993). — DOLD-SAMPLONIUS, YVONNE: "Bartel Leendert van der Waerden befragt von Yvonne Dold-Samplonius." *NTM* (2) **2** (1994), 129–147. Italian trans.: *Lettera Matematica P.RI.ST.EM* **15** (1995). English trans.: *Notices AMS* **44** (1997), 313–320. Japanese trans.: *Sugaku Seminar* **11** (1997), 2–8, **12** (1997), 62–66. — HOGENDIJK, JAN P.: "B. L. van der Waerden's Detective Work in Ancient and Medieval Mathematical Astronomy." *Nieuw Arch. Wiskd.* (4) **12** (1994), 145–158. — HLAWKA, EDMUND: "Bartel Leendert Van der Waerden." *Österreichische Akademie der Wissenschaften. Almanach* 1995/96. 146. Jahrgang, 399–405. — SCRIBA, CHRISTOPH J.: "Bartel Leendert van der Waerden (2. Februar 1903 – 12. Januar 1996)." *Ber. Wiss.Gesch.* **19** (1996), 245–251. — "Fra storia e memoria, Bartel Leendert van der Waerden" (Italian). *Lettera Matematica P.RI.ST.EM* **22** (1996), 31. — SPRINGER, T. A.: "Bartel Leendert van der Waerden (2 februari 1903 – 12 januari 1996)." *Koninklijke Nederlandse Akademie van Wetenschappen. Levensberichten en herdenkingen*, 1997, 45–50. — VAN DER WAERDEN, BARTEL LEENDERT: "Meine Göttinger Lehrjahre" (Gastvortrag in der Algebravorlesung, gehalten am 26.1.1979 in Heidelberg; mit einem Nachwort von PETER ROQUETTE). *Mitt. DMV*, no. 2 (1997), 20–27. — DOLD-SAMPLONIUS, YVONNE: *In Memoriam:* Bartel Leendert van der Waerden (1903–1996). *HM* **24** (1997), 125–130. Persian trans. in the Persian trans. of B. L. VAN DER WAERDEN: *A History of Algebra*, Tehran 1998, 351–359. E.N.

Van Ortroy, Fernand Gratien (* March 6, 1856, Aalst, Belgium; † August 22, 1934, Blankenberge, Belgium). VAN ORTROY was professor of geography at the University of Ghent. Among his areas of special interest were the history of cartography in the 16th and 17th centuries.

Works: *Bibliographie de l'œuvre de Pierre Apian* (1902, reprint 1963). — *Biobibliographie de Gemma Frisius, fondateur de l'école belge de géographie, de son fils Corneille et de ses neveux les Arsenius* (1920, reprint 1966).

Secondary literature: VERLINDEN, C., in: *Rijksuniversiteit te Gent. Liber Memorialis 1913–1960.* Part IV: *Faculteit der Wetenschappen. Faculteit der Toegepaste Wetenschappen.* Gent 1960, 53–55 (with portrait). P.B.

van Schooten Jr., Frans (* ca. 1615, Leiden, Netherlands; † May 29, 1660 Leiden, Netherlands). VAN SCHOOTEN studied at the University of Leiden, where his father, FRANS VAN SCHOOTEN THE ELDER, lectured at the university's engineering school. After his father's death in 1645, VAN SCHOOTEN took over his father's academic duties. He also prepared an edition of the collected mathematical writings of FRANÇOIS VIÈTE, and he published a Latin translation of DESCARTES's *Géométrie.* At VAN SCHOOTEN's burial, a funeral oration was delivered by SAMUEL TENNULIUS [TENNULIUS 1668].

Secondary literature: *May* 367. *DSB* **12**, 205–207. *P* II, 837. — *NNBW* **7**, 1110–1114. — HOFMANN, JOSEPH EHRENFRIED: *Frans van Schooten der Jüngere* (= *Boethius* **2**). Wiesbaden 1962 (with portrait). P.B.

Vasconcellos, Fernando de Almeida Loureiro e (* March 26, 1874, Chaves, Portugal; † November 1, 1944, Lisbon, Portugal). FERNANDO VASCONCELLOS, like PEDRO JOSÉ DA CUNHA, began his advanced studies at Lisbon's Polytechnic School, after which he went on to the Army School from which he graduated in military engineering. He not only followed a career in the army, attaining the post of colonel in 1926, but he also pursued at the same time an academic career which he started as a lecturer at Lisbon's Polytechnic School (after the Republican Revolution of 1910, the school was renamed the Faculty of Sciences of Lisbon). Later he joined the Advanced Institute for Agronomy as a full professor. He was the first president of the Lisbon Section of the Portuguese Group for History of Science, and in 1934 he was president of the Executive Commission of the Third International Congress of History of Science, which took place in the cities of Lisbon, Coimbra, and Oporto.

Secondary literature: *GEPB* **34**, 283–284 (with portrait). L.S.

Vashchenko-Zakharchenko, Mikhail Egorovich (* November 12, 1825, Malievka village, Poltava Region, Russian Empire; † August 27, 1912, Kiev, Russian Empire). MIKHAIL VASHCHENKO-ZAKHARCHENKO began teaching at the University of Kiev in 1863, and was promoted to the rank of professor in 1868. His major areas of research included the history of geometry in antiquity and the Middle Ages.

Works: *Euclid's Elements, with an Introduction and Commentaries of Professor M. E. Vashchenko-Zakharchenko.* Kiev 1880 (in Russian). — *History of Mathematics*, vol. 1. Kiev 1883 (in Russian).

Secondary literature: *May* 368. — GRATSIANSKAYA, L. N.: "Mikhail Egorovich Vashchenko-Zakharchenko." *IMI* **14** (1961), 441–464 (in Russian, with bibliography and portrait). — [BORODIN/BUGAI 1979]. — [BOGOLYUBOV 1983]. S.S.D.

Vasiliev, Aleksandr Vasilievich (* August 5, 1853, Kazan, Russian Empire; † October 9, 1929, Moscow, USSR). First *Privatdozent* (beginning in 1874) and then a professor at the University of Kazan (1874–1906), ALEKSANDR VASILIEV also taught as professor at the University of St. Petersburg/Petrograd University (1907–1923). His research interests included the history of mathematics in Russia, the life and works of N. I. LOBACHEVSKIĬ, and the history of the concept of the integers.

Works: *Integer. Historical essay.* Petrograd 1922 (in Russian). — *Nikolai Ivanovich Lobachevskiĭ.* Moscow 1992 (in Russian).

Secondary literature: *May* 368. *P* III, 1419; IV, 1599; V, 1337; VI, 2812. — [KUROSH 1959]. — [BORODIN/BUGAI 1979]. — [BOGOLYUBOV 1983]. — BAZHANOV, V. A., and A. P. YOUSHKEVICH: "A. V. Vasiliev as a Scientist and Social Activist." In: VASILIEV, A. V.: *Nikolai Ivanovich Lobachevskiĭ.* Moscow 1992, 221–226 (in Russian). — Portrait in [SHTOKALO 1966/70, vol. 4, book 2, p. 453]. S.S.D.

Velsius, Justus (* ca. 1510, The Hague, Netherlands; † after 1581 at an unknown location). VELSIUS graduated as *doctor medicinae* at Bologna in 1538 and at Leuven in 1541, where in 1542 he gave public lessons on EUCLID. Shortly afterwards, fear of the Inquisition caused him to flee to Strasbourg. He later spent time in Cologne, Marburg, London, Groningen (where he spent years in prison), and other places in the Netherlands. VELSIUS published the Greek text of PROCLUS's *De Motu* (On Motion), along with a Latin translation ([VELSIUS 1545]).

Secondary literature: FEIST-HIRSCH, E.: "The Strange Career of a Humanist. The Intellectual Development of Justus Velsius (1502–1582)." In: *Aspects de la progagande religieuse.* Genève 1957, 308–324. — DENIS, PHILIPPE: "Justus Velsius (Welsens)." *Bibliotheca dissidentium* **1** (1980), 49–95. — VANDEN BROECKE, STEVEN: "Humanism, Philosophy and the Teaching of Euclid at a Northern University. The Oration on the Various Uses and Dignities of the Mathematical Disciplines, 1544, of Justius Velsius (c. 1510/15 – after 1580)." *Lias* **25** (1998), 43–68. P.B.

Vera Fernández de Córdoba, Francisco (* February 26, 1888, Alconchel (Badajoz), Spain; † July 31, 1967, Buenos Aires, Argentina). A doctor of science and a professor of mathematics, VERA FERNÁNDEZ DE CÓRDOBA went into exile after the Spanish Civil War (1936–39), first in France, then the Dominican Republic, Colombia, and finally in Argentina. His main works in history of mathematics are an unfinished *History of Mathematics in Spain* in four volumes (1929–33) and *Historians of Spanish Mathematics* (1935).

Secondary literature: PELLECÍN, MANUEL: *Francisco Vera Fernández de Córdoba, Matemático e Historiador de la Ciencia.* Badajoz 1988. E.A.; M.H.

Ver Eecke, Paul Louis (* February 13, 1867, Menen, Belgium; † October 14, 1959, Berchem near Antwerp, Belgium). PAUL VER EECKE was a mining engineer who worked for the Belgian Ministry of Labor. He published numerous translations of classical Greek texts into French.

Secondary literature: *DSB* **13**, 615–616. *May* 369. — *BNB* **30**, 267–269. — PELSENEER, JEAN: "Paul Ver Eecke." *Osiris* **8** (1948), 5–11 (with bibliography and portrait). — MOGENET, JOSEPH: "Paul Ver Eecke (1867–1959)." *Arch. int. Hist. Sci.* **12** (1959), 296–297. — ITARD, JEAN: "Paul Ver Eecke (1867–1959)." *Rev. Hist. Sci.* **13** (1960), 141–143. — ROME, ADOLPHE: "Paul-Louis Ver Eecke." *Isis* **51** (1960), 202–203. P.B.

Veselovskiĭ, Ivan Nikolaevich (* November 26, 1892, Moscow, Russian Empire; † June 24, 1977, Moscow, USSR). Beginning in 1921, IVAN VESELOVSKIĬ taught in the Moscow Higher Technical School, where he was named professor in 1951. His areas of research interest were the mathematics of ancient Babylonia and Greece.

Main works: Russian translations of the works of ARCHIMEDES (1962), DIOPHANTUS (1974), and PTOLEMY (1997).

Secondary literature: [KUROSH 1959]. — [FOMIN/SHILOV 1969]. — Portrait in [SHTOKALO 1966/70, vol. 4, book 2, p. 488]. S.S.D.

Vetter, Quido Karl Ludwig (* June 5, 1881, Prague, Czechoslovakia; † October 20, 1960, Prague, Czechoslovakia). Following a maternal family tradition, QUIDO VETTER studied at the Technical University of Prague, but after two years turned to the study of mathematics and descriptive geometry at the Czech University of Prague, where he received his doctorate in 1913. His historical interests were inspired upon reading MORITZ CANTOR's monumental *Vorlesungen über Geschichte der Mathematik* (Lectures on History of Mathematics). Beginning in 1914, VETTER taught mathematics at a secondary school in Prague. At the time his two main fields of interest began to take shape: the history of mathematics and the methodology of mathematical instruction. Thus began a long stream of publications that resulted in one book and more than 250 articles.

Following his *Habilitation* in 1919, VETTER was the first scholar to receive permission to teach history of mathematics in Czechoslovakia. Some years later, in 1924, he was promoted to an extraordinary professorship, while at the same time he also served as director of a middle school at Humpolec. In 1945 VETTER resumed his lectures on the history and didactics of mathematics at the University of Prague.

VETTER's numerous papers are based on a sound familiarity with the historico-mathematical literature of his day. He had a vast knowledge of foreign languages: besides Russian, English, French, and German, he was able to read publications in Italian, Spanish, and Polish, among others. In 1926 he collected the articles he had published on ancient mathematics in various foreign journals into a book published in Czech: *Jak se počitalo a měřilo na insvitě kultury* (How One Counted and Measured at the Dawn of Culture) (1926). Another focus of his research was the development of the exact sciences in the Czech countries, from their earliest beginnings to the 17th century.

Among historians of mathematics, VETTER valued above all the works of GUSTAF ENESTRÖM and GINO LORIA. He always emphasized the importance of understanding mathematics as a general component of cultural history. He believed that the historian of mathematics must apply the methods developed by historians of general history, but adapted to the needs of writing the history of mathematics.

VETTER also took an active part in the educational reform of secondary schools in his country in the 1930s. He published articles on the application of synthetic and analytic methods in teaching mathematics, and wrote comparative reports about mathematical instruction in different European countries as well as the state of teaching in Czechoslovakia.

VETTER also played an important part in the foundation of national and international organizations for the history of science, for which his numerous international contacts were of considerable significance. He was among the small circle of scholars who were inspired to found the *Comité international d'histoire des sciences*, the forerunner of the *l'Académie internationale d'histoire des sciences*. VETTER, who had already served as president of the *Comité* in 1928, was actively involved in the work of the Academy, and in 1934 became vice-president, subsequently serving as its (fourth) president (1934–1937). In Czechoslovakia he founded (in 1928) the *Volné sdružení pro dějiny reálnych věd* (Free Association for the History of Science); later it was renamed the National Committee for the History of Science. It was under his presidency that the *IVth International Congress for History of Science* was organized in Prague in 1937 (see [FOLTA/NOVÝ 1973]).

Secondary literature: *May* 369. *P* VIIb(8), 5734. — NOVÝ, LUBOŠ: "Le Professeur Quido Vetter." *Arch. int. Hist. Sci.* **13** (1960), 270–273. — Portrait in *Dějiny věd a techniky* **23** (1990), 131.

L.N.

Vivanti, Giulio (* May 24, 1859, Mantua, Italy; † November 19, 1949, Milan, Italy). When VIVANTI died at the age of 90, he had long since been the dean of Italian mathematicians. He obtained his degree (*laurea*) first in engineering in 1881 from the University of Turin and then, two years later, in mathematics from the University of Bologna. After becoming a *libero docente*, he taught at the *Scuola Normale* in Pavia from 1892 to 1895, when he became extraordinary professor of calculus at the University of Messina. In 1908 he returned to Pavia and eventually to Milan, where he was among the founders of the local university in 1924. Years later, when he retired, he was named professor *emeritus*. VIVANTI's more than 170 scientific publications are devoted primarily to the field of mathematical analysis (and complex function theory in particular), the calculus of variations, and set theory. His major works include *Teoria delle funzioni analitiche* (Theory of Analytic Functions) (1901; German trans. 1906), and *Funzioni analitiche e funzioni trascendenti intere* (Analytical Functions and Entire Transcendental Functions) (1928). VIVANTI also devoted a number of papers to the history of mathematics. His major contribution to this field was a long article, "Infinitesimalrechnung, 1759–1799" (Infinitesimal Calculus, 1759–1799), included in CANTOR's *Geschichte der Mathematik*, (vol. 4, 639–869).

Secondary literature:: *May* 371. *P* IV, 1571; V, 1314; VI, 2766; VIIb(8), 5781. — CINQUINI, S.: "Giulio Vivanti." *Rend. Ist. Lomb. Sci. Lett. Parte Generale* (3) **83** (1950), 185–205 (with bibliography and portrait). U.B.

Vogel, Kurt (* September 30, 1888, Altdorf (near Nürnberg), Germany; † October 27, 1985, Munich, Germany). KURT VOGEL's father was a *Seminardirektor* (director of a teachers training college). KURT VOGEL passed the school-leaving examination in 1907 in Ansbach, at the same school where the historians of mathematics GOTTFRIED FRIEDLEIN and SIEGMUND GÜNTHER had taught. VOGEL then studied mathematics and physics at Erlangen and Göttingen from 1907 to 1911, *inter alia* under EILHARD WIEDEMANN, MAX NOETHER, PAUL GORDAN, ERHARD SCHMIDT, FELIX KLEIN, DAVID HILBERT, and OTTO TOEPLITZ. In 1911 he passed the *Lehramtsprüfung* (teaching examination) with a *Zulassungsarbeit* (qualifying essay) on FREDHOLM's integral equations. After World War I, VOGEL taught mathematics in Munich, first at the *Ludwigs-Realschule* (from 1920 to 1927) and then at the *Maximilians-Gymnasium*, one of the most traditional Bavarian *Gymnasiums* (from 1927 to his retirement in 1954).

Even during his school years, VOGEL was interested in the history of mathematics. He acquainted himself with problems which came from the Egyptian *Rhind papyrus*, and in order to study these texts in the original language, he began to teach himself Egyptian as a student in Göttingen. In 1927 he began to study ancient Egyptian with WILHELM SPIEGELBERG (1870–1930) at the University of Munich, and in 1929 he received his Ph.D. for a dissertation on the $(2 : n)$-table of the Rhind papyrus. One of the referees was HEINRICH WIELEITNER, whom VOGEL only met after his examination. WIELEITNER became a staunch supporter of VOGEL's in the academic world until WIELEITNER's early death two years later.

VOGEL was also interested in Greek science, especially practical mathematics — the so-called *logistic.* As his *Habilitationsschrift* (1933), he published *Beiträge zur Geschichte der griechischen Logistik* (Contributions to History of Greek Logistic), based on Greek and Byzantine sources, and his *Habilitation* was also on history of mathematics. In 1936 VOGEL became *Privatdozent* for history of mathematics at the University of Munich, and in 1940 he was promoted to *ausserplanmässiger Professor.*

Throughout his career, VOGEL hoped to establish the history of mathematics as an institute at the University of Munich. He first succeeded in founding an "Abteilung" (Section) within the Mathematical Institute for history of mathematics; its library was originally WIELEITNER's private library. About 1957 this "Abteilung" merged with the *Seminar für Geschichte der Naturwissenschaften* (Seminar for the History of Sciences), with VOGEL as director. Eventually, he was responsible for founding the *Institut für Geschichte der Naturwissenschaften* (Institute for the History of Sciences) at the University of Munich in 1963. VOGEL continued to teach at the University until the 1970s. In his seminars and lectures original texts were studied — specialists from other faculties, including assyriologists and classicists, also took part. Above all, VOGEL tried to interest future schoolteachers in history of mathematics.

The main subjects of VOGEL's scholarly output may be seen chronologically: he began, about 1930, with the then brand-new Egyptian and Babylonian texts. In the interpretation of Babylonian texts VOGEL, along with NEUGEBAUER and THUREAU-DANGIN, was a pioneer: in critical discussions (by correspondence) with OTTO NEUGEBAUER, VOGEL advanced several conclusions which are still considered valid today. There followed (before World War II) works on elementary Greek arithmetic. After the War VOGEL began studying Southern German texts of the 15th and 16th centuries, many of which were manuscripts in the Bavarian State Library. He later extended his research to general West European mathematics from the Middle Ages to the 16th century. He thus continued the research of EMIL WAPPLER, PETER TREUTLEIN, MAXIMILIAN CURTZE, and others, who had worked at the turn of the century. Because of VOGEL's research, relations between the 15th-century mathematical centers in Southern Germany and Austria, especially those at Vienna, Regensburg, and Leipzig, are now better understood.

VOGEL saw clearly that to make progress in history of mathematics, both generally for the medieval and early modern periods, texts had to be edited and studied critically. He therefore published editions (with translations and commentaries) of a series of texts, primarily on elementary arithmetic and geometry. The most important of these are: *Algorismus Ratisbonensis* (about 1450), two Byzantine *Rechenbücher* (reckoning books) (14th and 15th century), and the *Chiu Chang Suan Shu* (*Jiuzhang suanshu* = The Nine Chapters). To study this Chinese mathematical text, VOGEL taught himself Chinese at an advanced age. A *leitmotif* that runs throughout VOGEL's work is the occurrence of the same problems, typically in recreational mathematics, in different cultures, e.g. as found in Chinese, Indian, Greek, Arabic, Latin, Babylonian, and Egyptian sources. He carefully studied the differences in the various versions of such problems and drew insightful conclusions about how these problems moved from one culture to another.

VOGEL's encyclopedic knowledge of mathematics in various cultures and his ability to read and speak numerous languages made him a particularly well-suited editor for JOHANNES TROPFKE's *Geschichte der Elementarmathematik* (History of Elementary Matematics). VOGEL had assisted TROPFKE in the early 1930s in publishing the first three volumes of the third edition, and had himself brought out volume 4 in 1940. In the 1960s he began, with HELMUTH GERICKE and KARIN REICH, preparation of a fourth edition. Volume 1 (which comprises volumes 1 to 3 of the third edition) appeared in 1980. Subsequently VOGEL began reworking volume 4 of the third edition (devoted to plane geometry), on which he was still working the day before he died in Munich on October 27, 1985.

VOGEL's contributions to scholarship were internationally recognized. As early as 1931 he became a corresponding member of the *Académie Internationale d'Histoire des Sciences*; later he became an effective member and then vice president (1977–1981). In 1969 he received the Sarton Medal of the History of Science Society. He was a member of numerous German and international societies, including the *Deutsche Akademie der Naturforscher Leopoldina* in Halle.

KURT VOGEL had the ability not only to investigate problems in detail and to establish critical editions, but also to deduce relationships between different cultures. Here he was helped enormously by his outstanding linguistic talents. He used these to great advantage not only in his published works, but also in his lectures, both in the university and at international congresses.

Autobiography: VOGEL, KURT: "Rückschau auf 40 Jahre Mathematikgeschichtsforschung." In: STICKER, BERNHARD, and FRIEDRICH KLEMM (Eds.): *Wege zur Wissenschaftsgeschichte.* Wiesbaden 1969, 145–153 (with portrait and selected bibliography).

Main works: *Die Grundlagen der ägyptischen Arithmetik in ihrem Zusammenhang mit der* 2 : *n-Tabelle des Papyrus Rhind* (diss. Munich, 1929). — *Beiträge zur griechischen Logistik. Erster Teil* (Habilitationsschrift, 1936). — *Die Practica des Algorismus Ratisbonensis* (1954). — *Vorgriechische Mathematik.* 2 vols. (1958–1959). — *Ein byzantinisches Rechenbuch des 15. Jahrhunderts* (with HERBERT HUNGER, 1963). — *Chiu Chang Suan Shu. Neun Bücher arithmetischer Technik* (1968). — *Ein byzantinisches Rechenbuch des frühen 14. Jahrhunderts* (1968). — *Ein italienisches Rechenbuch aus dem 14. Jahrhundert (Columbia X 511 A 13)* (1977). — *Johannes Tropfke: Geschichte der Elementarmathematik, 4. Auflage, Bd. 1* (with HELMUTH GERICKE and KARIN REICH, 1980).

Secondary literature: *P* VI, 2769; VIIa(4,II), 777. — SCHNEIDER, IVO: "Ein Leben für die Wissenschaftsgeschichte: Kurt Vogel." *Beiträge zur Geschichte der Arithmetik von Kurt Vogel.* München 1978, 7–18. — FOLKERTS, MENSO: "Kurt Vogel: Biographie und Bibliographie." *HM* **10** (1983), 261–273. — SCRIBA, CHRISTOPH J.: "Kurt Vogel (1888–1985)." *Arch. int. Hist. Sci.* **85** (1985), 418–420. — FOLKERTS, MENSO: "Kurt Vogel (1888–1985)." *HM* **13** (1986), 98–105. — FOLKERTS, MENSO: "Kurt Vogel †." *Nbl. DGGMNT* **36** (1986), 10–12. — SCHNEI-

DER, IVO, and MICHAEL S. MAHONEY: "Kurt Vogel, 30 September 1888 – 27 October 1985." *Isis* **77** (1986), 667–669. — FOLKERTS, MENSO (Ed.): *Kurt Vogel, Kleinere Schriften zur Geschichte der Mathematik*, 2 vols. (= *Boethius* **20**). Stuttgart 1988 (In **1**, xiv–xlv: bibliography, papers, university lectures, éloges). — GUPTA, RADHA CHARAN: "Kurt Vogel (1888–1985), the Veteran German Historian of Mathematics." *Gaṇita Bhāratī* **10** (1988), 16–20 (with bibliographical note and portrait). — [VOGEL/YOUSHKEVICH 1997, with portrait]. M.F.

Vollgraff, Johan Adriaan (* April 28, 1877, Haarlem, Netherlands; † June 24, 1965, Leiden, Netherlands). VOLLGRAFF studied physics and mathematics at the University of Leiden, where he lectured on the history of mathematics as an external lecturer beginning in 1911; in 1916 he was appointed professor at the Flemish University of Ghent. After World War I, he returned to Holland where he lived and worked as a private scholar. In addition to editing volumes 16 to 20 of HUYGENS's *Œuvres*, he also published FREDERICUS RISNER's work on optics, including annotations made by WILLEBRORD SNELLIUS (1918), and the works of the 18th-century Dutch mathematician NICOLAAS STRUYCK (1912).

Secondary literature: *P* V, 1318; VI, 2773; VIIb(9), 5798. — *BWN* **2**, 587. — VAN PROOSDIJ, B. A.: "Publications de J. A. Vollgraff." *Janus* **46** (1957), 224–228. — STRUIK, DIRK J.: "Johan Adriaan Vollgraff (1877–1965)." *Isis* **57** (1966), 84. — WITTOP KONING, D. A.: "Johan Adriaan Vollgraff (1877–1965)." *Rev. Hist. Sci.* **19** (1966), 270. P.B.

Vorstermann von Oyen, George Auguste (* July 11, 1836, Gilze en Rijen, Netherlands; † October 13, 1915, Aardenburg, Netherlands). After teaching in a primary school at Nieuw-Loosdrecht, VORSTERMANN VON OYEN went on to teach mathematics and physics at the *Gymnasium* of Winschoten before his appointment, in 1860, as headmaster of the public school at Aardenburg. He wrote a number of handbooks on arithmetic, algebra, and geometry, and several articles and books on agriculture and horticulture. Among his writings devoted to history of mathematics are a study on LUDOLF VAN CEULEN and an article on 16th- and 17th-century Dutch surveyors, published in BONCOMPAGNI's *Bullettino*.

Secondary literature: *BWN* **1**, 629–630. P.B.

Vossius, Gerardus Johannes (* March or April 1577, Heidelberg, Germany; † March 17, 1649, Amsterdam, Netherlands). A humanist, classical scholar, and theologian, VOSSIUS was rector of the Latin School in Dordrecht (1600–1614) when he was appointed regent of the Theological College at Leiden. In 1619 he was dismissed because of supposed sympathy for the Remonstrance, a theological movement among Dutch Protestants, dissenting from stricter Calvinism. Three years later he became professor of eloquence and chronology, and subsequently of Greek as well, at the University of Leiden. In 1631, he moved to Amsterdam to become professor of ecclesiastical history in the newly-founded *Athenaeum Illustre*. His manuscript *De universae mathesios natura et constitutione liber* (Book on the

Nature and Constitution of the Universe of Mathematics), was published by his nephew FRANCISCUS JUNIUS in 1650 ([VOSSIUS 1650]). This book, dealing with the various disciplines of the mathematical sciences, listed numerous authors whose writings had contributed to each discipline over the centuries.

Secondary literature: *P* II, 1235. — RADEMAKER, CORNELIS SIMON M.: *Life and work of Gerardus Joannes Vossius (1577–1649).* Assen 1981 (with portrait).
P.B.

Vygodskiĭ, Mark Yakovlevich (* October 2, 1898, Minsk, Russian Empire; † September 26, 1965, Tula, USSR). MARK VYGODSKIĬ's father was a chemical engineer and his mother a professor of music. In 1916 he graduated from the *Gymnasium* in Baku, whereupon he became a member of the Bolshevik party and participated in the civil war. After the Russian Revolution, he studied at the University of Rostov-on-Don, where his teacher was the mathematician and historian of mathematics D. D. MORDUKHAĬ-BOLTOVSKOĬ. In 1923 VYGODSKIĬ completed his studies at the University of Moscow, where he became a postgraduate student of the history of mathematics in 1926. In 1931 he was promoted to the rank of professor, and together with S. A. YANOVSKAYA, he struggled to provide a Marxist education for his students. Soon VYGODSKIĬ was one of the most powerful figures in the Soviet mathematical community — he was appointed director of the Institute of Mathematics at the University of Moscow, served as vice-president of the Moscow Mathematical Society, and was editor in chief of the State Publishing House for Theoretical Technical Literature. As one of the principal ideologists in Soviet mathematics, he participated in the campaign directed against "reactionary professors," and he helped to criticize such mathematicians as D. F. EGOROV. At the same time, as a first-rate scientist and an excellent teacher, VYGODSKIĬ contributed substantially to the expansion of mathematics in the USSR. In 1933, again with S. A. YANOVSKAYA, he founded a seminar which is still in existence at the University of Moscow devoted to the history of mathematics. That same year due to internal political struggles he became an object of persecution and was expelled from the Communist Party. After World War II, he worked at the University of Moscow from 1945 until 1948, and from 1950 he also held a position at the Polytechnic and Pedagogical Institutes in Tula, where he died on September 26, 1965.

In the 1930s, together with YANOVSKAYA, VYGODSKIĬ undertook the creation of a specifically Marxist history of mathematics. According to VYGODSKIĬ, this meant interpreting the development of mathematics from a Marxist point of view, searching for causes, i.e. the forces which have served to determine the course of the history of mathematics. He posed the following three general problems for the history of mathematics:

1. The interaction between mathematics and culture;
2. The "class nature" of mathematics;
3. The historical interpretation of mathematics.

The first period of VYGODSKIĬ's scientific activity is reflected in the titles of his works: *Plato as a Mathematician* (1926) (a critical appreciation of the influence of PLATO and his followers in mathematics), *The Problem of Number and its Development* (1929), and *The Problems of History and Marxist Methodology* (1930). In the 1930s VYGODSKIĬ concentrated his attention on mathematics in antiquity. Beginning in 1940 he published his research on Babylonian mathematics, including his principal work, *Arithmetic and Algebra in Antiquity.* This book is characterized by its original view of ancient mathematics and by important results on the history of Babylonian mathematics. VYGODSKIĬ also studied questions concerning the development of mathematics in Russia, the history of mathematical analysis, and of differential geometry. He devoted considerable effort to the edition of classic mathematical works by EUCLID, KEPLER, MONGE, and EULER, among others. Together with S. A. YANOVSKAYA and A. P. YOUSHKEVICH, VYGODSKIĬ is considered one of the co-founders of the Soviet School of the History of Mathematics.

Secondary literature: *May* 374. — YOUSHKEVICH, ADOLF P.: "Mark Yakovlevich Vygodsky." *Arch. int. Hist. Sci.* **19** (1966), 271–272. — ROSENFELD, BORIS A.: "Mark Yakovlevich Vygodskiĭ and his Works on the History of Mathematics." In: VYGODSKIĬ, M. YA.: *Arithmetic and Algebra in Antiquity.* 2nd ed. Moscow 1967, 350–362 (in Russian). — Portrait in [SHTOKALO 1966/70, vol. 4, book 2, p. 461]. S.S.D.

Wallis, John (* November 23 (December 3), 1616, Ashford (Kent), England; † November 8, 1703, Oxford, England). WALLIS was named Savilian Professor of Geometry at Oxford University in 1649, and was one of the founding members of the Royal Society. He studied theology at Cambridge University, was ordained in 1640, and in 1644 appointed one of two secretaries to the Assembly of Divines at Westminster. During the Civil War he developed great skill in deciphering captured code letters for the Parliamentarians. In his mathematical works that later had a strong influence on NEWTON, he made free use of the method of indivisibles. Applying algebraic considerations, he derived an infinite product for π, but was criticized by FERMAT and others for his daring extrapolations. His *Treatise of Algebra, Both Historical and Practical* (1685, Latin edition in his *Opera Mathematica*, vol. II, 1693), was widely read; it marks the beginning of the serious study of the history of mathematics in England. WALLIS also edited several classical Greek texts on mathematics, astronomy, and music by ARCHIMEDES, PTOLEMY, ARISTARCHUS, PORPHYRY, and PAPPUS.

Autobiography: SCRIBA, CHRISTOPH J.: "The Autobiography of John Wallis, F.R.S." *Notes Rec. R. Soc. Lond.* **25** (1970), 17–46.

Secondary literature: *DSB* **14**, 146–155. *May* 377. *P* II, 1254. — PRAG, ADOLF: "John Wallis, 1616–1703. Zur Ideengeschichte der Mathematik im 17. Jahrhundert." *Q. St. G. Math. Astron. Phys.*, Abt. B **1** (1931), 381–412. — SCOTT, JOHN FREDERICK: *The Mathematical Work of John Wallis, D.D., F.R.S. (1616–1703).* London 1938; repr. New York 1981 (with portrait). — SCRIBA, CHRISTOPH

J.: *Studien zur Mathematik des John Wallis (1616–1703)* (= *Boethius* **6**). Wiesbaden 1966 (with portrait). C.J.S.

Wallis, Peter John (* June 4, 1918, London, England; † August 24, 1992, Newcastle-upon-Tyne, Scotland). PETER WALLIS was an historian and bibliographer of mathematics, along with his wife RUTH. He was especially interested in book subscription lists, historical bibliography, and Newtoniana. He edited *An Index of British Mathematicians: a check list* (1976), and produced a *Biobliography of British Mathematics and its Applications* (1986–1993, in three parts). I.G.G.

Wappler, Hermann Emil (* June 20, 1852, Bernsbach, Saxony, Germany; † October 6, 1899, Zwickau, Germany). WAPPLER was a German teacher of mathematics in Schkeuditz (1878) and Zwickau (1879–1899). He was mainly interested in the history of algebra in the 15th century.

Main works: "Zur Geschichte der deutschen Algebra im 15. Jahrhundert." *Programm Gymnasium Zwickau* 1887. — "Beitrag zur Geschichte der Mathematik." *Abh. Gesch. Math.* **5** (1890), 147–168. — "Bemerkungen zur Rhythmomachie." *Z. Math. Phys., Hist.-.lit. Abt.* **37** (1892), 1–17. — "Zur Geschichte der deutschen Algebra." *Abh. Gesch. Math.* **9** (1899), 537–554.

Secondary literature: *May* 377. *P* IV, 1597. — ENESTRÖM, GUSTAF: "Hermann Emil Wappler †." *Bibl. Math.* (3) **1** (1900), 225 (with portrait). M.F.

Weil, André (* May 6, 1906, Paris, France; † August 6, 1998, Princeton, NJ, USA). ANDRÉ WEIL was a child prodigy. He was educated at the *Ecole Normale Supérieure* and passed his *agrégation* examination in 1925, after which he studied in Rome, Göttingen, and Stockholm. He was awarded the D.Sc. degree by the University of Paris in 1928. After completing a year of military service in 1928–29, he went to India where he taught as professor of mathematics at Aligarh Muslim University from 1930–32. He then taught briefly at Marseilles before accepting a position in Strasbourg, where he remained until 1939. Meanwhile, in 1933 WEIL was one of a group of mathematicians who formed the *Séminaire Julia* in Paris, which after the War became the famous *Séminaire Bourbaki.*

Apparently WEIL was deeply impressed by the bitter results of World War I, when the notion of egalitarian conscription resulted in the decimation of an entire generation of young French mathematicians. Having made up his mind to leave France, he accepted an invitation from LARS AHLFORS to visit Finland, where in 1939 he was jailed as a spy. Due to efforts by ROLF NEVANLINNA, WEIL was deported to Sweden, and eventually returned to France. While imprisoned in Rouen, he was unusually productive, and proved the Riemann hypothesis for curves over finite fields, edited proof sheets for a book on integration (*L'intégration dans les groupes topologiques et ses applications* (Integration on Topological Groups and their Applications) (1940)), and drafted a volume devoted to the integral that later appeared in the Bourbaki series. He also began his *Foundations of Algebraic Geometry* (published in 1946). WEIL was eventually tried, found guilty of failure to report for duty, whereupon he asked to rejoin the army, for which permission was granted. But shortly thereafter, as Germany occupied much of France, WEIL was evacuated to England, and eventually found his way back to Marseilles via North Africa. In Marseilles he was demobilized, and learning that he had the offer of a position in the United States, prepared to leave France. His admirers, it has been said, found his record between 1939 and 1945 "an immense embarrassment," particularly in contrast to that of his sister, SIMONE WEIL, who was a social philosopher, religious mystic, and a heroine of the French Resistance (see "André Weil. Obituary," *The Economist* **348** (no. 8082) (August 22, 1998), p. 70).

Back in the United States, WEIL returned to Princeton, New Jersey, where he had spent a year at the Institute for Advanced Study in 1936–37. Now, with support from the Rockefeller Foundation, he spent some time in Princeton, and then from 1941–43, taught briefly at Haverford College, followed by a year at Lehigh University.

Subsequently, a Guggenheim Fellowship enabled him to go to Brazil, where he taught at Saõ Paolo from 1945–47. When he returned to the United States, he accepted a position at the University of Chicago, where he taught mathematics for more than a decade before accepting a permanent position at the Institute for Advanced Study in 1958. WEIL retired from the Institute in 1976.

It was at the end of his career that WEIL turned his attention increasingly to history. He edited three volumes in the Bernoulli edition: "Der Briefwechsel von Jacob Bernoulli" (with contributions by CLIFFORD TRUESDELL and FRITZ NAGEL; 1993); *Die Werke von Jacob Bernoulli*, vol. 4 "Reihenlehre" (1993), and vol. 5 "Differentialgeometrie" (together with M. MATTMÜLLER; 1999). He also published two important works, one devoted to an historical study of elliptic functions, the other a history of number theory from HAMMURAPI to LEGENDRE.

Main works: WEIL's own commentaries on his published papers may be found in his *Œuvres Scientifiques. Collected Papers*, in 3 vols. New York 1979. — WEIL's historical publications include: *Essais historiques sur la théorie des nombres (= Monographies de l'Enseignement mathématique* **22**). Genève 1975. — *Elliptic Functions According to Eisenstein and Kronecker.* Berlin/New York 1976. — *Number Theory: An Approach Through History from Hammurapi to Legendre.* Boston 1984; German trans. *Zahlentheorie: ein Gang durch die Geschichte; von Hammurapi bis Legendre.* Basel and Boston 1992.

Autobiography: *Souvenirs d'apprentissage.* Basel and Boston 1991; English trans. by JENNIFER GAGE: *The Apprenticeship of a Mathematician.* Basel and Boston 1992; German trans. *Lehr- und Wanderjahre eines Mathematikers.* Basel and Boston 1993.

Secondary literature: SERRE, JEAN-PIERRE, and GORO SHIMURA (Eds.): *Geometry and Number Theory. A Volume in Honor of André Weil.* Baltimore 1983. — BURKHART, FORD N.: "André Weil, Who Reshaped Mathematics, is Dead at 92." *New York Times* (August 10, 1998), B7, with portrait. — LANGLANDS, ROBERT P.: "André Weil (1906–98)." *Nature* **395** (no. 6705) (October 29, 1998), 848 (with portrait). — A special issue of the *Notices of the AMS* **46** (no. 4) (April, 1999), is devoted in part to articles about WEIL, and its cover carries a color portrait. The articles devoted to WEIL in this issue of the *Notices* include KNAPP, ANTHONY W.: "André Weil. A Prologue." 434–439; "André Weil. Reminiscences by Armand Borel, Pierre Cartier, Komaravolu Chandrasekharan, Shiing-Shen Chern, and Sho-Kichi Iyanaga." 440–447, including portraits; and "The Apprenticeship of a Mathematician — Autobiography of André Weil" (Book Review), 448–456 (with portraits). — CARTAN, HENRI: "André Weil: Memories of a Long Friendship." *Notices of the AMS* **46** (no. 6) (June/July, 1999), 633–636 (with portraits). — DIGNE, FRANÇOIS (Ed.): "André Weil (1906–1998)." *Gazette des mathématiciens* **80**, numéro spécial. Paris: Soc. Mathématique de France 1999. — SERRE, JEAN-PIERRE: "La vie et l'œuvre d'André Weil." *Enseign. Math.* (2) **45** (1999), 5–16. J.W.D.

Wertheim, Gustav (* June 9, 1843, Imbshausen near Northeim, Germany; † August 31, 1902, Frankfurt (Main), Germany). WERTHEIM was briefly a private tutor (1866–1870) who then taught at the Jewish "Realschule" in Frankfurt (1872–1900). His major research interest was number theory, but he also translated a number of modern mathematical and physical works, including ones by J. A. SERRET, P. G. TAIT, and W. THOMSON, into German. Among his historical works are *Die Arithmetik des Elia Misrachi* (1893, 2nd ed. 1896), "Die Berechnung der irrationalen Quadratwurzeln und die Erfindung der Kettenbrüche" (The Computation of Surds and the Invention of Continued Fractions), *Abh. Gesch. Math.* **8** (1898), 147–160; and "Die Algebra des Johann Heinrich Rahn (1659) und die englische Übersetzung derselben" (The Algebra of J. H. Rahn (1659) and Its English Translation), *Bibl. Math.* (3) **3** (1902), 113–126. WERTHEIM also produced a translation of DIOPHANTUS (1890).

Secondary literature: *May* 380. *P* IV, 1621; V, 1356. — [*Acta Math.* Tab. Gen. **1–35** (1913), 122; portrait 177]. — ENESTRÖM, GUSTAF: "Gustav Wertheim." *Bibl. Math.* (3) **3** (1902), 395–402 (with portrait). — Anonymous: "Gustav Wertheim †." *ZMNU* **34** (1903), 75–77 (with bibliography). M.F.

Whewell, William (* May 24, 1794, Lancaster, England; † March 6, 1866, Cambridge, England). Master of Trinity College, Cambridge, WHEWELL taught mineralogy and moral philosophy, but is best known as a philosopher of science with a special interest in methodology. Among his important works are a *History of the Inductive Sciences, from the Earliest to the Present Time* (1837), *History of Scientific Ideas* (3rd ed. 1858), *Novum Organon Renovatum* (1858), and *On the Philosophy of Discovery* (1860).

Secondary literature: *DSB* **14**, 292–295. *May* 381. *P* II, 1310; III, 1437. — BUTTS, R. E. (Ed.): *William Whewell's Theory of Scientific Method* (1968). — FISCH, M.: *William Whewell.* Oxford 1991 (with portrait). I.G.G.

Whittaker, Sir Edmund Taylor (* October 24, 1873, Birkdale, Lancashire, England; † March 24, 1956, Edinburgh, Scotland). In addition to a distinguished career in mathematics and mathematical physics, WHITTAKER made many contributions to the history of science, including unusually fine obituaries and centenary articles for major figures of the past. He is best remembered as an historian for *A History of the Theories of Aether and Electricity* (1911), although his treatment of relativity theory in the second edition (2 vols., 1951) has been criticized.

Secondary literature: *DSB* **14**, 316–318. *May* 382. *P* IV, 1627; V, 1361; VI, 2869; VIIb(9), 5977. — TEMPLE, GEORGE: "E. T. Whittaker." *Biogr. Mem. FRS* (2) **2** (1956), 299–325 (with portrait). — MCCREA, WILLIAM H.: "E. T. Whittaker." *J. Lond. Math. Soc.* (2) **32** (1957), 234–256. I.G.G.

Wiedemann, Eilhard (* August 1, 1852, Berlin, Germany; † January 7, 1928, Erlangen, Germany). A *Privatdozent* in physics at Leipzig in 1876, WIEDEMANN was made an extraordinary professor at Leipzig in 1878, and ordinary professor of physics at Darmstadt (1886) and Erlangen (1886–1926). His areas of research interest included physics and history of Arabic science.

Main works: "Beiträge zur Geschichte der Naturwissenschaften" I–LXXVIII. 78 contributions, in *Sb. Phys.-Med. Soc. Erlangen* **34** (1902) to **60** (1928). — *Aufsätze zur arabischen Wissenschaftsgeschichte.* 2 vols. Hildesheim 1970. — *Gesammelte Schriften zur arabisch-islamischen Wissenschaftsgeschichte.* Frankfurt am Main 1984.

Secondary literature: *May* 382. *P* III, 1441; IV, 1631; V, 1364.; VI, 2874. — SCHMIDT, GERHARD C.: "Eilhard Wiedemann. Geb. 1. August 1852 – gest. 7. Januar 1928." *Physik. Zs.* **29** (1928), 185–190 (with portrait). — SEEMANN, H. J.: "Eilhard Wiedemann, geb. 1. August 1852 – gest. 7. Januar 1928." *Isis* **14** (1930), 166–186 (with portrait). M.F.

Wieleitner, Heinrich (* October 31, 1874, Wasserburg (near Munich), Germany; † December 27, 1931, Munich, Germany). WIELEITNER came from humble origins, but his intellectual gifts were early recognized and he received a scholarship to the *Bischöfliches Seminar* in Scheyern (near Pfaffenhofen, Bavaria). After graduating from the Freising *Gymnasium* in 1893, he studied mathematics from 1893–1895 at the University of Munich, where he was very active in the Mathematical Society (*Mathematischer Verein*). His first lecture (1895) as a member of this society was on squarable *lunulae*, which so impressed his contemporary EDMUND LANDAU that he later made it the starting point of a new direction of his own mathematical investigations. Through his works on the lunes WIELEITNER also came in contact with JOHANNES TROPFKE, and the two became lifelong friends. In 1897 WIELEITNER passed the *Staatsexamen*, thus qualifying for a license to teach mathematics and physics at "Gymnasien." After a short time as an assistant to W. VON DYCK and S. FINSTERWALDER at the *Technische Hochschule* (1897–1898), WIELEITNER became a teacher at several *Gymnasien*: in Speyer (1898–1909 and 1915–1920), Pirmasens (1909–1915), and Augsburg (1920–1926). In 1926 he became director of the *Neues Realgymnasium* in Munich (1926–1931), a position he held until his premature death in 1931.

WIELEITNER was awarded his doctorate in 1901 with a dissertation *Über die Flächen 3. Ordnung mit Ovalpunkten* (On Surfaces of Third Order with Oval Points). He continued to study higher curves over the next decade. In 1903 he began writing a textbook on plane algebraic curves which was published in 1905, *Theorie der ebenen algebraischen Kurven höherer Ordnung* (Theory of Higher Order Plane Algebraic Curves); another substantial volume was *Spezielle ebene Kurven* (Special Plane Curves) (1908). In 1904 WIELEITNER attended the Third International Congress of Mathematicians in Heidelberg, where he became acquainted with the Italian geometers. Among his special interest was the teaching of mathematics in schools. In 1909 WIELEITNER became a "Beauftragter" (delegate) of the International Commission on Mathematical Education (ICME) *(Internationale Mathematische Unterrichtskommission (IMUK))* and thereby learned a great deal about how mathematics was taught in different schools.

Beginning around 1910, history of mathematics was the subject of most of WIELEITNER's publications. He came indirectly to the subject through ANTON VON BRAUNMÜHL: when the latter died in 1908, SIEGMUND GÜNTHER asked WIELEITNER to take over BRAUNMÜHL's part of the *Geschichte der Mathematik* (History of Mathematics), a two-volume series intended to update CANTOR's *Vorlesungen* and give teachers and students of mathematics a manual on history of mathematics. Volume 2, on which BRAUNMÜHL had worked, was to cover the period from DESCARTES to the end of the 18th century. WIELEITNER published the first part of this volume (on arithmetic, algebra, and calculus) in 1911, and the second (on geometry and trigonometry) in 1921. In 1927–1929 he published a four-fascicle set, *Mathematische Quellenbücher* (Sourcebooks on History of Mathematics).

WIELEITNER also worked on numerous special problems, in particular the prehistory of analytic geometry, graphs, perspective, and the treatment of the law of free fall according to the later scholastics. As early as 1912 he published several works on the mathematics of NICOLE ORESME and other scholastics, for which he sought out manuscripts in various libraries. WIELEITNER was one of the first to study the theory of the latitude of forms and their geometrical representation — a subject of particular interest to him because of its importance for the history of analytic geometry. He also wrote on infinite series in the Middle Ages.

In 1909 WIELEITNER was introduced to JULIUS RUSKA. Together they wrote a substantial treatise on inheritance problems in AL-KHWĀRIZMĪ's "Algebra" (published in 1922). After GÜNTHER's death (1923), WIELEITNER assumed responsibility for abstracts of new publications in history of mathematics and science for the *Mitteilungen zur Geschichte der Medizin und der Naturwissenschaften.* He also contributed to the sections for abstracts in *Isis* and *Archeion.* At the suggestion of ARNOLD SOMMERFELD, WIELEITNER became *Privatdozent* for history of mathematics in Munich (1928), and in 1930 he became honorary professor. He died only one year later, on December 27, 1931.

WIELEITNER was one of the leading historians of mathematics in his day. His writings have a characteristic precision, he always went straight to the sources, and he strove to write the history of mathematical ideas in which the question of development was central. He was the originator of this approach to history of mathematics, which even today has its adherents.

Main works: *Geschichte der Mathematik. 2. Teil: Von Cartesius bis zur Wende des 18. Jahrhunderts* (Sammlung Schubert, 2 vols. 1911/1921). — *Geschichte der Mathematik.* 2 vols. (Sammlung Göschen, 1922/1923). — *Die Geburt der modernen Mathematik.* 2 vols. (1924/1925). — *Der Gegenstand der Mathematik im Lichte ihrer Entwicklung* (1925). — *Mathematische Quellenbücher. I: Rechnen und Algebra, II: Geometrie und Trigonometrie, III: Analytische und synthetische Geometrie, IV: Infinitesimalrechnung* (1927/1929).

Secondary literature: *DSB* **14**, 336–337. *May* 382. *P* V, 1367; VI, 2877. — BORTOLOTTI, ETTORE: "Heinrich Wieleitner". *Boll. UMI* **11** (1932), 63–64. — HOFMANN, JOSEPH EHRENFRIED: "Heinrich Wieleitner (31. Oktober 1874 bis 27. Dezember 1931)." *Natur und Kultur* **29** (1932), 95–97. — HOFMANN, JOSEPH EHRENFRIED: "Heinrich Wieleitner. (31. 10. 1874 – 27. 12. 1931.). *Unterrichtsbl. Math. Nat.* **38** (1932), 58–62. — HOFMANN, JOSEPH EHRENFRIED: "Heinrich Wieleitner. (Geb. 31. X. 1874 zu Wasserburg am Inn, gest. 27. XII. 1931 zu München.)." *Jber. DMV* **42** (1933), 199–223 (with bibliography and portrait). — HOFMANN, JOSEPH EHRENFRIED: "Heinrich Wieleitner – das Wirken eines Wissenschaftshistorikers." *Das Weltall* **34** (1934/35), 5–8. — HUBER, L.: "Dr. Heinrich Wieleitner. Sein Leben und sein wissenschaftliches Werk;" SIEBAUER: "Wieleitner als Lehrer, Vorstand und Mensch." *Neues Land* **13** (1932), 53–54; 54–55. — LIETZMANN, WALTER: "Heinrich Wieleitner †." *ZMNU* **63** (1932), 91–93. — LOREY, WILHELM: "Heinrich Wieleitner zum Gedächtnis." *A.V.Z. Math. Naturw. Blätter* **26** (1932), 19–23. — RUSKA, JULIUS: "Heinrich Wieleitner (geb. 31. X. 1874, gest. 27. XII. 1931)." *Isis* **18** (1932), 150–165 (with bibliography and portrait). — TROPFKE, JOHANNES: "Heinrich Wieleitner. 31. Oktober 1874 – 27. Dezember 1931." *MGMNT* **31** (1932), 97–101. — VOGEL, KURT: "Heinrich Wieleitner (1874–1931)." *Archeion* **14** (1932), 112–115. — VOGEL, KURT: "Heinrich Wieleitner †." *Bayer. Bl. Gymn.* **68** (1932), 90–94 (with portrait). M.F.

Woepcke, Franz (* May 6, 1826, Dessau, Germany; † March 24, 1864, Paris, France). WOEPCKE pursued Arabic studies in Bonn (1848), Leiden, and Paris (1850), became *Privatdozent* in Bonn (1850), and later taught mathematics and physics in Berlin (1856–1858). Subsequently, he studied manuscripts in Paris and Rome (1858–1864). He was mainly interested in Arabic mathematics, and in Greek works preserved in Arabic sources.

Main works: Editions of OMAR KHAYYAM's (*i.e.* ʿUMAR KHAYYĀM's) *Algebra* (1851) and AL-KARAJĪ's *Algebra* (1853). — Reconstruction of writings by EUCLID (1851) and APOLLONIUS (1856). — Works on the transmission of the Hindu-Arabic numerals (1859) and on THĀBIT B. QURRA, ABŪ 'L-WAFĀʾ, IBN KHALDŪN, AL-QALAṢĀDĪ, AND LEONARDO FIBONACCI. — *Etudes sur les mathématiques arabo-islamiques.* Frankfurt am Main 1986.

Secondary literature: *DSB* **14**, 471–473. *May* 385. *P* II, 1353; III, 1458. — NARDUCCI, ENRICO: "Intorno alla vita ed agli scritti di Francesco Woepcke." *Bull. Boncompagni* **2** (1869), 119–152 (with bibliography). — BIERMANN, KURT-R.: "F. Woepckes Beziehungen zur Berliner Akademie." *Mber. DAW Berlin* **2** (1960), 240–249. M.F.

Wolf, Rudolf (1816–1893) (* July 7, 1816, Fällanden, Kanton Zürich, Switzerland; † December 6, 1893, Zürich, Switzerland). RUDOLF WOLF is best known as a pioneer in solar research, although he was no less successful as a university professor, science organizer, historian of science, and librarian. Never married, WOLF devoted much of his spare time to innumerable experiments in empirical probability. The son of a minister of the Protestant Reformed Church of Zürich, he studied applied mathematics at the University of Zürich (with KARL HEINRICH GRÄFFE, JOSEPH LUDWIG RAABE, and the geodesist JOHANNES ESCHMANN); at the University of Vienna (with JOSEPH JOHANN VON LITTROW, JOSEPH PETZVAL, and ANDREAS FREIHERR VON ETTINGHAUSEN); and at the University of Berlin (with JAKOB STEINER, PETER GUSTAV LEJEUNE DIRICHLET, and JOHANN FRANZ ENCKE). From 1839 to 1855 he taught mathematics and physics, as well as sports, at a *Gymnasium* in Bern, but also lectured on mathematics and astronomy at the University, first in the capacity of reader, and then, after 1853, as a professor. Simultaneously, he acted as director of the observatory. From 1855 until his death he held a joint professorship of astronomy at the *Eidgenössisches Polytechnikum* (today: *Eidgenössische Technische Hochschule (ETH)*) and at the University of Zürich.

The combined influence of WOLF's own background, teachers, and friends helped to insure that WOLF's own inclination towards the history of culture and science soon crystallized into the main object of his life's work. He published numerous articles in the proceedings of the scientific societies of Bern and Zürich. Meanwhile, the *Schweizerische Naturforschende Gesellschaft* (Swiss Academy of Sciences) benefitted from his organizational interests, including the promotion of meteorology in Switzerland and its participation in the European meridian degree measuring project. In 1841 WOLF presented a program to the Swiss Scientific Society: to write, in his spare time, a "special Swiss history of natural sciences and their auxiliary disciplines." WOLF was convinced that by "ranking the human agent higher than is done by any other kind of history," this might be a contribution to the general history of science and would, at the same time, "reflect the spirit of entire centuries" [WOLF 1841, 203–204].

WOLF, of course, relied upon the standard works of his time, including ABRAHAM GOTTHELF KÄSTNER's *Geschichte der Mathematik* (1796–1800), JEAN ETIENNE MONTUCLA's *Histoire des mathématiques* (1799–1802) — which WOLF tried to improve in biographical matters ([WOLF 1993, 97/116]) —, WILLIAM WHEWELL's *History of the Inductive Sciences* (1840–1841), and GUILIELMO LIBRI's *Histoire des sciences mathématiques en Italie* (1838–1841). WOLF first became active as an historian of science when, from 1843 to 1855, he was editor of the *Mittheilungen der Naturforschenden Gesellschaft in Bern*, in which he published forty "Notizen zur Geschichte der Mathematik und Physik in der Schweiz" (Notices on the History of Mathematics and Physics in Switzerland). He also contributed numerous historical notes and several hundred excerpts from the correspondence of well-known scientists (which may still have some value as primary sources). Even the *Taschenbuch für Mathematik, Physik, Geodäsie und Astronomie* (Pocket Book for Mathematics, Physics, Geodesy, and Astronomy) (1852; 6th edition 1895), adopted for use in schools, contains an appendix of several pages giving the most important dates in the history of mathematics. Acting as archivist of the *Schweizerische Naturforschende Gesellschaft* in Bern, he was the founder of its collection of autographs (now in the *Burgerbibliothek* in Bern, with large numbers of specimens of WOLF's own correspondence with contemporary scientists).

WOLF's first major contribution to the history of science was four volumes of the *Biographien zur Culturgeschichte der Schweiz* (Biographies on the Cultural History of Switzerland) [WOLF 1858/62]. Comprising more than 1600 pages, these cover the lives of eighty notable scholars, among them several mathematicians including BÜRGI, EULER, and the BERNOULLIS. In addition, WOLF's footnotes contain nearly 3000 further biographical sketches. The *Vierteljahrsschrift der Naturforschenden Gesellschaft in Zürich*, edited by WOLF from 1856 to 1893, published his numerous articles on historical subjects and various aspects of science, along with astronomical and some mathematical papers. Beginning in 1861, WOLF continued the *Biographien* with 475 *Notizen zur schweizerischen Kulturgeschichte* (Notices on Swiss Cultural History), and in 1871 he began a series of *Astronomische Mitteilungen*, mainly comprising his sunspot observations as well as notes on his collections of historical instruments. The *Handbuch der Mathematik, Physik, Geodäsie und Astronomie* (first published in 1869/70, and in two volumes in 1872), was the "detailed continuation of the 'pocket-Wolf' " ([BILLWILLER 1894, 246]). Finally, he recast the material in his *Handbuch der Astronomie, ihrer Geschichte und Literatur* (Handbook of Astronomy, its History and Literature) [WOLF 1890/1893]. In 1877 WOLF contributed the volume on *Geschichte der Astronomie* (History of Astronomy) [WOLF 1877] to the series *Geschichte der Wissenschaften in Deutschland* (History of the Sciences in Germany) under the aegis of the *Historische Comission der kgl. Akademie der Wissenschaften in München.*

As an historical introduction to the publications of the *Schweizerische geodätische Commission*, he published his *Geschichte der Vermessungen in der Schweiz* (History of Geodetic Surveys in Switzerland) [WOLF 1879]. Meanwhile, having become librarian of the *Polytechnikum* in 1855, WOLF was careful to make sure that in addition to the technical and scientific literature of the day, the library also did its best to obtain material on the history of science from the 15th to the 18th centuries.

WOLF's *Biographien zur Culturgeschichte der Schweiz* [WOLF 1858/62], along with his other historical books (all of them exemplarily indexed), remain treasure-troves for the historian of mathematics. The same holds for his innumerable, unfortunately nowadays hardly-known articles, especially his *Notizen zur schweizerischen Kulturgeschichte*, and excerpts from letters of famous scientists. WOLF deserves lasting credit for his efforts to locate the correspondence of the BERNOULLIS, which was finally discovered in 1877 in Stockholm ([SPIESS 1955, 31–38], [DSB 1970/90, **14**, 481]. WOLF's final work, the *Handbuch der Astronomie, ihrer Geschichte und Literatur* (Handbook of Astronomy, its History and Literature) [WOLF 1890/1893], contains "important results of his own investigations" in the history of science, in particular his research "about the development of trigonometry and the invention of logarithms" [BILLWILLER 1894, 247] and [LUTSTORF/WALTER 1992]. In the capacity of librarian, WOLF was succeeded by FERDINAND RUDIO of Wiesbaden, Germany, professor of mathematics at the *Polytechnikum* who shared WOLF's interest in the history of mathematics and also became a well-known historian of mathematics, notably through the edition of EULER's *Opera omnia.*

WOLF's biography was updated on the occasion of the one-hundredth anniversary of his death [BOSSHARD 1993], [GLAUS 1993], [LUTSTORF 1993], [WOLF 1993]; his bibliography was listed in contemporary obituaries [WEILENMANN 1894], [GRAF 1893].

Further publications: "Anzeige einer Specialgeschichte der Naturlehre und ihrer Hülfswissenschaften." *Vh. Schw. Nat. Ges.* **26** (1841), 203–209. — "Jugendtagebuch 1835/1841" (1993, ed. VERENA LARCHER). *Schriftenreihe der ETH-Bibliothek Zürich* **30** (1993).

Secondary literature: *DSB* **14**, 480–481. *May* 385. *P* II, 1357; III, 1460; IV, 1662; VIIa, Suppl., 784.. — *HBLS* **7**, 583. — *SL* VI, 686. — BILLWILLER, ROBERT: "† Prof. Dr. Rudolf Wolf. Nekrolog." *Vh. Schw. Nat. Ges.* **77** (1894), 237–249. — GRAF, JOHANN HEINRICH: "Prof. Dr. Rudolf Wolf." *Mitt. Nat. Ges. Bern aus dem Jahre 1893*, Fasc. 1329–1331 (1894), 193–231 (with bibliography). — WEILENMANN, AUGUST: "Nekrolog auf Prof. Dr. Johann Rudolf Wolf." *Vj. Nat. Ges. Zür.* **39** (1894), 1–64 (with bibliography). — [SPIESS 1955]. — LUTSTORF, HEINZ, and MAX WALTER: "Jost Bürgi's "Progress Tabulen" (Logarithmen)." *Schriftenreihe der ETH-Bibliothek* **28** (1992). — [BOSSHARD 1993]. — GLAUS, BEAT: "Rudolf Wolf: Lehrer, Forschungsorganisator und Wissenschaftshistoriker. Zu seinem 100. Todesjahr." *Gesnerus* **50** (1993), 223–241. — LUTSTORF, HEINZ: "Professor Rudolf Wolf und seine Zeit." *Schriftenreihe der ETH-Bibliothek Zürich* **31** (1993). — [FREI/STAMMBACH 1994, 13]. — Portrait: [BÖLLING 1994, 9.1]. B.G.

Wylie, Alexander (* April 6, 1815, London, England; † February 6, 1887, Hampstead, London, England). WYLIE, apprenticed to a cabinet-maker, studied Latin and Chinese on his own. Having become sufficiently competent, in 1847 he was sent by the London Missionary Society to supervise its publications in Shanghai. There he learned a number of additional European and Asian languages, and also acquired a special interest in Chinese mathematics. In 1852 he called attention to the fact that WILLIAM HORNER's method of solving algebraic equations had been anticipated by Chinese mathematicians in the 14th century. He was the first to introduce Western readers to the analysis of indeterminate equations due to QIN JIUSHAO (1202–1261). Some of WYLIE's publications later served as the basis of BIERNATZKI's article, "Die Arithmetik der Chinesen" in *Crelles Journal* (1856), which introduced a wider public to Chinese arithmetic (see BIERNATZKI's biography). Among WYLIE's numerous publications (both in English and Chinese) was a Chinese translation (1865) made with the assistance of the Chinese mathematician LI SHANLAN (1811–1882) of books VII–XV of the *Elements* of EUCLID. This was a continuation of the translation begun by the Jesuit missionary MATTEO RICCI in 1607 (see Section 17.1). — In 1877, when his eyesight began to fail, WYLIE returned to England. His library became the foundation of the *Bibliotheca Sinica*.

Main works: *Chinese Researches*. Shanghai 1897; repr. Peking 1936 and Taipei 1966. — BIERNATZKI, KARL L.: "Die Arithmetik der Chinesen." *Crelles J.* **52** (1856), 59–94; repr. Walluf near Wiesbaden 1973; French trans. by O. TERQUEM in *Nouvelles Annales de Mathématiques* **1** (1862), 35–44, and 529–540.

Secondary literature: CORDIER, HENRI: *Life and Labours of Alexander Wylie, Agent of the British and Foreign Bible Society in China.* [n.p.] 1887. C.J.S.

Yanovskaya, Sof'ya Aleksandrovna (* January 31, 1896, Odessa, Russian Empire; † October 24, 1966, Moscow, USSR). SOF'YA YANOVSKAYA was a professor and member of the Faculty of Mathematics and Mechanics of the University of Moscow (1931–1966). A mathematician and philosopher of mathematics, she was also a co-founder (with M. VYGODSKIĬ and A. YOUSHKEVICH) of the Soviet School of the history of mathematics. Among her areas of research interest were history of the axiomatic method, history of mathematical logic and the foundations of mathematics, and history of mathematics in Russia.

Main works: "From the History of the Teaching of Mathematics in the University of Moscow ." In: *IMI* **8** (1955), 127–480 (with I. I. LIKHOLETOV) (in Russian). — *Methodological Problems of Science* (in Russian). Moscow 1972 (with portrait).

Secondary literature: *May* 387. — [KUROSH 1959]. — BASHMAKOVA, IZABELLA G., and ADOLF-ANDREJ YOUSHKEVICH: "Sofia Alexandrovna Yanovskaya (1896–1966." *Arch. int. Hist. Sci.* **20** (1967), 105–107. — [FOMIN/SHILOV 1969]. — [BORODIN/BUGAI 1979]. — [BOGOLYUBOV 1983]. — ANELLIS, IRVING H.: "The Heritage of S. A. Janovskaya." *Hist. Phil. Logic* **8** (1987), 45–56. — ANELLIS, IRVING H.: "Sof'ya Aleksandrovna Yanovskaya's Contributions to Logic and History of Logic." *Modern Logic* **6** (1996), 7–36. — BASHMAKOVA, IZABELLA G., S. S. DEMIDOV, and V. A. USPENSKII: "Sof'ya Aleksandrovna Yanovskaya." *Modern Logic* **6** (1996), 357–372 (in Russian). — BASHMAKOVA, IZABELLA G., *et al.*: "A Passion for Clarity." *Vopr. Ist. Estest. Tekhn.* **49** (1996), 108–119, 173 (in Russian, English summary). — KUSHNER, B. A. : "Some Reminiscences About Sof'ya Aleksandrovna Yanovskaya." *Vopr. Ist. Estest. Tekhn.* **49** (1996), 119–123 (in Russian, translated from English). — "On the Centenary of the Birth of Sof'ya Alexandrovna Yanovskaya." Special section in *IMI* (2) **2 (37)**, (1997), 105–127 (in Russian). S.S.D.

Youshkevich, Adolf-Andreĭ Pavlovich (* July 15, 1906, Odessa, Russian Empire; † July 17, 1993, Moscow, Russia). The son of a well-known Russian philosopher and man of letters, PAVEL SOLOMONOVICH YOUSHKEVICH, ADOLF-ANDREĬ YOUSHKEVICH attended high school in St. Petersburg (1915–1917) and Odessa. In 1923 he entered the University of Moscow from which he graduated in 1929, having studied with both D. F. EGOROV and N. N. LUZIN. The areas of their research — set theory and theory of real functions — to a considerable extent predetermined YOUSHKEVICH's own future interests. The final choice of his academic specialty, however, was strongly influenced by his father who had received his own mathematical education at the Sorbonne in Paris, and was also greatly interested in the history and philosophy of science.

At this same time two young professors at the University of Moscow had begun to study and teach the history of mathematics — SOF'YA A. YANOVSKAYA (head of the seminar on the history and philosophy of science), and MARK YA. VYGODSKIĬ. YOUSHKEVICH formed friendships with both of them and later was invited to share direction of the seminar on history of mathematics which they organized in 1933.

The problems of the foundations of mathematics had always attracted interest in Moscow. They were discussed for example in two seminars, in the Communist Academy in ALEKSANDR YA. KHINCHIN's seminar on the methodology and philosophy of science, and in a seminar on the history and philosophy of science in the K. A. Timiryazev Institute. YOUSHKEVICH participated in both of these. Coupled with his own enthusiasm for the theory of real functions, this led to his choice of topics for his earliest research — the history of foundations of mathematical analysis. This choice was especially timely since MARX's mathematical manuscripts, being studied and actively discussed in Moscow during this period, were mostly devoted to the foundations of analysis. Consequently, study of the history of analysis and its foundations received a degree of ideological protection. These circumstances proved very useful for YOUSHKEVICH. His first publication was an article on LAZARE CARNOT's philosophy of mathematics (1929). Despite his scientific talents, YOUSHKEVICH was unable to become a post-graduate student for political reasons — because his father was considered an adherent of ERNST MACH, LENIN's ideological opponent.

In 1930 YOUSHKEVICH began work at the Supreme Technical School in Moscow. In 1938 he was awarded his first scientific degree, candidate of sciences, without any defense. He received a second degree, doctor of science, from the University of Moscow in 1940, for a dissertation he wrote on the history of mathematics in Russia in the 18th century. That same year he became professor of mathematics at Moscow Supreme Technical School, where he obtained a chair in 1941. In 1952, known in Soviet history as the year of the struggle against cosmopolitanism (*i.e.* what was considered to be a hostile ideology of the decadent West), a year marked by a wave of repression which led to a campaign directed first of all against Jews, YOUSHKEVICH was removed from his chair at the Supreme Technical School. Only with very great difficulty was he able to maintain his position at the Institute for the History of Science and Technology of the U.S.S.R. Academy of Sciences in Moscow, where he worked from 1945 until the end of his life. In the 1970s he headed a special group at the Institute on the history of mathematics.

During his long scientific career, YOUSHKEVICH wrote nearly 400 scientific papers on different problems of the history of science. Although the scope of his work is remarkable, it is possible to identify several main areas of his research interests:

- The origins and development of the principal notions of mathematical analysis;
- The history of mathematics in the 18th century, above all EULER's works;
- Medieval mathematics in Europe and the Orient;
- History of mathematics in Russia.

On the first of these subjects YOUSHKEVICH worked his entire life. The studies he devoted to the development of the notion of function from antiquity up to the end of the 19th century are widely known and were conveniently summarized in a paper published in the *Archive for History of Exact Sciences* (1976). His commentaries accompanying editions of classic works of mathematicians of the 17th and 18th centuries — those of DESCARTES, NEWTON, LEIBNIZ, and LAZARE CARNOT, for example — deserve mention, along with chapters he contributed on the development of mathematical analysis for the *History of Mathematics from Ancient Times up to the Beginning of the 19th Century* (1970–1972), a collaborative work in three volumes directed by YOUSHKEVICH [YOUSHKEVICH 1970/72].

There is also an important paper on the development of the notion of limit up to the contributions of KARL WEIERSTRASS (1986), as well as YOUSHKEVICH's research on LEONHARD EULER. YOUSHKEVICH was especially interested in EULER's concept of function, in his generalization of the notion of the sum of a series, and in his famous trilogy on mathematical analysis. For many years YOUSHKEVICH was occupied with the edition of EULER's works, specifically the fourth series of EULER's *Opera Omnia* (published with the collaboration of VLADIMIR I. SMIRNOV, EDUARD WINTER, ASHOT T. GRIGORIAN, and RENÉ TATON, among others). It is important to emphasize that YOUSHKEVICH was very much attracted to the 18th century — the Century of Reason —that was most congenial with his own rational mind.

Another focus of his research was medieval mathematics. His book *The History of Mathematics in the Middle Ages* (1961) has been translated into six languages and has become a basic handbook for historians of medieval science [YOUSHKEVICH 1961]. This work contains serious arguments in support of his views on medieval mathematics which he first advanced in 1951. According to YOUSHKEVICH, medieval mathematics has its own specific character which shares certain characteristics in common with various mathematical cultures of Europe and Asia. Here, as in his other works, YOUSHKEVICH shows a tendency to advance philosophical generalizations and to find general patterns in what at first sight may only seem to be heterogeneous systems. Successive translations of his book have included the author's additions, whereby he has incorporated the results of the most recent scientific research. In the Soviet Union YOUSHKEVICH was the founder of a school which soon became internationally known of historians of medieval Arabic mathematics that included BORIS A. ROZENFELD, GALINA P. MATVIEVSKAYA, MIRIAM M. ROZHANSKAYA, and others.

YOUSHKEVICH began his studies on mathematics in Russia prior to World War II. In addition to his doctoral dissertation mentioned above, he also devoted several special studies to S. E. GUR'EV, NIKOLAI I. LOBATCHEVSKIĬ, and MIKHAIL V. OSTROGRADSKIĬ, among others, which eventually resulted in a major study, a *History of Mathematics in Russia Before 1917* [YOUSHKEVICH 1968]. In this book, YOUSHKEVICH surveys the history of mathematics in Russia not only in the context of its social and cultural history, but as a natural part of the general process of the development of world science as well.

YOUSHKEVICH also devoted considerable time to editorial activities. He directed the monumental *History of Mathematics from Ancient Times up to the Beginning of the 19th Century*, which appeared in three volumes [YOUSHKEVICH 1970/72]. He founded the series *Mathematics of the 19th Century*, which he co-directed with the mathematician ANDREI N. KOLMOGOROV. He also directed and actively participated in the edition of a *Source Book on the History of Mathematics* (1976–1977) which has been widely utilized in education [YOUSHKEVICH 1976/77]. One of the most important contributions YOUSHKEVICH made to the history of mathematics was the creation, in 1948, along with GEORGII F. RYBKIN, of the journal *Istoriko-Matematicheskie Issledovaniya* (Research on the History of Mathematics), of which 35 volumes were published under his editorship. YOUSHKEVICH served as editor in chief until his death, and many of the most important studies on the history of mathematics produced in the Soviet Union were published in this journal. At the same time he was head (at first together with VYGODSKIĬ and YANOVSKAYA, later with the collaboration of IZABELLA G. BASHMAKOVA and KONSTANTIN A. RYBNIKOV) of the oldest seminar on the history of mathematics in the country, held at the University of Moscow. For all of these reasons, A. P. YOUSHKEVICH has exercised a strong influence on virtually all Soviet historians of mathematics. Among hallmarks of his style, he always demanded exactness in the matter of quotations and clarity in the interpretation of sources, as well as rationally well founded conclusions. YOUSHKEVICH had many students, among whom was FËDOR A. MEDVEDEV.

Among the Soviet scientists who began to participate actively in the international Scientific Community at the end of the 1950s, YOUSHKEVICH was one of the first to set about helping to restore the connections severed by the tragic events associated with the Cold War that had divided the world essentially into two blocks. He rapidly earned an international reputation, was elected into various academies and scientific societies, and was also awarded several prizes, including the *Alexandre Koyré Medal*, the *George Sarton Medal*, and the *Kenneth O. May Medal.* From 1965–1968 he served as president of the International Academy of the History of Science.

Further publication: *Matematika v ee Istorii.* Moskva 1996.

Secondary literature: *May* 388. — CHEMLA, KARINE: "Adolf Andrei Pavlovitch Youschkevitch. Interviewed by Karine Chemla." *NTM* **28** (1991), 1–11. — ZAITSEV, E.: "Adolf Pavlovich Yuskevich." *Stud. Leibnitiana* **25** (1993), 129–131. — BASHMAKOVA, ISABELLA, *et al.*: "*In memoriam:* Adolph Andrei Pavlovich Yushkevich (1906–1993)." *HM* **22** (1995), 113–118. — Bibliography: *IMI* **21** (1976), 312–327; **30** (1986), 352–357; 2nd series **1** (36), no. 1 (1995), 13–19. — Memorial articles: *IMI*, 2nd series **1** (36), no. 1 (1995), 7–13, 19–39 (with portrait). — [VOGEL/YOUSHKEVICH 1997, with portrait]. S.S.D.

Zeuthen, Hieronymus Georg (* February 15, 1839, Grimstrup, Denmark; † January 6, 1920, Copenhagen, Denmark). ZEUTHEN's scholarship in the history of mathematics is characterized by its penetrating analysis of the mathematical structure of the classics and of the logical connections between these works. He always stressed that he made his contributions to this field not as an historian, but as a professional mathematician.

HIERONYMUS ZEUTHEN was a son of the minister FREDERIK LUDVIG BANG ZEUTHEN and MAGDALENE HEDEVIG JOHANNE LAUB, who together had five children. When ZEUTHEN was ten years old the family moved to Sorø where in school he met his later colleague JULIUS PETERSEN (1839–1910). As a result of their common interest, they both went on to study mathematics at the University of Copenhagen (ZEUTHEN in 1857). A year after ZEUTHEN graduated in 1862, he had shown such talent in geometry that he obtained a scholarship to go to Paris to study enumerative methods with the master, MICHEL CHASLES. On his return to Copenhagen, ZEUTHEN defended his doctoral thesis on *Systems of Conics Which are Subject to Four Conditions* (1865), the methods of which he later generalized to surfaces of the second degree. He soon became known (together with HERMANN SCHUBERT) as an expert on enumerative geometry, and was chosen to write the survey of this subject for the *Encyklopädie der mathematischen Wissenschaften* (1906). Eight years later he completed his scientific career with the publication of his classic *Lehrbuch der abzählenden Methoden der Geometrie* (Textbook on Enumerative Geometry) (1914). He taught this subject as well as many other subjects, including history of mathematics, at the University of Copenhagen first as a *privatdocent* (beginning in 1866), then as a *docent* (from 1871), and finally as a professor from 1883 until his retirement in 1910. He was elected a member of the Royal Danish Academy of Sciences and Letters in 1872, and served as its secretary for 39 years. ZEUTHEN was married three times; in 1867 to JULIE HENRIETTE JESPERSEN who died nine years later; in 1879 to her older sister, LOUISE MARIE CHRISTINE, who died in 1886 and with whom he had one son; and in 1887 to SOPHIE CHRISTINE FREDERIKKE LAWAETZ, with whom he had two sons and a daughter.

ZEUTHEN's geometric research gave him a Platonic view of mathematics that was at the very heart of his historical work. According to ZEUTHEN, mathematics deals with unchanging ideas which can be expressed in different forms. When doing mathematics he wanted to find proofs which most clearly reflected the ideal core of the matter. Such proofs would often be close to the intuitive argument by which the theorem had been found. For this reason ZEUTHEN did not agree with those of his contemporaries who argued that a proof had to be arithmetical in order to be rigorous. He preferred geometric proofs of geometric theorems.

When doing history of mathematics he wanted to uncover the ideas and motives of the ancient masters. These ideas, he argued, were usually formulated in an unfamiliar language, but since the ideas themselves had not changed over time, it was possible for a modern mathematician to appreciate the work of a colleague 2000 years earlier. Still, ZEUTHEN repeatedly underlined that one cannot evaluate or understand mathematics of an earlier period on the basis of the mathematics of today. On the contrary, he thought it indispensable to be acquainted with the techniques and symbolism of former times in order to contrast those tools and what they could be used for with what they had actually been used for.

When ZEUTHEN wrote his first series of papers on the history of mathematics (1876–1883) entitled *From the History of Mathematics* (in Danish), he explicitly stated that his purpose was not historical but didactic and mathematical. He hoped that he could develop a mathematical sense in his mathematics students by showing them how a subject had previously been treated. In later publications he also emphasized that his approach to the history of mathematics could help research mathematicians by calling attention to forgotten procedures which, though less general or automatic than modern streamlined methods, might lead to a deeper understanding of particular cases and open up new areas of research. ZEUTHEN soon discovered that his approach also allowed him to contribute substantially to the history of mathematics itself, in particular to the history of ancient Greek mathematics. Indeed, the ancient masterpieces were built up according to a logically deductive plan which does not correspond to the way in which the results had originally been obtained. A critical mathematical analysis combined with a study of scattered evidence in the later commentators could be used to reveal the process of discovery and the more intuitive ideas behind the rigorous façade. This in turn might allow the reconstruction of lost works, the interpretation of foggy hints, and the connection of seemingly unconnected ideas.

ZEUTHEN demonstrated the value of this approach in his first major work, *The Theory of Conic Sections in Antiquity* (in Danish 1884 and German 1886). The synthetic style of APOLLONIUS's books had caused speculation about the original discovery of the results. Several 19th-century mathematicians had advanced the idea that the Greeks had used analytic geometry to find these results, but that they had kept the methods secret in order to impress their readers. ZEUTHEN found this hypothesis "tasteless." Instead, he insisted that the Greeks possessed a method which in many ways was equivalent to analytic geometry but which had a different form. Finding this method, which he called geometric algebra, was one of his most important contributions to interpreting the history of mathematics. In 1882 PAUL TANNERY had pointed out that EUCLID's book VI contains a geometric solution of quadratic equations, and he conjectured that the Pythagoreans had already derived this treatment from an arithmetic procedure, in order to circumvent the problem of incommensurable quantities. ZEUTHEN went further. In book II of the *Elements* he found a theory developed entirely deductively which in geometric language expressed many of the relations that are now expressed in arithmetic language, such as $(a+b)^2 = a^2 + 2ab + b^2$ (EUCLID II.4).

Geometric algebra provided ZEUTHEN with a means of analyzing APOLLONIUS's *Conics*. Indeed, he observed that after APOLLONIUS had derived the symptoms of the conic sections (i.e. the equivalents of their equations), the rest of the work was based entirely on these symptoms. ZEUTHEN's analysis extended and even contradicted the standard interpretation of APOLLONIUS. For example, the commentator GEMINUS had claimed that APOLLONIUS was the first to see that arbitrary sections in arbitrary cones would yield the same conic sections that earlier mathematicians had obtained through sections perpendicular to a generator in a right cone. ZEUTHEN refuted this by calling attention to the fact that ARCHIMEDES explicitly mentioned, as something well-known, that all sections of a cone whose planes meet all generators are ellipses or circles. With the kind of mathematical argument typical of ZEUTHEN, he then concluded that it was unthinkable that ARCHIMEDES and his predecessors could have known this without knowing similar facts about the parabola and hyperbola. Instead, ZEUTHEN interpreted GEMINUS's statement to concern the names of the curves only. Another example is ZEUTHEN's attempt to make sense of APOLLONIUS's claim that his third book contained a solution of the so-called problem of the three and four line loci. By means of an intricate analysis, ZEUTHEN tried to explain that APOLLONIUS's book three implicitly contains a solution, why EUCLID might have failed to solve the problem, and how EUCLID's lost *Porisms* might have been developed in this connection. These few examples suffice to show the strength of ZEUTHEN's approach to the history of mathematics. By entering into the mathematical spirit of the time, he was able to find mathematically plausible reconstructions of non-surviving theories, to evaluate surviving sources, and to show that some commentators (GEMINUS, for example), were probably mistaken.

ZEUTHEN's work on conic sections forced him to embark on a comprehensive study of Greek mathematics. Over the following years he continued to investigate further traces of the use of geometric algebra as well as the use of infinitesimal methods in Greek texts. These investigations resulted in his book *Lectures on the History of Mathematics; Antiquity and Middle Ages* (Danish 1893; German 1896; French 1902), which contained several new penetrating observations. For example, ZEUTHEN interpreted Greek constructions as existence proofs and the parallel postulate as a postulate securing the existence of the intersection of two lines. After 1893 he extended his studies to the 16th and 17th centuries, and was particularly interested in determining how geometric algebra was transformed into analytic geometry. Therefore he studied VIÈTE, FERMAT, and DESCARTES with special care. The development of the infinitesimal calculus also caught his interest. In particular, he re-evaluated BARROW's insights into the inverse nature of tangent constructions and the determination of areas.[8] ZEUTHEN published the results of his research first in a series of papers (1893–1900) in French, and then in his third historical monograph *Lectures on the History of Mathematics: 16th and 17th Centuries* (1903 in Danish and German). ZEUTHEN continued to publish new contributions to the history of mathematics until the year before he died. Among his major works are: *The Development of Mathematics as an Exact Science until the End of the 18th Century* (1896) and *How Mathematics in the Period from Plato to Euclid Became a Rational Science* (1917), both in Danish.

It is probable that it was CHASLES who aroused ZEUTHEN's interest in the history of mathematics when ZEUTHEN was in Paris. In fact, ZEUTHEN's first historical paper dealt with one of CHASLES's favorite subjects: Hindu mathematics. But it was another thirteen years before ZEUTHEN published anything of an historical nature. His interest in the history of conic sections was a natural result of his mathematical research in this area, but according to a letter to TANNERY, it was the publication of the first volume of MORITZ CANTOR's *Vorlesungen über Geschichte der Mathematik* that convinced him to pursue and publish his ideas:

> But it was the book of M. Cantor which showed me, by its mistakes in this respect, how necessary it was to make a profound analysis of the ancient theory of conic sections [TANNERY, *Mémoires Scientifiques* **16**, 645].

When CANTOR published the second edition of his *Vorlesungen* in 1894 and failed to mention ZEUTHEN's work on conic sections, ZEUTHEN complained in a rather aggressive note, "M. Maurice Cantor et la géométrie supérieure de l'antiquité," (M. Moritz Cantor and the Advanced Geometry in Antiquity) (*Bull. Sci. Math.* **18** (1894), 163–169). In an immediate reply, CANTOR elegantly succeeded in conveying his opposition to ZEUTHEN's approach in the title of his article, "M. Zeuthen et sa géométrie supérieure de l'antiquité" (M. Zeuthen and His Advanced Geometry in Antiquity), (*Bull. Sci. Math.* **19** (1895), 64–69). These two publications inaugurated a long dispute. In a later review of ZEUTHEN's *Lectures*, CANTOR made it even clearer that although ZEUTHEN's analyses might be called *geistvoll* (clever), CANTOR did not consider them as contributions to the history of mathematics. But others thought highly of ZEUTHEN's historical research; TANNERY, for example, who wrote laudatory reviews of ZEUTHEN's books and had the *Lectures on History of Mathematics: Antiquity and Middle Ages* translated into French. As one might expect, ZEUTHEN was also highly esteemed by mathematicians, and he was chosen (probably by FELIX KLEIN) to write the sections on *Die Mathematik im Alterthum und im Mittelalter* (Mathematics in Antiquity and the Middle Ages) for the collection *Kultur der Gegenwart.*

[8]In a recent paper SKULI SIGURDSSON has analyzed ZEUTHEN's comparison of the methods of NEWTON and LEIBNIZ, and he concludes that it reflects ZEUTHEN's elitist conception of mathematics [Sigurdsson 1992].

The Danish philologist JOHAN LUDVIG HEIBERG, who gained fame through his authoritative editions of the works of Greek mathematicians, was probably the person whom ZEUTHEN consulted most often in connection with his historical research. The achievements of these two connoisseurs of Greek mathematics stand out as remarkable examples of a fruitful collaboration between a scientist and a humanist. Many letters are preserved in which ZEUTHEN explains technical matters to HEIBERG. Considering that the two experts met at least every two weeks in the Royal Danish Academy, there is no doubt that their collaboration was much more intense than the written evidence shows. They published two joint papers (1906 and 1907), the first of which created a great sensation. It contained ARCHIMEDES's so-called *Method*, thought to be lost, but which HEIBERG had found as a palimpsest in Constantinople. HEIBERG provided a translation of the text and ZEUTHEN wrote the mathematical commentary. The content particularly pleased ZEUTHEN, for it showed that the ancients had been in possession of an intuitive infinitesimal method by which they first found the results that they later proved by the method of exhaustion. This was what ZEUTHEN and others had guessed on the basis of mathematical analyses of the surviving sources. Thus, as ZEUTHEN himself pointed out, HEIBERG's discovery brilliantly supported ZEUTHEN's approach to the history of mathematics.

Secondary literature: *DSB* **14**, 618–619. *May* 389. *P* III, 1482; IV, 1689; V, 1410; VI, 2963. — [*Acta Math.*, Tab. Gen. **1–35** (1913), 124; portrait 178]. — NOETHER, MAX: "Hieronymus Georg Zeuthen." *Math. Ann.* **83** (1921), 1–23. — KLEIMAN, STEVEN L.: "Hieronymus Georg Zeuthen (1839–1920)." *Contemp. Math.* **123** (1991), 1–13. — SIGURDSSON, SKULI: "Equivalence, Pragmatic Platonism, and Discovery of the Calculus." In: *The Invention of Physical Science*, ed. M. J. NYE, *et al.*, Dordrecht 1992, 97–116. — [LÜTZEN/PURKERT 1994]. — Portrait: [BÖLLING 1994, 1.1]. J.L.

Zubov, Vasiliĭ Pavlovich (* August 1, 1899, Aleksandrov, Vladimir Region, Russian Empire; † April 8, 1963, Moscow, USSR). VASILIĬ ZUBOV was a researcher in the Institute for the History of Science and Technology of the USSR Academy of Sciences in Moscow from 1946. He was an expert on the history of medieval European science, and published (with commentaries) the arithmetic text of KIRIK FROM NOVGOROD.

Works: The publication and the commentaries of the arithmetic text of KIRIK FROM NOVGOROD. In: *IMI* **6** (1953), 174–214 (in Russian). — "Quelques observations sur l'auteur du traite anonyme *Utrum alicuius quadrati sit commensurabilis costae eiusdem*." *Isis* **50** (1959), 130–134.

Secondary literature: GRIGORIAN, ASHOT T., B. G. KUZNETSOV, and A. P. YOUSCHKEVITCH: "Vassili Pavlovitch Zoubov." *Arch. int. Hist. Sci.* **16** (1963), 305–306. — [KUROSH 1959]. — [FOMIN/SHILOV 1969]. — [BORODIN/BUGAI 1979]. — [BOGOLYUBOV 1983]. — Portrait in [Shtokalo 1966/70, vol. 4, book 2, p. 493]. S.S.D.

Part III

Abbreviations, Bibliography and Index

Abbreviations

Abbreviation	Journal/Work of Reference
Abh. Bayer. Akad. der Wiss., math.-nat. Kl.	Abhandlungen der Bayerischen Akademie der Wissenschaften, mathematisch-naturwissenschaftliche Klasse
Abh. Braunschw. Wiss. Ges.	Abhandlungen der Braunschweigischen Wissenschaftlichen Gesellschaft
Abh. dt. Akad. Wiss., Kl. Math., Phys. u. Techn.	Abhandlungen der Deutschen Akademie der Wissenschaften zu Berlin, Klasse für Mathematik, Physik und Technik
Abh. Gesch. Math.	Abhandlungen zur Geschichte der Mathematik
Abh. Gesch. math. Wiss.	Abhandlungen zur Geschichte der mathematischen Wissenschaften mit Einschluss ihrer Anwendungen
Abh. Gesch. Med. Nat.	Abhandlungen zur Geschichte der Medizin und der Naturwissenschaften
Abh. Gesch. Nat. Techn.	Abhandlungen zur Geschichte der Naturwissenschaften und der Technik
Abh. Kön. Böhm. Ges. Wiss.	Abhandlungen der königlichen böhmischen Gesellschaft der Wissenschaften
Abh. math. Sem. Univ. Hamburg	Abhandlungen aus dem mathematischen Seminar der Universität Hamburg
Acta hist. rer. nat. techn.	Acta historiae rerum naturalium necnon technicarum (Prague)
Acta Math.	Acta Mathematica
ADB	Allgemeine Deutsche Biographie
AMM	The American Mathematical Monthly
Ann. Sci.	Annals of Science
Ann. Univ. Paris	Annales de l'Université de Paris
Arch. Gesch. Math. Nat. Techn.	Archiv für Geschichte der Mathematik, der Naturwissenschaften und der Technik
Arch. Gesch. Nat. Techn.	Archiv für die Geschichte der Naturwissenschaften und der Technik
Arch. Hist. ex. Sci.	Archive for History of Exact Sciences
Arch. int. Hist. Sci.	Archives internationales d'histoire des Sciences

Abbreviation	Journal/Work of Reference
Arch. Stor. Sci.	Archivio di Storia della Scienza
Atti Accad. Pont.	Atti Accademia Pontaniana
Atti Adunanze Ist. Veneto Sci. Lett. Arti	Atti Adunanze d'Istituto Veneto di Scienze Lettere ed Arti
Atti Ist. Veneto Sci. Lett. Arti	Atti Reale Istituto Veneto di Scienze Lettere ed Arti
Atti Mem. Accad. Sci. Modena	Atti e Memorie Accademia delle Scienze di Modena
A. V. Z. Math. Naturw. Blätter	A. V. Z. Mathematisch-Naturwissenschaftliche Blätter. Zeitschrift des Arnstädter Verbandes Mathematischer und Naturwissenschaftlicher Verbindungen an deutschen Hochschulen im Deutschen Wissenschafter-Verbande
Bad. Biogr.	Badische Biographien
Bayer. Bl. Gymn.	Bayerische Blätter für das Gymnasialschulwesen
BBKl	Biographisch-bibliographisches Kirchenlexikon
BCJ	De Backer, Augustin, and Aloys Carlos Sommervogel: Bibliothèque de la Compagnie de Jésus. 11 vols. Brussels: Oscar Schepens, and Paris: Alphonse Picard 1890–1932. Repr. Heverlee-Leuven: Editions de la Bibliothèque S.J., 1960. 12 vols.
Ber. Wiss. Gesch.	Berichte zur Wissenschaftsgeschichte
Bibl. diss.	Bibliotheca dissidentium. Répertoire des non-conformistes religieux des seizième et dix-septième siècles, André Séguenny (Ed.). Baden-Baden: Éditions Valentin Koerner, 1980–1999. 20 vols. published
Bibl. Math. (1)	Bibliotheca Mathematica, First Series (1884–1887)
Bibl. Math. (2)	Bibliotheca Mathematica, Second Series (1887–1899)
Bibl. Math. (3)	Bibliotheca Mathematica, Third Series (1900–1914)
Biogr. hervorr. Nat. Techn. Med.	Biographien hervorragender Naturwissenschaftler, Techniker und Mediziner
Biogr. Mem. FRS	Biographical Memoirs of Fellows of the Royal Society
Biogr. Mem. INSA	Biographical Memoirs of Fellows of the Indian National Science Academy
Biogr. Mem. NAS	Biographical Memoirs of the National Academy of Sciences (USA)
BJ	Biographisches Jahrbuch und Nekrolog
BNB	Biographie Nationale, publiée par l'Académie Royale des Sciences, des Lettres et des Beaux-Arts de Belgique. 49 vols. Brussels: 1866–1986
Bol. Acad. Nac. Cie. Cordoba	Bollettino de la Academia Nacional de Ciencias ... Cordoba
Bol. Arch. Gen. Nac.	Boletín del Archivo General de la Nación

Abbreviation	Journal/Work of Reference
Boll. Loria	Bollettino di bibliografia e storia delle scienze matematiche (LORIA)
Boll. Mat.	Bollettino di Matematica
Boll. UMI	Bollettino dell'Unione Matematica Italiana
Brit. J. Hist. Phil. Sci.	The British Journal for the History and Philosophy of Science
Brit. J. Hist. Sci.	British Journal of the History of Science
BUAM	Biographie Universelle (MICHAUD) ancienne et moderne. 2nd. ed. 45 vols. Paris: C. Desplaces, and Leipzig: F.A. Brockhaus, 1854
Bull. AMS	Bulletin of the American Mathematical Society
Bull. Boncompagni	Bullettino di bibliografia e di storia delle scienze matematiche e fisiche (BONCOMPAGNI)
Bull. LMS	Bulletin of the London Mathematical Society
Bull. Math.	Bulletin mathématique
Bull. Sci. Math.	Bulletin des sciences mathématiques
Bull. Soc. R. Sci. Liège	Bulletin de la Société royale des sciences de Liège
BWN	Biografisch Woordenboek van Nederland. 4 vols. published. The Hague: Instituut voor Nederlandse Geschiedenis, 1979–1994.
Byzantium	Byzantium. Revue internationale des études Byzantines
Cas. Pest. Mat. Fys.	Časopis pro pěstování mathematiky a fysiky
Centralbl. Bibl.	Centralblatt für Bibliothekswesen
Class. Philol.	Classical Philology
Contemp. Math.	Contemporary Mathematics
CR	Comptes rendus des séances hebdomadaires de l'Académie des sciences
Crelles J.	Crelles Journal für die reine und angewandte Mathematik
DAB	Dictionary of American Biography
DBI	Dizionario Biografico degli Italiani, Roma, since 1960
DBJ	Deutsches Biographisches Jahrbuch, 1917–1920. Stuttgart 1928
DBL	Dansk Biografisk Leksikon, 3rd edition, 16 vols., København: Gyldendal, 1979–1984
DGB	Deutsches Geschlechterbuch
DNB	Dictionary of National Biography (Oxford University Press, 1885)/Dictionary of National Biography from the Earliest Times to 1900 (London, 1885), with Supplements to 1990, publ. 1996
DSB	Dictionary of Scientific Biography
Dt. Math.	Deutsche Mathematik

Abbreviation	Journal/Work of Reference
EI	The Encyclopaedia of Islam. New Edition. Leiden: E. J. Brill, since 1996
EIT	Enciclopedia Italiana Treccani
Elem. Math.	Elemente der Mathematik
Enseign. Math.	L'Enseignement mathématique
Euclides	Euclides (Groningen)
FF	Forschungen und Fortschritte
Gaceta Mat.	Gaceta Matemática
Gaṇita-Bhāratī	Gaṇita-Bhāratī. Bulletin of the Indian Society for History of Mathematics
GEPB	Grande Enciclopédia Portuguesa e Brasileira
HBLS	Historisch-Biographisches Lexikon der Schweiz. 7 vols., 1 suppl. Neuenburg: Attinger, 1921–1934
Hist. Phil. Logic	History and Philosophy of Logic
Hist. Sci.	History of Science
Hist. Techn.	History and Technology
HM	Historia Mathematica
HSPS	Historical Studies in the Physical Sciences
Hum. Techn.	Humanismus und Technik
IMI	Istoriko-Matematicheskie Issledovaniya
Isis	Isis
JASB	Journal of the Asiatic Society of Bengal
Jber. DMV	Jahresbericht der Deutschen Mathematiker-Vereinigung
Jber. Fortschr. klass. Altertumswiss.	Jahresbericht über die Fortschritte der klassischen Altertumswissenschaft
J. Comb. Theory	Journal of Combinatorial Theory
JHA	Journal for the History of Astronomy
JHBS	Journal of the History of Behavioral Science
JHM	Journal for the History of Medicine
J. Lond. Math. Soc.	Journal of the London Mathematical Society
JS	Journal des savants
Lias	Lias. Sources and Documents Relating to the Early Modern History of Ideas
Llull	Llull. Revista de la Sociedad Española de Historia de las Ciencias y de las Técnicas (until 1985: Boletín de la Sociedad Española de Historia de las Ciencias)
MARS	Mémoires de l'Académie royale des sciences, Paris
Math. Ann.	Mathematische Annalen
Math. Didact.	Mathematica Didactica
Math. Gaz.	Mathematical Gazette
Math. Intell.	Mathematical Intelligencer
Math. Nachr.	Mathematische Nachrichten

Abbreviation	Journal/Work of Reference
Math.-phys. Semesterber.	Mathematisch-Physikalische Semesterberichte zur Pflege des Zusammenhangs von Schule und Universität
Math. Rev.	Mathematical Reviews
Mathesis	Mathesis (Mexico)
May	MAY, KENNETH O.: Bibliography and Research Manual of the History of Mathematics. University of Toronto Press, 1973
Mber. DAW Berlin	Monatsberichte der Deutschen Akademie der Wissenschaften zu Berlin
Mém. Acad. Sci. Danemark	Mémoires de l'Académie des sciences de Danemark
Mem. Litt. Acad. Sci. Lisboa	Memorias de Litteratura da Academia das Sciencias de Lisboa
MGMN	Mitteilungen zur Geschichte der Medizin und der Naturwissenschaften
MGMNT	Mitteilungen zur Geschichte der Medizin, der Naturwissenschaften und der Technik
Mh. Math. Phys.	Monatshefte für Mathematik und Physik
Mitt. DMV	Mitteilungen der Deutschen Mathematiker-Vereinigung
Mitt. Math. Ges. Hamburg	Mitteilungen der Mathematischen Gesellschaft in Hamburg
Mitt. Nat. Ges. Bern	Mittheilungen der Naturforschenden Gesellschaft in Bern
Mitt. Österr. Ges. Wiss.-Gesch.	Mitteilungen der Österreichischen Gesellschaft für Wissenschaftsgeschichte
M. N. Bl.	Mathematisch-naturwissenschaftliche Blätter
MNU	Der Mathematische und Naturwissenschaftliche Unterricht
Naturw. Rdsch.	Naturwissenschaftliche Rundschau
NBG	HOEFER, J. CH. F.: Nouvelle biographie générale. 46 vols. Paris: Firmin Didot, 1852–1866
NBL	Norsk Biografisk Leksikon, 19 vols. Kristiania/Oslo: Aschehoug, 1921–1971
Nbl. DGGMNT	Nachrichtenblatt der Deutschen Gesellschaft für Geschichte der Medizin, Naturwissenschaft und Technik e.V.
NBW	Nationaal Biografisch Woordenboek. 15 vols. published. Brussels: Koninklijke Vlaamse Academiën van België, 1964–1996
NDB	Neue Deutsche Biographie
Nieuw Arch. Wiskd.	Nieuw Archief voor Wiskunde
NNBW	Nieuw Nederlands Biografisch Woordenboek. 10 vols. Leiden: A. W. Sijthoff, 1910–1937

Abbreviation	Journal/Work of Reference
Normat	Nordisk Matematisk Tidsskrift, later Normat
Norsk Geogr. Selsk. Årbog	Norske Geografiske Selskab. Oslo. Årbog
Notes Rec. R. Soc. Lond.	Notes and Records of the Royal Society of London
Notices AMS	Notices of the American Mathematical Society
NTM	NTM. Schriftenreihe zur Geschichte der Naturwissenschaften, Technik und Medizin (Leipzig), *continued since 1993 as* NTM. Internationale Zeitschrift für Geschichte und Ethik der Naturwissenschaften, Technik und Medizin/International Journal for History of Natural Sciences, Technology and Medicine (Basel)
Obituary Notices Fellows RS	Obituary Notices of Fellows of the Royal Society
Österreich. Akad. Wiss., Math.-Natur. Kl., Sber. II	Österreichische Akademie der Wissenschaften, Mathematisch-Naturwissenschaftliche Klasse, Sitzungsberichte II
Ostw. Klass.	Ostwalds Klassiker der exakten Wissenschaften
P	POGGENDORFF, JOHANN CHRISTIAN: Biographisch-literarisches Handwörterbuch zur Geschichte der exakten Wissenschaften
Period. Mat.	Periodico di Matematiche
PGM	Petermanns geographische Mitteilungen
Physik. Bl.	Physikalische Blätter
Physik. Zs.	Physikalische Zeitschrift
Prax. Math.	Praxis der Mathematik. Monatsschrift der reinen und angewandten Mathematik im Unterricht
Proc. APS	Proceedings of the American Philosophical Society
Proc. BA	Proceedings of the British Academy
Proc. Edinburgh Math. Soc.	Proceedings of the Edinburgh Mathematical Society
Proc. London Math. Soc.	Proceedings of the London Mathematical Society
Quaderni P.RI.ST.EM	Quaderni *Progetto di Ricerche Storiche e Epistemologiche*
Q. St. G. Math. Astron. Phys.	Quellen und Studien zur Geschichte der Mathematik, [Astronomie und Physik]
Rend. Accad. Naz. Lincei, Classe Sci. Fis. Mat. Nat.	Rendiconti dell'Accademia Nazionale dei Lincei, Classe di Scienze Fisiche, Matematiche e Naturali
Rend. Accad. Sci. Ist. Bologna	Rendiconti dell'Accademia delle Scienze dell'Istituto di Bologna
Rend. Ist. Lomb. Sci. Lett. Parte Generale	Rendiconti dell'Istituto Lombardo di Scienze e Lettere. Parte Generale
Rev. Hist. Math.	Revue d'histoire des mathématiques
Rev. Hist. Sci.	Revue d'histoire des sciences

Abbreviation	Journal/Work of Reference
Rev. Quest. sci.	Revue des questions scientifiques
Sb. Akad. Heid., Math.-nat. Kl.	Sitzungsberichte der Heidelberger Akademie der Wissenschaften, Mathematisch-naturwissenschaftliche Klasse
Sb. Berliner Math. Ges.	Sitzungsberichte der Berliner Mathematischen Gesellschaft
SBL	Svenskt Biografiskt Lexikon, Stockholm, since 1918
Sb. Phys.-Med. Soc. Erlangen	Sitzungsberichte der Physikalisch-Medizinischen Sozietät zu Erlangen
Sci. Persp.	Science in Perspective
Scientia	Scientia, Rivista di scienza
Scripta Math.	Scripta Mathematica
Semester-Ber.	Semester-Berichte zur Pflege des Zusammenhangs von Universität und Schule im mathematischen Unterricht
SL	Schweizer Lexikon. 6 vols. Luzern 1991–1993
Stud. Leibnitiana	Studia Leibnitiana
Stud. Mater. Dziej. Nauki. Pol.	Studia i materiały z dziejów nauki polskiej, Series C.
Sudh. Arch.	Sudhoffs Archiv
Techn. Culture	Technology and Culture
TIIET	Trudy Instituta Istorii Estestvoznaniya i Tekhniki
Unterrichtsbl. Math. Nat.	Unterrichtsblätter für Mathematik und Naturwissenschaften
VdAa	VAN DER AA, A. J.: Biographisch Woordenboek der Nederlanden. Haarlem: J.J. van Brederode, 1852–1878. Repr. Amsterdam: B.M. Israel 1969
Vh. Nat. Ges. Basel	Verhandlungen der Naturforschenden Gesellschaft in Basel
Vh. Schw. Nat. Ges.	Verhandlungen der Schweizerischen Naturforschenden Gesellschaft
Vj. Nat. Ges. Zür.	Vierteljahrsschrift der Naturforschenden Gesellschaft in Zürich
Vopr. Ist. Estest. Techn.	Voprosy Istorii Estestvoznaniya i Tekhniki
Wiad. Mat.	Wiadomości Matematiczne
Wiss. Z. Univ. Greifswald	Wissenschaftliche Zeitschrift der Universität Greifswald
Math.-Nat. Reihe	Mathematisch-Naturwissenschaftliche Reihe
Z. dt. MG.	Zeitschrift der Deutschen Morgenländischen Gesellschaft
Z. Math. Phys.	Zeitschrift für Mathematik und Physik
Z. Math. Phys., Hist.-lit. Abt.	Zeitschrift für Mathematik und Physik, Historisch-literarische Abtheilung
ZMNU	Zeitschrift für mathematischen und naturwissenschaftlichen Unterricht

Bibliography

[ABREU 1813] ABREU, JOÃO MANUEL DE: "Nota sobre varios logares da censura dos redatores do *Edinburg Review* aos Principios Mathematicos do dr. José Anastácio da Cunha." *O Investigador Portuguez em Inglaterra* **8** (1813), 235–249, 442–455, 612–623.

[ABU-LUGHOD 1965] ABU-LUGHOD, IBRAHIM: *Arab Discovery of Europe. A Study in Cultural Encounters.* Princeton, NJ: Princeton University Press 1965.

[*Acta Math.*, Tab. Gen. **1–35** (1913)] *Acta Math. 1882–1912: Table Générale des Tomes 1–35*, RIESZ, MARCEL (Ed.) (with biographies and portraits). Berlin *et al.* 1913.

[ADAMS/NEUGEBAUER 1955a] ADAMS, C. R., and OTTO NEUGEBAUER: "Obituary. Raymond Clare Archibald. *In Memoriam.*" *AMM* **62** (1955), 743–745.

[ADAMS/NEUGEBAUER 1955b] ADAMS, C. R., and OTTO NEUGEBAUER: "Raymond Clare Archibald. A Minute." *Scripta Math.* **21** (1955), 293–295.

[ALLMAN 1877/85] ALLMAN, GEORGE JOHNSTON: *Greek Geometry from Thales to Euclid.* Dublin: Hodges, Figgis and Co.; London: Longmans, Green and Co. 1877–1885; 2nd ed. 1889.

[AMMA 1979] AMMA, T. A. SARASVATI: *Geometry in Ancient and Medieval India.* Delhi: Motilal Banarsidass 1979.

[*Anales* 1969/79] *Anales de la Sociedad Mexicana de Historia de la Ciencia y la Tecnología.* 5 vols. México: SMHC 1969–1979.

[ANDERSEN 1980] ANDERSEN, KIRSTI: "An Impression of Mathematics in Denmark in the Period 1600–1800." *Centaurus* **24** (1980), 316–334.

[ANDERSEN/BANG 1983] ANDERSEN, KIRSTI, and THØGER BANG: "Matematik." In: *Københavns Universitet 1479–1979*, vol. 12, 113–199. København: Gads Forlag 1983.

[ANDERSEN/DYBDAHL 1995] ANDERSEN, KIRSTI, and METTE DYBDAHL (Eds.): *World Directory of Historians of Mathematics.* Aarhus: International Commission on History of Mathematics, 3rd. ed. 1995.

[ANDRES 1918] ANDRES, HANS JACOB, *et al.*: *Zur Erinnerung an Herrn Professor Dr. J. H. Graf. Abschiedsworte gesprochen bei der Leichenfeier in der Johanneskirche, Donnerstag, den 20. Juni 1918.* Bern: Buchdruckerei Steiger 1918.

[ARBOLEDA 1985] ARBOLEDA, LUIS CARLOS: "Dificultades estructurales de la profesionalización de las matemáticas en Colombia." In: PESET, JOSÉ LUIZ (Ed.): *La Ciencia Moderna y el Nuevo Mundo.* Madrid: CSIC/SLAHCT 1985, 27–38.

[ARNETH 1852] ARNETH, ARTHUR: *Die Geschichte der reinen Mathematik in ihrer Beziehung zur Geschichte der Entwicklung des menschlichen Geistes.* Stuttgart: Franck 1852 (= special edition of *Neue Encyclopädie für Wissenschaften und Künste*).

[ARRIGHI 1989] ARRIGHI, GINO: "Baldassarre Boncompagni e la matematica medioevale." In: [BARBIERI/DEGANI 1989], 23–46.

[ASCHER 1981] ASCHER, MARCIA: *Code of the Qipu. A Study in Media, Mathematics and Culture.* Ann Arbor: University of Michigan Press 1981.

[ASCHER 1991] ASCHER, MARCIA: *Ethnomathematics: A Multicultural View of Mathematical Ideas.* Pacific Grove, CA: Brooks/Cole Publishing Co. 1991.

[AVEZ 1993] AVEZ, RENAUD: *L'Institut Français de Damas au Palais Azem (1922–1946)* (= *Collection Témoignages et Documents* **1**). Damas: L'Institut Français de Damas 1993.

[AVITAL 1995] AVITAL, SHMUEL: "History of Mathematics Can Help Improve Instruction and Learning." In: SWETZ, FRANK, *et al.* (Eds.): *Learn from the Masters!* Washington, D.C.: Mathematical Association of America 1995, 3–12.

[AZEVEDO 1994] AZEVEDO, FERNANDO DE (Ed.): *As Ciências no Brasil.* Rio de Janeiro: Editora UFRJ 1994 (original ed. 1955).

[BABINI/MIELI/PAPP 1945] BABINI, JOSÉ; ALDO MIELI, and DESIDERIO PAPP: *Panorama general de historia de la sciencia.* 10 vols. Buenos Aires and Mexico: Espasa-Calpe 1945.

[BABINI/REY PASTOR 1951] BABINI, JOSÉ, and JULIO REY PASTOR: *Historia de la Matemática.* Buenos Aires: Espasa-Calpe Argentina S.A. 1951.

[BACHELARD 1975] BACHELARD, GASTON: *Le nouvel esprit scientifique.* Paris: PUF 1975.

[BAG 1979] BAG, AMULYA KUMAR: *Mathematics in Ancient and Medieval India.* Varanasi: Chaukhamba 1979.

[BAGHERI 1995] BAGHERI, MOHAMMAD: "History of Mathematics in Iran." *History and Pedagogy of Mathematics Newsletter* **34** (1995), 6–7.

[BALDI 1998] BALDI, BERNARDINO: *Vite de' matematici,* ELIO NENCI (Ed.). Milano: Franco Angeli 1998.

[BANFI 1930] BANFI, ANTONIO: *Vita di Galileo Galilei.* Milano-Roma: Società Editrice La Cultura 1930; repr. Milano: Feltrinelli 1962.

[BANFI 1949] BANFI, ANTONIO: *Galileo Galilei.* Milano: Ambrosiana 1949; repr. Il Saggiatore 1961.

[BARANIECKI 1884] BARANIECKI, ALEKSANDER M.: "Krótki rys rozwoju arytmetyki i o jej nauczaniu w Polsce" (A Short Outline of the Development of Arithmetic and on Teaching It in Poland). In: *Arytmetyka,* pp. XIII–LVI. Warsaw 1884.

[BARBIERI/DEGANI 1989] BARBIERI, FRANCESCO, and FRANCA CATTELANI DEGANI (Eds.): *Pietro Riccardi (1828–1898) e la storiografia delle matematiche in Italia.* Modena: Università degli Studi di Modena, Dipartimento di Matematica Pura ed Applicata "G. Vitali" 1989.

[BARROW-GREEN 2000] BARROW-GREEN, JUNE: "Web Historical Resources for the Mathematics Teacher." In: *History in Mathematics Education. The ICMI Study*, FAUVEL, JOHN, and JAN VAN MAANEN (Eds.). Dordrecht: Kluwer 2000, 362–370.

[BASHMAKOVA 1972] BASHMAKOVA, IZABELLA G.: *Diofant i diofantovy uravneniya.* Moscow: Nauka 1972. (German trans.: *Diophant und diophantische Gleichungen.* Berlin: VEB Deutscher Verlag der Wissenschaften 1974; English trans.: *Diophantus and Diophantine Equations* (= Dolciani Math. Exposition, no. 20). New York: The Mathematical Association of America 1997.)

[BASHMAKOVA/SLAVUTIN 1984] BASHMAKOVA, IZABELLA G., and EVGENIĬ I. SLAVUTIN: *Istoriya Diofantova analiza ot Diofanta do Ferma.* Moscow: Nauka 1984.

[BASHMAKOVA/VESELOVSKIĬ 1974] BASHMAKOVA, IZABELLA G., and IVAN N. VESELOVSKIĬ (Eds.): *Diofant Aleksandriiskiĭ. Arifmetika.* Trans. by I. N. VESELOVSKIĬ, ed. with commentary by I. G. BASHMAKOVA. Moscow: Nauka 1974.

[BASTOS 1858] BASTOS, FRANCISCO ANTÓNIO MARTINS: "José Monteiro da Rocha." *O Instituto* **6** (1858), 261–262.

[BAUMGART 1989] BAUMGART, JOHN K., *et al.*: *Historical Topics for the Mathematics Classroom*, 2nd rev. ed. Washington, D.C. 1989.

[BAYS 1927a] BAYS, SÉVERIN: "Discours d'ouverture du Président annuel [sur les mathématiciens suisses du passé]." *Vh. Schw. Nat. Ges.* **107**. Jahresversammlung vom 30. August bis 1. September 1926 in Freiburg, II. Teil, 11–55.

[BAYS 1927b] BAYS, SÉVERIN: "Mathématiciens fribourgeois." *Bulletin de la Société Fribourgeoise des Sciences Naturelles* **28** (1927), 165–184.

[BEAULIEU 1994] BEAULIEU, LILIANE: "Dispelling a Myth: Questions and Answers About Bourbaki's Early Works, 1934–1944." In: SASAKI CHIKARA; SUGIURA MITSUO, and JOSEPH W. DAUBEN (Eds.): *The Intersection of History and Mathematics.* Basel, Boston, Berlin: Birkhäuser 1994, 241–252.

[BECHER 1980] BECHER, HARVEY: "William Whewell and Cambridge Mathematics." *HSPS* **11** (1980), 3–48.

[BECKER/HOFMANN 1951] BECKER, OSKAR, and JOSEPH EHRENFRIED HOFMANN: *Geschichte der Mathematik.* Bonn: Athenäum-Verlag 1951 (see esp. "Entwicklung der Mathematik-Geschichte," 254–261).

[BEČVÁŘ 1995] BEČVÁŘ, JINDŘICH: *Eduard Weyr, 1852–1903.* Praha: Prometheus 1995.

[BEHNKE 1949] BEHNKE, HEINRICH: "Otto Toeplitz zum Gedächtnis." *Math.-phys. Semesterber.* **1** (1949), 89–96.

[BELHOSTE/DAHAN/PICON 1994] BELHOSTE, BRUNO; AMY DAHAN DALMÉDICO, and ANTOINE PICON: *La formation polytechnicienne 1794–1994.* Paris: Dunod 1994.

[BELL 1937] BELL, ERIC T.: *Men of Mathematics.* New York: Simon and Schuster 1937.

[BELL 1940] BELL, ERIC T.: *Development of Modern Mathematics.* New York: Mc-Graw Hill 1940; 2nd ed. 1945.

[BELL 1955] BELL, ERIC T.: "Autobiographical Essay." *Twentieth Century Authors*, first supplement. In: KUNITZ, STANLEY J. (Ed.): *A Biographical Dictionary of Modern Literature.* New York: Wilson 1955, 70–71.

[BELTRAMI 1889] BELTRAMI, EUGENIO: "Un precursore italiano di Legendre e Lobachevskji." *Rend. Accad. Naz. Lincei* **5** (1st sem., 1889), 441–448.

[BELTRAMI 1919] BELTRAMI, LUCA: *Documenti e memorie riguardanti la vita e le opere di Leonardo da Vinci.* Milano: Fratelli Treves 1919.

[BELTRÁN 1964] BELTRÁN, ENRIQUE (Ed.): *Memorias del Primer Coloquio Mexicano de Historia de la Ciencia.* 2 vols. México: SMHN 1964.

[BERËZKINA 1957] BERËZKINA, EL'VIRA I.: "O 'Matematike v devyati knigakh.' " Trans. and commentary by E. I. BERËZKINA. *IMI* (1) **10** (1957), 427–584.

[BERG/SCHIFFERS 1992] BERG, HANS VAN DER, and NORBERT SCHIFFERS (Eds.): *La Cosmovisión Aymara.* La Paz: UCB/hisbol 1992.

[BERTRAND 1869] BERTRAND, JOSEPH: *Académie des sciences et les académiciens de 1666 à 1793.* Paris: J. Hetzel 1869.

[BESSMERTNY 1934] BESSMERTNY, BERTHA: "Savérien historien des sciences." *Archeion* **15** (1933), no. 3–4 [1934], 369–378.

[BETH 1929] BETH, HERMANUS JOHANNES ELISA: *Inleiding in de niet-Euclidische meetkunde op historischen grondslag.* Groningen: P. Noordhoff 1929.

[BETH 1932] BETH, HERMANUS JOHANNES ELISA: *Newton's "Principia."* 2 vols. Groningen-Batavia: P. Noordhoff 1932.

[BIARD 1997] BIARD, AGNÉS: *Henri Berr et la culture du XXe siècle.* Paris: Albin Michel 1997.

[BIEBERBACH 1934] BIEBERBACH, LUDWIG: "Persönlichkeitsstruktur und mathematisches Schaffen." *FF* **10** (1934), 235–237.

[BIELIŃSKI 1890] BIELIŃSKI, JÓZEF: "Stan nauk matematyczno-fizycznych za czasów Wszechnicy Wileńskiej" (The State of Mathematical and Physical Sciences in the Times of the Vilnius University). *Prace Matematyczno-Fizyczne* **2** (1890), 265–433.

[BIERENS DE HAAN 1883] BIERENS DE HAAN, DAVID: *Bibliographie Néerlandaise historique-scientifique, des ouvrages importants dont les Auteurs sont nés aux 16e, 17e et 18e siècles, sur les sciences mathématiques et physiques, avec leurs applications.* Rome: Imprimerie des sciences mathématiques et physiques 1883; repr. Nieuwkoop: B. de Graaf 1960.

[BIERENS DE HAAN 1891] BIERENS DE HAAN, DAVID: "Bibliographie de l'histoire des sciences mathématiques aux Pays-Bas." *Bibl. Math.* (2) **5** (1891), 13–22.

[BIERMANN 1983] BIERMANN, KURT-R.: "Aus der Vorgeschichte der Euler-Ausgabe 1783–1907." In: BURCKHARDT, JOHANN JAKOB; EMIL A. FELLMANN, and WALTER HABICHT (Eds.): *Leonhard Euler 1707–1783. Beiträge zu Leben und Werk. Gedenkband des Kantons Basel-Stadt.* Basel: Birkhäuser Verlag 1983, 489–500.

[BILINSKI 1977] BILINSKI, BRONISLAW: "Prolegomena alle *Vite de' matematici* di Bernardino Baldi (1587–1596)." *Accademia Polacca delle Scienze. Biblioteca e Centro di Studi a Roma, Conferenze e Studi* **71** (1977), 1–135.

[BINDER 1994] BINDER, CHRISTA: "Austria and Hungary." In: [Grattan-Guinness] 1994, 1457–1464].

[BINDER 1996] BINDER, CHRISTA: "Die erste Wiener mathematische Schule (Johannes von Gmunden, Georg Peuerbach)." In: GEBHARDT, RAINER, and HELMUTH ALBRECHT (Eds.): *Rechenmeister und Cossisten der frühen Neuzeit* (= *Schriften des Adam-Ries-Bundes Annaberg-Buchholz* **7**) (1996), 3–18.

[BINDER 1998] BINDER, CHRISTA: "Die zweite Wiener mathematische Schule." In: RÖTTEL, KARL (Ed.): *Ad Fontes Arithmeticae et Algebrae. Festschrift zum 70. Geburtstag von Wolfgang Kaunzner* (= *Staffelsteiner Schriften* **7**). Buxheim Eichstädt: Polygon-Verlag 1998, 60–66.

[BINDER 1999] BINDER, CHRISTA: "Georg Tannstetter (Collimitius (1482–1535))." In: *Rechenbücher und Mathematische Texte der frühen Neuzeit* (= *Schriften des Adam-Ries-Bundes Annaberg-Buchholz* **11**) (1999), 29–36.

[BIRKENMAJER/BIRKENMAJER 1918] BIRKENMAJER, LUDWIK ANTONI, and ALEKSANDER BIRKENMAJER: "Najważniejsze dezyderaty nauki polskiej w zakresie historii nauk matematycznych" (The Most Important Desiderata of Polish Science in the Field of Mathematical Science). *Nauka Polska* **1** (1918), 87–106.

[BIRKENMAJER/DICKSTEIN 1933] BIRKENMAJER, LUDWIK ANTONI, and SAMUEL DICKSTEIN: "Coup d'oeil sur l'histoire des sciences exactes an Pologne." Cracow: Polish Academy of Arts 1933.

[BKOUCHE 1994] BKOUCHE, RUDOLPH: "Desargues au XIXième siècle." In: DHOMBRES, JEAN, and JOËL SAKAROVITCH (Eds.): *Desargues et son temps.* Paris: Librairie Albert Blanchard 1994, 207–217.

[BÖLLING 1994] BÖLLING, REINHARD: *Das Fotoalbum für Weierstraß/A Photo Album for Weierstrass.* Braunschweig and Wiesbaden: Vieweg 1994.

[BOGOLYUBOV 1972] BOGOLYUBOV, ALEKSEĬ N.: "Razvitie idei operatsionnogo (simvolicheskogo) ischisleniya." In: SHTOKALO, I. Z.: *Operatsionnoe ischislenie.* Kiev: Naukova Dumka 1972.

[BOGOLYUBOV 1983] BOGOLYUBOV, ALEKSEĬ N.: *Mathematics. Mechanics. Biographical Dictionary* (in Russian). Kiev: Naukova Dumka 1983.

[BOLZANO 1882] BOLZANO, BERNARD: "Ryze analytický dükaz poučky [...] " (Purely Analytical Proof that between Two Values that Produce an Opposite Result at least one Real Root of the Function Exists) (trans. from German by F. J. STUDNIČKA). *Cas. Pest. Mat. Fys.* **9** (1882), 1–38.

[BONJOUR 1960] BONJOUR, EDGAR: *Die Universität Basel von den Anfängen bis zur Gegenwart, 1460–1960.* Basel: Helbing und Lichtenhahn 1960; 2nd rev. ed. Basel: Helbing und Lichtenhahn 1971.

[BORGATO 1989] BORGATO, MARIA TERESA: "La storia delle matematiche nell'opera di Lagrange." In: [BARBIERI/DEGANI 1989], 107–131.

[BORGATO 1992] BORGATO, MARIA TERESA: "On the History of Mathematics in Italy Before Political Unification." *Arch. int. Hist. Sci.* **42** (1992), 121–136.

[BORODIN/BUGAI 1979] BORODIN, A. I., and A. S. BUGAI (Eds.): *Biographical Dictionary of Mathematicians* (in Russian). Kiev: Radyanska Shkola 1979.

[BORTOLOTTI 1922a] BORTOLOTTI, ETTORE: Review of G. LORIA: *L'opera geometrica di Evangelista Torricelli.* In: *Period. Mat.* (4) **2** (1922), 274–279.

[Bortolotti 1922b] Bortolotti, Ettore: "Lettera del prof. Bortolotti al Direttore del Periodico prof. Enriques." *Period. Mat.* (4) **2** (1922), 368–369.

[Bortolotti 1931] Bortolotti, Ettore: *Elenco delle pubblicazioni nel 41º anno d'insegnamento.* Bologna: Società Tipografica già Compositori 1931.

[Bortolotti 1932] Bortolotti, Ettore: Nachruf auf Wieleitner. *Boll. UMI* **11** (1932), 63f.

[Bortolotti 1939] Bortolotti, Ettore: "Il primato dell'Italia nel campo della matematica." *Atti della Società Italiana per il Progresso delle Scienze* (Riunione XXVII, Bologna 4–11 Settembre 1938), **3** (1939), 598–614.

[Bortolotti 1947] Bortolotti, Ettore: *La storia della matematica nella università di Bologna.* Bologna: Zanichelli 1947.

[Bos/Mehrtens/Schneider 1981] Bos, Henk; Herbert Mehrtens, and Ivo Schneider: *Social History of Nineteenth Century Mathematics.* Boston, Basel, Stuttgart: Birkhäuser 1981.

[Bosshard 1993] Bosshard, Hans Heinrich (Ed.): [Festschrift zum 100. Todestag von Rudolf Wolf mit Beiträgen von J. J. Burckhardt, V. Larcher, H. Balmer, H. Lutstorf, T. K. Friedli, R. Ineichen.] *Vj. Nat. Ges. Zür.* **138** no. 4 (1993), 225–298.

[Bossut 1800] Bossut, Charles: *Essay sur l'histoire générale des mathématiques.* 2 vols. Paris: Louis 1800. Engl. trans., London: J. Johnson 1803; German trans., with comments and annotations by N. Th. Reimer: *Versuch einer allgemeinen Geschichte der Mathematik.* 2 vols. Hamburg: B. G. Hoffmann 1804.

[Bourbaki 1948] Bourbaki, Nicolas: "L'architecture des mathématiques." In: Le Lionnais, François (Ed.): *Les grands courants de la pensée mathématique.* Paris: Cahier du Sud 1948, 35–47.

[Bourbaki 1969] Bourbaki, Nicolas: *Eléments d'histoire des mathématiques.* 2nd ed., revised, corrected, augmented. Paris: Hermann 1969. (First ed.: Paris 1948).

[Boyer 1939] Boyer, Carl B.: *The Concepts of the Calculus; a Critical and Historical Discussion of the Derivative and the Integral.* New York: Columbia University Press 1939; repr. New York: Hafner Publishing Co. 1949; repr. as *The History of the Calculus and its Conceptual Development. (The Concepts of the Calculus).* New York: Dover 1959.

[Braunmühl 1895] Braunmühl, Anton von: "Der Unterricht in der Geschichte der Mathematik an der k. technischen Hochschule zu München." *Bibl. Math.* (2) **9** (1895), 89–90.

[Braunmühl 1897] Braunmühl, Anton von: "Mathematisch-historische Vorlesungen und Seminarübungen an der technischen Hochschule zu München." *Bibl. Math.* (2) **11** (1897), 113–115.

[Braunmühl 1900/03] Braunmühl, Anton von: *Vorlesungen über Geschichte der Trigonometrie.* 2 vols. Leipzig: B. G. Teubner 1900–1903; repr. Niederwalluf: Sändig 1971; Schaan/Liechtenstein: Sändig 1981.

[Braunmühl 1902] Braunmühl, Anton von: "Mathematisch-historische Vorlesungen und Seminarübungen an der technischen Hochschule in München 1897–1902." *Bibl. Math.* (3) **3** (1902), 403–404.

[BRAUNMÜHL 1905] BRAUNMÜHL, ANTON VON: "Le séminaire d'histoire des mathématiques à l'école polytechnique de Munich." *Enseign. Math.* **7** (1905), 65–66.

[BRENTJES 1992] BRENTJES, SONJA: "Historiographie der Mathematik im Islamischen Mittelalter." *Arch. int. Hist. Sci.* **42** (1992), 27–63.

[BRETSCHNEIDER 1870] BRETSCHNEIDER, CARL ANTON: *Die Geometrie und die Geometer vor Euklides. Ein historischer Versuch.* Leipzig: B. G. Teubner 1870.

[BRIOSCHI 1852] BRIOSCHI, FRANCESCO: "Prefazione a una memoria di G. Piola." *Memorie dell'Istituto Lombardo di Scienze e Lettere* (1) **3** (1852), 282–298. (Repr. in: BRIOSCHI: *Opere matematiche* **1**, Milano: Ulrico Hoepli 1901, 119–135.)

[BRIOSCHI 1863] BRIOSCHI, FRANCESCO: "Sulla risolvente di Malfatti per le equazioni del quinto grado." *Annali di matematica pura e applicata* (1) **5** (1863), 233–250. (Repr. in: BRIOSCHI: *Opere matematiche* **2**, Milano: Ulrico Hoepli 1902, 39–56.)

[BRUN 1962] BRUN, VIGGO: *Regnekunsten i det gamle Norge.* Oslo and Bergen: Universitetsforlaget 1962.

[BRUNETTI 1984] BRUNETTI, FRANZ: "De la mécanique à l'histoire." *Dix-huitième siècle* **16** (1984), 123–136.

[BRUNSCHVICG 1981] BRUNSCHVICG, LÉON: *Les étapes de la philosophie mathématique*, nouveau tirage augmenté d'une Préface de M. JEAN-TOUSSAINT DESANTI. Paris: A. Blanchard 1981 (1st ed.: Paris: F. Alcan 1912).

[BSHM 1992] The British Society for the History of Mathematics: *Newsletter* no. 21 (1992): "The First 21 Years: 1971–1992." (Contains an overview by IVOR GRATTAN-GUINNESS and a list of all meetings and lectures held.)

[BUCCIANTINI 1985] BUCCIANTINI, MASSIMO: "Favaro, Antonio." In: *Dizionario Biografico degli Italiani.* Roma: Istituto della Enciclopedia Italiana **30** (1985), 441–444.

[BUFFON 1740] BUFFON, GEORGES de, trad.: ISAAC NEWTON: *La méthode des fluxions et des suites infinies.* Paris 1740; repr. Paris: A. Blanchard 1966.

[BURCKHARDT 1980] BURCKHARDT, JOHANN JAKOB: *Die Mathematik an der Universität Zürich 1916–1950 unter den Professoren R. Fueter, A. Speiser, P. Finsler* (= Beihefte zur Zeitschrift *Elem. Math.* **16**). Basel: Birkhäuser 1980.

[BURCKHARDT 1983] BURCKHARDT, JOHANN JAKOB: "Die Euler-Kommission der Schweizerischen Naturforschenden Gesellschaft — ein Beitrag zur Editionsgeschichte." In: BURCKHARDT, JOHANN JAKOB; EMIL A. FELLMANN, and WALTER HABICHT (Eds.): *Leonhard Euler 1707–1783. Beiträge zu Leben und Werk. Gedenkband des Kantons Basel-Stadt.* Basel: Birkhäuser 1983, 501–509.

[BURZIO 1942] BURZIO, FILIPPO: *Lagrange.* Torino: UTET 1942; repr. with a preface by LUIGI PEPE: Torino: UTET 1993.

[CAJORI 1893] CAJORI, FLORIAN: *History of Mathematics.* New York: The Macmillan Company 1893; 2nd ed., revised and enlarged, New York: The Macmillan Company 1919.

[CAJORI 1899] CAJORI, FLORIAN: *A History of Physics in its Elementary Branches, Including the Evolution of Physical Laboratories.* New York: The Macmillan Company 1899.

[CAJORI 1920/21] CAJORI, FLORIAN: "Moritz Cantor, the Historian of Mathematics." *Bull. AMS* **27** (1920/21), 21–28.

[CAJORI 1928/29] CAJORI, FLORIAN: *History of Mathematical Notations.* 2 vols. Chicago, IL: The Open Court Publishing Company 1928–1929.

[CANDOLLE 1915] CANDOLLE, AUGUSTIN DE, *et al.* (Eds.): "Centenaire de la Société Helvétique des Sciences Naturelles. Notices historiques et documents réunis par la Commission historique instituée à l'occasion de la session annuelle de Genève (12–15 septembre 1915)." *Neue Denkschriften der Schweizerischen Naturforschenden Gesellschaft* **50** (1915).

[CANO PAVÓN 1993] CANO PAVÓN, JOSÉ MARÍA: *La Ciencia en Sevilla.* Sevilla: Servicio de Publicaciones de la Universidad de Sevilla 1993.

[CANTOR 1880/1908] CANTOR, MORITZ: *Vorlesungen über Geschichte der Mathematik.* 4 vols. Leipzig and Berlin: B. G. Teubner 1880–1908. (vol. 1: 1880, 2nd ed. 1894, 3rd ed. 1907; vol. 2: 1892, 2nd ed. 1899/1900; vol. 3: 1898, 2nd ed. 1900/1901; vol. 4: 1908; several reprints).

[CANTOR 1900] CANTOR, MORITZ: "C. I. Gerhardt." *Jber. DMV* **8** (1900), 28–30.

[CARRUCCIO 1958] CARRUCCIO, ETTORE: *Matematica e logica nella storia del pensiero contemporaneo.* Torino: Gheroni 1958. English trans. by ISABEL QUIGLY: *Mathematics and Logic in History and in Contemporary Thought.* London: Faber and Faber 1964.

[CASINI 1998] CASINI, PAOLO: *L'antica sapienza italica. Cronistoria di un mito.* Bologna: Il Mulino 1998.

[CASTELLI 1998] CASTELLI GATTINARA, ENRICO (Ed.): *Les inquiétudes de la raison. Epistémologie et histoire en France dans l'entre-deux-guerres.* Paris: Ed. de l'EHESS/Libr. J. Vrin 1998.

[CASTELNUOVO 1938] CASTELNUOVO, GUIDO: *Le origini del calcolo infinitesimale nell'era moderna.* Bologna: Zanichelli 1938; repr. Milano: Feltrinelli 1962.

[*Catálogo* 1992] *Catálogo de textos marginados novohispanos. Inquisición: siglo XVIII y XIX.* México: Archivo General de la Nación, UNAM y El Colegio de México 1992.

[CAVAILLÈS 1962] CAVAILLÈS, JEAN: *Philosophie mathématique* (= Collection *Histoire de la pensée* **6**). Paris: Hermann 1962.

[CAVERNI 1891–1898] CAVERNI, RAFFAELO: *Storia del metodo sperimentale in Italia*, 5 vols. Firenze: Stabilimento R. Civelli Editore 1891–1898.

[CHACE 1927/29] CHACE, ARNOLD BUFFUM: *The Rhind Mathematical Papyrus.* 2 vols. Oberlin, OH: The Mathematical Association of America 1927–1929; repr. Reston, VA: The National Council of Teachers of Mathematics [1979].

[CHANDLER/MAGNUS 1982] CHANDLER, BRUCE, and WILHELM MAGNUS: *The History of Combinatorial Group Theory. A Case Study in the History of Ideas.* New York: Springer 1982.

[CHARETTE 1995] CHARETTE, FRANÇOIS: "Orientalisme et histoire des sciences: l'historiographie européenne des sciences islamiques et hindoues, 1784–1900." Ph.D. thesis presented to the Faculty of Advanced Studies at the University of Montreal in May, 1995.

[CHASLES 1837] CHASLES, MICHEL: *Aperçu historique sur l'origine et le développement des méthodes en géométrie, particulièrement de celles qui se rapportent à la géométrie moderne, suivi d'un mémoire de géométrie sur deux principes généraux de la science, la dualité et l'homographie* (= Mémoires couronnés par l'Académie de Bruxelles **11** (1837); Brussels: M. Hayez 1837; repr. Paris: Editions Jacques Gabay 1989. — Repr. Paris: Gauthier-Villars et fils 1875 and 1888. — Partial German trans. by L. SOHNCKE, Halle 1839.

[CHASLES 1860] CHASLES, MICHEL: *Les trois livres de Porismes d'Euclide, rétablis pour la première fois, d'apres la notice et les lemmes de Pappus, et conformément au sentiment de R. Simson sur la forme des énoncés de ces propositions.* Paris: Mallet-Bachelier 1860.

[CHASLES 1870] CHASLES, MICHEL: *Rapport sur les progrès de la Géométrie.* Publication faite sour les auspices du Ministère de l'Instruction Publique. Paris: Imprimerie Nationale 1870.

[CHILD 1920] CHILD, JAMES MARC: *The Early Mathematical Manuscripts of Leibniz.* Chicago and London: Open Court 1920.

[CHRISTENSEN 1895] CHRISTENSEN, SOPHUS ANDREAS: *Matematikens Udvikling i Danmark og Norge i det XVIII Aarhundrede.* Odense: Hempelske Boghandels Forlag 1895.

[CHRISTENSEN/HEIBERG 1889] CHRISTENSEN, SOPHUS ANDREAS, and JOHAN LUDVIG HEIBERG: "Bibliographische Notiz über das Studium der Geschichte der Mathematik in Dänemark." *Bibl. Math.* (2) **3** (1889), 75–83.

[CHRISTIE 1990] CHRISTIE, JOHN: "The Development of the Historiography of Science." In: OLBY, R. C., *et al.* (Eds.): *Companion to the History of Modern Science.* London: Routledge 1990, 5–22.

[CHRYSOSTOMOS 1821] CHRYSOSTOMOS: "The Church and the Greek Education During the Years after the Capture." In: *Panegyric Speeches of Academicians.* Athens: Academy of Athens. Foundation Kostas and Eleni Ourani 1977, 98–136.

[CIFOLETTI 1996] CIFOLETTI, GIOVANNA: "The Creation of the History of Algebra in the Sixteenth Century." In: GOLDSTEIN, CATHERINE; JEREMY GRAY, and JIM RITTER (Eds.): *Mathematical Europe.* Paris: Edition de la Maison des Sciences de l'Homme 1996, 123–142.

[COBO 1964] COBO, BERNABÉ: *Historia del Nuevo Mundo* (1653). Madrid: Atlas 1964.

[COLEBROOKE 1817] COLEBROOKE, HENRY THOMAS: *Algebra with Arithmetic and Mensuration from the Sanskrit of Brahmegupta and Bháscara.* London: Murray 1817; repr. Wiesbaden: Sändig 1973.

[COLLOT D'ESCURY 1824/44] COLLOT D'ESCURY, HENDRIK: *Holland's roem in kunsten en wetenschappen.* 7 vols. s'Gravenhage and Amsterdam: Gebroeders van Cleef 1824–1844.

[COOLIDGE 1940] COOLIDGE, JULIAN LOWELL: *A History of Geometrical Methods.* Oxford: The Clarendon Press 1940; repr. 1947, and New York: Dover 1963.

[COSSALI 1797/99] COSSALI, PIETRO: *Origine, trasporto in Italia, primi progressi in essa dell'Algebra. Storia critica di nuove disquisizioni analitiche e metafisiche arrichita.* 2 vols. Parma: Reale Tipographice 1797–1799.

[COSSALI 1996] COSSALI, PIETRO: *La storia del caso irriducibile*, GATTO, ROMANO (Ed.). Venezia: Istituto Veneto di Scienze Lettere ed Arti (*Memorie Classe di Scienze Fisiche, Matematiche, Naturali* **36** (1996)).

[COSTA 1920] COSTA, MANOEL AMOROSO: *As Idéias Fundamentais da Matemática*. Rio de Janeiro: Pimenta de Mello and C. 1920.

[COSTABEL 1988] COSTABEL, PIERRE: "L'Académie et ses Secrétaires perpétuels: un aspect méconnu de l'histoire." *La vie des sciences*. Série générale, vol. 5, no. 2, Mars-Avril 1988, 155–168.

[COUMET 1981] COUMET, ERNEST: "Paul Tannery: L'organisation de l'enseignement de l'histoire des sciences." *Revue de Synthése* **101–102** (1981), 87–123.

[CRELIER 1918] CRELIER, LOUIS: "Professeur Dr J. H. Graf (1852–1918)." *Vh. Schw. Nat. Ges.* Jahrgang 1918, 105–115.

[CROMBIE 1994] CROMBIE, ALISTAIR: *Styles of Scientific Thinking in the European Tradition*. 3 vols. London: Duckworth 1994.

[CUNHA 1929a] CUNHA, PEDRO JOSÉ DA: *Bosquejo historico das mathematicas em Portugal*. Exposição Portuguesa em Sevilha 1929.

[CUNHA 1929b] CUNHA, P. J. DA: *A Astronomia, a Nautica, e as Sciencias afins*. Exposição Portuguesa em Sevilha 1929.

[CUNHA 1930] CUNHA, P. J. DA: *Nota ao bosquejo historico das matematicas em Portugal*. Lisboa: Imprensa Nacional 1930.

[CUNHA 1937] CUNHA, P. J. DA: *A Escola Politecnica de Lisboa — Breve noticia historica*. Lisboa: Faculdade de Ciencias de Lisboa 1937.

[CUNHA 1940a] CUNHA, P. J. DA: "Como se introduziram e desenvolverem entre nós as Noções de Limite e de Infinitamente Pequeno." *Congresso do Mundo Português* **12**, vol. 1 (1940), 35–57.

[CUNHA 1940b] CUNHA, P. J. DA: "O Cálculo Infinitesimal e a Escola Politécnica de Lisboa." *Congresso do Mundo Português* **12**, vol. 1 (1940), 59–78.

[CUNHA 1940c] CUNHA, P. J. DA: "As Matemáticas em Portugal no Século XVII." *Memorias da Academia das Ciências de Lisboa*, Classe de Ciências, **3** (separata) 1940.

[DAHLBO 1897] DAHLBO, J.: *Uppränning till Matematikens Historia i Finland frän äldsta Tider til stora Ofreden*. Nikolaistad: Wasa Tryckeri 1897.

[DAHLIN 1875] DAHLIN, ERNST MAURITZ: *Bidrag till de matematiska vetenskapernes historia i Sverige före 1670*. Uppsala Universitet Årsskrift 1875.

[D'ALEMBERT 1751] D'ALEMBERT, JEAN LE ROND: "Discours préliminaire." *Encyclopédie* I. Paris: André-François Le Breton et ses associés: Antoine-Claude Briasson, Michel Antoine David et Laurent Durand 1751.

[DAMARDĀSH 1965] DAMARDĀSH, AḤMAD SAʿĪD: *Maḥmūd Ḥamdī al-Falakī*. Cairo 1965.

[D'AMBROSIO 1988] D'AMBROSIO, UBIRATAN: "Ethnomathematics: A Research Program in the History of Ideas and in Cognition." *International Study Group on Ethnomathematics Newsletter* **4** (1988), 5–8.

[D'AMBROSIO 1994] D'AMBROSIO, UBIRATAN: "O Seminário Matemático e Físico da Universidade de São Paulo. Uma Tentativa de Institucionalização na Década de Trinta." *Temas e Debates* **VII**, n° 4 (1994), 20–27.

[DATTA 1927/29] DATTA, BIBHUTIBHUSAN: "Hindu Contribution to Mathematics." *Bulletin of the Allahabad University Mathematical Association* **1** (1927), 49–73; **2** (1929), 1–36.

[DATTA 1932] DATTA, BIBHUTIBHUSAN: *The Science of the Śulba.* Calcutta: University of Calcutta 1932; repr. Calcutta 1991.

[DATTA/SINGH 1935/38] DATTA, BIBHUTIBHUSAN, and AVADHES NARAYAN SINGH: *History of Hindu Mathematics. A Source Book.* Part I (1935), II (1938). Lahore: Motilal Banarsidass; repr. in one volume, Bombay: Asia Publishing House 1962. Part III was edited by KRIPA SHANKAR SHUKLA and published in a series of articles in the *Indian J. Hist. Sci.*, starting with vol. 15 (1980) up to vol. 28 (1993).

[DAUBEN 1985] DAUBEN, JOSEPH WARREN (Ed.): *The History of Mathematics from Antiquity to the Present: A Selective Annotated Bibliography.* New York: Garland Press 1985. (For CD-ROM edition, see [Dauben/Lewis 2000]).

[DAUBEN 1994] DAUBEN, JOSEPH WARREN: "Mathematics: an Historians Perspective." In: SASAKI CHIKARA; MITSUO SUGIURA, and JOSEPH W. DAUBEN (Eds.): *The Intersection of History and Mathematics.* Basel: Birkhäuser 1994, 1–13.

[DAUBEN 1995] DAUBEN, JOSEPH WARREN: *Abraham Robinson. The Creation of Nonstandard Analysis. A Personal and Mathematical Odyssey.* Princeton: Princeton University Press 1995 (paperback ed. 1998).

[DAUBEN 1998] DAUBEN, JOSEPH WARREN: "*Historia Mathematicae*: Journals of the History of Mathematics," International Symposium on Journals and History of Science, in honor of the 10th anniversary of the founding of the Italian journal *Nuncius* (Firenze, June 5–6, 1997). In: *Journals and History of Science*, BERETTA, MARCO; CLAUDIO POGLIANO, and PIETRO REDONDI (Eds.). Florence: Olschki Editore 1998, 1–30.

[DAUBEN 1999] DAUBEN, JOSEPH WARREN: "*Historia Mathematica*: 25 Years/Context and Content." *HM* **26** (1999), 1–28.

[DAUBEN/LEWIS 2000] LEWIS, ALBERT C. (Ed.): Revised edition on CD-ROM of [Dauben 1985]. Providence, RI: American Mathematical Society 2000.

[DAVIS 1790] DAVIS, SAMUEL: "On the Astronomical Computations of the Hindus." *Asiatick Researches* **2** (1790), 225–287.

[DAVIS 1994] DAVIS, CHANDLER: "Where Did 20th-Century Mathematics Go Wrong?" In: SASAKI CHIKARA; MITSUO SUGIURA, and JOSEPH W. DAUBEN (Eds.): *The Intersection of History and Mathematics.* Basel: Birkhäuser 1994, 129–142.

[DE GORTARI 1963] DE GORTARI, ELI: *La Ciencia en la Historia de México.* México: FCE 1963.

[DE MORGAN 1847] DE MORGAN, AUGUSTUS: *Arithmetical Books.* London: Taylor and Walton 1847; repr. London: H. K. Elliot (n. d.).

[DE SANTILLANA 1955] DE SANTILLANA, GIORGIO: *The Crime of Galileo.* Chicago: Chicago University Press 1955.

[De Vries 1926/40] De Vries, Hendrik: *Historische Studiën.* 3 vols. Groningen: P. Noordhoff 1926–1940.

[Dehn 1943/44] Dehn, Max: "Mathematics, 400 B. C. — 300 B. C.," *AMM* **50** (1943), 411–414; "Mathematics, 300 B. C. — 200 B. C.," *AMM* **51** (1944), 25–31; "Mathematics, 200 B. C. — 600 A. D.," *AMM* **51** (1944), 149–157.

[Delambre 1817] Delambre, Jean-Baptiste: *Histoire de l'astronomie*, I: *Astronomie ancienne.* Paris: V. Courcier 1817.

[Delambre 1989] Delambre, Jean-Baptiste: *Rapports à l'Empereur sur le progrès des sciences, des lettres et des arts depuis 1789.* In: Dhombres, Jean (Ed.): *Sciences mathématiques*, vol.I. Paris: Belin 1989.

[Delone 1947] Delone, Boris N.: *Peterburgskaya shkola teorii chisel.* Moscow and Leningrad: Izdatel'stvo AN SSSR 1947.

[Demidov 1992] Demidov, Sergeĭ S.: "Historiographie des Mathématiques en Russie et en URSS avant 1941." *Arch. Int. Hist. Sci.* **42** (1992), 94–113.

[Demidov 1993a] Demidov, Sergeĭ S.: "The Moscow School of the Theory of Functions in the 1930s." In: Zdravkovska, S., and P. L. Duren (Eds.): *Golden Years of Moscow Mathematics.* Providence, R. I.: AMS 1993, 35–54.

[Demidov 1993b] Demidov, Sergeĭ S.: "Brief Survey of Literature on the Development of Mathematics in the USSR." In: Zdravkovska, S., and P. L. Duren (Eds.).: *Golden Years of Moscow Mathematics.* Providence, R. I.: AMS 1993, 245–271.

[Demidov/Folkerts 1992] Demidov, Sergeĭ S., and Menso Folkerts: "Historiography and the History of Mathematics." *Arch. int. Hist. Sci.* **42** (1992), 1–144.

[Demidov/Ford 1996] Demidov, Sergeĭ S., and Charles E. Ford: "N. N. Luzin and the Affair of the 'National Fascist Center.' " In: Dauben, J. W.; M. Folkerts, E. Knobloch, and H. Wussing (Eds.): *History of Mathematics: States of the Art.* San Diego: Academic Press 1996, 137–148.

[Depman 1960] Depman, Ivan Ya.: "S.-Peterburgskoe matematicheskoe obshchestvo." *IMI* (1) **13** (1960), 11–106.

[Dhombres 1989] Dhombres, Nicole, and Jean Dhombres: *Naissance d'un nouveau pouvoir: sciences et savants en France 1793–1824.* Paris: Payot 1989.

[Di Sieno 1998] Di Sieno, Simonetta: "Storia e didattica." In: Di Sieno, Simonetta; Angelo Guerraggio, and Pietro Nastasi (Eds.): *La matematica italiana dopo l'unità. Gli anni tra le due guerre mondiali.* Milano: Marcos y Marcos editore 1998, 765–816.

[Dianni 1963] Dianni, Jadwiga: *Studium matematyki na Uniwersytecie Jagiellońskim do połowy XIX wieku* (Study of Mathematics at the Jagiellonian University until the First Half of the 19th Century). Cracow: Jagiellonian University 1963.

[Dianni 1974] Dianni, Jadwiga: "Zarys historyczny geometrii wykreślnej i jej recepcji w Polsce do końca XIX w" (An Historical Outline of Descriptive Geometry and of its Tradition in Poland in the 19th Ccentury). *Studia i Materiały z Dziejów Nauki Polskiej* S.C. **19** (1974), 91–129.

[Dianni/Wachułka 1963] Dianni, Jadwiga, and Adam Wachułka: *Tysiac lat polskiej myśli matematycznej* (A Thousand Years of Polish Mathematics). Warsaw: Państwowe Zakłady Wydawnictwo Szkolnych (PZWS) 1963.

[DICK 1970] DICK, AUGUSTE: *Emmy Noether (1882–1935)* (= Beihefte zur Zeitschrift *Elemente der Mathematik*, vol. 13). Basel und Stuttgart: Birkhäuser 1970.

[DICKSTEIN 1896] DICKSTEIN, SAMUEL: *Hoene-Wroński. Jego życie i prace* (Hoene-Wroński, his Life and Works). Cracow: Academy of Arts 1896.

[DICKSTEIN 1901/02] DICKSTEIN, SAMUEL: "Korespondencja Kochańskiego i Leibniza" (Letters of Kochański and Leibniz). *Prace Matematyczno-Fizyczne* **12** (1901), 225–278, **13** (1902) 237–283.

[DIJKSTERHUIS 1929/30] DIJKSTERHUIS, EDUARD JAN: *De Elementen van Euclides*. 2 vols. Groningen: P. Noordhoff 1929–1930.

[DIJKSTERHUIS 1938] DIJKSTERHUIS, EDUARD JAN: *Archimedes. Eerste deel*. Groningen: P. Noordhoff 1938. (Vol. II was never published.)

[DIJKSTERHUIS 1956] DIJKSTERHUIS, EDUARD JAN: *Archimedes*. Copenhagen: Munksgaard 1956 (Engl. trans. of [DIJKSTERHUIS 1938] and additional material).

[DIKSHIT 1896] DIKSHIT or DĪKṢITA, SAṄKARA BĀLAKṚṢṆA: *Indian Astronomy* (in Marathi). Poona: Aryabhusana Press 1896. English translation by R. V. VAIDYA. Calcutta: Meteorological Department. Two Parts 1969, 1981.

[DIVINE 1993] DIVINE, ROBERT A.: *The Sputnik Challenge*. New York: Oxford University Press 1993.

[DJEBBAR 1998] DJEBBAR, AHMAD: "A Panorama of Research on Mathematics in Andalus and in Maghreb between the IXth and XVIth Century." *Conference: New Perspectives on Science in Medieval Islam. (November 6–8, 1998). Manuscripts.* Cambridge, Mass.: The Dibner Institute, MIT 1998.

[DOBRZYCKI 1971] DOBRZYCKI, STANISŁAW: *Wydział Matematyczno-Fizyczny Szkoły Głównej Warszawskiej. Sekcja Matematyczna* (The Faculty of Mathematics and Physics at the University of Warsaw. Mathematical Department). Wrocław: Ossolineum 1971.

[DOMORADZKI 1992a] DOMORADZKI, STANISŁAW: "Czasopiśmiennictwo matematyczne polskie XIX i początku XX wieku" (Polish Mathematical Journals in the 19th Century and at the Beginning of the 20th Century). In: *Matematyka przełomu XIX u XX wieku* (Mathematics at the Turn of the 19th and 20th Century). Katowice: The Silesia University 1992, 61–74.

[DOMORADZKI 1992b] DOMORADZKI, STANISŁAW: "Samuel Dickstein (1851–1939) w świetle korespondencji z Władysławem Natansonem i Marianem Smoluchowskim. Problemy nurtujące naukę polską" (Samuel Dickstein in the Light of the Correspondence with Wł. Natanson and M. Smoluchowski. Problems of Polish Science). *Ibidem*, 84–103.

[DOMORADZKI 1995] DOMORADZKI, STANISŁAW: "Uwagi o literaturze matematycznej polskiej w latach 1851–1920" (Remarks about Polish Mathematical Literature, 1851–1920). In: *Matematyka Polska w stuleciu 1851–1950* (Polish Mathematics, 1851–1950). Uniwersytet Szczeciński. Materiały Konferencje **16** (1995), 161–180.

[DOROFEEVA 1961] DOROFEEVA, A. V.: "Razvitie variatsionnogo ischisleniya kak ischisleniya variatsiĭ. 2-ya polovina 18 — pervaya polovina 19 veka." *IMI* (1) **14** (1961), 101–180.

[Dow 1991] Dow, Peter B.: *Schoolhouse Politics: Lessons from the Sputnik Era.* Cambridge, MA: Harvard University Press 1991.

[Drake 1978] Drake, Stillman, *et al.*: "A Memorial Tribute to Kenneth O. May." *HM* **5** (1978), 3–12.

[Droysen 1800] Droysen, Johann Friedrich: *Rede von den Verdiensten der schwedischen Gelehrten um die Mathematik und Physik, zur Feyer des hohen Geburtsfestes unsers allerdurchlauchtigsten grossmächtigsten Königs und Herrn Gustav IV. Adolphs,* [...]. Greifswald: Mauritius 1800.

[DSB 1970/90] Gillispie, Charles C. (Ed.): *Dictionary of Scientific Biography.* 18 vols. New York: Charles Scribner's Sons 1970–1990.

[Du Mans 1890] Du Mans, P. Raphael: *Estat de la Perse en 1660.* Publié avec notes et appendice par Ch. Schefer. Paris: Ernest Leroux 1890.

[Duda 1996] Duda, Roman: "Fundamenta Mathematicae, Studia Mathematica, Acta Arithmetica — trzy pierwsze specialistyczne czasopisma matematyczne" (Fundamenta Mathematicae, Studia Mathematica, Acta Arithmetica — the First Three Topic-Oriented Mathematical Journals). *Zeszyty Naukowe Politechniki Śląskiej. Matematyka-Fizyka* **76** (1996), 47–80.

[Dunsch/Müller 1989] Dunsch, Lothar, and Hella Müller: *Ein Fundament zum Gebäude der Wissenschaften. Einhundert Jahre Ostwalds Klassiker der exakten Wissenschaften* (= *Ostwalds Klassiker der exakten Wissenschaften,* Sonderband). Leipzig: Akademische Verlagsgesellschaft Geest und Portig 1989.

[Duren/Askey/Merzbach 1988/89] Duren, Peter L.; Richard A. Askey, and Uta C. Merzbach (Eds.): *A Century of Mathematics in America.* Providence, R.I.: American Mathematical Society, vol. 1, 1988; vol. 2, 1989; vol. 3, 1989.

[Dvivedi 1890] Dvivedi, Sudhakara: *Gaṇaka Taraṅgiṇī* (in Sanskrit). Edited and published by his son Padmakara Dvivedi. Benares 1933.

[Dvivedi 1910] Dvivedi, Sudhakara: *A History of Mathematics. Part I* (in Hindi). Benares: Khelarilal 1910.

[Echegaray 1866] Echegaray, José: "Historia de las Matemáticas puras en nuestra España: Discurso de Recepción en la RACEFNM, 1866." In: *La Polémica de la Ciencia Española.* Madrid: Alianza Editorial 1970, 161–190.

[*EI*] *The Encyclopaedia of Islam.* New Edition. Leiden: E. J. Brill, since 1986.

[Einhorn 1985] Einhorn, Rudolf: *Vertreter der Mathematik und Geometrie an den Wiener Hochschulen 1900–1940.* (Dissertationen der Technischen Universität Wien, vol. 34, parts I and II.) Wien: VWGÖ 1985.

[Eisele 1985] Eisele, Carolyn: *Historical Perspectives on Peirce's Logic of Science. A History of Science.* 2 vols. Berlin: Mouton Publishers 1985.

[Ekama 1809] Ekama, Cornelius: *Oratio de Frisia ingeniorum mathematicorum in primis fertili.* Leeuwarden: D. van der Sluis 1809.

[Ekama 1823] Ekama, Cornelius: "Verhandeling over Gemma Frisius, den eersten grondlegger tot het bepalen van de lengte op zee." *Verhandelingen der Eerste Klasse van het Koninklijk Nederlandsch Instituut van Wetenschappen, Letterkunde en Schone Kunsten te Amsterdam* **7** (1823), 215–260.

[ELFVING 1981] ELFVING, GUSTAV: *The History of Mathematics in Finland 1828–1918.* Helsinki: Societas Scientiarum Fennica 1981.

[ENESTRÖM 1889] ENESTRÖM, GUSTAF: "Bibliographie suédoise de l'histoire des mathématiques 1667–1888". *Bibl. Math.* (2) **3** (1889), 1–14.

[ENESTRÖM 1901] ENESTRÖM, GUSTAF: "Account of *R. Guimarães. Les Mathématiques en Portugal au XIXe siècle.*" *Bibl. Math.* (3) **2** (1901), 168–169.

[ENESTRÖM 1910/13] ENESTRÖM, GUSTAF: "Verzeichnis der Schriften Leonhard Eulers." *Jber. DMV, Ergänzungsband* **4** (1910–1913).

[ENRIQUES 1915–1934] ENRIQUES, FEDERIGO: *Lezioni sulla teoria geometrica delle equazioni e delle funzioni algebriche, pubblicate a cura del dott. Oscar Chisini*: 4 vols. Bologna: Zanichelli 1915–1934; repr. Bologna: Zanichelli 1985.

[ENRIQUES 1922] ENRIQUES, FEDERIGO: "Presentazione di Einstein a Bologna." *Period. Mat.* (4) **2**, 76–79.

[ENRIQUES 1924–27] ENRIQUES, FEDERIGO: *Questioni riguardanti le matematiche elementari.* 4 vols. Bologna: Zanichelli 1924–1927; repr. Bologna: Zanichelli 1983.

[ENRIQUES 1934] ENRIQUES, FEDERIGO: *Signification de l'histoire de la pensée scientifique.* Paris: Hermann 1934.

[ENRIQUES 1936] ENRIQUES, FEDERIGO: *Il significato della storia del pensiero scientifico.* Italian trans. of [Enriques 1934]. Bologna: Zanichelli 1936.

[ENRIQUES 1938] ENRIQUES, FEDERIGO: *Le matematiche nella storia e nella cultura.* Bologna: Zanichelli 1938; repr. Bologna: Zanichelli 1971.

[ENRIQUES/SANTILLANA 1932] ENRIQUES, FEDERIGO, and GIORGIO DE SANTILLANA: *Storia del pensiero scientifico. Il mondo antico.* Bologna: Zanichelli 1932.

[ENRIQUES/SANTILLANA 1937] ENRIQUES, FEDERIGO, and GIORGIO DE SANTILLANA: *Compendio del pensiero scientifico dall'antichità fino ai tempi moderni.* Bologna: Zanichelli 1937; repr. Bologna: Zanichelli 1972.

[EPPLE 2000] EPPLE, MORITZ: "Genies, Ideen, Institutionen, mathematische Werkstätten: Formen der Mathematikgeschichte. Ein metahistorischer Essay." *Mathematische Semesterberichte* **47** (2000), 131-163.

[EVES 1964] EVES, HOWARD: *An Introduction to the History of Mathematics.* New York: Holt, Rinehart and Winston, rev. ed. 1964.

[EWING 1994] EWING, JOHN: *A Century of Mathematics: Through the Eyes of the "Monthly."* Washington, D.C.: Mathematical Association of America 1994.

[FÄH 1913] FÄH, HERMANN, *et al.*: "Prof. Dr. Hermann Kinkelin. 1832–1913." *Vh. Schw. Nat. Ges. 96. Jahresversammlung vom 7.–10. Sept. 1913 in Frauenfeld*, I. Teil, Anhang: Nekrologe und Biographien verstorbener Mitglieder, 34–48.

[FARRER 1907] FARRER, J. A.: *Literary Forgeries.* London: Longmans, Green, and Co. 1907.

[FAUSTMANN 1994] FAUSTMANN, GERLINDE: *Österreichische Mathematiker um 1800, unter besonderer Berücksichtigung ihrer logarithmischen Werke.* (Dissertationen der Technischen Universität Wien, vol. 59). Wien: Österreichischer Kunst- und Kulturverlag 1994.

[FAUVEL/VAN MAANEN 2000] FAUVEL, JOHN, and JAN VAN MAANEN (Eds.): *History in Mathematics Education* (= ICMI Studies Series, Vol. 6). Dordrecht: Kluwer Academic Publishers 2000.

[FAVARO 1888] FAVARO, ANTONIO: *Per la edizione nationale delle opere di Galileo Galilei.* Firenze: Esposizione e disegno 1888.

[FEINGOLD 1990] FEINGOLD, MORDECHAI (Ed.): *Before Newton. The Life and Times of Isaac Barrow.* Cambridge: Cambridge University Press 1990.

[FELLMANN/IM HOF 1993] FELLMANN, EMIL A., and HANS CHRISTOPH IM HOF: "Die Euler-Ausgabe — Ein Bericht zu ihrer Geschichte und ihrem aktuellen Stand." *Jahrbuch Überblicke Mathematik* 1993, 185–198.

[FERNÁNDEZ DEL CASTILLO 1953] FERNÁNDEZ DEL CASTILLO, FRANCISCO: *La Facultad de Medicina.* México: UNAM 1953.

[FERNÁNDEZ DEL CASTILLO 1982] FERNÁNDEZ DEL CASTILLO, FRANCISCO: *Libros y Libreros en el Siglo XVI.* México: FCE 1982.

[FERNÁNDEZ DE NAVARRETE 1846] FERNÁNDEZ DE NAVARRETE, MARTÍN: *Disertación sobre la historia de la náutica y de las ciencias matemáticas que han contribuído a sus progresos entre los españoles.* Madrid: Viuda de Calero 1846.

[FERRIER 1996] FERRIER, RONALD W. (Ed.): *A Journey to Persia. Jean Chardin's Portrait of a Seventeenth-Century Empire.* London and New York: I. B. Tauris Publishers 1996.

[FICHANT/PÊCHEUX 1974] FICHANT, MICHEL, and MICHEL PÊCHEUX: *Sur l'histoire des sciences.* Paris: F. Maspero 1974.

[FOLTA 1982] FOLTA, JAROSLAV: *Česká geometrická škola, historická analýza* (The Czech Geometric School, Historical Analysis). Prague: Academia 1982.

[FOLTA/NOVÝ 1973] FOLTA, JAROSLAV, and LUBOŠ NOVÝ: "The Fourth International Congress of the History of Sciences, Prague 1937." *Acta hist., Special Issue* **6** (1973). (Edition of Papers).

[FOMIN/SHILOV 1969] FOMIN, S. V., and G. E. SHILOV (Eds.): *Matematika v SSSR. 1958–1967.* 2 vols. (in Russian). Moscow: Nauka 1969.

[FOURCY 1828] FOURCY, AMBROISE L.: *Histoire de l'Ecole Polytechnique.* Paris 1828; repr. Paris: Belin 1987, with notes by JEAN DHOMBRES.

[FRAJESE 1964] FRAJESE, ATTILIO: *Galileo matematico.* Roma: Studium 1964.

[FRANCI 1989] FRANCI, RAFFAELLA: "Pietro Cossali: storico dell'algebra." In: [BARBIERI/DEGANI 1989], 199–217.

[FRANKE 1879] FRANKE, JAN N.: *Maciej Głoskowski, matematyk polski XVII wieku* (Maciej Głoskowski, a Polish Mathematician of the 17th Century). Cracow: Academy of Arts 1879.

[FRANKE 1884] FRANKE, JAN N.: *Jan Brożek, akademik krakowski (1585–1652)* (Jan Brożek, a Cracovian Academician). Cracow: Wdrukarni Uniwersytatu Jagiellońskiego 1884.

[FREI/STAMMBACH 1994] FREI, GÜNTHER, and URS STAMMBACH: *Die Mathematiker an den Zürcher Hochschulen.* Basel: Birkhäuser 1994.

[FREIRE 1872] FREIRE, FRANCISCO DO CASTRO: *Memória Histórica da Faculdade de Matemática nos cem annos decorridos desde a reforma da Universidade em 1772 até ao presente.* Coimbra: Imprensa da Universidade 1872.

[FREIRE 1884] FREIRE, FRANCISCO DO CASTRO: "A mathemática nas duas primeiras dinastias." *O Instituto* (2) **31** (1884), 405–410.

[FRIEDLEIN 1861] FRIEDLEIN, GOTTFRIED: *Gerbert, die Geometrie des Boethius und die indischen Ziffern.* Erlangen: Blaesing 1861.

[FRIEDLEIN 1869] FRIEDLEIN, GOTTFRIED: *Die Zahlzeichen und das elementare Rechnen der Griechen und Römer und des christlichen Abendlandes vom 7. bis 13. Jahrhundert.* Erlangen: Deichert 1869; repr. Wiesbaden: Sändig 1968.

[FROBESIUS 1750] FROBESIUS, JOHANN NIKOLAUS: *Historica et dogmatica ad Mathesin introductio, qua succincta matheseos historia cum ceteris ejusdem praecognitis, nec non systematis mathematici delineatio continentur.* Helmstedt: Weygand 1750.

[FROBESIUS 1751/54] FROBESIUS, JOHANN NIKOLAUS: *Rudimenta biographiae mathematicorum.* 3 parts. Helmstedt: Schnorr 1751–1754.

[FRYDE 1990] FRYDE, EDMUND B.: "Historiography and Historical Methodology." *The New Encyclopædia Britannica, Macropaedia* **8**. Chicago: Encyclopaedia Britannica 1990.

[FU 1990] FU ZUOHUA: "A Study on *Chouren zhuan.*" In: MEI RONGZHAO (Ed.): *Ming Qing shuxueshi lun wenji* (Collected Papers on the History of Mathematics in the Ming and Qing Periods). Nanjing: Jiangsu Education Press 1990, 235–237.

[FUETER 1941] FUETER, EDUARD: *Geschichte der exakten Wissenschaften in der Schweizerischen Aufklärung (1680–1780)* (= *Veröffentlichungen der Schweizerischen Gesellschaft für Geschichte der Medizin und der Naturwissenschaften* **12**). Aarau and Leipzig: Sauerländer 1941.

[FUJIWARA 1936] FUJIWARA MATSUSABURŌ: "Obituary Note, Hayashi Tshuruichi," *Tōhoku Mathematical Journal* **41** (1936), 265–289; repr. in [Hayashi 1937], vol. 1, 570–589; repr. again, Tokyo: Hohukan Shokan 1985.

[FUJIWARA 1942] FUJIWARA MATSUSABURŌ: "A Brief Sketch of Wazan. The Mathematics of the Old Japanese School," *Journal of the Sendai International Cultural Society* (1942), 64–84.

[FUJIWARA 1949a] FUJIWARA MATSUSABURŌ: "Yo no Wasan-shi Kenkyū" (My Study in History of Japanese Mathematics). *Kagakushi Kenkyū* **11** (1949), 79–86.

[FUJIWARA 1949b] FUJIWARA MATSUSABURŌ: "The List of Mathematical Papers by Prof. M. Fujiwara," *Tōhoku Mathematical Journal* **49** (1949), 133–138.

[FUJIWARA 1954/60] FUJIWARA MATSUSABURŌ: *Meiji-zen Nippon Sūgaku-shi* (History of Mathematics in the pre-Meiji Era), HIRAYAMA AKIRA (Ed.), 5 vols., published for the Japan Academy. Tokyo: Iwanami Shoten 1954–1960; repr. (with some corrections) Tokyo: Iwanami Shoten 1983.

[GÅRDING 1994] GÅRDING, LARS: *Matematik och Matematiker. Matematiken i Sverige före 1950.* Lund: Lund University Press 1994. English ed.: *Mathematics and Mathematicians: Mathematics in Sweden Before 1950.* Lund: Lund University Press 1997.

[GARCÍA DE GALDEANO 1921] GARCÍA DE GALDEANO, ZOEL: *La Matemática en España: Conferencia dada en Abril de 1921 en la Facultad de Ciencias.* Zaragoza: University of Zaragoza 1921.

[GARCÍA DE ZUÑIGA 1992] GARCÍA DE ZUÑIGA, EDUARDO: *Lecciones de Historia de las Matemáticas,* (MARIO H. OTERO, Ed.). Montevideo: Facultad de Humanidades y Ciencias de la Educación 1992.

[GASSER 1979] GASSER, ADOLF: "Prof. Dr. Joachim Otto Fleckenstein (7. Juli 1914 bis 21. Februar 1980)." *Vh. Nat. Ges. Basel* **90** (1979), 151–156.

[GEMIGNANI 1993] GEMIGNANI, GIUSEPPE (Ed.): *Contributi alla storia delle matematiche. Scritti in onore di Gino Arrighi.* Accademia Nazionale di Scienze, Lettere e Arti. Modena: Mucchi 1993.

[GERDES 1998] GERDES, PAULUS: *Women, Art and Geometry in Southern Africa.* Trenton, NJ: Africa World Press 1998.

[GERICKE 1966] GERICKE, HELMUTH: "Aus der Chronik der Deutschen Mathematiker-Vereinigung." *Jber. DMV* **68** (1966), 46–70.

[GERICKE 1968] GERICKE, HELMUTH: "Wilhelm Süss, der Gründer des Mathematischen Forschungsinstitutes Oberwolfach." *Jber. DMV* **69** (1967/68), 161-183.

[GERICKE 1984] GERICKE, HELMUTH: *Mathematik in Antike und Orient.* Berlin and Heidelberg: Springer 1984.

[GERICKE 1990] GERICKE, HELMUTH: *Mathematik im Abendland: Von den römischen Feldmessern bis zu Descartes.* Berlin and Heidelberg: Springer 1990.

[GEYMONAT 1947] GEYMONAT, LUDOVICO: *Storia e filosofia dell'analisi infinitesimale.* Torino: Levrotto e Bella 1947.

[GEYMONAT 1957] GEYMONAT, LUDOVICO: *Galileo Galilei.* Torino: Einaudi 1957.

[GILLIS 1973] GILLIS, JAN: "Paul Mansion en George Sarton." *Mededelingen Koninklijke Academie voor Wetenschappen, Letteren en Schone Kunsten van België, Klasse der Wetenschappen* **35**, no. 2 (1973).

[GINSBURG 1932] GINSBURG, JEKUTHIEL: "The Policy of *Scripta Mathematica.*" *Scripta Math.* **1** (1932), 1–2.

[GISPERT 1999] GISPERT, HÉLÈNE: "Les débuts de l'histoire des mathématiques sur les scènes internationales et le cas de l'entreprise encyclopédique de Felix Klein et Jules Molk." *HM* **26** (1999), 344–360.

[GJERTSEN 1986] GJERTSEN, DEREK: *The Newton Handbook.* London and New York: Routledge and Kegan Paul 1986.

[GLASS 1998] GLASS, DAGMAR: "Der Missionar Cornelius van Dyck (1818–1895) als Lehrbuchautor und Förderer des arabischen Wissenschaftsjournalismus." In: PREISSLER, HOLGER, and HEIDI STEIN (Eds.): *Annaeherung an das Fremde. XXVI. Deutscher Orientalistentag vom 25. bis 29. 9. 1995 in Leipzig (= Z. dt. MG., Supplement* **11** (1998), 187–198.

[GNEDENKO 1946] GNEDENKO, BORIS V.: *Ocherki po istorii matematiki v Rossii.* Moscow and Leningrad: Gostekhizdat 1946.

[GODEAUX 1943] GODEAUX, LUCIEN: *Esquisse d'une histoire des sciences mathématiques en Belgique.* Bruxelles: Office de Publicité 1943.

[GÖÇEK 1987] GÖÇEK, FATMA MÜGE: *East Encounters West. France and the Ottoman Empire in the Eighteenth Century.* New York and Oxford: Oxford University Press 1987.

[GOŁĄB 1964] GOŁĄB, STANISŁAW: "Zarys dziejów matematyki w Uniwersytecie Jagiellońskim w XX wieku" (An Outline of the History of Mathematics at the Jagiellonian University in the 20th Century). In: *Studia z dziejów Katedr Wydziału Matematyki, Fizyki i Chemii Uniwersytetu Jagiellońskiego* (Studies on History of the Departments of the Faculty of Mathematics, Physics and Chemistry at the Jagiellonian University). Cracow: Jagiellonian University 1964, 75–86. (Universitas Jagellonica Opera ab Universitate Jagellonica sexta solemnia saecularia celebrante edita **15**.)

[GOLDFARB 1994] GOLDFARB, JOSÉ LUIZ: *Voar também é com os homens. O Pensamento de Mário Schemberg.* São Paulo: Edusp. 1994.

[GOLDSTEIN/GRAY/RITTER 1996] GOLDSTEIN, CATHERINE; JEREMY GRAY, and JIM RITTER (Eds.): *L'Europe Mathématique. Histoire, Mythes, Identités.* Paris: Éditions de la Maison des sciences de l'homme 1996.

[GOLIUS 1669] GOLIUS, JACOB: *Muhammedis Fil. Ketiri Ferganensis, qui vulgo Alfraganus dicitur, Elementa astronomica, arabicè et latinè. Cum notis ad res exoticas sive orientales, quae in iis occurrunt. Opera Jacobi Golii.* Amsterdam: Johannes Jansonius van Waesberge 1669.

[GONÇALVES 1940] GONÇALVES, JOSÉ VICENTE: "Análise do Livro VIIII dos Principios Mathemáticos de José Anastácio da Cunha." *Congresso do Mundo Português* **12**, vol. 1 (1940), 123–140.

[GONÇALVES 1958] GONÇALVES, JOSÉ VICENTE: "O Professor Mira Fernandes." *Boletim da Academia das Ciências de Lisboa* **30** (1958), 168–171.

[GONÇALVES 1966] GONÇALVES, JOSÉ VICENTE: "Elogio Histórico de Pedro José da Cunha." *Memórias da Academia das Ciências de Lisboa*, Classe de Ciências, **9** (1966), 93–111.

[GONÇALVES 1971] GONÇALVES, JOSÉ VICENTE: "Aureliano de Mira Fernandes, Investigador e Ensaísta." Preface to *Obras Completas de Aureliano de Mira Fernandes*, vol. 1 (1971), pp. v–xxi.

[GRABINER 1975] GRABINER, JUDITH: "The Mathematician, the Historian and the History of Mathematics," *HM* **2** (1975), 439–447.

[GRAF 1886] GRAF, JOHANN HEINRICH: "Die Naturforschende Gesellschaft in Bern vom 18. Dezember 1786 bis 18. Dezember 1886. Ein Rückblick auf die Geschichte dieses Vereins bei Anlass der Feier des 100jährigen Bestehens." *Mitt. Nat. Ges. Bern*, Jahrgang 1886, 91–177.

[GRAF 1888/90] GRAF, JOHANN HEINRICH: *Geschichte der Mathematik und der Naturwissenschaften in bernischen Landen vom Wiederaufblühen der Wissenschaften bis in die neuere Zeit*, 3 Hefte in insgesamt 4 Teilen. Bern and Basel: Wyss 1888–1890.

[GRAF 1897] GRAF, JOHANN HEINRICH: "Verzeichnis der gedruckten mathematischen, astronomischen und physikalischen Doktor-Dissertationen der schweizer. Hochschulen bis zum Jahre 1896." *Mitt. Nat. Ges. Bern*, Jahrgang 1897, 53–60.

[GRAF 1902] GRAF, JOHANN HEINRICH: "Daniel Huber's trigonometrische Vermessung des Kantons Basel (1813–1824)." *Mitt. Nat. Ges. Bern*, Jahrgang 1902, 1–79.

[GRAF 1906] GRAF, JOHANN HEINRICH: "Zur Geschichte der mathematischen Wissenschaften an der ehemaligen Akademie und der Hochschule Bern." *Mitt. Nat. Ges. Bern*, Jahrgang 1906, 63–81.

[GRAF-STUHLHOFER 1996] GRAF-STUHLHOFER, FRANZ: *Humanismus zwischen Hof und Universität, Georg Tannstetter (Collimitius) und sein wissenschaftliches Umfeld im Wien des frühen 16. Jahrhunderts.* Wien: WUV Universitätsverlag 1996.

[GRATTAN-GUINNESS 1989] GRATTAN-GUINNESS, IVOR: "Some Remarks on the Recognition of Arabic Mathematics in the Writings of Montucla and Delambre." *Gaṇita-Bhāratī* **11** (1989), 12–17.

[GRATTAN-GUINNESS 1990] GRATTAN-GUINNESS, IVOR: *Convolutions in French Mathematics, 1800–1840. Fron the Calculus and Mechanics to Mathematical Analysis and Mathematical Physics* (= *Science Networks* **2–4**). 3 vols. Basel, Boston, Berlin: Birkhäuser 1990.

[GRATTAN-GUINNESS 1994] GRATTAN-GUINNESS, IVOR (Ed.): *The Companion Encyclopedia of the History and Philosophy of the Mathematical Sciences.* 2 vols. London and New York: Routledge 1994.

[GREENAWAY 1996] GREENAWAY, FRANK: *Science International. A History of the International Council of Scientific Unions.* Cambridge: Cambridge University Press 1996.

[GRINSTEIN/LIPSEY 2001] GRINSTEIN, LOUISE S., and SALLY I. LIPSEY (Eds.): *Encyclopedia of Mathematics Education.* New York: RoutledgeFalmer 2001.

[GRÖSSING 1983] GRÖSSING, HELMUTH: *Humanistische Naturwissenschaft. Zur Geschichte der Wiener mathematischen Schulen des 15. und 16. Jahrhunderts* (= *Saecula Spiritualia,* **8**). Baden-Baden: Verlag Valentin Loerner 1983.

[GRUBER 1996] GRUBER, MARIA: *Die Mathematik in Österreich im 17. Jahrhundert — anhand der Biographien zweier Benediktiner.* Dissertation Technische Universität Wien 1996.

[GUERRAGGIO 1989] GUERRAGGIO, ANGELO: "I contributi di Vailati alla storia della matematica." In: [BARBIERI/DEGANI 1989], 255–268.

[GUIMARÃES 1900] GUIMARÃES, RODOLFO: *Les Mathématiques en Portugal au XIX*[e] *Siècle.* Coimbra: Imprimerie de l'Université 1900.

[GUIMARÃES 1909] GUIMARÃES, R.: *Les Mathématiques en Portugal.* Coimbra: Imprimerie de l'Université 1909.

[GUIMARÃES 1911] GUIMARÃES, R.: *Les Mathématiques en Portugal. Appendice II.* Coimbra: Imprimerie de l'Université 1911.

[GUIMARÃES 1914/15] GUIMARÃES, R.: "Sur la vie et l'œuvre de Pedro Nunes." *Annaes Scientificos da Academia Polytechnica do Porto* **9** (1914), 54–64, 96–117, 152–167, 210–227, and **10** (1915), 20–36.

[GUIMARÃES 1915a] GUIMARÃES, R.: "Decreto ordenando a impressão das obras de Pedro Nunes." *O Instituto* (3) **67** (1915), 347–352.

[GUIMARÃES 1915b] GUIMARÃES, R.: "Bosquejo histórico sobre a historiografia das Matemáticas." *Boletim da 2ª Classe da Academia das Sciências de Lisboa* **9** (1915), 438–460.

[GUIMARÃES 1915c] GUIMARÃES, R.: "Programa de um Curso Universitário de História das Matemáticas." *Revista da Universidade de Coimbra* **4** (1915), 699–704.

[GUIMARÃES 1916a] GUIMARÃES, R.: "A edição de 1546 do livro de Pedro Nunes: De arte atque ratione navigandi." *Boletim bibliográfico da biblioteca da Universidade de Coimbra* **3** (1916), 28–36.

[GUIMARÃES 1916b] GUIMARÃES, R.: "Um opúsculo raríssimo de Pedro Nunes." *Boletim bibliográfico da biblioteca da Universidade de Coimbra* **3** (1916), 268–289.

[GUIMARÃES 1917] GUIMARÃES, R.: "O livro de Vernier." *Boletim bibliográfico da biblioteca da Universidade de Coimbra* **4** (1917), 85–121, 224–255.

[GUIMARÃES 1921] GUIMARÃES, R.: "Os Professores Monteiro da Rocha e Anastácio da Cunha." *Jornal de sciências matemáticas, físicas e naturais* (3) **2** (1921), 67–87.

[GUO 1993] GUO SHUCHUN: *Zhongguo gudai kexue jishu dianji tonghui. Shuxue juan* (Complete Edition of Ancient Books on Chinese Science and Technology. Mathematics Section). 5 vols. Zhengzhou: Henan Education Press 1993.

[GUPTA 1979] GUPTA, RADHA CHARAN: "Prabodh Chandra Sengupta (1876–1962), Historian of Indian Astronomy and Mathematics." *Gaṇita Bhāratī* **1** (1979), 31–35.

[GUPTA 1980] GUPTA, RADHA CHARAN: "Bibhutibhusan Datta (1888–1958), Historian of Indian Mathematics." *HM* **7** (1980), 126–133 (with bibliography and portrait).

[GUPTA 1990a] GUPTA, RADHA CHARAN: "The Chronic Problem of Ancient Indian Chronology." *Gaṇita Bhāratī* **12** (1990), 17–26.

[GUPTA 1990b] GUPTA, RADHA CHARAN: "Sudhākara Dvivedī (1855–1910), Historian of Indian Astronomy and Mathematics." *Gaṇita Bhāratī* **12** (1990), 83–96 (with bibliography and portrait).

[GURJAR 1947] GURJAR, LAXMAN VASUDEVA: *Ancient Indian Mathematics and Vedha.* Poona: Ideal Book Service 1947.

[HADJIDAKIS 1884] HADJIDAKIS, IOANNIS: "Inaugural Speech (pronounced in 7. 12. 1881)." *Parnassos* **8** (1884), 803–811.

[HADJIDAKIS 1885] HADJIDAKIS, IOANNIS: "History of Mathematics in Ancient Greece." *Parnassos* **9** (1885), 7–20.

[HADJIDAKIS 1899] HADJIDAKIS, NIKOLAOS: "Mathematics in Scandinavia." *Athena* **11** (1899), 582–587.

[HADJIDAKIS 1901/02] HADJIDAKIS, NIKOLAOS: "Sur l'Etat actuel des Mathématiques Supérieures en Grèce." *Enseign. Math.* **3** (1901/02), 397–400.

[HADJIDAKIS 1910] HADJIDAKIS, NIKOLAOS: "Mathematics in the Intellectual Life." *Panathinea* **19** (1910), 279–289, 320–324.

[HANKEL 1874] HANKEL, HERMANN: *Zur Geschichte der Mathematik in Alterthum und Mittelalter.* Leipzig: B. G. Teubner 1874; repr., with preface and index by J. E. HOFMANN: Hildesheim: Olms 1965.

[HARA 1987] HARA, KOKITI: "Necrologie: Koshiro Nakamura (1901–1986)." *Historia Scientiarum* (Tokyo) **32** (1987), 147–150.

[HARDY 1934] HARDY, GODEFREY HAROLD: "The J-type and the S-type among Mathematicians." *Nature* **134** (August 18, 1934), 250; repr. in *Math. Intell.* **6**, no. 3 (1984), 7.

[HARLFINGER 1989] HARLFINGER, DIETER (Ed.): *Graecogermania: Griechischstudien deutscher Humanisten. Die Editionstätigkeit der Griechen in der italienischen Renaissance (1469–1523)* (= *Ausstellungskataloge der Herzog August Bibliothek* **59**). Weinheim: VCH 1989.

[HARTING 1868] HARTING, PIETER: *Christiaan Huygens in zijn leven en werken geschetst.* Groningen: Gebroeders Hoitsema 1868.

[HARTSHORNE 1938] HARTSHORNE, E. Y.: "The German Universities and the Government." *The Annals of the American Academy of Political and Social Science* **200** (1938), 210–234.

[HAYASHI 1910] HAYASHI TSURUICHI: "The *Fukudai* and Determinants in Japanese Mathematics." *Proceedings of the Tokyo Mathematico-Physical Society* **5** (1910), 254–271.

[HAYASHI 1937] HAYASHI TSURUICHI: *Wasan Kenkyū Shūroku* (Collected Papers on Japanese Mathematics), 2 vols., FUJIWARA MATSUSABURŌ, *et al.* (Eds.). Tokyo: Tokyo Kaiseikan 1937; repr. Tokyo: Hohukan Shokan, 1985.

[HEIBERG 1883/86] HEIBERG, JOHAN LUDVIG: *Euclidis opera omnia. Elementa.* 4 vols. Leipzig: Teubner 1883–1886.

[HEILBRONNER 1739] HEILBRONNER, JOHANN CHRISTOPH: *Versuch einer mathematischen Historie [...].* Frankfurt and Leipzig: Wohler 1739.

[HEILBRONNER 1742] HEILBRONNER, JOHANN CHRISTOPH: *Historia matheseos universae a mundo condito ad saeculum p[ost] C[hristum] n[atum] XVI. Praecipuorum mathematicorum vitas, dogmata, scripta et manuscripta complexa, accedit recensio elementorum compendiorum et operum mathematicorum atque Historia Arithmetices ad nostra tempore.* Leipzig: Impensis Joh. Friderici Gleditschii 1742.

[HELFGOTT 1995] HELFGOTT, MICHEL: "Improved Teaching of the Calculus Through the Use of Historical Materials." In: SWETZ, FRANK, *et al.* (Eds.): *Learn from the Masters!* Washington, D.C.: Mathematical Association of America 1995, 135–144.

[HELLINGER/TOEPLITZ 1927] HELLINGER, ERNST, and OTTO TOEPLITZ: "Integralgleichungen und Gleichungen mit unendlichvielen Unbekannten." In: *Encyklopädie der mathematischen Wissenschaften*, Part II.3.2, C13 (1927), 1335–1597.

[HIERONYMUS 1992] HIERONYMUS, FRANK: *En Basileia polei tes Germanias: Griechischer Geist aus Basler Pressen.* Ausstellungskatalog Basel/Berlin/Mainz (= *Publikationen der Universitätsbibliothek Basel* **15**). Basel: Universitätsbibliothek 1992.

[HILL 1896] HILL, GEORGE WILLIAM: "Remarks on the Progress of Celestial Mechanics Since the Middle of the Century." *Bull. AMS* **2** (1896), 125–136.

[HIRAYAMA/ŌYA/SHIMODAIRA 1984] HIRAYAMA AKIRA; ŌYA SHIN'ICHI, and SHIMODAIRA KAZUO: *Bunkashi-jō yori mitaru Nippon no Sūgaku* (Japanese Mathematics from the Viewpoint of Cultural History). Tokyo: Kōseisha Kōseikaku 1984 (original edition 1922).

[HÖFLECHNER 1993] HÖFLECHNER, WALTER: "Wissenschaftsgeschichte in Österreich." *Mitt. Österr. Ges. Wiss.Gesch.* **13** (1993), 82–102.

[HOFER 1985] HOFER, HANS: *Querschnitt durch die Mathematik im Dritten Reich.* Dissertation Universität Wien 1985.

[HOFMANN 1940] HOFMANN, JOSEPH EHRENFRIED: "Über Ziele und Wege mathematikgeschichtlicher Forschung." *Dt. Math.* **5** (1940), 150–157.

[HOFMANN 1948a] HOFMANN, JOSEPH EHRENFRIED: "Geschichte der Mathematik." In: *FIAT Review of German Science 1939–1946. Vol. I: Pure Mathematics*, SÜSS, WILHELM (Ed.), Part I. German edition: *Naturforschung und Medizin in Deutschland*, Band I: *Reine Mathematik.* Teil I, 1–9. Wiesbaden: Dieterich'sche Verlagsbuchhandlung 1948.

[HOFMANN 1948b] HOFMANN, JOSEPH EHRENFRIED: "Über Wesen, Methoden und Aufgabe der Mathematikgeschichte und ihre Bedeutung für die Geschichte der Naturwissenschaften." *MNU* **1** (1948/49), 63–67.

[HOFMANN 1953/57/63] HOFMANN, JOSEPH EHRENFRIED: *Geschichte der Mathematik.* Vol. I: *Von den Anfängen bis zum Auftreten von Fermat und Descartes.* Berlin: De Gruyter 1953; 2nd. ed. 1963). Vol. II: *Von Fermat und Descartes bis zur Erfindung des Calculus [...]* Berlin: Walter de Gruyter 1957. Vol. III: *Von den Auseinandersetzungen um den Calculus bis zur Französischen Revolution.* Berlin: Walter de Gruyter 1957 (= *Sammlung Göschen*, vols. 226/226a, 875, 882).

[HOFMANN 1970] HOFMANN, JOSEPH EHRENFRIED: "*In memoriam* Siegfried Heller (1876–1970)." *Arch. int. Hist. Sci.* **23** (1970), 214–216.

[HOLST 1889] HOLST, ELLING: "Bibliographische Notiz über das Studium der Geschichte der Mathematik in Norwegen." *Bibl. Math.* (2) **3** (1889), 97–103.

[HUA 1983] HUA LUOGENG (L. K. HUA): *Qian Baocong kexueshi lunwen xuanji* (Selected Essays of Qian Baocong on the History of Science). Beijing: Science Press 1983.

[HULTSCH 1876/78] HULTSCH, FRIEDRICH: *Pappi alexandrini collectionis quae supersunt.* 3 vols. Berlin: Weidmann 1876–1878.

[HUNTER 1996] HUNTER, PATTI W.: "Drawing the Boundaries: Mathematical Statistics in 20th-Century America." *HM* **23** (1996), 7–30.

[HYPPOLITE 1964] HYPPOLITE, JEAN: "L'épistémologie de G. Bachelard." *Rev. Hist. Sci.* **17** (1964), 1–12.

[ISELY 1901] ISELY, LOUIS: *Histoire des sciences mathématiques dans la Suisse française.* Neuchâtel: Imprimerie nouvelle 1901.

[ISRAEL 1990] ISRAEL, GIORGIO: "On Correspondence Between B. Boncompagni and A. Genocchi." *HM* **17** (1990), 48–54.

[IWIŃSKI 1975] IWIŃSKI, TADEUSZ: *Ponad pół wieku działalności matematyków polskich* (Over Half a Century of Polish Mathematicians' Activity). Warsaw: Państwowe Wydawnictwo Naukowe 1975.

[JAMI 1994] JAMI, CATHERINE: "Learning Mathematical Sciences During the Early and Mid-Ch'ing." In: ELMAN, B. A., and A. WOODSIDE (Eds.): *Education and Society in Late Imperial China.* Berkeley: University of California Press 1994, 223–256.

[JARNÍK 1981] JARNÍK, VOJTĚCH: *Bernard Bolzano and the Foundation of Mathematical Analysis.* (Collection of articles, in Czech and English, originally published 1922–1961). Prague: Academia 1981.

[JAYAWARDENE 1983] JAYAWARDENE, S. A.: "Mathematical Sciences." In: CORSI, PIETRO, and PAUL WEINDLING (Eds.): *Information Sources in the History of Science and Medicine.* London: Butterworth 1983, 259–284 (with extensive bibliography).

[JOHNSON 1989] JOHNSON, WILLIAM: "Charles Hutton, 1737–1823." *Journal of Mechanical Working Technology* **18** (1989), 195–230.

[JOHNSON 1996] JOHNSON, WILLIAM: "Isaac Todhunter (1820–1884)." *International Journal of Mechanical Science* **38** (1996), 1231–1270.

[JONES 1976] JONES, PHILLIP S.: "Louis Charles Karpinski, Historian of Mathematics and Astronomy." *HM* **3** (1976), 185–202.

[JONES/ENROS 1984] JONES, CHARLES V., and PHILIP C. ENROS: "Kenneth O. May, 1915-1977. His Early Life to 1946." *HM* **11** (1984), 359–379.

[KÄSTNER 1796/1800] KÄSTNER, ABRAHAM GOTTHELF: *Geschichte der Mathematik seit der Wiederherstellung der Wissenschaften bis an das Ende des 18. Jahrhunderts.* 4 vols. Göttingen: Johann Georg Rosenbusch's Wittwe 1796–1800; repr., with introduction and index of names, by JOSEPH EHRENFRIED HOFMANN. 4 vols. Hildesheim and New York: Georg Olms Verlag 1970.

[KAGAN 1907] KAGAN, VENIAMIN F.: *Osnovaniya geometrii.* Part 2: *Istoricheskiĭ ocherk razvitiya ucheniya ob osnovaniyakh geometrii.* Odessa: "Ekonomicheskaya" Tipografiya 1907.

[KAISER/NÖBAUER 1984] KAISER, HANS, and WILFRIED NÖBAUER: *Geschichte der Mathematik für den Schulunterricht.* Wien: Hölder-Pichler-Tempsky 1984.

[KARAGIANNIDES 1893] KARAGIANNIDES, A.: *Die Nicht-Euklidische Geometrie vom Alterthum bis zur Gegenwart.* Berlin: Mayer und Müller 1893.

[KAWAHARA/YANO 1996] KAWAHARA HIDEKI, and YANO MICHIO: "Japanese Contributions to the History of Chinese Science." *Historia Scientiarum* (Tokyo) **6** (1996), 123–158.

[KAYE 1925] KAYE, GEORGE RUSBY: *Indian Mathematics.* Calcutta: Thacker and Spink 1925.

[KENNEDY 1980] KENNEDY, HUBERT: *Peano. Life and Works of Giuseppe Peano.* Dordrecht and Boston: Reidel 1980.

[KLEIN 1926/27] KLEIN, FELIX: *Vorlesungen über die Entwicklung der Mathematik im 19. Jahrhundert.* COURANT, RICHARD, and OTTO NEUGEBAUER (Eds.), 2 vols. Berlin: Julius Springer 1926–1927; repr. in one vol. 1979.

[KLINE 1953] KLINE, MORRIS: *Mathematics in Western Culture.* New York: Oxford University Press 1953; repr. 1964.

[KLINE 1962] KLINE, MORRIS: *Mathematics: a Cultural Approach.* Reading, MA: Addison-Wesley Pub. Co. 1962.

[KLINE 1972] KLINE, MORRIS: *Mathematical Thought from Ancient to Modern Times.* New York: Oxford University Press 1972; repr. 1990.

[KLINE 1976] KLINE, MORRIS: "Carl B. Boyer (1906–1976)." *HM* **3**, 387–401.

[KLÜGEL 1803/1831] KLÜGEL, GEORG SIMON: *Mathematisches Wörterbuch oder Erklärung der Begriffe, Lehrsätze, Aufgaben und Methoden der Mathematik mit den nöthigen Beweisen und literarischen Nachrichten begleitet in alphabetischer Ordnung.* 5 vols. in 6 parts (vol. 4 ed. by CARL BRANDAN MOLLWEIDE, vol. 5 (in 2 parts) by JOHANN AUGUST GRUNERT). Leipzig: Schwickert 1803/31. Supplements (2 vols.) ed. by J. A. GRUNERT. Leipzig: Schwickert 1833–1836.

[KOLMOGOROV 1938] KOLMOGOROV, ANDREĬ N.: "Matematika." In: *Bol'shaya Sovetskaya Entsiklopediya.* 1st ed. Moscow: Izdatel'stvo Sovetskaya Entsiklopediya **38**, 359–401.

[KOLMOGOROV 1946] KOLMOGOROV, ANDREĬ N.: "N'yuton i sovremennoe matematicheskoe myshlenie." In: *Moskovskiĭ universitet — pamyati Isaaka N'yutona (1643–1943).* Moscow: Izdatel'stvo Moskovskogo Universiteta 1946, 27–42.

[KOLMOGOROV/YOUSHKEVICH 1978/87] KOLMOGOROV, ANDREĬ N., and ADOL'F P. YOUSHKEVICH (Eds.): *Matematika XIX veka. Matematicheskaya logika. Algebra. Teoriya chisel. Teoriya veroyatnosteĭ.* Moscow: Nauka 1978. (English trans.: *Mathematics of the 19th Century. Mathematical Logic. Algebra. Number Theory. Probability Theory.* Basel: Birkhäuser 1992.)

KOLMOGOROV, ANDREĬ N., and ADOL'F P. YOUSHKEVICH (Eds.): *Matematika XIX veka. Geometriya. Teoriya analiticheskikh funktsiĭ.* Moscow: Nauka 1981. (English trans.: *Mathematics of the 19th Century. Geometry. Analytic Function Theory.* Basel: Birkhäuser 1996.)

KOLMOGOROV, ANDREĬ N., and ADOL'F P. YOUSHKEVICH (Eds.): *Matematika XIX veka. Chebyshevskoe napravlenie v teorii funktsiĭ. Obyknovennye differentsial'nye uravneniya. Variatsionnoe ischislenie. Teoriya konechnykh raznosteĭ.* Moscow: Nauka, 1987. (English trans.: *Mathematics of the 19th Century. Function Theory According to Chebyshev. Ordinary Differential Equations. Calculus of Variations. Theory of Finite Differences.* Basel: Birkhäuser 1998.)

[KONDŌ 1946] KONDŌ YŌITSU: *Kikagaku Shisō-shi* (The History of Geometrical Thought). Tokyo: Itō Shoten 1946. Second revised edition under the new title *Shin Kikagaku Shisō-shi* (The New History of Geometrical Thought). Tokyo: San'ichi Shobō 1966.

[KONDŌ 1947] KONDŌ YŌITSU: *Sūgaku Shisōshi Josetsu* (An Introduction to the History of Mathematical Thought). Kyoto: San-ichi Shobō 1947.

[KONDŌ 1994] KONDŌ YŌITSU: *Sōgakushi Chosakushō* (Collected Works on History of Mathematics, in 5 vols.), SASAKI CHIKARA (Ed.). Tokyo: Nihon Hyoron-sha 1994.

[KORDOS 1994] KORDOS, MAREK: *Wykłady z historii matematyki* (Lectures on History of Mathematics). Warsaw: Wydawnictwa Szkolne i Pedagogiczne 1994.

[KOYRÉ/COHEN 1972] *Isaac Newton's Philosophiae naturalis principia mathematica.* The Third Edition (1726), with variant readings, assembled and edited by ALEXANDRE KOYRÉ and I. BERNARD COHEN, with the assistance of ANNE WHITMAN. 2 vols. Cambridge, Mass.: Harvard University Press 1972.

[KRAFFT 1753] KRAFFT, GEORG WOLFGANG: *Institutiones geometriae sublimioris.* Tübingen: Berger 1753.

[KRAGH 1987] KRAGH, HELGE: *An Introduction to the Historiography of Science.* Cambridge, Mass.: Cambridge University Press 1987.

[KRAUS 1938] KRAUS, PAUL: "Julius Ruska." *Osiris* **5** (1938), 5–40.

[KROPP 1948] KROPP, GERHARD: *Beiträge zur Philosophie, Pädagogik und Geschichte der Mathematik.* (With an appendix: "Die mathematikgeschichtliche Forschung," 80–88.) Berlin: Fr. K. Koetschau Verlag 1948.

[KRYLOV 1916] KRYLOV, ALEKSEĬ N. (Ed.): *N'yuton I. Matematicheskie Nachala Natural'noĭ Filosofii.* Trans. by A. N. KRYLOV. *Izvestiya Nikolaevskoĭ Morskoĭ Akademii.* St. Peterburg: Nikolaevskaya Morskaya Akademiya 1916.

[KURATOWSKI 1973] KURATOWSKI, KAZIMIERZ: *Pół wieku matematyki polskiej 1920–1970.* Warsaw: Omega. Wiedza Powszechna 1973. English ed.: *A Half Century of Polish Mathematics.* Warsaw: PWN and Oxford: Pergamon Press 1980.

[KUROSH 1959] KUROSH, ALEKSANDR G. (Ed.): *Matematika v SSSR za 40 let. 1917–1957.* 2 vols. Moscow: Fizmatgiz 1959.

[KUROSH/MARKUSHEVICH/RASHEVSKIĬ 1948] KUROSH, ALEKSANDR G.; ALEKSEĬ I. MARKUSHEVICH, and PËTR K. RASHEVSKIĬ (Eds.): *Matematika v SSSR za 30 let. 1917–1947.* Moscow and Leningrad: Gostekhizdat 1948.

[LAGRANGE 1788] LAGRANGE, JOSEPH LOUIS: *Mécanique analytique.* Paris: Desaint 1788; 2nd ed., 2 vols., Paris: Courcier 1811–1815, in *Œuvres de Lagrange*, M. J.-A. SERRET (Ed.). Paris: Gauthier-Villars 1867–1892, **11–12**; repr. Hildesheim: Georg Olms 1973.

[LAUDAN 1993] LAUDAN, RACHEL: "Histories of the Sciences and their Uses: a Review to 1913." *Hist. Sci.* **31** (1993), 1–34.

[LE PAIGE 1884] LE PAIGE, CONSTANTIN: "Correspondance de René-François de Sluse publiée pour la première fois et précédée d'une introduction par M. C. Le Paige." *Bull. Boncompagni* **17** (1884), 427–554, 603–726. Separately published Rome: Imprimerie des sciences mathématiques et physiques 1885.

[LE PAIGE 1888] LE PAIGE, CONSTANTIN: "Notes pour servir à l'histoire des mathématiques dans l'ancien pays de Liège." *Bulletin de l'Institut archéologique liégeois* **21** (1888), 457–565. Separately published Liège: L. De Thier, 1890.

[LEIBNIZ 1714] LEIBNIZ, GOTTFRIED WILHELM: "Historia et origo calculi differentialis." In: GERHARDT, CARL IMMANUEL (Ed.): *G. W. Leibniz: Mathematische Schriften,* vol. 5, 392–410. Halle: Schmidt 1858; repr. Hildesheim: Olms 1962.

[LEJEUNE 1956] LEJEUNE, ALBERT: *L'Optique de Claude Ptolémée dans la version latine d'après l'arabe de l'émir Eugène de Sicile. Édition critique et exégétique.* Louvain: Publications Universitaire de Louvain 1956.

[LEJEUNE 1957] LEJEUNE, ALBERT: *Recherches sur la catoptrique grecque d'après les sources antiques et médiales.* Bruxelles: Palais des Académies 1957.

[LENIN 1963] LENIN, VLADIMIR I.: *Polnoe Sobranie Sochineniĭ.* 5th ed., vol. 29. Moskva: Politizdat 1963.

[LEONARDO DA VINCI 1942] LEONARDO DA VINCI: *I libri di meccanica nella ricostruzione ordinata di Arturo Uccelli, preceduti da un'introduzione critica e da un esame delle fonti.* Milano: Ulrico Hoepli 1942.

[LEVERE 1988] LEVERE, TREVOR H.: "The History of Science in Canada." *Brit. J. Hist. Phil. Sci.* **21** (1988), 419–425.

[LEVI 1947] LEVI, BEPPO: *Leyendo a Euclides.* Rosario: Editorial Rosario S.A. 1947.

[LHUILIER 1782] LHUILIER, SIMON: "Algebra dla szkół narodowych" (Algebra for National Schools). Marywil: Michat Gröll 1782.

[LI/DU 1987] LI YAN and DU SHIRAN: *Chinese Mathematics: A Concise History.* JOHN N. CROSSLEY and ANTHONY W.-C. LUN (Trans.). Oxford: Clarendon Press 1987.

[LIBRI 1838/41] LIBRI (actually LIBRI-CARUCCI DALLA SOMMAJA), GUGLIELMO BRUTUS ICILIUS TIMOLEON: *Histoire des sciences mathématiques en Italie, depuis la renaissance des lettres jusqu'à la fin du dix-septième siècle.* 4 vols. Paris: Jules Renouard 1838–1841; 2nd ed., 4 vols., Halle: Schmidt 1865; repr. New York: Johnson Reprint 1966; repr. Hildesheim: Olms 1967; repr. Bologna: Foreni 1989.

[LIKHOLETOV/YANOVSKAYA 1955] LIKHOLETOV, IVAN I., and SOF'YA A. YANOVSKAYA: "Iz istorii prepodavaniya matematiki v Moskovskom universitete (1804–1860)." *IMI* (1) **8** (1955), 127–480.

[LINDNER 1980] LINDNER, HELMUT: " 'Deutsche' und 'gegentypische' Mathematik. Zur Begründung einer 'arteigenen' Mathematik im Deutschen Reich durch Ludwig Bieberbach." In: *Naturwissenschaft, Technik und NS-Ideologie. Beiträge zur Wissenschaftsgeschichte des Dritten Reiches*, MEHRTENS, HERBERT, and STEFFEN RICHTER (Eds.). Frankfurt am Main: Suhrkamp 1980, 88–115.

[LIU 1986] LIU DUN: "Mei Wending — the Great Master of Astronomy and Mathematics in the Early Qing Period" (in Chinese). *Ziran bianzhengfa tongxun* (Journal of Dialectics of Nature) **8** no. 1 (1986), 52–64.

[LIU 1991] LIU DUN: "Transformation of Nationalism in the Early Qing and Its Influence on Astronomy and Mathematics of the Qing Dynasty" (in Chinese). *Ziran bianzhengfa tongxun* (Journal of Dialectics of Nature) **13** no. 3 (1981), 42–52.

[LIU 1994] LIU DUN: "400 Years of the History of Mathematics in China — An Introduction to the Major Historians of Mathematics since 1592." *Historia Scientiarum* (Tokyo), **4** (1994), 103–111.

[LOMBARDI 1827–30] LOMBARDI, ANTONIO: *Storia della letteratura italiana nel secolo XVIII*, 4 vols. Modena: Tipografia Camerale 1827–30.

[LORENTE PÉREZ 1921] LORENTE PÉREZ, Mª JOSÉ: "Biografía y análisis de las obras de matemática pura de P. Sánchez Ciruelo." In: *Publicaciones del Laboratorio y Seminario Matemático.* Madrid: JAE **3** (1921), 259–349.

[LOREY 1916] LOREY, WILHELM: *Das Studium der Mathematik an den deutschen Universitäten seit Anfang des 19. Jahrhunderts.* Leipzig and Berlin: Teubner 1916.

[LORIA 1892] LORIA, GINO: *Nicola Fergola e la scuola di matematici che lo ebbe a duce.* Atti dell'Università di Genova 1892.

[LORIA 1899] LORIA, GINO: "Il *Giornale de' Letterati d'Italia* di Venezia e la *Raccolta Calogerà* come fonti per la storia delle matematiche nel secolo XVIII." *Abh. Gesch. Math.* **9** (1899), 241–274.

[LORIA 1913] LORIA, GINO: "Lagrange e la storia delle matematiche." *Bibl. Math.* (3) **12** (1913), 333–338.

[LORIA 1914] LORIA, GINO: *Le scienze esatte nell'antica Grecia.* Milano: Ulrico Hoepli 1914; repr. Milano: Ulrico Hoepli 1987.

[LORIA 1916] LORIA, GINO: *Guida allo studio della storia delle matematiche.* Milano: U. Hoepli 1916; 2nd, revised ed. Milano: U. Hoepli 1946.

[LORIA 1918] LORIA, GINO: "Guglielmo Libri come storico della scienza." *Atti della Società Ligustica di Scienze Naturali e Geografiche* **28** (1918). Repr. in: [Loria 1937], 345–370.

[LORIA 1922] LORIA, GINO: "A proposito dell'edizione faentina delle 'Opere' di Evangelista Torricelli. L'imputato risponde." *Period. Mat.* (4) **2**, 364–368.

[LORIA 1929a] LORIA, GINO: "Inauguration solennelle du Comité international d'Histoire des Sciences et Commémoration de Paul Tannery." *Archeion* **11** (1929), lxxv–cviii.

[LORIA 1929/33] LORIA, GINO: *Storia delle matematiche dall'alba della civiltà al tramonto del secolo XIX.* 3 vols. Milano: Ulrico Hoepli 1929–1933. 2nd revised ed.: Milano: Ulrico Hoepli 1950; repr. Milano: Cisalpino-Goliardica 1982.

[LORIA 1932] LORIA, GINO: "L'ininterrotta continuità del pensiero matematico italiano." *Period. Mat.* (4) **12**, 1–16.

[LORIA 1937] LORIA, GINO: *Scritti, conferenze, discorsi sulla storia delle matematiche.* Padova: Cedam 1937.

[LORIA 1938] LORIA, GINO: *Galileo Galilei.* Milano: Ulrico Hoepli 1938.

[LORIA 1946] LORIA, GINO: *Guida allo studio della storia delle matematiche.* 2nd ed. of [Loria 1916]. Milano: Ulrico Hoepli 1946.

[LORIA 1950] LORIA, GINO: *Storia delle matematiche dall'alba della civiltà al tramonto del secolo XIX.* 3 vols. 2nd revised ed.: Milano: Ulrico Hoepli 1950; repr. Milano: Cisalpino-Goliardica 1982.

[LORIA/GAMBIOLI/VOLTERRA 1911–1912] LORIA, GINO; DIONISIO GAMBIOLI, and VITO VOLTERRA (Eds.): *Opere matematiche del marchese G. C. de' Toschi di Fagnano, pubblicate sotto gli auspici della Società Italiana per il Progresso delle Scienze dai soci V. Volterra, G. Loria, D. Gambioli.* Milano, Roma, Napoli: Società Editrice Dante Alighieri 1911 (vols. 1 and 2); 1912 (vol. 3).

[LORTIE 1955] LORTIE, LÉON: "Les mathématiques de nos ancêtres." *Memoirs of the Royal Society of Canada* **49**, 3rd Series, First Section (1955), 31–45.

[LUCKEY 1999] LUCKEY, PAUL: *Die Schrift des Ibrāhīm b. Sinān b. Ṯābit über die Schatteninstrumente* (1944), JAN P. HOGENDIJK (Ed.) (= *Publications of the Institute for the History of Arabic-Islamic Science* **101**). Frankfurt am Main: Institute for the History of Arabic-Islamic Science at the Johann Wolfgang Goethe University, 1999.

[LÜTZEN/PURKERT 1994] LÜTZEN, JESPER, and WALTER PURKERT: "Conflicting Tendencies in the Historiography of Mathematics: M. Cantor and H. G. Zeuthen." In: KNOBLOCH, EBERHARD, and DAVID E. ROWE (Eds.): *The History of Modern Mathematics* **3**, 1–42, 2 portraits. Boston: Academic Press 1994.

[MACFARLANE 1916] MACFARLANE, ALEXANDER: *Lectures on Ten British Mathematicians of the Nineteenth Century.* New York: Wiley 1916.

[MADDISON 1977] MADDISON, FRANCIS, *et al.* (Eds.): *Linacre Studies.* Oxford: Clarendon Press 1977.

[MAISTROV 1967] MAĬSTROV, LEONID E.: *Teoriya veroyatnosteĭ. Istoricheskiĭ ocherk.* Moskva: Nauka 1967. Engl. trans.: *Probability Theory: A Historical Sketch.* New York: Academic Press 1974.

[MANSION 1875] MANSION, P.: Review: "Zur Geschichte der Mathematik in Altertum und Mittelalter von Dr. Hermann Hankel [...]." *Bull. Boncompagni* **8** (1875), 185–220.

[MANSION 1888] MANSION, PAUL: "Sur le cours d'histoire des mathématiques de l'Université de Gand." *Bibl. Math.* (2) **2** (1888), 33–35.

[MANSION 1900] MANSION, PAUL: "Programme du Cours d'histoire des mathématiques de l'Université de Gand." *Bibl. Math.* (3) **1** (1900), 232–236.

[MARCZEWSKI 1948] MARCZEWSKI, EDWARD: *Rozwój matematyki w Polsce* (The Development of Mathematics in Poland). Cracow: Polish Academy of Arts 1948.

[MARIE 1883/88] MARIE, MAXIMILIEN: *Histoires des sciences mathématiques et physiques.* 12 vols. Paris: Gauthier-Villars 1883–1888.

[MARKUSHEVICH 1951] MARKUSHEVICH, ALEKSEĬ I.: *Ocherki po istorii teorii analiticheskikh funktsiĭ.* Moskva: Gostekhizdat 1951. German trans.: *Skizzen zur Geschichte der analytischen Funktionen.* Berlin: VEB Deutscher Verlag der Wissenschaften 1955.

[MARS 1666/99] *Mémoires de l'Académie royale des Sciences depuis 1666 jusqu'à 1699.* 11 vols. Paris: Jean Boudot 1729–1734.

[MARTÍNEZ 1994] MARTÍNEZ, M. L.: *Fondo Eduardo García de Zuñiga.* Montevideo: Galileo 1994. Segunda poca, no. 10.

[MARTZLOFF 1987] MARTZLOFF, JEAN-CLAUDE: *Histoire des mathématiques chinoises.* Paris: Masson 1987.

[MARTZLOFF 1997] MARTZLOFF, JEAN-CLAUDE: *A History of Chinese Mathematics.* STEPHEN S. WILSON (Trans.). Berlin: Springer 1997.

[MATVIEVSKAYA 1962] MATVIEVSKAYA, GALINA P.: *K istorii matematiki Sredneĭ Azii IX–XV vekov.* Tashkent: Izdatel'stvo AN UzSSR 1962.

[MATVIEVSKAYA 1967] MATVIEVSKAYA, GALINA P.: *Uchenie o chisle na srednevekovom Blizhnem i Srednem Vostoke.* Tashkent: FAN 1967.

[MATVIEVSKAYA/ROZENFELD 1985] MATVIEVSKAYA, GALINA P., and BORIS A. ROZENFELD: *Matematiki i astronomy musul'manskogo srednevekov'ja i ich trudy (VIII–XVII vv.).* 3 vols. Moscow: Nauka 1985.

[MAY 1968] MAY, KENNETH OWNSWORTH: "Growth and Quality of the Mathematical Literature." *Isis* **59** (1968), 363–371.

[MAY 1970] MAY, KENNETH OWNSWORTH: "E. T. Bell." *DSB* **1** (1970), 584–585.

[MAY 1972] MAY, KENNETH OWNSWORTH: "C. F. Gauss." *DSB* **5** (1972), 298–315.

[MAY 1973] MAY, KENNETH OWNSWORTH: *Bibliography and Research Manual of the History of Mathematics.* Toronto: University of Toronto Press 1973.

[May/Gardner 1972] May, Kenneth Ownsworth and Constance Moore Gardner: *World Directory of Historians of Mathematics.* Toronto: University of Toronto Press 1972.
2nd ed. by Kenneth Ownsworth May and Laura Roebuck. Toronto: University of Toronto Press 1978.
Supplement to the 2nd ed. by Christoph J. Scriba and Nancy Sacksteder. Hamburg: Institute for History of Science (IGN) 1985.
3rd ed. by Kirsti Andersen and Mette Dybdahl. Aarhus: DFI Print University of Aarhus 1995.

[McArthur 1990] McArthur, Charles W.: *Operations Analysis in the U.S. Army Eighth Air Force in World War II* (= *History of Mathematics*, **4**). Providence, R.I.: American Mathematical Society 1990.

[McCrea 1957] McCrea, William H.: Obituary of E. T. Whittaker. *J. Lond. math. Soc.* (2) **32** (1957), 234–256.

[Medvedev 1965] Medvedev, Fëdor A.: *Razvitie teorii mnozhestv v 19 veke.* Moskva: Nauka 1965.

[Medvedev 1974] Medvedev, Fëdor A.: *Razvitie ponyatiya integrala.* Moskva: Nauka 1974.

[Medvedev 1975] Medvedev, Fëdor A.: *Ocherki istorii teorii funktsii deĭstvitel'nogo peremennogo.* Moskva: Nauka 1975. Engl. trans.: *Scenes from the History of Real Functions.* Basel: Birkhäuser 1991.

[Mehrtens 1987a] Mehrtens, Herbert: "Ludwig Bieberbach and 'Deutsche Mathematik'." In: *Studies in the History of Mathematics*, Esther Phillips (Ed.). Washington, D.C.: The Mathematical Association of America 1987, 199–241.

[Mehrtens 1987b] Mehrtens, Herbert: "The Social System of Mathematics and National Socialism: a Survey." *Sociological Inquiry* **57** (1987), 159–182.

[Mehrtens 1990] Mehrtens, Herbert: *Moderne – Sprache – Mathematik.* Frankfurt (Main): Suhrkamp 1990.

[Mehrtens 1994] Mehrtens, Herbert: "Irresponsible Purity: the Political and Moral Structure of Mathematical Sciences in The National Socialist State." In: [Renneberg/Walker 1994], 324–338.

[Mei 1672] Mei Wending: "To Fang Zhongtong" (1672). In: Mei Juecheng (Ed.): *Jixuetang shichao* (Selection of Mei Wending's Poems). Qing edition 1757.

[Meinel 1998] Meinel, Christoph: "German History of Science Journals and the German History of Science Community." In: *Journals and History of Science*, Beretta, Marco; Claudio Pogliano, and Pietro Redondi (Eds.). Florence: Olschki 1998, 77–96.

[*Memoria* 1953/54] *Memoria del Congreso Científico Mexicano.* (IV Centenario de la Universidad de México (1551–1951)). 15 vols. México: UNAM 1953–1954.

[Merian 1830] Merian, Peter: "Daniel Huber, Professor der Mathematik und Bibliothekar zu Basel." *Verhandlungen der allgemeinen schweizerischen Gesellschaft für die gesammten Naturwissenschaften in ihrer sechzehnten Jahresversammlung zu St. Gallen den 26., 27. und 28. Juli 1830 [...]*, 145–152.

[MERIAN 1838] MERIAN, PETER: "Darstellung der Leistungen der Schweizer im Gebiete der Naturwissenschaften, seit der Zeit der Wiederherstellung der Wissenschaften bis gegen das Ende des vorigen Jahrhunderts." Eröffnungsrede bei der 23ten Jahresversammlung der allgemeinen schweizerischen Gesellschaft für die gesammten Naturwissenschaften. *Vh. Schw. Nat. Ges.* 23. Versammlung zu Basel den 12., 13. und 14. September 1838, pp. 1–33.

[MERIAN 1860] MERIAN, PETER: *Die Mathematiker Bernoulli.* Jubelschrift zur vierten Säcularfeier der Universität Basel. Basel: Schweighauser'sche Universitäts-Buchdruckerei 1860.

[MERIAN 1867] MERIAN, PETER: "Geschichte der Naturforschenden Gesellschaft in Basel während der ersten fünfzig Jahre ihres Bestehens." In: *Festschrift herausgegeben von der Naturforschenden Gesellschaft in Basel zur Feier des fünfzigjährigen Bestehens 1867.* Basel: Schultze 1867, 1–52.

[MERZ 1904/1912] MERZ, JOHN THEODORE: *A History of European Thought in the Nineteenth Century.* 4 vols. Edinburgh and London: William Blackwood 1904–1912; repr. Gloucester, Mass.: Peter Smith 1976.

[MERZBACH 1989] MERZBACH, UTA C.: "The Study of the History of Mathematics in America: A Centennial Sketch." In: [DUREN/ASKEY/MERZBACH 1988/89, vol. 3, 639–666].

[MERZBACH 1991] MERZBACH, UTA C.: "Aspects of the 20th Century." In: BOYER, CARL: *A History of Mathematics.* 2nd. rev. ed. 1991, 616–632 (Original: New York: John Wiley 1968, first rev. ed. New York: John Wiley 1989). .

[METHÉE 1991] METHÉE, PIERRE-DENIS: *Les mathématiques à l'Académie et à la Faculté des sciences de l'Université de Lausanne (= Études et Documents pour servir à l'histoire de l'Université de Lausanne* **29**). Lausanne: Université de Lausanne 1991.

[METT 1996] METT, RUDOLF: *Regiomontanus. Wegbereiter des neuen Weltbildes.* Stuttgart and Leipzig: Teubner; Zürich: vdf Hochschulverlag 1996.

[METZGER 1987] METZGER, HÉLÈNE: *La méthode philosophique en histoire des sciences, textes 1914–1939,* GAD FREUDENTHAL (Ed.). Corpus des œuvres de philosophie en langue française. Paris: Fayard 1987.

[MICHEL 1939] MICHEL, HENRI: *Introduction à l'étude d'une collection d'instruments anciens de mathématiques.* Antwerp: De Sikkel 1939.

[MICHEL 1947] MICHEL, HENRI: *Traité d'astrolabe.* Paris: Gauthier-Villars 1947; 2nd ed. Paris: Alain Brieux 1976.

[MIELI 1920] MIELI, ALDO: "L'opera di Raffaello Caverni come storico (Cenni preliminari)." *Arch. Stor. Sci.* **1** (1920), 262–266.

[MIELI 1926] MIELI, ALDO: "La storia della scienza in Italia." *Arch. Stor. Sci.* **7** (1926), 36–48.

[MIKAMI 1913] MIKAMI YOSHIO: *The Development of Mathematics in China and Japan.* Leipzig: Teubner 1913; repr. New York: Chelsea 1974.

[MIKAMI/SMITH 1914] MIKAMI YOSHIO and DAVID E. SMITH: *A History of Japanese Mathematics.* Chicago: Open Court 1914.

[MIODUSZEWSKI 1996] MIODUSZEWSKI, JERZY: "Mikołaj Kopernik (1473–1543) — matematyk" (Mikołaj Kopernik (1473–1543) — Mathematician). *Matematyka* **3** (1996), 131–140.

[MOGENET 1950] MOGENET, JOSEPH: *Autolicus de Pitane. Histoire du texte suivie de l'édition critique des traités de la sphère en mouvement et des levers et des couchers.* Louvain: Publications Universitaires de Louvain 1950.

[MOGENET 1985] MOGENET, JOSEPH: *Le "Grand Commentaire" de Théon d'Alexandrie aux Tables Faciles de Ptolémée. Livre I: Histoire du texte. Édition critique, Traduction. Revue et complétées par Anne Tihon. Commentaire par Anne Tihon.* Città del Vaticano: Bibliotheca Apostolica Vaticana 1985.

[MONTEVERDE 1853] MONTEVERDE, MANUEL: *Inmenso desarrollo que desde el siglo XVII han recibido las Matemáticas, manifestando su íntima asociación con la Física e indicando los trabajos de las Academias* (= *Memorias de la RACEFNM*, (1) **2**). Madrid 1853.

[MONTMORT 1713/14] MONTMORT, PIERRE RÉMOND DE: *Essai d'analyse sur les jeux du hasard.* Paris: Jacques Quillau; 2nd ed. 1713–1714; repr. New York: Chelsea 1979.

[MONTUCLA 1758] MONTUCLA, JEAN ÉTIENNE: *Histoire des mathématiques.* 2 vols. Paris: Ch. A. Jombert 1758. New edition completed by JÉROME DE LALANDE, 4 vols., Paris: Henri Agasse 1799–1802; repr. Paris: A. Blanchard 1960.

[MONTUCLA 1782/1804] MONTUCLA, JEAN ETIENNE: *Historie der Wiskunde sedert haaren oorsprong tot in onzen tyd; door den Heere Montucla [...]. Uit het Fransch vertaald, en met eenige byvoegselen en ophelderende aanmerkingen vermeerderd; door Arnoldus Bastiaan Strabbe.* 4 vols. Amsterdam: vol. 1: Erven van F. Houttuyn 1782, vols. 2–4: P. G. Geysbeek 1787, 1797, 1804.

[MONTUCLA 1799/1802] MONTUCLA, JEAN ÉTIENNE: *Histoire des Mathématiques, [...]* nouvelle édition, considérablement augmentée, et prolongée jusque vers l'époque nouvelle. 4 vols., complétée par JÉRÔME DE LALANDE. Paris: Henri Agasse 1799–1802. Nouveau tirage augmenté d'un avant-propos de CH. NAUX, Paris: A. Blanchard 1960.

[MORAIS 1994] ABRÃO DE MORAIS: "A Astronomia no Brasil." In: FERNANDO DE AZEVEDO (Ed.): *As Ciências no Brasil.* 2nd ed.: Rio de Janeiro: Editora UFRJ 1994, 97–189 (original edition 1955).

[MORDUKHAĬ-BOLTOVSKOĬ 1948/50] MORDUKHAĬ-BOLTOVSKOĬ, DMITRIĬ D. (Ed.): *Evklid. Nachala.* Trans. and commentary by D. D. MORDUKHAĬ-BOLTOVSKOĬ, in collaboration with M. YA. VYGODSKIĬ and I. N. VESELOVSKIĬ. 3 vols. Moscow and Leningrad: Gostekhizdat 1948/50.

[MORENO CORRAL 1992] MORENO CORRAL, MARCO A.: "Libros de Matemáticas llegados a America durante los Siglos XVI y XVII." *Mathesis* **8** (1992), 331–344.

[MORENO DE LOS ARCOS 1986] MORENO DE LOS ARCOS, ROBERTO: *Ensayos de Historia de la Ciencia y la Tecnología en México.* México: UNAM 1986.

[MOSCHEO 1988] MOSCHEO, ROSARIO: *Francesco Maurolico tra Rinascimento e scienza galileiana. Materiali e ricerche.* Messina: Società messinese di storia patria 1988.

[MÜLLER 1903] MÜLLER, CONRAD H.: *Studien zur Geschichte der Mathematik, insbesondere des mathematischen Unterrichts, an der Universität Göttingen im 18. Jahrhundert. Mit einer Einleitung: Über Charakter und Umfang historischer Forschung in der Mathematik.* Leipzig: Teubner 1904.

[MÜLLER 1909] MÜLLER, FELIX: *Führer durch die mathematische Literatur, mit besonderer Berücksichtigung der historisch wichtigen Schriften* (= *Abh. Gesch. math. Wiss.* **27**). Leipzig: Teubner 1909.

[MUIR 1906] THOMAS MUIR: *The Theory of Determinants in the Historical Order of Development.* vol. 1. London: Macmillan 1890; 2nd ed. 1906.

[MUNBY 1968] MUNBY, ALAN NOEL L.: *The History and Bibliography of Science in England: the First Phase, 1833–1845.* Berkeley: University of California Press 1968.

[MURAWSKI 1986] MURAWSKI, ROMAN: *Filozofia matematyki* (Mathematical Philosophy). Poznań: University of Adam Mickiewicz 1986.

[MUSIELAK 1993] MUSIELAK, JULIAN: "On the History of Functional Analysis." *Opuscula Mathematica* **13** (1993), 27–36 (= *Scientific Bulletins of the University of Mining and Metallurgy* 1522. Cracow 1993.)

[NAAS/SCHMIDT 1961] NAAS, JOSEF, and HERMANN LUDWIG SCHMIDT (Eds.): *Mathematisches Wörterbuch.* 2 vols. Berlin: Akademie Verlag, and Stuttgart: B. G. Teubner Verlag 1961. Several reprints.

[NADAL 1959] NADAL, ANDRÉ: "Gaston Milhaud (1858–1918)." *Rev. Hist. Sci.* **12** (1959), 97–110.

[NAGEL 1993] NAGEL, FRITZ: "Die Bernoulli-Edition Basel: Geschichte, gegenwärtiger Stand und Zukunftsperspektiven." *Bulletin: Pro Saeculo XVIII° Societas Helvetica* **2** (1993), 24–31.

[NARDUCCI 1867] NARDUCCI, ENRICO: "Intorno alla vita del conte Giammaria Mazzucchelli e alla collezione dei suoi manoscritti posseduti dalla Biblioteca Vaticana." *Giornale Arcadico di Scienze, Lettere ed Arti* **198** (New Series, **52**) (1867), 1–79.

[NASTASI 1990] NASTASI, PIETRO: "Aspetti istituzionali della storia della scienza in Italia nel periodo tra le due guerre." In: *Scritti di storia della scienza in onore di Giovanni Marini-Bettólo*, BALLIO, A., and L. PAOLONI (Eds.). Roma: Accademia Nazionale delle Scienze detta dei XL, 409–444.

[NATUCCI 1923] NATUCCI, ALPINOLO: *Il concetto di numero e le sue estensioni.* Torino: Bocca 1923.

[NEEDHAM 1954] NEEDHAM, JOSEPH: *Science and Civilisation in China*, vol. 1: *Introductory Orientations.* Cambridge: Cambridge University Press 1954.

[NEEDHAM 1959] NEEDHAM, JOSEPH: *Science and Civilisation in China*, vol. 3: *Mathematics and the Sciences of the Heavens and the Earth.* Cambridge: Cambridge University Press 1959.

[NESSELMANN 1842] NESSELMANN, GEORG HEINRICH FERDINAND: *Versuch einer kritischen Geschichte der Algebra. Erster Theil: Die Algebra der Griechen. Nach den Quellen bearbeitet.* Berlin: G. Reimer 1842; repr. Frankfurt: Minerva 1969. [No further parts were published].

[NEUENSCHWANDER 1993] NEUENSCHWANDER, ERWIN: "Der Nachlass von Erich Bessel-Hagen im Archiv der Universität Bonn." *HM* **20** (1993), 382–414.

[NEUENSCHWANDER 1999] NEUENSCHWANDER, ERWIN: "Zur Historiographie der Mathematik in der Schweiz." *Arch. int. Hist. Sci.* **49** (1999), 369–399.

[NEUGEBAUER/PINGREE 1970/71] NEUGEBAUER, OTTO, and DAVID PINGREE, (Eds. and Trans.): *Pañcasiddhāntikā of Varāhamihira.* 2 parts. Copenhagen: Munksgaard 1970/71.

[NEUMANN 1893] NEUMANN, THEODOR: *Das moderne Ägypten.* Leipzig: Duncker und Humblot 1893.

[NEWTON 1959/77] TURNBULL, HERBERT WESTREN, *et al.* (Eds.): *The Correspondence of Isaac Newton.* 7 vols. Cambridge: Cambridge University Press 1959–1977.

[NEWTON 1967/81] WHITESIDE, DEREK THOMAS (Ed.): *The Mathematical Papers of Isaac Newton.* 8 vols. Cambridge: Cambridge University Press 1967–1981.

[NIELSEN 1910/12] NIELSEN, NIELS: *Matematiken i Danmark 1801–1908.* København and Kristiania: Gyldendalske Boghandel and Nordisk Forlag 1910. *Matematiken i Danmark 1528–1800.* København and Kristiania: Gyldendalske Boghandel and Nordisk Forlag 1912.

[NOVÝ 1961] NOVÝ, LUBOŠ (Ed.): *Dějiny exaktních věd v českých zemích do honce 19. století* (History of Science in the Bohemian Lands up to the End of the 19th Century). Praha: ČSAV 1961.

[NOVÝ 1968] NOVÝ, LUBOŠ (Ed.): "La révolution scientifique du 17e siècle et les sciences mathématiques et physiques." *Acta hist. Special Issue* **4** (1968).

[NOVÝ 1973] NOVÝ, LUBOŠ: *Origins of Modern Algebra.* Prague: Academia 1973.

[NOVÝ 1982] NOVÝ, LUBOŠ (Ed.): "Bernard Bolzano (1781–1848). Bicentenary. Impact of Bolzano's Epoch on the Development of Science (Conference Papers)." *Acta hist., Special Issue* **13** (1982).

[OBENRAUCH 1897] OBENRAUCH, FERDINAND: *Geschichte der darstellenden und projektiven Geometrie mit besonderer Berücksichtigung ihrer Begründung in Frankreich und Deutschland und ihrer wissenschaftlichen Pflege in Österreich.* Brünn: Carl Winkler 1897.

[O'GORMAN 1939] O'GORMAN, EDMUNDO: "Bibliotecas y Librerías Coloniales." *Boletín del Archivo General de la Nación* **10**, no. 4 (1939), 662–1006.

[OGURA 1951] OGURA KINNOSUKE: *Life and Works of Dr. Yoshio Mikami* (in Japanese). *Kagakushi Kenkyū* **18** (1951), 1-11.

[OGURA 1974–75] OGURA KINNOSUKE: *Chosaku-shū* (Collected works). 8 vols. Tokyo: Keisō Shobō 1974–75.

[OLIVEIRA CASTRO 1994] OLIVEIRA CASTRO, FRANCISCO M. DE: "A Matemática no Brasil." In: FERNANDO DE AZEVEDO (Ed.): *As Ciências no Brasil.* 2nd ed.: Rio de Janeiro: Editora UFRJ 1994, 55–96 (original edition 1955).

[OPIAL 1964/65] OPIAL, ZDZISŁAW: "Zarys dziejów matematyki w Uniwersytecie Jagiellońskim w drugiej połowie XIX wieku" (An Outline of the History of Mathematics at the Jagiellonian University in the Second Half of the 19th Century).

In: *Studia z dziejów Katedr Wydziału Matematyki, Fizyki, Chemii Uniwersytetu Jagiellońskiego* (Studies on the History of the Departments of the Faculty of Mathematics, Physics and Chemistry at the Jagiellonian University). Cracow: Jagiellonian University 1964, 59–74 (= *Universitas Jagellonica Opera ab Universitate Jagellonica sexta solemnia saecularia celebrante edita* **15**, and *Wiad. Mat.* **8** (1965), 25–39).

[OPIAL 1966] OPIAL, ZDZISŁAW: "Dzieje nauk matematycznych w Polsce" (History of the Mathematical Sciences in Poland). In: *Studia i Materiały z Dziejów Nauki Polskiej,* series C **10** (1966), 137–166.

[ORELLANA 1985] ORELLANA, CARLOS GONZÁLEZ: *Historia de la Educación en Guatemala.* Guatemala: Editorial Universitaria 1985.

[ORELLANA 1991] ORELLANA C., MAURÍCIO: *Resumen de las Clases del Curso de Historia de la Matemática en América Latina y Venezuela.* Caracas: Universidad Pedagògica Experimental Libertador 1991 (mimeographed).

[*Ostwalds Klassiker* 1889] *Ostwalds Klassiker der exakten Wissenschaften.* Founded by WILHELM OSTWALD. [The series still continues; vol. 286 appeared in 2000, but the series does not have a continuous publishing history.] Leipzig: Wilhelm Engelmann 1889–1915; due to World War I, nothing was published between 1915–1922. When publication resumed in 1922, it was with a new publisher, Leipzig: Akademische Verlagsgesellschaft, 1922–1938. World War II again interrupted publication, and nothing appeared between 1938–1952. Then publication resumed in the former DDR, Leipzig: Akademische Verlagsgesellschaft Geest und Portig; 1952–1989. Meanwhile, a new series was published briefly in West Germany, a *Neue Folge,* vols. 1–6, Frankfurt (Main): Akademische Verlagsgesellschaft, 1965–1970. The reunification of Germany again disturbed publication, with no volumes of *Ostwalds Klassiker* appearing from Leipzig after 1989. The series was revived, beginning with vol. 280, in 1995, and since then has been published in Frankfurt (Main) and Thun: Harri Deutsch 1995.

[OTTOWITZ 1992] OTTOWITZ, NIKOLAUS: *Der Mathematikunterricht an der Technischen Hochschule in Wien 1815–1918.* (Dissertationen der Technischen Universität Wien, vol. 52, parts I and II). Wien: VWGÖ 1992.

[OZHIGOVA 1972] OZHIGOVA, ELENA P.: *Razvitie teopii chisel v Rossii.* Leningrad: Nauka 1972.

[PAGE 1985] PAGE, WARREN (Ed.): *American Perspectives on the 5th International Congress on Mathematical Education. ICME 5.* Washington, D.C.: Mathematical Association of America 1985.

[PALLADINO 1989] PALLADINO, FRANCO: "La storia delle scienze matematiche a Napoli tra Ottocento e Novecento: il contributo di Federico Amodeo." In: [BARBIERI/DEGANI 1989], 269–296.

[PAPLAUSKAS 1966] PAPLAUSKAS, AL'GIRDAS B.: *Trigonometricheskie ryady ot Eĭlera do Lebega.* Moskva: Nauka 1966.

[PARODI 1919] PARODI, DOMINIQUE: *La philosophie contemporaine en France.* Paris: Alcan 1919.

[PARSHALL/ROWE 1994] PARSHALL, KAREN, and DAVID ROWE: *The Emergence of the American Mathematical Research Community, 1876–1900.* Providence, RI: The American Mathematical Society, and London: London Mathematical Society 1994.

[PAUL 1976] PAUL, HARRY W.: "Scholarship and Ideology: The Chair of the General History of Science at the Collège de France, 1892–1913." *Isis* **67** (1976), 376–397.

[PAUL 1980] PAUL, CHARLES B.: *Science and Immortality. The Eloges of the Paris Academy of Sciences (1699–1791).* Berkeley: University of California Press 1980.

[PAWLIKOWSKA 1964/66] PAWLIKOWSKA, ZOFIA: "Z historii polskiej terminologii matematycznej" (On the History of Polish Mathematical Terminology). *Wiad. Mat.* **7** (1964), 165–190; **8** (1965), 41–64; **9** (1966), 23–43.

[PAWLIKOWSKA-BROŻEK 1983] PAWLIKOWSKA-BROŻEK, ZOFIA: "Matematyka" (Mathematics). In: *Zarys dziejów nauk przyrodniczych w Polsce* (An Outline of the History of Natural Sciences in Poland). Warsaw: Wiedza Powszechna 1983, 155–215.

[PAWLIKOWSKA-BROŻEK 1996] PAWLIKOWSKA-BROŻEK, ZOFIA: "On Mathematical Life in Poland," in [GOLDSTEIN/GRAY/RITTER 1996, 289–301].

[PELCZAR 1996] PELCZAR, ANDRZEJ: "Matematyka w Polsce u początków PTM (i nieco wcześniej)" (Mathematics in Poland at the Beginning of the Polish Mathematical Society (and Somewhat Earlier)). *Wiad. Mat.* **32** (1996), 137–152.

[PELZEL 1773/82] PELZEL, MARTIN (Ed.): *Abbildungen böhmischer und mährischer Gelehrten und Künstler nebst kurzen Nachrichten von ihrem Leben und Wirken.* 4 vols. Prague: W. Gerle 1773–1782. (Two volumes were also published in Latin by ADAUCTUS VOIGT under the title: *Effigies virorum eruditorum atque artificium Bohemiae et Moraviae una cum brevi vitae operumque ipsorum enumeratione.* Prague: W. Gerle 1773–1774.)

[PEPPENAUER 1953] PEPPENAUER, HELGA: *Geschichte des Studienfaches Mathematik an der Universität Wien von 1848 bis 1900.* Dissertation Universität Wien 1953.

[PERRIN 1983] PERRIN, FERDINAND: *La vie et l'œuvre de Charles Dupin (1784–1873), mathématicien, ingénieur et éducateur.* Paris 1983.

[PESET 1985] PESET, JOSÉ LUIZ (Ed.): *La Ciencia Moderna y el Nuevo Mundo.* Madrid: CSIC/SLAHCT 1985, 27–38.

[PETIT 1994] PETIT, ANNIE: "L'impérialisme des géomètres à l'Ecole polytechnique. Les critiques d'Auguste Comte," in [BELHOSTE/DAHAN/PICON 1994], 59–75.

[PETIT 1995] PETIT, ANNIE: "L'héritage du positivisme dans la création de la chaire d'histoire générale des sciences au Collège de France," *Rev. Hist. Sci.* **48** (1995), 521–556.

[PETROVA 1965] PETROVA, SVETLANA S.: "Printsip Dirikhle v rabotakh Rimana." *IMI* (1) **16** (1965), 295–310. (See also: "De l'histoire du principe variationnel de Dirichlet." In: DEMIDOV, S. S.; M. FOLKERTS, D. ROWE, and C. J. SCRIBA (Eds.): *Amphora.* Basel: Birkhäuser Verlag 1992, 539–550.)

[PETROVA 1996] PETROVA, SVETLANA S.: "La reforme de Kolmogorov de l'enseignement des mathématiques en Union sovietique." In: BELHOSTE, B.; HELÈNE GISPERT, and N. HULIN (Eds.): *Les sciences au lycee. Un siècle de reformes des mathématiques et de la physique en France et a l'etranger.* Paris: Librairie Vuibert 1996, 311–318.

[PEYRARD 1814/18] PEYRARD, FRANÇOIS: *Euclidis quae supersunt. Les Œuvres d'Euclide, en grec, en latin, et en français.* D'après un manuscrit très ancien, qui était resté inconnu jusqu'à nos jours. 3 vols. Paris: M. Patris 1814–1818.

[PHILI 1996] PHILI, CHRISTINE: "La reconstruction des mathématiques en Grèce: l'apport de Ioannis Carandinos (1784–1834)." In: [GOLDSTEIN/GRAY/RITTER 1996, 303–319].

[PHILI 1999] PHILI, CHRISTINE: "Kyparissos Stephanos and his Paper on Quaternions." *Acta hist.*, New Series **3** (1999), 35–46.

[PHILI 2000] PHILI, CHRISTINE: "Some Aspects of Scientific Society in Athens at the End of the XIXth Century: Mathematics and Mathematicians." *Arch. int. Hist. Sci.* **50** (2000), 302–320.

[PICATOSTE 1891] PICATOSTE, FELIPE: *Apuntes para una biblioteca científica española del siglo XVI.* Madrid: Manuel Tello 1891.

[PICCARD 1973] PICCARD, SOPHIE: "Le professeur Séverin Bays (1885–1972)." *Bulletin de la Société Fribourgeoise des Sciences Naturelles* **62** (1973), 86–95.

[PIEPER 1987] PIEPER, HERBERT (Ed.): *Briefwechsel zwischen Alexander von Humboldt und C. G. Jacob Jacobi (= Beiträge zur Alexander-von-Humboldt-Forschung 11).* Berlin: Akademie-Verlag 1987.

[PINL 1969] PINL, MAX: "Kollegen in einer dunklen Zeit." I. Teil. *Jber. DMV* **71** (1969), 167–228.

[PINL 1971] PINL, MAX: "Kollegen in einer dunklen Zeit." II. Teil. *Jber. DMV* **72** (1971), 165–189.

[PINL 1972] PINL, MAX: "Kollegen in einer dunklen Zeit." III. Teil. *Jber. DMV* **73** (1972), 153–208.

[PINL 1974] PINL, MAX (in collaboration with AUGUSTE DICK): "Kollegen in einer dunklen Zeit; Schluss." *Jber. DMV* **75** (1974), 166–208.

[PINL 1976] PINL, MAX (in collaboration with AUGUSTE DICK): "Kollegen in einer dunklen Zeit; Nachtrag und Berichtigung." *Jber. DMV* **77** (1976), 161–164.

[PINL/FURTMÜLLER 1973] PINL, MAX, and LUX FURTMÜLLER: "Mathematicians under Hitler." In *Publications of the Leo Baeck Institute, Year Book XVIII.* London: Secker and Warburg 1973, 129–182.

[PITCHER 1988] PITCHER, EVERETT: *A History of the Second Fifty Years. American Mathematical Society, 1939–1988.* Providence, Rhode Island: American Mathematical Society 1988.

[PLANCHEREL 1960] PLANCHEREL, MICHEL: "Mathématiques et mathématiciens en Suisse (1850–1950)." *L'Enseign. Math.* (2) **6** (1960), 194–218.

[POGREBYSSKIĬ 1966] POGREBYSSKIĬ, IOSIF B.: *Ot Lagranzha k Einshteinu.* Moscow: Nauka 1966.

[POIRIER 1926] POIRIER, RENÉ: *Philosophes et savants français du XXe siècle* II: La philosophie de la science. Paris: F. Alcari 1926.

[POMIAN 1975] POMIAN, KRISTOF: "L'histoire de la science et l'histoire de l'histoire." *Annales. Economies, Sociétés, Civilisations* **30** (1975), 935–952.

[Poppe 1797] Poppe, Johann Heinrich Moritz: *Geschichte der Entstehung und der Fortschritte der theoretischen und practischen Uhrmacherkunst* (1797). Leipzig: Gleditsch 1800.

[Poppe 1802] Poppe, Johann Heinrich Moritz: *Ausführliche Geschichte der Anwendung aller krummen Linien in mechanischen Künsten und in der Architektur, seit den ältesten Zeiten bis zu Anfange des neunzehnten Jahrhunderts.* Nürnberg: Raspe 1802.

[Poppe 1828] Poppe, Johann Heinrich Moritz: *Geschichte der Mathematik seit der ältesten bis auf die neueste Zeit.* Tübingen: Osiander 1828.

[Poudra 1864a] Poudra, Noël-Germinal (Ed.): *Œuvres de Desargues, précédées d'une nouvelle biographie de l'auteur.* 2 vols. Paris: Leiber 1864.

[Poudra 1864b] Poudra, Noël-Germinal: *Histoire de la perspective ancienne et moderne [...].* Paris: J. Couéard 1864.

[Poulitsas 1957] Poulitsas, Panagiotis: "What Did They Repay Us?" [Official speech of Oct. 28, 1957]. In: *The 28 of October 1940. Panegyric speeches of Academicians.* Athens: Academy of Athens, Foundation Kostas and Heleni Ourani 1978, 213–254.

[Powell/Frankenstein 1997] Powell, Arthur B., and Marilyn Frankenstein (Eds.): *Ethnomathematics: Challenging Eurocentrism in Mathematics Education.* Albany, NY: State University of New York Press 1997.

[Prag 1929] Prag, Adolf: "John Wallis, 1616–1703. Zur Ideengeschichte der Mathematik im 17. Jahrhundert." *Q. St. G. Math. Astron. Phys., Abt. B* **1** (1929), 381–412.

[Prasad 1933/34] Prasad, Ganesh: *Some Great Mathematicians of the Nineteenth Century: Their Lives and Their Works.* 2 vols. Benares: Benares Mathematical Society 1933/34.

[Prieto 1921/23] Prieto, Sotero: "La teoría de la relatividad." *El Maestro* (1921–1923). (A facsimile edition is included, as a supplement, in [Prieto 1991]).

[Prieto 1991] Prieto, Sotero: *Historia de las Matemáticas.* México: IMC Ediciones 1991.

[Procissi 1989] Procissi, Angiolo: "Guglielmo Libri come storico della matemática e amico di gioventù di Gino Capponi." In: [Barbieri/Degani 1989], 171–180.

[Pyenson 1985] Pyenson, Lewis: *Cultural Imperialism in the Exact Sciences. German Expansion Overseas 1900–1930.* New York: Peter Lang 1985.

[Pyenson 1993] Pyenson, Lewis: *Civilizing Mission: Exact Sciences and French Expansion, 1830–1940.* Baltimore and London: The Johns Hopkins University Press 1993.

[Quetelet 1864] Quetelet, Adolphe: *Histoire des sciences mathématiques et physiques chez les Belges.* Bruxelles: M. Hayez 1864.

[Qian 1800] Qian Daxin: "Preface" (in Chinese). In: Jiao Xun: *Tianyuanyi shi* (Explanation of the Tian Yuan Method). Qing edition, ca. 1800.

[Qian 1963] Qian Baocong: *Suanjing shishu* (Ten Books of Mathematical Classics). 2 vols. Beijing: Zuonghua Shuju 1963.

[RAINA 1999] RAINA, DHRUV: *Nationalism, Institutional Science, and Politics of Representation: Ancient Indian Astronomy in the Landscape of French Enlightenment Historiography.* Ph.D. Thesis, Faculty of Arts, Göteborg University 1999.

[RAJAGOPALAN 1979] RAJAGOPALAN, K. R., *et al.*: *History of Indian Mathematics* (in Tamil). Madras: Christian Literature Society 1979.

[RAMOS s/d] RAMOS, GERARDO: "El desarrollo de la Matemática en el Perú." In: YEPES, ERNESTO (Ed.): *Algunos aportes para el estudio de la historia de la ciencia en el Peru.* Lima: CONCYTEC s/d [undated], 15–19.

[RASHED 1984] RASHED, ROSHDI: *Entre arithmétique et algèbre. Recherches sur l'histoire des mathématiques arabes.* Paris: Les Belles Lettres 1984, 300–318.

[REDONDI 1986] REDONDI, PIETRO (Ed.): *Alexandre Koyré. De la mystique à la science. Cours, conférences et documents 1922–1962.* Paris: Ed. de l'EHESS 1986.

[REID 1993] REID, CONSTANCE: *The Search for E. T. Bell, also known as John Taine.* Washington, D.C.: The Mathematical Association of America 1993.

[REMMERT 1999] REMMERT, VOLKER R.: "Mathematicians at War. Power Struggles in Nazi Germany's Mathematical Community: Gustav Doetsch and Wilhelm Suess." *Rev. Hist. Math.* **5** (1999), 7–59.

[RENNEBERG/WALKER 1994] RENNEBERG, MONIKA, and MARK WALKER (Eds.): *Science, Technology, and National Socialism.* Cambridge: Cambridge University Press 1994.

[REY PASTOR 1913] REY PASTOR, JULIO: *Los matemáticos españoles en el siglo XVI. Discurso Inaugural del Curso Académico 1912–1913 de la Universidad de Oviedo.* Oviedo: Universidad de Oviedo 1913.

[REY PASTOR 1915] REY PASTOR, JULIO: "El progreso de España en las Ciencias y el progreso de las Ciencias en España: Discurso inaugural de la Sección Primera, Ciencias Matemáticas." In: *Asociación Española para el Progreso de las Ciencias. Quinto Congreso celebrado en Valladolid del 17 al 22 de octubre de 1915.* Madrid: Imprenta de Eduardo Arias, 1915, vol. 1 (2nd part), 11–25.

[REY PASTOR 1942] REY PASTOR, JULIO: *La Ciencia y la Técnica en el Descubrimiento de América.* Buenos Aires: Espasa-Calpe Argentina S.A. 1942.

[REYES PRÓSPER 1911] REYES PRÓSPER, VENTURA: "Juan Martínez Silíceo." *Revista de la Sociedad Matemática Española* **1** (1911), 153–156.

[RICCARDI 1897] RICCARDI, PIETRO: "Contributo degl'Italiani alla storia della matematica." *Memorie della R. Accademia delle Scienze dell'Istituto di Bologna* (V) **6** (1896–97), 755–775; **7** (1897), 371–425.

[RICE 1996] RICE, ADRIAN: "Augustus de Morgan: Historian of Science." *Hist. Sci.* **34** (1996), 201–240.

[RICHARD 1645] RICHARD, CLAUDE: *Euclidis Elementorum geometricorum Libros tredecim Isidorum et Hypsiclem et Recentiores de Corporibus Regularibus, et Procli Propositiones geometricas Immissionemque duarum rectarum linearum continué proportionalium inter duas rectas, tàm secundùm Antiquos, quàm secundùm Recentiores Geometras, novis ubique feré demonstrationibus illustravit, et multis definitionibus, axiomatibus, propositionibus, corollariis, et animadversionibus, ad Geometriam rectè intelligendam, necessariis, locupletavit Claudius Richardus e Societate Iesu Sacerdos, patria Ornacensis in libero Comitatu Burgundiæ, et Regius Mathematicarum Professor.* Antwerp: Hieronymus Verdussen 1645.

[RICHARD 1655] RICHARD, CLAUDE: *Apollonii Pergæi Conicorum libri IV. Cum Commentariis R. P. Claudii Richardi, E Societate Iesu Sacerdotis.* Antwerp: Hieronymus and Joannes Baptista Verdussen 1655.

[RICHESON 1934] RICHESON, ALLIE W.: "Courses on the History of Mathematics in the United States." *Scripta Math.* **2** (1934), 161–165.

[RICKEY 1995] RICKEY, FRED: "My Favorite Ways of Using History in Teaching Calculus." In: SWETZ, FRANK, *et al.* (Eds.): *Learn from the Masters!* Washington, D.C.: Mathematical Association of America 1995, 123–134.

[RIVOLO 1989] RIVOLO, MARIA TERESA: "Contributi di Pietro Franchini alla storia delle matematiche." In: [BARBIERI/DEGANI 1989], 231–238 (with list of publications).

[RODRIGUES 1813] RODRIGUES, ANASTÁCIO JOAQUIM: "Reflexões em defeza dos Principios Mathematicos do dr. José Anastácio da Cunha, censurados na Revista d'Edinburgo." *O Investigador Portuguez em Inglaterra* **8** (1813), 21–45.

[RÖTH 1846/58] RÖTH, EDUARD MAXIMILIAN: *Geschichte unserer abendländischen Philosophie.* Mannheim: Bassermann. vol. 1: 1846, vol. 2: 1858; repr. 1862.

[ROGER 1995] ROGER, JACQUES: "Pour une histoire historienne des sciences." In: CLAUDE BLANCKAERT (Ed.): *Pour une histoire des sciences à part entière.* Paris: Albin Michel 1995.

[ROMANUS 1597] ROMANUS, ADRIANUS: *In Archimedis Circuli dimensionem Expositio & Analysis. Apologia pro Archimede ad Clariss. virum Iosephum Scaligerum. Exercitationes Cyclicae contra Iosephum Scaligerum, Orontium Finæum, & Raymarum Ursum, in decem Dialogos distinctæ.* Würzburg 1597 [actually printed at Geneva by Pyrame de Candolle].

[ROMANUS ca. 1602] ROMANUS, ADRIANUS: *In Mahumedis Arabis Algebram Prolegomena.* Würzburg: Georg Fleischmann ca. 1602.

[ROUSE BALL 1927] ROUSE BALL, WILLIAM W.: *Compendio di storia delle matematiche,* trans. and ed. by GIULIO PULITI and DIONISIO GAMBIOLI. Bologna: Zanichelli 1927.

[ROYAL SOCIETY 1940] ROYAL SOCIETY OF LONDON: *Record of the Royal Society.* London: Royal Society, 4th ed. 1940; *Suppl. for the Years 1940–1989*: 1992.

[ROZENFELD 1976] ROZENFELD, BORIS A.: *Istoriya neevklidovoĭ geometrii.* Moscow: Nauka 1976. Engl. trans.: *A History of Non-Euclidean Geometry.* Berlin: Springer 1988.

[ROZENFELD/YOUSHKEVICH 1965] ROZENFELD, BORIS A., and ADOLF P. YOUSHKEVICH: *Omar Khaiyam.* Moscow: Nauka 1965.

[ROZHANSKAYA 1976] ROZHANSKAYA, MIRIAM M.: *Mechanika na srednevekovom Vostoke.* Moscow: Nauka 1976.

[RUAN 1799] RUAN YUAN: *Chouren zhuan* (Biographies of Astronomers and Mathematicians). Qing edition, 1799.

[RUDIO 1896] RUDIO, FERDINAND: "Die naturforschende Gesellschaft in Zürich 1746–1896." *Vj. Nat. Ges. Zür.* **41** (1896), Teil 1, 1–274 (= Part I of the Society's *Festschrift*). Zürich: Zürcher und Furrer 1896.

[RUDIO 1911] RUDIO, FERDINAND: "Vorwort zur Gesamtausgabe der Werke von Leonhard Euler." In: *Leonhardi Euleri Opera omnia*, ser. I, vol. 1, pp. ix–xli. Leipzig and Berlin: Teubner 1911.

[RUDIO 1923] RUDIO, FERDINAND: "Paul Stäckels Verdienste um die Gesamtausgabe der Werke von Leonhard Euler." *Jahresber. DMV* **32** (1923), 13–32.

[RUFINI 1926] RUFINI, ENRICO: *Il 'metodo' di Archimede e le origini del calcolo infinitesimale nell'Antichità.* Roma: Stock 1926; repr. Milano: Feltrinelli 1961.

[RUSKA 1925] RUSKA, JULIUS: *Erster Jahresbericht des Instituts für Geschichte der Naturwissenschaft zu Heidelberg.* Heidelberg: Carl Winter's Universitätsbuchhandlung 1925 (4 pp.).

[RUSKA 1926] RUSKA, JULIUS: *Zweiter Jahresbericht des Instituts für Geschichte der Naturwissenschaften zu Heidelberg.* Heidelberg: Carl Winter's Universitätsbuchhandlung 1926 (8 pp.).

[RUSKA 1928] RUSKA, JULIUS: *Forschungsinstitut für Geschichte der Naturwissenschaften in Berlin. Erster Jahresbericht. Mit einer wissenschaftlichen Beilage: Über die Aufgaben eines Forschungsinstituts für Geschichte der Naturwissenschaften.* Berlin: Julius Springer 1928 (24 pp.).

[RUSKA 1929] RUSKA, JULIUS: *Forschungs-Institut für Geschichte der Naturwissenschaften in Berlin. Zweiter Jahresbericht. Mit einer wissenschaftlichen Beilage: Aufgaben der Chemiegeschichte.* Berlin: Julius Springer 1929 (38 pp.).

[RUSKA 1930] RUSKA, JULIUS: *Forschungs-Institut für Geschichte der Naturwissenschaften in Berlin. Dritter Jahresbericht. Mit einer wissenschaftlichen Beilage: Der Zusammenbruch der Dschābir-Legende.* Berlin: Julius Springer 1930 (38 pp.).

[RYBKIN/YOUSHKEVICH 1948/66] RYBKIN, GEORGIĬ F., and ADOL'F P. YOUSHKEVICH (Eds.): *IMI.* Ser. 1. Vols. 1–17. Moscow: Fizmatgiz 1948–1966.

[RYBNIKOV 1949] RYBNIKOV, KONSTANTIN A.: "Pervye etapy razvitiya variatsionnogo ischisleniya." *IMI* (1) **2** (1949), 355–498.

[RYBNIKOV 1960/63] RYBNIKOV, KONSTANTIN A.: *Istoriya Matematiki.* 2 vols. Moscow: Izdatel'stvo MGU 1960/63; 2nd ed. 1974; 3d ed. 1994.

[RYBNIKOV 1966/89] RYBNIKOV, KONSTANTIN A. (Ed.): *Istoriya i metodologiya estestvennykh nauk.* Series *Matematika i Mekhanika.* Vols. 5, 9, 11, 14, 16, 20, 25, 29, 31, 32, 36. Moscow: Izdatel'stvo MGU 1966/89.

[RYCHLÍK 1962] RYCHLÍK, KAREL: *Theorie der reellen Zahlen im Bolzanos handschriftlichem Nachlasse.* Prag: Tschechoslowakische Akademie der Wissenschaften 1962.

[SA^cID AL-ANDALUSI 1935] SA^cID AL-ANDALUSI: *Kitâb Tabaqât al-Umam (Livre des Catégories des Nations). Traduction avec notes et indices, précedé d'une introduction par Régis Blachère.* Paris: Larose Editeurs 1935.

[SAMSÓ 1992] SAMSÓ, JULIO: *Las ciencias de los antiguos en al-Andalus.* Madrid: Mapfre 1992.

[SANTALÓ 1970] SANTALÓ, LUIS A.: *La Matemática en la Facultad de Ciencias Exactas y Naturales de la Universidad de Buenos Aires en el Período 1865–1930.* Bol. Acad. Nac. Cie. Cordoba **48** (1970), 255–273.

[SANTALÓ 1972] SANTALÓ, LUIS A.: *Evolución de las Ciencias en la Republica Argentina 1923–1972.* Vol. I: *Matemática.* Buenos Aires: Sociedad Científica Argentina 1972.

[SANTOS 1806a] SANTOS, ANTÓNIO RIBEIRO DOS: "Memoria da vida e escriptos de Don Francisco de Mello." *Mem. Litt. Acad. Sci. Lisboa* **7** (1806), 237–249.

[SANTOS 1806b] SANTOS, ANTÓNIO RIBEIRO DOS: "Memoria da vida e escriptos de Pedro Nunes." *Mem. Litt. Acad. Sci. Lisboa* **7** (1806), 250–283.

[SANTOS 1812] SANTOS, ANTÓNIO RIBEIRO DOS: "Memorias historicas sobre alguns mathemáticos portugueses e estrangeiros domiciliados em Portugal." *Mem. Litt. Acad. Sci. Lisboa* **8** (1812), 148–220.

[SARKAR 1918] SARKAR, BENOY KUMAR: *Hindu Achievements in Exact Sciences.* London: Longmans 1918.

[SARTON 1935] SARTON, GEORGE: "Preface to Volume XXIII of *Isis* (Quetelet)." *Isis* **23** (1935), 6–24.

[SARTON 1936a] SARTON, GEORGE: *The Study of the History of Mathematics.* Cambridge (Mass.): Harvard University Press 1936; repr. New York: Dover Publications, 1955. (Includes references on historians of mathematics.)

[SARTON 1936b] SARTON, GEORGE: *The Study of the History of Science.* Cambridge (Mass.): Harvard University Press 1936; cf. [SARTON 1955].

[SARTON 1936c] SARTON, GEORGE: "Montucla (1725–1799). His Life and Works." *Osiris* **1** (1936), 519–567, 2 pl.

[SARTON 1947] SARTON, GEORGE: "Paul, Jules, and Marie Tannery" (with a note on Grégoire Wyrouboff). *Isis* **38** (1947), 33–51.

[SARTON 1951] SARTON, GEORGE: "Acta atque Agenda" [on CANTOR, TANNERY, HEIBERG, DUHEM, HEATH, and MIELI]. *Arch. int. Hist. Sci.* **30** (1951), 323–356.

[SARTON 1952] SARTON, GEORGE: *Horus: A Guide to the History of Science.* Waltham (Mass.): Chronica Botanica, and New York: The Ronald Press Company 1952.

[SARTON 1955] SARTON, GEORGE: *The Study of the History of Mathematics* and *The Study of the History of Science* (Two volumes bound as one). Repr. of [SARTON 1936a] and [SARTON 1936b]. New York: Dover 1955.

[SASAKI 1994] SASAKI CHIKARA: "Kondō Yōitsu: The Birth of an Historian of Mathematics." In: [Kondo 1994], vol. 1, 397–440.

[SASAKI 1996] SASAKI CHIKARA: "Historians of Mathematics in Modern Japan." *Historia Scientiarum* (Tokyo) **6** (1996), 67–78.

[ŚĀSTRĪ 1858] ŚĀSTRĪ, BAPU DEVA: "Bhāskara's Knowledge of the Differential Calculus." *Journal of the Asiatic Society of Bengal* **27** (1858), 213–216.

[SCHINZEL 1995] SCHINZEL, ANDRZEJ: "Teoria liczb w Polsce w latach 1851–1950" (Number Theory in Poland in the Period 1851–1950). In: *Matematyka polska w stuleciu 1851–1950* (Polish Mathematics, 1851–1950). *Wiad. Mat.* **30** no. 1 (1993), 19–50. Uniwersytet Szczeciński. Materiały Konferencje **16** (1995), 115–160.

[SCHNEIDER 1914] SCHNEIDER, FELIX: "Prof. Dr. Fritz Burckhardt. 1830–1913." *Vh. Schw. Nat. Ges.* 1914, I. Teil, Anhang: Nekrologe und Biographien verstorbener Mitglieder, 1–13.

[SCHNEIDER 1992a] SCHNEIDER, IVO: "Hintergrund und Formen der Mathematikgeschichte des 18. Jahrhunderts." *Arch. int. Hist. Sci.* **42** (1992), 64–75.

[SCHNEIDER 1992b] SCHNEIDER, IVO: "The History of Mathematics: Aims, Results, and Future Prospects." In: SERGEI S. DEMIDOV *et al.*, (Eds.): *Amphora. Festschrift für Hans Wussing zu seinem 65. Geburtstag.* Basel: Birkhäuser 1992, 619–629.

[SCHÖNBECK 1994] SCHÖNBECK, JÜRGEN; HORST STRUVE, and KLAUS VOLKERT (Eds.): *Der Wandel im Lehren und Lernen von Mathematik und Naturwissenschaften.* Band I: *Mathematik* (= Schriftenreihe Pädagogische Hochschule Heidelberg **18**). Weinheim: Deutscher Studien Verlag 1994.

[SCHOLZ 1990] SCHOLZ, ERHARD: Review of JEAN DIEUDONNÉ: *A History of Algebraic and Differential Topology, 1900–1969.* Boston and Basel: Birkhäuser 1989. In: *Arch. int. Hist. Sci.* **40** (1990), 375–378.

[SCHUBRING 1985] SCHUBRING, GERT: "Die Entwicklung des Mathematischen Seminars der Universität Bonn 1864–1929." *Jahresber. DMV* **87** (1985), 139–163.

[SCHUBRING 2000] SCHUBRING, GERT: "Recent Research on Institutional History of Science and its Application to Islamic Civilization." In: *Science in Islamic Civilisation,* ISAHNOĞLU, EKMELEDDIN, and FEZA GÜNERGUN (Eds.). Istanbul: Organisation of the Islamic Conference, Research Centre for Islamic History, Art and Culture, Ircica 2000, 19–36.

[SCRIBA 1967] SCRIBA, CHRISTOPH J.: "Über Aufgaben und Probleme mathematikhistorischer Forschung." In: BARON, WALTER (Ed.): *Beiträge zur Methodik der Wissenschaftsgeschichte* (= *Beiträge zur Geschichte der Wissenschaft und der Technik* **9**), 54–80. Wiesbaden: F. Steiner 1967.

[SCRIBA 1968] SCRIBA, CHRISTOPH J.: "Geschichte der Mathematik." *Überblicke Mathematik* **1** (1968), 9–33.

[SCRIBA 1970] SCRIBA, CHRISTOPH J.: "Geschichtsschreibung der Mathematik." *Gießener Universitätsblätter* **3**, no. 2 (1970), 44–51.

[SCRIBA 1977] SCRIBA, CHRISTOPH J.: "Carl Friedrich Gauss in der Wissenschaftsgeschichte." *Abh. Braunschw. Wiss. Ges.* **27** (1977), 39–56.

[SCRIBA 1980] SCRIBA, CHRISTOPH J.: "Viggo Brun: In Memoriam." *HM* **7** (1980), 1–6.

[SCRIBA 1985] SCRIBA, CHRISTOPH J.: "Thirty Years of the 'History of Mathematics' at Oberwolfach. In remembrance of J. E. Hofmann (1900–1973)." *HM* **12** (1985), 369–371.

[SCRIBA 1993] SCRIBA, CHRISTOPH J.: "The Beginnings of the International Congresses of the History of Science." *XVIIIth International Congress of History of Science. Hamburg–Munich, 1st to 9th August, 1989. Final Report (= Sudhoffs Archiv),* Beiheft **30**, 3–10.

[SEGAL 1980] SEGAL, S. L.: "Helmut Hasse in 1934." *HM* **7** (1980), 46–56.

[SEGRE 1902/03] SEGRE, CORRADO: "Congetture intorno all'influenza di Girolamo Saccheri sulla formazione della geometria non euclidea." *Atti della R. Accademia delle Scienze di Torino* **38** (1902/03), 535–547.

[SEMPILIUS 1635] SEMPILIUS, HUGO: *De Mathematicis Disciplinis Libri duodecim.* Antwerp: Balthasar Moretus 1635.

[SEN/SHUKLA 1985] SEN, SAMARENDRA NATH, and KRIPA SHANKAR SHUKLA (Eds.): *History of Astronomy in India.* New Delhi: Indian National Science Academy 1985.

[SEQUENZ 1965] SEQUENZ, HEINRICH (Ed.): *150 Jahre Technische Hochschule Wien, Band 1: Geschichte und Ausstrahlungen.* Wien: Technische Hochschule Wien 1965.

[SERGESCU 1951] SERGESCU, PIERRE: *Coup d'œil sur les origines de la science exacte moderne* (= coll. "Esprit et Méthode" = 14 causeries faites en 1950 à l'Heure de culture française à la radio). Paris: Société d'Edition d'Enseignement supérieur 1951, 206.

[SHEYNIN 1965] SHEYNIN, OSKAR B.: "O rabotakh Roberta Edreina po teorii oshibok i eë prilozheniyam." *IMI* (1) **16** (1965), 325–336.

[SHEYNIN 1966] SHEYNIN, OSKAR B.: "Origin of the Theory of Errors." *Nature* **211**: 5052 (1966), 1003–1004.

[SHTOKALO 1959/63] SHTOKALO, IOSIF Z. (Ed.): *Istoryko-Matematychnyĭ Zbirnyk.* 4 vols. Kiev: Izdatel'stvo AN USSR 1955/63.

[SHTOKALO 1966/70] SHTOKALO, IOSIF Z.: *Istoriya Otechestvennoĭ Matematiki.* 4 vols. Kiev: Naukova dumka 1966/70.

[SICIAK 1993] SICIAK, JÓZEF: "Kilka uwag do dziejów analizy zespolonej w Krakowie" (A Few Remarks Concerning the History of Complex Analysis in Cracow). *Opuscula Mathematica* **13** (1993), 129–136 (= *Scientific Bulletins of the University of Mining and Metallurgy* 1522. Cracow 1993).

[SIEGEL 1965] SIEGEL, CARL LUDWIG: *Zur Geschichte des Frankfurter Mathematischen Seminars* (= *Frankfurter Universitätsreden*, no. 36.) Frankfurt am Main: Vittorio Klostermann 1965.

[SIEGMUND-SCHULTZE 1993] SIEGMUND-SCHULTZE, REINHARD: *Mathematische Berichterstattung in Hitlerdeutschland* (= *Studien zur Wissenschafts-, Sozial- und Bildungsgeschichte der Mathematik* **9**). Göttingen: Vandenhoeck und Ruprecht 1993.

[SIEGMUND-SCHULTZE 1998] SIEGMUND-SCHULTZE, REINHARD: *Mathematiker auf der Flucht vor Hitler: Quellen und Studien zur Emigration einer Wissenschaft* (= *Dokumente zur Geschichte der Mathematik* **10**). Braunschweig: Vieweg 1998.

[SIGURDSSON 1992] SIGURDSSON, SKULI: "Equivalence, Pragmatic Platonism, and Discovery of the Calculus." In: NYE, MARY JOE *et al.* (Eds.): *The Invention of Physical Science.* Dordrecht: Kluwer Academic Publishers 1992, 97–116.

[SILVA 1860/84] SILVA, INOCÊNCIO FRANCISCO DA: *Diccionario Bibliographico Portuguez.* 12 vols. Lisboa: Imprensa Nacional, 1860–1884.

[SILVA 1992] Silva, Clóvis Pereira da: *A Matemática no Brasil. Uma História do seu Desenvolvimento.* Curitiba: Editora da Universidade do Paraná 1992.

[SIMONOV 1957] SIMONOV, NIKOLAĬ I.: *Prikladnye metody analiza u Eĭlera.* Moscow: Gostekhizdat 1957.

[SIMONOV 1977] SIMONOV, REM A.: *Matematicheskaya mysl' Drevneĭ Rusi.* Moscow: Nauka 1977.

[SINACEUR 1994] SINACEUR, HOURYA: *Jean Cavaillès. Philosophie mathématique*. Paris: PUF, collection "Philosophies," 1994.

[SINHA 1974] SINHA, SASADHAR: *Asutosh Mookerjee*. New Delhi: Publication Division of Government of India 1974.

[SMIRNOV/YOUSHKEVICH 1974] SMIRNOV, VLADIMIR I., and ADOL'F P. YOUSHKEVICH (Eds.): *Leonard Eĭler. Perepiska. Annotirovannyĭ Ukazatel'*. Leningrad: Nauka 1974.

[SMITH 1930] SMITH, DAVID E.: "Florian Cajori." *Bull. AMS* **36** (1930), 777–780.

[SMOLÍK 1864] SMOLÍK, JOSEF: *Matematikové v Čechách od založeni university* (Mathematicians in Bohemia from the Foundation of the University of Prague). Prague: privately published, 1864.

[SNELLIUS 1608] SNELLIUS, WILLEBRORD: *Apollonius batavus, seu Exsuscitata Apollonii Pergæi Περὶ διωρισμένης τομ ῆς Geometria*. Leiden: Johannes van Dorp 1608.

[SPEISER 1939] SPEISER, ANDREAS: *Die Basler Mathematiker. 117. Neujahrsblatt. Hrsg. von der Gesellschaft zur Beförderung des Guten und Gemeinnützigen*. Basel: Helbing und Lichtenhahn 1939.

[SPEZIALI 1959] SPEZIALI, PIERRE: *Gabriel Cramer (1704–1752) et ses correspondants* (= *Les Conférences du Palais de la Découverte, Série D* **59**). Paris: Palais de la Découverte 1959.

[SPIESS 1929] SPIESS, OTTO: *Leonhard Euler. Ein Beitrag zur Geistesgeschichte des XVIII. Jahrhunderts*. Frauenfeld and Leipzig: Huber 1929.

[SPIESS 1955] SPIESS, OTTO: "Die Basler Mathematiker des achtzehnten Jahrhunderts. Plan zu einer Gesamtausgabe ihrer Werke." In: *Der Briefwechsel von Johann Bernoulli*, vol. 1, 9–85. Basel: Birkhäuser 1955.

[SRINIVASIENGAR 1967] SRINIVASIENGAR, C. N.: *The History of Ancient Indian Mathematics*. Calcutta: World Press 1967.

[STADLER 1997] STADLER, FRIEDRICH: *Studien zum Wiener Kreis*. Frankfurt am Main: Suhrkamp 1997.

[STÄCKEL/AHRENS 1908] STÄCKEL, PAUL, and WILHELM AHRENS: *Der Briefwechsel zwischen C. G. J. Jacobi und P. H. von Fuss über die Herausgabe der Werke Leonhard Eulers*. Leipzig: Teubner 1908.

[STAEHELIN 1957] STAEHELIN, ANDREAS: *Geschichte der Universität Basel, 1632–1818* (= *Studien zur Geschichte der Wissenschaften in Basel*, vol. IV/V). Basel: Helbing und Lichtenhahn 1957.

[STÄHELIN/SPEISER 1989/90] STÄHELIN, HARTMANN, and DAVID SPEISER: "55 Jahre Herausgabe der Bernoulli-Werke. Verein zur Förderung der Bernoulli-Edition." *Vh. Nat. Ges. Basel* **100** (1989/90), 5–8.

[STAMM 1935] STAMM, EDWARD: *Z historii matematyki w XVII wieku w Polsce* (From the History of Mathematics in the 17th-Century in Poland). Appendix to the *Wiadomośct Matematyczne*. Warsaw 1935.

[STEENSTRA 1759] STEENSTRA, PYBO: *Intree-rede over de Opkomst en den Voortgang der Meetkonst*. Leiden: Samuel and Johannes Luchtmans 1759.

[STEINSCHNEIDER 1893/1901] STEINSCHNEIDER, MORITZ: "Mathematik bei den Juden." A series of articles, originally published in *Bibl. Math.* (2) (1893–1899), (3) **2** (1901), and in *Abh. Gesch. Math.* **9** (1899). Repr. Hildesheim: Olms 1964.

[STEPHANIDES 1932] STEPHANIDES, MICHAEL: "Les savants byzantins et la science moderne. Renaissance et Byzance." *Archeion* **14** (1932), 492–496.

[STEVIN 1585] STEVIN, SIMON: *L'Arithmetique: Contenant les computations des nombres arithmetiques ou vulgaires: Aussi l'Algebre, avec les equations de cinc quantitez. Ensemble les quatre premiers livres d'Algebre de Diophante d'Alexandrie, maintenant premierement traduicts en François. Encore un livre particulier de la Pratique d'Arithmetique; Et un traicté des Incommensurables grandeurs: Avec l'explication du Dixièsme Livre d'Euclide.* Leiden: Christophle Plantin 1585.

[STOCKLER 1805a] STOCKLER, FRANCISCO DE BORJA GARÇÃO: "Elogio de João Le Rond d'Alembert." In: *Obras de Francisco de Borja Garção Stockler*, vol. 1, 1–128. Lisboa: Academia Real das Sciencias de Lisboa 1805.

[STOCKLER 1805b] STOCKLER, FRANCISCO DE BORJA GARÇÃO: "Elogio de José Joaquim Soares de Barros e Vasconcellos." In: *Obras de Francisco de Borja Garção Stockler*, vol. 1, 185–224. Lisboa: Academia Real das Sciencias de Lisboa 1805.

[STOCKLER 1805c] STOCKLER, FRANCISCO DE BORJA GARÇÃO: "Memória sobre a Originalidade dos Descobrimentos Marítimos dos Portugueses no Século Decimoquinto." In: *Obras de Francisco de Borja Garção Stockler*, vol. 1, 345–388. Lisboa: Academia Real das Sciencias de Lisboa 1805.

[STOCKLER 1805d] STOCKLER, FRANCISCO DE BORJA GARÇÃO: *Obras de Francisco de Borja Garção Stockler*, vol. 1. Lisboa: Academia Real das Sciencias de Lisboa 1805.

[STOCKLER 1819] STOCKLER, FRANCISCO DE BORJA GARÇÃO: *Ensaio Historico sobre as Origens e Progressos das Mathematicas em Portugal.* Paris: P. N. Rougeron 1819.

[STOLZ 1881] STOLZ, OTTO: "B. Bolzano's Bedeutung in der Geschichte der Infinitesimalrechnung." *Math. Ann.* **18** (1881), 255–279.

[STRUIK 1980] STRUIK, DIRK J.: "The Historiography of Mathematics from Proklos to Cantor." *NTM* **17**, no. 2 (1980), 1–22.

[STRUIK 1981] STRUIK, DIRK J.: *Matematica: un profilo storico*, BOTTAZZINI, U. (Ed.). Bologna: Il Mulino 1981.

[STUBHAUG 2000a] STUBHAUG, ARILD: *Niels Henrik Abel and his Times. Called Too Soon by Flames Afar.* Berlin: Springer 2000. 1st ed. in Norwegian: *Et foranskutt lyn. Niels Henrik Abel og hans tid.* Oslo: Aschehoug 1996.

[STUBHAUG 2000b] STUBHAUG, ARILD: *Dat war mine tankers Djervhet. Matematikeren Sophus Lie* (It was the Outspokeness of my Thoughts. The Mathematician Sophus Lie). Oslo: Aschehoug 2000.

[STUDNIČKA 1876] STUDNIČKA, FRANTIŠEK JOSEF: "A. L. Cauchy als formaler Begründer der Determinantentheorie." *Abh. Kön. Böhm. Ges. Wiss.* (6) **6** (1876), 1–40 (also published separately).

[STUDNIČKA 1886] STUDNIČKA, FRANTIŠEK JOSEF (Ed.): *Tycho Brahe Triangulorum planorum et sphaericorum praxis arithmetice [...]*. Prague: privately published, 1886.

[STUDNIČKA 1903] STUDNIČKA, FRANTIŠEK JOSEF (Ed.): *Tycho Brahe Brevissimum planimetriae compendium.* Prague: privately published, 1903.

[STULOFF 1988/91] STULOFF, NIKOLAI: *Die Entwicklung der Mathematik.* 4 vols. Mainz: Johannes Gutenberg Universität, Fachbereich 17, Mathematik, 1988–1991.

[STYAZHKIN 1967] STYAZHKIN, NIKOLAĬ I.: *Formirovanie matematicheskoĭ logiki.* Moscow: Nauka 1967. Engl. trans.: *History of Mathematical Logic from Leibniz to Peano.* Cambrige, MA.: MIT Press 1969.

[SUTER 1890] SUTER, HEINRICH: "Bibliographische Notiz über die mathematisch-historischen Studien in der Schweiz." *Bibl. Math.* (2) **4** (1890), 97–106.

[SUTER 1893] SUTER, HEINRICH: "Der V. Band des Katalogs der arabischen Bücher der vicekoniglichen Bibliothek in Kairo." *Z. Math. Phys.* **38** (1893), 1–24, 41–57, 161–184.

[SWERDLOW 1993a] SWERDLOW, NOEL M.: "Otto E. Neugebauer." *Proc. APS* **137** (1993), 137–165.

[SWERDLOW 1993b] SWERDLOW, NOEL M.: "Otto E. Neugebauer (26 May 1899 – 19 February 1990)." *JHA* **24** (1993), 289–299.

[SWERDLOW 1993c] SWERDLOW, NOEL M.: "Montucla's Legacy: The History of the Exact Sciences." *Journal of the History of Ideas* **54** (1993), 299–328.

[SZÉNÁSSY 1992] SZÉNÁSSY, BARNA: *History of Mathematics in Hungary until the 20th Century.* Berlin *et al.*: Springer 1992.

[ŚNIADECKI 1815] ŚNIADECKI, JAN: "O J. L. Lagrange, pierwszym geometrze naszego wieku" (On J. L. Lagrange, the First Geometrician of our Century). *Dziennik Wileński* **2**, 479–500, 641–689. Wilno 1815.

[ŚNIADECKI 1837a] ŚNIADECKI, JAN: "O rachunku losów (1817)" (On the Calculus of Chances). In: *Dzieła Jana Śniadeckiego* (Works of Jan Śniadecki), vol. 4. Warsaw 1837, and in: *Pisma filozoficzne* (Philosophical Papers). Warsaw: Państwowe Wydawnictwo Naukowe 1958, 99–116.

[ŚNIADECKI 1837b] ŚNIADECKI, JAN: "O rozumowaniu rachunkowym (1818)" (On Mathematical Reasoning). In: *Dzieła Jana Śniadeckiego* (Works of Jan Śniadecki), vol. 4. Warsaw 1837, and in: *Ibidem*, 117–140.

[TACQUET 1654] TACQUET, ANDREAS: *Elementa Geometriæ planæ ac solidæ quibus accedunt selecta ex Archimede Theoremata.* Antwerp: Jacob Meursius 1654.

[TANNERY 1912/50] TANNERY, PAUL: *Mémoires scientifiques*, HEIBERG, J.-L., and H.-G. ZEUTHEN (Eds.), 17 vols. Toulouse et Paris: E. Privat & Gauthier-Villars 1912–1950; repr. Paris: Editions Jacques Gabais 1995–1996.

[TATARKIEWICZ 1995] TATARKIEWICZ, KRZYSZTOF: "Uwagi o historii dwóch wieków mechaniki w Polsce (1795–1995)" (A Few Remarks on the History of Mechanics in Poland). In: "Matematyka Polska w stuleciu 1851–1950" (Polish Mathematics, 1851–1950). Uniwersytet Szczeciński. Materiały Konferencje **16** (1995), 115–160.

[TATON 1987] TATON, RENÉ: "Pierre Sergescu (1893–1954): son œuvre en histoire des sciences et son action pour la renaissance des *Archives internationales d'histoire des sciences.*" *Arch. int. Hist. Sci.* **37** (1987), 104–119.

[TEIXEIRA 1865] TEIXEIRA, ANTÓNIO JOSÉ: *Estudos sobre a doutrina da proporcionalidade,* etc. Coimbra: Imprensa da Universidade 1865.

[TEIXEIRA 1869] TEIXEIRA, ANTÓNIO JOSÉ: "Questão entre Jose Anastácio da Cunha e José Monteiro da Rocha." *Jornal Litterário* (1869), 97–100, 105–112, 125–127, 129–136, 139–142, 147–150, 156–159, 165–166.

[TEIXEIRA 1881a] TEIXEIRA, FRANCISCO GOMES: "Prelecção sobre a origem e sobre os princípios do cálculo infinitesimal, feita aos alunos da Universidade de Coimbra". *Jornal de Sciencias Mathematicas e Astronomicas* **3** (1881), 21–45.

[TEIXEIRA 1881b] TEIXEIRA, FRANCISCO GOMES: "Sobre a história do nonius." *Jornal de Sciencias Mathematicas e Astronomicas* **3** (1881), 73.

[TEIXEIRA 1890] TEIXEIRA, FRANCISCO GOMES: "Sur les écrits d'histoire des mathématiques publiés au Portugal." *Bibl. Math.* (1) **4** (1890), 91–92.

[TEIXEIRA 1902] TEIXEIRA, FRANCISCO GOMES: "Apontamentos Biográficos sobre Daniel Augusto da Silva." *Boletim da Direcção Geral de Instrucção Pública* **1** (1902), 829–840.

[TEIXEIRA 1905/06] TEIXEIRA, FRANCISCO GOMES: "Sobre uma questão entre Monteiro da Rocha e Anastácio da Cunha." *Annaes Scientificos da Academia Polytechnica do Porto* **1** (1905–1906), 7–15.

[TEIXEIRA 1923] TEIXEIRA, FRANCISCO GOMES: "Les Mathématiques en Portugal." *Enseign. Math.*, 23ᵉ année, **3–4** (1923), 137–142.

[TEIXEIRA 1925] TEIXEIRA, FRANCISCO GOMES: *Panegíricos e Conferências.* Coimbra: Imprensa da Universidade 1925.

[TEIXEIRA 1934] TEIXEIRA, FRANCISCO GOMES: *História das Matemáticas em Portugal.* Lisboa: Academia das Ciências 1934.

[TEMPLE 1956] TEMPLE, GEORGE: "Obituary of E. T. Whittaker." *Biographical Memoirs of Fellows of the Royal Society* (2) **2** (1956), 299–325.

[TENNULIUS 1668] TENNULIUS, SAMUEL: *Jamblichus Chalcidensis ex Cœlo-Syria in Nicomachi Geraseni Arithmeticam introductionem, et De Fato. Nunc primum editus, in Latinum sermonem conversus, notis perpetuis illustratus à Samuele Tennulio. Accedit Joachimi Camerarii Explicatio in duos libros Nicomachi.* Arnhem: Joh. Fridericus Hagius 1668.

[THACKRAY/MERTON 1972] THACKRAY, ARNOLD, and ROBERT K. MERTON: "On Discipline-Building: The Paradoxes of George Sarton." *Isis* **63** (1972), 473–495.

[THIBAUT 1984] THIBAUT, GEORGE: *Mathematics in the Making in Ancient India.* (Posthumous collection of his writings on the *Śulba Sūtras.*) Calcutta: K. P. Bagchi 1984.

[TIBENSKÝ 1976] TIBENSKÝ, JÁN: *Biobibliografia prírodných, lékařských a technických vied na Slovensku do roku 1850* (Biobibliography of Science, Medicine and Technology in Slovakia till 1850). 2 vols. Martin: Matica Slovenská 1976.

[TIBENSKÝ 1979] TIBENSKÝ, JÁN: *Dejiny vēdy a techniky na Slovensku* (History of Science and Technology in Slovakia). Bratislava: Osveta 1979.

[TIMCHENKO 1899] TIMCHENKO, IVAN YU.: *Osnovaniya teorii analiticheskikh funktsiĭ.* Part 1. *Istoricheskie svedeniya o razvitii ponyatiĭ i metodov, lezhashchikh v osnovanii teorii analiticheskikh funktsiĭ.* Odessa: Mathesis 1899.

[TIMCHENKO 1917] TIMCHENKO, IVAN YU. (Ed.): KÈDZHORI F. *Istoriya elementarnoĭ matematiki.* Trans., ed. and commentary with additions by I. YU. TIMCHENKO. Odessa: Mathesis 1917.

[TIRABOSCHI 1772–1795] TIRABOSCHI, GIROLAMO: *Storia della letteratura italiana.* 11 vols., vol. 6 in 2 parts, vol. 7 in 3 parts. Modena: Società Tipografica 1772–1795.

[TOEPLITZ 1927] TOEPLITZ, OTTO: "Das Problem der Universitätsvorlesungen über Infinitesimalrechnung und ihrer Abgrenzung gegenüber der Infinitesimalrechnung an den höheren Schulen." *Jahresber. DMV* **36** (1927), 88–100.

[TOMEK 1849] TOMEK, WÁCLAV WLADIVOJ: *Geschichte der Prager Universität.* Prag: Universität Prag 1849.

[TRABULSE 1974] TRABULSE, ELÍAS: *Ciencia y Religión en el Siglo XVII.* México: El Colegio de México 1974.

[TRABULSE 1983/89] TRABULSE, ELÍAS: *Historia de la Ciencia en México.* 5 vols. Mexico: Conacyt/FCE 1983–1989.

[TRABULSE 1984] TRABULSE, ELÍAS: *El Círculo Roto (= Col. Lecturas Mexicanas* no. 54). México: SEP/FCE 1984.

[TREMBLEY 1987] TREMBLEY, JACQUES (Ed.): *Les savants genevois dans l'Europe intellectuelle du XVII*[e] *au milieu du XIX*[e] *siècle.* Genève: Editions du Journal de Genève 1987.

[TROPFKE 1899] TROPFKE, JOHANNES: *Erstmaliges Auftreten der einzelnen Bestandtheile unserer Schulmathematik* (Schulprogramm). Berlin: Weidmann 1899.

[TRUESDELL 1984] TRUESDELL, CLIFFORD A.: *An Idiot's Fugitive Essays on Science.* New York: Springer 1984.

[ṬŪQĀN 1963] ṬŪQĀN, QADRĪ ḤĀFIẒ: *Turāth al-ʿarab al-ʿilmī fi'l-riyāḍiyyāt wa'l-falak.* Cairo: Dār al-shurūq 1941; repr. 1963.

[TURNBULL 1939] TURNBULL, HERBERT WESTREN (Ed.): *James Gregory Tercentenary Memorial Volume.* London: G. Bell and Sons 1939.

[VAN DER WAERDEN 1950] VAN DER WAERDEN, BARTEL LEENDERT: *Ontwakende Wetenschap. Egyptische, Babylonische en Griekse Wiskunde.* Groningen: P. Noordhoff 1950. English trans. *Science Awakening*, vol. 1. Groningen: P. Noordhoff 1950; repr. New York: John Wiley and Sons 1963.

[VAN DER WAERDEN 1983] VAN DER WAERDEN, BARTEL LEENDERT: *Zur algebraischen Geometrie. Selected Papers.* Berlin, *et al.*: Springer 1983 (with bibliography to 1972).

[VAN KAMPEN 1826] VAN KAMPEN, NICOLAAS GODFRIED: *Beknopte Geschiedenis der Letteren en Wetenschappen in de Nederlanden, van de vroegste tijden af, tot op het begin der negentiende eeuw.* Part 3, Delft: Weduwe J. Allart 1826.

[van Raemdonck 1869] van Raemdonck, Jan Hubert: *Gerard Mercator. Sa vie et ses œuvres.* Sint-Niklaas: E. Dalschaert-Praet 1869.

[Van Schooten 1657] Van Schooten, Frans: *Exercitationum Mathematicarum Libri quinque. I. Propositionum arithmeticarum et geometricarum centuria. II. Constructio problematum simplicium geometricorum. III. Apollonii Pergaei Loca Plana Restituta. IV. Organica conicarum sectionum in plano descriptio. V. Sectiones miscellaneae triginta. Quibus accedit Christiani Hugenii tractatus, de Ratiociniis in Aleae Ludo.* Leiden: Johannes Elsevier 1657.

[Vasconcellos 1925] Vasconcellos, Fernando de Almeida Loureiro e: *História das Matemáticas na Antiguidade.* Lisboa: Aillaud & Bertrand 1925.

[Vasconcelos 1926] Vasconcelos, José: *La Raza Cósmica.* México and Paris: Agencia Mundial de Libreria 1926.

[Vasil'ev 1919] Vasil'ev, Aleksandr V.: *Tseloe chislo. Istoricheskiĭ ocherk.* Petrograd: Nauchnoe Knigoizdatel'stvo 1919; 2nd ed. 1922.

[Velamazán 1994] Velamazán, Mª Angeles: *La enseñanza de las matemáticas en las Academias militares en España en el siglo XIX.* Zaragoza: Seminario de Historia de la Ciencia y de la Técnica de Aragón, Universidad de Zaragoza 1994.

[Velsius 1544] Velsius, Justus: *De Mathematicarum Disciplinarum vario usu dignitateque: quique ad harum comparandam cognitionem adhibendi sint: deque optima horum explicandorum ratione Iusti Velsij Hagani Medici Oratio.* Strasbourg: Crato Mylius 1544.

[Velsius 1545] Velsius, Justus: *Procli Diadochi De Motu libri duo, nunc primum latinitate donati, Iusto Velsio Hagano Medico interprete.* Basel: Joannes Hervagius 1545.

[Ver Eecke, Paul 1926] Ver Eecke, Paul: *Diophant d'Alexandrie. Les six Livres arithmétiques et le Livre des nombres polygones.* Bruges: Desclée, de Brouwer et Cie. 1926; repr. Paris: Albert Blanchard 1959.

[Veselovskiĭ 1962] Veselovskiĭ, Ivan N. (Ed.): *Arkhimed. Sochineniya.* Trans., with commentary, by I. N. Veselovskiĭ. Moscow: Fizmatgiz 1962.

[Vetter 1919] Vetter, Quido: "O metodice dějin matematiky." (On the Methodology of the History of Mathematics). *Věstník Královské české společnosti nauk, Třída matematicko-přírodovědecká. Ročník 1918.* Praha: Královská česká společnost nauk 1919, vol. III, 1–52.

[Vetter 1924] Vetter, Quido: "La Boemia nelle Storia della Matematica." (Bibliography = "Supplemento alla Guida allo studio della storia delle Matematiche del Prof. Gino Loria"). *Boll. Loria* 1924, 1–31.

[Vetter 1958/61] Vetter, Quido: "Esquisse du progrès des mathématiques dans les régions tchèques jusq'à la bataille de Bièla Gora." *IMI* **11** (1958), 461–514; "Esquisse du progrès des mathématiques dans les régions tchèques depuis la bataille de Bièla Gora jusqu'a la fin du XVII-ème siècle." *Ibid.* **14** (1961), 491–516.

[Vicente 1991] Vicente Maroto, Mª Isabel, and Mariano Esteban Piñeiro: *Aspectos de la ciencia aplicada en la España del Siglo de Oro.* Valladolid: Junta de Castilla y León 1991.

[VILLICUS 1897] VILLICUS, FRANZ: *Die Geschichte der Rechenkunst vom Alterthume bis zum XVIII. Jahrhundert*, 3rd enlarged ed. Wien: Gerold Verlag 1897.

[VISCHER 1860] VISCHER, WILHELM: *Geschichte der Universität Basel von der Gründung 1460 bis zur Reformation 1529.* Basel: Georg 1860.

[VITRAC 1996] VITRAC, BERNARD: "Mythes (et réalités?) dans l'histoire des mathématiques grecques anciennes." In: [GOLDSTEIN/GRAY/RITTER 1996, 31–51].

[VOGEL 1965] VOGEL, KURT: "L'historiographie mathématiques avant Montucla." In: *Actes du XIe Congrès International d'Histoire des Sciences, Varsovie–Cracovie 24–31 Août 1965*, vol. III. Wroclaw, Varsovie, Cracovie 1968, 179–184.

[VOGEL/YOUSHKEVICH 1997] FOLKERTS, MENSO; MIRIAM M. ROŽANSKAJA, and IRINA LUTHER (Eds.): *Mathematikgeschichte ohne Grenzen. Die Korrespondenz zwischen K. Vogel und A. P. Juschkewitsch (= Algorismus*, no. 22; with portrait). München: Institut für Geschichte der Naturwissenschaften 1997. Russian ed.: *Istoria Matematiki bez Granits.* Moscow: Ianus-K 1997 (with portrait).

[VOSSIUS 1650] VOSSIUS, GERARDUS JOANNES: *De universae mathesios natura et constitutione liber. Cui subjungitur Chronologia Mathematicorum* (= vol. 4 of *De quatuor artibus popularibus*). Amsterdam: Joannes Blaeu 1650; 2nd ed. 1660.

[VYDRA 1778] VYDRA, STANISLAW: *Historia matheseos in Bohemia et Moravia cultae.* Prague: W. Gerle 1778.

[VYGODSKIĬ 1941] VYGODSKIĬ, MARK YA.: *Arifmetika i algebra v drevnem mire.* Moscow and Leningrad: GITTL 1941; 2nd ed. 1967.

[VYGODSKIĬ 1948] VYGODSKIĬ, MARK YA.: "Matematika i eë deyateli v Moskovskom universitete vo vtoroĭ polovine XIX veka." *IMI* (1) **1** (1948), 141–183.

[WALLIS 1975] WALLIS, PETER, and RUTH WALLIS: *Eighteenth-Century British Historians of Mathematics*, 1975 (manuscript).

[WANG 1996] WANG QINGJIAN: "The Path of the Historian of Mathematics Prof. Liang Zongju" (in Chinese). *Zhonghuo keji shiliao* (Chinese Historical Materials on Science and Technology) **17** no. 4 (1996), 39–47.

[WEHRLI 1955] WEHRLI, FRITZ: *Die Schule des Aristoteles. Heft 4. Eudemus of Rhodos.* Basel: Schwabe und Co. 1955; 2nd ed.: *Heft 8. Eudemus of Rhodos*: 1969.

[WHISH 1835] WHISH, CHARLES M.: "On the Hindu Quadrature of the Circle and the Infinite Series of the Proportion of the Circumference to the Diameter Exhibited in the Four Śāstras, etc." *Transaction of the Royal Society of Great Britain and Ireland* **3** (1835), 509–523.

[WHITROW 1975] WHITROW, GERALD JOHN: "Henry Thomas Colebrooke (1765–1837) and Hindu mathematics." *Bulletin of the Institute of Mathematics and its Applications* **11** (1975), 147–154.

[WIELEITNER 1923] WIELEITNER, HEINRICH: "Siegmund Günther †." *MGMN* **22** (1923), 1–2.

[WIELEITNER 1927/29] WIELEITNER, HEINRICH: *Mathematische Quellenbücher.* 4 vols. Berlin: Verlag Salle 1927–1929.

[WIĘSŁAW 1997] WIĘSŁAW, WITOLD: *Matematyka i jej historia* (Mathematics and Its History). Opole: Novik 1997.

[WILDER 1981] WILDER, RAYMOND L.: *Mathematics as a Cultural System.* Oxford: Pergamon Press 1981.

[WOLF 1858/62] WOLF, RUDOLF: *Biographien zur Kulturgeschichte der Schweiz.* 4 vols. Zürich: Orell Füssli 1858–1862.

[WOLF 1877] WOLF, RUDOLF: *Geschichte der Astronomie* (= *Geschichte der Wissenschaften in Deutschland* **16**). München: Oldenbourg 1877; repr. London: Johnson 1965, and Osnabrück: R. Kuballe 1984.

[WOLF 1879] WOLF, RUDOLF: *Geschichte der Vermessungen in der Schweiz.* Zürich: S. Höhr 1879.

[WOLF 1890/1893] WOLF, RUDOLF: *Handbuch der Astronomie, ihrer Geschichte und Literatur.* 2 vols. Zürich: Schulthess 1890–1893; repr. Hildesheim: Olms 1973.

[WOLFF 1710/11] WOLFF, CHRISTIAN: *Die Anfangsgründe aller mathematischen Wissenschaften.* Vol. 5: *Kurtzer Unterricht von den vornehmsten mathematischen Schriften.* Halle: Renger 1710–1711; 2nd ed. 1750–1757.

[WOLFF 1716] WOLFF, CHRISTIAN: *Vollständiges Mathematisches Lexicon, darinnen all Kunst-Wörter und Sachen, welche in der erwegenden und ausübenden Mathesi vorzukommen pflegen, deutlich erkläret, über all aber Zur Historie der Mathematischen Wissenschaften dienliche Nachrichten eingestreuet, und die besten und auserlesensten Schrifften, welche jede Materie gründlich abgehandelt, angeführet [...].* Leipzig: Joh. Friedrich Gleditschens seel. Sohn 1716; 2nd ed. [not by WOLFF] Leipzig: Gleditsch 1734. Repr. of the 1st ed. (with an introduction and additional index of names and publications by JOSEPH EHRENFRIED HOFMANN) Hildesheim and New York: Olms 1965.

[WOLFF 1741] WOLFF, CHRISTIAN: *De Praecipuis Scriptis Mathematicis Brevis Commentatio.* Halle: Renger 1741.

[YAN/MEI 1990] YAN DUNJIE and MEI RONGZHAO: "Cheng Dawei and His Mathematical Works" (in Chinese). In: MEI RONGZHAO (Ed.): *Ming Qing shuxueshi lunwenji* (Collected Papers on the History of Mathematics in the Ming and Qing Periods). Nanjing: Jiangsu Education Press 1990, 26–52.

[YANOVSKAYA 1968] YANOVSKAYA, SOF'YA A. (Ed.): *K. Marks. Matematicheskie rukopisi.* Moscow: Nauka 1968.

[YOUSHKEVICH 1948a] YOUSHKEVICH, ADOL'F P.: "Istoriya matematiki." In: ALEKSANDR G. KUROSH; ALEKSEĬ I. MARKUSHEVICH, and PËTR K. RASHEVSKIĬ (Eds.): *Matematika v SSSR za 30 let. 1917–1947.* Moscow and Leningrad: Gostekhizdat 1948, 993–1023.

[YOUSHKEVICH 1948b] YOUSHKEVICH, ADOL'F P.: "Matematika v Moskovskom Universitete za pervye sto let ego sushchestvovaniya." *IMI* (1) **1** (1949), 43–140.

[YOUSHKEVICH 1959] YOUSHKEVICH, ADOL'F P.: "Istoriya matematiki." In: ALEKSANDR G. KUROSH (Ed.): *Matematika v SSSR za 40 let. 1917–1957.* 2 vols. Moscow: Fizmatgiz 1959, 953–987.

[YOUSHKEVICH 1961] YOUSHKEVICH, ADOL'F P.: *Istoriya matematiki v srednie veka.* Moscow: Fizmatgiz 1961. German trans.: *Geschichte der Mathematik im Mittelalter.* Leipzig: Teubner 1964.

[YOUSHKEVICH 1968] YOUSHKEVICH, ADOL'F P.: *Istoriya matematiki v Rossii do 1917 goda.* Moscow: Nauka 1968.

[YOUSHKEVICH 1970/72] YOUSHKEVICH, ADOL'F P. (Ed.): *Istoriya matematiki s drevneĭshikh vremën do nachala XIX stoletiya.* 3 vols. Moscow: Nauka 1970–1972.

[YOUSHKEVICH 1971] YOUSHKEVICH, ADOL'F P.: "The concept of function up to the middle of the 19th century." *Arch. Hist. ex. Sci.* **16** (1971), 37–85.

[YOUSHKEVICH 1973/94] YOUSHKEVICH, ADOL'F P. (Ed.): *IMI.* Ser. 1, vols. 18–35. Moscow: Nauka 1973/94.

[YOUSHKEVICH 1976/77] YOUSHKEVICH, ADOL'F P. (Ed.): *Khrestomatiya po istorii matematiki. Arifmetika i algebra. Teoriya chisel. Geometriya.* Moscow: Prosveshchenie 1976; *Matematicheskiĭ analiz. Teoriya veroyatnosteĭ.* Moscow: Prosveshchenie 1977.

[YOUSHKEVICH 1979] YOUSHKEVICH, ADOL'F P.: "Sovetskie issledovaniya po istorii matematiki za 60 let." *IMI* (1) **24** (1979), 9–87. (English trans. in: *Acta hist.* **18** (1982), 9–113.)

[YOUSHKEVICH 1983] YOUSHKEVICH, ADOL'F P.: "A. N. Kolmogorov o predmete matematiki i eë istorii." *VIET* **3** (1983), 67–74. (English trans. under the title "A. N. Kolmogorov: Historian and Philosopher of Mathematics on the Occasion of his 80th Birthday" in: *HM* **10** (1983), 383–395.)

[YOUSHKEVICH 1994] YOUSHKEVICH, ADOL'F P.: "A. N. Kolmogorov o sushchnosti matematiki i periodizacii eë istorii."*IMI* (1) **35** (1994), 8–16.

[YOUSHKEVICH/WINTER 1959/76] JUŠKEVIČ, ADOLF P., and EDUARD WINTER (Eds.): *Die Berliner und die Petersburger Akademie der Wissenschaften im Briefwechsel Leonhard Eulers.* 3 vols. Berlin: Akademie-Verlag 1959–1976.

[ZASLAVSKY 1973] ZASLAVSKY, CLAUDIA: *Africa Counts. Number and Pattern in African Culture.* Boston: Prindle, Weber and Schmidt 1973; 2nd ed. Brooklyn, N.Y.: Laurence Hill Books 1990.

[ZEA 1968] ZEA, LEOPOLDO: *El Positivismo en México: Nacimiento, Apogeo y Decadencia.* México: FCE 1968.

[ZEUTHEN 1905] ZEUTHEN, HIERONYMUS G.: "L'œuvre de Paul Tannery comme historien des mathématiques." *Bibl. Math.* (3) **6** (1905), 257–304.

[ŻEBRAWSKI 1873/86] ŻEBRAWSKI, TEOFIL: *Bibliografia piśmiennictwa polskiego z działu matematyki i fizyki oraz ich zastosowań* (Bibliography of Polish Literature on Mathematics and Physics and Their Applications). Cracow: Biblioteka Kórnicka 1886; new photo-reproduced edition Warsaw: Instytut Historii Nauki, Oświaty i Techniki PAN 1992.

[ZINNER 1990] ZINNER, ERNST: *Regiomontanus: His Life and Work* (trans. by EZRA BROWN). Amsterdam, New York, Oxford: North Holland 1990. (Original German ed.: *Leben und Wirken des Joh. Müller von Königsberg genannt Regiomontanus.* München 1938; 2nd corrected and expanded edition, Osnabrück: Otto Zeller 1968.)

[ZUBOV 1953] ZUBOV, VASILIĬ P. (Ed.): "Kirik Novgorodets. Uchenie im zhe vedati cheloveku chisla vsekh let" (with commentary). *IMI* (1) **6** (1953), 174–212.

[ZUÑIGA 1995] ZUÑIGA, ANGEL RUIZ (Ed.): *Historia de las Matemáticas en Costa Rica.* San José: Editorial de la Universidad de Costa Rica 1995.

Index

Aaboe, Asger (*1922), 158
Abbagnano, Nicola (1901–1990), 435
ʿAbd al-Ḥamīd b. Turk (early 9th century), 518
ʿAbd al-Raḥmān al-Akhdarī (†1510), 320
ʿAbd al-Raḥmān al-Sayyid, (1855–1902; Shaykh), 321
Abel, Niels Henrik (1802–1829), 23, 151–153, 314, 333, 364, 384, 449–450, 467, 500, 531, 534
Abel Edition, 467, 534
Abetti, Giorgio (1882–1982), 426
Abraham ibn Ezra (ca. 1090–1167), 119
Abraham Savasorda (ca. 1070–ca. 1136), 119–121, 403
Abrams, John (1914–1981), 286
Abū Kāmil (9th/10th century), 147, 270
Abū l-Rayḥān al-Bīrūnī, *see* al-Bīrūnī, Abū l-Rayḥān
Abū ʾl-Wafāʾ (940–998), 16–17, 132–133, 392
Abū Naṣr Manṣūr b. ʿAlī b. ʿIrāq (ca. 1000), 462
Abū Sahl Wayjan b. Rustam al-Kūhī (10th century), 518
Académie Internationale d'Histoire des Sciences (AIHS), see International Academy of the History of Science
Adam, 534
Adam, Charles (1857–1940), 26, 538
Ādityadāsa (ca. 500 A.D.), 308
Adnan, Abdülhak Adivar (1882–1955), 327
Agasse, Louis (18th century), 490
Agnesi, Maria Gaetana (1718–1799), 89
Agostini, Amedeo (1892–1958), **B**, 90–91, 93–94, 375, 423
Agudo, Fernando Roldão Dias (*1925), 246
Aguilon, Franciscus de (1566–1617), 47
Ahlfors, Lars (1907–1996), 561
Aḥmad ibn Yūsuf (†912/13), 391
Ahrens, Wilhelm (1872–1927), 129
Airy, George Biddell (1801–1892), 169
Aiton, Eric John (1920–1991), **B**, 407
Akhiezer, Naum I. (1901–1980), 190–192
al-Khwārizmī, Muḥammad b. Mūsā, *see* Muḥammad b. Mūsā al-Khwārizmī
Albèri, Eugenio (1807–1878), 85
Alberti, Leon Battista (1404–1472), 483
Albertus Magnus (1208?–1280), 448
Albuquerque, Luis de (1917–1992), 246
Aleksandrov, Aleksandr D. (1912–1999), 189, 261n
Aleksandrov, Pavel S. (1896–1982), 186–187
Aleksandrova, Nadezhda V. (*1932), 192
Alembert, *see* d'Alembert, Jean Baptiste le Rond
Alhazen, *see* Ibn al-Haytham
ʿAlī al-Qalaṣādī (†1486), 320, 324
Allman, George Johnston (1824–1904), 171, 178, 418, 535
Alphonse X, the Wise (1221–1284; King of Castille from 1252), 235
Alpoim, José Fernandes Pinto (1695–1765), 251
Alsina, Juan (†1807), 251
Amaldi, Ugo (1875–1957), 84, 423
al-ʿĀmilī, Bahāʾ al-Dīn, *see* Bahāʾ al-Dīn al-ʿĀmilī
Amma, T. A. Sarasvati (20th century), 315
Amodeo, Federico (1859–1946), **B**, 87–88, 355
Ampère, André Marie (1775–1836), 437, 482
al-Āmulī, Muḥammad b. Maḥmūd, *see* Muḥammad b. Maḥmūd al-Āmulī
Anbūba, ʿĀdil (20th century), 327
al-Andalusi, Saʿid, 231
Andersen, Kirsti (*1941), 151, 161
Anderson, Alexander (1582–1620?), 47

ANDRES, GIOVANNI (1740–1817), 67–68
ANDRÈS, JUAN, *see* ANDRES, GIOVANNI
ANDRONICUS II (1259/60–1332; Emperor 1282–1328), 223
ANFOSI, AGUSTÍN (1889–1966), 259
ANSELME OF CANTERBURY (1033–1109), 460
ANTHEMIUS OF TRALLEIS (†534), 52
ANTINORI, VINCENZO (1792–1865), 85
ANTIPHON (ca. 430 B.C.), 222
ANTONIO DE' MAZZINGHI (2nd half of 14th century), 370
ANTROPOVA, VARVARA I. (1924–1991), 189
APIANUS, PETER (1495–1552), 52, 551
APOKIN, IGOR A. (*1936), 193, 476
APOLLONIUS OF PERGE (ca. 260–ca. 190), 5, 46–47, 52, 64–65, 83, 118, 123, 139, 156–157, 162, 172, 228, 304, 389, 404, 441–443, 459, 509, 525, 529, 537, 575–576
Arab countries, Turkey, and Iran
  American University in Beirut, 322, 324, 514
  Arabic College in Jerusalem, 513
  Catholic University St. Joseph in Beirut, 324, 326
  Committee for Research into Turkish History in Ankara, 324
  Egyptian Academy in Cairo, 323
  Egyptian Institute in Alexandria, 322
  Egyptian Institute in Cairo, 15
  Egyptian Society for the History of Science in Cairo, 324
  Egyptian University in Cairo, 322
  First Muslim printing press in Constantinople, 318
  French Institute in Cairo, 326
  French Institute in Damascus, 326
  Geographical Society in Cairo, 322
  Institute for the History of Arabic Science in Aleppo, 328
  Maʿhad Mawlāy al-Ḥasan li'l-abḥāth in Tétouan, 326
  National Library of Egypt in Cairo, 321
  Naval School in Alexandria, 322
  School of Engineering in Cairo, 322
  School of Engineering and Artillery in Constantinople, 319
  School of Translation in Cairo, 321
  Science Heritage Center in Cairo, 328
  Society for Research into Turkish History in Ankara, 324n
  Society for the History of Mathematics (Algeria), 328
  Society for the Iranian Language in Teheran, 323
  Society for Turkish History in Ankara, 324n
  Society of Turkish Librarians, 517
  Syrian Protestant College in Beirut, 322
  Turkish Historical Society, 517
  Turkish Linguistic Society in Ankara, 323
  Turkish Society for Scientific and Technological Research, 517
  University of Abū Dīs, 514
  University of Aleppo, 328
  University of Alexandria, 324
  University of Ankara, 324, 516
  University of Bagdad, 383
  University of Cairo, 328
  University of Istanbul, 328
  University of Izmir, 427
  University of Jordan, 514
  University of Khartoum, 514
  University of Tehran, 324
  University of Tunis, 324
ARAGO, FRANÇOIS (1786–1853), 17, 20, 116, 322, 467
ARATUS (ca. 310–ca. 245), 63, 357
ARBOGAST, LOUIS FRANÇOIS ANTOINE (1759–1803), 12–13, 35, 76
ARCHIBALD, RAYMOND CLARE (1875–1955), **B**, 267–269, 271, 274, 285
ARCHIMEDES (287?–212), 13, 46, 52, 63–64, 66–68, 71, 83, 90, 92, 98–99, 120, 123, 131, 133–134, 139, 157, 172, 191, 197, 209, 218, 222–223, 228, 304, 389, 398, 404, 411, 424, 429, 441, 443, 446, 448, 459, 474–475, 502, 508–509, 529, 536, 542, 553, 560, 576–578
Archimedes Edition, 443, 529
ARCHYTAS (428–365), 47, 90, 222, 536
ARGAND, JEAN-ROBERT (1768–1822), 99
ARISTAEUS (4th century B.C.), 65
ARISTARCHUS OF SAMOS (3rd century B.C.), 63, 398, 441, 529, 560
ARISTOTLE (384–322), 83, 98, 172, 221–222, 357, 442–444
ARNETH, ARTHUR (1802–1858), **B**, 114, 116–117, 123, 387, 535
ARNOLD, IGOR' V. (1900–1948), 184
ARNOLD, VLADIMIR I. (*1937), 188
ARON, RAYMOND (*1905), 396
ARRHENIUS, SVANTE (1859–1927), 273
ARTIN, EMIL (1898–1962), 547–549

Arvesen, Ole Peder (1895–1991), **B**, 159
Āryabhaṭa I (*476 A.D.), 311
Ascher, Marcia (*1935), 282
Assanis, Spyridon (1756–1833), 225
Aubrac, Lucie (*1912), 395
Auger, Léon (1886–1964), 35, 37
Ausejo, Elena (*1961), 231
Austria
Benedictine Monastery of St. Peter in Salzburg, 389
Collegium poetarum in Vienna, 215
Deutsche Gesellschaft der Naturforscher und Ärzte in Vienna (1894), 217
First Viennese Mathematical School, 215
Kommission für Geschichte der Mathematik, Naturwissenschaften und Medizin, 219
Österreichische Akademie der Wissenschaften in Vienna, 216–219, 391
Österreichische Gesellschaft für Wissenschaftsgeschichte in Vienna, 219
Schola Sancti Petri in Salzburg, 213
Second Viennese Mathematical School, 215
University of Budapest, Hungary, 500
University of Innsbruck, 216
University of Vienna, 214, 218, 500, 505, 530, 567
Wiener Kreis (Vienna Circle), 219, 434
Autolycus of Pitane (ca. 330 B.C.), 64, 120, 404, 487
Ayyangar, A. A. Krishnaswami (1892–1953), 315
Azevedo, Fernando de (1894–1974), 254

Babbage, Charles (1792–1871), 163–164, 169, 407
Babini, José (1897–1983), **B**, 252–253, 484, 500
Bachelard, Gaston (1884–1962), **B**, 32, 35–36, 38, 43, 385
Bachet de Méziriac, Claude Gaspard (1581–1638), 5, 48
Bachmann, Paul (1837–1920), **B**, 128
Bacon, Francis (1561–1626), 167
Bag, Amulya Kumar (*1937), 315
Bagheri, Mohammed (*1951), 317n
Bahā᾽ al-Dīn al-ʿĀmilī (1547–1622), 115, 123, 320, 495
Baillet, Adrien (1649–1706), 5
Bailly, Jean Sylvain (1736–1793), 15, 310
Bailly, Francis (1774–1844), 168
Baire, Louis René (1874–1932), 413
Bakhmutskaya, Esfir' Ya. (1916–1972), 185
Balada, František (1902–1961), 210
Balbín, Bohuslav (1621–1688), 206n
Balbin, Valentin (1851–1901), 252
Balboa, Vasco Nuñez de (ca. 1475–1519), 250
Baldi, Bernardino (1533–1617), **B**, 63–64, 81
Baldini, Ugo (*1944), 400
Ball, Walter William Rouse, *see* Rouse Ball, Walter William
Banfi, Antonio (1886–1957), 86
Banū Mūsā (ca. 875 A.D.), 121, 402
Barajas, Alberto (*1913), 259n
Baraniecki, Aleksander M. (1848–1895), 200
Barlow, Peter (1776–1862), 164
Barner, Martin (*1921), xxvii
Barrau, Johan Anthony (1873–1953), 410
Barreda, Gabino (1818–1881), 257
Barrow, Isaac (1630–1677), 167, 577
Bartolache, José Ignacio (1739–1790), 256
Bashmakova, Izabella G. (*1921), 187–188, 191, 573
Bastos, Francisco António Martins (1799–1868), 245
Battaglini, Giuseppe (1826–1894), 83, 471, 477
Baumgart, John K. (*1916), 335
Bayer, Raymond (1898–1959), 38
Bays, Séverin (1885–1972), 101
Beaugrand, Jean (ca. 1595–ca. 1640), 409
Beaulieu, Armand (*1909), 26
Becker, Friedrich (1900–1985), 135
Becker, Oskar (1889–1964), **B**, 135, 139, 150, 448, 548–549
Bečvář, Jindřich (*1947), 211
Beekman, Isaac (1588–1637), 409
Behā᾽ al-Dīn al-ʿĀmilī, *see* Bahā᾽ al-Dīn al-ʿĀmilī
Behaim, Martin (ca. 1459–1507), 244
Belgium, *see* Benelux
Bell, Eric Temple (1883–1960), **B**, 272–273
Bellavitis, Giusto (1803–1880), 425
Bellone, Enrico (*1938), 435

Bellyustin, Vsevolod K. (1865–1925), 182
Beltrami, Eugenio (1835–1900), 23, 79, 402, 433, 469, 471, 474, 477
Beltrami, Luca (1854–1933), 87
Belyĭ, Yuriĭ A. (*1925), 190, 193
Beman, Wooster Woodruff (1850–1922), 270, 524
Benelux
  Amsterdam Municipal University, 54–56, 59
  Athenaeum Illustre in Amsterdam, 530, 558
  Athénée in Luxembourg, 437
  Collège Saint-Michel in Brussels, 376
  Comité Belge d'Histoire des Sciences, 58
  Delft Technical University, 56
  Ecole Normale des Sciences in Ghent, 51
  Flemish University of Ghent, 558
  Free University Amsterdam, 55
  Genootschap der Mathematische Weetenschappen in Amsterdam, 532
  Genootschap voor de Geschiedenis der Geneeskunde, Wiskunde, Natuurwetenschappen en Techniek (Gewina) in Leiden, 57–59
  Jesuit College in Antwerp, 534
  Jesuit College in Leuven, 534
  Musée des Sciences et des Lettres in Brussels, 50
  National Committee for Logic, History and Philosophy of Science, 58
  Polytechnical School in Delft, 376, 408
  Royal Academy of Brussels, 20
  Royal Academy of Science in Amsterdam, 53, 55, 58, 412
  Royal Military Academy in Breda, 376
  Royal Military Academy in Brussels, 438, 504
  Teacher Training College in Deventer, 54
  Theological College in Leiden, 558
  University of Amsterdam, 383, 408–409, 411, 429, 460, 547–548
  University of Brussels, 51–52, 502
  University of Franeker, 49, 57, 416
  University of Ghent, 49–52, 476, 504, 515, 551
  University of Groningen, 54–56, 410, 547
  University of Leiden, 47, 49–50, 54, 363, 411–412, 417, 438, 487, 525, 530, 551, 558, 566
  University of Leuven, 46, 49–52, 436, 486, 509, 552
  University of Liège, 49, 51–52, 438, 463
  University of Louvain-la-Neuve
    Centre d'Histoire des Sciences Grecques et Byzantines, 52
  University of Louvain-la-Neuve, 104
  University of Nijmegen, 540
  University of Utrecht, 50, 54–57, 59, 412, 429, 487–488
  Zuidnederlands Genootschap voor de Geschiedenis van Geneeskunde, Wiskunde en Natuurwetenschappen, 58–59
Benjamin of Lesbos (1762–1824), 224
Bensaúde, Joaquim (1859–1952), 244
Bentley, John (ca. 1800), 310
Berëzkina, El'vira I. (*1931), 131, 138, 192
Bergson, Henri (1859–1941), 31, 460
Bernal, John D. (1901–1971), 261n, 479
Bernoulli, Daniel I (1700–1782), 10, 39, 73, 527–528, 568
Bernoulli, Jakob I (1654–1705), 10, 13, 39, 74, 99, 104, 181, 428, 448, 568
Bernoulli, Johann I (1667–1748), 10, 13, 39, 73–74, 97, 99, 103–104, 401, 428, 448, 526–527, 568–569
Bernoulli, Johann II (1710–1790), 100–101, 526, 568–569
Bernoulli, Johann III (1744–1807), 526
Bernoulli, Nicolaus I (1687–1759), 6, 111
Bernoulli Edition, 103–104, 106, 401, 428, 526–528, 562
Bernshteĭn, Sergeĭ N. (1880–1968), 186
Berr, Henri (1863–1954), **B**, 28, 33–34, 37–39, 506
Berry, Arthur (1862–1929), 172, 178
Berthollet, Claude-Louis (1748–1822), 15
Bertini, Eugenio (1846–1933), 469
Bertrand, Joseph (1822–1900), **B**, 17–18, 20, 22
Bertrand, Louis (1731–1812), 99, 520
Berzolari, Luigi (1863–1949), 81–82, 375, 472

BESPAMYATNYKH, NIKIFOR D. (1910–1987), 193
BESSARION (1403–1472), Cardinal, 214
BESSEL, FRIEDRICH WILHELM (1784–1846), 128–129
BESSEL-HAGEN, ERICH (1898–1946), **B**, 134–135, 548
BETH, HERMANUS J. E. (1880–1950), **B**, 55, 58
BETTI, ENRICO (1823–1892), 77, 82, 421, 477
BEZOLD, WILHELM VON (1837–1907), 81
BEZOUT, ETIENNE (1730–1783), 73
BHASKARA (1114–ca. 1185), 117
BHĀU, DĀJI (1821–1874), 311
BIADIEGO, GIAMBATTISTA (1850–1923), 81
BIANCANI, GIUSEPPE (1566?–1624), 65
BIANCHI, CELESTINO (1817–1885), 85
BIANCHI, LUIGI (1856–1928), 421
BIEBERBACH, LUDWIG (1886–1982), 141–142
BIELIŃSKI, JÓZEF (1848–1926), 200
BIERENS DE HAAN, DAVID (1822–1895), **B**, 53–54, 81, 333, 418
BIERMANN, KURT-R. (*1919), 144, 191
BIERNATZKI, KARL L. (1814–1899), **B**, 570
BINDER, CHRISTA (*1947), 213
BIOCHE, CHARLES (1859–1950/51), 31
BIOT, EDOUARD-CONSTANT (1803–1850), **B**, 18
BIOT, JEAN BAPTISTE (1784–1862), **B**, 17–20, 116, 363, 520
BIRKELAND, OLAF KRISTIAN (1867–1917), 383
BIRKENMAJER, ALEKSANDER (1890–1967), 200
BIRKENMAJER, LUDWIK ANTONI (1855–1929), 200, 202
BIRKHOFF, GARRETT D. (1911–1996), **B**, 260n
BIRKHOFF, GEORGE DAVID (1884–1944), 260n, 364
AL-BĪRŪNĪ, ABŪ 'L-RAYḤĀN (973–1048), 133, 323, 327
BJERKNES, CARL ANTON (1825–1903), **B**, 23, 153, 418, 534
BJØRNBO, AXEL ANTHON (1874–1911), **B**, 118, 126, 132, 157, 403
BLOCH, MARC (1886–1944), 5
BOBYNIN, VIKTOR V. (1849–1919), **B**, 180–182, 185, 196–197, 389, 418
BOCHNER, SALOMON (1899–1982)), 141
BOCKSTAELE, PAUL (*1920), 45, 52
BÖHME, JAKOB (1575–1624), 460
BOETHIUS, ANICIUS MANLIUS TORQUATUS SEVERINUS (ca. 480–524), 21, 46, 61, 75, 119–120, 123, 213, 388, 537
BOFFITO, GIUSEPPE (1869–1944), 86
BOGGIO, TOMMASO (1877–1963), 393
BOGOLYUBOV, ALEKSEĬ N. (*1911), 189–192
BOGOMOLOV, STEPAN A. (1877–1965), 182
Bohemian Countries
Academy of Sciences in Prague, 211
Bolzano Commission, 209
Czech Polytechnic in Prague, 211
Czech University in Prague, 207–209, 553
Czechoslovakian Academy of Sciences in Prague, 210, 212
Fourth International Congress of the History of Science in Prague (1937), 554
Fourth International Congress of the History of Science in Prague (1937), 209, 212
Free Association for the History of Science in Prague, 209
German University in Prague, 207
Karl-Ferdinand-Universität (1622–1882), *see German University, Czech University, University of Prague*
National Comittee for the History of Science in Prague, 209, 554
National Technical Museum in Prague, 211
Slovakian Academy of Sciences in Bratislava, 210, 212
Technical University of Prague, 209, 553
University of Brno, 210
University of Prague, 206–208, 210–211, 214, 530, 554
BOHL, PIRS G. (1865–1921), 194
BOHR, HARALD (1887–1951), 496, 498
BOIVIN DE VILLE-NEUVE, JEAN (1666–1726), 5
BOIVIN, JEAN (*1945), 451
BOLKHOVITINOV, EVGENIĬ (1767–1837), 179
BÓLYAI, FARKAS (WOLFGANG) (1775–1856), 128, 217
BÓLYAI, JÁNOS (JOHANN) (1802–1860), 23, 79, 113, 128, 217, 372, 480
BOLZA, OSKAR (1857–1942), 128
BOLZANO, BERNARD (1781–1848), 33, 208–209, 211–212, 216
Bolzano Edition, 209

Bombelli, Rafael (1526–1572/73), 70, 90, 92, 257n, 374–375, 424, 448
Bonatti, Guido (1200/1220–1298?), 369
Bonaventura, Tommaso (1675–1731), 76
Boncompagni, Baldassarre (1821–1894), **B**, 19, 25, 30, 46n, 51, 63–64, 66, 70, 75, 79–81, 85–86, 89, 118, 122, 125, 283n, 358, 389, 400, 425, 432, 470–471
Boncompagni, Ugo (1502–1585; Pope Gregory XIII from 1572), 368
Bonnycastle, John (1750–1821), 13, 164, 376
Bonola, Roberto (1874–1911), **B**, 84, 89, 423, 471
Bopp, Karl (1877–1934), **B**, 130
Borel, Emile (1871–1956), 32, 413
Borel, Jean, *see* Borelli, Giovanni Alfonso
Borgato, Maria Teresa (*1950), 71
Borghetti, Smeraldo (fl. late 16th century), 80
Borrel, Jean, *see* Buteo, Jean
Borrelli, Giovanni Alfonso (1608–1679), 64–65
Borromeo, Carlo (1538–1584; Cardinal), 357
Borsuk, Karol (1905–1982), 201
Bortolotti, Ettore (1866–1947), **B**, 61, 67, 84–85, 88–93, 95, 137, 351, 423, 467, 469, 471
Bos, Henk J. M. (*1940), 56, 336
Bosmans, Henri (1852–1928), **B**, 46n, 52, 58, 390
Bosscha, Johannes (1831–1911), **B**, 54
Bossut, Charles (1730–1814), **B**, 9, 13, 23, 70–72, 114, 164
Bottazzini, Umberto (*1947), 61
Bourbaki, N(icolas) (20th century), **B**, 39–43, 74, 332, 336, 395, 410, 561
Boutroux, Emile (1845–1921), 31–32, 379, 384, 506, 535
Boutroux, Pierre (1880–1922), **B**, 31–32, 42, 384
Bowditch, Nathaniel (1773–1838), **B**, 264
Boyer, Carl B. (1906–1976), **B**, 277–279, 281, 416, 491
Bradwardine, Thomas (1290?–1349), 215, 402, 448
Brahe, Tycho (1546–1601), 17, 47, 207, 257n
Brahmā (Indian god), 307
Brand, Eugène (1861–1936), 51
Braudel, Fernand (1902–1985), 38
Braunmühl, Anton von (1853–1908), **B**, 124–127, 130–133, 136, 365, 389–390, 565
Bréhier, Emile (1876–1952), 395
Brendel, Martin (1862–1939), 128–129
Brentjes, Sonja (*1951), 15, 317
Bresciani, Benedetto (1658–1740), 76
Bretschneider, Carl Anton (1808–1878), **B**, 535
Brewster, David (1781–1868), 166, 170, 407
Brianchon, Charles Julien (1783–1864), 408
Briggs, Henry (1561–1630), 47, 163, 174
Brioschi, Francesco (1824–1897), 77–79, 82, 314, 402
British Isles, *see* Great Britain
Brocard, Henri (1845–1922), 24, 418
Broch, Ole Jacob (1818–1889), **B**, 154, 534
Brouncker, William (Lord, 1620?–1684), 448
Brouwer, Luitzen Egbertus Jan (1881–1966), 33, 429, 547
Brown, Ernest William (1866–1938), 274
Brown, Jack (around 1900), 174
Brożek, Jan (1585–1652), 199
Bruhat, François (*1929), 41
Bruins, Evert M. (1909–1990), **B**, 56–57
Brun, Viggo (1885–1978), **B**, 153, 159
Brunet, Pierre (1893–1950), 34–35, 37, 39
Bruno, Giordano (1548–1600), 257, 483
Brunschvicg, Léon (1869–1944), **B**, 31–32, 42, 395, 460, 493
Bruschi, Angelo (1858–1941), 426
Bryson (ca. 410 B.C.), 222
Bubnov, Nikolaĭ M. (1858–1943), **B**, 118, 120, 123, 180, 388
Buchan, Earl of (1742–1829), 163
Bueno, Cosme (1711–1798), 251
Bürger, Herman Carel (1893–1965), 132
Bürgi, Jost (1552–1632), 104, 568
Buffon, George Louis Leclerc Comte de (1707–1788), 11, 253
Bukharin, Nikolaĭ I. (1888–1938), 183
Bulmer Thomas, Ivor, *see* Thomas, Ivor
Bunyakovskiĭ, Viktor Ya. (1804–1889), 179
Burckhardt, Fritz (1830–1913), 101
Burckhardt, Johann Jakob (*1903), 104–105
Burgess, Ebenezer (1805–1870), 311

BURNET, JOHN (1863–1928), 171, 178
BUSARD, HUBERTUS L. L. (*1923), 56
BUTEO, JEAN (1492–1572), 4

CABRAL, PEDRO ALVARES (ca. 1467–1520?), 250
CAJORI, FLORIAN (1859–1930), **B**, 34, 84, 124, 181, 264–267, 269, 272–274, 281, 389, 542
CALANDRINI, JEAN-LOUIS (1703–1758), 99
CALVIN, JOHANNES (1509–1564), 98
CAMARERO, ERNESTO GARCÍA (20th century), 507
CAMPANUS, JOHANNES, of Novara (13th century), 4, 63
CAMPBELL, LEWIS (1830–1908), 169
CAMPEDELLI, LUIGI (1903–1978), 423
Canada
  Acadia University in Wolfville, Nova Scotia, 285, 287, 353
  Canadian Society for the History and Philosophy of Mathematics (CSHPM), 286–287
  Laval University in Québec, Québec, 287
  McMaster University in Hamilton, Ontario, 287
  Mount Allison Ladies College in Sackville, New Brunswick, 285, 353
  Ontario College of Education in Toronto, 285
  Ontario Education Communication Authority in Toronto, 480
  Simon Fraser University in Burnaby, British Columbia, 287
  Université de Québec in Montréal, Québec, 15n, 287
  University of Toronto, 260n, 286–287, 480–481
  York University in Downsview, Ontario, 287
CĀṆAKYA, *see* KAUṬILYA
CANDRAGUPTA MAURYA (Indian king about 300 B.C.), 308
CANGUILHEM, GEORGES (1904–1995), 36, 396
CANTOR, GEORG (1845–1918), 32–33, 142, 395, 418, 480, 482, 536
CANTOR, MORITZ BENEDIKT (1829–1920), **B**, 22, 25, 51, 81–82, 109–110, 116, 120–124, 127, 136, 149, 155–156, 169–171, 180, 213, 266, 283, 314, 368, 371–374, 386, 402, 418–419, 448, 491, 533, 536, 541, 553–555, 577
CARANDINOS, IOANNIS (1784–1834), 225–226
CARATHÉODORY, CONSTANTIN (1873–1950), 228, 456
CARDANO, GIROLAMO (1501–1576), 69–70, 73, 83, 90, 94, 257n, 374–375, 388, 391, 400, 438, 500
CÁRDENAS, FRANCISCO (1898–1969), 258
CÁRDENAS, LAZARO (1895–1970; President of Mexico 1934–1940), 260n
CARLI, ALARICO (1824–1900), 86
CARLYLE, THOMAS (1795–1881), 314
CARNOT, LAZARE NICOLAS M. (1753–1823), 16, 22, 186, 387, 397, 571–572
CARNOT, SADI (1796–1832), 102, 252
CARRA DE VAUX, BERNARD (1867–1953), **B**, 18
CARROLL, LEWIS, *see* DODGSON, CHARLES L.
CARRUCCIO, ETTORE (1908–1980), **B**, 92–95, 375, 424
CARSLAW, HORATIO S. (1870–1954), 372
CARTAN, ELIE (1869–1951), 41, 438
CARTAN, HENRI (*1904), 395
Cartan Edition, 41
CARUS, PAUL (1852–1919), 270
CARUSI, ENRICO (1878–1945), 86, 426
CARVALHO E MELO, SEBASTIÃO JOSÉ DE (1699–1782; Marquis of Pombal), 239–240, 250
CASORATI, FELICE (1835–1890), 78–79, 82, 469
CASPAR, MAX (1880–1956), **B**, 446
CASSINA, UGO (1897–1964), **B**, 82, 93–94
CASSINI, JEAN-DOMINIQUE (1625–1712), 15, 68
CASSIRER, ERNST (1874–1945), 395, 489
CASTELLI, BENEDETTO (1577–1644), 66–68
CASTELNUOVO, GUIDO (1865–1952), 81, 84, 92–93, 392, 421–424
CASTRO, FRANCISCO M. DE OLIVEIRA (1902–1993), 254
CASTRO FREIRE, FRANCISCO DE (1811–1884), **B**, 239–246
CATALDI, PIETRO ANTONIO (1548–1626), 90, 374–375
CAUCHY, AUGUSTIN LOUIS (1789–1857), 24, 39, 78, 181, 189, 216, 226, 314, 362, 374, 413, 465
Cauchy Edition, 24
CAUSSIN DE PERCEVAL, JEAN-JACQUES (1759–1835), 15

CAVAILLÈS, JEAN (1903–1944), **B**, 32–33, 385
CAVALIERI, BONAVENTURA (1598–1647), 68, 76–77, 90, 186, 375
CAVERNI, RAFFAELLO (1837–1900), 87
CAYLEY, ARTHUR (1821–1895), 314, 542
CECCO D'ASCOLI (1269–1327), 436
CELTIS, KONRAD (†1508), 215
CERRUTI, VALENTINO (1850–1909), 374, 477
CERVIÑO, PEDRO (†1816), 251
CESI, FEDERICO (1585–1630), 369
CEULEN, LUDOLPH VAN (1540–1610), 47
CEVA, TOMMASO (1649–1736), 64
CHACE, ARNOLD BUFFUM (1845–1932), **B**, 138, 268–269, 281, 354
CHAĬKOVSKIĬ, NIKOLAĬ A. (1887–1970), 183
CHAMBERS, EPHRAIM (1680?–1740), 163
CHARLES III (1718–1788; King of Spain from 1759), 234, 250
CHASLES, MICHEL (1793–1880), **B**, 17–23, 27, 42, 75, 88, 120, 123, 156, 267, 362, 370, 387, 437, 471, 491, 503, 520, 535, 574, 577
CHEBOTARËV, NIKOLAĬ G. (1894–1947), 183
CHEBYSHEV, PAFNUTIĬ L. (1821–1894), 181, 193–194, 196, 476
CHEIKHO, LOUIS (1889–1927), 327
CHEN JIXIN (late 18th century), 303
CHENG DAWEI (1533–1606), 297–298
CHESEAUX, JEAN-PHILIPPE LOYS DE (1718–1751), 99
CHILD, JAMES MARC (1871–1960), 170, 446
CHILDE, GORDON (1892–1957), 261n
China
- *Beijing University*, 306
- *Central University*, 504
- *Chinese Academy of Sciences (Academia Sinica)* in Beijing, 305–306, 465, 504
- *Chinese Astronomical Society*, 504
- *Chinese Mathematical Society*, 306, 504
- *Chinese Science Society*, 504
- *Chinese Society for Academic Studies and the Arts*, 504
- *Chinese Society for History of Mathematics*, 306
- *Chinese Society for the History of Science and Technology*, 306
- *Dalian University*, 306
- *Hangzhou University*, 306
- *Inner Mongolia Normal University*, 306
- *Institute for the History of Natural Sciences* in Beijing, 305, 465
- *Nankai University*, 504
- *Qian-Jia School*, 297, 300, 302–303
- *Qufu University*, 306
- *Shanghai University*, 306
- *Suzhou Industrial College*, 504
- *Suzhou Railway School*, 504
- *Tang Imperial College*, 301–302
- *Tangshan Railway and Mining College*, 464
- *Tianjin University*, 306
- *University of Hong Kong*, 544
- *University of Science and Technology* in Beijing, 305
- *Wuhan University*, 306
- *Xi'an University*, 306
- *Xuzhou University*, 306
- *Zhejiang University*, 504

CHIÒ, FELICE (1813–1871), 432
CHISINI, OSCAR (1889–1967), 84, 423
CHRISTENSEN, RICHARD (†1876), 364
CHRISTENSEN, SOPHUS ANDREAS (1861–1943), **B**, 153
CHUAQUI, ROLANDO (1935–1994), 254
CHUTHAN HSITA, *see* HSITA, CHUTHAN
CISCAR, GABRIEL (1759–1829), 427
CLAEYS, ARTHUR (1875–1949), 51
CLAGETT, MARSHALL (*1916), 282
CLAIRAUT, ALEXIS CLAUDE (1713–1765), 70
CLAVIUS, CHRISTOPH (1537–1612), 47, 65, 299, 357
CLAVUS, CLAUDIUS (fl. 1426), 366
CLEBSCH, ALFRED (1833–1872), 81, 455, 476
CLEOMEDES (1st century A.D.), 404
CLEOPATRA (69–30 B.C.; Egyptian Queen), 397
CLERSELIER, CLAUDE (1614–1686), 5
CLIO (Greek Muse of history), 307
COBO, BERNABÉ (17th century), 250
CODAZZA, GIOVANNI (1816–1877), 370
COHEN, I. BERNARD (*1914), 39, 364, 461
COHN-VOSSEN, STEFAN (1902–1936), 456
COIGNET, MICHEL (1549?–1623), 376
COLEBROOKE, HENRY THOMAS (1765–1837), **B**, 16, 166, 178, 310, 397
COLLALTO, ANTONIO (†1820), 67
COLLIMITIUS, *see* TANNSTETTER, GEORG
COLLOT D'ESCURY, HENDRIK (1773–1845), 56
COLUMBUS, CHRISTOPHER (1451–1506), 215

Comité International d'Histoire des Sciences, see International Academy of the History of Science
COMMANDINO, FEDERICO (1509–1575), 63–64, 357
COMTE, AUGUSTE (1789–1857), 14, 25, 28–30, 41, 257, 535
CONDORCET, JEAN ANTOINE CARITAT DE (1773–1791), 6–9, 14, 541
CONDUITT, JOHN (1688–1737), 162n
CONFORTO, FABIO (1909–1954), 424
CONTI, ALBERTO (1873–1940), 89, 471
COOLEY, HOLLIS R. (1899–1987), 279, 457
COOLIDGE, JULIAN LOWELL (1873–1954), **B**, 267, 272, 281
COPERNICUS, NICOLAUS (1473–1543), 12, 126, 147, 200–202, 257n, 351, 389–391, 402, 460, 476, 498, 517
Copernicus Edition, 147, 402, 428
CORREIA DA SERRA, JOSÉ FRANCISCO (1750–1823), 240
CORTESAO, ARMANDO (1891–1977), 37
COSSALI, PIETRO (1748–1815), **B**, 62, 69–72, 75, 89, 369–370
COSTA, MANOEL AMOROSO (1885–1928), 254
COSTABEL, PIERRE (1912–1989), **B**, 36–39
COTES, ROGER (1682–1716), 167, 448
COURANT, RICHARD (1888–1972), 127, 141, 279, 456, 496–497
COUSIN, VICTOR (1792–1867), 26, 538
COUTURAT, LOUIS (1868–1914), **B**, 29, 32, 82, 545
CRAMER, GABRIEL (1704–1752), 73, 97, 99–100
CREMONA, LUIGI (1830–1903), 77, 314, 402, 421, 474, 477
CROCE, BENEDETTO (1866–1952), 422
CROMBIE, ALISTAIR CAMERON (1915–1996), 177
CROUSAZ, JEAN-PIERRE DE (1663–1750), 99
CUESTA DUTARI, NORBERTO (1907–1989), 236n
CURTZE, MAXIMILIAN (1837–1903), **B**, 77, 81, 120–121, 124, 130, 366, 371, 388–390, 436, 556
CUSANUS, NICOLAUS, *see* NICHOLAS OF CUSA
CYRIL V (†1775; Patriarch of Constantinople), 224
Czechoslovakia, *see* Bohemian Countries
CZUBER, EMANUEL (1851–1925), 217
CZWALINA, ARTHUR (1884–1964), **B**, 139
D'ALEMBERT, JEAN BAPTISTE LE ROND (1717–1783), 8–9, 22, 68, 70, 73–74, 240, 251, 362, 376, 470, 489, 531, 541
D'AMBROSIO, UBIRATAN (*1932), 249
D'ETAPLES, JACQUES LEFÈVRE, *see* LEFÈVRE D'ETAPLES, JACQUES
D'HERBELOT, BARTHÉLÉMI (1625–1695), 14
D'OCAGNE, MAURICE, *see* OCAGNE, MAURICE D'
D'ORESME, NICOLE, *see* NICOLE ORESME
D'OVIDIO, ENRICO (1843–1933), 469–470
DA CUNHA, JOSÉ ANASTÁCIO (1744–1787), 241–242, 245–246, 540
DA CUNHA, PEDRO JOSÉ (1867–1945), **B**, 239, 241, 243–244, 246
DA MONTEFELTRO, FEDERICO (1422–1482; Duke of Urbino), 358
DA SILVA, DANIEL AUGUSTO (1814–1878), 242–244, 540
DA SILVA, LUCIANO PEREIRA (1864–1926), 244
DAHLBO, J. (19th century), 153
DAHLIN, ERNST MAURITZ (1843–1929), **B**, 153
DAI ZHEN (1724–1777), 300–303
AL-DAMARDĀSH, AḤMAD SAʿĪD (20th century), 324, 327
DAMOISEAU, MARIE-CHARLES-THÉODORE (1768–1846), 17
DAMPIER-WHETHAM, WILLIAM C. D. (1867–1952), 175
DANIËLS, CAREL EDOUARD (1839–1920), 57
DANNEMANN, FRIEDRICH (1859–1936), 227
DARBOUX, GASTON (1842–1917), 23, 32, 314, 373, 486
DARNTON, ROBERT ((*1939), 489
DAS, SUKUMAR RANJAN (1st half of 20th century), 315
DATTA, BIBHUTIBHUSAN (1888–1958), **B**, 314–315, 523
DAUBEN, JOSEPH W. (*1944), xxi, 263, 287, 297, 329, 337
DAVID, FLORENCE NIGHTINGALE (*1909), 175
DAVIET DE FONCENEX, FRANÇOIS (1733/34–1799), 79, 433
DAVIS, SAMUEL (†1819), 310
DE BEAUNE, FLORIMOND (1601–1652), 401, 409
DE GUA DE MALVES, JEAN PAUL, *see* GUA DE MALVES, JEAN PAUL DE
DE LA CAILLE, NICOLAS (1713–1762), 225
DE LA PEÑA, ÁNGEL (1837–1906), 258

DE L'HÔPITAL, *see* L'HÔSPITAL, GUILLAUME F. A. DE
DE MARCHI, LUIGI (1857–1936), 418
DE MELLO, DON FRANCISCO (1490–1536), 240
DE MORGAN, AUGUSTUS (1806–1871), **B**, 167–170, 178, 304, 467
DE PAOLIS, RICCARDO (1854–1892), 421
DE SACY, ANTOINE ISAAC SILVESTRE, *see* SILVESTRE DE SACY, ANTOINE ISAAC
DE SLUSE, RENÉ FRANÇOIS, *see* SLUSE, RENÉ FRANÇOIS DE
DE TRAVESEDO, FRANCISCO (1786–1861), 235
DE VRIES, HENDRIK (1867–1954), **B**, 54, 547–548
DE WAARD, CORNELIS (1879–1963), **B**, 26, 54
DEBYE, PETER (1884–1966), 456, 529
DEDEKIND, RICHARD (1831–1916), 92, 270, 395, 413, 424, 482, 547
DEDRON, PIERRE (†1970), 38
DEE, JOHN (1527–1608), 63
DEHN, MAX (1878–1952), **B**, 136, 141, 275–276, 445
DEL MONTE, GUIDUBALDO, *see* MONTE, GUIDUBALDO DEL
DELAMBRE, JEAN-BAPTISTE J. (1749–1822), **B**, 15–16, 21n, 42, 72, 487, 490, 520, 537
DELISLE, JOSEPH-NICOLAS (1688–1768), 15
DELLA FAILLE, JEAN CHARLES (1597–1652), 376
DELLA ROVERE, FRANCESCO MARIA II (1549–1631; DUKE OF URBINO), 358
DELONE, BORIS N. (1890–1980), 187
DELORME, SUZANNE (*1913), 37
DELUC, JEAN-ANDRÉ (1763–1847), 99
DEMIDOV, SERGEI S. (*[illegible]), 179
DEMOCRITUS OF ABDERA (late 5th century B.C.), 90
DENJOY, ARNAUD (1884–1974), 413
Denmark, *see* Scandinavia
DEPMAN, IVAN YA. (1885–1970), **B**, 190, 192–194
DESANTI, JEAN-TOUSSAINT (*1914), 32
DESARGUES, GIRARD (1591–1661), 21–22, 397, 408–409, 503
Desargues Edition, 503
DESBOVES, ADOLPHE (1818–1888), 24
DESCARTES, RENÉ (1596–1650), 5, 7, 10, 13, 21, 26, 31, 42, 48, 72, 122, 136, 162, 176, 186, 294, 303, 351, 401, 409, 448, 452, 460, 473, 486, 489–491, 537–538, 551, 572, 577
Descartes Edition, 26, 401, 537–538
DESCHALES, CLAUDE FRANÇOIS MILLIET (1621–1678), 5, 66, 112
DEVAKṚṢṆA, PANDIT (19th century), 312
ḌHUṆḌHIRĀJA MIŚRA (19th century), 312
DIANNI, JADWIGA (1886–1981), 201
DÍAZ, GUSTAVO (1911–1979; President of Mexico 1964–1970), 261
DÍAZ, PORFIRIO (1830–1915), 258
DICK, AUGUSTE (1910–1993), 218
DICKSTEIN, SAMUEL (1851–1939), **B**, 200, 202
DIDEROT, DENIS (1713–1784), 489
DIDYMUS OF ALEXANDRIA (1st century B.C.), 52
DIELS, HERMANN (1848–1922), 511, 537
DIEPGEN, PAUL (1878–1966), 132
DIEUDONNÉ, JEAN (1906–1992), **B**, 40–41, 378, 413
DÍEZ, JUAN (1480–1549), 269
DIHKHUDĀ, ʿALĪ AKBAR (1879–1955), 323
DIJKSTERHUIS, EDUARD JAN (1892–1965), **B**, 55–56, 58–59, 362
DIKSIT, *see* DĪKṢITA
DĪKṢITA, SAṄKARA BĀLAKṚṢṆA (1853–1898), 313
DING FUBAO (1874–1952), 304
DINGLER, HUGO (1881–1954), 137, 149
DINI, ULISSE (1845–1918), 421, 484
DIOGENES LAERTIUS (3rd century A.D.), 64
DIONÍSIO, JOSÉ JOAQUIM (1924–1999), 246
DIOPHANTUS OF ALEXANDRIA (ca. 250 A.D.), 4–5, 25–26, 30, 46–48, 52, 69, 98, 115–117, 119, 139, 171, 188, 191, 197, 214, 223, 228, 245, 404, 441–442, 529, 537, 553, 563
Diophantus Edition, 537
DIRICHLET, PETER GUSTAV LEJEUNE (1805–1859), 33, 186–188, 387, 482, 488, 567
DJEBBAR, AHMED (*1941), 323
DOBROVOL'SKIĬ, VYACHESLAV A. (*1919), 189–190
DOBRZYCKI, STANISŁAW (1905–1989), 201
DODGSON, CHARLES L. (alias: LEWIS CARROLL, 1832–1892), 450
DODT VAN FLENSBURG, JOHANNES JACOBUS (1800–1847), **B**, 53
DOLD-SAMPLONIUS, YVONNE (*1937), 337
DOM MANUEL (1469–1521; King of Portugal from 1495), 250

DOMINICUS DE CLAVASIO (14th century), 403
DOMORADZKI, STANISŁAW (*1958), 199, 201
DONDER, THÉOPHILE DE (1872–1957), 51, 502
DOROFEEVA, ALLA V. (*1935), 187–188
DOUMER, PAUL (1857–1932; President of the French Republic from 1931), 34
DRAKE, STILLMAN (1910–1993), 64
DROSTE, JOHANNES (1886–1963), 487
DROYSEN, JOHANN FRIEDRICH (1770–1814), **B**, 152
DU BOIS REYMOND, PAUL (1831–1889), 395, 538
DU HAMEL, JEAN-BAPTISTE (1623–1706), 7
DU SHIRAN (*1929), 305, 465
DUBBEY, JOHN (*1934), 178
DUDA, ROMAN (*1935), 201
DUERER, ALBRECHT (1471–1528), 391, 448
DUGAC, PIERRE (1926–2000) **B**, 41
DUGAS, RENÉ (1897–1957), **B**, 35, 37
DUHEM, PIERRE (1861–1916), **B**, 29–30, 37, 42, 87, 410, 477
DULONG, PIERRE LOUIS (1785–1838), 116
DUMAS, JEAN-BAPTISTE-ANDRÉ (1800–1884), 22, 116
DUNCAN, A. M. (1926–1999), 352
DUNNINGTON, GUY WALDO (1906–1974), **B**, 129
DUPIN, CHARLES (1784–1873), 225–226
DVIVEDI, SUDHAKARA (1855–1910/11), 311–313
DYCK, CORNELIUS VAN (1818–1895), 322
DYCK, WALTHER VON (1856–1934), 446, 564

ECHEGARAY, JOSÉ (1833–1916), **B**, 232, 233n, 236
EDLESTON, JOSEPH (1816–1895), 167
EECKE, PAUL VER, *see* VER EECKE, PAUL
EGOROV, DMITRIĬ F. (1869–1931), 184, 194, 559, 571
Egypt, *see* Arab Countries
EHRESMANN, CHARLES (1905–1979), 395–396
EIBE, THYRA (1866–1955), **B**, 157
EINSTEIN, ALBERT (1879–1955), 93, 173
EISELE, CAROLYN (1902–2000), **B**, 264, 282
EISENLOHR, AUGUST (1832–1902), **B**, 138
EISENSTEIN, GOTTHOLD (1823–1852), 511
EKAMA, CORNELIS (1773–1826), **B**, 49
ELFVING, GUSTAV (1908–1984), 153
ELIA MISRACHI, *see* MISRACHI, ELIA
ENCKE, JOHANN FRANZ (1791–1865), 567
END, WILHELM (1864–1922), 126
ENDŌ TOSHISADA (1843–1915), 289–290, 485
ENESTRÖM, GUSTAF HJALMAR (1852–1923), **B**, 24, 29, 51, 103, 124–125, 129–130, 137, 146, 154–157, 159, 180, 208, 243–244, 283, 388–390, 455, 471, 554
ENGEL, FRIEDRICH (1861–1941), **B**, 79, 127
ENRIQUES, FEDERIGO (1871–1946), **B**, 35, 61, 81–85, 90–95, 351, 372, 392, 435, 438, 545–546
EPAPHRODITUS (1st century A.D.), 123, 389
EPPLE, MORITZ (*1960), 143, 337
EPSTEIN, PAUL (1871–1939), 136
ERASMUS OF ROTTERDAM (ca. 1466–1536), 98
ERATOSTHENES (3rd century B.C.), 384
ERSCH, JOHANN SAMUEL (1766–1828), 119
ESCHERICH, GUSTAV VON (1849-1935), 217–218
ESCHMANN, JOHANNES (1808–1852), 567
ETTINGHAUSEN, ANDREAS FREIHERR VON (1796–1878), 567
EUCLID (ca. 365–ca. 300), 4, 13, 21–22, 30, 46–48, 52, 55, 61, 63–65, 79, 83, 92, 98, 118–119, 123, 126, 131–132, 139–140, 144, 147, 156–157, 163, 171–172, 179–180, 191, 197, 213–214, 222–223, 228, 233, 253, 257n, 294, 297–299, 304, 313, 320, 366, 388–389, 397–398, 402, 410–411, 415, 423, 429, 433, 441–443, 452, 459, 493, 502, 508–509, 529, 536–537, 545, 549, 552, 560, 570, 576–577
Euclid Edition, 536
EUDEMUS (4th century B.C.), 222
EUDOXOS (4th century B.C.), 90, 446, 536
EULER, LEONHARD (1707–1783), 39, 73–74, 90, 97, 101–106, 126–127, 129, 144–145, 163, 183, 186, 189–191, 194, 196, 228, 252, 254, 282, 292, 351–352, 373, 417, 428, 441, 446, 448–449, 470, 510–512, 522–523, 526–528, 544, 560, 568–569, 572
Euler Edition, 102–104, 106, 127, 129–130, 352, 418, 428, 446, 510–512, 544, 569, 572
EUPHORBUS (4th century B.C.), 358
EUTOCIUS (6th century A.D.), 63, 223
EVES, HOWARD (*1911), 277–278

FABBRONI, ANGELO (1732–1803), 68

FABER, GEORG (1877–1966), 446
FAGNANO, GIULIO CARLO (1682–1766), 66, 82, 374, 473
Fagnano Edition, 374, 473
FAILLE, JEAN CHARLES DELLA, *see* DELLA FAILLE, JEAN CHARLES
FAIRON, JOSEPH (1863–1925), 51
FANTUZZI, GIOVANNI (1718–1799), 436
AL-FARGHĀNĪ, ABŪ 'L-ʿABBĀS AḤMAD IBN MUḤAMMAD (9th century), 47
FARREN, EDWARD JAMES (19th. cent.), 168
FARRINGTON, BENJAMIN (1891–1974), 175, 178
FATIO DE DUILLIER, NICOLAS (1664–1753), 99
FAUVEL, JOHN (1947-2001), 161, 284–285, 335
FAVARO, ANTONIO (1847–1922), **B**, 77, 80–81, 85–88, 371, 418, 477
FEBVRE, LUCIEN (1878–1956), 35, 43, 506
FEHR, HENRI (1870–1954), 24
FELLMANN, EMIL A. (*1927), 103–104, 191
FENG ZHENG (19th century), 304
FERDINAND MAXIMILIAN JOSEPH (1832–1867; Emperor of Mexico from 1863), 257
FERDINAND VI (1713–1759; King of Spain from 1746), 234
FERGOLA, NICOLÒ (NICOLA) (1753–1824), 88, 471
FERMAT, PIERRE DE (1601–1665), 5, 10, 13, 24, 42, 54, 76, 90, 102, 119, 139, 294, 409, 441, 448, 452, 473, 476, 491, 537, 560, 576
Fermat Edition, 26, 54, 76, 537–538
FERMI, ENRICO (1901–1954), 426
FERNÁNDEZ DE NAVARRETE, MARTÍN (1765–1844), **B**, 234
FERRARI, LUDOVICO (1522–1565), 73, 375, 436, 438
FERRO, SCIPIONE DAL (1465–1526), 90, 374, 436
FETTWEIS, EWALD (1881–1967), **B**, 139
FIBONACCI, *see* LEONARDO OF PISA
FIELD, JUDITH V. (*1943), 352
FIENUS, THOMAS (fl. 1619), 406
Finland, *see* Scandinavia
FINSTERWALDER, SEBASTIAN (1862–1951), 564
FISHER, RONALD A. (1890–1962), 479
FLAMSTEED, JOHN (1646–1719), 167–168
FLATT, ROBERT (1863–1955), 102
FLAVIUS JOSEPHUS (37/39–after 93), 47
FLECKENSTEIN, JOACHIM OTTO (1914–1980), **B**, 104, 106
FOLKERTS, MENSO (*1943), 117, 145–147, 337n
FOLQUE, PHILLIPE (1800–1874), 242
FOLTA, JAROSLAV (*1933), 211
FONCENEX, FRANÇOIS DAVIET DE, *see* DAVIET DE FONCENEX, FRANÇOIS
FONTAINE DES BERTINS, ALEXIS (1705-1771), 71
FONTANA, GAETANO (1645–1719), 64
FONTANA, GREGORIO (1735–1803), 70–71, 78
FONTANA, MARIANO (1746–1808), 71
FONTENELLE, BERNARD LE BOVIER (1657–1757), 6–8, 10, 162n
FORBES, ROBERT JAMES (1900–1973), 412
FORSYTH, ANDREW RUSSELL (1858-1942), 172n
FORTI, UMBERTO (*1901), 92, 374
FOUCAULT, LÉON (1819–1868), 437
FOUCHY, JEAN-PAUL GRANDJEAN DE (1744–1776), 8
FOURCY, AMBROISE (1748–1842), 11
FOURIER, JEAN BAPTISTE JOSEPH (1768–1830), 15, 293, 466, 490
FRAENKEL, ABRAHAM ADOLF (1891–1965), 128, 274, 395
FRAJESE, ATTILIO (1902–1986), **B**, 86, 92–93, 424
France
  Académie des Sciences in Paris, 7, 15, 17–29, 41–42, 76, 362, 376, 384, 387, 401, 405, 410, 413–414, 422, 466, 490, 499, 520, 535
  Académie Française in Paris, 8, 323, 363
  Académie Internationale d'Histoire des Sciences (AIHS), *see* *International Academy of the History of Science* (under I)
  Académie royal des Sciences, *see* *Académie des Sciences* in Paris
  Association des Collaborateurs de Nicolas Bourbaki, 379
  Bibliothèque Nationale in Paris, 16, 19, 467
  Bureau of Longitudes in Paris, 16, 42, 520
  Center for Research on the History of Science and Technology in Paris, 191, 401
  Center for Research on the History of Science and Technology in Paris, 38–39, 461

Center for the History of Arabic Science and Philosophy in Paris, 39
Centre International de Synthèse (CIS) in Paris, 33–34, 38, 361, 493
Centre Koyré, *see Center for Research on the History of Science and Technology* in Paris
Centre National de la Recherche Scientifique (CNRS) in Paris, 38, 41, 509
Collège de France in Paris, 15, 18, 29, 31, 34, 363, 380, 401, 466, 535
Collège de Lyon, 463
Collège du Plessis in Paris, 405
Collège Henri IV in Paris, 520
Collège Royal in Paris, 4, 463, 489
Collège Saint-Louis in Paris, 520
Comité International d'Histoire des Sciences, *see International Academy of the History of Science* (under I)
Commission on Epistemology and History of Mathematics of IREM, 41
Confédération Générale du Travail (CGT) in Paris, 451
Dépôt d'Artillerie in Paris, 540
Ecole d'Application d'Artillerie in Metz, 478
Ecole d'Application des Manufactures de l'État in Paris, 535
Ecole d'Etat-Major in Paris, 503
Ecole des Chartes in Paris, 466–467
Ecole des Langues Orientales Vivantes in Paris, 519
Ecole du Génie in Mézières, 376
Ecole Militaire in Paris, 394
Ecole Normale des Instituteurs in Paris, 451
Ecole Normale Supérieure (ENS) in Paris, 24, 30–32, 379, 385, 394–395, 401, 409, 414, 450, 486, 561
Ecole Polytechnique in Paris, 11, 13–14, 16, 20, 24, 30, 226, 363, 396, 414, 478, 499, 502–503, 519, 534–535, 540
Ecole Pratique des Hautes Etudes (EPHE) in Paris, 38, 451, 460
Encyclopedias, 6–9, 13–15, 35, 42, 68, 376, 380, 463, 487–491, 506
Faculty of Sciences in Paris, 12
French Committee of Historians of Sciences, 34
French Corps of Civil Engineers, 25
French Mathematical Society, 41, 227
French National Committee of Mathematics, 41
Institut Catholique de Paris, 391
Institut de France in Paris, 363, 399, 405, 463, 490, 541
  Leibniz Committee, 32
Institut des Hautes Etudes Scientifiques in Bures-sur-Yvette, 410
Institute Henri Poincaré in Paris
  Seminar for History of Mathematics, 37, 41, 413
Institute Supérieur Ouvrier in Paris, 451
Institutes for Research on Mathematics Education (IREM), 41
International Congress of Comparative History in Paris (1900), 28
International Congress of Mathematicians in Paris (1900), 24, 82, 389
International Congress of Philosophy in Paris (1900), 538, 545
International Congress of the History of Science in Paris (1929), 34, 484, 538
International Congress of the History of Science in Paris (1968), 286, 481
International Congress on Bibliography of the Mathematical Sciences in Paris (1889), 24, 227
Jesuit College in Lyon, 509
Louvre in Paris, 376
Lycée Bonaparte in Paris, 502
Lycée de Mayence, 540
Lycée in Amiens, 395
Lycée in Bordeaux, 394
Lycée Louis-le-Grand in Paris, 394
Mazarine Library in Paris, 37
Ministry of Public Instruction, 26
Observatory in Paris, 504
Research Group for Epistemological and Historical Studies on the Exact Sciences and Scientific Institutions (REHSEIS) in Paris, 41
Séminaire Bourbaki in Paris, 377, 561

Séminaire Julia in Paris, 561
Society for the Physical and Natural Sciences in Bordeaux, 25, 535
Sorbonne, *see University of Paris*
The Worker's University, 480
Universal Exhibition in Paris (1900), 243, 246
University of Bordeaux, 22, 414, 450
University of Caen, 401
University of Dijon, 356
University of Grenoble, 24
University of Le Havre, 486
University of Lille, 414
University of Marseilles, 451
University of Montpellier, 486
University of Nancy, 409, 487
University of Nice, 410
University of Paris, 33–36, 227–228, 240, 356, 363, 384, 395–396, 401, 429–430, 436–438, 450, 460, 466, 480, 483, 486–487, 501, 506, 509, 561, 566, 571, 574
University of Paris VI, 41, 413
University of Paris VII, 41
University of Poitiers, 380
University of Rennes, 409, 414
University of Strassbourg, 395, 561
University of Toulouse, 489
Francesco V of Austria-Este (1819–1875; Duke of Modena), 508
Franchini, Pietro (1768–1837), 71–72
Franci, Raffaella (*1940), 69
Franco of Liège (ca. 1050), 537
Francœur, Louis Benjamin (1773–1849), 13, 394
François I (1494–1547; King of France from 1515), 4
Frank, Philip (1884–1966), 261n
Franke, Jan N. (1848–1918), 200
Frankl', Feliks I. (1905–1961), 189
Fraser, Craig (*1951), 285
Frechet, Maurice René (1878–1973), 35, 37, 261n
Frederick II (1194–1250; Emperor from 1220), 331
Frege, Gottlob (1848–1925), 538
Frénicle de Bessy, Bernard (ca. 1605–1675), 409
Freudenthal, Hans (1905–1990), **B**, 56, 59, 130, 384, 488
Fricke, Robert (1861–1930), 218
Friedlein, Gottfried (1828–1875), **B**, 119–120, 123, 388, 555
Frobenius, Georg (1849–1917), 526
Frobesius, Johann Nikolaus (1701–1756), **B**, 111–112
Fryde, Edmund Boleslav (*1923), 329
Frye, Richard N. (*1920), 517
Fryer, John (1839–1928), 304
Fubini, Guido (1879–1943), 434
Fueter, Rudolf (1880–1950), 105, 548
FUJISAWA Rikitaro (1861–1933), 430, 440, 485
FUJIWARA Matsusaburō (1881–1946), **B**, 291–293, 295, 440
Funk, Paul (1886–1969), 218
Furtwängler, Philipp (1869–1940), 217
Fuss, Paul Heinrich (1798–1855), 129

Gadamer, Hans-Georg (1900–2002), 548
Gaĭduk, Yuriĭ M. (1914-1993), 190
Galilei, Galileo (1564–1642), 12, 48, 51, 62, 68, 71, 74–77, 80, 85–87, 90, 92–93, 126, 257, 309, 389, 406, 410, 424–426, 429, 436, 460, 466, 474, 476–477, 480, 483
Galileo Edition, 71, 76, 85–86, 371, 374, 426
Galle, Andreas Wilhelm Gottfried (1858–1943), 128
Galluzzi, Paolo (*1942), 426
Galois, Evariste (1811–1832), 183, 186, 374, 480
Galton, Francis (1822–1911), 174
Galvani, Luigi (1737–1798), 436
Gambioli, Dionisio (1858–1941), 82, 473
GAN De (mid-4th century B.C.), 505
Gandz, Solomon (1883–1954), 497
Ganguli, Sarada Kanta (*1881), 315
Ganitanad (pseudonym), *see* Gupta, Radha Charan
Gans, David (1907–1999), 279, 457
Garbasso, Antonio (1871–1931), 426
Garbers, Karl (1898–1990), 133
García, Bacca David (1901–1992), 261n
García de Galdeano, Zoel (1846–1924), **B**, 232–233, 235–237, 506
García de Zuñiga, Eduardo (1867–1951), **B**, 253
Garciadiego, Alejandro R. (*1953), 256, 260n
Gårding, Lars (*1919), 153
Garnett, William (1850–1932), 169
Gassendi, Petrus (1592–1655), 509
Gattinara, Enrico Castelli (20th century), 30
Gatto, Romano (*1944), 70, 400
Gauss, Carl Friedrich (1777–1855), 78, 113, 127–129, 131, 135, 142,

144–145, 267, 286, 293, 314, 387, 391, 414, 424, 448, 449, 456, 480, 528
GAUSS, JOSEPH (1806–1873), 129
Gauss Edition, 127–129, 135, 456
GAUTIER, ALFRED (1793–1881), 12
GEIGY, JOHANN RUDOLF, 103
GEISER, CARL FRIEDRICH (1843–1934), 510–511
GELBART, ABE (1913–1994), 274
GELCICH, EUGEN (1854–1917), 418
GELFOND, ALEKSANDR O. (1906–1968), 186–187
GEMINUS (1st century B.C.), 576
GEMMA FRISIUS, REINER (1508–1555), 49–50, 52, 333, 417, 551
GENOCCHI, ANGELO (1817–1889), **B**, 79, 370, 470
GENTILE, GIOVANNI (1875–1944), 92, 422–424
GEORGIOS PACHYMERES (1242–ca. 1310), 223, 538
GEPPERT, HARALD ALOYSIUS (1902–1945), 128
GERARD OF CREMONA (1114?–1187), 63, 366, 369
GERBERT OF AURILLAC (ca. 940–1003; Pope SYLVESTER II from 999), 21, 120, 385, 389, 537
Gerbert Edition, 385
GERCKE, ALFRED (1860–1922), 444
GERHARDT, CARL IMMANUEL (1816–1899), **B**, 121–122, 449
GERICKE, HELMUTH (*1909), 138, 146, 557
GERLING, CHRISTIAN LUDWIG (1788–1864), 129
Germany
    Akademie der Wissenschaften in Berlin, 72, 129, 139, 145, 240, 447, 492, 531
    Akademie der Wissenschaften in Göttingen, 128
    Akademie der Wissenschaften in Heidelberg, 391
    Alexander-von-Humboldt-Forschungsstelle in Berlin, 144
    Bayerische Akademie der Wissenschaften in München, 568
    Bayerische Staatsbibliothek in München, 556
    Coppernicus-Verein für Wissenschaft und Kunst in Thorn, 402
    Deutsche Akademie der Naturforscher Leopoldina in Halle, 143, 557
    Deutsche Gesellschaft für Geschichte der Medizin und Naturwissenschaft, 109, 132
    Deutsche Mathematiker-Vereinigung, 129, 145, 148, 456
    Deutsche Morgenländische Gesellschaft, 517
    Encyclopedias, 114–116, 119–120, 150
    Fachsektion für Geschichte, Philosophie und Logik der Mathematik, 145
    Fachsektion "Geschichte der Mathematik" in der Deutschen Mathematiker-Vereinigung, 148
    Free University of Berlin, 427
    Gauss-Gesellschaft in Göttingen, 414
    Herzog August Bibliothek in Wolfenbüttel, 98n, 389
    Herzogliche Bibliothek in Gotha, 103, 526
    Institute for History of Medicine and Science in Leipzig, 143–144
    Institute for History of Medicine in Leipzig, 132–133, 146
    Institute for History of Science and Medicine in Berlin, 132–133, 143, 513
    Institute for History of Science and Technology in Berlin, 146
    Institute for History of Science in Berlin, 513
    Institute for History of Science in Frankfurt, 146
    Institute for History of Science in Hamburg, 146
    Institute for History of Science in Heidelberg, 132
    Institute for History of Science in Jena, 143
    Institute for History of Science in Munich, 146
    Institute of the History of Arabic-Islamic Sciences in Frankfurt, 147, 533
    International Congress of Mathematicians in Heidelberg (1904), 27, 473, 565
    Königliche Bibliothek in Berlin, 118, 530
    Kulturwissenschaftliche Bibliothek Warburg in Hamburg, 176
    Mathematisch-Physikalischer Salon in Dresden, 143
    Mathematische Gesellschaft der DDR, 145, 148

Mathematisches Forschungsinstitut Oberwolfach, 145, 148, 191, 447
Max Planck Institute for the History of Science in Berlin, 148
Niedersächsische Landesbibliothek in Hannover, 32, 121, 401, 447, 545
Pädagogische Hochschule in Aachen, 427
Physikalisch-medizinische Gesellschaft zu Erlangen, 533
Research Institute for History of Science in Berlin, 132
Staatsbibliothek in Berlin, 395
Technical University in Berlin, 147
Technical University in Darmstadt, 427, 446, 564
Technical University in Hannover, 493, 528
Technical University in Karlsruhe, 447, 528
Technical University in Munich, 104, 125–126, 365, 382, 427, 446, 455, 468, 564
Technical University in Munich, 439
University of Berlin, 117, 134, 227–228, 294, 353, 387, 395, 404, 429–431, 443, 445, 447, 474, 487, 510, 518, 526, 529–530, 533, 567
University of Bielefeld, 147
University of Bonn, 134–136, 358, 362, 455, 474, 519, 542, 566
University of Breslau, 356, 445, 518
University of Erfurt, 144
University of Erlangen, 122, 132, 363, 439, 455, 555, 564
University of Frankfurt (Main), 133, 136, 141, 275, 405, 445, 469, 519
University of Freiburg, 135, 293, 358, 395, 447, 475
University of Giessen, 420
University of Göttingen, 54, 113, 127–128, 134, 141, 228, 362, 384, 387, 395, 405, 430, 438, 440, 445, 450, 454–455, 460, 468, 493, 496, 505, 542, 547, 555, 561
University of Greifswald, 144, 148, 152, 402, 413, 420, 476, 540
University of Halle, 111, 362, 430, 528
University of Hamburg, 147, 395, 505, 547
University of Heidelberg, 102, 116, 123, 131, 354, 373, 387, 416, 445, 468, 474, 513, 528
University of Helmstedt, 112, 430
University of Kiel, 134–135, 363, 445, 519, 528, 542
University of Königsberg, 115, 495, 505, 528
University of Leipzig, 122, 144, 382, 420, 439, 443–444, 454–455, 467–469, 484, 530, 548, 564
University of Mainz, 147, 427
University of Marburg (Lahn), 111, 275, 404, 445, 474, 476, 505
University of Münster, 134, 356, 519
University of Munich, 136–137, 146–147, 365, 445–446, 495, 555–556, 564–565
University of Strassburg, 353, 454, 522
University of Tübingen, 112–113, 122, 439, 447, 474
University of Würzburg, 509
GERONO, CAMILLE CHRISTOPHE (1799–1892), 24, 540
GERSTINGER, HANS (1885–1971), 138
GESSNER, JOHANNES (1709–1790), 100
GEYMONAT, LUDOVICO (1908–1991), **B**, 86, 94–95, 425
GHANNŪN, ʿABDALLĀH (†1989), 323
GHERARDI, SILVESTRO (1802–1879), **B**, 77
GHERARDO DA SABBIONETA (13th century), 369
GHETALDI, MARINO (1566–1626), 65
GHIYĀTH AL-DĪN JAMSHĪD AL-KĀSHĪ, *see* AL-KĀSHĪ, JAMSHĪD B. MASʿŪD
GIBSON, GEORGE (1858–1930), 174
GIGLI, DUILIO (1878–1933), 82
GILBERT, PHILIPPE (1832–1892), **B**, 50–51
GINSBURG, JEKUTHIEL (1889–1957), **B**, 269–271, 274, 525
GIORDANI, ENRICO (19th century), 436
GIORDANI, VITALE (1633–1711), 65
GIRARD, ALBERT (1595–1632), 53, 70
GIUDICE, FRANCESCO (1855–1936), 83
GLAISHER, JAMES W. L. (1848–1928), 169
GLAREANUS (1488–1563), 98
GLEIZER, GERSH I. (1904–1967), 194
GLENNIE, JAMES (1730–1817), 163
GLIVENKO, VALERIĬ I. (1897–1940), 186
GLODEN, ALBERT (1901–1966), **B**, 53
GMUNDEN, JOHANNES VON (ca.1380–1442), 214–216
GNEDENKO, BORIS V. (1912–1995), **B**, 186–187, 192–193
GODEAUX, LUCIEN (1887–1975), **B**, 52
GÖDEL, KURT (1906–1978), 261n, 396
GOETHE, JOHANN WOLFGANG VON (1749–1832), 106

GOŁĄB, STANISŁAW (1902–1980), 201
GOLIUS, JACOB (1596–1667), **B**, 47
GOLUBEV, VLADIMIR V. (1884–1954), 187
GONÇALVES, JOSÉ VICENTE (1896–1985), 241, 246
GONSETH, FERDINAND (1890–1975), 105–106
GONZAGA, DUKE FERRANTE II (1550–1605), 357
GORDAN, PAUL (1837–1912), 555
GORTARI, ELI DE (1918–1991), 261
GOSSELIN, GUILLAUME (fl. 1577–1583), 4
GOTAMA SIDDHA, *see* HSITA, CHUTHAN
GOVI, GILBERTO (1826–1889), 418
GOVINDA VIṬṬHALA, *see* VIṬṬHALA, GOVINDA
GOW, JAMES (1854–1923), 171, 178
GRAEF, CARLOS (1911–1988), 259
GRAEFE, FRIEDRICH (1855–1918), 102
GRÄFFE, KARL HEINRICH (1799–1873), 567
GRAF, JOHANN HEINRICH (1852–1918), 101, 104, 106
GRANDI, LUIGI GUIDO (1671–1742), 68, 76, 89
GRANT, ROBERT (1814–1892), 167, 178
GRASSMANN, HERMANN GÜNTHER (1809–1877), 127, 420, 457
Grassmann Edition, 127, 457
GRATTAN-GUINNESS, IVOR (*1941), 161, 336n
GRAVE, DMITRIĬ A. (1863–1939), 183
GRAVELAAR, NICOLAAS LAMBERTUS W. A. (1851–1913), **B**, 54
GRAVES, ROBERT (1810–1893), 169
GRAY, GEORGE JOHN (1863–1934?), 170
Great Britain
- *Analytical Society*, 164
- *Bodleian Library* in Oxford, 162
- *British Academy* in London, 442
- *British Association for the Advancement of Science (BAAS)*, 164, 169–170
- *British Museum* in London, 268–269, 276, 513
- *British Society for the History of Mathematics*, 178, 287, 352
- *British Society for the History of Science*, 176–177
- *British Society for the Philosophy of Science*, 177
- *Cambridge University*, 167, 169–172, 175, 290, 406, 494, 503, 510, 542, 560
- *Chartered Accountants of Scotland*, 174
- *Didsbury College of Education*, 351
- *East India Company*, 166
- *Encyclopedias*, 163–164, 167–168, 173, 178, 406
- *Francis Galton Laboratory for National Eugenics*, 174
- *Imperial College London*, 177
- *International Congress of the History of Science* in Edinburgh (1977), 481
- *London Mathematical Society (LMS)*, 283
- *London School of Economics*, 479
- *Manchester Polytechnic*, 351, 504
- *Mathematical Association*, 352, 442
- *Museum of the History of Science* in Oxford, 175
- *Open University* in Milton Keynes, 178, 338
- *Oxford University*, 162, 443, 560
- *Royal Asiatic Society*, 399
- *Royal Astronomical Society*, 399
- *Royal Military Academy* at Woolwich, 163
- *Royal Observatory* at Edinburgh, 407
- *Royal Society of Edinburgh*, 174, 275, 494
- *Royal Society of London*, 11, 163–166, 168–171, 176, 240, 407, 442, 494, 544, 560
- *Society for the Diffusion of Useful Knowledge*, 168
- *Thomas Harriot Seminar*, 177
- *Trinity College, Cambridge*, 440, 453, 510, 563
- *University College, London*, 169–176, 406, 501
- *University College, Swansea*, 175
- *University of Birmingham*, 504
- *University of Edinburgh*, 173, 310
- *University of Glasgow*, 173–174
- *University of London*, 176, 351, 407, 479
- *University of Manchester*, 351
- *University of St. Andrews*, 174, 337, 544
- *Warburg Institute* in London, 176

Greece
- *Academy "Assured"* in Corfu, 225
- *Academy "Errants"* in Corfu, 225
- *Academy "Fertiles"* in Corfu, 225
- *Academy of Athens*, 227
- *Athonian Academy* at Mount Athos, 224

Encyclopedia Suda (10th century), 223
French Library in Corfu, 225
Ionian Academy in Corfu, 225–226
Mount Athos, 224
National Technical University in Athens, 226
School for Greek Culture at Mount Athos, 224
Scientific Society, 228
University of Alexandria, 224
University of Athens, 226–229, 529, 530
Vatopediou Convent at Mount Athos, 224
Greenstreet, William John (1861–1930), 176
Greenwood, Major, the younger (1880–1949), 175
Gregor XIII, Pope, *see* Boncompagni, Ugo
Gregorius a Sancto Vincentio (1584–1667), 373, 376, 448
Gregory, James (1638–1675), 139, 176, 275–276, 405, 448, 544
Grigoryan, Ashot Tigranovich (*1910), 572
Gruber, Johann Gottfried (1774–1851), 119
Grunert, Johann August (1797–1872), 402
Grynaeus, Simon (1493–1541), 98
Gua de Malves, Jean Paul de (ca. 1712–1786), 70, 489
Günther, Adam Wilhelm Siegmund (1848–1923), **B**, 25, 81, 123–127, 130, 133, 136, 371, 388–389, 418, 555, 565
Guglielmini, Domenico (1655–1710), 64, 68
Guglielmini, Giovanni Battista (1763–1817), 71
Guimarães, Rodolfo Ferreira Dias (1866–1918), **B**, 239, 243–246
Guizot, François Pierre Guillaume (1787–1874), 467
Gunter, Edmund (1581–1626), 47
Gunther, Robert T. (1869–1940), 175
GUO Shoujing (1231–1316), 300
GUO Shuchun (*1941), 304–305
GUO Zhangfa (18th century), 303
Gupta, Radha Charan (*1935), 307
Gur'ev, Semën E. (1766–1813), 179, 573
Gurjar, Laxman Vasudeva (1909–1982), 315
Gusak, Alekseĭ A. (*1927), 192
Gussov, Viktor V. (*1911), 192
Guter, Rafail S. (1919–1978), 193

Habicht, Walter (*1915), 103
Hadamard, Jacques (1865–1963), 35, 261n, 438
Hadjidakis, Ioannis (1844–1921), 227–229
Hadjidakis, Nikolaos (1873–1941), 228–229
Hagstroem, Karl Gustav (20th century), 419
Hájek von Hájek, Thaddäus (1525–1600), 46, 206
al-Ḥājjī (ca. 786–ca. 835), 140
Ḥājjī Khalīfa (†1657), 319
Hallam, Henry (1777–1859), 167
Halley, Edmund (1656–1743), 162, 166, 175
Halma, Nicolas (1755–1828), 13
Halstead, George Bruce (1853–1922), 372, 485
Hamilton, William Rowan (1805–1865), 169, 480
Hammurapi (1728–1686; Egyptian King), 562
Hankel, Hermann (1839–1873), **B**, 81, 122–123, 371, 535
Hannequin, Arthur (1856–1905), 26, 29–31
HARA Kokiti (*1918), 294
Harding, Karl Ludwig (1765–1834), 129
Hardy, Godfrey Harold (1877–1947), 498
Haridatta (ca. 683 A.D.), 309
Harig, Gerhard (1902–1966), 143–144
Harkness, James (1864–1923), 218
Haro, Guillermo (1913–1988), 261
Harriot, Thomas (ca. 1560–1621), 89, 159, 162, 177, 469
Harris, John (1667–1719), 163
Harting, Pieter (1813–1885), 57
Hartner, Willy (1905–1981), 191
Harvey, William (1578–1657), 261n
Hasse, Helmut (1898–1979), 519
Hausdorff, Felix (1868–1942), 145, 147
Hausdorff Edition, 147
HAYASHI Tsuruichi (1873–1935), **B**, 291–292, 430
HE Shaogeng (*1939), 305
Heath, Archibald Edward (1887–1961), 170
Heath, Sir Thomas Little (1861–1940), **B**, 26, 171, 173, 176–178, 273–274

HECKE, ERICH (1887–1947), 547
HEE, LOUIS VAN (1873–1951), 302
HEEGAARD, POUL (1871–1948), 275, 405
HEGEL, GEORG WILHELM FRIEDRICH (1770–1831), 116, 184, 461
HEGELER, EDWARD C. (1835–1910), 270
HEIBERG, JOHAN LUDVIG (1854–1928), **B**, 25, 84, 92, 118, 120, 154–157, 365, 415, 418, 441, 536, 578
HEIDEGGER, MARTIN (1889–1976), 135, 358, 395
HEILBRONNER, JOHANN CHRISTOPH (1706–1747), **B**, 111–112
HEIM, ALBERT (1849–1937), 511
HEINRICH VON LANGENSTEIN (ca.1340–1397), 216
HEISENBERG, WERNER (1901–1976), 548
HELLER, SIEGFRIED (1876–1970), **B**, 134, 448
HELLINGER, ERNST (1883–1950), **B**, 136, 141, 275–276, 405
HELLMAN, C. DORIS (1910–1973), 416
HENRION, DENIS (†1632?), 48
HENRY THE NAVIGATOR (1394–1460), 241
HENRY, CHARLES (1859–1926), 24, 26, 54, 76, 81, 537
HENRY, JOSEPH (1797–1878), 170
HERBART, JOHANN FRIEDRICH (1776–1841), 293
HERDER, JOHANN GOTTFRIED VON (1744–1803), 115
HERMANN OF CARINTHIA (12th century), 214
HERMELINK, HEINRICH (1920–1978), **B**, 133
HERMITE, CHARLES (1822–1901), 314, 476
HERODOTUS (484–ca. 425), 307
HERON OF ALEXANDRIA (ca. 75 A.D.), 5, 63–64, 71, 120, 126, 138, 157, 357, 392, 443, 509, 511, 536
HERSCHEL, JOHN (1792–1871), 164
HESSE, LUDWIG OTTO (1811–1874), 81
HILBERT, DAVID (1862–1943), 79, 131, 147, 196, 395, 405, 433, 445, 460, 547, 555
Hilbert Edition, 147
HILL, GEORGE WILLIAM (1838–1914), 269
HIPPARCHUS OF NIKAIA (180?–125?), 47, 441, 537
HIPPOKRATES OF CHIOS (ca. 440 B.C.), 71, 99, 222, 536
HIRAYAMA AKIRA (1904–1998), 290–292
HITLER, ADOLF (1889–1945), 140, 143
HJELMSLEV, JOHANNES (1873–1950), **B**, 158
HLAWKA, EDMUND (* 1916), 218
HOEFER, FERDINAND (1811–1878), 23
HOENE-WROŃSKI, JÓZEF (1776–1853), 200, 409
HOFMANN, JOSEPH EHRENFRIED (1900–1973), **B**, 112, 130, 135–140, 142–143, 145–146, 149–150, 191, 445, 476
HOFMANN, JOSEPHA (1912–1986), 139, 142
HOGBEN, LANCELOT (1895–1975), 175
HOGENDIJK, JAN (*1955), 56
HOLMBOE, BERNT MICHAEL (1795–1850), **B**, 153
HOLST, ELLING (1849–1915), **B**, 153–154, 531, 534
HOLTZMANN, WILHELM, *see* XYLANDER
HOOKE, ROBERT (1635–1703), 175
HOOYKAAS, REIJER (1906–1994), 55
HOPF, HEINZ (1894–1971), 294
HOPPE, EDMUND (1854–1928), **B**, 126
HOREM, NICOLAUS, *see* NICOLE ORESME
HORMIGON, MARIANO (*1946), 231
HORNER, WILLIAM GEORGE (1786–1837), 569
HORSBURGH, ELLICE MARTIN (1870–1935), 174
HORSLEY, SAMUEL (1733–1806), 163
HOÜEL, JULES (1823–1886), **B**, 20, 22–23, 25, 42
HOUZEL, CHRISTIAN (*1937), 41
HSITA, CHUTHAN (ca. 718 A.D.), 309
HUA HENGFANG (1833–1902), 304, 464
HUA LUOGENG (L. K. HUA) (1910–1985), 304
HUANG ZHONGJUN ((late 19th century), 303
HUBER, DANIEL (1768–1829), 100–101
HUDDE, JAN (1628–1704), 448
HUG, JOHANN CASPAR (1821–1884), 102
HUISMAN, ANDRÉ (20th century), 451
HULTSCH, FRIEDRICH (1833–1906), **B**, 25, 119–120, 443, 536
HUMBOLDT, ALEXANDER VON (1769–1859), 118, 128, 251, 448
HUMBOLDT, WILHELM VON (1767–1835), 114
HUND, FRIEDRICH (1896–1997), 548
Hungary, *see* Austria
HUNRATH, KARL (*1847), 418
HURWITZ, ADOLF (1859–1919), 511
HUSSERL, EDMUND (1859–1938), 135, 293, 358, 395, 460–461, 475, 538
HUTTON, CHARLES (1737–1823), 163–164
AL-HUWĀRĪ (13th century), 324
HUYGENS, CHRISTIAAN (1629–1695), 10, 51, 54, 57, 74, 333, 376, 448, 460, 558

Huygens Edition, 53–54, 558
HYPPOLITE, JEAN (*1907), 36
HYPSICLES OF ALEXANDRIA (ca. 180 B.C.), 47, 509

IAMBLICHUS (ca. 285–ca. 330), 47, 536
IBN ABĪ UṢAYBIʿA (†1270), 319
IBN AL-BANNĀʾ (ca. 1256–ca. 1321), 324–327
IBN AL-HAYTHAM (965–ca. 1040), 16, 81, 324–327, 448, 469
IBN KHALLIKĀN (†1282), 319
IBN AL-QĀḌĪ (†1616/17), 317
IBN SĪNĀ (†1036), 323, 517
IBN AL-ṬAḤḤĀNA (19th century), 319
IBN-YŪNUS (10th century), 15
IBRĀHĪM B. SINĀN B. THĀBIT B. QURRA (908–946), 474
IM HOF, HANS-CHRISTOPH (*1944), 103
India
- *Aligarh Muslim University*, 561
- *Allahabad University Mathematical Association*, 314
- *Allahabad University*, 503
- *Asiatic Society* in Calcutta, 310
- *Banaras Mathematical Society*, 503, 523
- *Bharata Ganita Parisad*, 523
- *Calcutta Mathematical Society*, 314
- *Central Hindu College* in Banaras (now Varanasi), 503
- *College of Fort Williams* in Calcutta, 399
- *Government Sanskrit College* in Banaras, 311–312
- *Hindu University* in Banaras, 503, 523
- *Late Aryabhaṭa School of South India*, 309
- *Lucknow University*, 315
- *Marathi School* in Nagpur, 312
- *Muir Central College* in Allahabad, 503
- *Queen's College* in Banaras (now Varanasi), 503
- *Sehore Sanskrit School*, 312
- *University of Bombay*, 311
- *University of Calcutta*, 311, 313–314, 405, 503, 523
- *University of Lucknow*, 523
- *University of Madras*, 311
- *University of Nālandā*, 311
- *University of Odantapura*, 311
- *University of Takṣilā*, 311

INITIUS ALGEBRAS (fictitious author, early 16th century), 121, 403
International Academy of the History of Science, 34, 191, 209, 227–228, 351, 381, 387, 393, 401, 454, 484–485, 501, 530, 538, 550, 554, 557, 574
International Commission of Mathematics Instruction (ICMI), 335, 355, 456, 524, 565
International Commission on Mathematical Education, *see* *International Commission of Mathematics Instruction (ICMI)*
International Commission on the History of Mathematics (ICHM), 282, 287, 480
International Commission on the Teaching of Mathematics, *see* *International Commission of Mathematics Instruction (ICMI)*
International Congresses in Philosophy, 27
International Congresses of Comparative History, 27–28
International Congresses of Mathematicians, 27, 82, 89, 124, 283, 472–473, 511, 565
International Congresses of the History of Science, 34, 209, 212, 286, 481, 484, 538, 551, 554
International Study Group on the Relations between History and Pedagogy of Mathematics, *see* *International Commission of Mathematics Instruction (ICMI)*
International Union of the History and Philosophy of Science (IUHPS), 37, 58, 521
Internationale Mathematische Unterrichtskommission (IMUK), *see* *International Commission of Mathematics Instruction (ICMI)*
Iran, *see* Arab countries, Turkey and Iran
ISAAK ARGYROS (1310?–1371), 223
ISELY, LOUIS (1854–1916), 105
ISIDORUS OF MILETUS (ca. 520), 47, 223
Italy
- *Academia pro Interlingua* of PEANO, 394
- *Academy of Sciences* in Modena, 507

Academy of Sciences in Turin, 72, 89, 391
Accademia dei Lincei in Rome, 369, 422, 546
Accademia del Cimento in Florence, 76
Accademia Pontificia dei Nuovi Lincei in Rome, 369
Ambrosiana in Milan, 370
Collegio Romano in Rome, 65, 369
Congress of the Italian Mathematical Union (1940), 93
Encyclopedias, 62, 65–66, 82, 92, 375, 424, 467, 472
International Congress of Comparative History in Rome (1903), 28, 538
International Congress of Mathematicians in Bologna (1928), 209n, 373–375
International Congress of Mathematicians in Rome (1908), 82, 89, 456, 473
International Congress of Philosophy in Bologna (1911), 422
Istituto Nazionale delle Assicurazioni (INA) in Rome, 91
Istituto Superiore di Magistero Femminile in Rome, 372
Istituto Tecnico in Rome, 477
Istituto Veneto di Scienze in Venice, 426, 471
Italian Society for the Advancement of Science, 90
Laurenziana Library in Florence, 13n
Liceo in Modica (Sicily), 373
Liceo of Lucca, 71
Mathesis (Italian Society for Teachers of Mathematics), 91, 423
Military Academy in Caserta, 393
Military School in Verona, 400
Moreniana Library, 467
National Institute for the History of Mathematical and Physical Sciences in Rome, 91–92
National Institute for the History of Science in Rome, 92–93
Naval Academy in Livorno, 94, 351, 393
Polytechnic in Milan, 94
Postgraduate School in History of Science in Rome, 92–93
Reale Accademia di Scienze, Lettere ed Arte in Padua, 417
Royal Vinciana Commission, 86, 426
S. Uffizio (Holy Office), 76–77, 436
Scuola di applicazioni per l'ingegneria in Bologna, 508
Scuola Normale in Palermo, 372
Scuola Normale in Pavia, 372, 555
Scuola Normale Superiore in Pisa, 94, 421
Sixth Congress of Italian Scientists in Milan (1844), 76
Società Filosofica Italiana, 422
Società Italiana di Scienze, 400
Technical Institute of Bologna, 373
Technical Institute of Naples, 353
Third Congress of Italian Scientists in Florence (1841), 85
Unione Matematica Italiana (UMI), 95, 394
University of Bologna, 71, 77, 90, 94–95, 351, 372–375, 392, 421–423, 435–436, 438, 552, 555
University of Genoa, 88, 469, 474, 545
University of Messina, 477, 555
University of Milan, 94–95, 393, 427, 434, 555
University of Modena, 71, 94, 373, 392, 508
University of Naples, 88, 353, 477
University of Padua, 69, 224, 357, 399–400, 425–426
University of Parma, 69, 400
University of Pavia, 78, 94, 393
University of Piacenza, 432
University of Pisa, 91, 94, 351, 421, 465, 484
University of Rome, 35, 54, 91–92, 94–95, 392, 421–424, 428, 477, 484, 545, 561
University of Turin, 82–83, 94, 392–393, 399, 425, 427, 432–435, 469, 545–546, 555
Vatican Library in Rome, 81, 366
ITARD, GILLES (*1936), 451
ITARD, JEAN (1902–1979), **B**, 37–38
IWIŃSKI, TADEUSZ (1906–1993), 201
IYANAGA SHOKICHI (*1906), 294

JACOBI, CARL GUSTAV JACOB (1804–1851), 115, 129, 142, 293, 314, 495
JALLABERT, JEAN (1712–1768), 99
JAMĀL AL-DĪN AL-AFGHĀNĪ (†1897), 322
JAMĀL AL-DĪN HUMĀʾĪ (1900–1980), 327
JAMI, CATHERINE (*1961), 494
JAMSHĪD B. MĀSʿŪD AL-KĀSHĪ, *see* AL-KĀSHĪ, JAMSHĪD B. MĀSʿŪD
JANISZEWSKI, ZYGMUNT (1888–1920), 202

Japan
History of Mathematics Society, 295
History of Science Society of Japan, 295
Imperial University of Tokyo, 290–291, 430, 440
Japan Academy, 291–292, 485
Japan Imperial Academy, 290, 430
Kyoto Imperial University, 292–293, 295, 440
Tōhoku Imperial University in Sendai, 291–293, 430, 485
Tokyo Mathematical Society, 290
Tokyo School of Physics, 485
University of Tokyo, 289, 295, 485
JARNÍK, VOJTĚCH (1897–1970), 209
JAŠEK, MARTIN (*1879), 209n
Jesuits, 9, 15, 46n, 52, 58, 65–68, 76, 216, 241, 250, 285, 297, 299, 304, 369, 376, 399, 425, 433, 463, 488, 507–508, 521, 534, 570
JIAO XUN (1763–1820), 302
JIAQING (1760–1820; Emperor of China from 1796), 300
JOÃO III, D. (1502–1557; King of Portugal from 1521), 241
JODE, CORNELIUS DE (1568–1600), 52
JODE, GERARD DE (1521–1591), 52
JOHANNES CAMPANUS, *see* CAMPANUS, JOHANNES of Novara
JOHANNES DE LINERIIS (14th century), 403
JOANNES DE MURIS (1290?–1360?), 215, 403
JOHANNES DE SACROBOSCO, *see* SACROBOSCO
JOHANNES PHILOPONUS (490–566), 223, 537
JOHN OF HOLYWOOD or HALIFAX, *see* SACROBOSCO
JOHN OF SEVILLE (13th century), 371
JOLIVET, JEAN (*1925), 39
JOMBERT, CHARLES-ANTOINE (1712–1784), 489
JONES, CHARLES V. (*1939), 286
JONES, PHILLIP S. (*1912), 270–271
JONES, WILLIAM (1746–1794), 310
JONQUIÈRES, JEAN P. ERNEST DE (1820–1901), 474
JORDAN, CAMILLE (1838–1922), 32, 373, 401
JORDANUS DE NEMORE (first half of 13th century), 30, 46, 120–121, 270, 366, 374, 402, 414, 543
JØRGENSEN, JØRGEN (1894–1969), **B**, 158
JOSÉ I (1714–1777; King of Portugal from 1750), 239, 241, 250
JOSEPHUS FLAVIUS, *see* FLAVIUS JOSEPHUS
JOURDAIN, PHILIP (1879–1919), **B**, 130, 170
Journals & Series
Abhandlungen über den mathematischen Unterricht in Deutschland, 456
Abhandlungen zur Geschichte der mathematischen Wissenschaften, 125, 149, 283, 334, 388
Abhandlungen zur Geschichte der Naturwissenschaften und der Medizin, 132–133
Acta Eruditorum, 431
Acta historiae rerum naturalium necnon technicarum, Special Issues, 210
Acta historica scientarum naturalium et medicinalium, 159
Acta Mathematica, 418, 470
Actualités scientifiques et industrielles, 35
Akademicheskie Izvestiya, 179
al-Abḥāth, 324
al-Mashriq, 324
al-Muqtaṭaf, 322
Algorismus, 146
Allgemeine deutsche Biographie, 391
American Mathematical Monthly, 275, 354, 524
Anais da Faculdade de Ciencias do Porto, 243
Annaes Scientificos da Academia Polytechnica do Porto, 243
Annales, 35
Annales de Gergonne, 431
Annales de l'Université d'Ankara, 517
Annali di scienze matematiche e fisiche, 369, 432
Arbor scientiarum, 146
Archeion, 34, 91, 253, 355, 484, 565
Archimede, 89
Archiv der Mathematik und Physik, 402
Archiv for Mathematik og Naturvidenskab, 467
Archive for History of Exact Sciences, 277, 282, 381, 459, 544, 548, 572
Archivio di Storia della Scienza, 91–92, 484
Archivo de Matemáticas, 507
Asiatick Researches, 310
Athena, 228

Atti dell'Accademia Nuovi Lincei, 371
Biblioteka Matematyczno-Fizyczna, 202
Bibliotheca Mathematica, 24, 51, 124–125, 154–155, 157, 159, 180, 208n, 244, 283, 388–390, 417–419, 455, 470–471, 545
Bibliothèque britannique, 12
Bibliothèque orientale, 14
Bibliothèque universelle, 12
Biographie universelle, 17
Biographien hervorragender Naturwissenschaftler, Techniker und Mediziner, 145
Biometrika, 174, 501
Boethius, 146, 337n
Boletin Informativo de la FEPAI, 252
Bollettino dell'Unione Matematica Italiana, 95
Bollettino di Bibliografia e Storia delle Scienze Matematiche (LORIA), 83, 89, 208n, 283, 334, 372, 471
Bollettino di Matematica, 89, 334, 471
Bollettino di Storia delle Scienze Matematiche, 95
Buletim da Academia Real das Sciencias de Lisboa, 241
Bulletin de bibliographie, d'histoire et de biographie mathématiques, 334, 540
Bulletin des sciences mathématiques, 23, 450
Bulletin of the American Mathematical Society, 524
Bullettino di Bibliografia e di Storia delle Scienze Matematiche e Fisiche (BONCOMPAGNI), 19, 24, 51, 53, 66, 79–81, 85, 122, 125, 283, 334, 358, 371, 388, 425–426, 432, 470–471, 558
Cahiers du séminaire d'histoire des mathématiques, 41, 413
Centaurus, 159
Colloquium Mathematicum, 201
Companion of the British Almanac, 168
Comptes rendus de l'Académie des Sciences de Paris, 17, 21, 251, 520
Crelles Journal für die Reine und Angewandte Mathematik, 431
De Gids, 411
Dějiny věd a techniky, 210
Deutsche Mathematik, 140, 142
Development of Physical and Mathematical Sciences, 367
Dictionary of Scientific Biography, 286, 381, 514
Dokumente zur Geschichte der Mathematik, 146
El Progreso Matemático, 232, 235, 507
Elemente der Mathematik, 104, 428
Eléments de mathématique, 377–379
Encyklopädie der mathematischen Wissenschaften, 28, 81, 83, 172–173, 217–219, 275–276, 379, 405, 418, 422, 432, 456, 472, 574
Enseignement des sciences, 451
Etudes socialistes, 451
Evolution de l'humanité, 506
FIAT Review of German Science 1939-1946, 140n
Fisiko-Matematicheskie Nauki v Khode ikh Razvitiya, 180
Fisiko-Matematicheskie Nauki v ikh Nastoyashchem i Proshedshem, 180
Gaṇita Bhāratī, 315
Geschichte der Wissenschaften in Deutschland, 568
Giornale de' Letterati d'Italia, 64, 357
Giornale di Matematiche, 83, 334, 471
Histoire de l'Académie royale des sciences – avec les mémoires de mathématiques et de physique, 7
Historia de la Ciencia, 252
Historia Mathematica, 277, 282, 286, 337, 381, 459, 481
Historia Scientiarum (Tokyo), 295
Historical Studies in the Physical Sciences, 355
Historische Bibliotheek voor de Exacte Wetenschappen, 55, 362
History of Mathematics Series, 334
International Catalogue of Scientific Literature, 170
Isis, 57, 273–274, 284, 459, 484, 515, 517, 565
Istoriko - Matematicheskie Issledovaniya (IMI), 187, 194–196, 573
Istoriya i Metodologiya Estestvennykh Nauk, 195

Istoryko-Matematychnyĭ Zbirnyk, 195
Izvestiya Severo-Kavkazskogo Universiteta, 182
Jahrbuch für Philosophie und phänomenologische Forschung, 475
Jahrbuch über die Fortschritte der Mathematik, 130, 417
Jahresbericht der Deutschen Mathematiker-Vereinigung, 456
Janus, 57, 384
Japanese Studies in the History of Science, 295
Jenaer Literaturzeitung, 391
Jornal de Sciencias Mathematicas, Physicas e Naturaes, 241
Jornal de Sciencias Mathematicas e Astronomicas, 243, 539
Journal asiatique, 16, 18n, 19, 363, 520
Journal des savants, 22, 361, 431
Journal of Cuneiform Studies, 513
Journal of the American Oriental Society, 517
Journal of the Institute of Arabic Manuscripts, 323
Kultur der Gegenwart, 456, 577
Kurze Mathematiker-Biographien, 104
Kwartalnik Historii Nauki i Techniki, 201
L'Enseignement mathématique, 24
L'Enseignement scientifique, 451
La France libre, 396
Literarisches Zentralblatt, 391
Llull, 236
Münchner allgemeine Zeitung, 391
Matematika v Shkole, 195
Matematyka, 201
Mathematica Notae, 253
Mathematical Intelligencer, 277
Mathematical Reviews, 158, 498
Mathematische Annalen, 217, 421, 455, 470, 507
Mathesis, 262
Mémoires de la société des sciences physiques et naturelles de Bordeaux, 23
Memoirs of the Analytical Society, 164
Memorias da Academia Real das Sciencias de Lisboa, 240–241, 531
Memorias de Litteratura da Academia das Sciencias de Lisboa, 240
Memorie dell'Accademia dei Lincei, 473
Memorie dell'Accademia delle Scienze di Torino, 470
Messenger of Mathematics, 169
Mitteilungen der Österreichischen Gesellschaft für Wissenschaftsgeschichte, 219
Mitteilungen zur Geschichte der Medizin und der Naturwissenschaften, 565
Mittheilungen der Naturforschenden Gesellschaft in Bern, 568
Monist, 270
Muslim World, 517
Narysy z Istorii Pryrodoznavsta i Tekhniky, 195
Nieuw Tijdschrift voor Wiskunde, 54
Nordisk Försäkringstidskrift, 417
Notes and Records of the Royal Society, 176
Notices et extraits des manuscrits de la Bibliothèque royale, 16, 520
Nouvelles annales de mathématiques, 24, 334, 540
Nuovo Cimento, 477
O Instituto, 242–243, 394
Obituary Notices of Fellows of the Royal Society, 176
Osiris, 273, 487, 509, 515
Ostwalds Klassiker der exakten Wissenschaften, 130, 144, 404, 529
Periodico di Matematiche, 90–92, 423
Philosophical Transactions of the Royal Society, 168
Physical and Mathematical Sciences in the Past and Present, 367
Poggendorff: Literarisch-Biographisches Handwörterbuch, 143
Poradnik dla Samouków, 202
Pour la révolution constructive, 451
Prace Matematyczno-Fizyczne, 200, 203, 409
Proceedings of the Canadian Society for History and Philosophy of Mathematics, 287
Proceedings of the Royal Society, 176, 251
Programy Wykładów i Składy Osobove Wydziatów, 200

Quarterly Journal of Education, 168
Quarterly Journal of Pure and Applied Mathematics, 169
Quellen und Studien zur Geschichte der Mathematik, Astronomie und Physik, 135, 140, 276, 496–497, 543, 548
Quellen und Studien zur Geschichte der Medizin und Naturwissenschaften, 132
Quipu, 254, 355
Revista Brasileira de História da Matemática, 255
Revista de la Sociedad Matemática Española, 232, 514
Revista de los Progresos de las Ciencias Exactas, Físicas y Naturales, 232
Revista Matemática Hispano-Americana, 232, 506, 514
Revue d'histoire des mathématiques, 41
Revue d'histoire des sciences, 36–37
Revue de synthèse, 33–34, 361
Revue de synthèse historique, 28, 33, 361
Revue des deux mondes, 22, 76
Revue des questions scientifiques, 390
Rivista di Matematica, 470, 546
Rivista di Scienza (Scientia), 84, 422
Rivista di Storia delle Scienze Mediche e Naturali, 484
Sbornik pro dějiny přirodnich věd a techniky, 210
Schola et vita, 394
Schriftenreihe zur Geschichte der Naturwissenschaften, Technik und Medizin (NTM), 144
Science Networks, xxxiv, 145, 334
Scientia, 484
Scientiarum Historia, 58–59
Scripta Mathematica, 271, 274–275, 381, 437, 523, 525
Séminaire Bourbaki, 377
Shuxueshi yanjiu wenji (Collection of Studies on History of Mathematics), 306
Sitzungsberichte der Physikalisch-Medizinischen Sozietät zu Erlangen, 132
Sources in the History of Mathematics and Physical Sciences, 334
Sprawozdania Dyrekcji Gimnazjów i Liceów, 200
Studien zur Wissenschafts-, Sozial- und Bildungsgeschichte der Mathematik, 146
Sūgakushi Kenkzū (The Journal of History of Mathematics, Japan), 295
Teubner-Archiv zur Mathematik, 144
Thalès, 35, 506
The Economist, 562
The History of Modern Mathematics, 334
The Records of the Royal Society, 176
The Royal Society Catalogue of Scientific Papers, 170
Tijdschrift voor de Geschiedenis der Geneeskunde, Natuurwetenschappen, Wiskunde en Techniek, 57–59
Tōhoku Mathematical Journal, 291, 440
Tōhoku Mathematical Paper, 417
Transactions of the Mathematical Section of the New Russian Society of Naturalists, 541
Un savant, une époque, 334
Uspekhi Matematicheskikh Nauk, 186
Vierteljahrsschrift der Naturforschenden Gesellschaft in Zürich, 511, 568
Vita Mathematica, 104, 334
Voprosy Istorii Estestvoznaniya i Tekhniki, 195
Wiadomości Matematyczne, 200–201, 203, 409
World Directory of Historians of Mathematics, 287, 481
Yearbook of Uppsala University, 417
Z dejín vied a techniky na Slovensku, 210
Zeitschrift für Geschichte der Arabisch-Islamischen Wissenschaften, 147
Zeitschrift für Mathematik und Physik, 124, 334, 388
Zeitschrift für Mathematik und Physik, Historisch-kritische Abteilung, 388
Zentralblatt für Mathematik und ihre Grenzgebiete, 130, 135, 158, 498, 530
Zhongguo keji shiliao (Chinese Historical Materials on Science and Technology), 306

Ziran bianzhengfa tongxun (Journal of the Dialectics of Nature), 306
Ziran kexueshi yanjiu (Studies in the History of Natural Sciences), 306
JULIEN, STANISLAS (1797–1873), 20
JUNG, CARL GUSTAV (1875–1961), 526
JUNGE, GUSTAV (1879–1959), **B**, 140
JUNGIUS, JOACHIM (1587–1657), 110
JUNIUS, FRANCISCUS (1624–1678), 48, 559
JUSCHKEWITSCH, ADOLF P., *see* YOUSHKEVICH, ADOLF-ANDREJ PAVLOVICH

KÄSTNER, ABRAHAM GOTTHELF (1719–1800), **B**, 113–114, 339, 373, 444, 449, 567
KAGAN, VENIAMIN F. (1869–1953), **B**, 181, 186–187, 192
KAHL, GUSTAV EMIL (*1827), 388
KAKEYA Sōichi (1886–1947), 417
KĀMIL, MUṢṬAFĀ PĀSHĀ (1874–1908), 322
KAMPEN, NICOLAAS GODFRIED VAN (1776–1839), 50
KANGXI (1654–1722; Emperor of China from 1661), 299
KANT, IMMANUEL (1724–1804), 23, 31, 168, 394, 429
AL-KARAJĪ (ca. 1000), 118
KARL IV (1316–1378; King of Bohemia 1347, Emperor 1355), 207
KARPINSKI, LOUIS C. (1878–1956), **B**, 269–271, 274–275, 525
KARY-NIYAZOV, TASHMUKHAMED N. (1896–1970), **B**, 190
AL-KĀSHĪ, JAMSHĪD B. MASʿŪD (†1429), 133, 320, 327, 475
KATSCHER, FRIEDRICH (*1923), 218
KAUṬILYA (ca. 300 B.C.), 308
KAYE, GEORGE RUSBY (1866–1926), 314
KEES, HERMANN (1886–1964), 496
KELVIN, LORD, *see* THOMSON, SIR WILLIAM
KEMAL, MUSTAFA ATATÜRK (1881–1938; President of Turkey from 1923), 324
KENNEDY, EDWARD S. (*1912), 326
KENNEDY, HUBERT (*1931), 393
KEPLER, JOHANNES (1571–1630), 39, 47, 147, 163, 186, 257n, 351–352, 393, 446–448, 452, 460, 469, 476, 560
Kepler Edition, 147, 393, 446
KERN, H. (1833–1917), 311
KEYSER, CASSIUS JACKSON (1862–1947), 274, 359
AL-KHAYYĀM, ʿUMAR, *see* ʿUMAR AL-KHAYYĀM
KHINCHIN, ALEKSANDR YA. (1894–1959), 186, 571
AL-KHWĀRIZMĪ, MUḤAMMAD B. MŪSĀ, *see* MUḤAMMAD B. MŪSĀ AL-KHWĀRIZMĪ
KIBRE, PEARL (1903–1985), 365
KIERKEGAARD, SØREN (1813–1855), 443
KIKUCHI DAIROKU (1855–1917), 290–291, 440, 485
AL-KINDĪ (ca. 800–ca. 873), 366
KINKELIN, HERMANN (1832–1913), 101, 526
KIRIK FROM NOVGOROD (12th century), 179, 193, 578
KIRO, SERGEĬ N. (1926–1990), 190
KISELËV, ANDREĬ A. (1916–1994), 190, 192
KLEIN, FELIX (1849–1925), **B**, 28, 81, 83, 109, 127–128, 131, 134, 141–142, 172, 177, 217, 227, 314, 362, 395, 420, 468, 524, 555, 577
KLEINER, ALFRED (1849–1916), 511
KLINE, MORRIS (1908–1992), **B**, 277–282, 305
KLINGENSTIERNA, SAMUEL (1698–1765), 152
KLÜGEL, GEORG SIMON (1739–1812), 114
KNESER, HELLMUTH (1898–1973), 547
KNOBLOCH, EBERHARD (*1943), 147
KNORR, WILBUR (1945–1997), **B**, 282
KNOTT, CARGILL GILSTON (1856–1922), 174
KOCHAŃSKI, ADAM (1631–1700), 200, 409
KOHEN, IGNACE (20th century), 451
KOHL, KARL (*1896), 133
KOHN, GUSTAV (1859–1921), 217
KOLLROS, LOUIS (1878–1959), 105
KOLMAN, ERNST (1892–1979), 184
KOLMOGOROV, ANDREĬ N. (1903–1987), **B**, 186–187, 195, 573
KOMMERELL, VIKTOR (1866–1948), 389
KONDŌ YŌITSU (1911–1979), 292–294
KONG JIHAN (1739–1784), 301
KORDOS, MAREK (*1940), 202
KORTEWEG, DIEDERIK JOHANNES (1848–1941), **B**, 54
KOUMAS, KONSTANTINOS (1777–1823), 225
KOUTSKÝ, KAREL (1897–1964), 210
KOVALEVSKAYA, SOF'YA V. (1850–1891), 144, 193–194, 197, 360
KOYRÉ, ALEXANDRE (1892–1964), **B**, 32, 34–35, 38–39, 191, 385, 412
KRAFFT, GEORG WOLFGANG (1701–1754), 111–112

Kragemo, Helge Bergh (1897–1968), **B**, 153
Kramar, Feodosiĭ D. (1911–1980), 192
Krause, Max (1909–1944) **B**, 133
Krauss, Paul (1904–1944), 326
Kravchuk, Mikhail F. (1892–1942), 183
Krazer, Adolf (1858–1926), 218, 511
Krbek, Franz von (1898–1984), 144
Kreck, Matthias (*1947), xxvii
Kronecker, Leopold (1823–1891), 33, 227, 314, 395
Kropp, Gerhard (1910–1974), 146
Kṛṣṇaśāstrī Goḍabole (1831–1886), 313
Krylov, Alekseĭ N. (1863–1945), **B**, 181
Kubera (Indian god of wealth), 307
al-Kūhī, *see* Abū Sahl Wayjan b. Rustam al-Kūhī
Kuhn, Thomas Samuel (1922–1995), 147
Kummer, Ernst Eduard (1810–1893), 227, 510
Kuratowski, Kazimierz (1896–1980), 201
Kūshyār b. Labbān (ca. 970–ca. 1030), 327
Kutta, Wilhelm (1867–1944), 126
Kuzicheva, Zinaida A. (*1933), 188
Kwietniewski, Stefan (1874–1940), 202

La Cour, Poul (1846–1908), **B**, 156
La Hire, Philippe de (1640–1718), 5
Lacaille, Nicolas Louis de, *see* de la Caille, Nicolas Louis
Lachelier, Jules (1832–1918), 32, 535
Lachmann, Karl (1793–1851), 117, 123, 389
Lacroix, Sylvestre François (1765–1843), **B**, 10, 12
Laertius, Diogenes, *see* Diogenes Laertius
Laffitte, Pierre (1823–1903), 29
Lagrange, Joseph Louis (1736–1813), 12, 41–42, 62, 68–69, 71–74, 78, 89, 183, 199, 270, 449, 473–474, 508
Laisant, Charles Auguste (1841–1920), 24
Lalande, Joseph Jérôme de (1732–1807), **B**, 9, 15, 67, 110, 376, 405, 489–490
Lalande, Pierre-André (1867–1963), 460
Lalla (8th century A.D.), 309
Lambert, Johann Heinrich (1728–1777), 100, 103, 106, 163, 373
Lambert Edition, 106
Lamé, Gabriel (1795–1870), 362
Landau, Edmund (1877–1938), 142, 496, 526, 564
Landen, John (1719–1790), 74
Lang, Arnold (1855–1914), 511
Langsdorff, Georg H. von (1774–1852), 251
Laplace, Pierre Simon de (1749–1827), 12, 15–16, 27, 42, 252, 264, 310, 380, 487
Laptev, Boris L. (1905–1989), 192
Lasserre, François (1919–1989), 105
Laue, Max von (1879–1960), 529
Lavisse, Ernest (1842–1922), 28
Le Breton, André François (1708–1779), 489
Le Gentil, Guillaume (1725–1792), 15, 310
Le Paige, Constantin (1852–1929), **B**, 51
Lebesgue, Henri Leon (1875–1941), 35, 294, 413
Leersum, E. C. van (1862–1938), 57
Lefèvre d'Etaples, Jacques (ca. 1455–1536), 4
Lefort, F. (19th century), 17
Lefranc, Georges (1904–1985), 451
Lefschetz, Solomon (1884–1972), 260n
Legendre, Adrien Marie (1752–1833), 79, 466, 562
Leibniz, Gottfried Wilhelm (1646–1716), 10, 21, 31–32, 64, 74, 83, 89–90, 99, 111, 118, 121, 131, 139, 147, 154, 162–163, 166–168, 200, 228, 291, 304, 332, 351–352, 392, 401, 406–407, 409, 428–429, 433, 446, 448–449, 476, 491, 493, 572
Leibniz Edition, 32, 139, 147, 401, 447–448, 476
Lejeune, Albert (1916–1988), **B**, 52
Lejeune-Dirichlet, Peter G., *see* Dirichlet, Peter Gustav Lejeune
Lemaître, Georges (1894–1966), 51
Lenin, Vladimir I. (1870–1924), 184, 571
Lenoble, Robert (1902–1959), **B**, 26, 37–38
Leonardo da Vinci (1452–1519), 75, 85–87, 389–391, 426, 466, 477–478, 483
Leonardo da Vinci Edition, 86–87, 426, 477
Leonardo Mainardi (fl. about 1488), 121, 403
Leonardo of Pisa [Fibonacci], ca. 1180–ca. 1250), 19, 69–71, 75, 79–80, 90, 118, 138, 162, 331, 369–370, 374, 400, 432, 502
Leonardo of Pisa Edition, 80, 370

LEOPARDI, GIACOMO (1798–1837), 68
LEVI, BEPPO (1875–1961), 253
LEVI-CIVITA, TULLIO (1873–1941), 61, 392, 498
L'HÔSPITAL, GUILLAUME F. A. DE (1661–1704), 13, 186, 527
L'HUILLIER, SIMON-ANTOINE (1750–1840), 97, 99
LI DI (*1927), 306
LI HUANG (†1812), 303
LI RUI (1768–1817), 302–303
LI SHANLAN (1811–1882), 464, 570
LI YAN (1892–1963), **B**, 297, 304–305, 504
LI ZHI (1192–1279), 302
LIANG ZHAOKENG (early 19th century), 304
LIANG ZONGJUS (1924–1995), 305
LIARD, LOUIS (1846–1917), 26, 32
LIBRI, GUGLIELMO BRUTO ICILIO TIMOLEON (1803–1869), **B**, 17, 20–21, 62, 74–76, 85–87, 89, 110, 120, 124, 369–370, 436, 520, 538, 567
LIBRI-CARUCCI DALLA SOMMAJA, *see* LIBRI, GUGLIELMO BRUTUS ICILIUS TIMOLEON
LICHTENSTEIN, LEON (1878–1933), 145
LIE, MARIUS SOPHUS (1842–1899), **B**, 126–127, 153, 333, 420, 455, 534
LIEBMANN, HEINRICH (1874–1939), **B**, 127
LIETZMANN, WALTER (1880–1959), **B**, 139
E LIGNE, D. JOÃO DE BRAGANÇA DE SOUSA (1719–1806; second Duke of Lafões), 240
LIN TINGKAN (1757–1809), 302
LINDEMANN, FERDINAND VON (1852–1939), 446
LIOUVILLE, JOSEPH (1809–1882), 118, 467
LISTING, JOHANN BENEDIKT (1808–1882), 186
LITTRÉ, EMILE (1801–1881), 29
LITTROW, JOSEPH JOHANN VON (1781–1840), 567
LIU DUN (*1957), 297
LIU DUO (late 19th century), 304
LIU HUI (late 3rd century A.D.), 300
LOBACHEVSKIĬ, NIKOLAĬ IVANOVICH (1792–1856), 23, 79, 113, 127, 144, 181, 186, 192–194, 196, 261n, 372, 420, 433, 482, 552, 573
LOHNE, JOHANNES AUGUST (1908–1993), **B**, 159
LOMBARDI, ANTONIO (1768–1847), 67
LOMBARDO RADICE, LUCIO (1916–1982), 95
LORENTE PÉREZ, JOSÉ M^a (20th century), 234n
LORENTZ, HENDRIK A. (1853–1928), 181
LOREY, WILHELM (1873–1955), **B**, 130, 177, 456
LORGNA, ANTONIO MARIA (1735–1796), 78, 400
LORIA, GINO (1862–1954), **B**, 25, 34, 67, 74, 80, 82–93, 95, 130, 208, 217, 274, 283, 334, 371–375, 389, 442, 466, 491, 546, 554
LORITI, HEINRICH, *see* GLAREANUS
LORTIE, LÉON (1902–1986), 285
LOUIS PHILIPPE D'ORLEANS (1773–1850; King of France 1830–1848), 466–467
LUCAS, EDOUARD (1842–1891), 24
LUCKEY, PAUL (1884–1949), **B**, 133
LUDOLF VAN CEULEN (1540–1610), 413, 525, 558
LÜTZEN, JESPER (*1951), 156
ŁUKASIEWICZ, JAN (1878–1956), 202
LUMISTE, YULO G. (*1929), 192
LUO SHILIN (1789–1853), 303
LUR'E, SOLOMON YA. (1891–1964), **B**, 183
Luxemburg, *see* Benelux
LUZIN, NIKOLAĬ N. (1883–1950), 186, 194, 571
LYAPUNOV, ALEKSANDR M. (1857–1918), 193–194, 523
LYUSTERNIK, LAZAR' A. (1899–1981), **B**, 187, 192–193

MABILLON, JEAN (1632–1707), 5
MACH, ERNST (1838–1916), 410, 571
MACKEY, JOHN STURGEON (1843–1914), 174
MACPIKE, EUGENE FAIRFIELD (*1870), 175
MAENNCHEN, PHILIPP (1869–1945), 128
MAGINI, GIOVANNI ANTONIO (1555–1617), 48
MAGNUS, WILHELM (1907–1990), 275
MAHĀVĪRA (ca. 850 A.D.), 309
MAHĀVĪRĀCĀRYA (9th cent.), 269
MAḤMŪD PĀSHĀ AL-FALAKĪ (1815–1885), 322
MAHNKE, DIETRICH (1884–1939), **B**, 447
MAIRAN, JEAN-JACQUES DORTOUS DE (1741–1743), 8
MAĬSTROV, LEONID E. (1920–1982), **B**, 188, 192–193
MAJUMDAR, NARENDRA KUMAR (1st half of 20th century), 315
MALEBRANCHE, NICOLA DE (1638–1715), 39, 401

Malebranche Edition, 401
MALFATTI, GIANFRANCESCO (1731–1807), 78, 81
MALLET, JACQUES-ANDRÉ (1740–1790), 99
MANDROU, ROBERT (*1921), 43
MANFREDI, GABRIELE (1681–1761), 68, 375
MANGIONE, CORRADO (*1930), 435
MAṆIBHADRA (Indian god), 307
MANIN, YURIĬ I. (*1937), 188
MANNHEIM, AMÉDÉ (1831–1906), 474
MANNOURY, GERRIT (1867–1956), 547
MANSION, PAUL (1844–1919), **B**, 51, 81, 122, 130, 418, 515
MANUEL MOSCHOPOULOS (1282–1328), 223, 537
MAO LI (18th century), 301
MARCHEVSKIĬ, MIKHAIL N. (1884–1974), 182
MARCI, MARCUS (1595–1667), 211
MARCOLONGO, ROBERTO (1862–1943), **B**, 87, 89, 91, 471
MARCZEWSKI, EDWARD (1907–1976), 201
MARIE, CHARLES FRANÇOIS MAXIMILIEN (1819–1891), **B**, 14, 20, 22, 110
MARINUS of Neapolis (Palestine) (2nd half of 5th century A.D.), 222
MARKUSHEVICH, ALEKSEĬ I. (1908–1979), **B**, 187, 192
MARRE, ARISTIDE (1823 or 1828–1918), 24, 81, 418
MARRE, ARISTIDE (1823–1918), 19
MARTIN, THÉODORE-HENRI (1813–1884), 19, 23
MARX, KARL (1818–1883), 182–186, 571
MASCHERONI, LORENZO (1750–1800), 408
MASERES, FRANCIS (1731–1824), 163
MASOTTI, ARNALDO (1902–1989), 94
MATTMÜLLER, MARTIN (*1957), 562
MATVIEVSKAYA, GALINA P. (*1930), 190–194, 573
MAUROLICO, FRANCESCO (1494–1575), 63–65, 71, 89
MAXIMILIAN I (1459–1519; Holy Roman Emperor from 1493), 215
MAXIMILIAN II (1527–1576; King of Bohemia 1562, Emperor 1564), 206
MAXIMOS PLANUDES (1255?–1310), 121, 223
MAXWELL, JAMES CLERK (1831–1879), 169
MAY, KENNETH OWNSWORTH (1915–1977), **B**, 262–263, 273, 282–283, 285–287, 360
MAY, SAMUEL CHESTER (1887–1955), 479
MAYER, ADOLF (1839–1908), 81
MAZARREDO, JOSÉ DE (1745–1812), 427
MAZURKIEWICZ, STEFAN (1888–1945), 202
MAZZINI, GIUSEPPE (1805–1872), 75
MAZZUCCHELLI, GIOVANNI MARIA (1707–1765), 66
MCCORMACK, THOMAS J. (1865–1932), 270
MEDICI, FERDINANDO II DE' (1621–1670), 76
MEDICI, LEOPOLDO DE' (1617–1675), 76
MEDVEDEV, FËDOR A. (1923–1993), **B**, 188–189, 193, 573
MEGERLIN, PETER (1623–1686), 99
MEHMED SAʿID EFENDI (†1761), 318
MEHMKE, RUDOLF (1857–1944), 388
MEHRTENS, HERBERT (*1946), 142, 147, 336
MEI RONGZHAO (*1935), 305
MEI WENDING (1633–1721), 299–300, 464
Meiji Emperor, *see* MUTSUHITO
MELETIOS (18th century; Abbot at Mount Athos), 224
MEL'NIKOV, IL'YA G. (1916–1979), 190
MENELAUS OF ALEXANDRIA (ca. 100 A.D.), 126, 132–133, 365–366, 462
MENGOLI, PIETRO (1625–1686), 375
MENNINGER, KARL (1898–1963), **B**, 139
MERCATOR, GERARD (1512–1594), 50, 52, 57, 333
MERCATOR, NICOLAUS (1620–1687), 139, 142, 448
MERIAN, PETER (1795–1883), 101
MERIMÉE, PROSPER (1803–1870), 467
MERSENNE, MARIN (1588–1648), 13, 26, 39, 42, 73, 409, 463, 509, 538
Mersenne Edition, 26, 409, 509, 538
MERZ, JOHN THEODORE (1840–1922), 172, 178, 339
MERZBACH, UTA C. (*1933), 364
METZGER, HÉLÈNE (1889–1944), 34–35, 39n
MEURICE, LÉON (1866–1943), 51
Mexico (Mexico City)
Asociación para la Historia, Filosofía y Pedagogía de las Ciencias Matemáticas, 262
Colegio de San Gregorio, 256n
Colegio de San Pablo, 256n
Colegio Mayor de Santa María de Todos los Santos, 256n
Congreso Científico Mexicano, 260
El Colegio de México, 260n
El Colegio Nacional, 260n
Escuela de Altos Estudios, 258
Escuela Nacional de Ingenieros, 259
Escuela Nacional Preparatoria, 258

Instituto de Matemáticas, 260, 262
Instituto Politécnico Nacional (IPN), 261
National University, 258
Seminario de Problemas Científicos y Filosóficos, 261
Sociedad Matemática Mexicana, 260, 262
Sociedad Mexicana de Historia Natural, 260n
Universidad Nacional Autónoma de México (UNAM), 260–263
Universidad Real y Pontificia de México, 256
University of Mexico, 258
MEYER, KIRSTINE (1861–1941), 415
MEYER, WILHELM FRANZ (1856–1934), 456
MICHEL, HENRI (1885–1981), **B**, 53
MICHEL, PAUL-HENRI (1894–1964), **B**, 37
MIELI, ALDO (1879–1950), **B**, 34–35, 37–39, 84, 91–92, 253, 356, 361, 485, 500, 521
MIKAMI YOSHIO (1875–1950), **B**, 269, 290–292, 295, 302, 431, 440, 525
MIKHAĬLOV, GLEB K. (*1929), 194
MILHAUD, GASTON (1858–1918), **B**, 29, 31, 42
MILLÁS VALLICROSA, JOSÉ MARIA (1897–1970), 147, 231n
MINICH, SERAFINO RAFFAELE (1808–1853), 425
MINTO, WALTER (1753–1796), 163
MIODUSZEWSKI, JERZY (*1927), 201
MIRA FERNANDES, AURELIANO DE (1884–1958), 246
MISES, RICHARD VON (1883-1952), 218
MISRACHI, ELIA (15th/16th century), 119, 563
MITTAG-LEFFLER, GÖSTA (1846–1927), 157, 314, 418–420
MÖBIUS, AUGUST FERDINAND (1790–1868), 408, 456
Möbius Edition, 456
MOGENET, JOSEPH (1913–1980), **B**, 52, 58
MOHR, GEORG (1640–1697), 158, 446
MOISEEV, NIKOLAĬ D. (1902–1955), 189
MOLK, JULES (1857–1914), **B**, 28, 155
MOLL, GERARD (1785–1838), **B**, 50
MOLLWEIDE, CARL BRANDAN (1774–1825), 114
MOLODSHIĬ, VLADIMIR N. (1906–1986), 192
MONGE, GASPARD (1746–1818), 15, 21–22, 159, 186, 226, 397, 408, 472–473, 560
MONNA, ANTONIE FRANS (1909–1995), **B**, 56
MONTE, GUIDOBALDO DEL (1545–1607), 48, 63, 357–358
MONTEIRO DA ROCHA, JOSÉ (1734–1819), 242, 245, 540
MONTESQUIEU, CHARLES DE SECONDAT, BARON DE (1689–1755), 9
MONTEVERDE, MANUEL († 1868), 234–235
MONTICELLI, ANGELO (ca. 1700), 358
MONTMORT, PIERRE RÉMOND DE (1678–1719), 6, 10, 111
MONTUCLA, JEAN ETIENNE (1725–1799), **B**, 3, 6, 9–11, 13, 23, 48–50, 67, 69–70, 72, 110, 114, 124, 179, 199, 240, 257, 376, 400, 444, 463, 532, 535, 567
MOOKERJEE, *see* MUKHERJEE
MOORE, ELIAKIM HASTINGS (1862–1932), 266
MOORE, JOHN HAMILTON († 1807), 380
MORAIS, ABRÃO DE (1916–1970), 254
MORDUKHAĬ-BOLTOVSKOĬ, DMITRIĬ D. (1876–1952), **B**, 182, 191, 559
MORENO, GABRIEL (1735–1809), 251
MORGADO, JOSÉ (*1921), 246
MORNET, DANIEL (1878–1954), 489
MOSER, CHRISTIAN (1861–1935), 511
MOSSOTTI, OTTAVIO (1791–1863), 225
MOUREY, C. V. (19th century), 470
MOUY, PAUL (1888–1946), **B**, 34
MOZZONI, ANDREA (1754–1842), 13
MUBĀRAK, ʿALĪ PĀSHĀ (1823–1893), 321
MÜLLER, ANTON (1798–1860), 102
MÜLLER, CONRAD HEINRICH (1878–1953), **B**, 109, 130
MÜLLER, EMIL (1861-1927), 217
MÜLLER, FELIX (1843–1928), **B**, 130
MÜLLER, JOHANNES, *see* REGIOMONTANUS
MÜTEFERRIKA, IBRAHIM (†1745), 318–319
MUḤAMMAD BARAKA AL-ʿABADĪ (13th century), 320
MUḤAMMAD AL-BIBLAWĪ (1863–1954), 321
MUḤAMMAD AL-FĀSĪ (20th century), 326
MUḤAMMAD B. MAḤMŪD AL-ĀMULĪ (†1352), 320
MUḤAMMAD B. MŪSĀ AL-KHWĀRIZMĪ (fl. ca. 780–830), 46–47, 63, 118, 137–138, 194, 270, 323–324, 327, 366, 371, 454, 509
MUḤAMMAD B. ZAKARIYĀʾ AL-RĀZĪ (10th century), 326
MUIR, THOMAS (1844–1934), **B**, 173, 178, 208n
MUKHERJEE, ASUTOSH (1864–1924), 313

MULLACH, FRIEDRICH WILHELM AUGUST (1807–1881), 537
MUNK, SALOMON (1805–1867), 17
MURAWSKI, ROMAN (20th century), 202
MURDOCH, JOHN (*1927), 459
MURSĪ, AḤMAD MUḤAMMAD (20th century), 324
MUSHARRAFA, ʿALĪ MUṢṬAFĀ (1898–1950), 324–327
MUSIELAK, JULIAN (*1928), 201
MUSSCHENBROEK, PIETER VAN (1692–1761), 48
MUTIS, JOSÉ CELESTINO (1732–1808), 251
MUTSUHITO (1852–1912; Emperor of Japan from 1867, later named MEIJI Emperor), 289

NACHBIN, LEOPOLDO (1922–1993), 254
NAFĪSĪ, SAʿĪD (1895–1967), 323
NAGEL, FRITZ (*1940), 562
NAKAMURA KOSHIRO (1901–1986), 294
NALLINO, CARLO A. (1872–1938), 322
NAPIER, JOHN (1550–1617), 47, 54, 163, 174, 438
NAPOLEON I BONAPARTE (1769–1821), 11–13, 15, 21, 45, 69, 400
NÁPOLES, ALFONSO (1897–1992), 258, 260
NARDUCCI, ENRICO (1832–1893), 64, 66, 81, 358, 372, 418
NARKIEWICZ, WŁADYSŁAW (20th century), 202
NAṢĪR AL-DĪN AL-ṬŪSĪ (†1274), 4, 147, 320, 323, 327
National Council of Teachers of Mathematics (NCTM), 335
NATUCCI, ALPINOLO (1883–1975), 94
NAUX, CHARLES (20th century), 37
AL-NAYRĪZĪ (†924), 121, 402
NAẒĪF, MUṢṬAFĀ (20th century), 324, 327
NECKER, LOUIS (1730–1804), 99
NEEDHAM, JOSEPH (1900–1995), **B**, 290, 302, 504
NELLI, GIOVANNI BATTISTA CLEMENTE DE' (1725–1793), 76
NENCI, ELIO (*1960), 64
NEPER, JOHN, *see* NAPIER, JOHN
NERNST, WALTER (1864–1941), 529
NESSELMANN, GEORG HEINRICH FERDINAND (1811–1881), **B**, 75, 112, 114–117, 123, 535–537
Netherlands, *see* Benelux
NETTO, EUGEN (1848–1919), 389, 418
NEUENSCHWANDER, ERWIN (1942), 97
NEUGEBAUER, OTTO (1899–1990), **B**, 127, 134–135, 138, 140–141, 158, 160, 276, 283, 384, 442, 456, 509, 513, 518, 543, 548, 556
NEVANLINNA, ROLF (1895–1980), **B**, 159, 561
NEWTON, HUBERT ANSON (1830–1896), 311
NEWTON, ISAAC (1642–1727), 10, 17, 21, 39, 55, 73–74, 90, 92, 131, 139, 144, 154, 162–163, 166–168, 170, 175–178, 181–183, 186–187, 224, 228, 251, 257, 267, 304, 351–352, 364, 386, 392, 406–407, 424, 428, 446–448, 452, 460–461, 469, 474, 491, 510, 545, 560, 572
Newton Edition, 461, 545
NI TINGMEI (late 8th century), 303
NICHOLAS OF CUSA (1401–1464), 139, 391, 448
NICOLAS RHABDAS († 1350), 537
NICOLE ORESME (1323?–1382), 121, 136, 215, 402, 565
NICOMACHUS OF GERASE (ca. 100 A.D.), 46–47, 223, 270, 326, 540
NIEBUHR, BARTHOLD GEORG (1776–1831), 115
NIELSEN, NIELS (1865–1931), **B**, 153–154, 158
NIKEPHOROS GREGORAS (1295–1359), 223
NIKOMACHUS OF GERASE, *see* NICOMACHUS OF GERASE
NĪLAKAṆṬHA, SOMAYĀJI (16th century), 309
NIMR, FĀRIS (1859–1951), 322
NIPSUS, M. IUNIUS (2nd century A.D.?), 123, 389
NISHIDA KITARO (1870–1945), 293
NÖBAUER, WILFRIED (1928–1988), 218
NØRLUND, NIELS ERIK (1885–1981), 158
NOETHER, EMMY (1882–1935), 218, 395, 496, 547–549
NOETHER, MAX (1844–1921), 457, 555
NORDEN, EDUARD (1868–1941), 444
NORTH, FREDERIC (1766–1827; fifth Earl of Guilford), 226
Norway, *see* Scandinavia
NOVÝ, LUBOŠ (*1929), 191, 205, 211
NUNES, PEDRO (1502–1578), 240–241, 244, 439, 540

OCAGNE, MAURICE D' (1862–1938), **B**, 37, 414
OCTAVIO DE TOLEDO, LUIS (1857–1934), 233
OETTINGEN, ARTHUR VON (1836–1920), 131
OGURA KINNOSUKE (1885–1962), 292–293
OLBERS, WILHELM (1758–1840), 128–129

OLDENBURG, HENRY (ca. 1620–1677), 51
OLIVEIRA, JOSÉ TIAGO DA FONSECA (1928–1992), 246
OMAR KHAYYAM, *see* ʿUMAR AL-KHAYYĀM
OPIAL, ZDZISŁAW (1930–1974), 201
ORE, ØYSTEIN (1899–1968), **B**, 153
ORESME, NICOLE, *see* NICOLE ORESME
ORTELIUS, ABRAHAM (1527–1598), 52
ORTHMANN, CARL (1839–1885), 130
ORTROY, FERNAND VAN (1856–1934), **B**, 52
OSGOOD, WILLIAM FOGG (1864–1943), 380
OSTROGRADSKIĬ, MIKHAIL V. (1801–1862), 193–194, 573
OSTROWSKI, ALEXANDER (1893–1986), 128
OSTWALD, WILHELM (1853–1932), 130, 484
OUGHTRED, WILLIAM (1575–1660), 266, 386
OWEN, GWILYM ELLIS LANE (*1922), 459
OZANAM, JACQUES (1640–1714), 448, 490
OZHIGOVA, ELENA P. (1923–1994), **B**, 190–192, 194

PACIOLI, LUCA (1445?–1514?), 63–64, 70, 400
PAGANI, GASPAR MICHEL (1796–1855), 437
PAIGE, CONSTANTIN LE, *see* LE PAIGE, CONSTANTINE
PANCKOUCKE, CHARLES-JOSEPH (1736–1798), 490
PĀṆINI (ca. 500 B.C.), 308
PAOLI, PIETRO (1759–1839), 78
PAOLO DELL'ABACO (†1375?), 370
PAPLAUSKAS, AL'GIRDAS B. (1931–1984), **B**, 188–189
PAPP, DESIDERIO (1895–1993), **B**, 253, 484
PAPPUS OF ALEXANDRIA (ca. 320), 5, 52, 63, 120–121, 132, 163, 222, 245, 357, 442–443, 509, 536, 560
Pappus Edition, 443, 536
PARAMEŚVARA (ca. 1440), 311
PAREDES, JOAQUÍN GREGORIO (1778–1839), 251
PARMENIDES (ca. 515–445), 93, 106
PARSHALL, KAREN HUNGER (*1955), 336
PARSHIN, ALEKSEĬ N. (*1942), 188
PASCAL, BLAISE (1623–1662), 13, 21–22, 51, 294, 362–363, 376, 408–409, 448, 452
Pascal Edition, 376
PASCH, MORITZ (1843–1930), 275
PASQUIER, LOUIS GUSTAVE DU (1876–1957), 105
PATERNÒ, EMANUELE (1847–1935), 484
PATRICIUS, FRANCIS (†1597), 406
PATY, MICHEL (*1938), 41
PAULY, AUGUST (1796–1845), 120
PAVLOV, IVAN P. (1849–1936), 261n
PAWLIKOWSKA-BROŻEK, ZOFIA (*1941), 199, 201
PEACOCK, GEORGE (1791–1858), 168, 406
PEANO, GIUSEPPE (1858–1932), 82–83, 91, 94, 393–394, 432–434, 470, 545–546
Peano Edition, 394
PEARSON, KARL (1857–1936), **B**, 174
PEDERSEN, OLAF (1920–1997), 158
PEDIASIMOS (ca. 1330), 119
PEET, THOMAS ERIC (1882–1934), 138
PEGOLOTTI, FRANCESCO BALDUCCI (ca. 1290–1347), 138
PEIFFER, JEANNE (*1948), xxi, 3, 161, 329
PEIPER, RUDOLF (1834–1898), 121
PEIRCE, BENJAMIN (1809–1880), 268
PEIRCE, CHARLES SANTIAGO SANDERS (1839–1914), **B**, 264, 281–282, 416, 480, 538
Peirce Edition, 416
PELCZAR, ANDRZEJ (*1937), 201
PELETIER, JACQUES (1517–1582), 4
PELSENEER, JEAN (1903–1985), **B**, 52, 58
PENHA, GUILHERME M. DE LA (1943–1996), 254
PESTRE, DOMINIQUE (*1950), 30
PETERSEN, CARL (1813–1880), 366
PETERSEN, JULIUS (1839–1910), 574
PETERSON, KARL M. (1828–1881), 192
PETROSYAN, GAREGIN B. (1902–1998), 185
PETROVA, SVETLANA S. (*1933), 187–188, 192, 475
PETRUS DE DACIA (13th century), 402
PETZVAL, JOSEPH (1807–1891), 567
PEUERBACH, GEORG VON (1423–1461), 46, 214–216, 352
PEYPERS, HENDRIK (1853–1904), 57
PEYRARD, FRANÇOIS (1760–1822), **B**, 13
PHILI, CHRISTINE (*1945), 221
PHILIP II (1527–1598; King of Spain from 1556), 45, 233
PICARD, CHARLES EMILE (1856–1941), 32, 373, 401, 422, 506
PICATOSTE RODRÍGUEZ, FELIPE (1834–1892), **B**, 233
PICCOLOMINI, ENEA SILVIO (1405–1464, Pope PIUS II from 1458), 98
PINCHERLE, SALVATORE (1853–1936), 81, 351, 373
PINGREE, DAVID (*1933), 496
PIOLA, GABRIO (1791–1850), 76–78
PITISCUS, BARTOLOMAEUS (1561–1613), 54, 438

PIUS II, Pope, *see* PICCOLOMINI, ENEA SILVIO
PIUS IX (1857–1939; Pope from 1922), 369
PLANA, GIOVANNI (1781–1864), 432
PLANCHEREL, MICHEL (1885–1967), 105
PLANCIUS, PETER (1552–1622), 52
PLANCK, MAX (1858–1947), 261n
PLATO (429–348), 106, 134, 142, 221–222, 224, 429, 536, 548–549, 577
PLATO OF TIVOLI (12th century), 63, 369
PLAYFAIR, JOHN (1748–1819), 163, 310
PLEDGE, HUMPHREY THOMAS (1903–1960), 175
PLIMPTON, GEORGE ARTHUR (1855–1936), 525
PLÜCKER, JULIUS (1801–1868), 81, 267, 408, 455–456
Plücker Edition, 456
PLUTARCH (ca. 46– after 120), 64
POGODA, ZDZISŁAW (*1955), 202
POGREBYSSKIĬ, IOSIF B. (1906–1972), **B**, 188–189
POINCARÉ, HENRI (1854–1912), 24, 32–33, 41, 141, 188, 194, 227, 273, 314, 373, 380, 413, 456, 485, 501
Poincaré Edition, 41
Poland
  Academy of Art in Warsaw, 409
  Adam Mickiewicz University in Poznań, 202
  Commission on History of Mathematics of the Polish Mathematical Society, 201–202
  Eleventh International Congress for the History of Science in Warsaw and Cracow (1965), 146
  Jagiellonian University in Cracow, 200–202
  Józef Mianowski Fund, 202
  Komisja Edukacji Narodowej, 199
  Polish Academy of Art in Warsaw, 409
  Polish Academy of Sciences (PAN) in Warsaw, 201
  Polish Mathematical Society, 201
  Polytechnic Institute in Warsaw, 492
  Royal University in Warsaw
    Department for the History of Mathematics, 199
  Schools of History of Mathematics (Conferences for mathematicians and historians of mathematics), 201
  University of Cracow, 214
  University of Warsaw, 182, 202, 409, 492
  University of Wrocław, 202
  Warsaw Society of Sciences, 409
POLISHCHUK, EFIM M. (1914–1988), 190
POLUBARINOVA-KOCHINA, PELAGEYA YA. (1899–1999), 197
PONCELET, JEAN VICTOR (1788–1867), 21, 362, 397, 450
POPOV, GEORGIĬ N. (1878–1930), 182
POPPE, JOHANN HEINRICH MORITZ (1776–1854), 113–114
PORPHYRY (234–ca. 305), 222, 560
Portugal
  Academy of Sciences in Lisbon, 240–243, 507, 531
  Advanced Institute for Agronomy in Lisbon, 551
  Army School in Lisbon, 244, 404, 439, 551
  International Congress of Bibliography of the Mathematical Sciences (1889), 243
  National Seminar for the History of Mathematics, 247
  O Instituto in Coimbra, 242, 394
  Polytechnic Academy in Oporto, 439
  Polytechnic School in Lisbon, 244, 404, 551
  Portuguese Group for History of Science, 551
  Third International Congress for the History of Science in Lisbon, Coimbra, and Oporto (1934), 35, 551
  University of Coimbra, 240–242, 246–247, 394, 507, 531, 539
  University of Lisbon, 245–247
  University of Minho, 247
  University of Oporto, 247
POSTEL, GUILLAUME (1510–1581), 4
POUDRA, NOËL-GERMINAL (1794–1894), **B**, 20, 22
POULIOT, ADRIEN (1896–1977), 285
POWELL, BADEN (1796–1860), 167
PRAG, ADOLF (*1906), 136, 139–141, 177
PRASAD, GANESH (1876–1935), **B**, 314
PRIETO, SOTERO (1884–1935), 256n, 258–259, 262
PRINGSHEIM, ALFRED (1850–1941), 432
PROCLUS (ca. 410–485), 46–47, 52, 98, 119, 161, 222, 509, 536, 552
PSELLOS, MICHAEL (1018–ca. 1078), 223, 444, 537

PTOLEMY (ca. 100–ca. 160), 13, 46–47, 52, 63, 98, 157, 196, 214, 245, 257n, 366, 441, 443, 463, 487, 509, 537, 553, 560
PUISEUX, VICTOR (1820–1883), 437
PYTHAGORAS (ca. 580–500 B.C.), 18n, 64, 67, 75, 221–222

AL-QALAṢĀDĪ, ʿALĪ, *see* ʿALĪ AL-QALAṢĀDĪ
QIAN BAOCONG (1892–1974), **B**, 297, 303–305, 464
QIAN DAXIN (1728–1804), 302
QIANLONG (1711–1799; Emperor of China from 1735), 300
QIN JIUSHAO (1202–1261), 302, 570
QUETELET, ADOLPHE (1796–1874), **B**, 50, 397
QURBĀNĪ, ABŪ L-QĀSIM (*1911), 327
QUSṬĀ IBN LŪQĀ (†912), 392

RAABE, JOSEPH LUDWIG (1801–1859), 567
RADAKOVIČ, MICHAEL (1866–1934), 495
RAEMDONCK, JAN HUBERT VAN (1817–1899), 57
RAHN, JOHANN HEINRICH (1622–1676), 563
RAJAGOPALAN, K. R. (20th century), 315
RAKHMANINOV, PËTR A. (†1813), 179
RAMANUJAN, SRINIVASA (1889–1920), 480, 531
RAMBAUD, ALFRED NICOLAS (1842–1905), 28
RAMEAU, JEAN PHILIPPE (1683–1764), 73
RAMÉE, PIERRE DE LA (1515–1572), 4, 48, 112, 388
RAMOS, SAMUEL (1897–1959), 261
RAMUS, PETRUS, *see* RAMÉE, PIERRE DE LA
RAṄGĀCĀRYA, M. MALUR (1861–1916), 269
RANKE, LEOPOLD VON (1795–1886), 115, 121
RAPHSON, JOSEPH (early 17th cent.), 162n
RASHED, ROSHDI (*1936), 39–41
RASHEVSKIĬ, PËTR K. (1907–1983), 184
RATDOLT, ERHARD (1447–1528?), 4
RAVENSTEIN, ERNEST GEORGE (1834–1913), 244
AL-RĀZĪ, MUḤAMMAD B. ZAKARĪYĀʾ, *see* MUḤAMMAD B. ZAKARĪYĀʾ AL-RĀZĪ
RECILLAS, FÉLIX (20th century), 260
RECORDE, ROBERT (1510–1558), 406
REGIOMONTANUS (1436–1476), 46, 110, 121, 214–216, 218, 403, 448
REICH, KARIN (*1941), 138, 557
REICHARDT, HANS (1908–1991), 144
REICHENBACH, HANS (1891–1953), 261n
REIDEMEISTER, KURT (1893–1971), **B**, 140
REIMER, NIKOLAUS THEODOR (1772–1832), 13, 114, 376
RELLICH, FRANZ (1906–1955), 548
REY, ABEL (1873–1940), **B**, 26, 31, 34–36, 43
REY PASTOR, JULIO (1888–1962), **B**, 233–237, 252–253, 355
REYES PRÓSPER, VENTURA (1863–1922), **B**, 232–233, 236
REYMOND, ARNOLD (1874–1958), 37
REZA PAHLEVI, KHAN (1878–1944; Shah of Iran from 1925), 525
RHIND, A. HENRY (1833–1863), 268
RIBEIRO DOS SANTOS, ANTÓNIO (1745–1818), **B**, 240
RICCARDI, PIETRO FRANCESCO (1828–1898), **B**, 66, 77, 80, 418, 471
RICCATI, VINCENZO (1707–1775), 68, 70
RICCI, MATTEO (1552–1610), 299, 570
RICCI, MICHELANGELO (1619–1682), 448
RICCIOLI, GIOVAMBATTISTA (1598–1671), 65
RICHARD, CLAUDE (1588/89–1664), **B**, 46
RICHELOT, FRIEDRICH JULIUS (1808–1875), 495
RICHESON, ALLIE W. (1897–1966), 281
RIEMANN, GEORG FRIEDRICH BERNHARD (1826–1866), 23, 81, 194, 293, 314, 457
Riemann Edition, 457
RIES, ADAM (1492–1555), 120, 144
RIGAUD, STEPHEN PETER (1774–1839), 166–167, 177
RIGNANO, EUGENIO (1870–1930), 422
RISNER, FRIEDRICH († ca. 1580), 558
RIVAUD, ALBERT (1876–1956), 32
ROBERT OF CHESTER (ca. 1140), 46–47, 270, 454
ROBERTUS ANGLICUS (13th century), 403
ROBERVAL, GILES PERSONE DE (1602–1675), 37, 90, 294, 409, 448, 452
ROCHOT, BERNARD (1900–1971), **B**, 26
RODET, LEON (*ca. 1850), 19
RODRÍGUEZ, FRIAR DIEGO (ca. 1596–1668), 256, 257n
RÖMER, OLE (1644–1710), 415
RÖTH, EDUARD MAXIMILIAN (1807–1858), 116, 149n
ROGER, JACQUES (1920–1990), 43
ROMANUS, ADRIANUS (1561–1615), **B**, 46–47, 376, 436

ROME, ADOLPHE (1889–1971), **B**, 52, 58, 486
ROSEN, FREDERIC (1805–1837), 118, 397
ROSENTHAL, ARTUR (1887–1959), 495
ROSS, DAVID (Sir, 1877–1971), 442
ROSSETTI, LUCIA (2nd half of 20th century), 426
ROUSE BALL, WALTER WILLIAM (1850–1925), **B**, 82, 169–170, 178
ROWE, DAVID (*1950), 147, 336
ROZENFELD, BORIS A. (*1917), 188–189, 192, 573
ROZHANSKAYA, MIRA M. (*1928), 192, 573
RUAN YUAN (1764–1849), 302–304
RUDIO, FERDINAND (1856–1929), **B**, 101–103, 106, 124, 129–130, 390, 569
RUDOLF II (1552–1612; King of Bohemia 1575, Emperor 1576), 206
RUFFINI, PAOLO (1765–1822), 67, 78, 90, 92, 374
Ruffini Edition, 374
RUFINI, ENRICO (1890–1924), 92
Rumania
  Cluj University, 521
  University of Bucharest, 521
RUSKA, JULIUS (1867–1949), **B**, 131–133, 137, 565
RUSSELL, BERTRAND (1872–1970), 395, 479–480, 538
Russia and the U.S.S.R.
  Academy of Pedagogical Sciences of the Russian Soviet Federate Socialist Republic (RSFSR), 478
  Academy of Pedagogical Sciences of the USSR, 478
  Academy of Sciences of the USSR, 460, 462, 475–476, 482, 500, 503, 523, 532, 572, 578
  Academy of Sciences of the Uzbek SSR, 455
  All-Union Conferences on the History of Physico-Mathematical Sciences, 194
  Chernovtsy University, 522
  Communist Academy, 571
  Don Polytechnic Institute in Novochenkassk, 492
  Ecole supérieure féminine in St. Petersburg, 385
  Encyclopedias, 186, 460, 542
  Financial Institute in Moscow, 476
  Gymnasium in St. Petersburg, 112
  Herzen Pedagogical Institute in Leningrad, 475
  Imperial Academy of St. Petersburg, 399
  Institute for the History of Science and Technology in Leningrad, 183
  Institute for the History of Science and Technology in Moscow, 187–189, 195
  Institute for the History of the Ukrainian Academy of Sciences in Kiev, 526
  Institute of Foreign Languages in Lvov, 475
  International Congress of the History of Science in Moscow (1971), 481
  Kaluga Pedagogical Institute, 482
  Kharkov Engineering-Economic Institute, 480
  Kiev Institute of Aviation Engineers, 522
  Leningrad Higher Artillery School, 500
  Leningrad Pedagogical Institute, 408
  Lomonosov University, *see University of Moscow*
  Lvov University, 475
  Military High School in Nizhniĭ Novgorod, 367
  Moscow Higher Technical School, *see Moscow Technical University*
  Moscow Mathematical Society, 559
  Moscow State University, *see University of Moscow*
  Moscow Supreme Technical School, *see Moscow Technical University*
  Moscow Technical University, 571
  Moscow Technological Institute, 522, 553
  Odessa Polytechnic Institute, 542
  Pyatigork Pedagogical Institute, 493
  Russian Academy of Sciences in St. Petersburg, 102–103, 112, 129, 391
  Samara University, 475
  Soviet Academy of Sciences in Leningrad, 475
  Soviet Academy of Sciences in Moscow, 183, 187
  Soviet National Committee of Historians of Science and Technology, 195

St. Petersburg Naval Academy, 462
St. Petersburg Technological Institute, 462
Timiryazev Institute, 571
Tula Pedagogical Institute, 559
Tula Polytechnic Institute, 559
Ukrainian Academy of Sciences in Kiev, 190, 437, 503, 526
University of Kazan', 181–183, 552
University of Kharkov, 183
University of Kiev, 179–180, 183, 385, 552
University of Leningrad, 183, 475, 496, 523, 532
University of Moscow, 180, 184, 186–188, 195, 367, 437, 454, 460, 475, 478, 559, 570–573
Department for the History of Mathematics and Mechanics, 187
University of Nizhnii Novgorod, 475
University of Odessa, 181, 183, 454, 541–542
University of Rostov-on-Don, 182–183, 492, 559
University of Saratov, 522
University of St. Petersburg, 385, 552
University of Tashkent, 190, 455
Vyatka Pedagogical Institute, 408
RUSSO, FRANÇOIS (1909–1998), 35, 37
RUTTEN, MARGUERITE (20th century), 383
RYBKIN, GEORGIĬ F. (1903–1972), 187, 573
RYBNIKOV, KONSTANTIN A. (*1913), 185–188, 195, 573
RYCHLÍK, KAREL (1885–1968), 209

ṢABRĀ, ʿABD AL-ḤAMĪD I. (*1924), 327, 459
SACCHERI, GIROLAMO (1667–1733), 67, 79, 83, 546
SACHS, ABRAHAM A. (1914–1983), **B**, 276
SACROBOSCO (1st half of 13th century), 121, 270, 402
SAʿĪDĀN, AḤMAD S. (1914–1991), **B**, 323, 327
SALADINI, GIROLAMO (1731–1813), 78
SALIÉ, HANS (1902–1978), 144
SÁNCHEZ PÉREZ, JOSÉ AUGUSTO (1882–1958), **B**, 236n
SANFORD, VERA (1891–1971), **B**, 274
SANTALÓ, LUIS (1911–2001), 253
SANTILLANA, GIORGIO DE (1902–1974), 86, 92–93, 424, 428
SARAIVA, LUIS M. R. (*1951), 239
SARKAR, BENOY KUMAR (ca. 1918), 315
ṢARRŪF, YAʿQŪB (1839–1912), 322
SARTON, GEORGE (1884–1956), **B**, 25, 29, 51n, 57, 110, 227, 273–274, 284, 329–330, 338, 484–485, 516
SASAKI CHIKARA (*1947), 289, 294
ŚASTRĪ, BĀPŪDEVA (1821–1900), 312
SAVASORDA, ABRAHAM, *see* ABRAHAM SAVASORDA
SAVÉRIEN, ALEXANDRE (*1720 or 1723), 8
SAVILE, HENRY (1549–1622), 162
SAYILI, AYDIN (1913–1993), **B**, 324–327
SCALIGER, JOSEPH (1540–1609), 46, 509
Scandinavia
Åbo Academy, 521
Askov Folk High School, South Jutland, 156, 462
Mittag-Leffler Institute in Djursholm, Sweden, 420
Norges Tekniske Högskole in Trondheim, 159, 384
Rask Ørsted Fond, 497
Royal Academy of Sciences in Stockholm, 152, 365
Royal Danish Academy of Sciences and Letters in Copenhagen, 152, 574, 578
Royal Danish Academy of Arts and Sciences in Copenhagen, 443
Royal Library in Copenhagen, 157, 366
Royal Library in Stockholm, 417
Royal Norwegian Academic Society of Sciences and Letters in Trondheim, 152
Royal Swedish Academy in Stockholm, 526
Stockholms Högskola, 420
Swedish University in Helsinki, 521
Technical University in Trondheim, 159, 275, 354
University of Åbo , 152
University of Aarhus, 159
University of Christiania, later University of Oslo, 151, 154, 364, 382, 449, 467, 531, 534
University of Copenhagen, 141, 152, 154, 157–158, 365, 398, 415, 442–443, 446, 453, 462, 496–497, 499, 518, 574
University of Helsinki, 499
University of Lund, 152, 417
University of Oslo, 159, 364, 384, 461, 469, 499
University of Stockholm, 561

University of Uppsala, 151, 158, 404, 417, 521
SCHÄFER, CLEMENS (1878–1968), 128
SCHEINER, CHRISTOPH (1575–1650), 47, 126
SCHEMBERG, MARIO (1916–1990), 254
SCHERING, ERNST C. J. (1833–1897), 81, 128
SCHEYNER, CHRISTOPH, *see* SCHEINER, CHRISTOPH
SCHINZEL, ANDRZEJ (*1937), 201
SCHLÄFLI, LUDWIG (1814–1895), 104
Schläfli Edition, 104
SCHLESINGER, LUDWIG (1864–1933), 128–129
SCHLICK, MORITZ (1882–1926), 434
SCHLÖMILCH, OSKAR XAVER (1823–1901), 124, 388
SCHLÜSSEL, CHRISTOPH, *see* CLAVIUS
SCHMIDT, ERHARD (1876–1959), 555
SCHMIDT, OLAF (1913–1996), **B**, 158
SCHMIDT, WILHELM (1862–1905), 120, 511
SCHNEIDER, IVO (*1938), 147, 336
SCHÖNE, HERMANN (1870–1941), 120
SCHOLZ, HEINRICH (1884–1956), **B**, 134, 445
SCHOOTEN, FRANS VAN, JR. (ca. 1615–1660), **B**, 5, 46, 448
SCHOOTEN, FRANS VAN, SR. (1581?–1645), 551
SCHOTT, ALBERT (*1901, † probably between 1941 and 1945), 135
SCHOTTKY, FRIEDRICH (1851–1935), 526
SCHOY, CARL (1877–1925), **B**, 133
SCHRECKER, PAUL (1889–1963), 35
SCHREIBER, PETER (*1938), 148
SCHREIER, OTTO (1901–1929), 547
SCHRÖDINGER, ERWIN (1887–1961), 529
SCHROETER, CARL (1855–1939), 511
SCHUBERT, HERMANN (1848–1911), 574
SCHUBRING, GERT (*1944), 147
SCHUMACHER, HEINRICH CHRISTIAN (1780–1850), 128–129
SCHWARZ, HERMANN AMANDUS (1843–1921), 526
SCINÀ, DOMENICO (1764–1837), 71
SCOTT, JOHN FREDERICK (1892–1971), 176
SCRIBA, CHRISTOPH J. (*1929), xxi, 109, 145–147, 191, 285–287, 329, 481
SÉDILLOT, JEAN-JACQUES E. (1777–1832), **B**, 16, 23, 520
SÉDILLOT, LOUIS PIERRE-EUGÈNE AMÉLIE (1808–1875), **B**, 16–21, 81, 363
SEGNER, JOHANN ANDREAS (1704–1777), 224–225
SEGRE, CORRADO (1863–1924), 79, 399, 421, 470, 474
SEIDEL, LUDWIG (1821–1896), 126
SEIDENBERG, ABRAHAM (1916–1988), **B**, 282
SEKI TAKAKAZU (†1708), 291, 485
SELENIUS, CLAS-OLOF (1922–1991), **B**, 158
SELLA, QUINTINO (1827–1884), 369
SEMPILIUS, HUGO, *see* SEMPLE, HUGH
SEMPLE, HUGH (1596–1654), **B**, 47–48
SEN, SAMARENDRA NATH (1918–1992), 315
SENGUPTA, PRABODH CHANDRA (1876–1962), 315
SERENUS OF ANTINOË (4th century A.D.), 52, 63, 157, 443
SERGESCU, PIERRE (1893–1954), **B**, 31, 35–38
SERRET, JOSEPH ALFRED (1819–1885), 563
SERRUS, CHARLES (1886–1946), 34
SETH, son of ADAM, 534
SETHE, KURT (1869–1934), 496
SEVĀRĀMA, PANDIT (19th century), 312
SEVERI, FRANCESCO (1879–1961), 392, 421–423, 438
SEZGIN, FUAT (*1924), 147
SHAFAREVICH, IGOR' R. (*1923), 188
SHAHRĪYĀRĪ, PARVĪZ (*1926), 327
SHEARMAN, ARTHUR THOMAS (1866–1937), 170
SHEĬNIN, OSKAR B. (*1925), 193
SHEREMETIEVSKIĬ, V. P. (†1919), 181
SHI SHEN (SHI SHENFU) (mid-4th century B.C.), 505
SHRAER, MOISEĬ G. (1918–1983), 189
SHTOKALO, IOSIF Z. (1897–1987), 190, 194
SHTYKAN, ABRAM B. (1906–1985), 193
SHUKLA, KRIPA SHANKAR (*1918), 315
SIBṬ AL-MĀRIDĪNĪ († ca. 1495), 320
SICIAK, JÓSEF (*1931), 201
SIDLER, GEORG JOSEPH (1831–1907), 102
SIEGEL, CARL LUDWIG (1896–1981), 134–136
SIEGMUND-SCHULTZE, REINHARD (*1953), 147
SIERPIŃSKI, WACŁAW (1882–1969), 202
SIERRA, JUSTO (1848–1912), 258
SILVESTRE DE SACY, ANTOINE ISAAC (1758–1838), 16, 18, 519
SIMART, GEORGES (*1846), 422
SIMON, MAX (1844–1918), **B**, 126
SIMONOV, NIKOLAĬ I. (1910–1979), **B**, 189, 194
SIMONOV, REM A. (*1929), 193
SIMONS, LAO G. (1870–1949), **B**, 274

SIMPLICIUS (ca. 520 A.D.), 222, 443, 511, 536
SIMSON, ROBERT (1687–1768), 21, 163, 171, 175
SINGH, AVADHES NARAYAN (1901–1954), **B**, 314–315
SIRAZHDINOV, SAGDY KH. (1921–1988), 190
ŚIVA (Indian god), 307
SIXTUS IV (1414–1484; Pope from 1471), 215
SLAVUTIN, EVGENIĬ I. (*1948), 188
ŚLEBODZIŃSKI, WŁADYSŁAW (1884–1972), 201
SLUSE, RENÉ FRANÇOIS DE (1622–1685), 51, 437, 463
SMEUR, ALPHONSE J. E. M. (*1925), 56
SMIRNOV, VLADIMIR I. (1887–1974), **B**, 189, 194, 572
SMITH, DAVID EUGENE (1860–1944), **B**, 137, 264–266, 269–270, 272–274, 290, 302, 386–387, 418, 437, 454, 464, 485
SMOLÍK, JOSEF (1832–1915), 207
SNELLIUS, WILLEBRORD (1580–1626), **B**, 46–47, 413, 558
ŚNIADECKI, JAN (1756–1830), 199–200
SOLOGUB, VLADIMIR S. (1927–1982), **B**, 189–190
SOLOV'ËV, ALEKSANDR D. (*1927), 188
SOMMERFELD, ARNOLD (1868–1951), 495, 565
SOMMERVILLE, DUNCAN MCLAREN YOUNG (1879–1934), 174
SOPPELSA, MARIA LAURA (*1947), 426
SOUISSI, MOHAMED (20th century), 324
South America
Colegio Nacional de Buenos Aires, 252
Consejo Nacional de Investigaciones Científicas (CONICET) in Argentina, 355
Escola Politécnica do Rio de Janeiro, 254
Fundación para el Estudio del Pensamiento Argentino e Iberoamericano (FEPAI), 252
Grupo Argentino de Historia de la Ciencia (Buenos Aires), 501
Grupo Chileno de Historia de la Ciencia (Santiago de Chile), 501
Pontificia Universidad Católica de Chile in Santiago, 254
Seminario Nacional de Historia da Matematica in Brazil, 255
Sociedad Científica Argentina, 254
Sociedad Latinoamericana de Historia de las Ciencias y la Tecnologia (SLHCT), 254–255
Sociedade Brasileira de História da Matemática (SBHMat), 255
Unión Matemática Argentina, 252
Universidad de Buenos Aires, 252–253, 355
Universidad de la Republica in Montevideo, 253–254, 431
Universidad de Rosario in Argentina, 253
Universidad de Saõ Paolo, 562
Universidad Nacional del Litoral in Santa Rosario, 484
Universidad Nacional del Litoral in Santa Fé, 252–253, 484
University of Buenos Aires, 506
SOUZA, JOAQUIM GOMES DE (1829–1863), 251
SOUZINHA, *see* SOUZA, JOAQUIM GOMES DE
Spain
Academia de Matemáticas in Madrid, 233
Asociación de Historiadores de la Ciencia Española, 236
Biblioteca del Escorial in Madrid, 514
Civil Engineering School in Madrid, 415
Colegio Imperial in Madrid, 509
Faculty of Philosophy in Madrid, 235
Faculty of Science in Madrid, 235
Institute for Islamic Studies in Madrid, 326
International Exhibition in Seville (1929), 244, 247
Junta para Ampliación de Estudios e Investigaciones Científicas (JAE), 233n, 415, 506
Real Academia de Ciencias Exactas, Físicas y Naturales (RACEFNM) in Madrid, 232, 234, 415, 417, 539
Sociedad Española de Historia de las Ciencias y de las Técnicas (SEHCYT) in Madrid, 236
Spanish Mathematical Society, 415, 507
Spanish school of Millás Vallicrosa in Barcelona, 231n, 236n
University of Madrid, 506
University of Oviedo, 233, 506
University of Salamanca, 256
University of Zaragoza, 431, 506

SPEISER, ANDREAS (1885–1970), 103, 106, 527
SPEISER, DAVID (*1926), 104
SPELUZZI, BERNARDINO (1835–1898), 251
SPEZIALI, PIERRE (1913–1995), 105
SPIEGELBERG, WILHELM (1870–1930), 555
SPIESS, OTTO (1878–1966), **B**, 102–104, 106, 428
SPINOZA, BARUCH DE (1632–1677), 53
SPRAT, THOMAS (1635–1713), 166
SPRINGER, FERDINAND (1881–1965), 158
ŚRĪDHARA (ca. 750 A.D.), 309
SRINIVASIENGAR, C. N. (1901–1972), 315
ST. RUPERT (6th century A.D.), 213
STABIUS, JOHANNES (†1522), 215–216
STÄCKEL, PAUL (1862–1919), **B**, 79, 103, 127–130, 420, 445, 511
STAIGER, ELISABETH, née KLEIN (1888–1968), 456
STALIN, IOSIF V. (1879–1953), 143, 183, 190
STAMATIS, EVANGELOS (1898–1990), **B**, 228–229
STAMM, EDWARD (1886–1940), 201
STAUDT, CHRISTIAN VON (1798–1867), 448
STEENSTRA, PIBO (†1788), **B**, 49
STEINER, JAKOB (1796–1863), 104, 267, 387, 408, 567
STEINITZ, ERNST (1871–1928), 547
STEINSCHNEIDER, MORITZ (1816–1907), **B**, 118–119, 418
STEKLOV, VLADIMIR A. (1864–1926), 182
STENZEL, JULIUS (1883–1935), 134–135, 276, 445, 496, 543, 548
STEPANOV, VYACHESLAV V. (1899–1950), 186
STEPHANIDES, MICHAEL (1868–1957), 227, 530
STEPHANOS, KYPARISSOS (1857–1917), 227
STERN, MORITZ ABRAHAM (1807–1894), 387, 511
STEVIN, SIMON (1548–1620), 46–47, 50, 53, 55, 57, 298, 333, 376, 406, 411–413, 438, 525
Stevin Edition, 55
STEWART, MATTHEW (1717–1785), 171
STIBORIUS, ANDREAS (ca. 1470– ca. 1515), 215–216
STIFEL, MICHAEL (ca. 1487–1567), 110, 388, 448
STIRLING, JAMES (1692–1770), 175
STOCKLER, FRANCISCO DE BORJA GARÇÃO (1759–1829), **B**, 239–246
STÖRMER, CARL (1874–1957), **B**, 153, 383, 534
STOLZ, OTTO (1842–1905), 216
STONE, EDMUND (1700?–1768), 163
STRABBE, ARNOLDUS BASTIAAN (1741–1805), **B**, 49
STRÖMER, M. (18th century), 152
STRUIK, DIRK JAN (1894–2000), **B**, 54–55, 61, 110, 113, 144, 260n, 282, 305, 456
STRUIK, RUTH (1896–1995), 61, 92, 423
STRUVE, VASILIĬ V. (1889–1965), **B**, 135, 183, 496–497
STRUYCK, NICOLAAS (1687–1769), 558
STUDNIČKA, FRANTIŠEK JOSEF (1836–1903), 207–208
STUDY, EDUARD (1862–1930), 399
STULOFF, NICOLAI N. (*1914), 147
STURM, JOHANN CHRISTOPH (1635–1703), 110
STYAZHKIN, NIKOLAĬ I. (1932–1986), 188
SUDHOFF, KARL FRIEDRICH JAKOB (1853–1938), 132
SÜSS, WILHELM (1895–1958), 140, 141n
SUNDMAN, KARL FRITHIOF (1873–1949), 477
SUSHKEVICH, ANTON K. (1889–1961), 183
SUTER, HEINRICH (1848–1922), **B**, 23n, 105, 124, 132, 321, 390, 418
SUZZI, GIUSEPPE (1701–1764), 224
SWAMI VIDYARANYA, *see* DATTA, BIBHUTIBHUSAN
Sweden, *see* Scandinavia
SWERDLOW, NOEL (*1941), 30, 498
SWETZ, FRANK J. (*1937), 336
Switzerland
Academy of Geneva, 97, 99
Academy of Lausanne, 99
Bernoulli Kommission, 103
Central Library in Zürich, 510
Ecole Polytechnique Fédérale de Lausanne, 107
Eidgenössische Polytechnische Hochschule in Zürich, *see* *Eidgenössische Technische Hochschule (ETH)* in Zürich
Eidgenössische Technische Hochschule (ETH) in Zürich, 97, 101–102, 105, 408, 425, 487, 510–511, 533, 550, 567–569
Euler Kommission, 103, 129, 511
GeP Society, 510
International Congress of Philosophy in Geneva (1904), 538
International Congresses of Mathematicians in Zürich, 97, 102, 472–473, 511

Naturforschende Gesellschaften:
Aargauische Naturforschende Gesellschaft, 100
Naturforschende Gesellschaft in Basel, 100–101, 103
Naturforschende Gesellschaft in Bern, 100–101
Naturforschende Gesellschaft in Zürich, 100–101, 510
Schweizerische Naturforschende Gesellschaft, 100–104, 129, 417, 511, 567–568
Société de physique et d'histoire naturelle de Genève (SPHN), 100
Société vaudoise des sciences naturelles, 100
Observatory in Geneva, 99
Otto-Spiess-Foundation in Basel, 104, 528
Schweizerische Akademie der Naturwissenschaften, 129
Schweizerische Akademie der Naturwissenschaften in Bern, 100
Schweizerische Bibliothek-Kommission, 101
Schweizerische Landesbibliothek, 101
Schweizerische Mathematische Gesellschaft, 104
Steiner-Schläfli Committee, 104
Swiss Academy of Sciences, 511, 567
Swiss National Science Foundation in Bern, 104
University of Basel, 97–102, 106, 427, 526–528
University of Bern, 98, 399, 511, 567
University of Fribourg (Freiburg), 98
University of Geneva, 98, 107
University of Lausanne, 98
University of Neuchâtel (Neuenburg), 98
University of Zürich, 55, 98, 102, 106–107, 294, 510–511, 533, 548, 567
SYLOW, LUDVIG (1832–1918), **B**, 153, 461, 467, 531
SYLVESTER, JAMES JOSEPH (1814–1897), 476, 542
SZARSKI, JACEK (1921–1980), 201

TACQUET, ANDREAS (1612–1660), **B**, 48–49, 112, 376
AL-ṬAHṬĀWĪ, RIFĀʿA (1801–1873), 321
TAINE, JOHN, *see* BELL, ERIC TEMPLE
TAIT, PETER GUTHRIE (1831–1901), 563
TAKAGI TEIJI (1875–1960), 430–431, 440
TAKEBE KATAHIRO (1661–1739), 291
TAMARKIN, JACOB DAVID (1888–1945), 498
TANABE HAJIME (1885–1962), 292–293
TANNER-YOUNG, CECILY (1900–1992), 177
TANNERY, JULES (1848–1910), 32, 401, 535
TANNERY, PAUL (1843–1904), **B**, 14, 23–31, 34, 36–38, 42–43, 54, 76, 118, 124, 137, 155, 390, 418, 426, 444, 474, 506, 576–577
TANNERY-PRISSET, MARIE (1856–1945), 26, 535, 538
TANNSTETTER, GEORG (1482–1535), 215–216
TARTAGLIA, NICCOLÒ (1499/1500–1557), 63, 69–70, 73, 90, 94, 257n, 374–375, 400, 436
TARTINI, GIUSEPPE (1692–1770), 73
TATARKIEWICZ, KRZYSZTOF (*1923), 201
TATON, RENÉ (*1915), 24, 35–39, 191, 286, 401, 438, 481, 572
TAYLOR, ALFRED EDWARD (1869–1945), 170
TAYLOR, BROOK (1685–1731), 73, 83
TAYLOR, EVA GERMAINE RIMINGTON (1879–1966), 177
TEDONE, ORAZIO (1870–1922), 81
TEIXEIRA, ANTÓNIO JOSÉ (1830–1900), 245
TEIXEIRA, FRANCISCO GOMES (1851–1933), **B**, 239, 242–247
TEKELI, SEVIM (*1924), 324
TENCA, LUIGI (1877–1960), 93
TENNULIUS, SAMUEL (*1635), **B**, 47, 551
TERQUEM, OLRY (1782–1862), **B**, 24, 334
TEUBNER, BENEDICTUS GOTTHELF (1784–1856), 418
THĀBIT B. QURRA (826–901), 132–133, 326, 366, 518
THAER, CLEMENS (1883–1974), **B**, 139
THALES (ca. 624–ca. 546), 4, 47, 221–222, 442
THEODORUS PRODROMUS (12th century), 537
THEODOSIUS OF PITANE (ca. 100 B.C.), 52, 63, 120, 157, 366, 404, 443
THEON OF ALEXANDRIA (ca. 370 A.D.), 4, 13, 48, 487, 509
THÉVENOT, MELCHISEDECH (1620 or 1621–1692), 5
THIBAUT, GEORGE FREDERICK WILLIAM (1848–1914), 311
THOMAS BRADWARDINE, *see* BRADWARDINE, THOMAS
THOMAS, IVOR (1905–1993), 176–178

THOMPSON, SYLVANUS PHILLIP (1851–1916), 169
THOMSON, JOSEPH JOHN (1856–1940), 169
THOMSON, SIR WILLIAM (LORD KELVIN) (1824–1907), 169, 173, 494, 563
THOMSON, THOMAS (1773–1852), 166, 178
THORNDIKE, LYNN (1882–1965), 365
THUREAU-DANGIN, FRANÇOIS (1872–1944), **B**, 556
TIBENSKÝ, JÁN (*1923), 210
TIETZE, HEINRICH (1880-1964), 217
TIKHOMIROV, VLADIMIR M. (*1934), 188
TIMCHENKO, IVAN YU. (1863–1939), **B**, 180–182, 196
TIETZE, HEINRICH (1873–1945), 456
TIRABOSCHI, GIROLAMO (1731–1794), 67, 75, 90
TOALDO, GIUSEPPE (1719–1797), 76
TODHUNTER, ISAAC (1820–1884), **B**, 169, 173–174, 178
TOEPLITZ, OTTO (1881–1940), **B**, 134–135, 141, 276, 362, 445, 496–497, 548–549, 555
TOMEK, V. V. (1818–1905), 207
TORRICELLI, EVANGELISTA (1608–1647), 68, 89–90, 375, 392, 473, 546
Torricelli Edition, 89, 375, 473, 546
TORROJA Y CABALLÉ, EDUARDO (1847–1918), 236
TORTOLINI, BARNABA (1808–1874), 369
TOTOK, WILHELM (*1921), 447
TRAIL, WILLIAM (1746–1831), 163, 175
TRAUBE, LUDWIG (1861–1907), 365
TREUTLEIN, PETER (1845–1912), **B**, 120, 130, 556
TREYTORRENS, FRANÇOIS-FRÉDÉRIC DE (1687–1737), 99
TREYTORRENS, LOUIS DE (1726–1794), 99
TROPFKE, JOHANNES (1866–1939), **B**, 130, 136–139, 418, 557, 564
TRUESDELL, CLIFFORD AMBROSE (1919–2000), **B**, 282, 562
TSCHIRNHAUS, EHRENFRIED WALTHER VON (1651–1708), 50, 448
TURAEV, BORIS A. (1868–1920), 183, 496–497
TURAZZA, DOMENICO (1813–1892), 425
TURGOT, ANNE ROBERT JACQUES, BARON DE AULNE (1727–1781), 6, 9
Turkey, *see* Arab countries, Turkey and Iran
TURNBULL, HERBERT WESTREN (1885–1961), **B**, 176–177
AL-ṬŪSĪ, NAṢĪR AL-DĪN, *see* NAṢĪR AL-DĪN AL-ṬŪSĪ
TUTANKHAMEN (Egyptian King 1347–1337 B.C.), 268
TWEEDIE, CHARLES (1867/68–1925), 175

U.S.S.R., *see* Russia
ULUG BEG (1394–1449), 190
ʿUMAR AL-KHAYYĀM (1048?–1131?), 19, 118, 323, 327, 525
UNESCO
Science Division, 494
United States of America
American Association for the Advancement of Science (AAAS), 283, 354, 381, 387, 523
American Historical Association (AHA), 265
American Mathematical Society (AMS), 268–269, 283–284, 353, 498, 524
American Philosophical Society (APS), 498
Black Mountain College in Black Mountain, NC, 275, 405
Brooklyn College in Brooklyn, NY, 380–381, 459
Brown University in Providence, RI, 141, 158, 268–269, 271, 276, 281, 285, 353, 396, 496–498, 513, 518, 544
Department for History of Mathematics, 276, 283
California Institute of Technology in Pasadena, CA, 359, 544
Carleton College in Northfield, MN, 286, 480
Carnegie Corporation in New York, NY, 498
Carnegie Institution in Washington, DC, 515
Catholic University of America in Washington, DC, 381
Centenary of the City of Chicago (1933), 271
Charles S. Peirce Society, 416
City University of New York (CUNY), 415, 459
Colorado College in Colorado Springs, CO, 265, 386
Columbia University in New York, NY, 270, 359, 380, 415, 437, 485, 515, 523
Dale Collection, 269
Plimpton Collection, 269–270
Smith Collection, 269–270, 525

Teachers College, 270, 274, 454, 524
Cornell University in Ithaca, NY, 454
Dartmouth College in Hanover, NH, 281
Ecole Libre des Hautes Etudes in New York, 461
Encyclopedias, 354
George Washington University in St. Louis, MO, 515
Harvard University in Cambridge, MA, 353, 359, 364, 399, 459–461, 502, 515–516
Haverford College in Haverford, PA, 562
History of Science Society (HSS), 227, 274, 283–284, 381, 387, 416, 454, 524, 557
Hunter College in New York, NY, 274, 281, 415, 523
Indiana University in Bloomington, IN, 544
Indiana University-Purdue University in Indianapolis, IN, 416
Institute for Advanced Study in Princeton, NJ, 276, 457, 459–461, 513, 562
Institute of Current World Affairs in New York, 479
International Exposition in New York (1939), 37
Johns Hopkins University in Baltimore, MD, 265, 386, 502, 513, 520, 544, 548
Lawrence Scientific School in Cambridge, MA, 502
Lehigh University in Bethlehem, PA, 416, 562
Massachusetts Institute of Technology (MIT) in Cambridge, MA, 55, 532, 544
Mathematical Association of America (MAA), 268, 271, 283–284, 338, 354, 360, 387, 396, 454, 524
National Academy of Sciences, 498
National Council of Teachers of Mathematics (NCTM), 283
National Education Association (NEA), 265
New York Academy of Sciences, 416
New York Historical Society, 269
New York Mathematical Society, 265, 524
New York University, 279, 457–459
Northwestern State University in Natchitoches, LA, 414
Northwestern University in Evanston, IL, 274–275, 445
Oregon State College in Corvallis, OR, 277
Princeton University in Princeton, NJ, 544
Radcliffe College in Cambridge, MA, 515
Rockefeller Foundation in New York, NY, 497–498, 513
Rutgers University in Rutgers, NJ, 381
Smithsonian Institution in Washington, DC, 170
Stanford University in Stanford, CA, 359, 459
State Normal College in Ypsilanti, MI, 524
State University of New York (SUNY) in Oneonta, NY, 515
Syracuse University in Syracuse, NY, 524
Texas Tech University in Lubbock, TX, 416
Tulane University in New Orleans, LA, 265, 386
U. S. Coast and Geodetic Survey in Washington, DC, 502
University of California in Berkeley, CA, 266, 281, 381, 386, 479–480, 520–521
University of Chicago in Chicago, IL, 266, 359, 415, 461, 513
University of Illinois in Chicago, IL, 515, 562
University of Kansas in Lawrence, KA, 381
University of Michigan in Ann Arbor, MI, 270, 381, 454
University of Northern Iowa in Cedar Falls, IA, 381
University of Washington in Seattle, WA, 359
University of Wisconsin in Madison, WI, 265, 386
Vassar College in Poughkeepsie, NY, 523
Western Reserve University in Cleveland, OH, 515
World's Fair of 1933 in Chicago, 272
Yale University in New Haven, CT, 153, 158, 311, 461, 499

Yeshiva University in New York, NY, 274, 381, 437
al-Uqlīdisī, Abū' l-Ḥasan Aḥmad b. Ibrāhīm (10th century), 514

Vacca, Giovanni (1872–1953), **B**, 61, 82–85, 89, 91–93, 392, 424, 471, 546
Vahlen, Theodor (1869–1945), 140–142
Vailati, Giovanni (1863–1909), **B**, 82–84, 87–89, 471
Vaiman, Aizik A. (*1922), 191
Valentin, Georg Hermann (1848–1926), 418
Vallée Poussin, Charles J. de la (1866–1962), 51
Valson, Claude-Alphonse (1826–1901), 24
Van Allen, James A. (*1914), 383
van der Waerden, Bartel Leendert (1903–1996), **B**, 55, 59, 105–106
van Maanen, Jan (*1953), 335
van Roomen, Adriaan, *see* Romanus, Adrianus
van Swinden, Jan Hendrik (1746–1823), 487
Vandermonde, Alexandre A. T. (1735–1796), 35, 208n
Varāhamihira (6th century A.D.), 308
Varignon, Pierre (1654–1722), 13, 401
Varignon Edition, 401
Vasconcellos, Fernando de Almeida Loureiro e (1874–1944), **B**, 245
Vasconcellos, José Joaquim Soares de Barros e (1721–1793), 531
Vashchenko-Zakharchenko, Mikhail E. (1825–1912), **B**, 179–180
Vasil'ev, Aleksandr V. (1853–1929), **B**, 181–182
Vassura, Giuseppe (1866–1949), 375, 473, 546
Veblen, Oswald (1880–1960), 498
Veda, 315
Velsius, Justus (ca. 1510–ca. 1581), **B**, 46
Venturi, Giovanni Battista (1746–1822), 71, 75
Ver Eecke, Paul (1867–1959), **B**, 52, 58, 537n
Vera Fernández de Córdoba, Francisco (1888–1967), **B**, 234n
Verbiest, Ferdinand (1623–1688), 376
Verley, Jean-Luc (*1939), 41
Veronese, Giuseppe (1854–1917), 421
Veselovskiĭ, Ivan N. (1892–1977), **B**, 185, 191
Vetter, Quido (1881–1960), **B**, 34, 191, 205n, 207n, 208–212
Vico, Giovanni Batista (1668–1744), 259
Victorius (ca. 450), 119
Viète, François (1540–1603), 5, 48, 406, 448–449, 551, 576
Viète Edition, 551
Vietoris, Leopold (*1891), 217
Villicus, Franz (19th century), 216
Vimercati, Cipriano (18th century), 427
Viṣṇugupta, *see* Kauṭilya
Vitali, Giuseppe (1875–1932), 84, 423
Vitruvius Pollio (1st century B.C.), 357
Vitruvius Rufus (1st century A.D.), 123, 389
Viṭṭhala, Govinda (19th century), 313
Vivanti, Giuseppe (1859–1949), **B**, 82, 390
Viviani, Vincenzo (1622–1703), 65, 76
Vizgin, Vladimir P. (*1936), 189
Vogel, Kurt (1888–1985), **B**, 112, 130–131, 136–138, 140, 142–143, 146, 149, 191
Vollers, Karl (1857–1909), 321
Vollgraff, Johan Adriaan (1877–1965), **B**, 54, 57
Volodarskiĭ, Aleksandr I. (*1938), 192
Volta, Alessandro (1745–1827), 436
Voltaire (François Marie Arouet) (1694–1778), 9
Volterra, Vito (1860–1940), 82–83, 392, 473, 546
von Escherich, Gustav, *see* Escherich, Gustav von
Voronoĭ, Georgiĭ F. (1868–1908), 194
Vorsterman van Oyen, George Auguste (1836–1915), **B**, 53
Voss, Aurel (1845–1931), 456
Vossius, Gerard Johann (1577–1649), **B**, 48, 112, 162, 331
Vossius, Gerardus Joannes, *see* Vossius, Gerard Johann
Voulgaris, Eugenios (1716–1806), 224–225, 229
Vrain-Lucas, Denis (*1818), 21
Vries, Hendrik De, *see* De Vries, Hendrick
Vydra, Stanislav (1741–1804), 206
Vygodskiĭ, Mark Ya. (1898–1965), **B**, 184–187, 191, 196, 570–573

Waard, Cornelis de, *see* de Waard, Cornelis
Wachułka, Adam (1909–1991), 201

WAERDEN, BARTEL LEENDERT VAN DER, *see* VAN DER WAERDEN, BARTEL LEENDERT
WAHLERT, HOWARD E. (1905–1975), 279, 457
WALLIS, JOHN (1616–1703), **B**, 8, 51, 70, 73, 112, 136, 162, 167, 172, 176, 448
WALLIS, PETER (1918–1992), **B**, 177
WALLIS, RUTH, 177, 561
WALLNER, CARL RAIMUND (1881–1934), 126, 390
WALUSINSKI, GILBERT (20th century), 451
WANG LING (1918–1994), 494
WAPPLER, HERMANN EMIL (1852–1899), **B**, 120, 556
WARING, EDWARD (1736–1798), 73
WEBER, AUFRECHT (mid-19th century), 311
WEBER, HEINRICH (1842–1913), 454, 456, 512, 547
WEBER, WILHELM (1804–1891), 387
WEIERSTRASS, KARL THEODOR WILHELM (1815–1897), 78, 126, 145, 216, 228, 314, 510, 572
WEIL, ANDRÉ (1906–1998), **B**, 378, 395
WEIL, SIMONE (1909–1943), 562
WEILL, E. (20th century), 451
WEITZENBÖCK, ROLAND (1885–1955), 218, 495, 547
WENDELEN, GODFRIED (1580–1667), 376, 463
WERNER, JOHANNES (1468–1528), 366
WERTHEIM, GUSTAV (1843–1902), **B**, 119, 124, 390
WESSEL, CASPAR (1745–1818), 153, 333, 467
WHEWELL, WILLIAM (1794–1866), **B**, 167–169, 178, 264, 406–407, 542, 567
WHISH, CHARLES M. (ca. 1835), 311
WHISTON, WILLIAM (1667–1752), 48
WHITEHEAD, ALFRED NORTH (1861–1947), 173, 479
WHITESIDE, DEREK THOMAS (*1932), 141, 177
WHITNEY, WILLIAM D. (1827–1894), 311
WHITROW, GERALD JAMES (1912–2000), 177–178
WHITTAKER, EDMUND (1873–1956), **B**, 173, 176–178
WIEDEMANN, EILHARD (1852–1928), **B**, 132, 555
WIELEITNER, HEINRICH (1874–1931), **B**, 125–126, 136–137, 140, 149, 418, 446, 555
WIENER, NORBERT (1894–1964), 260n
WIĘSŁAW, WITOLD (*1944), 202
WILDER, RAYMOND (1896–1982), 277, 282
WILKINSON, LANCELOT (19th century), 312
WILSON, D. K. (1st half of 20th cent.), 176
WINTER, EDUARD (1896–1956), 191, 572
WIRTINGER, WILHELM (1865–1945), 457
WIRTINGER, WILHELM (1865-1945), 217–218
WISSOWA, GEORG (1859–1931), 120
WITELO (1225?–1280?), 46
WOEPCKE, FRANZ (1826–1864), **B**, 18–19, 118, 147, 520
WOHLWILL, EMIL (1835–1912), 418
WOLF, ABRAHAM (1876–1948), 175, 178
WOLF, JOHANN RUDOLF (1816–1893), **B**, 101–102, 106, 418, 510
WOLFF, CHRISTIAN (1679–1754), 99, 111–112, 114, 257, 430, 448–449
WOODHOUSE, ROBERT (1773–1827), 164, 169
WROŃSKI, JÓZEF, *see* HOENE-WROŃSKI, JÓZEF
WURSTISEN, CHRISTIAN (1544–1588), 99
WUSSING, HANS (*1927), xxi, 143–144, 191, 329
WYDRA, STANISLAV, *see* VYDRA, STANISLAV
WYLIE, ALEXANDER (1815–1887), **B**, 363
WYROUBOFF, GRÉGOIRE (1843–1913), 29

XU GUANGQI (1562–1633), 299
XYLANDER (1532–1576), 46, 98

YAN DUNJIE (1917–1988), 304–305
YANG HUI (13th century), 303
YANOVSKAYA, SOF'YA A. (1896–1966), **B**, 184–185, 187–188, 196, 559–560, 571–573
YELDHAM, FLORENCE ANNIE (*1877), 176
YONGLE (1360–1424; Emperor of China from 1402), 301
YOSIDA MITUYOSI (1598–1672), 298
YOUNG, GRACE CHISOLM (1868–1944), 172, 177
YOUNG, WILLIAM HENRY (1863–1942), 172, 177
YOUSHKEVICH, ADOLF-ANDREJ PAVLOVICH (1906–1993), **B**, 144, 184–197, 286, 481–482, 560, 570
Yugoslavia
University of Ljubljana, 385

YUSHKEVICH, ADOLF P., *see* YOUSHKEVICH, ADOLF-ANDREJ PAVLOVICH
YUSUPOV, NURI (1888–1937?), 183

ZACUTO, ABRAHAM (ABRÃO) (ca. 1450–ca. 1522), 244
ZĀKĪ, AḤMAD PĀSHĀ (1867–1934), 322
ZAMBERTI, BARTOLOMEO (*1473?), 4, 63
ZARATHUSTRA (between 1000 and 600 BC; Persian prophet), 116
ZARCO DEL VALLE, ANTONIO REMÓN (1775–1866), 234–235
ZAREMBA, STANISŁAW (1863–1942), 202
ZARISKI, OSCAR (1899–1986), 92, 424
ZASLAVSKY, CLAUDIA (*1917), 282
ŻEBRAWSKI, TEOFIL (1800–1887), 200
ZELLER, EDUARD (1814–1908), 535
ZENO OF ELEA (ca. 495–ca. 430 B.C.), 486, 536, 548
ZENODORUS (2nd century B.C.), 387
ZERMELO, ERNST (1871–1953), 217, 395
ZERVOS, PANAGIOTIS (1878–1952), 229
ZEUTHEN, HIERONYMUS GEORG (1839–1920), **B**, 22, 25–26, 28, 84, 137, 154–158, 160, 331, 365, 418, 443–444, 456, 518
ZHANG DUNREN (1754–1834), 302
ZHANG QIUJIAN (5th century A.D.), 301
ZHAO SHUANG (3rd century A.D.?), 300
ZHOU YUNQING (mid 20th century), 304
ZHU KEBAO (1845–1903), 303
ZHU KEZHEN (1890–1974), 305
ZHU SHIJIE (ca. 1300 A.D.), 302–303
ZHU ZAIYU (1536–1611), 298
ZHUKOVSKIĬ, NIKOLAĬ E. (1847–1921), 194
ZINDLER, KONRAD (1866–1934), 218
ZOLOTARËV, EGOR I. (1847–1878), 186
ZOROASTER, *see* ZARATHUSTRA
ZUBIETA, FRANCISCO (*1911), 259
ZUBOV, VASILIĬ P. (1899–1963), **B**, 188, 191, 193
ZWINGLI, ULRICH (1484–1531), 98